Be sure to keep up-to

Register now and we'll notify yo
when the newest cost index is available

Please provide your name, address, and email below
and return this card by mail or
Fax: 1-800-632-6732.

To register online: www.rsmeans411.com/register

Name _____

email _____

Company _____

Street _____

City/Town _____ State/Prov. _____ Zip/Postal Code _____

RSMeans

RSMeans
Quarterly Updates
63 Smiths Lane
Kingston, MA 02364-3008
USA

$184.95 per copy (in United States)
Price is subject to change without prior notice.

0050

RSMeans Square Foot Costs

31ST ANNUAL EDITION

2010

RSMeans
A division of Reed Construction Data
Construction Publishers & Consultants
63 Smiths Lane
Kingston, MA 02364-3008
(781) 422-5000

Copyright©2009 by RSMeans
A division of Reed Construction Data
All rights reserved.

Printed in the United States of America
ISSN 1540-6326
ISBN 978-0-87629-829-9

Senior Editor
Barbara Balboni

Contributing Editors
Christopher Babbitt
Ted Baker
Robert A. Bastoni
John H. Chiang, PE
Gary W. Christensen
David G. Drain, PE
Cheryl Elsmore
Robert J. Kuchta
Robert C. McNichols
Melville J. Mossman, PE
Jeannene D. Murphy
Stephen C. Plotner
Eugene R. Spencer
Marshall J. Stetson
Phillip R. Waier, PE

Book Design
Norman R. Forgit

President & CEO
Iain Melville

**Vice President,
Product & Development**
Richard Remington

**Vice President of
Operations**
David C. Walsh

Vice President of Sales
Thomas Kesler

Director, Cost Products
Daimon Bridge

Marketing Director
John M. Shea

Engineering Manager
Bob Mewis, CCC

Cover Design
Paul Robertson
Wayne Engebretson

Production Manager
Michael Kokernak

Production Coordinator
Jill Goodman

Production
Paula Reale-Camelio
Sheryl A. Rose
Jonathan Forgit
Debbie Panarelli
Mary Lou Geary

Technical Support
A.K. Dash
Tom Dion
Roger Hancock
Gary L. Hoitt
Genevieve Medeiros
Kathryn S. Rodriguez
Sharon Proulx

Foreword

Our Mission

Since 1942, RSMeans has been actively engaged in construction cost publishing and consulting throughout North America.

Today, more than 65 years after RSMeans began, our primary objective remains the same: to provide you, the construction and facilities professional, with the most current and comprehensive construction cost data possible.

Whether you are a contractor, owner, architect, engineer, facilities manager, or anyone else who needs a reliable construction cost estimate, you'll find this publication to be a highly useful and necessary tool.

With the constant flow of new construction methods and materials today, it's difficult to find the time to look at and evaluate all the different construction cost possibilities. In addition, because labor and material costs keep changing, last year's cost information is not a reliable basis for today's estimate or budget.

That's why so many construction professionals turn to RSMeans. We keep track of the costs for you, along with a wide range of other key information, from city cost indexes . . . to productivity rates . . . to crew composition . . . to contractor's overhead and profit rates.

RSMeans performs these functions by collecting data from all facets of the industry and organizing it in a format that is instantly accessible to you. From the preliminary budget to the detailed unit price estimate, you'll find the data in this book useful for all phases of construction cost determination.

The Staff, the Organization, and Our Services

When you purchase one of RSMeans' publications, you are, in effect, hiring the services of a full-time staff of construction and engineering professionals.

Our thoroughly experienced and highly qualified staff works daily at collecting, analyzing, and disseminating comprehensive cost information for your needs. These staff members have years of practical construction experience and engineering training prior to joining the firm. As a result, you can count on them not only for accurate cost figures, but also for additional background reference information that will help you create a realistic estimate.

The RSMeans organization is always prepared to help you solve construction problems through its variety of data solutions, including online, CD, and print book formats, as well as cost estimating expertise available via our business solutions, training, and seminars.

Besides a full array of construction cost estimating books, RSMeans also publishes a number of other reference works for the construction industry. Subjects include construction estimating and project and business management, special topics such as green building and job order contracting, and a library of facility management references.

In addition, you can access all of our construction cost data electronically using *Means CostWorks*® CD or on the Web using **MeansCostWorks.com.**

What's more, you can increase your knowledge and improve your construction estimating and management performance with an RSMeans construction seminar or in-house training program. These two-day seminar programs offer unparalleled opportunities for everyone in your organization to become updated on a wide variety of construction-related issues.

RSMeans is also a worldwide provider of construction cost management and analysis services for commercial and government owners.

In short, RSMeans can provide you with the tools and expertise for constructing accurate and dependable construction estimates and budgets in a variety of ways.

Robert Snow Means Established a Tradition of Quality That Continues Today

Robert Snow Means spent years building RSMeans, making certain he always delivered a quality product.

Today, at RSMeans, we do more than talk about the quality of our data and the usefulness of our books. We stand behind all of our data, from historical cost indexes to construction materials and techniques to current costs.

If you have any questions about our products or services, please call us toll-free at 1-800-334-3509. Our customer service representatives will be happy to assist you. You can also visit our Web site at **www.rsmeans.com.**

Table of Contents

Related RSMeans Products and Services

The Engineers at RSMeans, suggest the following products and services as companion information resources to *RSMeans Square Foot Costs*:

Construction Cost Data Books
Building Construction Cost Data 2010
Assemblies Cost Data 2010
Light Commercial Cost Data 2010
Facilities Construction Cost Data 2010

Reference Books
Building Security: Strategies & Costs
Designing & Building with the IBC
Estimating Building Costs
Estimating Handbook
How to Estimate with Means Data and CostWorks
Project Scheduling and Management for Construction
Square Foot & Assemblies Estimating Methods

Seminars and In-House Training
Square Foot & Assemblies Estimating
Means CostWorks® Training
Means Data for Job Order Contracting (JOC)
Plan Reading & Material Takeoff
Scheduling & Project Management

RSMeans on the Internet
Visit RSMeans at **www.rsmeans.com** for a one-stop portal for the most reliable and current resources available on the market. Here you can request or download **FREE** estimating software demos, visit our bookstore for convenient ordering, and learn more about new publications and companion products.

Also visit **www.reedconstructiondata.com** for RSMeans business solutions, seminars, training, and more.

RSMeans Electronic Data
Get the information found in RSMeans cost books electronically on *Means CostWorks®* CD or on the Web at **MeansCostWorks.com**.

RSMeans Business Solutions
Cost engineers and analysts conduct facility life cycle and benchmark research studies and predictive cost modeling, as well as offer consultation services for conceptual estimating, real property management, and job order contracting. Research studies are designed with the application of proprietary cost and project data from Reed Construction Data and RSMeans' extensive North American databases. Analysts offer building product manufacturers qualitative and quantitative market research, as well as time/motion studies for new products, and Web-based dashboards for market opportunity analysis. Clients are from both the public and private sectors, including federal, state, and municipal agencies, corporations, institutions, construction management firms, hospitals, and associations.

RSMeans for Job Order Contracting (JOC)
Best practice job order contracting (JOC) services support the owner in the development of their program. RSMeans develops all required contracting documents, works with stakeholders, and develops qualified and experienced contractor lists. Means JOCWorks™ software is the best of class in the functional tools that meet JOC specific requirements. RSMeans data is organized to meet JOC cost estimating requirements.

Construction Costs for Software Applications
Over 25 unit price and assemblies cost databases are available through a number of leading estimating and facilities management software providers (listed below). For more information see the "Other RSMeans Products" pages at the back of this publication.

MeansData™ is also available to federal, state, and local government agencies as multi-year, multi-seat licenses.

- 4Clicks-Solutions, LLC
- ArenaSoft Estimating
- Beck Technology
- BSD – Building Systems Design, Inc.
- CMS – Construction Management Software
- Corecon Technologies, Inc.
- CorVet Systems
- Estimating Systems, Inc.
- Hard Dollar Corporation
- MC^2
- Parsons Corporation
- Sage Timberline Office
- UDA Technologies, Inc.
- US Cost, Inc.
- VFA – Vanderweil Facility Advisers
- WinEstimator, Inc.

How the Book Is Built: An Overview

A Powerful Construction Tool

You have in your hands one of the most powerful construction tools available today. A successful project is built on the foundation of an accurate and dependable estimate. This book will enable you to construct just such an estimate.

For the casual user the book is designed to be:

- quickly and easily understood so you can get right to your estimate.
- filled with valuable information so you can understand the necessary factors that go into the cost estimate.

For the regular user, the book is designed to be:

- a handy desk reference that can be quickly referred to for key costs.
- a comprehensive, fully reliable source of current construction costs for typical building structures when only minimum dimensions, specifications, and technical data are available. It is specifically useful in the conceptual or planning stage for preparing preliminary estimates which can be supported and developed into complete engineering estimates for construction.
- a source book for rapid estimating of replacement costs and preparing capital expenditure budgets.

To meet all of these requirements, we have organized the book into the following clearly defined sections.

How To Use the Book: The Details

This section contains an in-depth explanation of how the book is arranged and how you can use it to determine a reliable construction cost estimate. It includes information about how we develop our cost figures and how to completely prepare your estimate.

Residential Section

Model buildings for four classes of construction—economy, average, custom, and luxury—are developed and shown with complete costs per square foot.

Commercial, Industrial, Institutional Section

This section contains complete costs for 75 typical model buildings expressed as costs per square foot.

Assemblies Section

The cost data in this section has been organized in an "Assemblies" format. These assemblies are the functional elements of a building and are arranged according to the 7 divisions of the UNIFORMAT II classification system. The costs in this section include standard markups for the overhead and profit of the installing contractor. The assemblies costs are those used to calculate the total building costs in the Commercial/Industrial/Institutional section.

Reference Section

This section includes information on General Conditions, Location Factors, Historical Cost Indexes, a Glossary, and a listing of Abbreviations used in this book.

Location Factors: You can adjust the total project costs to over 900 locations throughout the U.S. and Canada by using the data in this section.

Historical Cost Indexes: These indexes provide you with data to adjust construction costs over time.

Glossary: A listing of common terms related to construction and their definitions is included.

Abbreviations: A listing of the abbreviations used throughout this book, along with the terms they represent, is included.

The Scope of This Book

This book is designed to be as comprehensive and as easy to use as possible. To that end we have made certain assumptions and limited its scope in two key ways:

1. We have established material prices based on a national average.
2. We have computed labor costs based on a 30-city national average of union wage rates.

All costs and adjustments in this manual were developed from the *RSMeans Building Construction Cost Data, Assemblies Cost Data, Residential Cost Data,* and *Light Commercial Cost Data.* Costs include labor, material, overhead and profit, and general conditions and architects' fees where specified. No allowance is made for sales tax, financing, premiums for material and labor, or unusual business and weather conditions.

What's Behind the Numbers? The Development of Cost Data

The staff at RSMeans continuously monitors developments in the construction industry in order to ensure reliable, thorough, and up-to-date cost information.

While overall construction costs may vary relative to general economic conditions, price fluctuations within the industry are dependent upon many factors. Individual price variations may, in fact, be opposite to overall economic trends. Therefore, costs are continually monitored and complete updates are published yearly. Also, new items are frequently added in response to changes in materials and methods.

Costs—$ (U.S.)

All costs represent U.S. national averages and are given in U.S. dollars. The RSMeans Location Factors can be used to adjust costs to a particular location. The Location Factors for Canada can be used to adjust U.S. national averages to local costs in Canadian dollars. No exchange rate conversion is necessary.

Factors Affecting Costs

Due to the almost limitless variation of building designs and combinations of construction methods and materials, costs in this book must be used with discretion. The editors have made every effort to develop costs for "typical" building types. By definition, a square foot estimate, due to the short time and limited amount of detail required, involves a relatively lower degree of accuracy than a detailed unit cost estimate, for example. The accuracy of an estimate will increase as the project is defined in more detailed components. The user should examine closely and compare the specific requirements of the project being estimated with the components included in the models in this book. Any differences should be accounted for in the estimate.

In many building projects, there may be factors that increase or decrease the cost beyond the range shown in this manual. Some of these factors are:

- Substitution of building materials or systems for those used in the model
- Custom-designed finishes, fixtures, and/or equipment
- Special structural qualities (allowance for earthquake, future expansion, high winds, long spans, and unusual shape)
- Special foundations and substructures to compensate for unusual soil conditions
- Inclusion of typical "owner-supplied" equipment in the building cost (special testing and monitoring equipment for hospitals, kitchen equipment for hotels, and offices that have cafeteria accommodations for employees, etc.)
- An abnormal shortage of labor or materials

- Isolated building site or rough terrain that would affect the transportation of personnel, material, or equipment
- Unusual climatic conditions during the construction process

We strongly urge that, for maximum accuracy, these factors be considered each time a structure is being evaluated.

If users require greater accuracy than this manual can provide, the editors recommend that the *RSMeans Building Construction Cost Data* or the *RSMeans Assemblies Cost Data* be consulted.

Final Checklist

Estimating can be a straightforward process provided you remember the basics. Here's a checklist of some of the steps you should remember to complete before finalizing your estimate.

Did you remember to . . .

- include the Location Factor for your city?
- mark up the entire estimate sufficiently for your purposes?
- include all components of your project in the final estimate?
- double check your figures for accuracy?
- call RSMeans if you have any questions about your estimate or the data you've found in our publications?

Remember, RSMeans stands behind its publications. If you have any questions about your estimate, about the costs you've used from our books, or even about the technical aspects of the job that may affect your estimate, feel free to call the RSMeans editors at 1-800-334-3509.

Residential Section

Table of Contents

Introduction to Residential Section

The residential section of this manual contains costs per square foot for four classes of construction in seven building types. Costs are listed for various exterior wall materials which are typical of the class and building type. There are cost tables for Wings and Ells with modification tables to adjust the base cost of each class of building. Non-standard items can easily be added to the standard structures.

Cost estimating for a residence is a three-step process:
  (1) Identification
  (2) Listing dimensions
  (3) Calculations

Guidelines and a sample cost estimating procedure are shown on the following pages.

Identification

To properly identify a residential building, the class of construction, type, and exterior wall material must be determined. Page 6 has drawings and guidelines for determining the class of construction. There are also detailed specifications and additional drawings at the beginning of each set of tables to further aid in proper building class and type identification.

Sketches for eight types of residential buildings and their configurations are shown on pages 8 and 9. Definitions of living area are next to each sketch. Sketches and definitions of garage types are on page 10.

Living Area

Base cost tables are prepared as costs per square foot of living area. The living area of a residence is that area which is suitable and normally designed for full time living. It does not include basement recreation rooms or finished attics, although these areas are often considered full time living areas by the owners.

Living area is calculated from the exterior dimensions without the need to adjust for exterior wall thickness. When calculating the living area of a 1-1/2 story, two story, three story or tri-level residence, overhangs and other differences in size and shape between floors must be considered.

Only the floor area with a ceiling height of six feet or more in a 1-1/2 story residence is considered living area. In bi-levels and tri-levels, the areas that are below grade are considered living area, even when these areas may not be completely finished.

Base Tables and Modifications

Base cost tables show the base cost per square foot without a basement, with one full bath and one full kitchen for economy and average homes, and an additional half bath for custom and luxury models. Adjustments for finished and unfinished basements are part of the base cost tables. Adjustments for multi family residences, additional bathrooms, townhouses, alternate roofs, and air conditioning and heating systems are listed in modification tables below the base cost tables.

Costs for other modifications, including garages, breezeways and site improvements, are in the modification tables on pages 58 to 61.

Listing of Dimensions

To use this section of the manual, only the dimensions used to calculate the horizontal area of the building and additions and modifications are needed. The dimensions, normally the length and width, can come from drawings or field measurements. For ease in calculation, consider measuring in tenths of feet, i.e., 9 ft. 6 in. = 9.5 ft., 9 ft. 4 in. = 9.3 ft.

In all cases, make a sketch of the building. Any protrusions or other variations in shape should be noted on the sketch with dimensions.

Calculations

The calculations portion of the estimate is a two-step activity:
  (1) The selection of appropriate costs from the tables
  (2) Computations

Selection of Appropriate Costs

To select the appropriate cost from the base tables, the following information is needed:
  (1) Class of construction
      (a) Economy
      (b) Average
      (c) Custom
      (d) Luxury
  (2) Type of residence
      (a) 1 story
      (b) 1-1/2 story
      (c) 2 story
      (d) 3 story
      (e) Bi-level
      (f) Tri-level
  (3) Occupancy
      (a) One family
      (b) Two family
      (c) Three family
  (4) Building configuration
      (a) Detached
      (b) Town/Row house
      (c) Semi-detached
  (5) Exterior wall construction
      (a) Wood frame
      (b) Brick veneer
      (c) Solid masonry
  (6) Living areas

Modifications are classified by class, type, and size.

Computations

The computation process should take the following sequence:
  (1) Multiply the base cost by the area.
  (2) Add or subtract the modifications.
  (3) Apply the location modifier.

When selecting costs, interpolate or use the cost that most nearly matches the structure under study. This applies to size, exterior wall construction, and class.

How to Use the Residential Square Foot Cost Pages

The following is a detailed explanation of a sample entry in the Residential Square Foot Cost Section. Each bold number below corresponds to the item being described on the facing page with the appropriate component of the sample entry following in parenthesis.

Prices listed are costs that include overhead and profit of the installing contractor. Total model costs include an additional mark-up for General Contractors' overhead and profit, and fees specific to class of construction.

RESIDENTIAL — Average **1** — 2 Story **2**

- Simple design from standard plans
- Single family — 1 full bath, 1 kitchen
- No basement
- Asphalt shingles on roof
- Hot air heat
- Gypsum wallboard interior finishes
- Materials and workmanship are average
- Detail specifications on p. 27

3

Note: The illustration shown may contain some optional components (for example: garages and/or fireplaces) whose costs are shown in the modifications, adjustments, & alternatives below or at the end of the square foot section.

Base cost per square foot of living area

Exterior Wall **4**	1000	1200	1400	1600	1800	Living Area **5** 2000	2200	2600	3000	3400	3800
Wood Siding - Wood Frame	119.55	108.20	102.90	99.30	95.55	91.50	88.85	83.75	78.85	76.55	74.50
Brick Veneer - Wood Frame	127.75	115.85	109.95	106.10	101.85	97.65	9**6**	89.00	83.70	81.20	78.95
Stucco on Wood Frame	115.90	104.85	99.75	96.30	92.70	88.70	**6**	81.30	76.60	74.45	72.55
Solid Masonry	143.00	129.90	123.15	118.60	113.65	109.00	105.45	98.80	92.75	89.75	87.10
Finished Basement, Add **7**	19.70	19.40	18.75	18.35	17.90	17.65	**8**	16.60	16.15	15.90	15.65
Unfinished Basement, Add	7.65	7.10	6.70	6.40	6.10	5.90		5.35	5.05	4.80	4.70

Modifications

Add to the total cost

Upgrade Kitchen Cabinets **9**	$	+ 3521
Solid Surface Countertops (Included)		
Full Bath - including plumbing, wall and floor finishes		+ 5993
Half Bath - including plumbing, wall and floor finishes		+ 3570
One Car Attached Garage		+ 12,676
One Car Detached Garage		+ 16,871
Fireplace & Chimney		+ 5836

Adjustments

For multi family - add to total cost

Additional Kitchen	$	+ 6796
Additional Bath		+ 5993
Additional Entry & Exit		+ 1640
Separate Heating		+ 1409
Separate Electric		+ 1915

For Townhouse/Rowhouse -
Multiply cost per square foot by **10**

Inner Unit	+ .90
End Unit	+ .95

Alternatives

Add to or deduct from the cost per square foot of living area

Cedar Shake Roof	+ 1.95
Clay Tile Roof	+ 2.25
Slate Roof	+ 3.30
Upgrade Walls to Skim Coat Plaster	+ .49
Upgrade Ceilings to Textured Finish	+ .48
Air Conditioning, in Heating Ductwork	+ 2.54
In Separate Ductwork	+ 6.36
Heating Systems, Hot Water	+ 2.01
Heat Pump	+ 3.11
Electric Heat	– .80
Not Heated	– 3.55

11

Additional upgrades or components

Kitchen Cabinets & Countertops	Page 58
Bathroom Vanities	59
Fireplaces & Chimneys	59
Windows, Skylights & Dormers	59
Appliances	60
Breezeways & Porches	60
Finished Attic	60
Garages	61
Site Improvements	61
Wings & Ells	37

12

Important: See the Reference Section for Location Factors (to adjust for your city) and Estimating Forms

1 Class of Construction (Average)

The class of construction depends upon the design and specifications of the plan. The four classes are economy, average, custom, and luxury.

2 Type of Residence (2 Story)

The building type describes the number of stories or levels in the model. The seven building types are 1 story, 1-1/2 story, 2 story, 2-1/2 story, 3 story, bi-level, and tri-level.

3 Specification Highlights (Hot Air Heat)

These specifications include information concerning the components of the model, including the number of baths, roofing types, HVAC systems, and materials and workmanship. If the components listed are not appropriate, modifications can be made by consulting the information shown lower on the page or the Assemblies Section.

4 Exterior Wall System (Wood Siding - Wood Frame)

This section includes the types of exterior wall systems and the structural frame used. The exterior wall systems shown are typical of the class of construction and the building type shown.

5 Living Areas (2000 SF)

The living area is that area of the residence which is suitable and normally designed for full time living. It does not include basement recreation rooms or finished attics. Living area is calculated from the exterior dimensions without the need to adjust for exterior wall thickness. When calculating the living area of a 1-1/2 story, 2 story, 3 story or tri-level residence, overhangs and other differences in size and shape between floors must be considered. Only the floor area with a ceiling height of six feet or more in a 1-1/2 story residence is considered living area. In bi-levels and tri-levels, the areas that are below grade are considered living area, even when these areas may not be completely finished. A range of various living areas for the residential model are shown to aid in selection of values from the matrix.

6 Base Costs per Square Foot of Living Area ($91.50)

Base cost tables show the cost per square foot of living area without a basement, with one full bath and one full kitchen for economy and average homes, and an additional half bath for custom and luxury models. When selecting costs, interpolate or use the cost that most nearly matches the residence under consideration for size, exterior wall system, and class of construction. Prices listed are costs that include overhead and profit of the installing contractor, a general contractor markup, and an allowance for plans that vary by class of construction. For additional information on contractor overhead and architectural fees, see the Reference Section.

7 Basement Types (Finished)

The two types of basements are finished and unfinished. The specifications and components for both are shown on the Building Classes page in the Introduction to this section.

8 Additional Costs for Basements ($17.65)

These values indicate the additional cost per square foot of living area for either a finished or an unfinished basement.

9 Modifications and Adjustments (Upgrade Kitchen Cabinets $3521)

Modifications and Adjustments are costs added to or subtracted from the total cost of the residence. The total cost of the residence is equal to the cost per square foot of living area times the living area. Typical modifications and adjustments include kitchens, baths, garages and fireplaces.

10 Multiplier for Townhouse/ Rowhouse (Inner Unit .90)

The multipliers shown adjust the base costs per square foot of living area for the common wall condition encountered in townhouses or rowhouses.

11 Alternatives (Skim Coat Plaster $.49)

Alternatives are costs added to or subtracted from the base cost per square foot of living area. Typical alternatives include variations in kitchens, baths, roofing, and air conditioning and heating systems.

12 Additional Upgrades or Components (Wings & Ells page 37)

Costs for additional upgrades or components, including wings or ells, breezeways, porches, finished attics, and site improvements, are shown in other locations in the Residential Square Foot Cost Section.

Building Classes

Economy Class

An economy class residence is usually built from stock plans. The materials and workmanship are sufficient to satisfy building codes. Low construction cost is more important than distinctive features. The overall shape of the foundation and structure is seldom other than square or rectangular.

An unfinished basement includes 7' high 8" thick foundation wall composed of either concrete block or cast-in-place concrete.

Included in the finished basement cost are inexpensive paneling or drywall as the interior finish on the foundation walls, a low cost sponge backed carpeting adhered to the concrete floor, a drywall ceiling, and overhead lighting.

Custom Class

A custom class residence is usually built from plans and specifications with enough features to give the building a distinction of design. Materials and workmanship are generally above average with obvious attention given to construction details. Construction normally exceeds building code requirements.

An unfinished basement includes a 7'-6" high 10" thick cast-in-place concrete foundation wall or a 7'-6" high 12" thick concrete block foundation wall.

A finished basement includes painted drywall on insulated 2" x 4" wood furring as the interior finish to the concrete walls, a suspended ceiling, carpeting adhered to the concrete floor, overhead lighting and heating.

Average Class

An average class residence is a simple design and built from standard plans. Materials and workmanship are average, but often exceed minimum building codes. There are frequently special features that give the residence some distinctive characteristics.

An unfinished basement includes 7'-6" high 8" thick foundation wall composed of either cast-in-place concrete or concrete block.

Included in the finished basement are plywood paneling or drywall on furring that is fastened to the foundation walls, sponge backed carpeting adhered to the concrete floor, a suspended ceiling, overhead lighting and heating.

Luxury Class

A luxury class residence is built from an architect's plan for a specific owner. It is unique both in design and workmanship. There are many special features, and construction usually exceeds all building codes. It is obvious that primary attention is placed on the owner's comfort and pleasure. Construction is supervised by an architect.

An unfinished basement includes 8' high 12" thick foundation wall that is composed of cast-in-place concrete or concrete block.

A finished basement includes painted drywall on 2" x 4" wood furring as the interior finish, suspended ceiling, tackless carpet on wood subfloor with sleepers, overhead lighting and heating.

Configurations

Detached House

This category of residence is a free-standing separate building with or without an attached garage. It has four complete walls.

Semi-Detached House

This category of residence has two living units side-by-side. The common wall is a fireproof wall. Semi-detached residences can be treated as a row house with two end units. Semi-detached residences can be any of the building types.

Town/Row House

This category of residence has a number of attached units made up of inner units and end units. The units are joined by common walls. The inner units have only two exterior walls. The common walls are fireproof. The end units have three walls and a common wall. Town houses/row houses can be any of the building types.

Building Types

One Story

This is an example of a one-story dwelling. The living area of this type of residence is confined to the ground floor. The headroom in the attic is usually too low for use as a living area.

One-and-a-half Story

The living area in the upper level of this type of residence is 50% to 90% of the ground floor. This is made possible by a combination of this design's high-peaked roof and/or dormers. Only the upper level area with a ceiling height of 6 feet or more is considered living area. The living area of this residence is the sum of the ground floor area plus the area on the second level with a ceiling height of 6 feet or more.

One Story with Finished Attic

The main living area in this type of residence is the ground floor. The upper level or attic area has sufficient headroom for comfortable use as a living area. This is made possible by a high peaked roof. The living area in the attic is less than 50% of the ground floor. The living area of this type of residence is the ground floor area only. The finished attic is considered an adjustment.

Two Story

This type of residence has a second floor or upper level area which is equal or nearly equal to the ground floor area. The upper level of this type of residence can range from 90% to 110% of the ground floor area, depending on setbacks or overhangs. The living area is the sum of the ground floor area and the upper level floor area.

Two-and-one-half Story

This type of residence has two levels of equal or nearly equal area and a third level which has a living area that is 50% to 90% of the ground floor. This is made possible by a high peaked roof, extended wall heights and/or dormers. Only the upper level area with a ceiling height of 6 feet or more is considered living area. The living area of this residence is the sum of the ground floor area, the second floor area and the area on the third level with a ceiling height of 6 feet or more.

Three Story

This type of residence has three levels which are equal or nearly equal. As in the 2 story residence, the second and third floor areas may vary slightly depending on setbacks or overhangs. The living area is the sum of the ground floor area and the two upper level floor areas.

Bi-level

This type of residence has two living areas, one above the other. One area is about 4 feet below grade and the second is about 4 feet above grade. Both areas are equal in size. The lower level in this type of residence is originally designed and built to serve as a living area and not as a basement. Both levels have full ceiling heights. The living area is the sum of the lower level area and the upper level area.

Tri-level

This type of residence has three levels of living area: one at grade level, one about 4 feet below grade, and one about 4 feet above grade. All levels are originally designed to serve as living areas. All levels have full ceiling heights. The living area is a sum of the areas of each of the three levels.

Garage Types

Attached Garage

Shares a common wall with the dwelling. Access is typically through a door between dwelling and garage.

Basement Garage

Constructed under the roof of the dwelling but below the living area.

Built-In Garage

Constructed under the second floor living space and above basement level of dwelling. Reduces gross square feet of living area.

Detached Garage

Constructed apart from the main dwelling. Shares no common area or wall with the dwelling.

Building Components

1. Excavation
2. Sill Plate
3. Basement Window
4. Floor Joist
5. Shoe Plate
6. Studs
7. Drywall
8. Plate
9. Ceiling Joists
10. Rafters
11. Collar Ties
12. Ridge Rafter
13. Roof Sheathing
14. Roof Felt
15. Roof Shingles
16. Flashing
17. Flue Lining
18. Chimney
19. Roof Shingles
20. Gutter
21. Fascia
22. Downspout
23. Shutter
24. Window
25. Wall Shingles
26. Building Paper
27. Wall Sheathing
28. Fire Stop
29. Dampproofing
30. Foundation Wall
31. Backfill
32. Drainage Stone
33. Drainage Tile
34. Wall Footing
35. Gravel
36. Concrete Slab
37. Column Footing
38. Pipe Column
39. Expansion Joint
40. Girder
41. Sub-floor
42. Finish Floor
43. Attic Insulation
44. Soffit
45. Ceiling Strapping
46. Wall Insulation
47. Cross Bridging
48. Bulkhead Stairs

Exterior Wall Construction

Typical Frame Construction

Typical wood frame construction consists of wood studs with insulation between them. A typical exterior surface is made up of sheathing, building paper and exterior siding consisting of wood, vinyl, aluminum or stucco over the wood sheathing.

Brick Veneer

Typical brick veneer construction consists of wood studs with insulation between them. A typical exterior surface is sheathing, building paper and an exterior of brick tied to the sheathing, with metal strips.

Stone

Typical solid masonry construction consists of a stone or block wall covered on the exterior with brick, stone or other masonry.

Residential Cost Estimate Worksheet

Worksheet Instructions

The residential cost estimate worksheet can be used as an outline for developing a residential construction or replacement cost. It is also useful for insurance appraisals. The design of the worksheet helps eliminate errors and omissions. To use the worksheet, follow the example below.

1. Fill out the owner's name, residence address, the estimator or appraiser's name, some type of project identifying number or code, and the date.
2. Determine from the plans, specifications, owner's description, photographs or any other means possible the class of construction. The models in this book use economy, average, custom and luxury as classes. Fill in the appropriate box.
3. Fill in the appropriate box for the residence type, configuration, occupancy, and exterior wall. If you require clarification, the pages preceding this worksheet describe each of these.
4. Next, the living area of the residence must be established. The heated or air conditioned space of the residence, not including the basement, should be measured. It is easiest to break the structure up into separate components as shown in the example. The main house (A), a one-and-one-half story wing (B), and a one story wing (C). The breezeway (D), garage (E), and open covered porch (F) will be treated differently. Data entry blocks for the living area are included on the worksheet for your use. Keep each level of each component separate, and fill out the blocks as shown.
5. By using the information on the worksheet, find the model, wing or ell in the following square foot cost pages that best matches the class, type, exterior finish and size of the residence being estimated. Use the *Modifications, Adjustments, and Alternatives* to determine the adjusted cost per square foot of living area for each component.
6. For each component, multiply the cost per square foot by the living area square footage. If the residence is a townhouse/rowhouse, a multiplier should be applied based upon the configuration.
7. The second page of the residential cost estimate worksheet has space for the additional components of a house. The cost for additional bathrooms, finished attic space, breezeways, porches, fireplaces, appliance or cabinet upgrades, and garages should be added on this page. The information for each of these components is found with the model being used, or in the *Modifications, Adjustments, and Alternatives* pages.
8. Add the total from page one of the estimate worksheet and the items listed on page two. The sum is the adjusted total building cost.
9. Depending on the use of the final estimated cost, one of the remaining two boxes should be filled out. Any additional items or exclusions should be added or subtracted at this time. The data contained in this book is a national average. Construction costs are different throughout the country. To allow for this difference, a location factor based upon the first three digits of the residence's zip code must be applied. The location factor is a multiplier that increases or decreases the adjusted total building cost. Find the appropriate location factor and calculate the local cost. If depreciation is a concern, a dollar figure should be subtracted at this point.
10. No residence will match a model exactly. Many differences will be found. At this level of estimating, a variation of plus or minus 10% should be expected.

Adjustments Instructions

No residence matches a model exactly in shape, material, or specifications.

The common differences are:

1. Two or more exterior wall systems:
    a. Partial basement
    b. Partly finished basement
2. Specifications or features that are between two classes
3. Crawl space instead of a basement

EXAMPLES

Below are quick examples. See pages 15-17 for complete examples of cost adjustments for these differences:

1. Residence "A" is an average one-story structure with 1,600 S.F. of living area and no basement. Three walls are wood siding on wood frame, and the fourth wall is brick veneer on wood frame. The brick veneer wall is 35% of the exterior wall area.

    Use page 28 to calculate the Base Cost per S.F. of Living Area.
    Wood Siding for 1,600 S.F. = $94.05 per S.F.
    Brick Veneer for 1,600 S.F. = $108.35 per S.F.
    .65 ($94.05) + .35 ($108.35) = $99.05 per S.F. of Living Area.

2. a. Residence "B" is the same as Residence "A"; However, it has an unfinished basement under 50% of the building. To adjust the $99.05 per S.F. of living area for this partial basement, use page 38.

    $99.05 + .50 ($9.50) = $103.80 per S.F. of Living Area.

b. Residence "C" is the same as Residence "A"; However, it has a full basement under the entire building. 640 S.F. or 40% of the basement area is finished.

Using Page 38:

$99.05 + .40 ($27.95) + .60 ($9.50) = $115.93 per S.F. of Living Area.

3. When specifications or features of a building are between classes, estimate the percent deviation, and use two tables to calculate the cost per S.F.

    A two-story residence with wood siding and 1,800 S.F. of living area has features 30% better than Average, but 70% less than Custom.

    From pages 30 and 42:

    Custom 1,800 S.F. Base Cost = $116.70 per S.F.
    Average 1,800 S.F. Base Cost = $95.55 per S.F.
    DIFFERENCE = $21.15 per S.F.

    Cost is $95.55 + .30 ($21.15) = $101.90 per S.F. of Living Area.

4. To add the cost of a crawl space, use the cost of an unfinished basement as a maximum. For specific costs of components to be added or deducted, such as vapor barrier, underdrain, and floor, see the "Assemblies" section, pages 229 to 450.

Model Residence Example

First Floor Plan

E

D

C

F

A

B

Second Floor Plan

A

B

A = Main House
B = 1-1/2 Story Wing
C = 1 Story Wing
D = Breezeway
E = Garage
F = Open Covered Porch

15

RESIDENTIAL COST ESTIMATE

OWNERS NAME: **Albert Westenberg**	APPRAISER:	**Nicole Wojtowicz**
RESIDENCE ADDRESS: **300 Sygiel Road**	PROJECT:	**# 55**
CITY, STATE, ZIP CODE: **Three Rivers, MA 01080**	DATE:	**Jan. 1, 2010**

CLASS OF CONSTRUCTION
- ☐ ECONOMY
- ☑ AVERAGE
- ☐ CUSTOM
- ☐ LUXURY

RESIDENCE TYPE
- ☐ 1 STORY
- ☐ 1 1/2 STORY
- ☑ 2 STORY
- ☐ 2 1/2 STORY
- ☐ 3 STORY
- ☐ BI-LEVEL
- ☐ TRI-LEVEL

CONFIGURATION
- ☑ DETACHED
- ☐ TOWN/ROW HOUSE
- ☐ SEMI-DETACHED

OCCUPANCY
- ☑ ONE FAMILY
- ☐ TWO FAMILY
- ☐ THREE FAMILY
- ☐ OTHER

EXTERIOR WALL SYSTEM
- ☑ WOOD SIDING - WOOD FRAME
- ☐ BRICK VENEER - WOOD FRAME
- ☐ STUCCO ON WOOD FRAME
- ☐ PAINTED CONCRETE BLOCK
- ☐ SOLID MASONRY (AVERAGE & CUSTOM)
- ☐ STONE VENEER - WOOD FRAME
- ☐ SOLID BRICK (LUXURY)
- ☐ SOLID STONE (LUXURY)

* LIVING AREA (Main Building)		
First Level	**1288**	S.F.
Second level	**1288**	S.F.
Third Level		S.F.
Total	**2576**	S.F.

* LIVING AREA (Wing or Ell)	(**B**)	
First Level	**360**	S.F.
Second level	**310**	S.F.
Third Level		S.F.
Total	**670**	S.F.

* LIVING AREA (Wing or Ell)	(**C**)	
First Level	**192**	S.F.
Second level		S.F.
Third Level		S.F.
Total	**192**	S.F.

* Basement Area is not part of living area.

MAIN BUILDING		COSTS PER S.F. LIVING AREA
Cost per Square Foot of Living Area, from Page **42**		$ 83.75
Basement Addition: _____ % Finished, **100** % Unfinished		+ 5.35
Roof Cover Adjustment: **Cedar Shake** Type, Page **42** (Add or Deduct)		( ) 1.95
Central Air Conditioning: ☐ Separate Ducts ☑ Heating Ducts, Page **42**		+ 2.54
Heating System Adjustment: _____ Type, Page _____ (Add or Deduct)		( ) _____
Main Building: Adjusted Cost per S.F. of Living Area		$ # 93.59

MAIN BUILDING TOTAL COST	$ **93.59** /S.F.	X	**2,576** S.F.	X	_____	=	$ **241,088**
	Cost per S.F. Living Area		Living Area		Town/Row House Multiplier (Use 1 for Detached)		TOTAL COST

WING OR ELL (B) **1 - 1/2** STORY		COSTS PER S.F. LIVING AREA
Cost per Square Foot of Living Area, from Page **56**		$ 78.20
Basement Addition: **100** % Finished, _____ % Unfinished		+ 24.75
Roof Cover Adjustment: _____ Type, Page _____ (Add or Deduct)		( ) —
Central Air Conditioning: ☐ Separate Ducts ☑ Heating Ducts, Page **42**		+ 2.54
Heating System Adjustment: _____ Type, Page _____ (Add or Deduct)		( ) _____
Wing or Ell (**B**): Adjusted Cost per S.F. of Living Area		$ 105.49

WING OR ELL (B) TOTAL COST	$ **105.49** /S.F.	X	**670** S.F.	=	$ **70,678**
	Cost per S.F. Living Area		Living Area		TOTAL COST

WING OR ELL (C) **1** STORY		COSTS PER S.F. LIVING AREA
Cost per Square Foot of Living Area, from Page # **56** (WOOD SIDING)		$ 118.75
Basement Addition: _____ % Finished, _____ % Unfinished		+ —
Roof Cover Adjustment: _____ Type, Page _____ (Add or Deduct)		( ) —
Central Air Conditioning: ☐ Separate Ducts ☐ Heating Ducts, Page _____		+ —
Heating System Adjustment: _____ Type, Page _____ (Add or Deduct)		( ) —
Wing or Ell (C) Adjusted Cost per S.F. of Living Area		$ 118.75

WING OR ELL (C) TOTAL COST	$ **118.75** /S.F.	X	**192** S.F.	=	$ **22,800**
	Cost per S.F. Living Area		Living Area		TOTAL COST

	TOTAL THIS PAGE	**334,566**

Page 1 of 2

RESIDENTIAL
COST ESTIMATE

					QUANTITY		UNIT COST			
Total Page 1									$	**334,566**
					QUANTITY		UNIT COST			
Additional Bathrooms:	**2**	Full,	**1**	Half	**2 @ 5993**	**1 @ 3570**			+	**15,556**
Finished Attic:	**N/A**	Ft. x		Ft.		S.F.				
Breezeway: ☑ Open ☐ closed		**12**	Ft. x	**12** Ft.	144	S.F.	23.52		+	**3,387**
Covered Porch: ☑ Open ☐ Enclosed		**18**	Ft. x	**12** Ft.	216	S.F.	34.85		+	**7,528**
Fireplace: ☑ Interior Chimney ☐ Exterior Chimney										
☑ No. of Flues (**2**) ☑ Additional Fireplaces			**1 - 2nd Story**						+	**9,853**
Appliances:									+	—
Kitchen Cabinets Adjustments:				(±)						—
☑ Garage ☐ Carport: **2** Car(s) Description **Wood, Attached**				(±)						**21,272**
Miscellaneous:									+	

ADJUSTED TOTAL BUILDING COST	$	**392,162**

REPLACEMENT COST		
ADJUSTED TOTAL BUILDING COST	$	**392,162**
Site Improvements		
(A) Paving & Sidewalks	$	
(B) Landscaping	$	
(C) Fences	$	
(D) Swimming Pools	$	
(E) Miscellaneous	$	
TOTAL	$	**392,162**
Location Factor	x	**1.040**
Location Replacement Cost	$	**407,848**
Depreciation	-$	**40,785**
LOCAL DEPRECIATED COST	$	**367,063**

INSURANCE COST		
ADJUSTED TOTAL BUILDING COST	$	
Insurance Exclusions		
(A) Footings, sitework, Underground Piping	-$	
(B) Architects Fees	-$	
Total Building Cost Less Exclusion	$	
Location Factor	x	
LOCAL INSURABLE REPLACEMENT COST	$	

SKETCH AND ADDITIONAL CALCULATIONS

1 Story

1-1/2 Story

2 Story

Bi-Level

Tri-Level

©Design Basics, Inc.

	Components
1 Site Work	Site preparation for slab or excavation for lower level; 4' deep trench excavation for foundation wall.
2 Foundations	Continuous reinforced concrete footing, 8" deep x 18" wide; dampproofed and insulated 8" thick reinforced concrete block foundation wall, 4' deep; 4" concrete slab on 4" crushed stone base and polyethylene vapor barrier, trowel finish.
3 Framing	Exterior walls—2 x 4 wood studs, 16" O.C.; 1/2" insulation board sheathing; wood truss roof 24" O.C. or 2 x 6 rafters 16" O.C. with 1/2" plywood sheathing, 4 in 12 or 8 in 12 roof pitch.
4 Exterior Walls	Beveled wood siding and #15 felt building paper on insulated wood frame walls; sliding sash or double hung wood windows; 2 flush solid core wood exterior doors. **Alternates:** • Brick veneer on wood frame, has 4" veneer of common brick. • Stucco on wood frame has 1" thick colored stucco finish. • Painted concrete block has 8" concrete block sealed and painted on exterior and furring on the interior for the drywall.
5 Roofing	20 year asphalt shingles; #15 felt building paper; aluminum gutters, downspouts, drip edge and flashings.
6 Interiors	Walls and ceilings—1/2" taped and finished drywall, primed and painted with 2 coats; painted baseboard and trim; rubber backed carpeting 80%, asphalt tile 20%; hollow core wood interior doors.
7 Specialties	Economy grade kitchen cabinets—6 L.F. wall and base with plastic laminate counter top and kitchen sink; 30 gallon electric water heater.
8 Mechanical	1 lavatory, white, wall hung; 1 water closet, white; 1 bathtub, enameled steel, white; gas fired warm air heat.
9 Electrical	100 Amp. service; romex wiring; incandescent lighting fixtures, switches, receptacles.
10 Overhead and Profit	General Contractor overhead and profit.

Adjustments

Unfinished Basement:
7' high 8" concrete block or cast-in-place concrete walls.

Finished Basement:
Includes inexpensive paneling or drywall on foundation walls. Inexpensive sponge backed carpeting on concrete floor, drywall ceiling, and lighting.

- **Mass produced from stock plans**
- **Single family — 1 full bath, 1 kitchen**
- **No basement**
- **Asphalt shingles on roof**
- **Hot air heat**
- **Gypsum wallboard interior finishes**
- **Materials and workmanship are sufficient to meet codes**
- **Detail specifications on page 19**

Note: The illustration shown may contain some optional components (for example: garages and/or fireplaces) whose costs are shown in the modifications, adjustments, & alternatives below or at the end of the square foot section.

©Home Planners, Inc.

Base cost per square foot of living area

Exterior Wall	Living Area										
	600	800	1000	1200	1400	1600	1800	2000	2400	2800	3200
Wood Siding - Wood Frame	113.20	102.25	94.05	87.30	81.50	77.80	75.95	73.45	68.45	64.80	62.30
Brick Veneer - Wood Frame	118.40	106.90	98.30	91.10	85.00	81.05	79.10	76.40	71.15	67.30	64.55
Stucco on Wood Frame	106.90	96.65	89.00	82.75	77.35	73.90	72.20	69.95	65.25	61.85	59.50
Painted Concrete Block	112.10	101.25	93.15	86.50	80.75	77.10	75.30	72.85	67.90	64.30	61.80
Finished Basement, Add	27.90	26.30	25.10	24.00	23.20	22.60	22.30	21.85	21.15	20.70	20.25
Unfinished Basement, Add	12.30	10.95	10.05	9.15	8.45	8.00	7.80	7.40	6.90	6.50	6.15

Modifications

Add to the total cost

Upgrade Kitchen Cabinets	$ + 803
Solid Surface Countertops	+ 576
Full Bath - including plumbing, wall and floor finishes	+ 4775
Half Bath - including plumbing, wall and floor finishes	+ 2845
One Car Attached Garage	+ 11,273
One Car Detached Garage	+ 14,630
Fireplace & Chimney	+ 5267

Adjustments

For multi family - add to total cost

Additional Kitchen	$ + 3709
Additional Bath	+ 4775
Additional Entry & Exit	+ 1640
Separate Heating	+ 1409
Separate Electric	+ 1066

For Townhouse/Rowhouse -
Multiply cost per square foot by

Inner Unit	+ .95
End Unit	+ .97

Alternatives

Add to or deduct from the cost per square foot of living area

Composition Roll Roofing	– 1.05
Cedar Shake Roof	+ 4.20
Upgrade Walls and Ceilings to Skim Coat Plaster	+ .65
Upgrade Ceilings to Textured Finish	+ .48
Air Conditioning, in Heating Ductwork	+ 4.05
In Separate Ductwork	+ 7.72
Heating Systems, Hot Water	+ 2.11
Heat Pump	+ 2.56
Electric Heat	– 1.81
Not Heated	– 4.56

Additional upgrades or components

Kitchen Cabinets & Countertops	Page 58
Bathroom Vanities	59
Fireplaces & Chimneys	59
Windows, Skylights & Dormers	59
Appliances	60
Breezeways & Porches	60
Finished Attic	60
Garages	61
Site Improvements	61
Wings & Ells	25

- **Mass produced from stock plans**
- **Single family — 1 full bath, 1 kitchen**
- **No basement**
- **Asphalt shingles on roof**
- **Hot air heat**
- **Gypsum wallboard interior finishes**
- **Materials and workmanship are sufficient to meet codes**
- **Detail specifications on page 19**

Note: The illustration shown may contain some optional components (for example: garages and/or fireplaces) whose costs are shown in the modifications, adjustments, & alternatives below or at the end of the square foot section.

Base cost per square foot of living area

Exterior Wall	Living Area										
	600	800	1000	1200	1400	1600	1800	2000	2400	2800	3200
Wood Siding - Wood Frame	129.30	107.15	95.65	90.25	86.40	80.55	77.65	74.60	68.45	66.00	63.50
Brick Veneer - Wood Frame	136.55	112.45	100.60	94.85	90.80	84.50	81.40	78.15	71.60	69.00	66.25
Stucco on Wood Frame	120.50	100.80	89.80	84.80	81.15	75.80	73.10	70.30	64.65	62.45	60.20
Painted Concrete Block	127.75	106.05	94.65	89.30	85.45	79.70	76.85	73.85	67.75	65.40	62.90
Finished Basement, Add	21.55	18.25	17.45	16.80	16.35	15.70	15.35	15.00	14.30	14.00	13.65
Unfinished Basement, Add	10.90	8.30	7.65	7.15	6.80	6.25	6.00	5.70	5.15	4.95	4.60

Modifications

Add to the total cost

Upgrade Kitchen Cabinets	$ + 803
Solid Surface Countertops	+ 576
Full Bath - including plumbing, wall and floor finishes	+ 4775
Half Bath - including plumbing, wall and floor finishes	+ 2845
One Car Attached Garage	+ 11,273
One Car Detached Garage	+ 14,630
Fireplace & Chimney	+ 5267

Adjustments

For multi family - add to total cost

Additional Kitchen	$ + 3709
Additional Bath	+ 4775
Additional Entry & Exit	+ 1640
Separate Heating	+ 1409
Separate Electric	+ 1066

For Townhouse/Rowhouse -
Multiply cost per square foot by

Inner Unit	+ .95
End Unit	+ .97

Alternatives

Add to or deduct from the cost per square foot of living area

Composition Roll Roofing	– .70
Cedar Shake Roof	+ 3.05
Upgrade Walls and Ceilings to Skim Coat Plaster	+ .66
Upgrade Ceilings to Textured Finish	+ .48
Air Conditioning, in Heating Ductwork	+ 3.03
In Separate Ductwork	+ 6.79
Heating Systems, Hot Water	+ 2.01
Heat Pump	+ 2.82
Electric Heat	– 1.44
Not Heated	– 4.20

Additional upgrades or components

Kitchen Cabinets & Countertops	Page 58
Bathroom Vanities	59
Fireplaces & Chimneys	59
Windows, Skylights & Dormers	59
Appliances	60
Breezeways & Porches	60
Finished Attic	60
Garages	61
Site Improvements	61
Wings & Ells	25

- **Mass produced from stock plans**
- **Single family — 1 full bath, 1 kitchen**
- **No basement**
- **Asphalt shingles on roof**
- **Hot air heat**
- **Gypsum wallboard interior finishes**
- **Materials and workmanship are sufficient to meet codes**
- **Detail specifications on page 19**

Note: The illustration shown may contain some optional components (for example: garages and/or fireplaces) whose costs are shown in the modifications, adjustments, & alternatives below or at the end of the square foot section.

Base cost per square foot of living area

Exterior Wall	Living Area										
	1000	1200	1400	1600	1800	2000	2200	2600	3000	3400	3800
Wood Siding - Wood Frame	101.65	91.80	87.20	84.10	80.90	77.30	74.90	70.40	66.05	64.05	62.30
Brick Veneer - Wood Frame	107.30	97.00	92.10	88.70	85.30	81.55	78.95	74.05	69.45	67.20	65.30
Stucco on Wood Frame	94.95	85.55	81.35	78.50	75.65	72.25	70.05	66.05	62.05	60.25	58.65
Painted Concrete Block	100.45	90.75	86.20	83.10	80.00	76.40	74.05	69.65	65.35	63.35	61.65
Finished Basement, Add	14.65	14.05	13.50	13.20	12.80	12.60	12.35	11.85	11.50	11.25	11.05
Unfinished Basement, Add	6.70	6.15	5.75	5.50	5.25	5.05	4.85	4.45	4.15	4.00	3.80

Modifications

Add to the total cost

Upgrade Kitchen Cabinets	$ + 803
Solid Surface Countertops	+ 576
Full Bath - including plumbing, wall and floor finishes	+ 4775
Half Bath - including plumbing, wall and floor finishes	+ 2845
One Car Attached Garage	+ 11,273
One Car Detached Garage	+ 14,630
Fireplace & Chimney	+ 5820

Adjustments

For multi family - add to total cost

Additional Kitchen	$ + 3709
Additional Bath	+ 4775
Additional Entry & Exit	+ 1640
Separate Heating	+ 1409
Separate Electric	+ 1066

For Townhouse/Rowhouse -
Multiply cost per square foot by

Inner Unit	+ .93
End Unit	+ .96

Alternatives

Add to or deduct from the cost per square foot of living area

Composition Roll Roofing	– .50
Cedar Shake Roof	+ 2.10
Upgrade Walls and Ceilings to Skim Coat Plaster	+ .67
Upgrade Ceilings to Textured Finish	+ .48
Air Conditioning, in Heating Ductwork	+ 2.45
In Separate Ductwork	+ 6.21
Heating Systems, Hot Water	+ 1.96
Heat Pump	+ 3
Electric Heat	– 1.26
Not Heated	– 3.96

Additional upgrades or components

Kitchen Cabinets & Countertops	Page 58
Bathroom Vanities	59
Fireplaces & Chimneys	59
Windows, Skylights & Dormers	59
Appliances	60
Breezeways & Porches	60
Finished Attic	60
Garages	61
Site Improvements	61
Wings & Ells	25

- **Mass produced from stock plans**
- **Single family — 1 full bath, 1 kitchen**
- **No basement**
- **Asphalt shingles on roof**
- **Hot air heat**
- **Gypsum wallboard interior finishes**
- **Materials and workmanship are sufficient to meet codes**
- **Detail specifications on page 19**

Note: The illustration shown may contain some optional components (for example: garages and/or fireplaces) whose costs are shown in the modifications, adjustments, & alternatives below or at the end of the square foot section.

Base cost per square foot of living area

Exterior Wall	Living Area										
	1000	1200	1400	1600	1800	2000	2200	2600	3000	3400	3800
Wood Siding - Wood Frame	94.20	84.90	80.75	77.85	75.10	71.70	69.60	65.55	61.60	59.80	58.25
Brick Veneer - Wood Frame	98.40	88.80	84.40	81.40	78.35	74.85	72.60	68.25	64.10	62.20	60.50
Stucco on Wood Frame	89.15	80.15	76.35	73.70	71.15	67.85	65.95	62.30	58.55	56.90	55.50
Painted Concrete Block	93.30	84.05	79.95	77.15	74.40	71.00	68.95	65.00	61.10	59.30	57.75
Finished Basement, Add	14.65	14.05	13.50	13.20	12.80	12.60	12.35	11.85	11.50	11.25	11.05
Unfinished Basement, Add	6.70	6.15	5.75	5.50	5.25	5.05	4.85	4.45	4.15	4.00	3.80

Modifications

Add to the total cost

Upgrade Kitchen Cabinets	$ + 803
Solid Surface Countertops	+ 576
Full Bath - including plumbing, wall and floor finishes	+ 4775
Half Bath - including plumbing, wall and floor finishes	+ 2845
One Car Attached Garage	+ 11,273
One Car Detached Garage	+ 14,630
Fireplace & Chimney	+ 5267

Adjustments

For multi family - add to total cost

Additional Kitchen	$ + 3709
Additional Bath	+ 4775
Additional Entry & Exit	+ 1640
Separate Heating	+ 1409
Separate Electric	+ 1066

For Townhouse/Rowhouse -
Multiply cost per square foot by

Inner Unit	+ .94
End Unit	+ .97

Alternatives

Add to or deduct from the cost per square foot of living area

Composition Roll Roofing	– .50
Cedar Shake Roof	+ 2.10
Upgrade Walls and Ceilings to Skim Coat Plaster	+ .63
Upgrade Ceilings to Textured Finish	+ .48
Air Conditioning, in Heating Ductwork	+ 2.45
In Separate Ductwork	+ 6.21
Heating Systems, Hot Water	+ 1.96
Heat Pump	+ 3
Electric Heat	– 1.26
Not Heated	– 3.96

Additional upgrades or components

Kitchen Cabinets & Countertops	Page 58
Bathroom Vanities	59
Fireplaces & Chimneys	59
Windows, Skylights & Dormers	59
Appliances	60
Breezeways & Porches	60
Finished Attic	60
Garages	61
Site Improvements	61
Wings & Ells	25

Important: See the Reference Section for Location Factors (to adjust for your city) and Estimating Forms

- **Mass produced from stock plans**
- **Single family — 1 full bath, 1 kitchen**
- **No basement**
- **Asphalt shingles on roof**
- **Hot air heat**
- **Gypsum wallboard interior finishes**
- **Materials and workmanship are sufficient to meet codes**
- **Detail specifications on page 19**

Note: The illustration shown may contain some optional components (for example: garages and/or fireplaces) whose costs are shown in the modifications, adjustments, & alternatives below or at the end of the square foot section.

©Design Basics, Inc.

Base cost per square foot of living area

Exterior Wall	Living Area										
	1200	1500	1800	2000	2200	2400	2800	3200	3600	4000	4400
Wood Siding - Wood Frame	86.90	79.75	74.35	72.15	69.00	66.45	64.50	61.80	58.55	57.45	55.00
Brick Veneer - Wood Frame	90.75	83.25	77.50	75.15	71.90	69.15	67.10	64.20	60.80	59.65	57.05
Stucco on Wood Frame	82.30	75.50	70.50	68.50	65.50	63.20	61.35	58.85	55.90	54.85	52.55
Solid Masonry	86.10	78.95	73.65	71.50	68.35	65.90	63.95	61.20	58.05	57.05	54.60
Finished Basement, Add*	17.50	16.70	16.00	15.70	15.40	15.05	14.80	14.45	14.10	13.95	13.70
Unfinished Basement, Add*	7.30	6.70	6.15	5.85	5.60	5.35	5.15	4.85	4.60	4.50	4.30

*Basement under middle level only.

Modifications

Add to the total cost

Upgrade Kitchen Cabinets	$ + 803
Solid Surface Countertops	+ 576
Full Bath - including plumbing, wall and floor finishes	+ 4775
Half Bath - including plumbing, wall and floor finishes	+ 2845
One Car Attached Garage	+ 11,273
One Car Detached Garage	+ 14,630
Fireplace & Chimney	+ 5267

Adjustments

For multi family - add to total cost

Additional Kitchen	$ + 3709
Additional Bath	+ 4775
Additional Entry & Exit	+ 1640
Separate Heating	+ 1409
Separate Electric	+ 1066

For Townhouse/Rowhouse -
Multiply cost per square foot by

Inner Unit	+ .93
End Unit	+ .96

Alternatives

Add to or deduct from the cost per square foot of living area

Composition Roll Roofing	– .70
Cedar Shake Roof	+ 3.05
Upgrade Walls and Ceilings to Skim Coat Plaster	+ .57
Upgrade Ceilings to Textured Finish	+ .48
Air Conditioning, in Heating Ductwork	+ 2.10
In Separate Ductwork	+ 5.78
Heating Systems, Hot Water	+ 1.88
Heat Pump	+ 3.11
Electric Heat	– 1.09
Not Heated	– 3.84

Additional upgrades or components

Kitchen Cabinets & Countertops	Page 58
Bathroom Vanities	59
Fireplaces & Chimneys	59
Windows, Skylights & Dormers	59
Appliances	60
Breezeways & Porches	60
Finished Attic	60
Garages	61
Site Improvements	61
Wings & Ells	25

1 Story — Base cost per square foot of living area

Exterior Wall	Living Area							
	50	100	200	300	400	500	600	700
Wood Siding - Wood Frame	157.10	119.20	102.75	84.95	79.60	76.35	74.25	74.80
Brick Veneer - Wood Frame	170.25	128.60	110.65	90.15	84.30	80.75	78.40	78.85
Stucco on Wood Frame	141.35	107.95	93.40	78.70	74.00	71.10	69.15	69.95
Painted Concrete Block	154.35	117.20	101.15	83.90	78.60	75.40	73.35	73.90
Finished Basement, Add	42.80	34.90	31.55	26.05	24.95	24.30	23.85	23.55
Unfinished Basement, Add	23.70	17.55	14.95	10.70	9.80	9.30	8.95	8.70

1-1/2 Story — Base cost per square foot of living area

Exterior Wall	Living Area							
	100	200	300	400	500	600	700	800
Wood Siding - Wood Frame	124.70	99.95	84.90	75.60	71.05	68.85	66.10	65.30
Brick Veneer - Wood Frame	136.45	109.35	92.75	81.70	76.70	74.15	71.05	70.25
Stucco on Wood Frame	110.60	88.70	75.50	68.25	64.35	62.50	60.05	59.30
Painted Concrete Block	122.15	98.00	83.20	74.35	69.90	67.75	64.95	64.20
Finished Basement, Add	28.55	25.25	23.00	20.60	19.95	19.50	19.10	19.05
Unfinished Basement, Add	14.50	11.95	10.25	8.35	7.80	7.50	7.15	7.15

2 Story — Base cost per square foot of living area

Exterior Wall	Living Area							
	100	200	400	600	800	1000	1200	1400
Wood Siding - Wood Frame	127.10	94.10	79.70	65.30	60.65	57.75	55.90	56.70
Brick Veneer - Wood Frame	140.25	103.50	87.55	70.55	65.35	62.20	60.10	60.75
Stucco on Wood Frame	111.35	82.85	70.35	59.05	55.00	52.55	50.90	51.85
Painted Concrete Block	124.35	92.10	78.10	64.20	59.65	56.90	55.05	55.85
Finished Basement, Add	21.45	17.45	15.85	13.10	12.55	12.20	12.00	11.80
Unfinished Basement, Add	11.90	8.80	7.50	5.35	4.95	4.70	4.45	4.35

Base costs do not include bathroom or kitchen facilities. Use Modifications/Adjustments/Alternatives on pages 58-61 where appropriate.

1 Story

1-1/2 Story

2 Story

2-1/2 Story

Bi-Level

Tri-Level

Components

1 Site Work

Site preparation for slab or excavation for lower level; 4' deep trench excavation for foundation wall.

2 Foundations

Continuous reinforced concrete footing, 8" deep x 18" wide; dampproofed and insulated 8" thick reinforced concrete foundation wall, 4' deep; 4" concrete slab on 4" crushed stone base and polyethylene vapor barrier, trowel finish.

3 Framing

Exterior walls—2 x 4 wood studs, 16" O.C.; 1/2" plywood sheathing; 2 x 6 rafters 16" O.C. with 1/2" plywood sheathing, 4 in 12 or 8 in 12 roof pitch, 2 x 6 ceiling joists; 1/2" plywood subfloor on 1 x 2 wood sleepers 16" O.C.; 2 x 8 floor joists with 5/8" plywood subfloor on models with more than one level.

4 Exterior Walls

Beveled wood siding and #15 felt building paper on insulated wood frame walls; double hung windows; 3 flush solid core wood exterior doors with storms.
Alternates:
• Brick veneer on wood frame, has 4" veneer of common brick.
• Stucco on wood frame has 1" thick colored stucco finish.
• Solid masonry has a 6" concrete block load bearing wall with insulation and a brick or stone exterior. Also solid brick or stone walls.

5 Roofing

25 year asphalt shingles; #15 felt building paper; aluminum gutters, downspouts, drip edge and flashings.

6 Interiors

Walls and ceilings—1/2" taped and finished drywall, primed and painted with 2 coats; painted baseboard and trim, finished hardwood floor 40%, carpet with 1/2" underlayment 40%, vinyl tile with 1/2" underlayment 15%, ceramic tile with 1/2" underlayment 5%; hollow core and louvered interior doors.

7 Specialties

Average grade kitchen cabinets—14 L.F. wall and base with plastic laminate counter top and kitchen sink; 40 gallon electric water heater.

8 Mechanical

1 lavatory, white, wall hung; 1 water closet, white; 1 bathtub, enameled steel, white; gas fired warm air heat.

9 Electrical

100 Amp. service; romex wiring; incandescent lighting fixtures, switches, receptacles.

10 Overhead and Profit

General Contractor overhead and profit.

Adjustments

Unfinished Basement:
7'-6" high cast-in-place concrete walls 8" thick or 8" concrete block.

Finished Basement:
Includes plywood paneling or drywall on furring on foundation walls, sponge backed carpeting on concrete floor, suspended ceiling, lighting and heating.

- **Simple design from standard plans**
- **Single family — 1 full bath, 1 kitchen**
- **No basement**
- **Asphalt shingles on roof**
- **Hot air heat**
- **Gypsum wallboard interior finishes**
- **Materials and workmanship are average**
- **Detail specifications on p. 27**

Note: The illustration shown may contain some optional components (for example: garages and/or fireplaces) whose costs are shown in the modifications, adjustments, & alternatives below or at the end of the square foot section.

●Home Planners, Inc.

Base cost per square foot of living area

Exterior Wall	Living Area										
	600	800	1000	1200	1400	1600	1800	2000	2400	2800	3200
Wood Siding - Wood Frame	135.60	122.45	112.70	104.85	98.30	94.05	91.75	88.85	83.15	78.95	76.10
Brick Veneer - Wood Frame	152.80	138.85	128.50	120.05	112.95	108.35	105.95	102.75	96.70	92.20	89.10
Stucco on Wood Frame	141.80	129.05	119.55	112.00	105.65	101.50	99.35	96.50	91.05	86.95	84.20
Solid Masonry	166.85	151.40	139.85	130.30	122.25	117.10	114.35	110.65	103.95	98.90	95.30
Finished Basement, Add	33.95	32.80	31.25	29.85	28.75	27.95	27.60	27.00	26.10	25.50	24.90
Unfinished Basement, Add	13.90	12.50	11.55	10.65	9.95	9.50	9.20	8.85	8.30	7.90	7.60

Modifications

Add to the total cost

Upgrade Kitchen Cabinets	$ + 3521
Solid Surface Countertops (Included)	
Full Bath - including plumbing, wall and floor finishes	+ 5993
Half Bath - including plumbing, wall and floor finishes	+ 3570
One Car Attached Garage	+ 12,676
One Car Detached Garage	+ 16,871
Fireplace & Chimney	+ 5234

Adjustments

For multi family - add to total cost

Additional Kitchen	$ + 6796
Additional Bath	+ 5993
Additional Entry & Exit	+ 1640
Separate Heating	+ 1409
Separate Electric	+ 1915

For Townhouse/Rowhouse -
Multiply cost per square foot by

Inner Unit	+ .92
End Unit	+ .96

Alternatives

Add to or deduct from the cost per square foot of living area

Cedar Shake Roof	+ 3.85
Clay Tile Roof	+ 4.55
Slate Roof	+ 6.65
Upgrade Walls to Skim Coat Plaster	+ .41
Upgrade Ceilings to Textured Finish	+ .48
Air Conditioning, in Heating Ductwork	+ 4.19
In Separate Ductwork	+ 8.02
Heating Systems, Hot Water	+ 2.16
Heat Pump	+ 2.66
Electric Heat	– 1.01
Not Heated	– 3.80

Additional upgrades or components

Kitchen Cabinets & Countertops	Page 58
Bathroom Vanities	59
Fireplaces & Chimneys	59
Windows, Skylights & Dormers	59
Appliances	60
Breezeways & Porches	60
Finished Attic	60
Garages	61
Site Improvements	61
Wings & Ells	37

- **Simple design from standard plans**
- **Single family — 1 full bath, 1 kitchen**
- **No basement**
- **Asphalt shingles on roof**
- **Hot air heat**
- **Gypsum wallboard interior finishes**
- **Materials and workmanship are average**
- **Detail specifications on p. 27**

Note: The illustration shown may contain some optional components (for example: garages and/or fireplaces) whose costs are shown in the modifications, adjustments, & alternatives below or at the end of the square foot section.

©By Designer

Base cost per square foot of living area

Exterior Wall	Living Area										
	600	800	1000	1200	1400	1600	1800	2000	2400	2800	3200
Wood Siding - Wood Frame	151.70	126.70	113.30	106.95	102.40	95.75	92.40	88.90	82.05	79.25	76.35
Brick Veneer - Wood Frame	162.30	134.35	120.45	113.60	108.75	101.55	97.85	94.10	86.60	83.60	80.35
Stucco on Wood Frame	146.95	123.20	110.15	103.95	99.55	93.15	89.95	86.60	79.95	77.30	74.50
Solid Masonry	181.90	148.55	133.55	125.90	120.45	112.20	107.95	103.70	95.10	91.65	87.80
Finished Basement, Add	27.80	24.25	23.20	22.40	21.85	20.95	20.45	20.10	19.15	18.80	18.30
Unfinished Basement, Add	12.10	9.45	8.70	8.20	7.85	7.30	7.00	6.75	6.20	5.90	5.65

Modifications

Add to the total cost

Upgrade Kitchen Cabinets	$ + 3521
Solid Surface Countertops (Included)	
Full Bath - including plumbing, wall and floor finishes	+ 5993
Half Bath - including plumbing, wall and floor finishes	+ 3570
One Car Attached Garage	+ 12,676
One Car Detached Garage	+ 16,871
Fireplace & Chimney	+ 5234

Adjustments

For multi family - add to total cost

Additional Kitchen	$ + 6796
Additional Bath	+ 5993
Additional Entry & Exit	+ 1640
Separate Heating	+ 1409
Separate Electric	+ 1915

For Townhouse/Rowhouse -
Multiply cost per square foot by

Inner Unit	+ .92
End Unit	+ .96

Alternatives

Add to or deduct from the cost per square foot of living area

Cedar Shake Roof	+ 2.80
Clay Tile Roof	+ 3.30
Slate Roof	+ 4.80
Upgrade Walls to Skim Coat Plaster	+ .47
Upgrade Ceilings to Textured Finish	+ .48
Air Conditioning, in Heating Ductwork	+ 3.19
In Separate Ductwork	+ 7.01
Heating Systems, Hot Water	+ 2.05
Heat Pump	+ 2.96
Electric Heat	- .91
Not Heated	- 3.67

Additional upgrades or components

Kitchen Cabinets & Countertops	Page 58
Bathroom Vanities	59
Fireplaces & Chimneys	59
Windows, Skylights & Dormers	59
Appliances	60
Breezeways & Porches	60
Finished Attic	60
Garages	61
Site Improvements	61
Wings & Ells	37

Important: See the Reference Section for Location Factors (to adjust for your city) and Estimating Forms

- Simple design from standard plans
- Single family — 1 full bath, 1 kitchen
- No basement
- Asphalt shingles on roof
- Hot air heat
- Gypsum wallboard interior finishes
- Materials and workmanship are average
- Detail specifications on p. 27

Note: The illustration shown may contain some optional components (for example: garages and/or fireplaces) whose costs are shown in the modifications, adjustments, & alternatives below or at the end of the square foot section.

Base cost per square foot of living area

Exterior Wall	Living Area										
	1000	1200	1400	1600	1800	2000	2200	2600	3000	3400	3800
Wood Siding - Wood Frame	119.55	108.20	102.90	99.30	95.55	91.50	88.85	83.75	78.85	76.55	74.50
Brick Veneer - Wood Frame	127.75	115.85	109.95	106.10	101.85	97.65	94.70	89.00	83.70	81.20	78.95
Stucco on Wood Frame	115.90	104.85	99.75	96.30	92.70	88.70	86.25	81.30	76.60	74.45	72.55
Solid Masonry	143.00	129.90	123.15	118.60	113.65	109.00	105.45	98.80	92.75	89.75	87.10
Finished Basement, Add	19.70	19.40	18.75	18.35	17.90	17.65	17.25	16.60	16.15	15.90	15.65
Unfinished Basement, Add	7.65	7.10	6.70	6.40	6.10	5.90	5.70	5.35	5.05	4.80	4.70

Modifications

Add to the total cost

Upgrade Kitchen Cabinets	$ + 3521
Solid Surface Countertops (Included)	
Full Bath - including plumbing, wall and floor finishes	+ 5993
Half Bath - including plumbing, wall and floor finishes	+ 3570
One Car Attached Garage	+ 12,676
One Car Detached Garage	+ 16,871
Fireplace & Chimney	+ 5836

Adjustments

For multi family - add to total cost

Additional Kitchen	$ + 6796
Additional Bath	+ 5993
Additional Entry & Exit	+ 1640
Separate Heating	+ 1409
Separate Electric	+ 1915

For Townhouse/Rowhouse -
Multiply cost per square foot by

Inner Unit	+ .90
End Unit	+ .95

Alternatives

Add to or deduct from the cost per square foot of living area

Cedar Shake Roof	+ 1.95
Clay Tile Roof	+ 2.25
Slate Roof	+ 3.30
Upgrade Walls to Skim Coat Plaster	+ .49
Upgrade Ceilings to Textured Finish	+ .48
Air Conditioning, in Heating Ductwork	+ 2.54
In Separate Ductwork	+ 6.36
Heating Systems, Hot Water	+ 2.01
Heat Pump	+ 3.11
Electric Heat	− .80
Not Heated	− 3.55

Additional upgrades or components

Kitchen Cabinets & Countertops	Page 58
Bathroom Vanities	59
Fireplaces & Chimneys	59
Windows, Skylights & Dormers	59
Appliances	60
Breezeways & Porches	60
Finished Attic	60
Garages	61
Site Improvements	61
Wings & Ells	37

- **Simple design from standard plans**
- **Single family — 1 full bath, 1 kitchen**
- **No basement**
- **Asphalt shingles on roof**
- **Hot air heat**
- **Gypsum wallboard interior finishes**
- **Materials and workmanship are average**
- **Detail specifications on p. 27**

Note: The illustration shown may contain some optional components (for example: garages and/or fireplaces) whose costs are shown in the modifications, adjustments, & alternatives below or at the end of the square foot section.

Base cost per square foot of living area

Exterior Wall	Living Area										
	1200	1400	1600	1800	2000	2400	2800	3200	3600	4000	4400
Wood Siding - Wood Frame	118.55	111.25	101.50	99.60	95.95	90.20	85.75	81.10	78.80	74.55	73.25
Brick Veneer - Wood Frame	127.20	119.10	108.75	106.90	102.80	96.40	91.65	86.50	83.85	79.25	77.85
Stucco on Wood Frame	114.60	107.75	98.25	96.35	92.90	87.45	83.10	78.75	76.45	72.45	71.20
Solid Masonry	143.45	133.70	122.15	120.35	115.45	107.85	102.55	96.35	93.20	88.05	86.25
Finished Basement, Add	16.70	16.30	15.75	15.70	15.25	14.65	14.35	13.85	13.60	13.30	13.15
Unfinished Basement, Add	6.35	5.80	5.50	5.40	5.15	4.75	4.55	4.25	4.10	3.90	3.85

Modifications

Add to the total cost

Upgrade Kitchen Cabinets	$ + 3521
Solid Surface Countertops (Included)	
Full Bath - including plumbing, wall and floor finishes	+ 5993
Half Bath - including plumbing, wall and floor finishes	+ 3570
One Car Attached Garage	+ 12,676
One Car Detached Garage	+ 16,871
Fireplace & Chimney	+ 6634

Adjustments

For multi family - add to total cost

Additional Kitchen	$ + 6796
Additional Bath	+ 5993
Additional Entry & Exit	+ 1640
Separate Heating	+ 1409
Separate Electric	+ 1915

For Townhouse/Rowhouse -
Multiply cost per square foot by

Inner Unit	+ .90
End Unit	+ .95

Alternatives

Add to or deduct from the cost per square foot of living area

Cedar Shake Roof	+ 1.70
Clay Tile Roof	+ 1.95
Slate Roof	+ 2.90
Upgrade Walls to Skim Coat Plaster	+ .47
Upgrade Ceilings to Textured Finish	+ .48
Air Conditioning, in Heating Ductwork	+ 2.31
In Separate Ductwork	+ 6.14
Heating Systems, Hot Water	+ 1.81
Heat Pump	+ 3.19
Electric Heat	– 1.40
Not Heated	– 4.15

Additional upgrades or components

Kitchen Cabinets & Countertops	Page 58
Bathroom Vanities	59
Fireplaces & Chimneys	59
Windows, Skylights & Dormers	59
Appliances	60
Breezeways & Porches	60
Finished Attic	60
Garages	61
Site Improvements	61
Wings & Ells	37

- **Simple design from standard plans**
- **Single family — 1 full bath, 1 kitchen**
- **No basement**
- **Asphalt shingles on roof**
- **Hot air heat**
- **Gypsum wallboard interior finishes**
- **Materials and workmanship are average**
- **Detail specifications on p. 27**

Note: The illustration shown may contain some optional components (for example: garages and/or fireplaces) whose costs are shown in the modifications, adjustments, & alternatives below or at the end of the square foot section.

Base cost per square foot of living area

Exterior Wall	Living Area										
	1500	1800	2100	2500	3000	3500	4000	4500	5000	5500	6000
Wood Siding - Wood Frame	109.85	99.65	95.10	91.60	85.00	82.05	77.95	73.60	72.25	70.55	68.85
Brick Veneer - Wood Frame	118.00	107.30	102.20	98.40	91.15	87.90	83.25	78.50	77.00	75.10	73.15
Stucco on Wood Frame	106.20	96.30	91.95	88.60	82.30	79.45	75.55	71.40	70.10	68.50	67.00
Solid Masonry	133.25	121.30	115.35	110.90	102.55	98.65	93.05	87.55	85.70	83.55	81.10
Finished Basement, Add	14.35	14.20	13.80	13.50	13.05	12.70	12.40	12.00	11.95	11.75	11.60
Unfinished Basement, Add	5.15	4.80	4.55	4.35	4.05	3.85	3.60	3.40	3.30	3.25	3.15

Modifications

Add to the total cost

Upgrade Kitchen Cabinets	$ + 3521
Solid Surface Countertops (Included)	
Full Bath - including plumbing, wall and floor finishes	+ 5993
Half Bath - including plumbing, wall and floor finishes	+ 3570
One Car Attached Garage	+ 12,676
One Car Detached Garage	+ 16,871
Fireplace & Chimney	+ 6634

Adjustments

For multi family - add to total cost

Additional Kitchen	$ + 6796
Additional Bath	+ 5993
Additional Entry & Exit	+ 1640
Separate Heating	+ 1409
Separate Electric	+ 1915

For Townhouse/Rowhouse -
Multiply cost per square foot by

Inner Unit	+ .88
End Unit	+ .94

Alternatives

Add to or deduct from the cost per square foot of living area

Cedar Shake Roof	+ 1.30
Clay Tile Roof	+ 1.50
Slate Roof	+ 2.20
Upgrade Walls to Skim Coat Plaster	+ .49
Upgrade Ceilings to Textured Finish	+ .48
Air Conditioning, in Heating Ductwork	+ 2.31
In Separate Ductwork	+ 6.14
Heating Systems, Hot Water	+ 1.81
Heat Pump	+ 3.19
Electric Heat	− 1.09
Not Heated	− 3.85

Additional upgrades or components

Kitchen Cabinets & Countertops	Page 58
Bathroom Vanities	59
Fireplaces & Chimneys	59
Windows, Skylights & Dormers	59
Appliances	60
Breezeways & Porches	60
Finished Attic	60
Garages	61
Site Improvements	61
Wings & Ells	37

- **Simple design from standard plans**
- **Single family — 1 full bath, 1 kitchen**
- **No basement**
- **Asphalt shingles on roof**
- **Hot air heat**
- **Gypsum wallboard interior finishes**
- **Materials and workmanship are average**
- **Detail specifications on p. 27**

Note: The illustration shown may contain some optional components (for example: garages and/or fireplaces) whose costs are shown in the modifications, adjustments, & alternatives below or at the end of the square foot section.

Base cost per square foot of living area

Exterior Wall	Living Area										
	1000	1200	1400	1600	1800	2000	2200	2600	3000	3400	3800
Wood Siding - Wood Frame	111.50	100.75	95.90	92.60	89.25	85.40	83.10	78.45	74.00	72.00	70.10
Brick Veneer - Wood Frame	117.65	106.45	101.25	97.70	94.00	90.00	87.50	82.45	77.70	75.45	73.45
Stucco on Wood Frame	108.75	98.20	93.50	90.30	87.05	83.35	81.15	76.75	72.35	70.40	68.70
Solid Masonry	129.05	117.00	111.05	107.10	102.80	98.50	95.60	89.80	84.45	81.85	79.60
Finished Basement, Add	19.70	19.40	18.75	18.35	17.90	17.65	17.25	16.60	16.15	15.90	15.65
Unfinished Basement, Add	7.65	7.10	6.70	6.40	6.10	5.90	5.70	5.35	5.05	4.80	4.70

Modifications

Add to the total cost

Upgrade Kitchen Cabinets	$ + 3521
Solid Surface Countertops (Included)	
Full Bath - including plumbing, wall and floor finishes	+ 5993
Half Bath - including plumbing, wall and floor finishes	+ 3570
One Car Attached Garage	+ 12,676
One Car Detached Garage	+ 16,871
Fireplace & Chimney	+ 5234

Adjustments

For multi family - add to total cost

Additional Kitchen	$ + 6796
Additional Bath	+ 5993
Additional Entry & Exit	+ 1640
Separate Heating	+ 1409
Separate Electric	+ 1915

For Townhouse/Rowhouse -
Multiply cost per square foot by

Inner Unit	+ .91
End Unit	+ .96

Alternatives

Add to or deduct from the cost per square foot of living area

Cedar Shake Roof	+ 1.95
Clay Tile Roof	+ 2.25
Slate Roof	+ 3.30
Upgrade Walls to Skim Coat Plaster	+ .45
Upgrade Ceilings to Textured Finish	+ .48
Air Conditioning, in Heating Ductwork	+ 2.54
In Separate Ductwork	+ 6.36
Heating Systems, Hot Water	+ 2.01
Heat Pump	+ 3.11
Electric Heat	– .80
Not Heated	– 3.55

Additional upgrades or components

Kitchen Cabinets & Countertops	Page 58
Bathroom Vanities	59
Fireplaces & Chimneys	59
Windows, Skylights & Dormers	59
Appliances	60
Breezeways & Porches	60
Finished Attic	60
Garages	61
Site Improvements	61
Wings & Ells	37

Important: See the Reference Section for Location Factors (to adjust for your city) and Estimating Forms

- **Simple design from standard plans**
- **Single family — 1 full bath, 1 kitchen**
- **No basement**
- **Asphalt shingles on roof**
- **Hot air heat**
- **Gypsum wallboard interior finishes**
- **Materials and workmanship are average**
- **Detail specifications on p. 27**

Note: The illustration shown may contain some optional components (for example: garages and/or fireplaces) whose costs are shown in the modifications, adjustments, & alternatives below or at the end of the square foot section.

eDesign Basics, Inc.

Base cost per square foot of living area

Exterior Wall	Living Area										
	1200	1500	1800	2100	2400	2700	3000	3400	3800	4200	4600
Wood Siding - Wood Frame	105.90	97.55	91.35	86.00	82.50	80.50	78.35	76.25	72.80	70.00	68.65
Brick Veneer - Wood Frame	111.55	102.70	95.95	90.20	86.45	84.25	81.90	79.70	76.05	73.00	71.55
Stucco on Wood Frame	103.35	95.25	89.25	84.10	80.75	78.80	76.75	74.70	71.40	68.60	67.35
Solid Masonry	122.00	112.15	104.55	97.95	93.70	91.30	88.55	86.10	82.00	78.55	76.90
Finished Basement, Add*	22.50	22.10	21.15	20.35	19.95	19.65	19.25	19.00	18.60	18.25	18.10
Unfinished Basement, Add*	8.40	7.75	7.20	6.70	6.40	6.25	6.00	5.80	5.55	5.35	5.25

*Basement under middle level only.

Modifications

Add to the total cost

Upgrade Kitchen Cabinets	$ + 3521
Solid Surface Countertops (Included)	
Full Bath - including plumbing, wall and floor finishes	+ 5993
Half Bath - including plumbing, wall and floor finishes	+ 3570
One Car Attached Garage	+ 12,676
One Car Detached Garage	+ 16,871
Fireplace & Chimney	+ 5234

Adjustments

For multi family - add to total cost

Additional Kitchen	$ + 6796
Additional Bath	+ 5993
Additional Entry & Exit	+ 1640
Separate Heating	+ 1409
Separate Electric	+ 1915

For Townhouse/Rowhouse -
Multiply cost per square foot by

Inner Unit	+ .90
End Unit	+ .95

Alternatives

Add to or deduct from the cost per square foot of living area

Cedar Shake Roof	+ 2.80
Clay Tile Roof	+ 3.30
Slate Roof	+ 4.80
Upgrade Walls to Skim Coat Plaster	+ .39
Upgrade Ceilings to Textured Finish	+ .48
Air Conditioning, in Heating Ductwork	+ 2.14
In Separate Ductwork	+ 6
Heating Systems, Hot Water	+ 1.94
Heat Pump	+ 3.23
Electric Heat	– .65
Not Heated	– 3.43

Additional upgrades or components

Kitchen Cabinets & Countertops	Page 58
Bathroom Vanities	59
Fireplaces & Chimneys	59
Windows, Skylights & Dormers	59
Appliances	60
Breezeways & Porches	60
Finished Attic	60
Garages	61
Site Improvements	61
Wings & Ells	37

- Post and beam frame
- Log exterior walls
- Simple design from standard plans
- Single family — 1 full bath, 1 kitchen
- No basement
- Asphalt shingles on roof
- Hot air heat
- Gypsum wallboard interior finishes
- Materials and workmanship are average
- Detail specification on page 27

Note: The illustration shown may contain some optional components (for example: garages and/or fireplaces) whose costs are shown in the modifications, adjustments, & alternatives below or at the end of the square foot section.

Base cost per square foot of living area

Exterior Wall	Living Area										
	600	800	1000	1200	1400	1600	1800	2000	2400	2800	3200
6" Log - Solid Wall	152.60	138.45	128.00	119.40	112.20	107.60	105.10	101.85	95.75	91.20	88.05
8" Log - Solid Wall	143.40	130.20	120.50	112.65	106.15	101.85	99.55	96.65	90.95	86.80	84.00
Finished Basement, Add	33.95	32.80	31.25	29.85	28.75	27.95	27.60	27.00	26.10	25.50	24.90
Unfinished Basement, Add	13.90	12.50	11.55	10.65	9.95	9.50	9.20	8.85	8.30	7.90	7.60

Modifications

Add to the total cost

Upgrade Kitchen Cabinets	$ + 3521
Solid Surface Countertops (Included)	
Full Bath - including plumbing, wall and floor finishes	+ 5993
Half Bath - including plumbing, wall and floor finishes	+ 3570
One Car Attached Garage	+ 12,676
One Car Detached Garage	+ 16,871
Fireplace & Chimney	+ 5234

Adjustments

For multi family - add to total cost

Additional Kitchen	$ + 6796
Additional Bath	+ 5993
Additional Entry & Exit	+ 1640
Separate Heating	+ 1409
Separate Electric	+ 1915

For Townhouse/Rowhouse -
Multiply cost per square foot by

Inner Unit	+ .92
End Unit	+ .96

Alternatives

Add to or deduct from the cost per square foot of living area

Cedar Shake Roof	+ 3.85
Air Conditioning, in Heating Ductwork	+ 4.19
In Separate Ductwork	+ 8
Heating Systems, Hot Water	+ 2.16
Heat Pump	+ 2.65
Electric Heat	– 1.04
Not Heated	– 3.80

Additional upgrades or components

Kitchen Cabinets & Countertops	Page 58
Bathroom Vanities	59
Fireplaces & Chimneys	59
Windows, Skylights & Dormers	59
Appliances	60
Breezeways & Porches	60
Finished Attic	60
Garages	61
Site Improvements	61
Wings & Ells	37

- Post and beam frame
- Log exterior walls
- Simple design from standard plans
- Single family — 1 full bath, 1 kitchen
- No basement
- Asphalt shingles on roof
- Hot air heat
- Gypsum wallborad interior finishes
- Materials and workmanship are average
- Detail specification on page 27

Note: The illustration shown may contain some optional components (for example: garages and/or fireplaces) whose costs are shown in the modifications, adjustments, & alternatives below or at the end of the square foot section.

Base cost per square foot of living area

Exterior Wall	Living Area										
	2000	1200	1400	1600	1800	2000	2200	2600	3000	3400	3800
6" Log-Solid	135.50	123.40	117.50	113.50	109.20	104.90	101.95	96.10	90.70	88.20	85.90
8" Log-Solid	125.50	114.20	108.85	105.25	101.45	97.40	94.85	89.70	84.75	82.55	80.50
Finished Basement, Add	19.70	19.40	18.75	18.35	17.90	17.65	17.25	16.60	16.15	15.90	15.65
Unfinished Basement, Add	7.65	7.10	6.70	6.40	6.10	5.90	5.70	5.35	5.05	4.80	4.70

Modifications

Add to the total cost

Upgrade Kitchen Cabinets	$ + 3521
Solid Surface Countertops (Included)	
Full Bath - including plumbing, wall and floor finishes	+ 5993
Half Bath - including plumbing, wall and floor finishes	+ 3570
One Car Attached Garage	+ 12,676
One Car Detached Garage	+ 16,871
Fireplace & Chimney	+ 5836

Adjustments

For multi family - add to total cost

Additional Kitchen	$ + 6796
Additional Bath	+ 5993
Additional Entry & Exit	+ 1640
Separate Heating	+ 1409
Separate Electric	+ 1915

For Townhouse/Rowhouse -
Multiply cost per square foot by

Inner Unit	+ .92
End Unit	+ .96

Alternatives

Add to or deduct from the cost per square foot of living area

Cedar Shake Roof	+ 2.80
Air Conditioning, in Heating Ductwork	+ 2.54
In Separate Ductwork	+ 6.36
Heating Systems, Hot Water	+ 2.01
Heat Pump	+ 3.11
Electric Heat	– .80
Not Heated	– 3.55

Additional upgrades or components

Kitchen Cabinets & Countertops	Page 58
Bathroom Vanities	59
Fireplaces & Chimneys	59
Windows, Skylights & Dormers	59
Appliances	60
Breezeways & Porches	60
Finished Attic	60
Garages	61
Site Improvements	61
Wings & Ells	37

1 Story — Base cost per square foot of living area

Exterior Wall	Living Area							
	50	100	200	300	400	500	600	700
Wood Siding - Wood Frame	178.60	136.90	118.75	99.85	93.95	90.35	88.00	88.90
Brick Veneer - Wood Frame	188.10	140.95	120.55	97.85	91.15	87.10	84.45	85.10
Stucco on Wood Frame	169.85	130.55	113.40	96.30	90.65	87.30	85.05	86.05
Solid Masonry	233.15	175.90	151.20	121.50	113.45	108.50	106.90	106.95
Finished Basement, Add	54.20	44.95	40.45	32.90	31.40	30.55	29.90	29.50
Unfinished Basement, Add	25.55	19.20	16.60	12.20	11.30	10.80	10.40	10.15

1-1/2 Story — Base cost per square foot of living area

Exterior Wall	Living Area							
	100	200	300	400	500	600	700	800
Wood Siding - Wood Frame	144.50	115.95	98.90	88.95	83.75	81.35	78.20	77.25
Brick Veneer - Wood Frame	198.15	147.90	122.45	106.90	99.25	95.15	90.70	89.05
Stucco on Wood Frame	173.40	128.05	105.95	94.10	87.40	83.95	80.10	78.60
Solid Masonry	229.75	173.15	143.55	123.40	114.45	109.45	104.25	102.50
Finished Basement, Add	36.60	33.10	30.15	26.80	25.95	25.30	24.75	24.65
Unfinished Basement, Add	16.00	13.30	11.60	9.60	9.10	8.75	8.40	8.40

2 Story — Base cost per square foot of living area

Exterior Wall	Living Area							
	100	200	400	600	800	1000	1200	1400
Wood Siding - Wood Frame	144.30	107.80	91.75	76.45	71.25	68.05	66.00	67.10
Brick Veneer - Wood Frame	201.90	140.70	112.80	90.45	82.90	78.25	75.25	75.70
Stucco on Wood Frame	174.20	120.90	96.25	79.50	73.00	69.10	66.50	67.25
Solid Masonry	237.35	165.95	133.90	104.50	95.50	90.05	86.50	86.55
Finished Basement, Add	29.05	24.45	22.15	18.45	17.65	17.25	16.95	16.75
Unfinished Basement, Add	13.00	9.80	8.45	6.25	5.80	5.55	5.40	5.25

Base costs do not include bathroom or kitchen facilities. Use Modifications/Adjustments/Alternatives on pages 58-61 where appropriate.

Important: See the Reference Section for Location Factors (to adjust for your city) and Estimating Forms.

1 Story

1-1/2 Story

2 Story

2-1/2 Story

Bi-Level

Tri-Level

	Components
1 Site Work	Site preparation for slab or excavation for lower level; 4' deep trench excavation for foundation wall.
2 Foundations	Continuous reinforced concrete footing, 8" deep x 18" wide; dampproofed and insulated 8" thick reinforced concrete foundation wall, 4' deep; 4" concrete slab on 4" crushed stone base and polyethylene vapor barrier, trowel finish.
3 Framing	Exterior walls—2 x 6 wood studs, 16" O.C.; 1/2" plywood sheathing; 2 x 8 rafters 16" O.C. with 1/2" plywood sheathing, 4 in 12, 6 in 12 or 8 in 12 roof pitch, 2 x 6 or 2 x 8 ceiling joists; 1/2" plywood subfloor on 1 x 3 wood sleepers 16" O.C.; 2 x 10 floor joists with 5/8" plywood subfloor on models with more than one level.
4 Exterior Walls	Horizontal beveled wood siding and #15 felt building paper on insulated wood frame walls; double hung windows; 3 flush solid core wood exterior doors with storms. **Alternates:** • Brick veneer on wood frame, has 4" veneer high quality face brick or select common brick. • Stone veneer on wood frame has exterior veneer of field stone or 2" thick limestone. • Solid masonry has an 8" concrete block wall with insulation and a brick or stone facing. It may be a solid brick or stone structure.
5 Roofing	30 year asphalt shingles; #15 felt building paper; aluminum gutters, downspouts and drip edge; copper flashings.
6 Interiors	Walls and ceilings—5/8" drywall, skim coat plaster, primed and painted with 2 coats; hardwood baseboard and trim; sanded and finished, hardwood floor 70%, ceramic tile with 1/2" underlayment 20%, vinyl tile with 1/2" underlayment 10%; wood panel interior doors, primed and painted with 2 coats.
7 Specialties	Custom grade kitchen cabinets—20 L.F. wall and base with plastic laminate counter top and kitchen sink; 4 L.F. bathroom vanity; 75 gallon electric water heater; medicine cabinet.
8 Mechanical	Gas fired warm air heat/air conditioning; one full bath including bathtub, corner shower, built in lavatory and water closet; one 1/2 bath including built in lavatory and water closet.
9 Electrical	100 Amp. service; romex wiring; incandescent lighting fixtures, switches, receptacles.
10 Overhead and Profit	General Contractor overhead and profit.

Adjustments

Unfinished Basement:
7'-6" high cast-in-place concrete walls 10" thick or 12" concrete block.

Finished Basement:
Includes painted drywall on 2 x 4 wood furring with insulation, suspended ceiling, carpeting on concrete floor, heating and lighting.

- **A distinct residence from designer's plans**
- **Single family — 1 full bath, 1 half bath, 1 kitchen**
- **No basement**
- **Asphalt shingles on roof**
- **Forced hot air heat/air conditioning**
- **Gypsum wallboard interior finishes**
- **Materials and workmanship are above average**
- **Detail specifications on page 39**

Note: The illustration shown may contain some optional components (for example: garages and/or fireplaces) whose costs are shown in the modifications, adjustments, & alternatives below or at the end of the square foot section.

©Design Basics, Inc.

Base cost per square foot of living area

Exterior Wall	Living Area										
	800	1000	1200	1400	1600	1800	2000	2400	2800	3200	3600
Wood Siding - Wood Frame	161.95	146.65	134.40	124.60	117.80	114.15	109.80	101.65	95.70	91.55	87.25
Brick Veneer - Wood Frame	180.80	164.70	151.65	141.15	133.95	130.10	125.30	116.75	110.45	105.90	101.30
Stone Veneer - Wood Frame	185.65	169.10	155.60	144.80	137.35	133.35	128.40	119.55	113.05	108.35	103.55
Solid Masonry	190.20	173.15	159.30	148.15	140.50	136.45	131.25	122.10	115.45	110.55	105.60
Finished Basement, Add	50.70	50.50	48.40	46.60	45.50	44.85	43.90	42.60	41.60	40.75	39.95
Unfinished Basement, Add	20.90	19.65	18.50	17.60	17.00	16.65	16.20	15.50	14.95	14.50	14.10

Modifications

Add to the total cost

Upgrade Kitchen Cabinets	$ + 1108
Solid Surface Countertops (Included)	
Full Bath - including plumbing, wall and floor finishes	+ 7102
Half Bath - including plumbing, wall and floor finishes	+ 4230
Two Car Attached Garage	+ 21,916
Two Car Detached Garage	+ 25,336
Fireplace & Chimney	+ 5760

Adjustments

For multi family - add to total cost

Additional Kitchen	$ + 14,160
Additional Full Bath & Half Bath	+ 11,332
Additional Entry & Exit	+ 1640
Separate Heating & Air Conditioning	+ 6271
Separate Electric	+ 1915

For Townhouse/Rowhouse -
Multiply cost per square foot by

Inner Unit	+ .90
End Unit	+ .95

Alternatives

Add to or deduct from the cost per square foot of living area

Cedar Shake Roof	+ 3.45
Clay Tile Roof	+ 4.10
Slate Roof	+ 6.20
Upgrade Ceilings to Textured Finish	+ .48
Air Conditioning, in Heating Ductwork	Base System
Heating Systems, Hot Water	+ 2.21
Heat Pump	+ 2.64
Electric Heat	− 2.95
Not Heated	− 4.77

Additional upgrades or components

Kitchen Cabinets & Countertops	Page 58
Bathroom Vanities	59
Fireplaces & Chimneys	59
Windows, Skylights & Dormers	59
Appliances	60
Breezeways & Porches	60
Finished Attic	60
Garages	61
Site Improvements	61
Wings & Ells	47

- **A distinct residence from designer's plans**
- **Single family — 1 full bath, 1 half bath, 1 kitchen**
- **No basement**
- **Asphalt shingles on roof**
- **Forced hot air heat/air conditioning**
- **Gypsum wallboard interior finishes**
- **Materials and workmanship are above average**
- **Detail specifications on page 39**

Note: The illustration shown may contain some optional components (for example: garages and/or fireplaces) whose costs are shown in the modifications, adjustments, & alternatives below or at the end of the square foot section.

•Donald A. Gardner Architects, Inc.

Base cost per square foot of living area

Exterior Wall	Living Area										
	1000	1200	1400	1600	1800	2000	2400	2800	3200	3600	4000
Wood Siding - Wood Frame	145.30	135.10	127.80	119.10	114.10	109.20	99.85	95.55	91.55	88.55	84.30
Brick Veneer - Wood Frame	154.60	143.80	136.15	126.65	121.25	115.90	105.80	101.25	96.85	93.65	89.00
Stone Veneer - Wood Frame	159.75	148.60	140.65	130.80	125.15	119.70	109.10	104.35	99.70	96.45	91.60
Solid Masonry	164.50	153.05	144.90	134.60	128.75	123.10	112.15	107.25	102.35	99.00	94.00
Finished Basement, Add	33.65	33.90	32.95	31.60	30.90	30.15	28.80	28.15	27.50	27.05	26.50
Unfinished Basement, Add	14.15	13.45	13.05	12.30	11.90	11.55	10.85	10.50	10.15	9.95	9.60

Modifications

Add to the total cost

Upgrade Kitchen Cabinets	$ + 1108
Solid Surface Countertops (Included)	
Full Bath - including plumbing, wall and floor finishes	+ 7102
Half Bath - including plumbing, wall and floor finishes	+ 4230
Two Car Attached Garage	+ 21,916
Two Car Detached Garage	+ 25,336
Fireplace & Chimney	+ 5760

Adjustments

For multi family - add to total cost

Additional Kitchen	$ + 14,160
Additional Full Bath & Half Bath	+ 11,332
Additional Entry & Exit	+ 1640
Separate Heating & Air Conditioning	+ 6271
Separate Electric	+ 1915

For Townhouse/Rowhouse -
Multiply cost per square foot by

Inner Unit	+ .90
End Unit	+ .95

Alternatives

Add to or deduct from the cost per square foot of living area

Cedar Shake Roof	+ 2.50
Clay Tile Roof	+ 2.95
Slate Roof	+ 4.50
Upgrade Ceilings to Textured Finish	+ .48
Air Conditioning, in Heating Ductwork	Base System
Heating Systems, Hot Water	+ 2.10
Heat Pump	+ 2.77
Electric Heat	– 2.60
Not Heated	– 4.40

Additional upgrades or components

Kitchen Cabinets & Countertops	Page 58
Bathroom Vanities	59
Fireplaces & Chimneys	59
Windows, Skylights & Dormers	59
Appliances	60
Breezeways & Porches	60
Finished Attic	60
Garages	61
Site Improvements	61
Wings & Ells	47

- A distinct residence from designer's plans
- Single family — 1 full bath, 1 half bath, 1 kitchen
- No basement
- Asphalt shingles on roof
- Forced hot air heat/air conditioning
- Gypsum wallboard interior finishes
- Materials and workmanship are above average
- Detail specifications on page 39

Note: The illustration shown may contain some optional components (for example: garages and/or fireplaces) whose costs are shown in the modifications, adjustments, & alternatives below or at the end of the square foot section.

Base cost per square foot of living area

Exterior Wall	Living Area										
	1200	1400	1600	1800	2000	2400	2800	3200	3600	4000	4400
Wood Siding - Wood Frame	135.85	127.55	121.75	116.70	111.20	103.20	96.50	92.00	89.40	86.55	84.20
Brick Veneer - Wood Frame	145.80	136.90	130.60	125.10	119.25	110.45	103.10	98.15	95.30	92.15	89.55
Stone Veneer - Wood Frame	151.25	141.95	135.45	129.60	123.65	114.45	106.70	101.55	98.60	95.20	92.55
Solid Masonry	156.35	146.70	140.00	133.90	127.70	118.10	110.05	104.70	101.65	98.05	95.30
Finished Basement, Add	27.05	27.25	26.50	25.75	25.35	24.25	23.40	22.85	22.45	22.00	21.75
Unfinished Basement, Add	11.30	10.80	10.45	10.05	9.85	9.25	8.80	8.50	8.35	8.10	7.95

Modifications

Add to the total cost

Upgrade Kitchen Cabinets	$ + 1108
Solid Surface Countertops (Included)	
Full Bath - including plumbing, wall and floor finishes	+ 7102
Half Bath - including plumbing, wall and floor finishes	+ 4230
Two Car Attached Garage	+ 21,916
Two Car Detached Garage	+ 25,336
Fireplace & Chimney	+ 6502

Adjustments

For multi family - add to total cost

Additional Kitchen	$ + 14,160
Additional Full Bath & Half Bath	+ 11,332
Additional Entry & Exit	+ 1640
Separate Heating & Air Conditioning	+ 6271
Separate Electric	+ 1915

For Townhouse/Rowhouse -
Multiply cost per square foot by

Inner Unit	+ .87
End Unit	+ .93

Alternatives

Add to or deduct from the cost per square foot of living area

Cedar Shake Roof	+ 1.70
Clay Tile Roof	+ 2.05
Slate Roof	+ 3.10
Upgrade Ceilings to Textured Finish	+ .48
Air Conditioning, in Heating Ductwork	Base System
Heating Systems, Hot Water	+ 2.05
Heat Pump	+ 3.08
Electric Heat	– 2.60
Not Heated	– 4.16

Additional upgrades or components

Kitchen Cabinets & Countertops	Page 58
Bathroom Vanities	59
Fireplaces & Chimneys	59
Windows, Skylights & Dormers	59
Appliances	60
Breezeways & Porches	60
Finished Attic	60
Garages	61
Site Improvements	61
Wings & Ells	47

- **A distinct residence from designer's plans**
- **Single family — 1 full bath, 1 half bath, 1 kitchen**
- **No basement**
- **Asphalt shingles on roof**
- **Forced hot air heat/air conditioning**
- **Gypsum wallboard interior finishes**
- **Materials and workmanship are above average**
- **Detail specifications on page 39**

Note: The illustration shown may contain some optional components (for example: garages and/or fireplaces) whose costs are shown in the modifications, adjustments, & alternatives below or at the end of the square foot section.

Base cost per square foot of living area

Exterior Wall	Living Area										
	1500	1800	2100	2400	2800	3200	3600	4000	4500	5000	5500
Wood Siding - Wood Frame	133.30	119.90	112.00	107.30	101.15	95.30	92.05	87.00	84.30	81.90	79.45
Brick Veneer - Wood Frame	143.65	129.40	120.50	115.35	108.85	102.30	98.65	93.10	90.10	87.45	84.70
Stone Veneer - Wood Frame	149.30	134.65	125.20	119.80	113.15	106.15	102.30	96.45	93.30	90.50	87.60
Solid Masonry	154.55	139.50	129.55	123.90	117.05	109.70	105.65	99.65	96.30	93.30	90.25
Finished Basement, Add	21.50	21.50	20.35	19.80	19.35	18.65	18.20	17.70	17.30	17.05	16.80
Unfinished Basement, Add	9.10	8.60	8.00	7.75	7.50	7.15	6.90	6.65	6.45	6.35	6.15

Modifications

Add to the total cost

Upgrade Kitchen Cabinets	$ + 1108
Solid Surface Countertops (Included)	
Full Bath - including plumbing, wall and floor finishes	+ 7102
Half Bath - including plumbing, wall and floor finishes	+ 4230
Two Car Attached Garage	+ 21,916
Two Car Detached Garage	+ 25,336
Fireplace & Chimney	+ 7342

Adjustments

For multi family - add to total cost

Additional Kitchen	$ + 14,160
Additional Full Bath & Half Bath	+ 11,332
Additional Entry & Exit	+ 1640
Separate Heating & Air Conditioning	+ 6271
Separate Electric	+ 1915

For Townhouse/Rowhouse -
Multiply cost per square foot by

Inner Unit	+ .87
End Unit	+ .94

Alternatives

Add to or deduct from the cost per square foot of living area

Cedar Shake Roof	+ 1.50
Clay Tile Roof	+ 1.80
Slate Roof	+ 2.70
Upgrade Ceilings to Textured Finish	+ .48
Air Conditioning, in Heating Ductwork	Base System
Heating Systems, Hot Water	+ 1.85
Heat Pump	+ 3.19
Electric Heat	– 4.56
Not Heated	– 4.16

Additional upgrades or components

Kitchen Cabinets & Countertops	Page 58
Bathroom Vanities	59
Fireplaces & Chimneys	59
Windows, Skylights & Dormers	59
Appliances	60
Breezeways & Porches	60
Finished Attic	60
Garages	61
Site Improvements	61
Wings & Ells	47

- **A distinct residence from designer's plans**
- **Single family — 1 full bath, 1 half bath, 1 kitchen**
- **No basement**
- **Asphalt shingles on roof**
- **Forced hot air heat/air conditioning**
- **Gypsum wallboard interior finishes**
- **Materials and workmanship are above average**
- **Detail specifications on page 39**

Note: The illustration shown may contain some optional components (for example: garages and/or fireplaces) whose costs are shown in the modifications, adjustments, & alternatives below or at the end of the square foot section.

Base cost per square foot of living area

Exterior Wall	Living Area										
	1500	1800	2100	2500	3000	3500	4000	4500	5000	5500	6000
Wood Siding - Wood Frame	132.75	119.45	112.95	107.60	99.40	95.25	90.20	84.90	83.00	80.95	78.80
Brick Veneer - Wood Frame	143.45	129.40	122.25	116.45	107.45	102.85	97.15	91.30	89.15	86.85	84.40
Stone Veneer - Wood Frame	149.35	134.85	127.35	121.30	111.90	107.00	100.95	94.80	92.60	90.15	87.45
Solid Masonry	154.80	139.90	132.05	125.85	115.95	110.90	104.45	98.05	95.75	93.20	90.30
Finished Basement, Add	18.85	18.85	18.10	17.60	16.85	16.40	15.80	15.30	15.15	14.90	14.65
Unfinished Basement, Add	7.95	7.50	7.20	6.90	6.50	6.30	6.00	5.75	5.65	5.55	5.35

Modifications

Add to the total cost

Upgrade Kitchen Cabinets	$ + 1108
Solid Surface Countertops (Included)	
Full Bath - including plumbing, wall and floor finishes	+ 7102
Half Bath - including plumbing, wall and floor finishes	+ 4230
Two Car Attached Garage	+ 21,916
Two Car Detached Garage	+ 25,336
Fireplace & Chimney	+ 7342

Adjustments

For multi family - add to total cost

Additional Kitchen	$ + 14,160
Additional Full Bath & Half Bath	+ 11,332
Additional Entry & Exit	+ 1640
Separate Heating & Air Conditioning	+ 6271
Separate Electric	+ 1915

For Townhouse/Rowhouse -
Multiply cost per square foot by

Inner Unit	+ .85
End Unit	+ .93

Alternatives

Add to or deduct from the cost per square foot of living area

Cedar Shake Roof	+ 1.15
Clay Tile Roof	+ 1.35
Slate Roof	+ 2.05
Upgrade Ceilings to Textured Finish	+ .48
Air Conditioning, in Heating Ductwork	Base System
Heating Systems, Hot Water	+ 1.85
Heat Pump	+ 3.19
Electric Heat	– 4.56
Not Heated	– 4.03

Additional upgrades or components

Kitchen Cabinets & Countertops	Page 58
Bathroom Vanities	59
Fireplaces & Chimneys	59
Windows, Skylights & Dormers	59
Appliances	60
Breezeways & Porches	60
Finished Attic	60
Garages	61
Site Improvements	61
Wings & Ells	47

Important: See the Reference Section for Location Factors (to adjust for your city) and Estimating Forms

- **A distinct residence from designer's plans**
- **Single family — 1 full bath, 1 half bath, 1 kitchen**
- **No basement**
- **Asphalt shingles on roof**
- **Forced hot air heat/air conditioning**
- **Gypsum wallboard interior finishes**
- **Materials and workmanship are above average**
- **Detail specifications on page 39**

Note: The illustration shown may contain some optional components (for example: garages and/or fireplaces) whose costs are shown in the modifications, adjustments, & alternatives below or at the end of the square foot section.

Base cost per square foot of living area

Exterior Wall	Living Area										
	1200	1400	1600	1800	2000	2400	2800	3200	3600	4000	4400
Wood Siding - Wood Frame	128.45	120.70	115.20	110.45	105.15	97.80	91.50	87.30	84.95	82.35	80.05
Brick Veneer - Wood Frame	135.85	127.65	121.80	116.75	111.25	103.20	96.40	91.95	89.40	86.55	84.15
Stone Veneer - Wood Frame	139.95	131.45	125.45	120.15	114.50	106.20	99.15	94.50	91.85	88.85	86.35
Solid Masonry	143.80	135.00	128.85	123.35	117.55	108.95	101.65	96.80	94.10	91.05	88.45
Finished Basement, Add	27.05	27.25	26.50	25.75	25.35	24.25	23.40	22.85	22.45	22.00	21.75
Unfinished Basement, Add	11.30	10.80	10.45	10.05	9.85	9.25	8.80	8.50	8.35	8.10	7.95

Modifications

Add to the total cost

Upgrade Kitchen Cabinets	$ + 1108
Solid Surface Countertops (Included)	
Full Bath - including plumbing, wall and floor finishes	+ 7102
Half Bath - including plumbing, wall and floor finishes	+ 4230
Two Car Attached Garage	+ 21,916
Two Car Detached Garage	+ 25,336
Fireplace & Chimney	+ 5760

Adjustments

For multi family - add to total cost

Additional Kitchen	$ + 14,160
Additional Full Bath & Half Bath	+ 11,332
Additional Entry & Exit	+ 1640
Separate Heating & Air Conditioning	+ 6271
Separate Electric	+ 1915

*For Townhouse/Rowhouse -
Multiply cost per square foot by*

Inner Unit	+ .89
End Unit	+ .95

Alternatives

Add to or deduct from the cost per square foot of living area

Cedar Shake Roof	+ 1.70
Clay Tile Roof	+ 2.05
Slate Roof	+ 3.10
Upgrade Ceilings to Textured Finish	+ .48
Air Conditioning, in Heating Ductwork	Base System
Heating Systems, Hot Water	+ 2.05
Heat Pump	+ 3.08
Electric Heat	– 2.60
Not Heated	– 4.03

Additional upgrades or components

Kitchen Cabinets & Countertops	Page 58
Bathroom Vanities	59
Fireplaces & Chimneys	59
Windows, Skylights & Dormers	59
Appliances	60
Breezeways & Porches	60
Finished Attic	60
Garages	61
Site Improvements	61
Wings & Ells	47

- **A distinct residence from designer's plans**
- **Single family — 1 full bath, 1 half bath, 1 kitchen**
- **No basement**
- **Asphalt shingles on roof**
- **Forced hot air heat/air conditioning**
- **Gypsum wallboard interior finishes**
- **Materials and workmanship are above average**
- **Detail specifications on page 39**

Note: The illustration shown may contain some optional components (for example: garages and/or fireplaces) whose costs are shown in the modifications, adjustments, & alternatives below or at the end of the square foot section.

eDesign Basics, Inc.

Base cost per square foot of living area

Exterior Wall	Living Area										
	1200	1500	1800	2100	2400	2800	3200	3600	4000	4500	5000
Wood Siding - Wood Frame	132.40	120.30	111.05	103.45	98.30	94.60	90.20	85.55	83.60	79.25	76.85
Brick Veneer - Wood Frame	139.80	127.00	117.10	108.90	103.45	99.55	94.80	89.80	87.75	83.10	80.45
Stone Veneer - Wood Frame	143.85	130.70	120.40	111.95	106.30	102.30	97.35	92.15	90.05	85.15	82.40
Solid Masonry	147.60	134.05	123.45	114.70	108.90	104.80	99.65	94.30	92.15	87.10	84.20
Finished Basement, Add*	33.80	33.65	32.25	31.05	30.30	29.80	29.10	28.40	28.15	27.50	27.05
Unfinished Basement, Add*	13.95	13.10	12.35	11.75	11.30	11.10	10.70	10.35	10.20	9.85	9.60

*Basement under middle level only.

Modifications

Add to the total cost

Upgrade Kitchen Cabinets	$ + 1108
Solid Surface Countertops (Included)	
Full Bath - including plumbing, wall and floor finishes	+ 7102
Half Bath - including plumbing, wall and floor finishes	+ 4230
Two Car Attached Garage	+ 21,916
Two Car Detached Garage	+ 25,336
Fireplace & Chimney	+ 5760

Adjustments

For multi family - add to total cost

Additional Kitchen	$ + 14,160
Additional Full Bath & Half Bath	+ 11,332
Additional Entry & Exit	+ 1640
Separate Heating & Air Conditioning	+ 6271
Separate Electric	+ 1915

For Townhouse/Rowhouse -
Multiply cost per square foot by

Inner Unit	+ .87
End Unit	+ .94

Alternatives

Add to or deduct from the cost per square foot of living area

Cedar Shake Roof	+ 2.50
Clay Tile Roof	+ 2.95
Slate Roof	+ 4.50
Upgrade Ceilings to Textured Finish	+ .48
Air Conditioning, in Heating Ductwork	Base System
Heating Systems, Hot Water	+ 1.99
Heat Pump	+ 3.22
Electric Heat	– 2.31
Not Heated	– 4.03

Additional upgrades or components

Kitchen Cabinets & Countertops	Page 58
Bathroom Vanities	59
Fireplaces & Chimneys	59
Windows, Skylights & Dormers	59
Appliances	60
Breezeways & Porches	60
Finished Attic	60
Garages	61
Site Improvements	61
Wings & Ells	47

1 Story   Base cost per square foot of living area

Exterior Wall	Living Area							
	50	100	200	300	400	500	600	700
Wood Siding - Wood Frame	210.95	164.35	144.15	122.55	115.95	111.95	109.35	110.25
Brick Veneer - Wood Frame	235.90	182.20	159.05	132.55	124.95	120.35	117.35	117.90
Stone Veneer - Wood Frame	249.65	192.00	167.20	137.95	129.80	124.90	121.65	122.05
Solid Masonry	262.40	201.15	174.80	143.05	134.35	129.20	125.75	126.00
Finished Basement, Add	80.90	69.05	62.60	51.90	49.75	48.50	47.60	47.00
Unfinished Basement, Add	59.60	40.75	32.00	24.30	22.25	21.00	20.20	19.60

1-1/2 Story   Base cost per square foot of living area

Exterior Wall	Living Area							
	100	200	300	400	500	600	700	800
Wood Siding - Wood Frame	169.35	139.35	121.15	110.40	104.85	102.25	98.90	97.85
Brick Veneer - Wood Frame	191.65	157.25	136.05	122.05	115.60	112.45	108.45	107.35
Stone Veneer - Wood Frame	203.95	167.05	144.25	128.40	121.50	118.00	113.65	112.55
Solid Masonry	215.30	176.15	151.85	134.30	126.95	123.10	118.60	117.35
Finished Basement, Add	54.05	50.10	45.80	41.05	39.80	38.95	38.10	38.00
Unfinished Basement, Add	35.35	26.65	22.50	19.05	17.80	17.00	16.30	16.05

2 Story   Base cost per square foot of living area

Exterior Wall	Living Area							
	100	200	400	600	800	1000	1200	1400
Wood Siding - Wood Frame	168.45	129.20	112.00	95.55	89.95	86.60	84.40	85.55
Brick Veneer - Wood Frame	193.45	147.10	126.90	105.50	98.90	95.00	92.35	93.25
Stone Veneer - Wood Frame	207.20	156.85	135.10	110.95	103.80	99.50	96.70	97.45
Solid Masonry	219.95	166.00	142.65	116.10	108.40	103.80	100.75	101.35
Finished Basement, Add	40.45	34.55	31.35	26.00	24.95	24.25	23.90	23.55
Unfinished Basement, Add	29.85	20.35	16.00	12.15	11.10	10.55	10.10	9.85

Base costs do not include bathroom or kitchen facilities. Use Modifications/Adjustments/Alternatives on pages 58-61 where appropriate.

Important: See the Reference Section for Location Factors (to adjust for your city) and Estimating Forms.

1 Story

1-1/2 Story

2 Story

2-1/2 Story

Bi-Level

Tri-Level

	Components
1 Site Work	Site preparation for slab or excavation for lower level; 4' deep trench excavation for foundation wall.
2 Foundations	Continuous reinforced concrete footing, 8" deep x 18" wide; dampproofed and insulated 12" thick reinforced concrete foundation wall, 4' deep; 4" concrete slab on 4" crushed stone base and polyethylene vapor barrier, trowel finish.
3 Framing	Exterior walls—2 x 6 wood studs, 16" O.C.; 1/2" plywood sheathing; 2 x 8 rafters 16" O.C. with 1/2" plywood sheathing, 4 in 12, 6 in 12 or 8 in 12 roof pitch, 2 x 6 or 2 x 8 ceiling joists; 1/2" plywood subfloor on 1 x 3 wood sleepers 16" O.C.; 2 x 10 floor joists with 5/8" plywood subfloor on models with more than one level.
4 Exterior Walls	Face brick veneer and #15 felt building paper on insulated wood frame walls; double hung windows; 3 flush solid core wood exterior doors with storms. **Alternates:** • Wood siding on wood frame, has top quality cedar or redwood siding or hand split cedar shingles or shakes. • Solid brick may have solid brick exterior wall or brick on concrete block. • Solid stone concrete block with selected fieldstone or limestone exterior.
5 Roofing	Red cedar shingles; #15 felt building paper; aluminum gutters, downspouts and drip edge; copper flashings.
6 Interiors	Walls and ceilings—5/8" drywall, skim coat plaster, primed and painted with 2 coats; hardwood baseboard and trim; sanded and finished, hardwood floor 70%, ceramic tile with 1/2" underlayment 20%, vinyl tile with 1/2" underlayment 10%; wood panel interior doors, primed and painted with 2 coats.
7 Specialties	Luxury grade kitchen cabinets—25 L.F. wall and base with plastic laminate counter top and kitchen sink; 6 L.F. bathroom vanity; 75 gallon electric water heater; medicine cabinet.
8 Mechanical	Gas fired warm air heat/air conditioning; one full bath including bathtub, corner shower, built in lavatory and water closet; one 1/2 bath including built in lavatory and water closet.
9 Electrical	100 Amp. service; romex wiring; incandescent lighting fixtures, switches, receptacles.
10 Overhead and Profit	General Contractor overhead and profit.

Adjustments

Unfinished Basement:
8" high cast-in-place concrete walls 12" thick or 12" concrete block.

Finished Basement:
Includes painted drywall on 2 x 4 wood furring with insulation, suspended ceiling, carpeting on subfloor with sleepers, heating and lighting.

- **Unique residence built from an architect's plan**
- **Single family — 1 full bath, 1 half bath, 1 kitchen**
- **No basement**
- **Cedar shakes on roof**
- **Forced hot air heat/air conditioning**
- **Gypsum wallboard interior finishes**
- **Many special features**
- **Extraordinary materials and workmanship**
- **Detail specifications on page 49**

Note: The illustration shown may contain some optional components (for example: garages and/or fireplaces) whose costs are shown in the modifications, adjustments, & alternatives below or at the end of the square foot section.

eHome Planners, Inc.

Base cost per square foot of living area

Exterior Wall	Living Area										
	1000	1200	1400	1600	1800	2000	2400	2800	3200	3600	4000
Wood Siding - Wood Frame	185.40	170.80	159.15	151.00	146.60	141.40	131.70	124.80	119.85	114.85	110.75
Brick Veneer - Wood Frame	194.65	179.10	166.70	158.10	153.45	147.80	137.55	130.25	124.85	119.50	115.15
Solid Brick	210.80	193.70	179.90	170.60	165.45	159.10	147.90	139.70	133.65	127.75	122.80
Solid Stone	208.45	191.55	178.05	168.75	163.70	157.40	146.35	138.35	132.35	126.50	121.70
Finished Basement, Add	48.80	52.45	50.30	48.85	48.00	46.90	45.30	44.05	42.95	42.00	41.25
Unfinished Basement, Add	21.40	20.05	18.90	18.25	17.80	17.20	16.40	15.80	15.20	14.70	14.30

Modifications

Add to the total cost

Upgrade Kitchen Cabinets	$ + 1462
Solid Surface Countertops (Included)	
Full Bath - including plumbing, wall and floor finishes	+ 8238
Half Bath - including plumbing, wall and floor finishes	+ 4907
Two Car Attached Garage	+ 26,381
Two Car Detached Garage	+ 30,256
Fireplace & Chimney	+ 8141

Adjustments

For multi family - add to total cost

Additional Kitchen	$ + 18,138
Additional Full Bath & Half Bath	+ 13,145
Additional Entry & Exit	+ 2342
Separate Heating & Air Conditioning	+ 6271
Separate Electric	+ 1915

For Townhouse/Rowhouse -
Multiply cost per square foot by

Inner Unit	+ .90
End Unit	+ .95

Alternatives

Add to or deduct from the cost per square foot of living area

Heavyweight Asphalt Shingles	– 3.45
Clay Tile Roof	+ .65
Slate Roof	+ 2.75
Upgrade Ceilings to Textured Finish	+ .48
Air Conditioning, in Heating Ductwork	Base System
Heating Systems, Hot Water	+ 2.39
Heat Pump	+ 2.86
Electric Heat	– 2.60
Not Heated	– 5.21

Additional upgrades or components

Kitchen Cabinets & Countertops	Page 58
Bathroom Vanities	59
Fireplaces & Chimneys	59
Windows, Skylights & Dormers	59
Appliances	60
Breezeways & Porches	60
Finished Attic	60
Garages	61
Site Improvements	61
Wings & Ells	57

Important: See the Reference Section for Location Factors (to adjust for your city) and Estimating Forms

- **Unique residence built from an architect's plan**
- **Single family — 1 full bath, 1 half bath, 1 kitchen**
- **No basement**
- **Cedar shakes on roof**
- **Forced hot air heat/air conditioning**
- **Gypsum wallboard interior finishes**
- **Many special features**
- **Extraordinary materials and workmanship**
- **Detail specifications on page 49**

Note: The illustration shown may contain some optional components (for example: garages and/or fireplaces) whose costs are shown in the modifications, adjustments, & alternatives below or at the end of the square foot section.

©Larry E. Belk Designs

Base cost per square foot of living area

Exterior Wall	Living Area										
	1000	1200	1400	1600	1800	2000	2400	2800	3200	3600	4000
Wood Siding - Wood Frame	172.45	159.95	151.00	140.65	134.60	128.70	117.65	112.65	107.80	104.25	99.30
Brick Veneer - Wood Frame	183.10	169.95	160.55	149.25	142.80	136.50	124.55	119.15	113.85	110.05	104.70
Solid Brick	201.80	187.50	177.25	164.50	157.20	150.20	136.65	130.60	124.40	120.30	114.15
Solid Stone	199.05	184.95	174.75	162.20	155.10	148.15	134.85	128.90	122.85	118.80	112.75
Finished Basement, Add	34.35	37.30	36.10	34.50	33.55	32.75	31.05	30.30	29.30	28.85	28.15
Unfinished Basement, Add	15.55	14.75	14.25	13.40	12.95	12.55	11.65	11.30	10.75	10.55	10.15

Modifications

Add to the total cost

Upgrade Kitchen Cabinets	$ + 1462
Solid Surface Countertops (Included)	
Full Bath - including plumbing, wall and floor finishes	+ 8238
Half Bath - including plumbing, wall and floor finishes	+ 4907
Two Car Attached Garage	+ 26,381
Two Car Detached Garage	+ 30,256
Fireplace & Chimney	+ 8141

Adjustments

For multi family - add to total cost

Additional Kitchen	$ + 18,138
Additional Full Bath & Half Bath	+ 13,145
Additional Entry & Exit	+ 2342
Separate Heating & Air Conditioning	+ 6271
Separate Electric	+ 1915

For Townhouse/Rowhouse -
Multiply cost per square foot by

Inner Unit	+ .90
End Unit	+ .95

Alternatives

Add to or deduct from the cost per square foot of living area

Heavyweight Asphalt Shingles	– 2.85
Clay Tile Roof	+ .50
Slate Roof	+ 2
Upgrade Ceilings to Textured Finish	+ .48
Air Conditioning, in Heating Ductwork	Base System
Heating Systems, Hot Water	+ 2.27
Heat Pump	+ 3.17
Electric Heat	– 2.60
Not Heated	– 4.80

Additional upgrades or components

Kitchen Cabinets & Countertops	Page 58
Bathroom Vanities	59
Fireplaces & Chimneys	59
Windows, Skylights & Dormers	59
Appliances	60
Breezeways & Porches	60
Finished Attic	60
Garages	61
Site Improvements	61
Wings & Ells	57

- **Unique residence built from an architect's plan**
- **Single family — 1 full bath, 1 half bath, 1 kitchen**
- **No basement**
- **Cedar shakes on roof**
- **Forced hot air heat/air conditioning**
- **Gypsum wallboard interior finishes**
- **Many special features**
- **Extraordinary materials and workmanship**
- **Detail specifications on page 49**

Note: The illustration shown may contain some optional components (for example: garages and/or fireplaces) whose costs are shown in the modifications, adjustments, & alternatives below or at the end of the square foot section.

Base cost per square foot of living area

Exterior Wall	Living Area										
	1200	1400	1600	1800	2000	2400	2800	3200	3600	4000	4400
Wood Siding - Wood Frame	159.45	149.60	142.50	136.45	129.95	120.50	112.65	107.30	104.25	100.90	98.05
Brick Veneer - Wood Frame	170.85	160.25	152.65	146.00	139.15	128.80	120.20	114.40	111.05	107.30	104.25
Solid Brick	190.90	178.95	170.50	162.80	155.40	143.40	133.50	126.85	123.10	118.65	115.25
Solid Stone	187.95	176.15	167.90	160.35	153.00	141.25	131.55	125.05	121.35	116.90	113.55
Finished Basement, Add	27.65	29.95	29.10	28.15	27.60	26.30	25.15	24.45	24.10	23.50	23.15
Unfinished Basement, Add	12.45	11.85	11.50	11.00	10.75	10.00	9.50	9.15	8.95	8.60	8.40

Modifications

Add to the total cost

Upgrade Kitchen Cabinets	$ + 1462
Solid Surface Countertops (Included)	
Full Bath - including plumbing, wall and floor finishes	+ 8238
Half Bath - including plumbing, wall and floor finishes	+ 4907
Two Car Attached Garage	+ 26,381
Two Car Detached Garage	+ 30,256
Fireplace & Chimney	+ 8924

Adjustments

For multi family - add to total cost

Additional Kitchen	$ + 18,138
Additional Full Bath & Half Bath	+ 13,145
Additional Entry & Exit	+ 2342
Separate Heating & Air Conditioning	+ 6271
Separate Electric	+ 1915

For Townhouse/Rowhouse -
Multiply cost per square foot by

Inner Unit	+ .86
End Unit	+ .93

Alternatives

Add to or deduct from the cost per square foot of living area

Heavyweight Asphalt Shingles	– 1.70
Clay Tile Roof	+ .35
Slate Roof	+ 1.40
Upgrade Ceilings to Textured Finish	+ .48
Air Conditioning, in Heating Ductwork	Base System
Heating Systems, Hot Water	+ 2.21
Heat Pump	+ 3.34
Electric Heat	– 2.34
Not Heated	– 4.54

Additional upgrades or components

Kitchen Cabinets & Countertops	Page 58
Bathroom Vanities	59
Fireplaces & Chimneys	59
Windows, Skylights & Dormers	59
Appliances	60
Breezeways & Porches	60
Finished Attic	60
Garages	61
Site Improvements	61
Wings & Ells	57

- **Unique residence built from an architect's plan**
- **Single family — 1 full bath, 1 half bath, 1 kitchen**
- **No basement**
- **Cedar shakes on roof**
- **Forced hot air heat/air conditioning**
- **Gypsum wallboard interior finishes**
- **Many special features**
- **Extraordinary materials and workmanship**
- **Detail specifications on page 49**

Note: The illustration shown may contain some optional components (for example: garages and/or fireplaces) whose costs are shown in the modifications, adjustments, & alternatives below or at the end of the square foot section.

©Larry W. Garnett & Associates, Inc

Base cost per square foot of living area

Exterior Wall	Living Area										
	1500	1800	2100	2500	3000	3500	4000	4500	5000	5500	6000
Wood Siding - Wood Frame	155.05	139.30	129.95	123.10	113.75	106.95	100.45	97.40	94.45	91.60	88.60
Brick Veneer - Wood Frame	166.90	150.25	139.70	132.30	122.10	114.55	107.50	104.05	100.80	97.65	94.40
Solid Brick	187.70	169.40	156.90	148.50	136.80	127.90	119.85	115.80	112.00	108.30	104.60
Solid Stone	184.65	166.60	154.25	146.10	134.65	126.00	118.05	114.05	110.30	106.70	103.15
Finished Basement, Add	22.05	23.75	22.40	21.65	20.60	19.70	19.05	18.65	18.30	17.90	17.65
Unfinished Basement, Add	10.05	9.50	8.80	8.45	7.90	7.45	7.15	6.95	6.75	6.55	6.40

Modifications

Add to the total cost

Upgrade Kitchen Cabinets	$	+ 1462
Solid Surface Countertops (Included)		
Full Bath - including plumbing, wall and floor finishes		+ 8238
Half Bath - including plumbing, wall and floor finishes		+ 4907
Two Car Attached Garage		+ 26,381
Two Car Detached Garage		+ 30,256
Fireplace & Chimney		+ 9765

Adjustments

For multi family - add to total cost

Additional Kitchen	$	+ 18,138
Additional Full Bath & Half Bath		+ 13,145
Additional Entry & Exit		+ 2342
Separate Heating & Air Conditioning		+ 6271
Separate Electric		+ 1915

For Townhouse/Rowhouse -
Multiply cost per square foot by

Inner Unit	+ .86
End Unit	+ .93

Alternatives

Add to or deduct from the cost per square foot of living area

Heavyweight Asphalt Shingles	– 1.50
Clay Tile Roof	+ .30
Slate Roof	+ 1.20
Upgrade Ceilings to Textured Finish	+ .48
Air Conditioning, in Heating Ductwork	Base System
Heating Systems, Hot Water	+ 2
Heat Pump	+ 3.44
Electric Heat	– 4.60
Not Heated	– 4.54

Additional upgrades or components

Kitchen Cabinets & Countertops	Page 58
Bathroom Vanities	59
Fireplaces & Chimneys	59
Windows, Skylights & Dormers	59
Appliances	60
Breezeways & Porches	60
Finished Attic	60
Garages	61
Site Improvements	61
Wings & Ells	57

Important: See the Reference Section for Location Factors (to adjust for your city) and Estimating Forms

- **Unique residence built from an architect's plan**
- **Single family — 1 full bath, 1 half bath, 1 kitchen**
- **No basement**
- **Cedar shakes on roof**
- **Forced hot air heat/air conditioning**
- **Gypsum wallboard interior finishes**
- **Many special features**
- **Extraordinary materials and workmanship**
- **Detail specifications on page 49**

Note: The illustration shown may contain some optional components (for example: garages and/or fireplaces) whose costs are shown in the modifications, adjustments, & alternatives below or at the end of the square foot section.

Base cost per square foot of living area

Exterior Wall	Living Area										
	1500	1800	2100	2500	3000	3500	4000	4500	5000	5500	6000
Wood Siding - Wood Frame	153.95	138.35	130.60	124.15	114.55	109.65	103.70	97.65	95.30	92.90	90.45
Brick Veneer - Wood Frame	166.25	149.75	141.25	134.35	123.75	118.35	111.65	105.00	102.40	99.75	96.90
Solid Brick	187.85	169.75	159.95	152.15	140.00	133.70	125.55	117.85	114.85	111.70	108.20
Solid Stone	184.60	166.80	157.15	149.55	137.55	131.40	123.55	115.95	113.05	110.00	106.50
Finished Basement, Add	19.35	20.80	19.95	19.35	18.40	17.85	17.10	16.50	16.25	16.00	15.60
Unfinished Basement, Add	8.85	8.30	7.90	7.60	7.15	6.85	6.55	6.15	6.10	5.95	5.75

Modifications

Add to the total cost

Upgrade Kitchen Cabinets	$ + 1462
Solid Surface Countertops (Included)	
Full Bath - including plumbing, wall and floor finishes	+ 8238
Half Bath - including plumbing, wall and floor finishes	+ 4907
Two Car Attached Garage	+ 26,381
Two Car Detached Garage	+ 30,256
Fireplace & Chimney	+ 9765

Adjustments

For multi family - add to total cost

Additional Kitchen	$ + 18,138
Additional Full Bath & Half Bath	+ 13,145
Additional Entry & Exit	+ 2342
Separate Heating & Air Conditioning	+ 6271
Separate Electric	+ 1915

For Townhouse/Rowhouse -
Multiply cost per square foot by

Inner Unit	+ .84
End Unit	+ .92

Alternatives

Add to or deduct from the cost per square foot of living area

Heavyweight Asphalt Shingles	– 1.15
Clay Tile Roof	+ .20
Slate Roof	+ .90
Upgrade Ceilings to Textured Finish	+ .48
Air Conditioning, in Heating Ductwork	Base System
Heating Systems, Hot Water	+ 2
Heat Pump	+ 3.44
Electric Heat	– 4.60
Not Heated	– 4.40

Additional upgrades or components

Kitchen Cabinets & Countertops	Page 58
Bathroom Vanities	59
Fireplaces & Chimneys	59
Windows, Skylights & Dormers	59
Appliances	60
Breezeways & Porches	60
Finished Attic	60
Garages	61
Site Improvements	61
Wings & Ells	57

- **Unique residence built from an architect's plan**
- **Single family — 1 full bath, 1 half bath, 1 kitchen**
- **No basement**
- **Cedar shakes on roof**
- **Forced hot air heat/air conditioning**
- **Gypsum wallboard interior finishes**
- **Many special features**
- **Extraordinary materials and workmanship**
- **Detail specifications on page 49**

Note: The illustration shown may contain some optional components (for example: garages and/or fireplaces) whose costs are shown in the modifications, adjustments, & alternatives below or at the end of the square foot section.

Base cost per square foot of living area

Exterior Wall	Living Area										
	1200	1400	1600	1800	2000	2400	2800	3200	3600	4000	4400
Wood Siding - Wood Frame	150.85	141.60	134.85	129.20	123.05	114.15	106.85	101.85	99.05	96.05	93.40
Brick Veneer - Wood Frame	159.40	149.55	142.45	136.35	129.95	120.40	112.55	107.20	104.20	100.85	98.05
Solid Brick	174.40	163.55	155.85	148.95	142.10	131.30	122.55	116.55	113.25	109.25	106.20
Solid Stone	172.20	161.50	153.90	147.15	140.30	129.70	121.05	115.20	111.85	108.05	105.00
Finished Basement, Add	27.65	29.95	29.10	28.15	27.60	26.30	25.15	24.45	24.10	23.50	23.15
Unfinished Basement, Add	12.45	11.85	11.50	11.00	10.75	10.00	9.50	9.15	8.95	8.60	8.40

Modifications

Add to the total cost

Upgrade Kitchen Cabinets	$ + 1462
Solid Surface Countertops (Included)	
Full Bath - including plumbing, wall and floor finishes	+ 8238
Half Bath - including plumbing, wall and floor finishes	+ 4907
Two Car Attached Garage	+ 26,381
Two Car Detached Garage	+ 30,256
Fireplace & Chimney	+ 8141

Adjustments

For multi family - add to total cost

Additional Kitchen	$ + 18,138
Additional Full Bath & Half Bath	+ 13,145
Additional Entry & Exit	+ 2342
Separate Heating & Air Conditioning	+ 6271
Separate Electric	+ 1915

For Townhouse/Rowhouse -
Multiply cost per square foot by

Inner Unit	+ .89
End Unit	+ .94

Alternatives

Add to or deduct from the cost per square foot of living area

Heavyweight Asphalt Shingles	– 1.70
Clay Tile Roof	+ .35
Slate Roof	+ 1.40
Upgrade Ceilings to Textured Finish	+ .48
Air Conditioning, in Heating Ductwork	Base System
Heating Systems, Hot Water	+ 2.21
Heat Pump	+ 3.34
Electric Heat	– 2.34
Not Heated	– 4.54

Additional upgrades or components

Kitchen Cabinets & Countertops	Page 58
Bathroom Vanities	59
Fireplaces & Chimneys	59
Windows, Skylights & Dormers	59
Appliances	60
Breezeways & Porches	60
Finished Attic	60
Garages	61
Site Improvements	61
Wings & Ells	57

- **Unique residence built from an architect's plan**
- **Single family — 1 full bath, 1 half bath, 1 kitchen**
- **No basement**
- **Cedar shakes on roof**
- **Forced hot air heat/air conditioning**
- **Gypsum wallboard interior finishes**
- **Many special features**
- **Extraordinary materials and workmanship**
- **Detail specifications on page 49**

Note: The illustration shown may contain some optional components (for example: garages and/or fireplaces) whose costs are shown in the modifications, adjustments, & alternatives below or at the end of the square foot section.

©Home Planners, Inc.

Base cost per square foot of living area

Exterior Wall	Living Area										
	1500	1800	2100	2400	2800	3200	3600	4000	4500	5000	5500
Wood Siding - Wood Frame	142.05	131.05	121.95	115.95	111.55	106.30	100.95	98.60	93.55	90.65	87.60
Brick Veneer - Wood Frame	149.75	138.00	128.30	121.85	117.25	111.60	105.85	103.35	97.90	94.80	91.50
Solid Brick	163.25	150.15	139.40	132.25	127.20	120.85	114.40	111.70	105.55	102.00	98.30
Solid Stone	161.25	148.30	137.75	130.75	125.70	119.50	113.10	110.45	104.45	100.90	97.30
Finished Basement, Add*	32.50	35.00	33.50	32.60	32.00	31.10	30.20	29.85	29.05	28.50	28.05
Unfinished Basement, Add*	14.25	13.35	12.65	12.20	11.85	11.35	10.95	10.75	10.30	10.05	9.80

*Basement under middle level only.

Modifications

Add to the total cost

Upgrade Kitchen Cabinets	$ + 1462
Solid Surface Countertops (Included)	
Full Bath - including plumbing, wall and floor finishes	+ 8238
Half Bath - including plumbing, wall and floor finishes	+ 4907
Two Car Attached Garage	+ 26,381
Two Car Detached Garage	+ 30,256
Fireplace & Chimney	+ 8141

Adjustments

For multi family - add to total cost

Additional Kitchen	$ + 18,138
Additional Full Bath & Half Bath	+ 13,145
Additional Entry & Exit	+ 2342
Separate Heating & Air Conditioning	+ 6271
Separate Electric	+ 1915

For Townhouse/Rowhouse -
Multiply cost per square foot by

Inner Unit	+ .86
End Unit	+ .93

Alternatives

Add to or deduct from the cost per square foot of living area

Heavyweight Asphalt Shingles	– 2.50
Clay Tile Roof	+ .50
Slate Roof	+ 2
Upgrade Ceilings to Textured Finish	+ .48
Air Conditioning, in Heating Ductwork	Base System
Heating Systems, Hot Water	+ 2.15
Heat Pump	+ 3.48
Electric Heat	– 2.06
Not Heated	– 4.40

Additional upgrades or components

Kitchen Cabinets & Countertops	Page 58
Bathroom Vanities	59
Fireplaces & Chimneys	59
Windows, Skylights & Dormers	59
Appliances	60
Breezeways & Porches	60
Finished Attic	60
Garages	61
Site Improvements	61
Wings & Ells	57

1 Story — Base cost per square foot of living area

Exterior Wall	Living Area							
	50	100	200	300	400	500	600	700
Wood Siding - Wood Frame	229.75	176.80	153.85	129.05	121.60	117.00	114.00	114.95
Brick Veneer - Wood Frame	258.50	197.30	170.95	140.45	131.85	126.60	123.15	123.75
Solid Brick	308.95	233.30	200.90	160.45	149.80	143.35	139.15	139.15
Solid Stone	301.50	228.00	196.50	157.55	147.20	140.90	136.80	136.85
Finished Basement, Add	87.35	79.10	70.90	57.25	54.60	52.90	51.85	51.05
Unfinished Basement, Add	45.65	35.00	30.50	23.15	21.65	20.75	20.15	19.75

1-1/2 Story — Base cost per square foot of living area

Exterior Wall	Living Area							
	100	200	300	400	500	600	700	800
Wood Siding - Wood Frame	183.15	149.05	128.30	115.85	109.60	106.60	102.70	101.55
Brick Veneer - Wood Frame	208.80	169.55	145.40	129.20	121.90	118.20	113.65	112.45
Solid Brick	253.80	205.55	175.45	152.60	143.45	138.60	132.95	131.60
Solid Stone	247.10	200.25	171.00	149.10	140.20	135.60	130.10	128.80
Finished Basement, Add	57.45	56.70	51.30	45.25	43.65	42.55	41.55	41.40
Unfinished Basement, Add	28.85	24.40	21.45	18.20	17.30	16.70	16.15	16.10

2 Story — Base cost per square foot of living area

Exterior Wall	Living Area							
	100	200	400	600	800	1000	1200	1400
Wood Siding - Wood Frame	180.20	135.40	115.90	96.80	90.45	86.60	84.05	85.30
Brick Veneer - Wood Frame	208.90	155.90	133.00	108.20	100.70	96.15	93.15	94.10
Solid Brick	259.35	191.90	163.00	128.20	118.65	112.95	109.15	109.50
Solid Stone	251.90	186.65	158.50	125.25	116.00	110.45	106.75	107.20
Finished Basement, Add	43.70	39.60	35.50	28.65	27.30	26.45	25.95	25.55
Unfinished Basement, Add	22.80	17.50	15.30	11.60	10.85	10.35	10.10	9.90

Base costs do not include bathroom or kitchen facilities. Use Modifications/Adjustments/Alternatives on pages 58-61 where appropriate.

Kitchen cabinets - Base units, hardwood *(Cost per Unit)*

	Economy	Average	Custom	Luxury
24" deep, 35" high,				
One top drawer,				
One door below				
12" wide	$ 213	$ 284	$ 378	$ 497
15" wide	221	295	392	516
18" wide	240	320	426	560
21" wide	251	335	446	586
24" wide	285	380	505	665
Four drawers				
12" wide	222	296	394	518
15" wide	225	300	399	525
18" wide	248	330	439	578
24" wide	278	370	492	648
Two top drawers,				
Two doors below				
27" wide	300	400	532	700
30" wide	326	435	579	761
33" wide	345	460	612	805
36" wide	360	480	638	840
42" wide	383	510	678	893
48" wide	409	545	725	954
Range or sink base				
(Cost per unit)				
Two doors below				
30" wide	278	370	492	648
33" wide	300	400	532	700
36" wide	315	420	559	735
42" wide	334	445	592	779
48" wide	349	465	618	814
Corner Base Cabinet				
(Cost per unit)				
36" wide	480	640	851	1120
Lazy Susan *(Cost per unit)*				
With revolving door	634	845	1124	1479

Kitchen cabinets - Wall cabinets, hardwood *(Cost per Unit)*

	Economy	Average	Custom	Luxury
12" deep, 2 doors				
12" high				
30" wide	$ 193	$ 257	$ 342	$ 450
36" wide	223	297	395	520
15" high				
30" wide	161	214	285	375
33" wide	212	283	376	495
36" wide	224	299	398	523
24" high				
30" wide	255	340	452	595
36" wide	278	370	492	648
42" wide	315	420	559	735
30" high, 1 door				
12" wide	180	240	319	420
15" wide	199	265	352	464
18" wide	218	290	386	508
24" wide	248	330	439	578
30" high, 2 doors				
27" wide	274	365	485	639
30" wide	281	375	499	656
36" wide	319	425	565	744
42" wide	349	465	618	814
48" wide	390	520	692	910
Corner wall, 30" high				
24" wide	225	300	399	525
30" wide	251	335	446	586
36" wide	281	375	499	656
Broom closet				
84" high, 24" deep				
18" wide	536	715	951	1251
Oven Cabinet				
84" high, 24" deep				
27" wide	788	1050	1397	1838

Kitchen countertops *(Cost per L.F.)*

	Economy	Average	Custom	Luxury
Solid Surface				
24" wide, no backsplash	92	123	164	215
with backsplash	100	133	177	233
Stock plastic laminate, 24" wide				
with backsplash	17	23	31	40
Custom plastic laminate, no splash				
7/8" thick, alum. molding	27	36	48	63
1-1/4" thick, no splash	36	48	64	84
Marble				
1/2" - 3/4" thick w/splash	48	64	86	113
Maple, laminated				
1-1/2" thick w/splash	72	95	127	167
Stainless steel				
(per S.F.)	130	173	230	303
Cutting blocks, recessed				
16" x 20" x 1" (each)	97	129	172	226

Vanity bases *(Cost per Unit)*

2 door, 30" high, 21" deep	Economy	Average	Custom	Luxury
24" wide	217	289	384	506
30" wide	251	335	446	586
36" wide	334	445	592	779
48" wide	398	530	705	928

Solid surface vanity tops *(Cost Each)*

Center bowl	Economy	Average	Custom	Luxury
22" x 25"	$ 261	$ 282	$ 305	$ 329
22" x 31"	298	322	348	376
22" x 37"	340	367	396	428
22" x 49"	420	454	490	529

Fireplaces & Chimneys *(Cost per Unit)*

	1-1/2 Story	2 Story	3 Story
Economy (prefab metal)			
Exterior chimney & 1 fireplace	$ 5266	$ 5819	$ 6382
Interior chimney & 1 fireplace	5046	5610	5869
Average (masonry)			
Exterior chimney & 1 fireplace	5233	5835	6633
Interior chimney & 1 fireplace	4902	5503	5992
For more than 1 flue, add	376	640	1073
For more than 1 fireplace, add	3710	3710	3710
Custom (masonry)			
Exterior chimney & 1 fireplace	5759	6501	7341
Interior chimney & 1 fireplace	5400	6112	6588
For more than 1 flue, add	451	782	1060
For more than 1 fireplace, add	4137	4137	4137
Luxury (masonry)			
Exterior chimney & 1 fireplace	8140	8923	9764
Interior chimney & 1 fireplace	7766	8490	8987
For more than 1 flue, add	672	1121	1566
For more than 1 fireplace, add	6416	6416	6416

Windows and Skylights *(Cost Each)*

	Economy	Average	Custom	Luxury
Fixed Picture Windows				
3'-6" x 4'-0"	$ 344	$ 458	$ 610	$ 803
4'-0" x 6'-0"	606	808	1075	1414
5'-0" x 6'-0"	677	902	1200	1579
6'-0" x 6'-0"	677	902	1200	1579
Bay/Bow Windows				
8'-0" x 5'-0"	860	1146	1525	2007
10'-0" x 5'-0"	916	1221	1625	2138
10'-0" x 6'-0"	1452	1936	2575	3388
12'-0" x 6'-0"	1889	2518	3350	4408
Palladian Windows				
3'-2" x 6'-4"		1466	1950	2566
4'-0" x 6'-0"		1710	2275	2993
5'-5" x 6'-10"		2236	2975	3914
8'-0" x 6'-0"		2650	3525	4638
Skylights				
46" x 21-1/2"	416	450	590	637
46" x 28"	453	490	625	675
57" x 44"	550	595	735	793

Dormers *(Cost/S.F. of plan area)*

	Economy	Average	Custom	Luxury
Framing and Roofing Only				
Gable dormer, 2" x 6" roof frame	$ 26	$ 29	$ 31	$ 51
2" x 8" roof frame	27	30	32	54
Shed dormer, 2" x 6" roof frame	16	18	20	33
2" x 8" roof frame	18	19	21	34
2" x 10" roof frame	19	21	22	35

Appliances *(Cost per Unit)*

	Economy	Average	Custom	Luxury
Range				
30" free standing, 1 oven	$ 440	$ 1345	$ 1798	$ 2250
2 oven	1575	1700	1763	1825
30" built-in, 1 oven	820	1435	1743	2050
2 oven	1900	2185	2328	2470
21" free standing				
1 oven	420	468	492	515
Counter Top Ranges				
4 burner standard	400	603	704	805
As above with griddle	1600	2713	3269	3825
Microwave Oven	228	452	564	675
Compactor				
4 to 1 compaction	665	858	954	1050
Deep Freeze				
15 to 23 C.F.	645	745	795	845
30 C.F.	990	1108	1167	1225
Dehumidifier, portable, auto.				
15 pint	170	196	209	221
30 pint	188	216	230	244
Washing Machine, automatic	540	1058	1317	1575
Water Heater				
Electric, glass lined				
30 gal.	680	828	902	975
80 gal.	1300	1638	1807	1975
Water Heater, Gas, glass lined				
30 gal.	1075	1300	1413	1525
50 gal.	1275	1575	1725	1875
Dishwasher, built-in				
2 cycles	375	503	567	630
4 or more cycles	445	550	988	1425
Dryer, automatic	625	1200	1488	1775
Garage Door Opener	500	605	658	710
Garbage Disposal	135	201	234	266
Heater, Electric, built-in				
1250 watt ceiling type	212	264	290	315
1250 watt wall type	277	301	313	325
Wall type w/blower				
1500 watt	277	319	340	360
3000 watt	490	564	601	637
Hood For Range, 2 speed				
30" wide	172	636	868	1100
42" wide	370	1298	1762	2225
Humidifier, portable				
7 gal. per day	182	210	224	237
15 gal. per day	218	251	267	283
Ice Maker, automatic				
13 lb. per day	1025	1179	1256	1333
51 lb. per day	1600	1840	1960	2080
Refrigerator, no frost				
10-12 C.F.	550	570	580	590
14-16 C.F.	585	740	818	895
18-20 C.F.	725	1038	1194	1350
21-29 C.F.	1100	2338	2957	3575
Sump Pump, 1/3 H.P.	281	398	457	515

Breezeway *(Cost per S.F.)*

Class	Type	Area (S.F.)			
		50	100	150	200
Economy	Open	$ 23.18	$ 19.70	$ 16.53	$ 16.26
	Enclosed	111.19	85.87	71.31	62.46
Average	Open	30.49	26.89	23.52	21.37
	Enclosed	119.08	88.74	72.58	63.89
Custom	Open	40.58	35.72	31.08	28.44
	Enclosed	166.68	124.00	101.37	89.12
Luxury	Open	42.69	37.42	33.72	32.75
	Enclosed	171.90	127.49	103.13	92.32

Porches *(Cost per S.F.)*

Class	Type	Area (S.F.)				
		25	50	100	200	300
Economy	Open	$ 66.45	$ 44.46	$ 34.75	$ 29.41	$ 25.09
	Enclosed	132.79	92.53	69.94	54.56	46.71
Average	Open	85.84	54.65	41.93	34.85	34.85
	Enclosed	155.77	105.54	79.83	61.81	52.37
Custom	Open	103.63	68.76	51.76	45.22	40.52
	Enclosed	206.04	140.77	107.12	83.42	71.87
Luxury	Open	113.02	73.88	54.57	49.03	43.72
	Enclosed	221.26	155.39	115.28	89.58	77.11

Finished attic *(Cost per S.F.)*

Class	Area (S.F.)				
	400	500	600	800	1000
Economy	$ 17.73	$ 17.14	$ 16.42	$ 16.15	$ 15.54
Average	26.66	26.08	25.44	25.12	24.42
Custom	33.03	32.29	31.56	31.08	30.44
Luxury	42.43	41.40	40.43	39.47	38.82

Alarm system *(Cost per System)*

	Burglar Alarm	Smoke Detector
Economy	$ 405	$ 67
Average	460	81
Custom	793	178
Luxury	1150	242

Sauna, prefabricated
(Cost per unit, including heater and controls—7' high)

Size	Cost
6' x 4'	$ 5250
6' x 5'	5850
6' x 6'	6225
6' x 9'	7625
8' x 10'	10,100
8' x 12'	11,700
10' x 12'	12,200

Garages *

(Costs include exterior wall systems comparable with the quality of the residence. Included in the cost is an allowance for one personnel door, manual overhead door(s) and electrical fixture.)

Class	Type									
	Detached			Attached			Built-in		Basement	
	One Car	Two Car	Three Car	One Car	Two Car	Three Car	One Car	Two Car	One Car	Two Car
Economy										
Wood	$14,630	$22,195	$29,761	$11,273	$19,305	$26,871	$-1477	$-2953	$1277	$1666
Masonry	19,785	28,647	37,510	14,499	23,828	32,690	-2110	-4220		
Average										
Wood	16,871	25,001	33,130	12,676	21,272	29,401	-1752	-3504	1472	2056
Masonry	20,570	29,630	38,689	14,991	24,516	33,576	-2206	-3652		
Custom										
Wood	16,887	25,336	33,785	12,844	21,916	30,365	-3574	-4365	1366	1845
Masonry	21,566	31,191	40,816	15,771	26,020	35,646	-4148	-5514		
Luxury										
Wood	19,802	30,256	40,710	15,304	26,381	36,835	-3723	-4664	2259	3216
Masonry	25,168	36,971	48,774	18,661	31,088	42,891	-4382	-5982		

*See the Introduction to this section for definitions of garage types.

Swimming pools *(Cost per S.F.)*

Residential	(includes equipment)
In-ground	$ 25.00 - 64.50
Deck equipment	1.30
Paint pool, preparation & 3 coats (epoxy)	4.75
Rubber base paint	4.03
Pool Cover	.88
Swimming Pool Heaters	
(not including wiring, external piping, base or pad)	
Gas	
155 MBH	$ 2525.00
190 MBH	3050.00
500 MBH	10,900.00
Electric	
15 KW 7200 gallon pool	2700.00
24 KW 9600 gallon pool	3125.00
54 KW 24,000 gallon pool	4850.00

Wood and coal stoves

Wood Only	
Free Standing (minimum)	$ 1825
Fireplace Insert (minimum)	1915
Coal Only	
Free Standing	$ 2164
Fireplace Insert	2369
Wood and Coal	
Free Standing	$ 4450
Fireplace Insert	4561

Sidewalks *(Cost per S.F.)*

Concrete, 3000 psi with wire mesh	4" thick	$ 3.51
	5" thick	4.27
	6" thick	4.80
Precast concrete patio blocks (natural)	2" thick	10.60
Precast concrete patio blocks (colors)	2" thick	11.20
Flagstone, bluestone	1" thick	15.15
Flagstone, bluestone	1-1/2" thick	20.15
Slate (natural, irregular)	3/4" thick	15.25
Slate (random rectangular)	1/2" thick	23.30
Seeding		
Fine grading & seeding includes lime, fertilizer & seed per S.Y.		2.44
Lawn Sprinkler System	per S.F.	.89

Fencing *(Cost per L.F.)*

Chain Link, 4' high, galvanized	$ 16.60
Gate, 4' high (each)	157.00
Cedar Picket, 3' high, 2 rail	12.25
Gate (each)	186.00
3 Rail, 4' high	15.30
Gate (each)	197.00
Cedar Stockade, 3 Rail, 6' high	15.45
Gate (each)	197.00
Board & Battens, 2 sides 6' high, pine	21.00
6' high, cedar	29.00
No. 1 Cedar, basketweave, 6' high	16.90
Gate, 6' high (each)	225.00

Carport *(Cost per S.F.)*

Economy	$ 8.24
Average	12.26
Custom	18.03
Luxury	20.86

Insurance Exclusions

Insurance exclusions are a matter of policy coverage and are not standard or universal. When making an appraisal for insurance purposes, it is recommended that some time be taken in studying the insurance policy to determine the specific items to be excluded. Most homeowners insurance policies have, as part of their provisions, statements that read "the policy permits the insured to exclude from value, in determining whether the amount of the insurance equals or exceeds 80% of its replacement cost, such items as building excavation, basement footings, underground piping and wiring, piers and other supports which are below the undersurface of the lowest floor or where there is no basement below the ground." Loss to any of these items, however, is covered.

Costs for Excavation, Spread and Strip Footings and Underground Piping

This chart shows excluded items expressed as a percentage of total building cost.

| Class | Number of Stories | | | |
	1	1-1/2	2	3
Luxury	3.4%	2.7%	2.4%	1.9%
Custom	3.5%	2.7%	2.4%	1.9%
Average	5.0%	4.3%	3.8%	3.3%
Economy	5.0%	4.4%	4.1%	3.5%

Architect/Designer Fees

Architect/designer fees as presented in the following chart are typical ranges for 4 classes of residences. Factors affecting these ranges include economic conditions, size and scope of project, site selection, and standardization. Where superior quality and detail is required or where closer supervision is required, fees may run higher than listed. Lower quality or simplicity may dictate lower fees.

The editors have included architect/designer fees from the "mean" column in calculating costs.

Listed below are average costs expressed as a range and mean of total building cost for 4 classes of residences.

Class		Range	Mean
Luxury	Architecturally designed and supervised	5%–15%	10%
Custom	Architecturally modified designer plans	$2000–$3500	$2500
Average	Designer plans	$800–$1500	$1100
Economy	Stock plans	$100–$500	$300

Depreciation as generally defined is "the loss of value due to any cause." Specifically, depreciation can be broken down into three categories: **Physical Depreciation, Functional Obsolescence,** and **Economic Obsolescence.**

Physical Depreciation is the loss of value from the wearing out of the building's components. Such causes may be decay, dry rot, cracks, or structural defects.

Functional Obsolescence is the loss of value from features which render the residence less useful or desirable. Such features may be higher than normal ceilings, design and/or style changes, or outdated mechanical systems.

Economic Obsolescence is the loss of value from external features which render the residence less useful or desirable. These features may be changes in neighborhood socio-economic grouping, zoning changes, legislation, etc.

Depreciation as it applies to the residential portion of this manual deals with the observed physical condition of the residence being appraised. It is in essence *"the cost to cure."*

The ultimate method to arrive at *"the cost to cure"* is to analyze each component of the residence.

For example:

    Component, Roof Covering – Cost = $2400
    Average Life = 15 years
    Actual Life = 5 years
    Depreciation = 5 ÷ 15 = .33 or 33%

    $2400 × 33% = $792 = Amount of Depreciation

The following table, however, can be used as a guide in estimating the % depreciation using the actual age and general condition of the residence.

Depreciation Table — Residential

Age in Years	Good	Average	Poor
2	2%	3%	10%
5	4	6	20
10	7	10	25
15	10	15	30
20	15	20	35
25	18	25	40
30	24	30	45
35	28	35	50
40	32	40	55
45	36	45	60
50	40	50	65

Commercial/Industrial/ Institutional Section

Table of Contents

Introduction to the Commercial/ Industrial/Institutional Section

General

The Commercial/Industrial/Institutional section of this manual contains base building costs per square foot of floor area for 75 model buildings. Each model has a table of square foot costs for combinations of exterior wall and framing systems. This table is supplemented by a list of common additives and their unit costs. A breakdown of the component costs used to develop the base cost for the model is on the opposite page. The total cost derived from the base cost and additives must be modified using the appropriate location factor from the Reference Section.

This section may be used directly to estimate the construction cost of most types of buildings knowing only the floor area, exterior wall construction and framing systems. To adjust the base cost for components which are different than the model, use the tables from the Assemblies Section.

Building Identification & Model Selection

The building models in this section represent structures by use. Occupancy, however, does not necessarily identify the building, i.e., a restaurant could be a converted warehouse. In all instances, the building should be described and identified by its own physical characteristics. The model selection should also be guided by comparing specifications with the model. In the case of converted use, data from one model may be used to supplement data from another.

Adjustments

The base cost tables represent the base cost per square foot of floor area for buildings without a basement and without unusual special features. Basement costs and other common additives are listed below the base cost table. Cost adjustments can also be made to the model by using the tables from the Assemblies Section. This table is for example only.

Dimensions

All base cost tables are developed so that measurement can be readily made during the inspection process. Areas are calculated from exterior dimensions and story heights are measured from the top surface of one floor to the top surface of the floor above. Roof areas are measured by horizontal area covered and costs related to inclines are converted with appropriate factors. The precision of measurement is a matter of the user's choice and discretion. For ease in calculation, consideration should be given to measuring in tenths of a foot, i.e., 9 ft. 6 in. = 9.5 ft., 9 ft. 4 in. = 9.3 ft.

Floor Area

The term "Floor Area" as used in this section includes the sum of floor plate at grade level and above. This dimension is measured from the outside face of the foundation wall. Basement costs are calculated separately. The user must exercise his/her own judgment, where the lowest level floor is slightly below grade, whether to consider it at grade level or make the basement adjustment.

How to Use the Commercial/ Industrial/Institutional Section

The following is a detailed explanation of a sample entry in the Commercial/Industrial/Institutional Square Foot Cost Section. Each bold number below corresponds to the item being described on the following page with the appropriate component or cost of the sample entry following in parenthesis.

Prices listed are costs that include overhead and profit of the installing contractor and additional mark-ups for General Conditions and Architects' Fees.

COMMERCIAL/INDUSTRIAL/INSTITUTIONAL	M.010 **1**	Apartment, 1-3 Story **2**

Costs per square foot of floor area

Exterior Wall **3**	S.F. Area	8000	12000	15000	19000	22500	25000	29000	32000	36000
	L.F. Perimeter	213	280	330	350	400	433	442	480	520
Face Brick with Concrete Block Back-up	Wood Joists	198.55	175.50	166.20	154.45	149.90	147.25	141.55	139.80	137.40
	Steel Joists	199.55	177.60	168.85	158.25	153.90	151.40	146.40	144.70	142.45
Stucco on Concrete Block	Wood Joists	176.80	155.80	147.35	137.75	133.60	131.25	126.80	125.15	123.10
	Steel Joists	189.00	167.85	159.30	149.65	145.50	143.10	138.60	136.95	134.90
Wood Siding	Wood Frame	176.20	155.25	146.80	137.35	133.15	130.80	126.45	124.80	122.75
Brick Veneer	Wood Frame	184.85	162.75	153.85	143.15	138.85	136.30	131.20	129.50	127.30
Perimeter Adj., Add or Deduct	Per 100 L.F.	20.40	13.60	10.95	8.60	7.30	6.60	5.65	5.05	4.55
Story Hgt. Adj., Add or Deduct **6**	Per 1 Ft.	3.85	3.35	3.25	2.70	2.55	2.55	2.20	2.15	2.10

4 **5** **7** **8**

For Basement, add $33.15 per square foot of basement area **9**

The above costs were calculated using the basic specifications shown on the facing page. These costs should be adjusted where necessary for design alternatives and owner's requirements. Reported completed project costs, for this type of structure, range from $55.30 to $205.70 per S.F.

Common additives

Description	Unit	$ Cost
Appliances		
Cooking range, 30" free standing		
1 oven	Each	460 - 2300
2 oven	Each	1600 - 1850
30" built-in		
1 oven	Each	845 - 2150
2 oven	Each	1950 - 2025
Counter top cook tops, 4 burner	Each	425 - 855
Microwave oven	Each	267 - 750
Combination range, refrig. & sink, 30" wide	Each	1625 - 5250
72" wide	Each	5625
Combination range, refrigerator, sink,		
microwave oven & icemaker	Each	6400
Compactor, residential, 4-1 compaction	Each	685 - 1075
Dishwasher, built-in, 2 cycles	Each	535 - 955
4 cycles	Each	605 - 1750
Garbage disposer, sink type	Each	201 - 330
Hood for range, 2 speed, vented, 30" wide	Each	279 - 1275
42" wide	Each	475 - 2400
Refrigerator, no frost 10-12 C.F.	Each	570 - 630
18-20 C.F.	Each	750 - 1400

10

Description	Unit	$ Cost
Closed Circuit Surveillance, One station		
Camera and monitor	Each	1875
For additional camera stations, add	Each	1025
Elevators, Hydraulic passenger, 2 stops		
2000# capacity	Each	63,100
2500# capacity	Each	64,300
3500# capacity	Each	67,600
Additional stop, add	Each	8250
Emergency Lighting, 25 watt, battery operated		
Lead battery	Each	287
Nickel cadmium	Each	845
Laundry Equipment		
Dryer, gas, 16 lb. capacity	Each	900
30 lb. capacity	Each	3600
Washer, 4 cycle	Each	1100
Commercial	Each	1450
Smoke Detectors		
Ceiling type	Each	233
Duct type	Each	525

1 Model Number (M.010)

"M" distinguishes this section of the book and stands for model. The number designation is a sequential number.

2 Type of Building (Apartment, 1-3 Story)

There are 45 different types of commercial/industrial/institutional buildings highlighted in this section.

3 Exterior Wall Construction and Building Framing Options (Face Brick with Concrete Block Back-up and Open Web Steel Bar Joists)

Three or more commonly used exterior walls, and in most cases, two typical building framing systems are presented for each type of building. The model selected should be based on the actual characteristics of the building being estimated.

4 Total Square Foot of Floor Area and Base Perimeter Used to Compute Base Costs (22,500 Square Feet and 400 Linear Feet)

Square foot of floor area is the total gross area of all floors at grade, and above, and does not include a basement. The perimeter in linear feet used for the base cost is for a generally rectangular economical building shape.

5 Cost per Square Foot of Floor Area ($153.90)

The highlighted cost is for a building of the selected exterior wall and framing system and floor area. Costs for buildings with floor areas other than those calculated may be interpolated between the costs shown.

6 Building Perimeter and Story Height Adjustments

Square foot costs for a building with a perimeter or floor to floor story height significantly different from the model used to calculate the base cost may be adjusted, add or deduct, to reflect the actual building geometry.

7 Cost per Square Foot of Floor Area for the Perimeter and/or Height Adjustment ($7.30 for Perimeter Difference and $2.55 for Story Height Difference)

Add (or deduct) $7.30 to the base square foot cost for each 100 feet of perimeter difference between the model and the actual building. Add (or deduct) $2.55 to the base square foot cost for each 1 foot of story height difference between the model and the actual building.

8 Optional Cost per Square Foot of Basement Floor Area ($33.15)

The cost of an unfinished basement for the building being estimated is $33.15 times the gross floor area of the basement.

9 Range of Cost per Square Foot of Floor Area for Similar Buildings ($55.30 to $205.70)

Many different buildings of the same type have been built using similar materials and systems. RSMeans historical cost data of actual construction projects indicates a range of $55.30 to $205.70 for this type of building.

10 Common Additives

Common components and/or systems used in this type of building are listed. These costs should be added to the total building cost. Additional selections may be found in the Assemblies Section.

How to Use the Commercial/ Industrial/Institutional Section (Continued)

The following is a detailed explanation of a specification and costs for a model building in the Commercial/Industrial/Institutional Square Foot Cost Section. Each bold number below corresponds to the item being described on the following page with the appropriate component of the sample entry following in parenthesis.

Prices listed are costs that include overhead and profit of the installing contractor.

Model costs calculated for a 3 story building with 10' story height and 22,500 square feet of floor area **1**

2 Apartment, 1-3 Story

				Unit	Unit Cost	Cost Per S.F.	% Of Sub-Total
A. SUBSTRUCTURE							
1010	Standard Foundations	Poured concrete; strip and spread footings		S.F. Ground	5.73	1.91	
1020	Special Foundations	N/A		—	—	—	
1030	Slab on Grade	4" reinforced concrete with vapor barrier and granular base		S.F. Slab	4.77	1.59	4.4%
2010	Basement Excavation	Site preparation for slab and trench for foundation wall and footing		S.F. Ground	.18	.06	
2020	Basement Walls	4' foundation wall		L.F. Wall	71	1.46	
B. SHELL							
B10 Superstructure							
1010	Floor Construction	Open web steel joists, slab form, concrete, interior steel columns		S.F. Floor	17.31	11.54	12.5%
1020	Roof Construction	Open web steel joists with rib metal deck, interior steel columns		S.F. Roof	8.16	2.72	
B20 Exterior Enclosure							
2010	Exterior Walls	Face brick with concrete block backup	88% of wall	S.F. Wall	22.65	10.63	
2020	Exterior Windows	Aluminum horizontal sliding	12% of wall	Each	497	2.12	11.5%
2030	Exterior Doors	Aluminum and glass		Each	1830	.32	
B30 Roofing							
3010	Roof Coverings	Built-up tar and gravel with flashing; perlite/EPS composite insulation		S.F. Roof	6.06	2.02	1.8%
3020	Roof Openings	N/A **6**		—	—	—	
C. INTERIORS							
1010	Partitions	Gypsum board and sound deadening board on metal studs	10 S.F. of Floor/L.F. Partition	S.F. Partition	6.11	5.43	**9**
1020	Interior Doors **3**	15% solid core wood, 85% hollow core wood	80 S.F. Floor/Door	Each	510	6.37	
1030	Fittings	Kitchen Cabinets **4**		S.F. Floor	2.83	2.83	
2010	Stair Construction	Concrete filled metal pan		Flight	12,275	2.73	24.7%
3010	Wall Finishes	70% paint, 25% vinyl wall covering, 5% ceramic tile **5**		S.F. Surface	**7**		
3020	Floor Finishes	60% carpet, 30% vinyl composition tile, 10% ceramic tile		S.F. Floor	5.?2	**8**	
3030	Ceiling Finishes	Painted gypsum board on resilient channels		S.F. Ceiling	.11		
D. SERVICES							
D10 Conveying							
1010	Elevators & Lifts	One hydraulic passenger elevator		Each	103,950	4.62	4.1%
1020	Escalators & Moving Walks	N/A		—	—	—	
D20 Plumbing							
2010	Plumbing Fixtures	Kitchen, bathroom and service fixtures, supply and drainage	1 Fixture/200 S.F. Floor	Each	2488	12.44	
2020	Domestic Water Distribution	Gas fired water heater		S.F. Floor	3.92	3.92	14.7%
2040	Rain Water Drainage	Roof drains		S.F. Roof	1.11	.37	
D30 HVAC							
3010	Energy Supply	Oil fired hot water, baseboard radiation		S.F. Floor	7.54	7.54	
3020	Heat Generating Systems	N/A		—	—	—	
3030	Cooling Generating Systems	Chilled water, air cooled condenser system		S.F. Floor	8.59	8.59	14.2%
3050	Terminal & Package Units	N/A		—	—	—	
3090	Other HVAC Sys. & Equipment	N/A		—	—	—	
D40 Fire Protection							
4010	Sprinklers	Wet pipe sprinkler system		S.F. Floor	3.10	3.10	2.7%
4020	Standpipes	N/A		—	—	—	
D50 Electrical							
5010	Electrical Service/Distribution	600 ampere service, panel board and feeders		S.F. Floor	2.55	2.55	
5020	Lighting & Branch Wiring	Incandescent fixtures, receptacles, switches, A.C. and misc. power		S.F. Floor	6.78	6.78	9.6%
5030	Communications & Security	Alarm systems, internet wiring, and emergency lighting		S.F. Floor	1.37	1.37	
5090	Other Electrical Systems	Emergency generator, 11.5 kW		S.F. Floor	.21	.21	
E. EQUIPMENT & FURNISHINGS							
1010	Commercial Equipment	N/A		—	—	—	
1020	Institutional Equipment	N/A		—	—	—	0.0%
1030	Vehicular Equipment	N/A		—	—	—	
1090	Other Equipment	N/A **10**		—	—	—	
F. SPECIAL CONSTRUCTION							
1020	Integrated Construction	N/A		—	—	—	0.0%
1040	Special Facilities	N/A		—	—	—	
G. BUILDING SITEWORK	N/A						
				Sub-Total		113.99	100%
	CONTRACTOR FEES (General Requirements: 10%, Overhead: 5%, Profit: 10%) **11**		**12**		25%	28.51	
	ARCHITECT FEES				8%	11.40	
				Total Building Cost		**153.90**	

70

1. Building Description
(Model costs are calculated for a three-story apartment building with 10' story height and 22,500 square feet of floor area)

The model highlighted is described in terms of building type, number of stories, typical story height, and square footage.

2. Type of Building
(Apartment, 1-3 Story)

3. Division C Interiors
(C1020 Interior Doors)

System costs are presented in divisions according to the 7-division UNIFORMAT II classifications. Each of the component systems are listed.

4. Specification Highlights
(15% solid core wood; 85% hollow core wood)

All systems in each subdivision are described with the material and proportions used.

5. Quality Criteria
(80 S.F. Floor/Door)

The criteria used in determining quantities for the calculations are shown.

6. Unit (Each)

The unit of measure shown in this column is the unit of measure of the particular system shown that corresponds to the unit cost.

7. Unit Cost ($510)

The cost per unit of measure of each system subdivision.

8. Cost per Square Foot ($6.37)

The cost per square foot for each system is the unit cost of the system times the total number of units divided by the total square feet of building area.

9. % of Sub-Total (24.7%)

The percent of sub-total is the total cost per square foot of all systems in the division divided by the sub-total cost per square foot of the building.

10. Sub-Total ($113.99)

The sub-total is the total of all the system costs per square foot.

11. Project Fees
(Contractor Fees) (25%)
(Architect Fees) (8%)

Contractor Fees to cover the general requirements, overhead and profit of the General Contractor are added as a percentage of the sub-total. Architect Fees, also as a percentage of the sub-total, are also added. These values vary with the building type.

12. Total Building Cost ($153.90)

The total building cost per square foot of building area is the sum of the square foot costs of all the systems plus the General Contractor's general requirements, overhead and profit, and the Architect fees. The total building cost is the amount which appears shaded in the Cost per Square Foot of Floor Area table shown previously.

Examples

Example 1

This example illustrates the use of the base cost tables. The base cost is adjusted for different exterior wall systems, different story height and a partial basement.

CII APPRAISAL FIELD DATA FORM

COMMERCIAL
INDUSTRIAL
INSTITUTIONAL

Use sheets A & D for
App. Form 2
Use sheets A,B & C for
App. Form 3

CII APPRAISAL

1. SUBJECT PROPERTY:	Westernberg Holdings, Inc.	5. DATE:	Jan. 1, 2010
2. BUILDING:	# 623	6. APPRAISER:	SJS
3. ADDRESS:	55 Pine wood Rd., Bondsville, N.J. 07410		
4. BUILDING USE:	Apartments	7. YEAR BUILT:	1990

N
191'

W
60'

70'

E

S

North wall - Face brick w/concrete block back-up
East, West & South walls - Decorative concrete block

8. EXTERIOR WALL CONSTRUCTION:
9. FRAME: **Steel**
10. GROUND FLOOR AREA: 60' x 191' 11,460 S.F.   11. GROSS FLOOR AREA (EXCL. BASEMENT): 80, 220 (7 Floors) S.F.
12. NUMBER OF STORIES: 7   13. STORY HEIGHT: 11 ' - 4"
14. PERIMETER: 502 L.F.   15. BASEMENT AREA: 4200 S.F. S.F.
16. GENERAL COMMENTS:

Example 1 (continued)

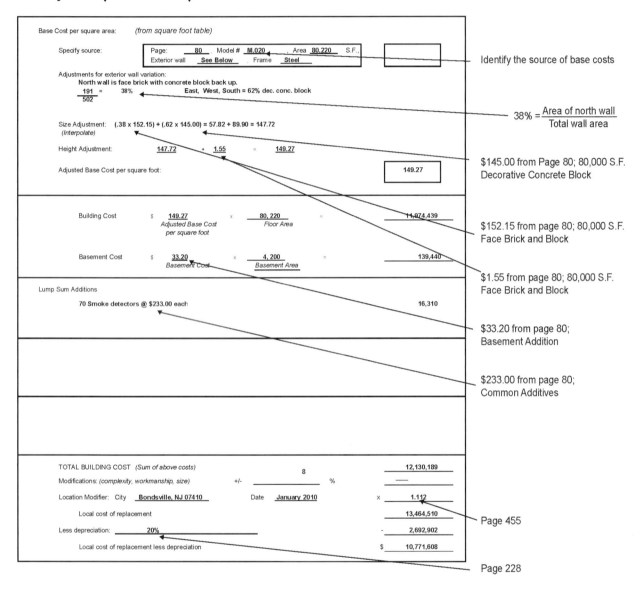

Base Cost per square area: *(from square foot table)*

Specify source: Page: __80__, Model # __M.020__, Area __80,220__ S.F., Exterior wall __See Below__, Frame __Steel__ — Identify the source of base costs

Adjustments for exterior wall variation:
North wall is face brick with concrete block back up.
$\frac{191}{502}$ = 38% **East, West, South = 62% dec. conc. block**

$38\% = \dfrac{\text{Area of north wall}}{\text{Total wall area}}$

Size Adjustment: (.38 x 152.15) + (.62 x 145.00) = 57.82 + 89.90 = 147.72
(Interpolate)

$145.00 from Page 80; 80,000 S.F. Decorative Concrete Block

Height Adjustment: __147.72__ + __1.55__ = __149.27__

Adjusted Base Cost per square foot: __149.27__

Building Cost $ __149.27__ x __80, 220__ = __11,974,439__
Adjusted Base Cost per square foot   *Floor Area*

$152.15 from page 80; 80,000 S.F. Face Brick and Block

Basement Cost $ __33.20__ x __4, 200__ = __139,440__
Basement Cost   *Basement Area*

$1.55 from page 80; 80,000 S.F. Face Brick and Block

Lump Sum Additions
70 Smoke detectors @ $233.00 each   16,310

$33.20 from page 80; Basement Addition

$233.00 from page 80; Common Additives

TOTAL BUILDING COST *(Sum of above costs)*   8   12,130,189

Modifications: *(complexity, workmanship, size)*   +/- _____ %   ——

Location Modifier: City __Bondsville, NJ 07410__   Date __January 2010__   x __1.112__

Local cost of replacement   13,464,510   Page 455

Less depreciation: __20%__   - 2,692,902

Local cost of replacement less depreciation $ __10,771,608__

Page 228

Example 2

This example shows how to modify a model building.

Model #M.020, 4–7 Story Apartment, page 80, matches the example quite closely. The model specifies a 6-story building with a 10' story height.

The following adjustments must be made:
- add stone ashlar wall
- adjust model to a 5-story building
- change partitions
- change floor finish

CII APPRAISAL

		5. DATE:	**Jan. 2, 2010**
1. SUBJECT PROPERTY:	**Westernberg Holdings, Inc.**	6. APPRAISER:	**SJS**
2. BUILDING:	**# 623**		
3. ADDRESS:	**55 Pinewood Rd., Bondsville, N.J. 07410**		
4. BUILDING USE:	**Apartments**	7. YEAR BUILT:	**1990**

N
170 '

W
64.7'

E

S

Two exterior wall types

North wall - 4" stone Ashlar w/8" block back up (split Face)

8. EXTERIOR WALL CONSTRUCTION: **East, West & South walls - 12" Decorative concrete block**

9. FRAME: **Steel**

10. GROUND FLOOR AREA:	**11,000**	11. GROSS FLOOR AREA (EXCL. BASEMENT):	**55,000**	S.F.
12. NUMBER OF STORIES:	**5**	13. STORY HEIGHT:	**11 Ft.**	
14. PERIMETER:	**469.4**	15. BASEMENT AREA:	None	S.F.

16. GENERAL COMMENTS:

Example 2 (continued)

Square foot area (excluding basement) from item 11 ___55,000___ S.F.
Perimeter from item 14 ___469.4___ Lin. Ft.
Item 17-Model square foot costs (from model sub-total) ___$123.97___

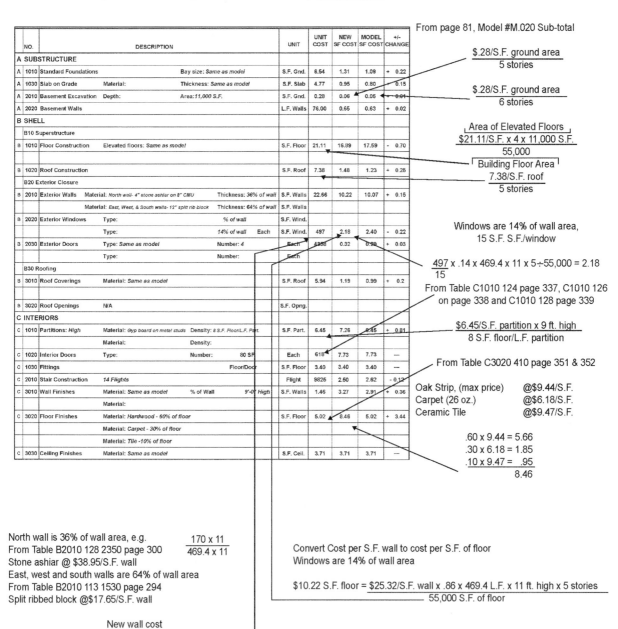

From page 81, Model #M.020 Sub-total

NO.	DESCRIPTION	UNIT	UNIT COST	NEW SF COST	MODEL SF COST	+/- CHANGE
A SUBSTRUCTURE						
A 1010 Standard Foundations	Bay size: *Same as model*	S.F. Gnd.	6.54	1.31	1.09	+ 0.22
A 1030 Slab on Grade	Material: Thickness: *Same as model*	S.F. Slab	4.77	0.95	0.80	0.15
A 2010 Basement Excavation	Depth: Area:*11,000 S.F.*	S.F. Gnd.	0.28	0.06	0.05	0.01
A 2020 Basement Walls		L.F. Walls	76.00	0.65	0.63	+ 0.02
B SHELL						
B10 Superstructure						
B 1010 Floor Construction	Elevated floors: *Same as model*	S.F. Floor	21.11	16.89	17.59	- 0.70
B 1020 Roof Construction		S.F. Roof	7.38	1.48	1.23	+ 0.25
B20 Exterior Closure						
B 2010 Exterior Walls	Material: *North wall- 4" stone ashlar on 8" CMU* Thickness: *36% of wall*	S.F. Walls	22.66	10.22	10.07	+ 0.15
	Material: *East, West, & South walls- 12" split rib block* Thickness: *64% of wall*	S.F. Walls				
B 2020 Exterior Windows	Type: % of wall	S.F. Wind.				
	Type: *14% of wall* Each	S.F. Wind.	497	2.18	2.40	- 0.22
B 2030 Exterior Doors	Type: *Same as model* Number: 4	Each	1358	0.32	0.29	+ 0.03
	Type: Number:	Each				
B30 Roofing						
B 3010 Roof Coverings	Material: *Same as model*	S.F. Roof	5.94	1.19	0.99	+ 0.2
B 3020 Roof Openings	N/A	S.F. Opng.				
C INTERIORS						
C 1010 Partitions: *High*	Material: *Gyp board on metal studs* Density: *8 S.F. Floor/L.F. Part.*	S.F. Part.	6.45	7.26	6.45	+ 0.81
	Material: Density:					
C 1020 Interior Doors	Type: Number: 80 SF	Each	618	7.73	7.73	—
C 1030 Fittings	Floor/Door	S.F. Floor	3.40	3.40	3.40	—
C 2010 Stair Construction	14 Flights	Flight	9825	2.50	2.62	- 0.12
C 3010 Wall Finishes	Material: *Same as model* % of Wall *9'-0" High*	S.F. Walls	1.46	3.27	2.91	+ 0.36
	Material:					
C 3020 Floor Finishes	Material: *Hardwood - 60% of floor*	S.F. Floor	5.02	8.46	5.02	+ 3.44
	Material: *Carpet - 30% of floor*					
	Material: *Tile -10% of floor*					
C 3030 Ceiling Finishes	Material: *Same as model*	S.F. Ceil.	3.71	3.71	3.71	—

$.28/S.F. ground area
─────────────────────
5 stories

$.28/S.F. ground area
─────────────────────
6 stories

Area of Elevated Floors
$21.11/S.F. x 4 x 11,000 S.F.
──────────────────────────────
55,000
Building Floor Area

7.38/S.F. roof
──────────────
5 stories

Windows are 14% of wall area,
15 S.F. S.F./window

497 x .14 x 469.4 x 11 x 5÷55,000 = 2.18
──
15

From Table C1010 124 page 337, C1010 126
on page 338 and C1010 128 page 339

$6.45/S.F. partition x 9 ft. high
─────────────────────────────────
8 S.F. floor/L.F. partition

From Table C3020 410 page 351 & 352

Oak Strip, (max price) @$9.44/S.F.
Carpet (26 oz.) @$6.18/S.F.
Ceramic Tile @$9.47/S.F.

.60 x 9.44 = 5.66
.30 x 6.18 = 1.85
.10 x 9.47 = .95
─────
8.46

North wall is 36% of wall area, e.g.
From Table B2010 128 2350 page 300 — 170 x 11 / 469.4 x 11
Stone ashiar @ $38.95/S.F. wall
East, west and south walls are 64% of wall area
From Table B2010 113 1530 page 294
Split ribbed block @$17.65/S.F. wall

New wall cost
(.36 x $38.95/S.F.) + (.64 x $17.65/S.F.) = $25.32/S.F. wall

Convert Cost per S.F. wall to cost per S.F. of floor
Windows are 14% of wall area

$10.22 S.F. floor = $25.32/S.F. wall x .86 x 469.4 L.F. x 11 ft. high x 5 stories
───
55,000 S.F. of floor

Example 2 (continued)

	NO.	SYSTEM/COMPONENT	DESCRIPTION	UNIT	UNIT COST	NEW S.F. COST	MODEL S.F. COST	+/- CHANGE
D	**SERVICES**							
	D10 Conveying							
D	1010	Elevators & Lifts 2	Type: *Same as model* Capacity: Stops:	Each	224,700	8.17	7.49	+ 0.68
D	1020	Escalators & Moving Walks Type:		Each				
	D20 Plumbing							
D	2010	Plumbing *Kitchen, bath, + service fixtures*	*1 Fixture/215 S.F. Floor*	Each	2240	11.35	11.35	---
D	2020	Domestic Water Distribution		S.F. Floor	3.79	3.79	3.79	---
D	2040	Rain Water Drainage		S.F. Roof	1.26	0.25	0.21	+ 0.04
	D30 HVAC							
D	3010	Energy Supply		S.F. Floor	6.65	6.65	6.65	---
D	3020	Heat Generating Systems Type: *Same as model*		S.F. Floor				
D	3030	Cooling Generating Systems Type: *Same as model*		S.F. Floor	8.03	8.03	8.03	---
D	3090	Other HVAC Sys. & Equipment		Each				
	D40 Fire Protection							
D	4010	Sprinklers *Same as model*		S.F. Floor	2.62	2.62	2.62	---
D	4020	Standpipes *Same as model*		S.F. Floor	0.46	0.46	0.46	---
	D50 Electrical							
D	5010	Electrical Service/Distribution *Same as model*		S.F. Floor	2.45	2.45	2.45	---
D	5020	Lighting & Branch Wiring		S.F. Floor	6.73	6.73	6.73	---
D	5030	Communications & Security		S.F. Floor	0.95	0.95	0.95	---
D	5090	Other Electrical Systems		S.F. Floor	0.18	0.18	0.18	---
E	**EQUIPMENT & FURNISHINGS**							
E	1010	Commercial Equipment		Each				
E	1020	Institutional Equipment		Each				
E	1030	Vehicular Equipment		Each				
E	1090	Other Equipment		Each				
F	**SPECIAL CONSTRUCTION**							
F	1020	Integrated Construction N/A		S.F.				
F	1040	Special Facilities N/A		S.F.				
G	**BUILDING SITEWORK**							

ITEM				
			Total Change	+ 5.32
17	Model 020 total	$ 117.89		
18	Adjusted S.F. cost $ **123.21** item 17 +/- changes			
19	Building area - from item 11 **55,000** S.F. x adjusted S.F. cost	$	6,776,550	
20	Basement area - from item 15 S.F. x S.F. cost $	$		
21	Base building sub-total - item 20 + item 19	$	6,776,550	
22	Miscellaneous addition (quality, etc.)	$		
23	Sub-total - item 22 + 21	$	6,776,550	
24	General conditions -25 % of item 23	$	1,694,138	
25	Sub-total - item 24 + item 23	$	8,470,688	
26	Architects fees **5.5** % of item 25	$	465,888	
27	Sub-total - item 26 + item 27	$	8,936,575	
28	Location modifier zip code 07410	x	1.11	
29	Local replacement cost - item 28 x item 27	$	9,919,599	
30	Depreciation **20** % of item 29	$	1,983,920	
31	Depreciated local replacement cost - item 29 less item 30	$	7,935,679	
32	Exclusions	$		
33	Net depreciated replacement cost - item 31 less item 32	$	7,935,679	

Costs per square foot of floor area

Exterior Wall	S.F. Area	8000	12000	15000	19000	22500	25000	29000	32000	36000
	L.F. Perimeter	213	280	330	350	400	433	442	480	520
Face Brick with Concrete Block Back-up	Wood Joists	198.55	175.50	166.20	154.45	149.90	147.25	141.55	139.80	137.40
	Steel Joists	199.55	177.60	168.85	158.25	153.90	151.40	146.40	144.70	142.45
Stucco on Concrete Block	Wood Joists	176.80	155.80	147.35	137.75	133.60	131.25	126.80	125.15	123.10
	Steel Joists	189.00	167.85	159.30	149.65	145.50	143.10	138.60	136.95	134.90
Wood Siding	Wood Frame	176.20	155.25	146.80	137.35	133.15	130.80	126.45	124.80	122.75
Brick Veneer	Wood Frame	184.85	162.75	153.85	143.15	138.85	136.30	131.20	129.50	127.30
Perimeter Adj., Add or Deduct	Per 100 L.F.	20.40	13.60	10.95	8.60	7.30	6.60	5.65	5.05	4.55
Story Hgt. Adj., Add or Deduct	Per 1 Ft.	3.85	3.35	3.25	2.70	2.55	2.55	2.20	2.15	2.10
For Basement, add $33.15 per square foot of basement area										

The above costs were calculated using the basic specifications shown on the facing page. These costs should be adjusted where necessary for design alternatives and owner's requirements. Reported completed project costs, for this type of structure, range from $55.30 to $205.70 per S.F.

Common additives

Description	Unit	$ Cost
Appliances		
Cooking range, 30" free standing		
1 oven	Each	460 - 2300
2 oven	Each	1600 - 1850
30" built-in		
1 oven	Each	845 - 2150
2 oven	Each	1950 - 2025
Counter top cook tops, 4 burner	Each	425 - 855
Microwave oven	Each	267 - 750
Combination range, refrig. & sink, 30" wide	Each	1625 - 5250
72" wide	Each	5625
Combination range, refrigerator, sink, microwave oven & icemaker	Each	6400
Compactor, residential, 4-1 compaction	Each	685 - 1075
Dishwasher, built-in, 2 cycles	Each	535 - 955
4 cycles	Each	605 - 1750
Garbage disposer, sink type	Each	201 - 330
Hood for range, 2 speed, vented, 30" wide	Each	279 - 1275
42" wide	Each	475 - 2400
Refrigerator, no frost 10-12 C.F.	Each	570 - 630
18-20 C.F.	Each	750 - 1400

Description	Unit	$ Cost
Closed Circuit Surveillance, One station		
Camera and monitor	Each	1875
For additional camera stations, add	Each	1025
Elevators, Hydraulic passenger, 2 stops		
2000# capacity	Each	63,100
2500# capacity	Each	64,300
3500# capacity	Each	67,600
Additional stop, add	Each	8250
Emergency Lighting, 25 watt, battery operated		
Lead battery	Each	287
Nickel cadmium	Each	845
Laundry Equipment		
Dryer, gas, 16 lb. capacity	Each	900
30 lb. capacity	Each	3600
Washer, 4 cycle	Each	1100
Commercial	Each	1450
Smoke Detectors		
Ceiling type	Each	233
Duct type	Each	525

Important: See the Reference Section for Location Factors

Model costs calculated for a 3 story building with 10' story height and 22,500 square feet of floor area

Apartment, 1-3 Story

				Unit	Unit Cost	Cost Per S.F.	% Of Sub-Total
A.	**SUBSTRUCTURE**						
1010	Standard Foundations	Poured concrete; strip and spread footings		S.F. Ground	5.73	1.91	
1020	Special Foundations	N/A		—	—	—	
1030	Slab on Grade	4" reinforced concrete with vapor barrier and granular base		S.F. Slab	4.77	1.59	4.4%
2010	Basement Excavation	Site preparation for slab and trench for foundation wall and footing		S.F. Ground	.18	.06	
2020	Basement Walls	4' foundation wall		L.F. Wall	71	1.46	
B.	**SHELL**						
	B10 Superstructure						
1010	Floor Construction	Open web steel joists, slab form, concrete, interior steel columns		S.F. Floor	17.31	11.54	12.5%
1020	Roof Construction	Open web steel joists with rib metal deck, interior steel columns		S.F. Roof	8.16	2.72	
	B20 Exterior Enclosure						
2010	Exterior Walls	Face brick with concrete block backup	88% of wall	S.F. Wall	22.65	10.63	
2020	Exterior Windows	Aluminum horizontal sliding	12% of wall	Each	497	2.12	11.5%
2030	Exterior Doors	Aluminum and glass		Each	1830	.32	
	B30 Roofing						
3010	Roof Coverings	Built-up tar and gravel with flashing; perlite/EPS composite insulation		S.F. Roof	6.06	2.02	1.8%
3020	Roof Openings	N/A		—	—	—	
C.	**INTERIORS**						
1010	Partitions	Gypsum board and sound deadening board on metal studs	10 S.F. of Floor/L.F. Partition	S.F. Partition	6.11	5.43	
1020	Interior Doors	15% solid core wood, 85% hollow core wood	80 S.F. Floor/Door	Each	510	6.37	
1030	Fittings	Kitchen Cabinets		S.F. Floor	2.83	2.83	
2010	Stair Construction	Concrete filled metal pan		Flight	12,275	2.73	24.7%
3010	Wall Finishes	70% paint, 25% vinyl wall covering, 5% ceramic tile		S.F. Surface	1.32	2.34	
3020	Floor Finishes	60% carpet, 30% vinyl composition tile, 10% ceramic tile		S.F. Floor	5.02	5.02	
3030	Ceiling Finishes	Painted gypsum board on resilient channels		S.F. Ceiling	3.41	3.41	
D.	**SERVICES**						
	D10 Conveying						
1010	Elevators & Lifts	One hydraulic passenger elevator		Each	103,950	4.62	4.1%
1020	Escalators & Moving Walks	N/A		—	—	—	
	D20 Plumbing						
2010	Plumbing Fixtures	Kitchen, bathroom and service fixtures, supply and drainage	1 Fixture/200 S.F. Floor	Each	2488	12.44	
2020	Domestic Water Distribution	Gas fired water heater		S.F. Floor	3.92	3.92	14.7%
2040	Rain Water Drainage	Roof drains		S.F. Roof	1.11	.37	
	D30 HVAC						
3010	Energy Supply	Oil fired hot water, baseboard radiation		S.F. Floor	7.54	7.54	
3020	Heat Generating Systems	N/A		—	—	—	
3030	Cooling Generating Systems	Chilled water, air cooled condenser system		S.F. Floor	8.59	8.59	14.2%
3050	Terminal & Package Units	N/A		—	—	—	
3090	Other HVAC Sys. & Equipment	N/A		—	—	—	
	D40 Fire Protection						
4010	Sprinklers	Wet pipe sprinkler system		S.F. Floor	3.10	3.10	2.7%
4020	Standpipes	N/A		—	—	—	
	D50 Electrical						
5010	Electrical Service/Distribution	600 ampere service, panel board and feeders		S.F. Floor	2.55	2.55	
5020	Lighting & Branch Wiring	Incandescent fixtures, receptacles, switches, A.C. and misc. power		S.F. Floor	6.78	6.78	9.6%
5030	Communications & Security	Alarm systems, internet wiring, and emergency lighting		S.F. Floor	1.37	1.37	
5090	Other Electrical Systems	Emergency generator, 11.5 kW		S.F. Floor	.21	.21	
E.	**EQUIPMENT & FURNISHINGS**						
1010	Commercial Equipment	N/A		—	—	—	
1020	Institutional Equipment	N/A		—	—	—	
1030	Vehicular Equipment	N/A		—	—	—	0.0%
1090	Other Equipment	N/A		—	—	—	
F.	**SPECIAL CONSTRUCTION**						
1020	Integrated Construction	N/A		—	—	—	
1040	Special Facilities	N/A		—	—	—	0.0%
G.	**BUILDING SITEWORK**	**N/A**					

						Sub-Total	113.99	**100%**
	CONTRACTOR FEES (General Requirements: 10%, Overhead: 5%, Profit: 10%)					25%	28.51	
	ARCHITECT FEES					8%	11.40	
						Total Building Cost	**153.90**	

Costs per square foot of floor area

Exterior Wall	S.F. Area	40000	45000	50000	55000	60000	70000	80000	90000	100000
	L.F. Perimeter	366	400	433	466	500	510	530	550	594
Face Brick with Concrete Block Back-up	Steel Frame	161.95	160.55	159.40	158.40	**157.70**	154.30	152.15	150.50	149.70
	R/Conc. Frame	171.20	169.70	168.40	167.35	166.50	162.65	160.20	158.15	157.35
Decorative Concrete Block	Steel Frame	153.30	152.00	151.00	150.15	149.50	146.80	145.00	143.60	142.95
	R/Conc. Frame	156.75	155.45	154.40	153.55	152.85	150.10	148.30	146.85	146.20
Precast Concrete Panels	Steel Frame	166.30	164.65	163.30	162.15	161.30	157.10	154.40	152.25	151.35
	R/Conc. Frame	169.40	167.80	166.40	165.25	164.45	160.15	157.45	155.35	154.45
Perimeter Adj., Add or Deduct	Per 100 L.F.	6.50	5.75	5.20	4.70	4.35	3.75	3.25	2.85	2.60
Story Hgt. Adj., Add or Deduct	Per 1 Ft.	2.10	2.10	2.00	2.00	1.90	1.75	1.55	1.35	1.40
For Basement, add $33.20 per square foot of basement area										

The above costs were calculated using the basic specifications shown on the facing page. These costs should be adjusted where necessary for design alternatives and owner's requirements. Reported completed project costs, for this type of structure, range from $66.25 to $188.35 per S.F.

Common additives

Description	Unit	$ Cost	Description	Unit	$ Cost
Appliances			Closed Circuit Surveillance, One station		
Cooking range, 30" free standing			Camera and monitor	Each	1875
1 oven	Each	460 - 2300	For additional camera stations, add	Each	1025
2 oven	Each	1600 - 1850	Elevators, Electric passenger, 5 stops		
30" built-in			2000# capacity	Each	158,200
1 oven	Each	845 - 2150	3500# capacity	Each	166,200
2 oven	Each	1950 - 2025	5000# capacity	Each	169,700
Counter top cook tops, 4 burner	Each	425 - 855	Additional stop, add	Each	14,000
Microwave oven	Each	267 - 750	Emergency Lighting, 25 watt, battery operated		
Combination range, refrig. & sink, 30" wide	Each	1625 - 5250	Lead battery	Each	287
72" wide	Each	5625	Nickel cadmium	Each	845
Combination range, refrigerator, sink,			Laundry Equipment		
microwave oven & icemaker	Each	6400	Dryer, gas, 16 lb. capacity	Each	900
Compactor, residential, 4-1 compaction	Each	685 - 1075	30 lb. capacity	Each	3600
Dishwasher, built-in, 2 cycles	Each	535 - 955	Washer, 4 cycle	Each	1100
4 cycles	Each	605 - 1750	Commercial	Each	1450
Garbage disposer, sink type	Each	201 - 330	Smoke Detectors		
Hood for range, 2 speed, vented, 30" wide	Each	279 - 1275	Ceiling type	Each	233
42" wide	Each	475 - 2400	Duct type	Each	525
Refrigerator, no frost 10-12 C.F.	Each	570 - 630			
18-20 C.F.	Each	750 - 1400			

Model costs calculated for a 6 story building with 10'-4" story height and 60,000 square feet of floor area

Apartment, 4-7 Story

				Unit	Unit Cost	Cost Per S.F.	% Of Sub-Total
A.	**SUBSTRUCTURE**						
1010	Standard Foundations	Poured concrete; strip and spread footings and 4' foundation wall		S.F. Ground	6.54	1.09	
1020	Special Foundations	N/A		–	–	–	
1030	Slab on Grade	4" reinforced concrete with vapor barrier and granular base		S.F. Slab	4.77	.80	2.2%
2010	Basement Excavation	Site preparation for slab and trench for foundation wall and footing		S.F. Ground	.28	.05	
2020	Basement Walls	4' foundation wall		L.F. Wall	76	.63	
B.	**SHELL**						
	B10 Superstructure						
1010	Floor Construction	Open web steel joists, slab form, concrete, steel columns		S.F. Floor	21.11	17.59	16.0%
1020	Roof Construction	Open web steel joists with rib metal deck, steel columns		S.F. Roof	7.38	1.23	
	B20 Exterior Enclosure						
2010	Exterior Walls	Face brick with concrete block backup	86% of wall	S.F. Wall	22.66	10.07	
2020	Exterior Windows	Aluminum horizontal sliding	14% of wall	Each	497	2.40	10.8%
2030	Exterior Doors	Aluminum and glass		Each	4358	.29	
	B30 Roofing						
3010	Roof Coverings	Built-up tar and gravel with flashing; perlite/EPS composite insulation		S.F. Roof	5.94	.99	0.8%
3020	Roof Openings	N/A		–	–	–	
C.	**INTERIORS**						
1010	Partitions	Gypsum board and sound deadening board on metal studs	8 S.F. Floor/L.F. Partitions	S.F. Partition	6.45	6.45	
1020	Interior Doors	15% solid core wood, 85% hollow core wood	80 S.F Floor/Door	Each	618	7.73	
1030	Fittings	Kitchen cabinets		S.F. Floor	3.40	3.40	
2010	Stair Construction	Concrete filled metal pan		Flight	9825	2.62	27.0%
3010	Wall Finishes	70% paint, 25% vinyl wall covering, 5% ceramic tile		S.F. Surface	1.46	2.91	
3020	Floor Finishes	60% carpet, 30% vinyl composition tile, 10% ceramic tile		S.F. Floor	5.02	5.02	
3030	Ceiling Finishes	Painted gypsum board on resilient channels		S.F. Ceiling	3.71	3.71	
D.	**SERVICES**						
	D10 Conveying						
1010	Elevators & Lifts	Two geared passenger elevators		Each	224,700	7.49	6.4%
1020	Escalators & Moving Walks	N/A		–	–	–	
	D20 Plumbing						
2010	Plumbing Fixtures	Kitchen, bathroom and service fixtures, supply and drainage	1 Fixture/215 S.F. Floor	Each	2440	11.35	
2020	Domestic Water Distribution	Gas fired water heater		S.F. Floor	3.79	3.79	13.0%
2040	Rain Water Drainage	Roof drains		S.F. Roof	1.26	.21	
	D30 HVAC						
3010	Energy Supply	Oil fired hot water, baseboard radiation		S.F. Floor	6.65	6.65	
3020	Heat Generating Systems	N/A		–	–	–	
3030	Cooling Generating Systems	Chilled water, air cooled condenser system		S.F. Floor	8.03	8.03	12.5%
3050	Terminal & Package Units	N/A		–	–	–	
3090	Other HVAC Sys. & Equipment	N/A		–	–	–	
	D40 Fire Protection						
4010	Sprinklers	Wet pipe sprinkler system		S.F. Floor	2.62	2.62	2.6%
4020	Standpipes	Standpipe		S.F. Floor	.46	.46	
	D50 Electrical						
5010	Electrical Service/Distribution	1600 ampere service, panel board and feeders		S.F. Floor	2.45	2.45	
5020	Lighting & Branch Wiring	Incandescent fixtures, receptacles, switches, A.C. and misc. power		S.F. Floor	6.73	6.73	8.7%
5030	Communications & Security	Alarm systems, internet wiring, emergency lighting, and intercom		S.F. Floor	.95	.95	
5090	Other Electrical Systems	Emergency generator, 11.5 kW		S.F. Floor	.18	.18	
E.	**EQUIPMENT & FURNISHINGS**						
1010	Commercial Equipment	N/A		–	–	–	
1020	Institutional Equipment	N/A		–	–	–	
1030	Vehicular Equipment	N/A		–	–	–	0.0 %
1090	Other Equipment	N/A		–	–	–	
F.	**SPECIAL CONSTRUCTION**						
1020	Integrated Construction	N/A		–	–	–	0.0 %
1040	Special Facilities	N/A		–	–	–	
G.	**BUILDING SITEWORK**	**N/A**					

			Sub-Total	117.89	**100%**
CONTRACTOR FEES (General Requirements: 10%, Overhead: 5%, Profit: 10%)		25%	29.49	
ARCHITECT FEES		7%	10.32	
		Total Building Cost	**157.70**	

Costs per square foot of floor area

Exterior Wall	S.F. Area	95000	112000	129000	145000	170000	200000	275000	400000	600000
	L.F. Perimeter	345	386	406	442	480	510	530	570	630
Ribbed Precast Concrete Panel	Steel Frame	220.60	216.15	211.05	208.70	204.75	200.35	191.50	184.30	178.85
	R/Conc. Frame	215.25	210.95	206.15	203.85	200.10	196.00	187.65	180.85	175.80
Face Brick with Concrete Block Back-up	Steel Frame	195.05	191.20	187.15	185.15	181.95	178.50	171.75	166.25	162.15
	R/Conc. Frame	205.95	202.20	198.15	196.10	192.90	189.50	182.75	177.20	173.15
Stucco on Concrete Block	Steel Frame	184.85	181.60	178.35	176.60	174.00	171.40	166.35	162.25	159.25
	R/Conc. Frame	195.85	192.60	189.30	187.55	185.00	182.40	177.35	173.25	170.15
Perimeter Adj., Add or Deduct	Per 100 L.F.	11.50	9.75	8.50	7.50	6.40	5.45	4.00	2.70	1.90
Story Hgt. Adj., Add or Deduct	Per 1 Ft.	3.65	3.50	3.20	3.05	2.80	2.55	1.95	1.40	1.05

For Basement, add $34.00 per square foot of basement area

The above costs were calculated using the basic specifications shown on the facing page. These costs should be adjusted where necessary for design alternatives and owner's requirements. Reported completed project costs, for this type of structure, range from $84.80 to $199.00 per S.F.

Common additives

Description	Unit	$ Cost
Appliances		
Cooking range, 30″ free standing		
1 oven	Each	460 - 2300
2 oven	Each	1600 - 1850
30″ built-in		
1 oven	Each	845 - 2150
2 oven	Each	1950 - 2025
Counter top cook tops, 4 burner	Each	425 - 855
Microwave oven	Each	267 - 750
Combination range, refrig. & sink, 30″ wide	Each	1625 - 5250
72″ wide	Each	5625
Combination range, refrigerator, sink,		
microwave oven & icemaker	Each	6400
Compactor, residential, 4-1 compaction	Each	685 - 1075
Dishwasher, built-in, 2 cycles	Each	535 - 955
4 cycles	Each	605 - 1750
Garbage disposer, sink type	Each	201 - 330
Hood for range, 2 speed, vented, 30″ wide	Each	279 - 1275
42″ wide	Each	475 - 2400
Refrigerator, no frost 10-12 C.F.	Each	570 - 630
18-20 C.F.	Each	750 - 1400

Description	Unit	$ Cost
Closed Circuit Surveillance, One station		
Camera and monitor	Each	1875
For additional camera stations, add	Each	1025
Elevators, Electric passenger, 10 stops		
3000# capacity	Each	430,000
4000# capacity	Each	433,000
5000# capacity	Each	437,000
Additional stop, add	Each	14,000
Emergency Lighting, 25 watt, battery operated		
Lead battery	Each	287
Nickel cadmium	Each	845
Laundry Equipment		
Dryer, gas, 16 lb. capacity	Each	900
30 lb. capacity	Each	3600
Washer, 4 cycle	Each	1100
Commercial	Each	1450
Smoke Detectors		
Ceiling type	Each	233
Duct type	Each	525

Important: See the Reference Section for Location Factors

Model costs calculated for a 15 story building with 10'-6" story height and 145,000 square feet of floor area

Apartment, 8-24 Story

				Unit	Unit Cost	Cost Per S.F.	% Of Sub-Total
A. SUBSTRUCTURE							
1010	Standard Foundations	CIP concrete pile caps		S.F. Ground	8.55	.57	
1020	Special Foundations	Steel H-piles, concrete grade beams		S.F. Ground	199	13.24	
1030	Slab on Grade	4" reinforced concrete with vapor barrier and granular base		S.F. Slab	4.77	.32	9.1%
2010	Basement Excavation	Site preparation for slab, piles and grade beam		S.F. Ground	.28	.02	
2020	Basement Walls	4' Foundation wall		L.F. Wall	71	.26	
B. SHELL							
	B10 Superstructure						
1010	Floor Construction	Open web steel joists, slab form, concrete, interior steel columns		S.F. Floor	19.18	17.90	11.6%
1020	Roof Construction	Open web steel joists with rib metal deck, interior steel columns		S.F. Roof	5.40	.36	
	B20 Exterior Enclosure						
2010	Exterior Walls	Ribbed precast concrete panel	87% of wall	S.F. Wall	45.54	19.02	
2020	Exterior Windows	Aluminum horizontal sliding	13% of wall	Each	497	2.07	15.1%
2030	Exterior Doors	Aluminum and glass		Each	3053	2.62	
	B30 Roofing						
3010	Roof Coverings	Built-up tar and gravel with flashing; perlite/EPS composite insulation		S.F. Roof	5.85	.39	0.2%
3020	Roof Openings	N/A		—	—	—	
C. INTERIORS							
1010	Partitions	Gypsum board on concrete block and metal studs	10 S.F. of Floor/L.F. Partition	S.F. Partition	13.06	13.06	
1020	Interior Doors	15% solid core wood, 85% hollow core wood	80 S.F. Floor/Door	Each	630	7.88	
1030	Fittings	Kitchen cabinets		S.F. Floor	3.51	3.51	
2010	Stair Construction	Concrete filled metal pan		Flight	9825	2.91	24.7%
3010	Wall Finishes	70% paint, 25% vinyl wall covering, 5% ceramic tile		S.F. Surface	1.44	2.87	
3020	Floor Finishes	60% carpet, 30% vinyl composition tile, 10% ceramic tile		S.F. Floor	5.02	5.02	
3030	Ceiling Finishes	Painted gypsum board on resilient channels		S.F. Ceiling	3.71	3.71	
D. SERVICES							
	D10 Conveying						
1010	Elevators & Lifts	Four geared passenger elevators		Each	483,938	13.35	8.5%
1020	Escalators & Moving Walks	N/A		—	—	—	
	D20 Plumbing						
2010	Plumbing Fixtures	Kitchen, bathroom and service fixtures, supply and drainage	1 Fixture/210 S.F. Floor	Each	2734	13.02	
2020	Domestic Water Distribution	Gas fired water heater		S.F. Floor	4.35	4.35	11.1%
2040	Rain Water Drainage	Roof drains		S.F Roof	2.55	.17	
	D30 HVAC						
3010	Energy Supply	Oil fired hot water, baseboard radiation		S.F.Floor	6.65	6.65	
3020	Heat Generating Systems	N/A		—	—	—	
3030	Cooling Generating Systems	Chilled water, air cooled condenser system		S.F. Floor	8.03	8.03	9.3%
3050	Terminal & Package Units	N/A		—	—	—	
3090	Other HVAC Sys. & Equipment	N/A		—	—	—	
	D40 Fire Protection						
4010	Sprinklers	Wet pipe sprinkler system		S.F. Floor	2.57	2.57	2.4%
4020	Standpipes	Standpipe		S.F. Floor	1.21	1.21	
	D50 Electrical						
5010	Electrical Service/Distribution	4000 ampere service, panel board and feeders		S.F. Floor	2.12	2.12	
5020	Lighting & Branch Wiring	Incandescent fixtures, receptacles, switches, A.C. and misc. power		S.F. Floor	7.31	7.31	7.9%
5030	Communications & Security	Alarm systems, internet wiring, emergency lighting, antenna, intercom and security television		S.F. Floor	2.80	2.80	
5090	Other Electrical Systems	Emergency generator, 80 kW		S.F. Floor	.19	.19	
E. EQUIPMENT & FURNISHINGS							
1010	Commercial Equipment	N/A		—	—	—	
1020	Institutional Equipment	N/A		—	—	—	
1030	Vehicular Equipment	N/A		—	—	—	0.0 %
1090	Other Equipment	N/A		—	—	—	
F. SPECIAL CONSTRUCTION							
1020	Integrated Construction	N/A		—	—	—	0.0 %
1040	Special Facilities	N/A		—	—	—	
G. BUILDING SITEWORK	**N/A**						

			Sub-Total	157.50	**100%**
CONTRACTOR FEES (General Requirements: 10%, Overhead: 5%, Profit: 10%)			25%	39.39	
ARCHITECT FEES			6%	11.81	

	Total Building Cost	**208.70**

Costs per square foot of floor area

Exterior Wall	S.F. Area	2000	3000	4000	5000	7500	10000	15000	25000	50000
	L.F. Perimeter	188	224	260	300	356	400	504	700	1200
Brick Veneer	Wood Frame	235.60	202.85	186.45	177.05	161.20	152.60	144.60	137.90	133.00
	Steel Frame	259.10	226.55	210.35	201.05	185.30	176.80	168.90	162.30	157.40
Brick Veneer on Block	Wood Truss	226.20	197.35	183.00	174.90	160.25	152.20	144.80	138.60	134.05
	Bearing Wall	270.40	236.15	219.05	209.30	192.55	183.35	174.85	167.75	162.60
Wood Siding	Wood Frame	221.90	191.65	176.55	167.85	153.55	145.90	138.70	132.65	128.30
Vinyl Siding	Wood Frame	220.60	190.90	176.05	167.45	153.65	146.15	139.20	133.45	129.15
Perimeter Adj., Add or Deduct	Per 100 L.F.	28.65	19.10	14.30	11.50	7.65	5.75	3.80	2.30	1.10
Story Hgt. Adj., Add or Deduct	Per 1 Ft.	3.60	2.80	2.45	2.30	1.80	1.55	1.30	1.05	0.90

For Basement, add $29.90 per square foot of basement area

The above costs were calculated using the basic specifications shown on the facing page. These costs should be adjusted where necessary for design alternatives and owner's requirements. Reported completed project costs, for this type of structure, range from $83.30 to $208.90 per S.F.

Common additives

Description	Unit	$ Cost
Appliances		
Cooking range, 30" free standing		
1 oven	Each	460 - 2300
2 oven	Each	1600 - 1850
30" built-in		
1 oven	Each	845 - 2150
2 oven	Each	1950 - 2025
Counter top cook tops, 4 burner	Each	425 - 855
Microwave oven	Each	267 - 750
Combination range, refrig. & sink, 30" wide	Each	1625 - 5250
60" wide	Each	4375
72" wide	Each	5625
Combination range, refrigerator, sink,		
microwave oven & icemaker	Each	6400
Compactor, residential, 4-1 compaction	Each	685 - 1075
Dishwasher, built-in, 2 cycles	Each	535 - 955
4 cycles	Each	605 - 1750
Garbage disposer, sink type	Each	201 - 330
Hood for range, 2 speed, vented, 30" wide	Each	279 - 1275
42" wide	Each	475 - 2400

Description	Unit	$ Cost
Appliances, cont.		
Refrigerator, no frost 10-12 C.F.	Each	570 - 630
14-16 C.F.	Each	610 - 945
18-20 C.F.	Each	750 - 1400
Laundry Equipment		
Dryer, gas, 16 lb. capacity	Each	900
30 lb. capacity	Each	3600
Washer, 4 cycle	Each	1100
Commercial	Each	1450
Sound System		
Amplifier, 250 watts	Each	2400
Speaker, ceiling or wall	Each	196
Trumpet	Each	375

Important: See the Reference Section for Location Factors

Model costs calculated for a 1 story building with 10' story height and 10,000 square feet of floor area

Assisted - Senior Living

				Unit	Unit Cost	Cost Per S.F.	% Of Sub-Total
A.	**SUBSTRUCTURE**						
1010	Standard Foundations	Poured concrete strip footings		S.F. Ground	1.85	1.85	
1020	Special Foundations	N/A		—	—	—	
1030	Slab on Grade	4" reinforced concrete with vapor barrier and granular base		S.F. Slab	4.77	4.77	9.2%
2010	Basement Excavation	Site preparation for slab and trench for foundation wall and footing		S.F. Ground	.52	.52	
2020	Basement Walls	4' foundation wall		L.F. Wall	76	3.02	
B.	**SHELL**						
	B10 Superstructure						
1010	Floor Construction	N/A		—	—	—	6.4 %
1020	Roof Construction	Plywood on wood trusses (pitched); fiberglass batt insulation		S.F. Roof	7.08	7.08	
	B20 Exterior Enclosure						
2010	Exterior Walls	Brick veneer on wood studs, insulated	80% of wall	S.F. Wall	23.03	7.37	
2020	Exterior Windows	Double hung wood	20% of wall	Each	592	2.97	10.6%
2030	Exterior Doors	Aluminum and glass; 18 ga. steel flush		Each	2237	1.34	
	B30 Roofing						
3010	Roof Coverings	Asphalt shingles with flashing (pitched); al. gutters & downspouts		S.F. Roof	2.63	2.63	2.4%
3020	Roof Openings	N/A		—	—	—	
C.	**INTERIORS**						
1010	Partitions	Gypsum board on wood studs	6 S.F. Floor/L.F. Partition	S.F. Partition	4.57	6.09	
1020	Interior Doors	Single leaf wood	80 S.F. Floor/Door	Each	576	6.78	
1030	Fittings	N/A		—	—	—	
2010	Stair Construction	N/A		—	—	—	29.6%
3010	Wall Finishes	Paint on drywall		S.F. Surface	.74	1.97	
3020	Floor Finishes	80% carpet, 20% ceramic tile		S.F. Floor	9.84	9.84	
3030	Ceiling Finishes	Gypsum board on wood furring, painted		S.F. Ceiling	7.88	7.88	
D.	**SERVICES**						
	D10 Conveying						
1010	Elevators & Lifts	N/A		—	—	—	0.0 %
1020	Escalators & Moving Walks	N/A		—	—	—	
	D20 Plumbing						
2010	Plumbing Fixtures	Kitchen, toilet and service fixtures, supply and drainage	1 Fixture/200 S.F. Floor	Each	1742	8.71	
2020	Domestic Water Distribution	Included with D2010		—	—	—	7.9%
2040	Rain Water Drainage	N/A		—	—	—	
	D30 HVAC						
3010	Energy Supply	N/A		—	—	—	
3020	Heat Generating Systems	N/A		—	—	—	
3030	Cooling Generating Systems	N/A		—	—	—	2.8 %
3050	Terminal & Package Units	Gas fired hot water boiler; terminal package heat/A/C units		S.F. Floor	3.08	3.08	
3090	Other HVAC Sys. & Equipment	N/A		—	—	—	
	D40 Fire Protection						
4010	Sprinklers	Wet pipe sprinkler system		S.F. Floor	3.16	3.16	2.9%
4020	Standpipes	N/A		—	—	—	
	D50 Electrical						
5010	Electrical Service/Distribution	600 ampere service, panel board, feeders and branch wiring		S.F. Floor	9.44	9.44	
5020	Lighting & Branch Wiring	Fluorescent & incandescent fixtures, receptacles, switches, A.C. and misc. power		S.F. Floor	10.74	10.74	26.8%
5030	Communications & Security	Alarm, fire detection, and communication system		S.F. Floor	9.12	9.12	
5090	Other Electrical Systems	Emergency generator, 7.5 kW		S.F. Floor	.20	.20	
E.	**EQUIPMENT & FURNISHINGS**						
1010	Commercial Equipment	Residential washers and dryers		S.F. Floor	.40	.40	
1020	Institutional Equipment	N/A		—	—	—	1.3%
1030	Vehicular Equipment	N/A		—	—	—	
1090	Other Equipment	Residential ranges and refrigerators		S.F. Floor	1.01	1.01	
F.	**SPECIAL CONSTRUCTION**						
1020	Integrated Construction	N/A		—	—	—	0.0 %
1040	Special Facilities	N/A		—	—	—	
G.	**BUILDING SITEWORK**	**N/A**					

		Sub-Total	109.97	**100%**
CONTRACTOR FEES (General Requirements: 10%, Overhead: 5%, Profit: 10%)			25%	27.51
ARCHITECT FEES			11%	15.12
		Total Building Cost	**152.60**	

Costs per square foot of floor area

Exterior Wall	S.F. Area	12000	15000	18000	21000	24000	27000	30000	33000	36000
	L.F. Perimeter	440	500	540	590	640	665	700	732	770
Face Brick with Concrete Block Back-up	Steel Frame	175.90	170.35	164.90	161.75	159.40	156.20	154.10	152.25	151.00
	Bearing Wall	169.85	164.85	159.90	157.10	155.00	152.10	150.25	148.55	147.40
Precast Concrete	Steel Frame	180.75	174.85	168.95	165.60	163.05	159.65	157.40	155.40	154.00
Decorative Concrete Block	Bearing Wall	152.35	148.85	145.45	143.50	142.05	140.10	138.85	137.70	136.95
Concrete Block	Steel Frame	157.40	154.50	151.75	150.10	148.95	147.35	146.35	145.40	144.75
	Bearing Wall	149.00	145.80	142.70	140.95	139.60	137.85	136.75	135.65	134.95
Perimeter Adj., Add or Deduct	Per 100 L.F.	12.40	9.90	8.30	7.05	6.15	5.45	4.90	4.50	4.10
Story Hgt. Adj., Add or Deduct	Per 1 Ft.	1.95	1.75	1.55	1.50	1.40	1.25	1.25	1.20	1.05
For Basement, add $27.15 per square foot of basement area										

The above costs were calculated using the basic specifications shown on the facing page. These costs should be adjusted where necessary for design alternatives and owner's requirements. Reported completed project costs, for this type of structure, range from $83.95 to $212.50 per S.F.

Common additives

Description	Unit	$ Cost
Closed Circuit Surveillance, One station		
Camera and monitor	Each	1875
For additional camera stations, add	Each	1025
Emergency Lighting, 25 watt, battery operated		
Lead battery	Each	287
Nickel cadmium	Each	845
Seating		
Auditorium chair, all veneer	Each	240
Veneer back, padded seat	Each	290
Upholstered, spring seat	Each	279
Classroom, movable chair & desk	Set	65 - 120
Lecture hall, pedestal type	Each	229 - 680
Smoke Detectors		
Ceiling type	Each	233
Duct type	Each	525
Sound System		
Amplifier, 250 watts	Each	2400
Speaker, ceiling or wall	Each	196
Trumpet	Each	375

Important: See the Reference Section for Location Factors

Model costs calculated for a 1 story building with 24′ story height and 24,000 square feet of floor area

Auditorium

			Unit	Unit Cost	Cost Per S.F.	% Of Sub-Total
A. SUBSTRUCTURE						
1010	Standard Foundations	Poured concrete; strip and spread footings	S.F. Ground	1.32	1.32	
1020	Special Foundations	N/A	—	—	—	
1030	Slab on Grade	6″ reinforced concrete with vapor barrier and granular base	S.F. Slab	5.79	5.79	7.8%
2010	Basement Excavation	Site preparation for slab and trench for foundation wall and footing	S.F. Ground	.18	.18	
2020	Basement Walls	4′ foundation wall	L.F. Wall	84	2.24	
B. SHELL						
	B10 Superstructure					
1010	Floor Construction	Open web steel joists, slab form, concrete (balcony)	S.F. Floor	20.32	2.54	8.0%
1020	Roof Construction	Metal deck on steel truss	S.F. Roof	7.25	7.25	
	B20 Exterior Enclosure					
2010	Exterior Walls	Precast concrete panel 80% of wall (adjusted for end walls)	S.F. Wall	41.29	21.14	
2020	Exterior Windows	Glass curtain wall 20% of wall	Each	38.85	4.98	23.0%
2030	Exterior Doors	Double aluminum and glass and hollow metal	Each	3793	1.89	
	B30 Roofing					
3010	Roof Coverings	Built-up tar and gravel with flashing; perlite/EPS composite insulation	S.F. Roof	5.60	5.60	4.8%
3020	Roof Openings	Roof hatches	S.F. Roof	.21	.21	
C. INTERIORS						
1010	Partitions	Concrete Block and toilet partitions 40 S.F. Floor/L.F. Partition	S.F. Partition	8.63	3.45	
1020	Interior Doors	Single leaf hollow metal 400 S.F. Floor/Door	Each	1001	2.50	
1030	Fittings	N/A	—	—	—	
2010	Stair Construction	Concrete filled metal pan	Flight	14,675	1.84	21.1%
3010	Wall Finishes	70% paint, 30% epoxy coating	S.F. Surface	5.01	4.01	
3020	Floor Finishes	70% vinyl tile, 30% carpet	S.F. Floor	9.60	9.60	
3030	Ceiling Finishes	Fiberglass board, suspended	S.F. Ceiling	4.31	4.31	
D. SERVICES						
	D10 Conveying					
1010	Elevators & Lifts	One hydraulic passenger elevator	Each	84,960	3.54	2.9%
1020	Escalators & Moving Walks	N/A	—	—	—	
	D20 Plumbing					
2010	Plumbing Fixtures	Toilet and service fixtures, supply and drainage 1 Fixture/800 S.F. Floor	Each	3760	4.70	
2020	Domestic Water Distribution	Gas fired water heater	S.F. Floor	.36	.36	4.9%
2040	Rain Water Drainage	Roof drains	S.F. Roof	.91	.91	
	D30 HVAC					
3010	Energy Supply	N/A	—	—	—	
3020	Heat Generating Systems	Included in D3050	—	—	—	
3030	Cooling Generating Systems	N/A	—	—	—	9.9%
3050	Terminal & Package Units	Single zone rooftop unit, gas heating, electric cooling	S.F. Floor	12.05	12.05	
3090	Other HVAC Sys. & Equipment	N/A	—	—	—	
	D40 Fire Protection					
4010	Sprinklers	Wet pipe sprinkler system	S.F. Floor	3.16	3.16	2.6%
4020	Standpipes	N/A	—	—	—	
	D50 Electrical					
5010	Electrical Service/Distribution	800 ampere service, panel board and feeders	S.F. Floor	2.40	2.40	
5020	Lighting & Branch Wiring	Fluorescent fixtures, receptacles, switches, A.C. and misc. power	S.F. Floor	11.41	11.41	15.1%
5030	Communications & Security	Alarm systems and emergency lighting, and public address system	S.F. Floor	3.27	3.27	
5090	Other Electrical Systems	Emergency generator, 100 kW	S.F. Floor	1.27	1.27	
E. EQUIPMENT & FURNISHINGS						
1010	Commercial Equipment	N/A	—	—	—	
1020	Institutional Equipment	N/A	—	—	—	
1030	Vehicular Equipment	N/A	—	—	—	0.0%
1090	Other Equipment	N/A	—	—	—	
F. SPECIAL CONSTRUCTION						
1020	Integrated Construction	N/A	—	—	—	0.0%
1040	Special Facilities	N/A	—	—	—	
G. BUILDING SITEWORK	**N/A**					

		Sub-Total	121.92	100%
CONTRACTOR FEES (General Requirements: 10%, Overhead: 5%, Profit: 10%)			25%	30.46
ARCHITECT FEES			7%	10.67

Total Building Cost	**163.05**

Costs per square foot of floor area

Exterior Wall	S.F. Area	2000	2700	3400	4100	4800	5500	6200	6900	7600
	L.F. Perimeter	180	208	236	256	280	303	317	337	357
Face Brick with Concrete Block Back-up	Steel Frame	241.15	226.00	217.00	209.35	204.65	201.05	196.85	194.35	192.30
	R/Conc. Frame	257.70	242.55	233.55	225.90	221.20	217.55	213.45	210.90	208.85
Precast Concrete Panel	Steel Frame	246.40	230.50	221.00	213.00	208.05	204.25	199.85	197.20	195.05
	R/Conc. Frame	263.00	247.05	237.60	229.55	224.65	220.75	216.40	213.75	211.60
Limestone with Concrete Block Back-up	Steel Frame	282.70	261.50	249.00	238.20	231.60	226.40	220.50	216.90	214.00
	R/Conc. Frame	299.25	278.10	265.55	254.70	248.15	243.00	237.05	233.40	230.50
Perimeter Adj., Add or Deduct	Per 100 L.F.	45.85	33.95	27.00	22.40	19.10	16.60	14.80	13.30	12.10
Story Hgt. Adj., Add or Deduct	Per 1 Ft.	4.35	3.75	3.35	3.00	2.85	2.65	2.50	2.35	2.25

For Basement, add $31.40 per square foot of basement area

The above costs were calculated using the basic specifications shown on the facing page. These costs should be adjusted where necessary for design alternatives and owner's requirements. Reported completed project costs, for this type of structure, range from $128.00 to $315.05 per S.F.

Common additives

Description	Unit	$ Cost
Bulletproof Teller Window, 44" x 60"	Each	4400
60" x 48"	Each	5375
Closed Circuit Surveillance, One station		
Camera and monitor	Each	1875
For additional camera stations, add	Each	1025
Counters, Complete	Station	5800
Door & Frame, 3' x 6'-8", bullet resistant steel		
with vision panel	Each	6700 - 8675
Drive-up Window, Drawer & micr., not incl. glass	Each	7475 - 12,400
Emergency Lighting, 25 watt, battery operated		
Lead battery	Each	287
Nickel cadmium	Each	845
Night Depository	Each	9300 - 14,000
Package Receiver, painted	Each	1825
stainless steel	Each	2825
Partitions, Bullet resistant to 8' high	L.F.	310 - 495
Pneumatic Tube Systems, 2 station	Each	31,500
With TV viewer	Each	58,000

Description	Unit	$ Cost
Service Windows, Pass thru, steel		
24" x 36"	Each	3350
48" x 48"	Each	3900
72" x 40"	Each	5550
Smoke Detectors		
Ceiling type	Each	233
Duct type	Each	525
Twenty-four Hour Teller		
Automatic deposit cash & memo	Each	51,500
Vault Front, Door & frame		
1 hour test, 32"x 78"	Opening	5375
2 hour test, 32" door	Opening	5825
40" door	Opening	6775
4 hour test, 32" door	Opening	6300
40" door	Opening	7525
Time lock, two movement, add	Each	1875

Important: See the Reference Section for Location Factors

Model costs calculated for a 1 story building with 14' story height and 4,100 square feet of floor area

Bank

				Unit	Unit Cost	Cost Per S.F.	% Of Sub-Total
A. SUBSTRUCTURE							
1010	Standard Foundations	Poured concrete; strip and spread footings		S.F. Ground	3.82	3.82	
1020	Special Foundations	N/A		—	—	—	
1030	Slab on Grade	4" reinforced concrete with vapor barrier and granular base		S.F. Slab	4.77	4.77	8.9%
2010	Basement Excavation	Site preparation for slab and trench for foundation wall and footing		S.F. Ground	.31	.31	
2020	Basement Walls	4' Foundation wall		L.F. Wall	76	5.66	
B. SHELL							
	B10 Superstructure						
1010	Floor Construction	Cast-in-place columns		L.F. Column	112	5.44	12.3%
1020	Roof Construction	Cast-in-place concrete flat plate		S.F. Roof	14.65	14.65	
	B20 Exterior Enclosure						
2010	Exterior Walls	Face brick with concrete block backup	80% of wall	S.F. Wall	30.30	21.19	
2020	Exterior Windows	Horizontal aluminum sliding	20% of wall	Each	557	6.49	17.9%
2030	Exterior Doors	Double aluminum and glass and hollow metal		Each	3020	1.47	
	B30 Roofing						
3010	Roof Coverings	Built-up tar and gravel with flashing; perlite/EPS composite insulation		S.F. Roof	6.99	6.99	4.3%
3020	Roof Openings	N/A		—	—	—	
C. INTERIORS							
1010	Partitions	Gypsum board on metal studs	20 S.F. of Floor/L.F. Partition	S.F. Partition	9.40	4.70	
1020	Interior Doors	Single leaf hollow metal	200 S.F. Floor/Door	Each	1001	5.01	
1030	Fittings	N/A		—	—	—	
2010	Stair Construction	N/A		—	—	—	14.0%
3010	Wall Finishes	50% vinyl wall covering, 50% paint		S.F. Surface	1.29	1.29	
3020	Floor Finishes	50% carpet tile, 40% vinyl composition tile, 10% quarry tile		S.F. Floor	5.31	5.31	
3030	Ceiling Finishes	Mineral fiber tile on concealed zee bars		S.F. Ceiling	6.51	6.51	
D. SERVICES							
	D10 Conveying						
1010	Elevators & Lifts	N/A		—	—	—	0.0%
1020	Escalators & Moving Walks	N/A		—	—	—	
	D20 Plumbing						
2010	Plumbing Fixtures	Toilet and service fixtures, supply and drainage	1 Fixture/580 S.F. Floor	Each	4449	7.67	
2020	Domestic Water Distribution	Gas fired water heater		S.F. Floor	1.08	1.08	6.0%
2040	Rain Water Drainage	Roof drains		S.F. Roof	.99	.99	
	D30 HVAC						
3010	Energy Supply	N/A		—	—	—	
3020	Heat Generating Systems	Included in D3050		—	—	—	
3030	Cooling Generating Systems	N/A		—	—	—	7.0%
3050	Terminal & Package Units	Single zone rooftop unit, gas heating, electric cooling		S.F. Floor	11.45	11.45	
3090	Other HVAC Sys. & Equipment	N/A		—	—	—	
	D40 Fire Protection						
4010	Sprinklers	Wet pipe sprinkler system		S.F. Floor	3.99	3.99	2.5%
4020	Standpipes	N/A		—	—	—	
	D50 Electrical						
5010	Electrical Service/Distribution	200 ampere service, panel board and feeders		S.F. Floor	2.41	2.41	
5020	Lighting & Branch Wiring	Fluorescent fixtures, receptacles, switches, A.C. and misc. power		S.F. Floor	10.24	10.24	13.0%
5030	Communications & Security	Alarm systems, internet and phone wiring, emergency lighting, and security television		S.F. Floor	7.99	7.99	
5090	Other Electrical Systems	Emergency generator, 15 kW, Uninterruptible power supply		S.F. Floor	.60	.60	
E. EQUIPMENT & FURNISHINGS							
1010	Commercial Equipment	Automatic teller, drive up window, night depository		S.F. Floor	7.52	7.52	
1020	Institutional Equipment	Closed circuit TV monitoring system		S.F. Floor	3.09	3.09	6.5%
1030	Vehicular Equipment	N/A		—	—	—	
1090	Other Equipment	N/A		—	—	—	
F. SPECIAL CONSTRUCTION							
1020	Integrated Construction	N/A		—	—	—	7.5%
1040	Special Facilities	Security vault door		S.F. Floor	12.17	12.17	
G. BUILDING SITEWORK	**N/A**						

			Sub-Total	162.81	**100%**
CONTRACTOR FEES (General Requirements: 10%, Overhead: 5%, Profit: 10%)			25%	40.70	
ARCHITECT FEES			11%	22.39	

Total Building Cost   **225.90**

Costs per square foot of floor area

Exterior Wall	S.F. Area	12000	14000	16000	18000	20000	22000	24000	26000	28000
	L.F. Perimeter	460	491	520	540	566	593	620	645	670
Concrete Block	Steel Roof Deck	161.10	158.00	155.60	153.45	151.80	150.65	149.60	148.65	147.90
Decorative Concrete Block	Steel Roof Deck	164.95	161.55	158.90	156.50	154.75	153.40	152.20	151.20	150.35
Face Brick on Concrete Block	Steel Roof Deck	173.05	168.90	165.70	162.80	160.65	159.05	157.60	156.40	155.35
Jumbo Brick on Concrete Block	Steel Roof Deck	169.25	165.45	162.50	159.80	157.85	156.35	155.05	153.90	153.00
Stucco on Concrete Block	Steel Roof Deck	164.20	160.85	158.20	155.85	154.10	152.85	151.70	150.70	149.85
Precast Concrete Panel	Steel Roof Deck	168.80	165.05	162.10	159.50	157.55	156.05	154.75	153.65	152.70
Perimeter Adj., Add or Deduct	Per 100 L.F.	4.75	4.15	3.60	3.25	2.95	2.55	2.40	2.25	2.05
Story Hgt. Adj., Add or Deduct	Per 1 Ft.	1.00	0.90	0.90	0.85	0.85	0.70	0.70	0.70	0.65

For Basement, add $30.95 per square foot of basement area

The above costs were calculated using the basic specifications shown on the facing page. These costs should be adjusted where necessary for design alternatives and owner's requirements. Reported completed project costs, for this type of structure, range from $56.60 to $139.10 per S.F.

Common additives

Description	Unit	$ Cost
Bowling Alleys, Incl. alley, pinsetter,		
scorer, counter & misc. supplies, average	Lane	60,500
For automatic scorer, add (maximum)	Lane	10,900
Emergency Lighting, 25 watt, battery operated		
Lead battery	Each	287
Nickel cadmium	Each	845
Lockers, Steel, Single tier, 60" or 72"	Opening	200 - 350
2 tier, 60" or 72" total	Opening	124 - 152
5 tier, box lockers	Opening	70 - 82.50
Locker bench, lam., maple top only	L.F.	22
Pedestals, steel pipe	Each	64
Seating		
Auditorium chair, all veneer	Each	240
Veneer back, padded seat	Each	290
Upholstered, spring seat	Each	279
Sound System		
Amplifier, 250 watts	Each	2400
Speaker, ceiling or wall	Each	196
Trumpet	Each	375

Model costs calculated for a 1 story building with 14' story height and 20,000 square feet of floor area

			Unit	Unit Cost	Cost Per S.F.	% Of Sub-Total
A. SUBSTRUCTURE						
1010	Standard Foundations	Poured concrete; strip and spread footings	S.F. Ground	1.12	1.12	
1020	Special Foundations	N/A	—	—	—	
1030	Slab on Grade	4" reinforced concrete with vapor barrier and granular base	S.F. Slab	4.77	4.77	7.5%
2010	Basement Excavation	Site preparation for slab and trench for foundation wall and footing	S.F. Ground	.52	.52	
2020	Basement Walls	4' Foundation wall	L.F. Wall	76	2.14	
B. SHELL						
	B10 Superstructure					
1010	Floor Construction	N/A	—	—	—	6.3 %
1020	Roof Construction	Metal deck on open web steel joists with columns and beams	S.F. Roof	7.17	7.17	
	B20 Exterior Enclosure					
2010	Exterior Walls	Concrete block 90% of wall	S.F. Wall	11.75	4.19	
2020	Exterior Windows	Horizontal pivoted 10% of wall	Each	1875	3.10	6.9%
2030	Exterior Doors	Double aluminum and glass and hollow metal	Each	2236	.57	
	B30 Roofing					
3010	Roof Coverings	Built-up tar and gravel with flashing; perlite/EPS composite insulation	S.F. Roof	5.36	5.36	4.8%
3020	Roof Openings	Roof hatches	S.F. Roof	.14	.14	
C. INTERIORS						
1010	Partitions	Concrete block 50 S.F. Floor/L.F. Partition	S.F. Partition	9.60	1.92	
1020	Interior Doors	Hollow metal single leaf 1000 S.F. Floor/Door	Each	1001	1	
1030	Fittings	N/A	—	—	—	
2010	Stair Construction	N/A	—	—	—	6.0%
3010	Wall Finishes	Paint and block filler	S.F. Surface	5.48	2.19	
3020	Floor Finishes	Vinyl tile 25% of floor	S.F. Floor	2.48	.62	
3030	Ceiling Finishes	Suspended fiberglass board 25% of area	S.F. Ceiling	4.31	1.08	
D. SERVICES						
	D10 Conveying					
1010	Elevators & Lifts	N/A	—	—	—	0.0 %
1020	Escalators & Moving Walks	N/A	—	—	—	
	D20 Plumbing					
2010	Plumbing Fixtures	Toilet and service fixtures, supply and drainage 1 Fixture/2200 S.F. Floor	Each	2970	1.35	
2020	Domestic Water Distribution	Gas fired water heater	S.F. Floor	.21	.21	1.9%
2040	Rain Water Drainage	Roof drains	S.F. Roof	.58	.58	
	D30 HVAC					
3010	Energy Supply	N/A	—	—	—	
3020	Heat Generating Systems	Included in D3050	—	—	—	
3030	Cooling Generating Systems	N/A	—	—	—	12.1 %
3050	Terminal & Package Units	Single zone rooftop unit, gas heating, electric cooling	S.F. Floor	13.70	13.70	
3090	Other HVAC Sys. & Equipment	N/A	—	—	—	
	D40 Fire Protection					
4010	Sprinklers	Sprinklers, light hazard	S.F. Floor	3.16	3.16	3.2%
4020	Standpipes	Standpipe, wet, Class III	S.F. Floor	.51	.51	
	D50 Electrical					
5010	Electrical Service/Distribution	800 ampere service, panel board and feeders	S.F. Floor	2.88	2.88	
5020	Lighting & Branch Wiring	High efficiency fluorescent fixtures, receptacles, switches, A.C. and misc. power	S.F. Floor	10.56	10.56	13.0%
5030	Communications & Security	Addressable alarm systems and emergency lighting	S.F. Floor	1.27	1.27	
5090	Other Electrical Systems	Emergency generator, 7.5 kW	S.F. Floor	.07	.07	
E. EQUIPMENT & FURNISHINGS						
1010	Commercial Equipment	N/A	—	—	—	
1020	Institutional Equipment	N/A	—	—	—	
1030	Vehicular Equipment	N/A	—	—	—	38.2 %
1090	Other Equipment	Bowling Alley, ballrack, automatic scorers	S.F. Floor	43.34	43.34	
F. SPECIAL CONSTRUCTION						
1020	Integrated Construction	N/A	—	—	—	0.0 %
1040	Special Facilities	N/A	—	—	—	
G. BUILDING SITEWORK	**N/A**					

		Sub-Total	113.52	100%
CONTRACTOR FEES (General Requirements: 10%, Overhead: 5%, Profit: 10%)		25%	28.35	
ARCHITECT FEES		7%	9.93	

	Total Building Cost	**151.80**

Costs per square foot of floor area

Exterior Wall	S.F. Area	6000	8000	10000	12000	14000	16000	18000	20000	22000
	L.F. Perimeter	320	386	453	520	540	597	610	660	710
Face Brick with Concrete Block Back-up	Bearing Walls	151.70	145.60	142.10	139.75	135.40	133.95	130.95	129.95	129.15
	Steel Frame	154.25	148.30	144.80	142.50	138.20	136.75	133.75	132.80	132.05
Decorative Concrete Block	Bearing Walls	143.00	137.75	134.70	132.65	129.05	127.80	125.40	124.55	123.90
	Steel Frame	145.60	140.40	137.40	135.40	131.90	130.70	128.25	127.45	126.85
Precast Concrete Panels	Bearing Walls	150.65	144.70	141.25	138.90	134.70	133.25	130.25	129.30	128.60
	Steel Frame	153.30	147.35	144.00	141.65	137.50	136.05	133.15	132.20	131.45
Perimeter Adj., Add or Deduct	Per 100 L.F.	13.10	9.90	7.85	6.55	5.60	4.90	4.35	3.95	3.60
Story Hgt. Adj., Add or Deduct	Per 1 Ft.	2.20	2.00	1.90	1.75	1.55	1.55	1.35	1.35	1.30
Basement—Not Applicable										

The above costs were calculated using the basic specifications shown on the facing page. These costs should be adjusted where necessary for design alternatives and owner's requirements. Reported completed project costs, for this type of structure, range from $68.40 to $164.90 per S.F.

Common additives

Description	Unit	$ Cost
Directory Boards, Plastic, glass covered		
30" x 20"	Each	605
36" x 48"	Each	1325
Aluminum, 24" x 18"	Each	585
36" x 24"	Each	685
48" x 32"	Each	975
48" x 60"	Each	2025
Emergency Lighting, 25 watt, battery operated		
Lead battery	Each	287
Nickel cadmium	Each	845
Benches, Hardwood	L.F.	107 - 195
Ticket Printer	Each	7125
Turnstiles, One way		
4 arm, 46" dia., manual	Each	1100
Electric	Each	2300
High security, 3 arm		
65" dia., manual	Each	5400
Electric	Each	6650
Gate with horizontal bars		
65" dia., 7' high transit type	Each	6125

Model costs calculated for a 1 story building with 14' story height and 12,000 square feet of floor area

				Unit	Unit Cost	Cost Per S.F.	% Of Sub-Total
A.	**SUBSTRUCTURE**						
1010	Standard Foundations	Poured concrete; strip and spread footings		S.F. Ground	2.36	2.36	
1020	Special Foundations	N/A		—	—	—	
1030	Slab on Grade	6" reinforced concrete with vapor barrier and granular base		S.F. Slab	7.26	7.26	12.7%
2010	Basement Excavation	Site preparation for slab and trench for foundation wall and footing		S.F. Ground	.31	.31	
2020	Basement Walls	4' Foundation wall		L.F. Wall	78	3.36	
B.	**SHELL**						
	B10 Superstructure						
1010	Floor Construction	N/A		—	—	—	
1020	Roof Construction	Metal deck on open web steel joists with columns and beams		S.F. Roof	5.48	5.48	5.2 %
	B20 Exterior Enclosure						
2010	Exterior Walls	Face brick with concrete block backup	70% of wall	S.F. Wall	30.31	12.87	
2020	Exterior Windows	Store front	30% of wall	Each	29.20	5.32	19.1%
2030	Exterior Doors	Double aluminum and glass		Each	5275	1.76	
	B30 Roofing						
3010	Roof Coverings	Built-up tar and gravel with flashing; perlite/EPS composite insulation		S.F. Roof	6.25	6.25	6.0%
3020	Roof Openings	N/A		—	—	—	
C.	**INTERIORS**						
1010	Partitions	Lightweight concrete block	15 S.F. Floor/L.F. Partition	S.F. Partition	9.60	6.40	
1020	Interior Doors	Hollow metal	150 S.F. Floor/Door	Each	1001	6.67	
1030	Fittings	N/A		—	—	—	
2010	Stair Construction	N/A		—	—	—	29.6%
3010	Wall Finishes	Glazed coating		S.F. Surface	1.79	2.39	
3020	Floor Finishes	Quarry tile and vinyl composition tile		S.F. Floor	8.94	8.94	
3030	Ceiling Finishes	Mineral fiber tile on concealed zee bars		S.F. Ceiling	6.51	6.51	
D.	**SERVICES**						
	D10 Conveying						
1010	Elevators & Lifts	N/A		—	—	—	
1020	Escalators & Moving Walks	N/A		—	—	—	0.0 %
	D20 Plumbing						
2010	Plumbing Fixtures	Toilet and service fixtures, supply and drainage	1 Fixture/850 S.F. Floor	Each	2907	3.42	
2020	Domestic Water Distribution	Electric water heater		S.F. Floor	.76	.76	4.6%
2040	Rain Water Drainage	Roof drains		S.F. Roof	.65	.65	
	D30 HVAC						
3010	Energy Supply	N/A		—	—	—	
3020	Heat Generating Systems	Included in D3050		—	—	—	
3030	Cooling Generating Systems	N/A		—	—	—	9.8 %
3050	Terminal & Package Units	Single zone rooftop unit, gas heating, electric cooling		S.F. Floor	10.26	10.26	
3090	Other HVAC Sys. & Equipment	N/A		—	—	—	
	D40 Fire Protection						
4010	Sprinklers	Wet pipe sprinkler system		S.F. Floor	4.11	4.11	3.9%
4020	Standpipes	N/A		—	—	—	
	D50 Electrical						
5010	Electrical Service/Distribution	400 ampere service, panel board and feeders		S.F. Floor	1.68	1.68	
5020	Lighting & Branch Wiring	Fluorescent fixtures, receptacles, switches, A.C. and misc. power		S.F. Floor	7.02	7.02	9.0%
5030	Communications & Security	Alarm systems and emergency lighting		S.F. Floor	.28	.28	
5090	Other Electrical Systems	Emergency generator, 7.5 kW		S.F. Floor	.41	.41	
E.	**EQUIPMENT & FURNISHINGS**						
1010	Commercial Equipment	N/A		—	—	—	
1020	Institutional Equipment	N/A		—	—	—	
1030	Vehicular Equipment	N/A		—	—	—	0.0 %
1090	Other Equipment	N/A		—	—	—	
F.	**SPECIAL CONSTRUCTION**						
1020	Integrated Construction	N/A		—	—	—	
1040	Special Facilities	N/A		—	—	—	0.0 %
G.	**BUILDING SITEWORK**	**N/A**					

		Sub-Total	104.47	100%
CONTRACTOR FEES (General Requirements: 10%, Overhead: 5%, Profit: 10%)		25%	26.14	
ARCHITECT FEES		7%	9.14	
	Total Building Cost		**139.75**	

Costs per square foot of floor area

Exterior Wall	S.F. Area	600	800	1000	1200	1600	2000	2400	3000	4000
	L.F. Perimeter	100	114	128	139	164	189	214	250	314
Brick with Concrete Block Back-up	Steel Frame	349.70	336.30	328.20	321.50	314.05	309.60	306.65	303.45	300.75
	Bearing Walls	344.50	331.05	322.95	316.25	308.80	304.35	301.45	298.15	295.45
Concrete Block	Steel Frame	310.15	302.50	297.80	294.00	289.70	287.20	285.50	283.65	282.10
	Bearing Walls	304.95	297.20	292.60	288.75	284.45	281.90	280.25	278.40	276.85
Galvanized Steel Siding	Steel Frame	311.00	304.30	300.25	296.90	293.15	290.95	289.50	287.85	286.50
Metal Sandwich Panel	Steel Frame	309.95	301.40	296.25	292.00	287.25	284.45	282.55	280.45	278.75
Perimeter Adj., Add or Deduct	Per 100 L.F.	92.60	69.40	55.55	46.25	34.70	27.80	23.15	18.45	13.85
Story Hgt. Adj., Add or Deduct	Per 1 Ft.	5.65	4.80	4.35	3.90	3.50	3.20	3.00	2.75	2.65
Basement—Not Applicable										

The above costs were calculated using the basic specifications shown on the facing page. These costs should be adjusted where necessary for design alternatives and owner's requirements. Reported completed project costs, for this type of structure, range from $122.75 to $348.65 per S.F.

Common additives

Description	Unit	$ Cost
Air Compressors, Electric		
1-1/2 H.P., standard controls	Each	1400
Dual controls	Each	1750
5 H.P., 115/230 volt, standard control	Each	3675
Dual controls	Each	3825
Emergency Lighting, 25 watt, battery operated		
Lead battery	Each	287
Nickel cadmium	Each	845
Fence, Chain link, 6' high		
9 ga. wire, galvanized	L.F.	22.50
6 ga. wire	L.F.	31.50
Gate	Each	335
Product Dispenser with vapor recovery		
for 6 nozzles	Each	24,800
Laundry Equipment		
Dryer, gas, 16 lb. capacity	Each	900
30 lb. capacity	Each	3600
Washer, 4 cycle	Each	1100
Commercial	Each	1450

Description	Unit	$ Cost
Lockers, Steel, single tier, 60" or 72"	Opening	200 - 350
2 tier, 60" or 72" total	Opening	124 - 152
5 tier, box lockers	Opening	70 - 82.50
Locker bench, lam. maple top only	L.F.	22
Pedestals, steel pipe	Each	64
Paving, Bituminous		
Wearing course plus base course	Sq. Yard	8.75
Safe, Office type, 1 hour rating		
30" x 18" x 18"	Each	2400
60" x 36" x 18", double door	Each	8975
Sidewalks, Concrete 4" thick	S.F.	4.09
Yard Lighting,		
20' aluminum pole with 400 watt high pressure sodium fixture	Each	2900

Important: See the Reference Section for Location Factors

Model costs calculated for a 1 story building with 12′ story height and 800 square feet of floor area

Car Wash

				Unit	Unit Cost	Cost Per S.F.	% Of Sub-Total
A.	**SUBSTRUCTURE**						
1010	Standard Foundations	Poured concrete; strip and spread footings		S.F. Ground	2.87	2.87	
1020	Special Foundations	N/A		—	—	—	
1030	Slab on Grade	5″ reinforced concrete with vapor barrier and granular base		S.F. Slab	6.33	6.33	7.9%
2010	Basement Excavation	Site preparation for slab and trench for foundation wall and footing		S.F. Ground	.96	.96	
2020	Basement Walls	4′ Foundation wall		L.F. Wall	67	9.60	
B.	**SHELL**						
	B10 Superstructure						
1010	Floor Construction	N/A		—	—	—	2.8 %
1020	Roof Construction	Metal deck, open web steel joists, beams, columns		S.F. Roof	7.12	7.12	
	B20 Exterior Enclosure						
2010	Exterior Walls	Face brick with concrete block backup	70% of wall	S.F. Wall	30.30	36.27	
2020	Exterior Windows	Horizontal pivoted steel	5% of wall	Each	741	7.04	23.8%
2030	Exterior Doors	Steel overhead hollow metal		Each	3298	16.48	
	B30 Roofing						
3010	Roof Coverings	Built-up tar and gravel with flashing; perlite/EPS composite insulation		S.F. Roof	7.96	7.96	3.2%
3020	Roof Openings	N/A		—	—	—	
C.	**INTERIORS**						
1010	Partitions	Concrete block	20 S.F. Floor/S.F. Partition	S.F. Partition	9.18	4.59	
1020	Interior Doors	Hollow metal	600 S.F. Floor/Door	Each	1001	1.67	
1030	Fittings	N/A		—	—	—	
2010	Stair Construction	N/A		—	—	—	2.9%
3010	Wall Finishes	Paint		S.F. Surface	1.12	1.12	
3020	Floor Finishes	N/A		—	—	—	
3030	Ceiling Finishes	N/A		—	—	—	
D.	**SERVICES**						
	D10 Conveying						
1010	Elevators & Lifts	N/A		—	—	—	0.0 %
1020	Escalators & Moving Walks	N/A		—	—	—	
	D20 Plumbing						
2010	Plumbing Fixtures	Toilet and service fixtures, supply and drainage	1 Fixture/160 S.F. Floor	Each	3139	19.62	
2020	Domestic Water Distribution	Gas fired water heater		S.F. Floor	38.23	38.23	24.5%
2040	Rain Water Drainage	Roof drains		S.F. Roof	3.87	3.87	
	D30 HVAC						
3010	Energy Supply	N/A		—	—	—	
3020	Heat Generating Systems	Unit heaters		Each	19.70	19.70	
3030	Cooling Generating Systems	N/A		—	—	—	7.8 %
3050	Terminal & Package Units	N/A		—	—	—	
3090	Other HVAC Sys. & Equipment	N/A		—	—	—	
	D40 Fire Protection						
4010	Sprinklers	N/A		—	—	—	0.0 %
4020	Standpipes	N/A		—	—	—	
	D50 Electrical						
5010	Electrical Service/Distribution	400 ampere service, panel board and feeders		S.F. Floor	22.19	22.19	
5020	Lighting & Branch Wiring	Fluorescent fixtures, receptacles, switches and misc. power		S.F. Floor	44.02	44.02	27.0%
5030	Communications & Security	N/A		—	—	—	
5090	Other Electrical Systems	Emergency generator, 7.5 kW		S.F. Floor	1.79	1.79	
E.	**EQUIPMENT & FURNISHINGS**						
1010	Commercial Equipment	N/A		—	—	—	
1020	Institutional Equipment	N/A		—	—	—	0.0 %
1030	Vehicular Equipment	N/A		—	—	—	
1090	Other Equipment	N/A		—	—	—	
F.	**SPECIAL CONSTRUCTION**						
1020	Integrated Construction	N/A		—	—	—	0.0 %
1040	Special Facilities	N/A		—	—	—	
G.	**BUILDING SITEWORK**	**N/A**					

					Sub-Total	251.43	100%
	CONTRACTOR FEES (General Requirements: 10%, Overhead: 5%, Profit: 10%)				25%	62.87	
	ARCHITECT FEES				7%	22	

	Total Building Cost	**336.30**

Costs per square foot of floor area

Exterior Wall	S.F. Area	2000	7000	12000	17000	22000	27000	32000	37000	42000
	L.F. Perimeter	180	340	470	540	640	740	762	793	860
Decorative Concrete Block	Wood Arch	239.45	180.55	167.70	158.25	154.65	152.35	147.95	145.00	143.70
	Steel Truss	239.30	180.40	167.60	158.10	154.50	152.25	147.80	144.85	143.55
Stone with Concrete Block Back-up	Wood Arch	294.50	210.30	191.70	177.65	172.40	169.15	162.55	158.10	156.20
	Steel Truss	286.90	202.70	184.10	170.05	164.80	161.55	154.95	150.45	148.60
Face Brick with Concrete Block Back-up	Wood Arch	269.90	197.00	180.95	**169.00**	164.50	161.65	156.00	152.25	150.65
	Steel Truss	262.25	189.40	173.35	161.40	156.90	154.05	148.40	144.60	143.05
Perimeter Adj., Add or Deduct	Per 100 L.F.	60.25	17.20	10.05	7.00	5.50	4.45	3.75	3.20	2.85
Story Hgt. Adj., Add or Deduct	Per 1 Ft.	3.50	1.85	1.55	1.20	1.10	1.15	0.95	0.75	0.80
For Basement, add $32.35 per square foot of basement area										

The above costs were calculated using the basic specifications shown on the facing page. These costs should be adjusted where necessary for design alternatives and owner's requirements. Reported completed project costs, for this type of structure, range from $78.35 to $320.40 per S.F.

Common additives

Description	Unit	$ Cost
Altar, Wood, custom design, plain	Each	2475
Deluxe	Each	12,800
Granite or marble, average	Each	13,600
Deluxe	Each	38,200
Ark, Prefabricated, plain	Each	10,100
Deluxe	Each	130,500
Baptistry, Fiberglass, incl. plumbing	Each	5525 - 8775
Bells & Carillons, 48 bells	Each	1,031,000
24 bells	Each	485,000
Confessional, Prefabricated wood		
Single, plain	Each	3525
Deluxe	Each	8625
Double, plain	Each	6625
Deluxe	Each	19,300
Emergency Lighting, 25 watt, battery operated		
Lead battery	Each	287
Nickel cadmium	Each	845
Lecterns, Wood, plain	Each	925
Deluxe	Each	5900

Description	Unit	$ Cost
Pews/Benches, Hardwood	L.F.	107 - 195
Pulpits, Prefabricated, hardwood	Each	1600 - 9475
Railing, Hardwood	L.F.	204
Steeples, translucent fiberglass		
30" square, 15' high	Each	7650
25' high	Each	8825
Painted fiberglass, 24" square, 14' high	Each	6650
28' high	Each	7100
Aluminum		
20' high, 3'- 6" base	Each	7800
35' high, 8'- 0" base	Each	30,200
60' high, 14'- 0" base	Each	65,000

Model costs calculated for a 1 story building with 24' story height and 17,000 square feet of floor area

Church

				Unit	Unit Cost	Cost Per S.F.	% Of Sub-Total
A.	**SUBSTRUCTURE**						
1010	Standard Foundations	Poured concrete; strip and spread footings		S.F. Ground	3.60	3.60	
1020	Special Foundations	N/A		—	—	—	
1030	Slab on Grade	4" reinforced concrete with vapor barrier and granular base		S.F. Slab	4.77	4.77	10.0%
2010	Basement Excavation	Site preparation for slab and trench for foundation wall and footing		S.F. Ground	.18	.18	
2020	Basement Walls	4' Foundation wall		L.F. Wall	76	3.60	
B.	**SHELL**						
	B10 Superstructure						
1010	Floor Construction	N/A		—	—	—	16.1%
1020	Roof Construction	Wood deck on laminated wood arches		S.F. Roof	18.32	19.55	
	B20 Exterior Enclosure						
2010	Exterior Walls	Face brick with concrete block backup	80% of wall (adjusted for end walls)	S.F. Wall	35.99	21.95	
2020	Exterior Windows	Aluminum, top hinged, in-swinging and curtain wall panels	20% of wall	S.F. Window	28.40	4.33	22.0%
2030	Exterior Doors	Double hollow metal swinging, single hollow metal		Each	1484	.52	
	B30 Roofing						
3010	Roof Coverings	Asphalt shingles with flashing; polystyrene insulation		S.F. Roof	5.48	5.48	4.5%
3020	Roof Openings	N/A		—	—	—	
C.	**INTERIORS**						
1010	Partitions	Plaster on metal studs	40 S.F. Floor/L.F. Partitions	S.F. Partition	14	8.40	
1020	Interior Doors	Hollow metal	400 S.F. Floor/Door	Each	1001	2.50	
1030	Fittings	N/A		—	—	—	
2010	Stair Construction	N/A		—	—	—	17.4%
3010	Wall Finishes	Paint		S.F. Surface	1.05	1.26	
3020	Floor Finishes	Carpet		S.F. Floor	8.98	8.98	
3030	Ceiling Finishes	N/A		—	—	—	
D.	**SERVICES**						
	D10 Conveying						
1010	Elevators & Lifts	N/A		—	—	—	0.0%
1020	Escalators & Moving Walks	N/A		—	—	—	
	D20 Plumbing						
2010	Plumbing Fixtures	Kitchen, toilet and service fixtures, supply and drainage	1 Fixture/2430 S.F. Floor	Each	3815	1.57	
2020	Domestic Water Distribution	Gas fired hot water heater		S.F. Floor	.28	.28	1.5%
2040	Rain Water Drainage	N/A		—	—	—	
	D30 HVAC						
3010	Energy Supply	Oil fired hot water, wall fin radiation		S.F. Floor	9.66	9.66	
3020	Heat Generating Systems	N/A		—	—	—	
3030	Cooling Generating Systems	N/A		—	—	—	16.1%
3050	Terminal & Package Units	Split systems with air cooled condensing units		S.F. Floor	9.95	9.95	
3090	Other HVAC Sys. & Equipment	N/A		—	—	—	
	D40 Fire Protection						
4010	Sprinklers	Wet pipe sprinkler system		S.F. Floor	3.16	3.16	2.6%
4020	Standpipes	N/A		—	—	—	
	D50 Electrical						
5010	Electrical Service/Distribution	400 ampere service, panel board and feeders		S.F. Floor	.78	.78	
5020	Lighting & Branch Wiring	Fluorescent fixtures, receptacles, switches, A.C. and misc. power		S.F. Floor	8.33	8.33	9.9%
5030	Communications & Security	Alarm systems, sound system and emergency lighting		S.F. Floor	2.82	2.82	
5090	Other Electrical Systems	Emergency generator, 7.5 kW		S.F. Floor	.12	.12	
E.	**EQUIPMENT & FURNISHINGS**						
1010	Commercial Equipment	N/A		—	—	—	
1020	Institutional Equipment	N/A		—	—	—	0.0%
1030	Vehicular Equipment	N/A		—	—	—	
1090	Other Equipment	N/A		—	—	—	
F.	**SPECIAL CONSTRUCTION**						
1020	Integrated Construction	N/A		—	—	—	0.0%
1040	Special Facilities	N/A		—	—	—	
G.	**BUILDING SITEWORK**	**N/A**					

				Sub-Total	121.79	100%
	CONTRACTOR FEES (General Requirements: 10%, Overhead: 5%, Profit: 10%)				25%	30.46
	ARCHITECT FEES				11%	16.75
				Total Building Cost		169

Costs per square foot of floor area

Exterior Wall	S.F. Area	2000	4000	6000	8000	12000	15000	18000	20000	22000
	L.F. Perimeter	180	280	340	386	460	535	560	600	640
Stone Ashlar Veneer On Concrete Block	Wood Truss	253.40	217.10	199.15	188.65	176.75	173.20	168.30	166.95	165.80
	Steel Joists	248.50	212.20	194.20	183.75	171.90	168.30	163.45	162.05	160.90
Stucco on Concrete Block	Wood Truss	228.65	197.85	183.55	175.40	166.25	163.40	159.75	158.75	157.80
	Steel Joists	224.25	193.45	179.15	170.95	161.80	158.95	155.30	154.30	153.35
Brick Veneer	Wood Frame	237.90	204.85	189.10	180.05	169.80	166.70	162.55	161.40	160.35
Wood Shingles	Wood Frame	226.40	195.95	181.90	173.90	164.90	162.15	158.60	157.55	156.65
Perimeter Adj., Add or Deduct	Per 100 L.F.	43.95	21.95	14.60	10.90	7.35	5.85	4.90	4.40	4.00
Story Hgt. Adj., Add or Deduct	Per 1 Ft.	4.90	3.85	3.05	2.60	2.15	1.90	1.75	1.65	1.55
For Basement, add $28.35 per square foot of basement area										

The above costs were calculated using the basic specifications shown on the facing page. These costs should be adjusted where necessary for design alternatives and owner's requirements. Reported completed project costs, for this type of structure, range from $77.50 to $257.85 per S.F.

Common additives

Description	Unit	$ Cost	Description	Unit	$ Cost
Bar, Front bar	L.F.	375	Sauna, Prefabricated, complete, 6' x 4'	Each	5875
Back bar	L.F.	300	6' x 9'	Each	8550
Booth, Upholstered, custom, straight	L.F.	212 - 390	8' x 8'	Each	10,100
"L" or "U" shaped	L.F.	220 - 370	10' x 12'	Each	14,000
Fireplaces, Brick not incl. chimney			Smoke Detectors		
or foundation, 30" x 24" opening	Each	2875	Ceiling type	Each	233
Chimney, standard brick			Duct type	Each	525
Single flue 16" x 20"	V.L.F.	94	Sound System		
20" x 20"	V.L.F.	109	Amplifier, 250 watts	Each	2400
2 flue, 20" x 32"	V.L.F.	166	Speaker, ceiling or wall	Each	196
Lockers, Steel, single tier, 60" or 72"	Opening	200 - 350	Trumpet	Each	375
2 tier, 60" or 72" total	Opening	124 - 152	Steam Bath, Complete, to 140 C.F.	Each	2075
5 tier, box lockers	Opening	70 - 82.50	To 300 C.F.	Each	2350
Locker bench, lam. maple top only	L.F.	22	To 800 C.F.	Each	6425
Pedestals, steel pipe	Each	64	To 2500 C.F.	Each	9600
Refrigerators, Prefabricated, walk-in			Swimming Pool Complete, gunite	S.F.	68.50 - 84.50
7'-6" High, 6' x 6'	S.F.	173	Tennis Court, Complete with fence		
10' x 10'	S.F.	136	Bituminous	Each	41,100 - 72,000
12' x 14'	S.F.	120	Clay	Each	41,100 - 61,000
12' x 20'	S.F.	106			

Important: See the Reference Section for Location Factors

Model costs calculated for a 1 story building with 12' story height and 6,000 square feet of floor area

Club, Country

				Unit	Unit Cost	Cost Per S.F.	% Of Sub-Total
A. SUBSTRUCTURE							
1010	Standard Foundations	Poured concrete; strip and spread footings		S.F. Ground	2.62	2.62	
1020	Special Foundations	N/A		—	—	—	
1030	Slab on Grade	4" reinforced concrete with vapor barrier and granular base		S.F. Slab	4.77	4.77	9.2%
2010	Basement Excavation	Site preparation for slab and trench for foundation wall and footing		S.F. Ground	.31	.31	
2020	Basement Walls	4' Foundation wall		L.F. Wall	76	5.56	
B. SHELL							
	B10 Superstructure						
1010	Floor Construction	N/A		—	—	—	
1020	Roof Construction	Wood truss with plywood sheathing		S.F. Roof	6.46	6.46	4.5 %
	B20 Exterior Enclosure						
2010	Exterior Walls	Stone ashlar veneer on concrete block	65% of wall	S.F. Wall	38.96	17.22	
2020	Exterior Windows	Aluminum horizontal sliding	35% of wall	Each	497	7.88	19.3%
2030	Exterior Doors	Double aluminum and glass, hollow metal		Each	2584	2.58	
	B30 Roofing						
3010	Roof Coverings	Asphalt shingles		S.F. Roof	16.14	3.10	2.2%
3020	Roof Openings	N/A		—	—	—	
C. INTERIORS							
1010	Partitions	Gypsum board on metal studs, load bearing	14 S.F. Floor/L.F. Partition	S.F. Partition	7.25	5.18	
1020	Interior Doors	Single leaf wood	140 S.F. Floor/Door	Each	576	4.11	
1030	Fittings	N/A		—	—	—	
2010	Stair Construction	N/A		—	—	—	16.7%
3010	Wall Finishes	40% vinyl wall covering, 40% paint, 20% ceramic tile		S.F. Surface	2.41	3.44	
3020	Floor Finishes	50% carpet, 30% hardwood tile, 20% ceramic tile		S.F. Floor	7.25	7.25	
3030	Ceiling Finishes	Gypsum plaster on wood furring		S.F. Ceiling	3.95	3.95	
D. SERVICES							
	D10 Conveying						
1010	Elevators & Lifts	N/A		—	—	—	0.0 %
1020	Escalators & Moving Walks	N/A		—	—	—	
	D20 Plumbing						
2010	Plumbing Fixtures	Kitchen, toilet and service fixtures, supply and drainage	1 Fixture/125 S.F. Floor	Each	2596	20.77	
2020	Domestic Water Distribution	Gas fired hot water		S.F. Floor	5.57	5.57	18.4%
2040	Rain Water Drainage	N/A		—	—	—	
	D30 HVAC						
3010	Energy Supply	N/A		—	—	—	
3020	Heat Generating Systems	Included in D3050		—	—	—	
3030	Cooling Generating Systems	N/A		—	—	—	20.5 %
3050	Terminal & Package Units	Multizone rooftop unit, gas heating, electric cooling		S.F. Floor	29.45	29.45	
3090	Other HVAC Sys. & Equipment	N/A		—	—	—	
	D40 Fire Protection						
4010	Sprinklers	Wet pipe sprinkler system		S.F. Floor	3.99	3.99	2.8%
4020	Standpipes	N/A		—	—	—	
	D50 Electrical						
5010	Electrical Service/Distribution	400 ampere service, panel board and feeders		S.F. Floor	3.43	3.43	
5020	Lighting & Branch Wiring	Fluorescent fixtures, receptacles, switches, A.C. and misc. power		S.F. Floor	5.25	5.25	6.5%
5030	Communications & Security	Alarm systems and emergency lighting		S.F. Floor	.39	.39	
5090	Other Electrical Systems	Emergency generator, 11.5 kW		S.F. Floor	.22	.22	
E. EQUIPMENT & FURNISHINGS							
1010	Commercial Equipment	N/A		—	—	—	
1020	Institutional Equipment	N/A		—	—	—	
1030	Vehicular Equipment	N/A		—	—	—	0.0 %
1090	Other Equipment	N/A		—	—	—	
F. SPECIAL CONSTRUCTION							
1020	Integrated Construction	N/A		—	—	—	0.0 %
1040	Special Facilities	N/A		—	—	—	
G. BUILDING SITEWORK	**N/A**						

					Sub-Total	143.50	**100%**
	CONTRACTOR FEES (General Requirements: 10%, Overhead: 5%, Profit: 10%)				25%	35.92	
	ARCHITECT FEES				11%	19.73	

	Total Building Cost	**199.15**

Costs per square foot of floor area

Exterior Wall	S.F. Area	4000	8000	12000	17000	22000	27000	32000	37000	42000
	L.F. Perimeter	280	386	520	585	640	740	840	940	940
Stone Ashlar on Concrete Block	Steel Joists	185.25	156.85	149.70	139.55	133.55	131.40	129.95	128.85	125.70
	Wood Joists	182.25	154.70	147.80	137.90	132.10	130.05	128.65	127.55	124.50
Face Brick on Concrete Block	Steel Joists	178.90	152.50	145.80	136.40	130.90	128.95	127.65	126.60	123.70
	Wood Joists	176.70	150.95	144.40	135.30	129.95	128.05	126.75	125.70	122.95
Decorative Concrete Block	Steel Joists	169.80	146.25	140.20	131.90	127.15	125.35	124.20	123.30	120.75
	Wood Joists	166.75	143.85	137.90	130.00	125.35	123.65	122.50	121.65	119.20
Perimeter Adj., Add or Deduct	Per 100 L.F.	24.70	12.40	8.25	5.80	4.50	3.70	3.10	2.70	2.35
Story Hgt. Adj., Add or Deduct	Per 1 Ft.	4.10	2.80	2.55	1.95	1.70	1.55	1.50	1.55	1.35
For Basement, add $28.90 per square foot of basement area										

The above costs were calculated using the basic specifications shown on the facing page. These costs should be adjusted where necessary for design alternatives and owner's requirements. Reported completed project costs, for this type of structure, range from $79.25 to $227.75 per S.F.

Common additives

Description	Unit	$ Cost	Description	Unit	$ Cost
Bar, Front bar	L.F.	375	Kitchen Equipment, cont.		
Back bar	L.F.	300	Freezer, 44 C.F., reach-in	Each	3250
Booth, Upholstered, custom, straight	L.F.	212 - 390	Ice cube maker, 50 lb. per day	Each	1875
"L" or "U" shaped	L.F.	220 - 370	Lockers, Steel, single tier, 60" or 72"	Opening	200 - 350
Emergency Lighting, 25 watt, battery operated			2 tier, 60" or 72" total	Opening	124 - 152
Lead battery	Each	287	5 tier, box lockers	Opening	70 - 82.50
Nickel cadmium	Each	845	Locker bench, lam. maple top only	L.F.	22
Flagpoles, Complete			Pedestals, steel pipe	Each	64
Aluminum, 20' high	Each	1700	Refrigerators, Prefabricated, walk-in		
40' High	Each	3600	7'-6" High, 6' x 6'	S.F.	173
70' High	Each	10,300	10' x 10'	S.F.	136
Fiberglass, 23' High	Each	1825	12' x 14'	S.F.	120
39'-5" High	Each	3475	12' x 20'	S.F.	106
59' High	Each	6625	Smoke Detectors		
Kitchen Equipment			Ceiling type	Each	233
Broiler	Each	4050	Duct type	Each	525
Coffee urn, twin 6 gallon	Each	3025	Sound System		
Cooler, 6 ft. long, reach-in	Each	5150	Amplifier, 250 watts	Each	2400
Dishwasher, 10-12 racks per hr.	Each	5075	Speaker, ceiling or wall	Each	196
Food warmer, counter 1.2 kw	Each	740	Trumpet	Each	375

Important: See the Reference Section for Location Factors

Model costs calculated for a 1 story building with 12' story height and 22,000 square feet of floor area

Club, Social

				Unit	Unit Cost	Cost Per S.F.	% Of Sub-Total
A.	**SUBSTRUCTURE**						
1010	Standard Foundations	Poured concrete; strip and spread footings		S.F. Ground	1.66	1.66	
1020	Special Foundations	N/A		—	—	—	
1030	Slab on Grade	4" reinforced concrete with vapor barrier and granular base		S.F. Slab	4.77	4.77	10.1%
2010	Basement Excavation	Site preparation for slab and trench for foundation wall and footing		S.F. Ground	.18	.18	
2020	Basement Walls	4' Foundation wall		L.F. Wall	76	3.52	
B.	**SHELL**						
	B10 Superstructure						
1010	Floor Construction	N/A		—	—	—	5.0 %
1020	Roof Construction	Metal deck on open web steel joists		S.F. Roof	4.97	4.97	
	B20 Exterior Enclosure						
2010	Exterior Walls	Stone, ashlar veneer on concrete block	65% of wall	S.F. Wall	38.96	8.84	
2020	Exterior Windows	Window wall	35% of wall	Each	43.60	5.32	14.9%
2030	Exterior Doors	Double aluminum and glass doors		Each	2584	.71	
	B30 Roofing						
3010	Roof Coverings	Built-up tar and gravel with flashing; perlite/EPS composite insulation		S.F. Roof	5.70	5.70	5.7%
3020	Roof Openings	N/A		—	—	—	
C.	**INTERIORS**						
1010	Partitions	Lightweight concrete block	14 S.F. Floor/L.F. Partition	S.F. Partition	10.64	7.60	
1020	Interior Doors	Single leaf wood	140 S.F. Floor/Door	Each	576	4.11	
1030	Fittings	N/A		—	—	—	
2010	Stair Construction	N/A		—	—	—	27.4%
3010	Wall Finishes	65% paint, 25% vinyl wall covering, 10% ceramic tile		S.F. Surface	1.63	2.33	
3020	Floor Finishes	60% carpet, 35% hardwood, 15% ceramic tile		S.F. Floor	6.76	6.76	
3030	Ceiling Finishes	Mineral fiber tile on concealed zee bars		S.F. Ceiling	6.51	6.51	
D.	**SERVICES**						
	D10 Conveying						
1010	Elevators & Lifts	N/A		—	—	—	0.0 %
1020	Escalators & Moving Walks	N/A		—	—	—	
	D20 Plumbing						
2010	Plumbing Fixtures	Kitchen, toilet and service fixtures, supply and drainage	1 Fixture/1050 S.F. Floor	Each	2898	2.76	
2020	Domestic Water Distribution	Gas fired hot water heater		S.F. Floor	.22	.22	3.5%
2040	Rain Water Drainage	Roof drains		S.F. Roof	.53	.53	
	D30 HVAC						
3010	Energy Supply	N/A		—	—	—	
3020	Heat Generating Systems	Included in D3050		—	—	—	
3030	Cooling Generating Systems	N/A		—	—	—	23.4 %
3050	Terminal & Package Units	Multizone rooftop unit, gas heating, electric cooling		S.F. Floor	23.40	23.40	
3090	Other HVAC Sys. & Equipment	N/A		—	—	—	
	D40 Fire Protection						
4010	Sprinklers	Wet pipe sprinkler system		S.F. Floor	3.16	3.16	3.2%
4020	Standpipes	N/A		—	—	—	
	D50 Electrical						
5010	Electrical Service/Distribution	400 ampere service, panel board and feeders		S.F. Floor	1.12	1.12	
5020	Lighting & Branch Wiring	Fluorescent fixtures, receptacles, switches, A.C. and misc. power		S.F. Floor	5.34	5.34	6.8%
5030	Communications & Security	Alarm systems and emergency lighting		S.F. Floor	.22	.22	
5090	Other Electrical Systems	Emergency generator, 11.5 kW		S.F. Floor	.12	.12	
E.	**EQUIPMENT & FURNISHINGS**						
1010	Commercial Equipment	N/A		—	—	—	
1020	Institutional Equipment	N/A		—	—	—	
1030	Vehicular Equipment	N/A		—	—	—	0.0 %
1090	Other Equipment	N/A		—	—	—	
F.	**SPECIAL CONSTRUCTION**						
1020	Integrated Construction	N/A		—	—	—	
1040	Special Facilities	N/A		—	—	—	0.0 %
G.	**BUILDING SITEWORK**	**N/A**					

			Sub-Total	99.85	**100%**
CONTRACTOR FEES (General Requirements: 10%, Overhead: 5%, Profit: 10%)		25%	24.96		
ARCHITECT FEES		7%	8.74		

Total Building Cost	**133.55**

Costs per square foot of floor area

Exterior Wall	S.F. Area	15000	20000	28000	38000	50000	65000	85000	100000	150000
	L.F. Perimeter	350	400	480	550	630	660	750	825	1035
Face Brick with Concrete Block Back-up	Steel Frame	208.50	198.80	190.45	183.90	179.40	174.70	171.85	170.55	167.70
	Bearing Walls	206.50	195.55	186.20	178.70	173.55	168.00	164.65	163.15	159.80
Decorative Concrete Block	Steel Frame	200.75	192.15	184.75	179.15	175.25	171.35	168.90	167.80	165.40
	Bearing Walls	198.70	188.90	180.50	173.95	169.40	164.65	161.70	160.40	157.50
Stucco on Concrete Block	Steel Frame	200.40	191.75	184.45	178.90	175.00	171.15	168.75	167.65	165.30
	Bearing Walls	198.35	188.60	180.20	173.70	169.15	164.50	161.55	160.25	157.40
Perimeter Adj., Add or Deduct	Per 100 L.F.	8.75	6.55	4.70	3.50	2.65	2.05	1.55	1.30	0.85
Story Hgt. Adj., Add or Deduct	Per 1 Ft.	2.15	1.85	1.60	1.40	1.20	0.95	0.80	0.75	0.65
For Basement, add $ 33.70 per square foot of basement area										

The above costs were calculated using the basic specifications shown on the facing page. These costs should be adjusted where necessary for design alternatives and owner's requirements. Reported completed project costs, for this type of structure, range from $ 114.65 to $ 269.45 per S.F.

Common additives

Description	Unit	$ Cost	Description	Unit	$ Cost
Carrels Hardwood	Each	605 - 1150	Lockers, Steel, single tier, 60" or 72"	Opening	200 - 350
Clock System			2 tier, 60" or 72" total	Opening	124 - 152
20 Room	Each	16,000	5 tier, box lockers	Opening	70 - 82.50
50 Room	Each	39,100	Locker bench, lam. maple top only	L.F.	22
Elevators, Hydraulic passenger, 2 stops			Pedestals, steel pipe	Each	64
1500# capacity	Each	60,900	Seating		
2500# capacity	Each	64,300	Auditorium chair, all veneer	Each	240
3500# capacity	Each	67,600	Veneer back, padded seat	Each	290
Additional stop, add	Each	8250	Upholstered, spring seat	Each	279
Emergency Lighting, 25 watt, battery operated			Classroom, movable chair & desk	Set	65 - 120
Lead battery	Each	287	Lecture hall, pedestal type	Each	229 - 680
Nickel cadmium	Each	845	Smoke Detectors		
Flagpoles, Complete			Ceiling type	Each	233
Aluminum, 20' high	Each	1700	Duct type	Each	525
40' High	Each	3600	Sound System		
70' High	Each	10,300	Amplifier, 250 watts	Each	2400
Fiberglass, 23' High	Each	1825	Speaker, ceiling or wall	Each	196
39'-5" High	Each	3475	Trumpet	Each	375
59' High	Each	6625	TV Antenna, Master system, 12 outlet	Outlet	320
			30 outlet	Outlet	207
			100 outlet	Outlet	199

Important: See the Reference Section for Location Factors

Model costs calculated for a 2 story building with 12' story height and 50,000 square feet of floor area

College, Classroom, 2-3 Story

				Unit	Unit Cost	Cost Per S.F.	% Of Sub-Total
A.	**SUBSTRUCTURE**						
1010	Standard Foundations	Poured concrete; strip and spread footings		S.F. Ground	1.20	.60	
1020	Special Foundations	N/A		—	—	—	
1030	Slab on Grade	4" reinforced concrete with vapor barrier and granular base		S.F. Slab	4.77	2.39	3.6%
2010	Basement Excavation	Site preparation for slab and trench for foundation wall and footing		S.F. Ground	.28	.14	
2020	Basement Walls	4' Foundation wall		L.F. Wall	71	1.46	
B.	**SHELL**						
	B10 Superstructure						
1010	Floor Construction	Open web steel joists, slab form, concrete		S.F. Floor	13.02	6.51	7.9%
1020	Roof Construction	Metal deck on open web steel joists, columns		S.F. Roof	6.92	3.46	
	B20 Exterior Enclosure						
2010	Exterior Walls	Decorative concrete block	65% of wall	S.F. Wall	14.35	2.82	
2020	Exterior Windows	Window wall	35% of wall	Each	36.95	3.90	5.8%
2030	Exterior Doors	Double glass and aluminum with transom		Each	5275	.64	
	B30 Roofing						
3010	Roof Coverings	Built-up tar and gravel with flashing; perlite/EPS composite insulation		S.F. Roof	5.56	2.78	2.2%
3020	Roof Openings	N/A		—	—	—	
C.	**INTERIORS**						
1010	Partitions	Concrete block	20 S.F. Floor/L.F. Partition	S.F. Partition	14.12	7.06	
1020	Interior Doors	Single leaf hollow metal	200 S.F. Floor/Door	Each	1001	5.01	
1030	Fittings	Chalkboards, counters, cabinets		S.F. Floor	4.75	4.75	
2010	Stair Construction	Concrete filled metal pan		Flight	14,675	2.94	27.4%
3010	Wall Finishes	95% paint, 5% ceramic tile		S.F. Surface	3.67	3.67	
3020	Floor Finishes	70% vinyl composition tile, 25% carpet, 5% ceramic tile		S.F. Floor	4.73	4.73	
3030	Ceiling Finishes	Mineral fiber tile on concealed zee bars		S.F. Ceiling	6.51	6.51	
D.	**SERVICES**						
	D10 Conveying						
1010	Elevators & Lifts	Two hydraulic passenger elevators		Each	79,000	3.16	2.5%
1020	Escalators & Moving Walks	N/A		—	—	—	
	D20 Plumbing						
2010	Plumbing Fixtures	Toilet and service fixtures, supply and drainage	1 Fixture/455 S.F. Floor	Each	6780	14.90	
2020	Domestic Water Distribution	Oil fired hot water heater		S.F. Floor	2.18	2.18	13.9%
2040	Rain Water Drainage	Roof drains		S.F. Roof	1.10	.55	
	D30 HVAC						
3010	Energy Supply	N/A		—	—	—	
3020	Heat Generating Systems	Included in D3050		—	—	—	
3030	Cooling Generating Systems	N/A		—	—	—	14.5%
3050	Terminal & Package Units	Multizone unit, gas heating, electric cooling		S.F. Floor	18.40	18.40	
3090	Other HVAC Sys. & Equipment	N/A		—	—	—	
	D40 Fire Protection						
4010	Sprinklers	Sprinklers, light hazard		S.F. Floor	2.86	2.86	2.5%
4020	Standpipes	Standpipe, dry, Class III		S.F. Floor	.29	.29	
	D50 Electrical						
5010	Electrical Service/Distribution	2000 ampere service, panel board and feeders		S.F. Floor	4.64	4.64	
5020	Lighting & Branch Wiring	High efficiency fluorescent fixtures, receptacles, switches, A.C. and misc. power		S.F. Floor	12.55	12.55	19.7%
5030	Communications & Security	Addressable alarm systems, internet wiring, communications systems and emergency lighting		S.F. Floor	7.06	7.06	
5090	Other Electrical Systems	Emergency generator, 100 kW		S.F. Floor	.68	.68	
E.	**EQUIPMENT & FURNISHINGS**						
1010	Commercial Equipment	N/A		—	—	—	
1020	Institutional Equipment	N/A		—	—	—	0.0%
1030	Vehicular Equipment	N/A		—	—	—	
1090	Other Equipment	N/A		—	—	—	
F.	**SPECIAL CONSTRUCTION**						
1020	Integrated Construction	N/A		—	—	—	0.0%
1040	Special Facilities	N/A		—	—	—	
G.	**BUILDING SITEWORK**	**N/A**					

	Sub-Total		126.64	100%
CONTRACTOR FEES (General Requirements: 10%, Overhead: 5%, Profit: 10%)		25%	31.68	
ARCHITECT FEES		7%	11.08	
	Total Building Cost		**169.40**	

Costs per square foot of floor area

Exterior Wall	S.F. Area	10000	15000	25000	40000	55000	70000	80000	90000	100000
	L.F. Perimeter	260	320	400	476	575	628	684	721	772
Face Brick with Concrete Block Back-up	R/Conc. Frame	221.80	205.10	189.00	177.90	173.60	170.05	168.85	167.50	166.80
	Steel Frame	232.20	215.55	199.45	188.35	184.10	180.45	179.30	178.00	177.20
Decorative Concrete Block	R/Conc. Frame	205.75	192.00	179.20	170.55	167.25	164.50	163.60	162.65	162.05
	Steel Frame	216.25	202.40	189.60	181.05	177.65	174.95	174.05	173.10	172.50
Precast Concrete Panels	R/Conc. Frame	227.10	209.50	192.35	180.35	175.80	171.85	170.65	169.15	168.35
	Steel Frame	237.55	219.95	202.75	190.80	186.25	182.30	181.05	179.65	178.80
Perimeter Adj., Add or Deduct	Per 100 L.F.	17.70	11.85	7.05	4.50	3.25	2.45	2.25	2.00	1.75
Story Hgt. Adj., Add or Deduct	Per 1 Ft.	3.40	2.80	2.10	1.60	1.40	1.15	1.10	1.05	1.00
For Basement, add $32.80 per square foot of basement area										

The above costs were calculated using the basic specifications shown on the facing page. These costs should be adjusted where necessary for design alternatives and owner's requirements. Reported completed project costs, for this type of structure, range from $74.05 to $218.25 per S.F.

Common additives

Description	Unit	$ Cost	Description	Unit	$ Cost
Carrels Hardwood	Each	605 - 1150	Kitchen Equipment		
Closed Circuit Surveillance, One station			Broiler	Each	4050
Camera and monitor	Each	1875	Coffee urn, twin 6 gallon	Each	3025
For additional camera stations, add	Each	1025	Cooler, 6 ft. long	Each	5150
Elevators, Hydraulic passenger, 2 stops			Dishwasher, 10-12 racks per hr.	Each	5075
2000# capacity	Each	63,100	Food warmer	Each	740
2500# capacity	Each	64,300	Freezer, 44 C.F., reach-in	Each	3250
3500# capacity	Each	67,600	Ice cube maker, 50 lb. per day	Each	1875
Additional stop, add	Each	8250	Range with 1 oven	Each	2700
Emergency Lighting, 25 watt, battery operated			Laundry Equipment		
Lead battery	Each	287	Dryer, gas, 16 lb. capacity	Each	900
Nickel cadmium	Each	845	30 lb. capacity	Each	3600
Furniture	Student	1900 - 3600	Washer, 4 cycle	Each	1100
Intercom System, 25 station capacity			Commercial	Each	1450
Master station	Each	2725	Smoke Detectors		
Intercom outlets	Each	174	Ceiling type	Each	233
Handset	Each	480	Duct type	Each	525
			TV Antenna, Master system, 12 outlet	Outlet	320
			30 outlet	Outlet	207
			100 outlet	Outlet	199

Model costs calculated for a 3 story building with 12' story height and 25,000 square feet of floor area

			Unit	Unit Cost	Cost Per S.F.	% Of Sub-Total
A. SUBSTRUCTURE						
1010	Standard Foundations	Poured concrete; strip and spread footings	S.F. Ground	5.67	1.89	
1020	Special Foundations	N/A	—	—	—	
1030	Slab on Grade	4" reinforced concrete with vapor barrier and granular base	S.F. Slab	4.77	1.59	3.4%
2010	Basement Excavation	Site preparation for slab and trench for foundation wall and footing	S.F. Ground	.18	.06	
2020	Basement Walls	4' Foundation wall	L.F. Wall	76	1.21	
B. SHELL						
	B10 Superstructure					
1010	Floor Construction	Concrete flat plate	S.F. Floor	23.15	15.43	13.9%
1020	Roof Construction	Concrete flat plate	S.F. Roof	12.69	4.23	
	B20 Exterior Enclosure					
2010	Exterior Walls	Face brick with concrete block backup 80% of wall	S.F. Wall	30.32	13.97	
2020	Exterior Windows	Aluminum horizontal sliding 20% of wall	Each	703	3.52	13.3%
2030	Exterior Doors	Double glass & aluminum doors	Each	5400	1.30	
	B30 Roofing					
3010	Roof Coverings	Single ply membrane, loose laid and ballasted; perlite/EPS composite insulation	S.F. Roof	5.22	1.74	1.2%
3020	Roof Openings	N/A	—	—	—	
C. INTERIORS						
1010	Partitions	Gypsum board on metal studs, concrete block 9 S.F. Floor/L.F. Partition	S.F. Partition	5.84	6.49	
1020	Interior Doors	Single leaf wood 90 S.F. Floor/Door	Each	576	6.40	
1030	Fittings	Closet shelving, mirrors, bathroom accessories	S.F. Floor	1.73	1.73	
2010	Stair Construction	Cast in place concrete	Flight	4775	2.48	21.9%
3010	Wall Finishes	95% paint, 5% ceramic tile	S.F. Surface	1.69	3.76	
3020	Floor Finishes	80% carpet, 10% vinyl composition tile, 10% ceramic tile	S.F. Floor	9.10	9.10	
3030	Ceiling Finishes	90% paint, 10% suspended fiberglass board	S.F. Ceiling	.93	.93	
D. SERVICES						
	D10 Conveying					
1010	Elevators & Lifts	One hydraulic passenger elevator	Each	110,250	4.41	3.1%
1020	Escalators & Moving Walks	N/A	—	—	—	
	D20 Plumbing					
2010	Plumbing Fixtures	Toilet and service fixtures, supply and drainage 1 Fixture/455 S.F. Floor	Each	8363	18.38	
2020	Domestic Water Distribution	Electric water heater	S.F. Floor	3.57	3.57	15.8%
2040	Rain Water Drainage	Roof drains	S.F. Roof	1.17	.39	
	D30 HVAC					
3010	Energy Supply	N/A	—	—	—	
3020	Heat Generating Systems	Included in D3050	—	—	—	
3030	Cooling Generating Systems	N/A	—	—	—	8.2%
3050	Terminal & Package Units	Rooftop multizone unit system	S.F. Floor	11.55	11.55	
3090	Other HVAC Sys. & Equipment	N/A	—	—	—	
	D40 Fire Protection					
4010	Sprinklers	Wet pipe sprinkler system	S.F. Floor	2.73	2.73	2.4%
4020	Standpipes	Standpipe, dry, Class III	S.F. Floor	.72	.72	
	D50 Electrical					
5010	Electrical Service/Distribution	800 ampere service, panel board and feeders	S.F. Floor	2.55	2.55	
5020	Lighting & Branch Wiring	High efficiency fluorescent fixtures, receptacles, switches, A.C. and misc. power	S.F. Floor	9.28	9.28	14.0%
5030	Communications & Security	Alarm systems, internet and phone wiring, communications systems and emergency lighting	S.F. Floor	7.93	7.93	
5090	Other Electrical Systems	Emergency generator, 7.5 kW	S.F. Floor	.08	.08	
E. EQUIPMENT & FURNISHINGS						
1010	Commercial Equipment	N/A	—	—	—	
1020	Institutional Equipment	N/A	—	—	—	2.8%
1030	Vehicular Equipment	N/A	—	—	—	
2020	Other Equipment	Dormitory furniture	S.F. Floor	3.90	3.90	
F. SPECIAL CONSTRUCTION						
1020	Integrated Construction	N/A	—	—	—	0.0%
1040	Special Facilities	N/A	—	—	—	
G. BUILDING SITEWORK	**N/A**					

			Sub-Total	141.32	**100%**
CONTRACTOR FEES (General Requirements: 10%, Overhead: 5%, Profit: 10%)			25%	35.31	
ARCHITECT FEES			7%	12.37	
		Total Building Cost		**189**	

Costs per square foot of floor area

Exterior Wall	S.F. Area	20000	35000	45000	65000	85000	110000	135000	160000	200000
	L.F. Perimeter	260	340	400	440	500	540	560	590	640
Face Brick with Concrete Block Back-up	R/Conc. Frame	214.85	199.65	195.70	187.20	183.50	179.60	176.70	174.95	173.05
	Steel Frame	224.85	209.65	205.70	197.25	193.55	189.70	186.75	184.95	183.10
Decorative Concrete Block	R/Conc. Frame	200.40	188.80	185.75	179.65	176.95	174.15	172.10	170.85	169.55
	Steel Frame	210.40	198.90	195.85	189.70	187.05	184.15	182.15	180.85	179.55
Precast Concrete Panels With Exposed Aggregate	R/Conc. Frame	220.90	204.20	199.80	190.35	186.25	181.90	178.65	176.65	174.55
	Steel Frame	230.80	214.10	209.75	200.30	196.25	191.90	188.65	186.65	184.55
Perimeter Adj., Add or Deduct	Per 100 L.F.	17.25	9.85	7.60	5.30	4.05	3.15	2.55	2.15	1.70
Story Hgt. Adj., Add or Deduct	Per 1 Ft.	3.45	2.60	2.35	1.80	1.55	1.30	1.15	0.95	0.85

For Basement, add $32.80 per square foot of basement area

The above costs were calculated using the basic specifications shown on the facing page. These costs should be adjusted where necessary for design alternatives and owner's requirements. Reported completed project costs, for this type of structure, range from $105.25 to $253.85 per S.F.

Common additives

Description	Unit	$ Cost	Description	Unit	$ Cost
Carrels Hardwood	Each	605 - 1150	Kitchen Equipment		
Closed Circuit Surveillance, One station			Broiler	Each	4050
Camera and monitor	Each	1875	Coffee urn, twin, 6 gallon	Each	3025
For additional camera stations, add	Each	1025	Cooler, 6 ft. long	Each	5150
Elevators, Electric passenger, 5 stops			Dishwasher, 10-12 racks per hr.	Each	5075
2000# capacity	Each	158,200	Food warmer	Each	740
2500# capacity	Each	162,200	Freezer, 44 C.F., reach-in	Each	3250
3500# capacity	Each	162,700	Ice cube maker, 50 lb. per day	Each	1875
Additional stop, add	Each	14,000	Range with 1 oven	Each	2700
Emergency Lighting, 25 watt, battery operated			Laundry Equipment		
Lead battery	Each	287	Dryer, gas, 16 lb. capacity	Each	900
Nickel cadmium	Each	845	30 lb. capacity	Each	3600
Furniture	Student	1900 - 3600	Washer, 4 cycle	Each	1100
Intercom System, 25 station capacity			Commercial	Each	1450
Master station	Each	2725	Smoke Detectors		
Intercom outlets	Each	174	Ceiling type	Each	233
Handset	Each	480	Duct type	Each	525
			TV Antenna, Master system, 12 outlet	Outlet	320
			30 outlet	Outlet	207
			100 outlet	Outlet	199

Model costs calculated for a 6 story building with 12' story height and 85,000 square feet of floor area

				Unit	Unit Cost	Cost Per S.F.	% Of Sub-Total
A. SUBSTRUCTURE							
1010	Standard Foundations	Poured concrete; strip and spread footings		S.F. Ground	8.34	1.39	
1020	Special Foundations	N/A		—	—	—	
1030	Slab on Grade	4" reinforced concrete with vapor barrier and granular base		S.F. Slab	4.77	.80	2.0%
2010	Basement Excavation	Site preparation for slab and trench for foundation wall and footing		S.F. Ground	.28	.05	
2020	Basement Walls	4' foundation wall		L.F. Wall	71	.62	
B. SHELL							
	B10 Superstructure						
1010	Floor Construction	Concrete slab with metal deck and beams		S.F. Floor	24.90	20.75	17.1%
1020	Roof Construction	Concrete slab with metal deck and beams		S.F. Roof	18.60	3.10	
	B20 Exterior Enclosure						
2010	Exterior Walls	Decorative concrete block	80% of wall	S.F. Wall	15.85	5.37	
2020	Exterior Windows	Aluminum horizontal sliding	20% of wall	Each	497	2.81	6.1%
2030	Exterior Doors	Double glass & aluminum doors		Each	3305	.31	
	B30 Roofing						
3010	Roof Coverings	Single ply membrane, loose laid and ballasted; perlite/EPS composite insulation		S.F. Roof	4.80	.80	0.6%
3020	Roof Openings	N/A		—	—	—	
C. INTERIORS							
1010	Partitions	Concrete block	9 S.F. Floor/L.F. Partition	S.F. Partition	9.59	10.65	
1020	Interior Doors	Single leaf wood	90 S.F. Floor/Door	Each	576	6.40	
1030	Fittings	Closet shelving, mirrors, bathroom accessories		S.F. Floor	1.42	1.42	
2010	Stair Construction	Concrete filled metal pan		Flight	14,675	3.10	26.2%
3010	Wall Finishes	95% paint, 5% ceramic tile		S.F. Surface	2.30	5.10	
3020	Floor Finishes	80% carpet, 10% vinyl composition tile, 10% ceramic tile		S.F. Floor	9.10	9.10	
3030	Ceiling Finishes	Mineral fiber tile on concealed zee bars, paint		S.F. Ceiling	.93	.93	
D. SERVICES							
	D10 Conveying						
1010	Elevators & Lifts	Four geared passenger elevators		Each	234,600	11.04	7.9%
1020	Escalators & Moving Walks	N/A		—	—	—	
	D20 Plumbing						
2010	Plumbing Fixtures	Toilet and service fixtures, supply and drainage	1 Fixture/390 S.F. Floor	Each	7141	18.31	
2020	Domestic Water Distribution	Oil fired water heater		S.F. Floor	1.58	1.58	14.4%
2040	Rain Water Drainage	Roof drains		S.F. Roof	1.86	.31	
	D30 HVAC						
3010	Energy Supply	Oil fired hot water, wall fin radiation		S.F. Floor	3.90	3.90	
3020	Heat Generating Systems	N/A		—	—	—	
3030	Cooling Generating Systems	Chilled water, air cooled condenser system		S.F. Floor	8.03	8.03	8.5%
3050	Terminal & Package Units	N/A		—	—	—	
3090	Other HVAC Sys. & Equipment	N/A		—	—	—	
	D40 Fire Protection						
4010	Sprinklers	Sprinklers, light hazard		S.F. Floor	2.72	2.72	2.4%
4020	Standpipes	Standpipe, dry, Class III		S.F. Floor	.57	.57	
	D50 Electrical						
5010	Electrical Service/Distribution	1200 ampere service, panel board and feeders		S.F. Floor	1.17	1.17	
5020	Lighting & Branch Wiring	High efficiency fluorescent fixtures, receptacles, switches, A.C. and misc. power		S.F. Floor	9.17	9.17	12.5%
5030	Communications & Security	Alarm systems, internet and phone wiring, communications systems and emergency lighting		S.F. Floor	6.91	6.91	
5090	Other Electrical Systems	Emergency generator, 30 kW		S.F. Floor	.21	.21	
E. EQUIPMENT & FURNISHINGS							
1010	Commercial Equipment	N/A		—	—	—	
1020	Institutional Equipment	N/A		—	—	—	
1030	Vehicular Equipment	N/A		—	—	—	2.3%
2020	Other Equipment	Built-in dormitory furnishings		S.F. Floor	3.19	3.19	
F. SPECIAL CONSTRUCTION							
1020	Integrated Construction	N/A		—	—	—	
1040	Special Facilities	N/A		—	—	—	0.0%
G. BUILDING SITEWORK	**N/A**						

		Sub-Total	139.81	100%
CONTRACTOR FEES (General Requirements: 10%, Overhead: 5%, Profit: 10%)		25%	35.01	
ARCHITECT FEES		7%	12.23	
	Total Building Cost		**187.05**	

Costs per square foot of floor area

Exterior Wall	S.F. Area	12000	20000	28000	37000	45000	57000	68000	80000	92000
	L.F. Perimeter	470	600	698	793	900	1060	1127	1200	1320
Face Brick with Concrete Brick Back-up	Steel Frame	285.90	233.05	208.95	193.85	186.10	178.55	172.55	167.85	165.05
	Bearing Walls	281.25	228.35	204.35	189.15	181.50	173.90	167.90	163.15	160.40
Decorative Concrete Block	Steel Frame	278.90	227.65	204.50	190.00	182.55	175.20	169.60	165.20	162.50
	Bearing Walls	274.40	223.20	200.05	185.50	178.05	170.75	165.10	160.70	158.05
Stucco on Concrete Block	Steel Frame	277.85	226.85	203.85	189.45	182.00	174.70	169.15	164.80	162.10
	Bearing Walls	273.40	222.40	199.35	185.00	177.50	170.25	164.70	160.35	157.65
Perimeter Adj., Add or Deduct	Per 100 L.F.	10.10	6.10	4.40	3.25	2.75	2.10	1.80	1.50	1.30
Story Hgt. Adj., Add or Deduct	Per 1 Ft.	1.80	1.35	1.20	0.95	0.95	0.90	0.75	0.70	0.70
For Basement, add $21.20 per square foot of basement area										

The above costs were calculated using the basic specifications shown on the facing page. These costs should be adjusted where necessary for design alternatives and owner's requirements. Reported completed project costs, for this type of structure, range from $151.60 to $283.10 per S.F.

Common additives

Description	Unit	$ Cost
Cabinets, Base, door units, metal	L.F.	268
Drawer units	L.F.	525
Tall storage cabinets, open	L.F.	500
With doors	L.F.	590
Wall, metal 12-1/2" deep, open	L.F.	200
With doors	L.F.	360
Carrels Hardwood	Each	605 - 1150
Countertops, not incl. base cabinets, acid proof	S.F.	48.50 - 61
Stainless steel	S.F.	121
Fume Hood, Not incl. ductwork	L.F.	855 - 2600
Ductwork	Hood	5125 - 8425
Glassware Washer, Distilled water rinse	Each	6900 - 14,200
Seating		
Auditorium chair, all veneer	Each	240
Veneer back, padded seat	Each	290
Upholstered, spring seat	Each	279
Classroom, movable chair & desk	Set	65 - 120
Lecture hall, pedestal type	Each	229 - 680

Description	Unit	$ Cost
Safety Equipment, Eye wash, hand held	Each	445
Deluge shower	Each	815
Sink, One piece plastic		
Flask wash, freestanding	Each	2375
Tables, acid resist. top, drawers	L.F.	195
Titration Unit, Four 2000 ml reservoirs	Each	6050

Important: See the Reference Section for Location Factors

Model costs calculated for a 1 story building with 12' story height and 45,000 square feet of floor area

				Unit	Unit Cost	Cost Per S.F.	% Of Sub-Total

A. SUBSTRUCTURE

				Unit	Unit Cost	Cost Per S.F.	% Of Sub-Total
1010	Standard Foundations	Poured concrete; strip and spread footings		S.F. Ground	2.84	2.84	
1020	Special Foundations	N/A		—	—	—	
1030	Slab on Grade	4" reinforced concrete with vapor barrier and granular base		S.F. Slab	4.77	4.77	10.4%
2010	Basement Excavation	Site preparation for slab and trench for foundation wall and footing		S.F. Ground	.18	.18	
2020	Basement Walls	4' foundation wall		L.F. Wall	71	5.88	

B. SHELL

B10 Superstructure

				Unit	Unit Cost	Cost Per S.F.	% Of Sub-Total
1010	Floor Construction	Metal deck on open web steel joists	(5680 S.F.)	S.F. Floor	13	1.69	4.0%
1020	Roof Construction	Metal deck on open web steel joists		S.F. Roof	3.58	3.58	

B20 Exterior Enclosure

2010	Exterior Walls	Face brick with concrete block backup	75% of wall	S.F. Wall	30.28	5.45	
2020	Exterior Windows	Window wall	25% of wall	Each	36.95	2.21	7.3%
2030	Exterior Doors	Glass and metal doors and entrances with transom		Each	4523	2.02	

B30 Roofing

3010	Roof Coverings	Built-up tar and gravel with flashing; perlite/EPS composite insulation		S.F. Roof	5.34	5.34	4.3%
3020	Roof Openings	Skylight		S.F. Roof	.32	.32	

C. INTERIORS

				Unit	Unit Cost	Cost Per S.F.	% Of Sub-Total
1010	Partitions	Concrete block	10 S.F. Floor/L.F. Partition	S.F. Partition	11.08	11.08	
1020	Interior Doors	Single leaf-kalamein fire doors	820 S.F. Floor/Door	Each	1000	1.22	
1030	Fittings	Lockers		S.F. Floor	.04	.04	
2010	Stair Construction	N/A		—	—	—	22.6%
3010	Wall Finishes	60% paint, 40% epoxy coating		S.F. Surface	2.73	5.45	
3020	Floor Finishes	60% epoxy, 20% carpet, 20% vinyl composition tile		S.F. Floor	5.49	5.49	
3030	Ceiling Finishes	Mineral fiber tile on concealed zee runners		S.F. Ceiling	6.51	6.51	

D. SERVICES

D10 Conveying

				Unit	Unit Cost	Cost Per S.F.	% Of Sub-Total
1010	Elevators & Lifts	N/A		—	—	—	0.0 %
1020	Escalators & Moving Walks	N/A		—	—	—	

D20 Plumbing

2010	Plumbing Fixtures	Toilet and service fixtures, supply and drainage	1 Fixture/260 S.F. Floor	Each	6666	25.64	
2020	Domestic Water Distribution	Gas fired water heater		S.F. Floor	2.43	2.43	21.7%
2040	Rain Water Drainage	Roof drains		S.F. Roof	.60	.60	

D30 HVAC

3010	Energy Supply	N/A		—	—	—	
3020	Heat Generating Systems	Included in D3050		—	—	—	
3030	Cooling Generating Systems	N/A		—	—	—	13.9 %
3050	Terminal & Package Units	Multizone unit, gas heating, electric cooling		S.F. Floor	18.40	18.40	
3090	Other HVAC Sys. & Equipment	N/A		—	—	—	

D40 Fire Protection

4010	Sprinklers	Sprinklers, light hazard		S.F. Floor	2.78	2.78	2.3%
4020	Standpipes	Standpipe, wet, Class III		S.F. Floor	.26	.26	

D50 Electrical

5010	Electrical Service/Distribution	1000 ampere service, panel board and feeders		S.F. Floor	2	2	
5020	Lighting & Branch Wiring	High efficiency fluorescent fixtures, receptacles, switches, A.C. and misc. power		S.F. Floor	10.46	10.46	12.3%
5030	Communications & Security	Addressable alarm systems, internet wiring, and emergency lighting		S.F. Floor	3.62	3.62	
5090	Other Electrical Systems	Emergency generator, 11.5 kW, Uninterruptible power supply		S.F. Floor	.21	.21	

E. EQUIPMENT & FURNISHINGS

				Unit	Unit Cost	Cost Per S.F.	% Of Sub-Total
1010	Commercial Equipment	N/A		—	—	—	
1020	Institutional Equipment	Cabinets, fume hoods, lockers, glassware washer		S.F. Floor	1.49	1.49	1.1 %
1030	Vehicular Equipment	N/A		—	—	—	
1090	Other Equipment	N/A		—	—	—	

F. SPECIAL CONSTRUCTION

1020	Integrated Construction	N/A		—	—	—	0.0%
1040	Special Facilities	N/A		—	—	—	

G. BUILDING SITEWORK   N/A

Sub-Total	131.96	**100%**

CONTRACTOR FEES (General Requirements: 10%, Overhead: 5%, Profit: 10%)	25%	33.04	
ARCHITECT FEES	10%	16.50	

Total Building Cost	**181.50**

Costs per square foot of floor area

Exterior Wall	S.F. Area	15000	20000	25000	30000	35000	40000	45000	50000	55000
	L.F. Perimeter	354	425	457	513	568	583	629	644	683
Brick Face with Concrete Block Back-up	Steel Frame	175.15	169.70	164.40	162.00	160.15	157.50	156.30	154.55	153.65
	R/Conc. Frame	166.05	160.65	155.25	152.90	151.00	148.35	147.20	145.45	144.50
Precast Concrete Panel	Steel Frame	180.50	174.65	168.65	166.00	163.95	160.90	159.65	157.60	156.60
	R/Conc. Frame	171.95	165.95	159.85	157.15	155.10	152.00	150.70	148.65	147.65
Limestone Face Concrete Block Back-up	Steel Frame	192.00	184.95	177.45	174.20	171.75	167.90	166.30	163.80	162.55
	R/Conc. Frame	182.90	175.80	168.35	165.10	162.65	158.75	157.15	154.65	153.40
Perimeter Adj., Add or Deduct	Per 100 L.F.	8.80	6.65	5.30	4.40	3.75	3.25	2.95	2.65	2.40
Story Hgt. Adj., Add or Deduct	Per 1 Ft.	2.20	2.00	1.75	1.60	1.50	1.35	1.30	1.20	1.15
For Basement, add $35.15 per square foot of basement area										

The above costs were calculated using the basic specifications shown on the facing page. These costs should be adjusted where necessary for design alternatives and owner's requirements. Reported completed project costs, for this type of structure, range from $124.20 to $253.40 per S.F.

Common additives

Description	Unit	$ Cost	Description	Unit	$ Cost
Carrels Hardwood	Each	605 - 1150	Lockers, Steel, Single tier, 60" or 72"	Opening	200 - 350
Elevators, Hydraulic passenger, 2 stops			2 tier, 60" or 72" total	Opening	124 - 152
2000# capacity	Each	63,100	5 tier, box lockers	Opening	70 - 82.50
2500# capacity	Each	64,300	Locker bench, lam. maple top only	L.F.	22
3500# capacity	Each	67,600	Pedestals, steel pipe	Each	64
Emergency Lighting, 25 watt, battery operated			Sound System		
Lead battery	Each	287	Amplifier, 250 watts	Each	2400
Nickel cadmium	Each	845	Speaker, ceiling or wall	Each	196
Escalators, Metal			Trumpet	Each	375
32" wide, 10' story height	Each	139,700			
20' story height	Each	168,000			
48" wide, 10' Story height	Each	148,200			
20' story height	Each	176,000			
Glass					
32" wide, 10' story height	Each	133,200			
20' story height	Each	161,000			
48" wide, 10' story height	Each	140,700			
20' story height	Each	170,000			

Model costs calculated for a 2 story building with 12' story height and 25,000 square feet of floor area

				Unit	Unit Cost	Cost Per S.F.	% Of Sub-Total
A.	**SUBSTRUCTURE**						
1010	Standard Foundations	Poured concrete; strip and spread footings		S.F. Ground	4.22	2.11	
1020	Special Foundations	N/A		—	—	—	
1030	Slab on Grade	4" reinforced concrete with vapor barrier and granular base		S.F. Slab	4.77	2.39	5.1%
2010	Basement Excavation	Site preparation for slab and trench for foundation wall and footing		S.F. Ground	.18	.09	
2020	Basement Walls	4' foundation wall		L.F. Wall	76	1.38	
B.	**SHELL**						
	B10 Superstructure						
1010	Floor Construction	Concrete flat plate		S.F. Floor	24.60	12.30	16.7%
1020	Roof Construction	Concrete flat plate		S.F. Roof	14.10	7.05	
	B20 Exterior Enclosure						
2010	Exterior Walls	Face brick with concrete block backup	75% of wall	S.F. Wall	30.30	9.97	
2020	Exterior Windows	Window wall	25% of wall	Each	36.95	4.05	12.4%
2030	Exterior Doors	Double aluminum and glass		Each	2440	.39	
	B30 Roofing						
3010	Roof Coverings	Built-up tar and gravel with flashing; perlite/EPS composite insulation		S.F. Roof	6	3	2.6%
3020	Roof Openings	Roof hatches		S.F. Roof	.10	.05	
C.	**INTERIORS**						
1010	Partitions	Gypsum board on metal studs	14 S.F. Floor/L.F. Partition	S.F. Partition	5.52	3.94	
1020	Interior Doors	Single leaf hollow metal	140 S.F. Floor/Door	Each	1001	7.15	
1030	Fittings	N/A		—	—	—	
2010	Stair Construction	Cast in place concrete		Flight	7300	1.16	22.6%
3010	Wall Finishes	50% paint, 50% vinyl wall covering		S.F. Surface	2.16	3.08	
3020	Floor Finishes	50% carpet, 50% vinyl composition tile		S.F. Floor	6.60	6.60	
3030	Ceiling Finishes	Suspended fiberglass board		S.F. Ceiling	4.31	4.31	
D.	**SERVICES**						
	D10 Conveying						
1010	Elevators & Lifts	One hydraulic passenger elevator		Each	79,000	3.16	2.7%
1020	Escalators & Moving Walks	N/A		—	—	—	
	D20 Plumbing						
2010	Plumbing Fixtures	Toilet and service fixtures, supply and drainage	1 Fixture/1040 S.F. Floor	Each	2860	2.75	
2020	Domestic Water Distribution	Gas fired water heater		S.F. Floor	.76	.76	3.4%
2040	Rain Water Drainage	Roof drains		S.F. Roof	.98	.49	
	D30 HVAC						
3010	Energy Supply	N/A		—	—	—	
3020	Heat Generating Systems	Included in D3050		—	—	—	
3030	Cooling Generating Systems	N/A		—	—	—	15.8%
3050	Terminal & Package Units	Multizone unit, gas heating, electric cooling		S.F. Floor	18.40	18.40	
3090	Other HVAC Sys. & Equipment	N/A		—	—	—	
	D40 Fire Protection						
4010	Sprinklers	Wet pipe sprinkler system		S.F. Floor	2.86	2.86	3.1%
4020	Standpipes	Standpipe, wet, Class III		S.F. Floor	.71	.71	
	D50 Electrical						
5010	Electrical Service/Distribution	600 ampere service, panel board and feeders		S.F. Floor	2.23	2.23	
5020	Lighting & Branch Wiring	High efficiency fluorescent fixtures, receptacles, switches, A.C. and misc. power		S.F. Floor	12.13	12.13	15.5%
5030	Communications & Security	Addressable alarm systems, internet wiring, communications systems and emergency lighting		S.F. Floor	3.43	3.43	
5090	Other Electrical Systems	Emergency generator, 11.5 kW		S.F. Floor	.15	.15	
E.	**EQUIPMENT & FURNISHINGS**						
1010	Commercial Equipment	N/A		—	—	—	
1020	Institutional Equipment	N/A		—	—	—	0.0 %
1030	Vehicular Equipment	N/A		—	—	—	
1090	Other Equipment	N/A		—	—	—	
F.	**SPECIAL CONSTRUCTION**						
1020	Integrated Construction	N/A		—	—	—	0.0 %
1040	Special Facilities	N/A		—	—	—	
G.	**BUILDING SITEWORK**	**N/A**					

			Sub-Total	116.09	**100%**
CONTRACTOR FEES (General Requirements: 10%, Overhead: 5%, Profit: 10%)			25%	29	
ARCHITECT FEES			7%	10.16	

Total Building Cost	**155.25**

Costs per square foot of floor area

Exterior Wall	S.F. Area	4000	6000	8000	10000	12000	14000	16000	18000	20000
	L.F. Perimeter	260	340	420	453	460	510	560	610	600
Face Brick with Concrete Block Back-up	Bearing Walls	163.50	155.20	151.05	144.50	138.35	136.55	135.20	134.20	130.75
	Steel Frame	158.25	151.00	147.45	141.85	136.50	134.95	133.80	132.90	129.95
Decorative Concrete Block	Bearing Walls	143.30	136.90	133.60	128.65	123.95	122.60	121.60	120.75	118.15
	Steel Frame	144.70	139.25	136.45	132.40	128.50	127.35	126.45	125.80	123.70
Tilt Up Concrete Wall Panels	Bearing Walls	147.25	141.10	137.95	133.25	128.75	127.45	126.45	125.75	123.25
	Steel Frame	142.05	136.85	134.30	130.50	126.90	125.90	125.00	124.40	122.50
Perimeter Adj., Add or Deduct	Per 100 L.F.	21.30	14.25	10.65	8.50	7.10	6.10	5.30	4.70	4.25
Story Hgt. Adj., Add or Deduct	Per 1 Ft.	2.65	2.30	2.15	1.90	1.60	1.55	1.45	1.35	1.20
For Basement, add $30.75 per square foot of basement area										

The above costs were calculated using the basic specifications shown on the facing page. These costs should be adjusted where necessary for design alternatives and owner's requirements. Reported completed project costs, for this type of structure, range from $75.80 to $239.45 per S.F.

Common additives

Description	Unit	$ Cost
Bar, Front bar	L.F.	375
Back bar	L.F.	300
Booth, Upholstered, custom straight	L.F.	212 - 390
"L" or "U" shaped	L.F.	220 - 370
Bowling Alleys, incl. alley, pinsetter		
Scorer, counter & misc. supplies, average	Lane	60,500
For automatic scorer, add	Lane	10,900
Emergency Lighting, 25 watt, battery operated		
Lead battery	Each	287
Nickel cadmium	Each	845
Kitchen Equipment		
Broiler	Each	4050
Coffee urn, twin 6 gallon	Each	3025
Cooler, 6 ft. long	Each	5150
Dishwasher, 10-12 racks per hr.	Each	5075
Food warmer	Each	740
Freezer, 44 C.F., reach-in	Each	3250
Ice cube maker, 50 lb. per day	Each	1875
Range with 1 oven	Each	2700

Description	Unit	$ Cost
Movie Equipment		
Projector, 35mm	Each	12,200 - 16,700
Screen, wall or ceiling hung	S.F.	8.50 - 12.90
Partitions, Folding leaf, wood		
Acoustic type	S.F.	73.50 - 122
Seating		
Auditorium chair, all veneer	Each	240
Veneer back, padded seat	Each	290
Upholstered, spring seat	Each	279
Classroom, movable chair & desk	Set	65 - 120
Lecture hall, pedestal type	Each	229 - 680
Sound System		
Amplifier, 250 watts	Each	2400
Speaker, ceiling or wall	Each	196
Trumpet	Each	375
Stage Curtains, Medium weight	S.F.	10.40 - 37
Curtain Track, Light duty	L.F.	80.50
Swimming Pools, Complete, gunite	S.F.	68.50 - 84.50

Community Center

Model costs calculated for a 1 story building with 12′ story height and 10,000 square feet of floor area

				Unit	Unit Cost	Cost Per S.F.	% Of Sub-Total
A. SUBSTRUCTURE							
1010	Standard Foundations	Poured concrete; strip and spread footings		S.F. Ground	2.10	2.10	
1020	Special Foundations	N/A		—	—	—	
1030	Slab on Grade	4" reinforced concrete with vapor barrier and granular base		S.F. Slab	4.77	4.77	13.8%
2010	Basement Excavation	Site preparation for slab and trench for foundation wall and footing		S.F. Ground	.31	.31	
2020	Basement Walls	4' foundation wall		L.F. Wall	71	7.49	
B. SHELL							
	B10 Superstructure						
1010	Floor Construction	N/A		—	—	—	4.7 %
1020	Roof Construction	Metal deck on open web steel joists		S.F. Roof	4.96	4.96	
	B20 Exterior Enclosure						
2010	Exterior Walls	Face brick with concrete block backup	80% of wall	S.F. Wall	30.31	13.18	
2020	Exterior Windows	Aluminum sliding	20% of wall	Each	590	2.01	15.2%
2030	Exterior Doors	Double aluminum and glass and hollow metal		Each	2367	.95	
	B30 Roofing						
3010	Roof Coverings	Built-up tar and gravel with flashing; perlite/EPS composite insulation		S.F. Roof	6.87	6.87	6.6%
3020	Roof Openings	Roof hatches		S.F. Roof	.09	.09	
C. INTERIORS							
1010	Partitions	Gypsum board on metal studs	14 S.F. Floor/L.F. Partition	S.F. Partition	7.59	5.42	
1020	Interior Doors	Single leaf hollow metal	140 S.F. Floor/Door	Each	1001	7.15	
1030	Fittings	Toilet partitions, directory board, mailboxes		S.F. Floor	2.03	2.03	
2010	Stair Construction	N/A		—	—	—	28.1%
3010	Wall Finishes	Paint		S.F. Surface	1.61	2.30	
3020	Floor Finishes	50% carpet, 50% vinyl tile		S.F. Floor	6.40	6.40	
3030	Ceiling Finishes	Mineral fiber tile on concealed zee bars		S.F. Ceiling	6.51	6.51	
D. SERVICES							
	D10 Conveying						
1010	Elevators & Lifts	N/A		—	—	—	0.0 %
1020	Escalators & Moving Walks	N/A		—	—	—	
	D20 Plumbing						
2010	Plumbing Fixtures	Kitchen, toilet and service fixtures, supply and drainage	1 Fixture/910 S.F. Floor	Each	2876	3.16	
2020	Domestic Water Distribution	Electric water heater		S.F. Floor	6.32	6.32	9.4%
2040	Rain Water Drainage	Roof drains		S.F. Roof	.47	.47	
	D30 HVAC						
3010	Energy Supply	N/A		—	—	—	
3020	Heat Generating Systems	Included in D3050		—	—	—	
3030	Cooling Generating Systems	N/A		—	—	—	8.9 %
3050	Terminal & Package Units	Single zone rooftop unit, gas heating, electric cooling		S.F. Floor	9.43	9.43	
3090	Other HVAC Sys. & Equipment	N/A		—	—	—	
	D40 Fire Protection						
4010	Sprinklers	Wet pipe sprinkler system		S.F. Floor	3.16	3.16	3.0%
4020	Standpipes	N/A		—	—	—	
	D50 Electrical						
5010	Electrical Service/Distribution	200 ampere service, panel board and feeders		S.F. Floor	.98	.98	
5020	Lighting & Branch Wiring	High efficiency fluorescent fixtures, receptacles, switches, A.C. and misc. power		S.F. Floor	5.56	5.56	7.7%
5030	Communications & Security	Addressable alarm systems and emergency lighting		S.F. Floor	1.43	1.43	
5090	Other Electrical Systems	Emergency generator, 15 kW		S.F. Floor	.18	.18	
E. EQUIPMENT & FURNISHINGS							
1010	Commercial Equipment	Freezer, chest type		S.F. Floor	.81	.81	
1020	Institutional Equipment	N/A		—	—	—	2.7%
1030	Vehicular Equipment	N/A		—	—	—	
1090	Other Equipment	Kitchen equipment, directory board, mailboxes, built-in coat racks		S.F. Floor	2.05	2.05	
F. SPECIAL CONSTRUCTION							
1020	Integrated Construction	N/A		—	—	—	0.0 %
1040	Special Facilities	N/A		—	—	—	
G. BUILDING SITEWORK	**N/A**						
				Sub-Total		106.09	**100%**
	CONTRACTOR FEES (General Requirements: 10%, Overhead: 5%, Profit: 10%)				25%	26.48	
	ARCHITECT FEES				9%	11.93	
				Total Building Cost		**144.50**	

Costs per square foot of floor area

Exterior Wall	S.F. Area	10000	12500	15000	17500	20000	22500	25000	30000	40000
	L.F. Perimeter	400	450	500	550	600	625	650	700	800
Brick Veneer	Steel Frame	305.35	302.15	300.00	298.40	297.30	295.50	294.10	292.00	289.35
Brick Veneer	Bearing Wall	316.90	312.55	309.60	307.50	305.95	303.50	301.50	298.70	295.05
Tilt Up Concrete	Steel Frame	297.50	294.30	292.15	290.55	289.45	287.65	286.25	284.15	281.50
Concrete Panels	Bearing Wall	327.00	321.65	318.05	315.50	313.55	310.55	308.20	304.65	300.20
EIFS	Steel Frame	301.90	299.30	297.55	296.20	295.35	293.85	292.70	291.00	288.85
EIFS	Bearing Wall	301.75	298.95	297.00	295.65	294.65	293.05	291.80	289.95	287.55
Perimeter Adj., Add or Deduct	Per 100 L.F.	7.85	6.30	5.20	4.55	3.90	3.55	3.15	2.60	2.00
Story Hgt. Adj., Add or Deduct	Per 1 Ft.	1.35	1.30	1.15	1.15	1.05	0.95	0.90	0.85	0.70
For Basement, add $33.80 per square foot of basement area										

The above costs were calculated using the basic specifications shown on the facing page. These costs should be adjusted where necessary for design alternatives and owner's requirements. Reported completed project costs, for this type of structure, range from $156.40 to $317.40 per S.F.

Common additives

Description	Unit	$ Cost	Description	Unit	$ Cost
Clock System			Smoke Detectors		
20 room	Ea.	16,000	Ceiling type	Ea.	233
50 room	Ea.	39,100	Duct type	Ea.	525
Closed Circuit Surveillance, one station			Sound System		
Camera and monitor	Total	1875	Amplifier, 250 watts	Ea.	2400
For additional camera stations, add	Ea.	1025	Speakers, ceiling or wall	Ea.	196
For zoom lens - remote control, add	Ea.	2975 - 10,000	Trumpets	Ea.	375
For automatic iris for low light, add	Ea.	2625			
Directory Boards, plastic, glass covered					
30" x 20"	Ea.	605			
36" x 48"	Ea.	1325			
Aluminum, 24" x 18"	Ea.	585			
36" x 24"	Ea.	685			
48" x 32"	Ea.	975			
48" x 60"	Ea.	2025			
Emergency Lighting, 25 watt, battery operated					
Lead battery	Ea.	287			
Nickel cadmium	Ea.	845			

Important: See the Reference Section for Location Factors

Model costs calculated for a 1 story building with 16'-6" story height and 22,500 square feet of floor area

Computer Data Center

				Unit	Unit Cost	Cost Per S.F.	% Of Sub-Total
A. SUBSTRUCTURE							
1010	Standard Foundations	Poured concrete; strip and spread footings		S.F. Ground	1.51	1.51	
1020	Special Foundations	N/A		—	—	—	
1030	Slab on Grade	4" concrete with vapor barrier and granular base		S.F. Slab	4.77	4.77	3.9%
2010	Basement Excavation	Site preparation for slab and trench for foundation wall and footing		S.F. Ground	.18	.20	
2020	Basement Walls	4' foundation wall		L.F. Wall	76	2.10	
B. SHELL							
	B10 Superstructure						
1010	Floor Construction	N/A		—	—	—	3.6 %
1020	Roof Construction	Steel joist, girders & deck on columns		S.F. Roof	7.98	7.98	
	B20 Exterior Enclosure						
2010	Exterior Walls	Brick veneer on metal studs		S.F. Wall	23.27	10.13	
2020	Exterior Windows	Aluminum sliding	5% of Wall	Each	557	.85	5.4%
2030	Exterior Doors	Aluminum and glass		Each	3482	.93	
	B30 Roofing						
3010	Roof Coverings	Single ply membrane, loose laid and ballasted; polyisocyanurate insulation		S.F. Roof	6.84	6.84	3.1%
3020	Roof Openings	Roof hatch		S.F. Roof	.09	.09	
C. INTERIORS							
1010	Partitions	Gypsum board on metal studs	13 S.F. Floor/L.F. Partition	S.F. Partition	4.33	4	
1020	Interior Doors	Solid core wood, metal frame	370 S.F. Floor/Door	Each	638	1.73	
1030	Fittings	Toilet partitions		S.F. Floor	.46	.46	
2010	Stair Construction	N/A		—	—	—	10.0%
3010	Wall Finishes	90% paint, 10% ceramic tile		S.F. Surface	1.88	3.47	
3020	Floor Finishes	65% carpet tile, 10% ceramic tile, 10% quarry tile		S.F. Floor	5.84	5.84	
3030	Ceiling Finishes	Mineral fiber tile on concealed zee bars		S.F. Ceiling	6.51	6.51	
D. SERVICES							
	D10 Conveying						
1010	Elevators & Lifts	N/A		—	—	—	0.0 %
1020	Escalators & Moving Walks	N/A		—	—	—	
	D20 Plumbing						
2010	Plumbing Fixtures	Toilet and service fixtures, supply and drainage	1 fixture/1180 S.F. Floor	Each	4236	3.59	
2020	Domestic Water Distribution	Gas fired water heater		S.F. Floor	.61	.61	2.3%
2040	Rain Water Drainage	Roof drains		S.F. Floor	.97	.97	
	D30 HVAC						
3010	Energy Supply	Hot water reheat		S.F. Floor	5.57	5.57	
3020	Heat Generating Systems	Boiler, oil fired hot water		Each	39,950	7.66	
3030	Cooling Generating Systems	Cooling tower		S.F. Floor	8.59	8.59	38.4%
3050	Terminal & Package Units	N/A		—	—	—	
3090	Other HVAC Sys. & Equipment	Heat exchanger, ductwork, air handling units, VAV terminals		S.F. Floor	63	63.05	
	D40 Fire Protection						
4010	Sprinklers	85% wet pipe sprinkler system, 15% preaction sprinkler system		S.F. Floor	3.71	3.71	1.9%
4020	Standpipes	Standpipes and hose systems		S.F. Floor	.54	.54	
	D50 Electrical						
5010	Electrical Service/Distribution	1200 ampere service, panel board and feeders		S.F. Floor	4.40	4.40	
5020	Lighting & Branch Wiring	High efficiency fluorescent fixtures, receptacles, A.C. and misc. power		S.F. Floor	15.57	15.57	30.4%
5030	Communications & Security	Addressable alarm & telephone systems, internet wiring and emergency lighting		S.F. Floor	32.20	32.20	
5090	Other Electrical Systems	Emergency generator, 350 kW; UPS system		S.F. Floor	15.10	15.10	
E. EQUIPMENT & FURNISHINGS							
1010	Commercial Equipment	N/A		—	—	—	
1020	Institutional Equipment	N/A		—	—	—	0.0 %
1030	Vehicular Equipment	N/A		—	—	—	
1090	Other Equipment	N/A		—	—	—	
F. SPECIAL CONSTRUCTION & DEMOLITION							
1020	Integrated Construction	Pedestal access floor	15% of Floor		1.96	1.96	0.9%
1040	Special Facilities	N/A		—	—	—	
G. BUILDING SITEWORK	**N/A**						

			Sub-Total	220.93	**100%**
CONTRACTOR FEES (General Requirements: 10%, Overhead: 5%, Profit: 10%)			25%	55.24	
ARCHITECT FEES			7%	19.33	

Total Building Cost	**295.50**

Costs per square foot of floor area

Exterior Wall	S.F. Area	16000	23000	30000	37000	44000	51000	58000	65000	72000
	L.F. Perimeter	597	763	821	968	954	1066	1090	1132	1220
Limestone with Concrete Block Back-up	R/Conc. Frame	214.25	205.50	195.95	193.25	186.50	184.95	181.70	179.55	178.65
	Steel Frame	213.40	204.60	195.10	192.40	185.65	184.10	180.85	178.70	177.80
Face Brick with Concrete Block Back-up	R/Conc. Frame	195.25	188.60	182.05	179.95	175.55	174.35	172.15	170.70	170.05
	Steel Frame	194.40	187.75	181.15	179.10	174.65	173.50	171.30	169.85	169.20
Stone with Concrete Block Back-up	R/Conc. Frame	199.70	192.50	185.30	183.10	178.05	176.80	174.35	172.70	172.05
	Steel Frame	198.90	191.75	184.50	182.30	177.30	176.05	173.60	171.95	171.20
Perimeter Adj., Add or Deduct	Per 100 L.F.	8.35	5.75	4.50	3.65	3.05	2.65	2.35	2.05	1.85
Story Hgt. Adj., Add or Deduct	Per 1 Ft.	3.00	2.65	2.25	2.15	1.80	1.70	1.55	1.45	1.40
For Basement, add $29.05 per square foot of basement area										

The above costs were calculated using the basic specifications shown on the facing page. These costs should be adjusted where necessary for design alternatives and owner's requirements. Reported completed project costs, for this type of structure, range from $133.45 to $248.60 per S.F.

Common additives

Description	Unit	$ Cost	Description	Unit	$ Cost
Benches, Hardwood	L.F.	107 - 195	Flagpoles, Complete		
Clock System			Aluminum, 20' high	Each	1700
20 room	Each	16,000	40' high	Each	3600
50 room	Each	39,100	70' high	Each	10,300
Closed Circuit Surveillance, One station			Fiberglass, 23' high	Each	1825
Camera and monitor	Each	1875	39'-5" high	Each	3475
For additional camera stations, add	Each	1025	59' high	Each	6625
Directory Boards, Plastic, glass covered			Intercom System, 25 station capacity		
30" x 20"	Each	605	Master station	Each	2725
36" x 48"	Each	1325	Intercom outlets	Each	174
Aluminum, 24" x 18"	Each	585	Handset	Each	480
36" x 24"	Each	685	Safe, Office type, 1 hour rating		
48" x 32"	Each	975	30" x 18" x 18"	Each	2400
48" x 60"	Each	2025	60" x 36" x 18", double door	Each	8975
Emergency Lighting, 25 watt, battery operated			Smoke Detectors		
Lead battery	Each	287	Ceiling type	Each	233
Nickel cadmium	Each	845	Duct type	Each	525

Important: See the Reference Section for Location Factors

Model costs calculated for a 1 story building with 14′ story height and 30,000 square feet of floor area

				Unit	Unit Cost	Cost Per S.F.	% Of Sub-Total
A. SUBSTRUCTURE							
1010	Standard Foundations	Poured concrete; strip and spread footings		S.F. Ground	1.57	1.57	
1020	Special Foundations	N/A		—	—	—	
1030	Slab on Grade	4″ reinforced concrete with vapor barrier and granular base		S.F. Slab	4.77	4.77	6.0%
2010	Basement Excavation	Site preparation for slab and trench for foundation wall and footing		S.F. Ground	.18	.18	
2020	Basement Walls	4′ foundation wall		L.F. Wall	76	2.21	
B. SHELL							
	B10 Superstructure						
1010	Floor Construction	Cast-in-place columns		L.F. Column	66	1.66	13.7%
1020	Roof Construction	Cast-in-place concrete waffle slab		S.F. Roof	18.35	18.35	
	B20 Exterior Enclosure						
2010	Exterior Walls	Limestone panels with concrete block backup	75% of wall	S.F. Wall	67	19.11	
2020	Exterior Windows	Aluminum with insulated glass	25% of wall	Each	703	2.92	15.4%
2030	Exterior Doors	Double wood		Each	2024	.47	
	B30 Roofing						
3010	Roof Coverings	Built-up tar and gravel with flashing; perlite/EPS composite insulation		S.F. Roof	5.34	5.34	3.7%
3020	Roof Openings	Roof hatches		S.F. Roof	.09	.09	
C. INTERIORS							
1010	Partitions	Plaster on metal studs	10 S.F. Floor/L.F. Partition	S.F. Partition	12.28	14.73	
1020	Interior Doors	Single leaf wood	100 S.F. Floor/Door	Each	576	5.76	
1030	Fittings	Toilet partitions		S.F. Floor	.61	.61	
2010	Stair Construction	N/A		—	—	—	33.4%
3010	Wall Finishes	70% paint, 20% wood paneling, 10% vinyl wall covering		S.F. Surface	2.35	5.65	
3020	Floor Finishes	60% hardwood, 20% carpet, 20% terrazzo		S.F. Floor	12.88	12.88	
3030	Ceiling Finishes	Gypsum plaster on metal lath, suspended		S.F. Ceiling	9.30	9.30	
D. SERVICES							
	D10 Conveying						
1010	Elevators & Lifts	N/A		—	—	—	0.0 %
1020	Escalators & Moving Walks	N/A		—	—	—	
	D20 Plumbing						
2010	Plumbing Fixtures	Toilet and service fixtures, supply and drainage	1 Fixture/1110 S.F. Floor	Each	4396	3.96	
2020	Domestic Water Distribution	Electric hot water heater		S.F. Floor	2.27	2.27	4.9%
2040	Rain Water Drainage	Roof drains		S.F. Roof	.96	.96	
	D30 HVAC						
3010	Energy Supply	N/A		—	—	—	
3020	Heat Generating Systems	Included in D3050		—	—	—	
3030	Cooling Generating Systems	N/A		—	—	—	12.6 %
3050	Terminal & Package Units	Multizone unit, gas heating, electric cooling		S.F. Floor	18.40	18.40	
3090	Other HVAC Sys. & Equipment	N/A		—	—	—	
	D40 Fire Protection						
4010	Sprinklers	Wet pipe sprinkler system		S.F. Floor	2.78	2.78	2.2%
4020	Standpipes	Standpipe, wet, Class III		S.F. Floor	.39	.39	
	D50 Electrical						
5010	Electrical Service/Distribution	400 ampere service, panel board and feeders		S.F. Floor	.95	.95	
5020	Lighting & Branch Wiring	High efficiency fluorescent fixtures, receptacles, switches, A.C. and misc. power		S.F. Floor	9.43	9.43	8.3%
5030	Communications & Security	Addressable alarm systems, internet wiring, and emergency lighting		S.F. Floor	1.64	1.64	
5090	Other Electrical Systems	Emergency generator, 11.5 kW		S.F. Floor	.13	.13	
E. EQUIPMENT & FURNISHINGS							
1010	Commercial Equipment	N/A		—	—	—	
1020	Institutional Equipment	N/A		—	—	—	0.0 %
1030	Vehicular Equipment	N/A		—	—	—	
1090	Other Equipment	N/A		—	—	—	
F. SPECIAL CONSTRUCTION							
1020	Integrated Construction	N/A		—	—	—	0.0 %
1040	Special Facilities	N/A		—	—	—	
G. BUILDING SITEWORK	**N/A**						

			Sub-Total	146.51	100%
CONTRACTOR FEES (General Requirements: 10%, Overhead: 5%, Profit: 10%)		25%	36.62		
ARCHITECT FEES		7%	12.82		
Total Building Cost				**195.95**	

Costs per square foot of floor area

Exterior Wall	S.F. Area	30000	40000	45000	50000	60000	70000	80000	90000	100000
	L.F. Perimeter	**410**	**493**	**535**	**533**	**600**	**666**	**733**	**800**	**795**
Limestone with Concrete Block Back-up	R/Conc. Frame	236.95	226.20	222.60	216.50	211.75	208.30	205.70	203.75	199.50
	Steel Frame	237.70	226.95	223.35	217.30	212.50	209.05	206.45	204.50	200.25
Face Brick with Concrete Block Back-up	R/Conc. Frame	219.05	210.00	207.00	202.50	198.60	195.80	193.65	192.10	189.05
	Steel Frame	219.85	210.85	207.85	203.35	**199.45**	196.65	194.50	192.90	189.85
Stone with Concrete Block Back-up	R/Conc. Frame	223.35	213.90	210.80	205.90	201.80	198.80	196.55	194.90	191.55
	Steel Frame	224.10	214.70	211.50	206.65	202.55	199.60	197.35	195.65	192.35
Perimeter Adj., Add or Deduct	Per 100 L.F.	12.40	9.25	8.30	7.45	6.20	5.30	4.70	4.15	3.75
Story Hgt. Adj., Add or Deduct	Per 1 Ft.	4.05	3.65	3.50	3.15	2.95	2.80	2.75	2.60	2.30
For Basement, add $29.65 per square foot of basement area										

The above costs were calculated using the basic specifications shown on the facing page. These costs should be adjusted where necessary for design alternatives and owner's requirements. Reported completed project costs, for this type of structure, range from $134.05 to $250.05 per S.F.

Common additives

Description	Unit	$ Cost	Description	Unit	$ Cost
Benches, Hardwood	L.F.	107 - 195	Emergency Lighting, 25 watt, battery operated		
Clock System			Lead battery	Each	287
20 room	Each	16,000	Nickel cadmium	Each	845
50 room	Each	39,100	Flagpoles, Complete		
Closed Circuit Surveillance, One station			Aluminum, 20' high	Each	1700
Camera and monitor	Each	1875	40' high	Each	3600
For additional camera stations, add	Each	1025	70' high	Each	10,300
Directory Boards, Plastic, glass covered			Fiberglass, 23' high	Each	1825
30" x 20"	Each	605	39'-5" high	Each	3475
36" x 48"	Each	1325	59' high	Each	6625
Aluminum, 24" x 18"	Each	585	Intercom System, 25 station capacity		
36" x 24"	Each	685	Master station	Each	2725
48" x 32"	Each	975	Intercom outlets	Each	174
48" x 60"	Each	2025	Handset	Each	480
Elevators, Hydraulic passenger, 2 stops			Safe, Office type, 1 hour rating		
1500# capacity	Each	60,900	30" x 18" x 18"	Each	2400
2500# capacity	Each	64,300	60" x 36" x 18", double door	Each	8975
3500# capacity	Each	67,600	Smoke Detectors		
Additional stop, add	Each	8250	Ceiling type	Each	233
			Duct type	Each	525

Model costs calculated for a 3 story building with 12' story height and 60,000 square feet of floor area

				Unit	Unit Cost	Cost Per S.F.	% Of Sub-Total
A. SUBSTRUCTURE							
1010	Standard Foundations	Poured concrete; strip and spread footings		S.F. Ground	3.57	1.19	
1020	Special Foundations	N/A		—	—	—	
1030	Slab on Grade	4" reinforced concrete with vapor barrier and granular base		S.F. Slab	4.77	1.59	2.5%
2010	Basement Excavation	Site preparation for slab and trench for foundation wall and footing		S.F. Ground	.18	.06	
2020	Basement Walls	4' foundation wall		L.F. Wall	71	.87	
B. SHELL							
	B10 Superstructure						
1010	Floor Construction	Concrete slab with metal deck and beams		S.F. Floor	25.82	17.21	15.3%
1020	Roof Construction	Concrete slab with metal deck and beams		S.F. Roof	16.89	5.63	
	B20 Exterior Enclosure						
2010	Exterior Walls	Face brick with concrete block backup	75% of wall	S.F. Wall	32.78	8.85	
2020	Exterior Windows	Horizontal pivoted steel	25% of wall	Each	707	7.07	10.9%
2030	Exterior Doors	Double aluminum and glass and hollow metal		Each	3218	.33	
	B30 Roofing						
3010	Roof Coverings	Built-up tar and gravel with flashing; perlite/EPS composite insulation		S.F. Roof	5.73	1.91	1.3%
3020	Roof Openings	N/A		—	—	—	
C. INTERIORS							
1010	Partitions	Plaster on metal studs	10 S.F. Floor/L.F. Partition	S.F. Partition	12.41	12.41	
1020	Interior Doors	Single leaf wood	100 S.F. Floor/Door	Each	576	5.76	
1030	Fittings	Toilet partitions		S.F. Floor	.30	.30	
2010	Stair Construction	Concrete filled metal pan		Flight	17,125	2.85	32.3%
3010	Wall Finishes	70% paint, 20% wood paneling, 10% vinyl wall covering		S.F. Surface	2.35	4.70	
3020	Floor Finishes	60% hardwood, 20% terrazzo, 20% carpet		S.F. Floor	12.88	12.88	
3030	Ceiling Finishes	Gypsum plaster on metal lath, suspended		S.F. Ceiling	9.30	9.30	
D. SERVICES							
	D10 Conveying						
1010	Elevators & Lifts	Five hydraulic passenger elevators		Each	166,920	13.91	9.3%
1020	Escalators & Moving Walks	N/A		—	—	—	
	D20 Plumbing						
2010	Plumbing Fixtures	Toilet and service fixtures, supply and drainage	1 Fixture/665 S.F. Floor	Each	2580	3.88	
2020	Domestic Water Distribution	Electric water heater		S.F. Floor	3.78	3.78	5.5%
2040	Rain Water Drainage	Roof drains		S.F. Roof	1.74	.58	
	D30 HVAC						
3010	Energy Supply	N/A		—	—	—	
3020	Heat Generating Systems	Included in D3050		—	—	—	
3030	Cooling Generating Systems	N/A		—	—	—	12.3 %
3050	Terminal & Package Units	Multizone unit, gas heating, electric cooling		S.F. Floor	18.40	18.40	
3090	Other HVAC Sys. & Equipment	N/A		—	—	—	
	D40 Fire Protection						
4010	Sprinklers	Wet pipe sprinkler system		S.F. Floor	2.81	2.81	2.1%
4020	Standpipes	Standpipe, wet, Class III		S.F. Floor	.36	.36	
	D50 Electrical						
5010	Electrical Service/Distribution	800 ampere service, panel board and feeders		S.F. Floor	1.20	1.20	
5020	Lighting & Branch Wiring	High efficiency fluorescent fixtures, receptacles, switches, A.C. and misc. power		S.F. Floor	9.57	9.57	8.4%
5030	Communications & Security	Addressable alarm systems, internet wiring, and emergency lighting		S.F. Floor	1.50	1.50	
5090	Other Electrical Systems	Emergency generator, 15 kW		S.F. Floor	.21	.21	
E. EQUIPMENT & FURNISHINGS							
1010	Commercial Equipment	N/A		—	—	—	
1020	Institutional Equipment	N/A		—	—	—	0.0 %
1030	Vehicular Equipment	N/A		—	—	—	
1090	Other Equipment	N/A		—	—	—	
F. SPECIAL CONSTRUCTION							
1020	Integrated Construction	N/A		—	—	—	0.0 %
1040	Special Facilities	N/A		—	—	—	
G. BUILDING SITEWORK	**N/A**						

Sub-Total		149.11	**100%**
CONTRACTOR FEES (General Requirements: 10%, Overhead: 5%, Profit: 10%)	25%	37.29	
ARCHITECT FEES	7%	13.05	
Total Building Cost		**199.45**	

Costs per square foot of floor area

Exterior Wall	S.F. Area	2000	5000	7000	10000	12000	15000	18000	21000	25000
	L.F. Perimeter	200	310	360	440	480	520	560	600	660
Tiltup Concrete Panel	Steel Joists	191.05	156.10	147.75	141.70	138.80	135.15	132.80	131.10	129.60
Decorative Concrete Block	Bearing Walls	186.55	152.05	143.75	137.80	134.95	131.35	128.95	127.30	125.80
Brick on Block	Steel Joists	233.90	191.65	181.00	173.50	169.65	164.85	161.60	159.35	157.30
Stucco on Concrete Block	Wood Truss	191.60	156.25	147.85	141.80	138.80	135.25	132.85	131.15	129.65
Brick Veneer	Wood Frame	201.45	164.30	155.20	**148.70**	145.45	141.40	138.70	136.80	135.15
Wood Siding	Wood Frame	183.80	154.35	147.25	142.15	139.70	136.65	134.65	133.20	132.00
Perimeter Adj., Add or Deduct	Per 100 L.F.	25.70	10.30	7.30	5.10	4.30	3.45	2.85	2.50	2.00
Story Hgt. Adj., Add or Deduct	Per 1 Ft.	2.55	1.60	1.30	1.15	1.05	0.90	0.80	0.75	0.70
For Basement, add $28.85 per square foot of basement area										

The above costs were calculated using the basic specifications shown on the facing page. These costs should be adjusted where necessary for design alternatives and owner's requirements. Reported completed project costs, for this type of structure, range from $57.80 to $210.65 per S.F.

Common additives

Description	Unit	$ Cost
Emergency Lighting, 25 watt, battery operated		
Lead battery	Each	287
Nickel cadmium	Each	845
Flagpoles, Complete		
Aluminum, 20' high	Each	1700
40' high	Each	3600
70' high	Each	10,300
Fiberglass, 23' high	Each	1825
39'-5" high	Each	3475
59' high	Each	6625
Gym Floor, Incl. sleepers and finish, maple	S.F.	13.15
Intercom System, 25 Station capacity		
Master station	Each	2725
Intercom outlets	Each	174
Handset	Each	480

Description	Unit	$ Cost
Lockers, Steel, single tier, 60" to 72"	Opening	200 - 350
2 tier, 60" to 72" total	Opening	124 - 152
5 tier, box lockers	Opening	70 - 82.50
Locker bench, lam. maple top only	L.F.	22
Pedestals, steel pipe	Each	64
Smoke Detectors		
Ceiling type	Each	233
Duct type	Each	525
Sound System		
Amplifier, 250 watts	Each	2400
Speaker, ceiling or wall	Each	196
Trumpet	Each	375

Important: See the Reference Section for Location Factors

Model costs calculated for a 1 story building with 12′ story height and 10,000 square feet of floor area

Day Care Center

				Unit	Unit Cost	Cost Per S.F.	% Of Sub-Total
A.	**SUBSTRUCTURE**						
1010	Standard Foundations	Poured concrete; strip and spread footings		S.F. Ground	1.91	1.91	
1020	Special Foundations	N/A		—	—	—	
1030	Slab on Grade	4" concrete with vapor barrier and granular base		S.F. Slab	4.29	4.29	9.0%
2010	Basement Excavation	Site preparation for slab and trench for foundation wall and footing		S.F. Ground	.31	.31	
2020	Basement Walls	4′ foundation wall		L.F. Wall	76	3.32	
B.	**SHELL**						
	B10 Superstructure						
1010	Floor Construction	Wood beams on columns		S.F. Floor	.15	.15	7.5%
1020	Roof Construction	Wood trusses		S.F. Roof	8.02	8.02	
	B20 Exterior Enclosure						
2010	Exterior Walls	Brick veneer on wood studs	85% of wall	S.F. Wall	23.31	10.46	
2020	Exterior Windows	Window wall	15% of wall	Each	43.60	3.46	16.5%
2030	Exterior Doors	Aluminum and glass; steel		Each	2132	4.06	
	B30 Roofing						
3010	Roof Coverings	Asphalt shingles, 9" fiberglass batt insulation, gutters and downspouts		S.F. Roof	4.08	4.08	3.7%
3020	Roof Openings	N/A		—	—	—	
C.	**INTERIORS**						
1010	Partitions	Gypsum board on wood studs	8 S.F. Floor/S.F. Partition	S.F. Partition	7.14	4.76	
1020	Interior Doors	Single leaf hollow metal	700 S.F. Floor/Door	Each	1001	2.67	
1030	Fittings	Toilet partitions		S.F. Floor	.55	.55	
2010	Stair Construction	N/A		—	—	—	15.3%
3010	Wall Finishes	Paint		S.F. Surface	1.40	1.87	
3020	Floor Finishes	5% quarry tile, 95% vinyl composition tile		S.F. Floor	2.51	2.51	
3030	Ceiling Finishes	Fiberglass tile on tee grid		S.F. Ceiling	4.31	4.31	
D.	**SERVICES**						
	D10 Conveying						
1010	Elevators & Lifts	N/A		—	—	—	0.0 %
1020	Escalators & Moving Walks	N/A		—	—	—	
	D20 Plumbing						
2010	Plumbing Fixtures	Toilet and service fixtures, supply and drainage	1 Fixture/455 S.F. Floor	Each	1648	9.26	
2020	Domestic Water Distribution	Electric water heater		S.F. Floor	2.22	2.22	10.5%
2040	Rain Water Drainage	N/A		—	—	—	
	D30 HVAC						
3010	Energy Supply	Oil fired hot water, wall fin radiation		S.F. Floor	10.63	10.63	
3020	Heat Generating Systems	N/A		—	—	—	
3030	Cooling Generating Systems	N/A		—	—	—	21.4%
3050	Terminal & Package Units	Split systems with air cooled condensing units		S.F. Floor	12.71	12.71	
3090	Other HVAC Sys. & Equipment	N/A		—	—	—	
	D40 Fire Protection						
4010	Sprinklers	Sprinkler, light hazard		S.F. Floor	3.16	3.16	2.9%
4020	Standpipes	N/A		—	—	—	
	D50 Electrical						
5010	Electrical Service/Distribution	200 ampere service, panel board and feeders		S.F. Floor	.98	.98	
5020	Lighting & Branch Wiring	Fluorescent fixtures, receptacles, switches, A.C. and misc. power		S.F. Floor	8.94	8.94	9.6%
5030	Communications & Security	Alarm systems and emergency lighting		S.F. Floor	.44	.44	
5090	Other Electrical Systems	Emergency generator, 15 kW		S.F. Floor	.08	.08	
E.	**EQUIPMENT & FURNISHINGS**						
1010	Commercial Equipment	N/A		—	—	—	
1020	Institutional Equipment	Cabinets and countertop		S.F. Floor	3.96	3.96	3.6 %
1030	Vehicular Equipment	N/A		—	—	—	
1090	Other Equipment	N/A		—	—	—	
F.	**SPECIAL CONSTRUCTION**						
1020	Integrated Construction	N/A		—	—	—	0.0%
1040	Special Facilities	N/A		—	—	—	
G.	**BUILDING SITEWORK**	**N/A**					

			Sub-Total	109.11	100%
CONTRACTOR FEES (General Requirements: 10%, Overhead: 5%, Profit: 10%)		25%	27.31		
ARCHITECT FEES		9%	12.28		

Total Building Cost	**148.70**

Costs per square foot of floor area

Exterior Wall	S.F. Area	12000	18000	24000	30000	36000	42000	48000	54000	60000
	L.F. Perimeter	460	580	713	730	826	880	965	1006	1045
Concrete Block	Steel Frame	128.10	121.40	118.30	114.05	112.60	110.95	110.10	109.00	108.05
	Bearing Walls	125.65	118.95	115.90	111.65	110.15	108.50	107.70	106.60	105.60
Precast Concrete Panels	Steel Frame	136.05	128.10	124.50	119.10	117.40	115.35	114.35	112.90	111.70
Insulated Metal Panels	Steel Frame	130.70	123.55	120.30	115.65	114.15	112.40	111.50	110.25	109.25
Face Brick on Common Brick	Steel Frame	147.00	137.20	132.90	126.05	123.95	121.30	120.05	118.15	116.65
Tilt-up Concrete Panel	Steel Frame	130.80	123.55	120.35	115.75	114.25	112.45	111.55	110.30	109.30
Perimeter Adj., Add or Deduct	Per 100 L.F.	5.10	3.35	2.55	2.05	1.70	1.50	1.30	1.10	1.00
Story Hgt. Adj., Add or Deduct	Per 1 Ft.	0.75	0.55	0.60	0.45	0.45	0.45	0.40	0.35	0.35
For Basement, add $29.40 per square foot of basement area										

The above costs were calculated using the basic specifications shown on the facing page. These costs should be adjusted where necessary for design alternatives and owner's requirements. Reported completed project costs, for this type of structure, range from $42.90 to $166.05 per S.F.

Common additives

Description	Unit	$ Cost	Description	Unit	$ Cost
Clock System			Dock Levelers, Hinged 10 ton cap.		
20 room	Each	16,000	6' x 8'	Each	6375
50 room	Each	39,100	7' x 8'	Each	6975
Dock Bumpers, Rubber blocks			Partitions, Woven wire, 10 ga., 1-1/2" mesh		
4-1/2" thick, 10" high, 14" long	Each	73	4' wide x 7' high	Each	180
24" long	Each	119	8' high	Each	188
36" long	Each	131	10' High	Each	246
12" high, 14" long	Each	121	Platform Lifter, Portable, 6'x 6'		
24" long	Each	137	3000# cap.	Each	9150
36" long	Each	157	4000# cap.	Each	11,300
6" thick, 10" high, 14" long	Each	120	Fixed, 6' x 8', 5000# cap.	Each	12,000
24" long	Each	150			
36" long	Each	194			
20" high, 11" long	Each	187			
Dock Boards, Heavy					
60" x 60" Aluminum, 5,000# cap.	Each	1275			
9000# cap.	Each	1575			
15,000# cap.	Each	1725			

Important: See the Reference Section for Location Factors

Model costs calculated for a 1 story building with 20′ story height and 30,000 square feet of floor area

				Unit	Unit Cost	Cost Per S.F.	% Of Sub-Total
A.	**SUBSTRUCTURE**						
1010	Standard Foundations	Poured concrete; strip and spread footings		S.F. Ground	1.29	1.29	
1020	Special Foundations	N/A		—	—	—	
1030	Slab on Grade	4″ reinforced concrete with vapor barrier and granular base		S.F. Slab	6.33	6.33	11.7%
2010	Basement Excavation	Site preparation for slab and trench for foundation wall and footing		S.F. Ground	.18	.18	
2020	Basement Walls	4′ foundation wall		L.F. Wall	76	2.20	
B.	**SHELL**						
	B10 Superstructure						
1010	Floor Construction	N/A		—	—	—	8.5 %
1020	Roof Construction	Metal deck, open web steel joists, beams and columns		S.F. Roof	7.25	7.25	
	B20 Exterior Enclosure						
2010	Exterior Walls	Concrete block	75% of wall	S.F. Wall	7.73	2.82	
2020	Exterior Windows	Industrial horizontal pivoted steel	25% of wall	Each	815	3.10	8.3%
2030	Exterior Doors	Double aluminum and glass, hollow metal, steel overhead		Each	2261	1.13	
	B30 Roofing						
3010	Roof Coverings	Built-up tar and gravel with flashing; perlite/EPS composite insulation		S.F. Roof	5.51	5.51	6.8%
3020	Roof Openings	Roof hatches		S.F. Roof	.32	.32	
C.	**INTERIORS**						
1010	Partitions	Concrete block	60 S.F. Floor/L.F. Partition	S.F. Partition	8.10	1.62	
1020	Interior Doors	Single leaf hollow metal and fire doors	600 S.F. Floor/Door	Each	1001	1.67	
1030	Fittings	Toilet partitions		S.F. Floor	1.20	1.20	
2010	Stair Construction	N/A		—	—	—	8.4%
3010	Wall Finishes	Paint		S.F. Surface	4.25	1.70	
3020	Floor Finishes	Vinyl composition tile	10% of floor	S.F. Floor	2.90	.29	
3030	Ceiling Finishes	Fiberglass board on exposed grid system	10% of area	S.F. Ceiling	6.51	.65	
D.	**SERVICES**						
	D10 Conveying						
1010	Elevators & Lifts	N/A		—	—	—	0.0 %
1020	Escalators & Moving Walks	N/A		—	—	—	
	D20 Plumbing						
2010	Plumbing Fixtures	Toilet and service fixtures, supply and drainage	1 Fixture/1000 S.F. Floor	Each	7720	7.72	
2020	Domestic Water Distribution	Gas fired water heater		S.F. Floor	.51	.51	10.6%
2040	Rain Water Drainage	Roof drains		S.F. Roof	.81	.81	
	D30 HVAC						
3010	Energy Supply	Oil fired hot water, unit heaters		S.F. Floor	9.17	9.17	
3020	Heat Generating Systems	N/A		—	—	—	
3030	Cooling Generating Systems	Chilled water, air cooled condenser system		S.F. Floor	10.49	10.49	23.1%
3050	Terminal & Package Units	N/A		—	—	—	
3090	Other HVAC Sys. & Equipment	N/A		—	—	—	
	D40 Fire Protection						
4010	Sprinklers	Sprinklers, ordinary hazard		S.F. Floor	3.66	3.66	4.9%
4020	Standpipes	Standpipe, wet, Class III		S.F. Floor	.49	.49	
	D50 Electrical						
5010	Electrical Service/Distribution	600 ampere service, panel board and feeders		S.F. Floor	1.33	1.33	
5020	Lighting & Branch Wiring	High intensity discharge fixtures, receptacles, switches, A.C. and misc. power		S.F. Floor	12.44	12.44	17.8%
5030	Communications & Security	Addressable alarm systems and emergency lighting		S.F. Floor	1.39	1.39	
5090	Other Electrical Systems	N/A		—	—	—	
E.	**EQUIPMENT & FURNISHINGS**						
1010	Commercial Equipment	N/A		—	—	—	
1020	Institutional Equipment	N/A		—	—	—	
1030	Vehicular Equipment	N/A		—	—	—	0.0 %
1090	Other Equipment	N/A		—	—	—	
F.	**SPECIAL CONSTRUCTION**						
1020	Integrated Construction	N/A		—	—	—	0.0 %
1040	Special Facilities	N/A		—	—	—	
G.	**BUILDING SITEWORK**	**N/A**					

		Sub-Total	85.27	100%
CONTRACTOR FEES (General Requirements: 10%, Overhead: 5%, Profit: 10%)		25%	21.32	
ARCHITECT FEES		7%	7.46	
	Total Building Cost		**114.05**	

Costs per square foot of floor area

Exterior Wall	S.F. Area	20000	30000	40000	50000	60000	70000	80000	90000	100000
	L.F. Perimeter	362	410	493	576	600	628	660	700	744
Face Brick Common Brick Back-up	Steel Frame	160.20	146.25	141.45	138.55	134.15	131.20	129.10	127.65	126.65
	Concrete Frame	155.90	141.95	137.15	134.25	129.85	126.90	124.80	**123.35**	122.35
Face Brick Concrete Block Back-up	Steel Frame	158.05	144.65	140.00	137.15	133.00	130.15	128.10	126.80	125.80
	Concrete Frame	153.75	140.35	135.70	132.85	128.70	125.85	123.80	122.45	121.50
Stucco on Concrete Block	Steel Frame	147.95	137.00	133.05	130.75	127.40	125.20	123.50	122.40	121.65
	Concrete Frame	143.65	132.70	128.75	126.45	123.10	120.85	119.20	118.10	117.35
Perimeter Adj., Add or Deduct	Per 100 L.F.	12.25	8.15	6.10	4.85	4.10	3.55	3.05	2.75	2.45
Story Hgt. Adj., Add or Deduct	Per 1 Ft.	3.40	2.60	2.30	2.15	1.90	1.70	1.55	1.50	1.45
For Basement, add $32.05 per square foot of basement area										

The above costs were calculated using the basic specifications shown on the facing page. These costs should be adjusted where necessary for design alternatives and owner's requirements. Reported completed project costs, for this type of structure, range from $42.90 to $166.10 per S.F.

Common additives

Description	Unit	$ Cost		Description	Unit	$ Cost
Clock System				Dock Levelers, Hinged 10 ton cap.		
20 room	Each	16,000		6' x 8'	Each	6375
50 room	Each	39,100		7' x 8'	Each	6975
Dock Bumpers, Rubber blocks				Elevator, Hydraulic freight, 2 stops		
4-1/2" thick, 10" high, 14" long	Each	73		3500# capacity	Each	146,700
24" long	Each	119		4000# capacity	Each	155,700
36" long	Each	131		Additional stop, add	Each	10,400
12" high, 14" long	Each	121		Partitions, Woven wire, 10 ga., 1-1/2" mesh		
24" long	Each	137		4' Wide x 7" high	Each	180
36" long	Each	157		8' High	Each	188
6" thick, 10" high, 14" long	Each	120		10' High	Each	246
24" long	Each	150		Platform Lifter, Portable, 6'x 6'		
36" long	Each	194		3000# cap.	Each	9150
20" high, 11" long	Each	187		4000# cap.	Each	11,300
Dock Boards, Heavy				Fixed, 6'x 8', 5000# cap.	Each	12,000
60" x 60" Aluminum, 5,000# cap.	Each	1275				
9000# cap.	Each	1575				
15,000# cap.	Each	1725				

Important: See the Reference Section for Location Factors

Model costs calculated for a 3 story building with 12' story height and 90,000 square feet of floor area

				Unit	Unit Cost	Cost Per S.F.	% Of Sub-Total
A.	**SUBSTRUCTURE**						
1010	Standard Foundations	Poured concrete; strip and spread footings		S.F. Ground	5.97	1.99	
1020	Special Foundations	N/A		—	—	—	
1030	Slab on Grade	4" reinforced concrete with vapor barrier and granular base		S.F. Slab	9.80	3.27	6.3%
2010	Basement Excavation	Site preparation for slab and trench for foundation wall and footing		S.F. Ground	.18	.06	
2020	Basement Walls	4' foundation wall		L.F. Wall	76	.59	
B.	**SHELL**						
	B10 Superstructure						
1010	Floor Construction	Concrete flat slab		S.F. Floor	21.95	14.63	20.9%
1020	Roof Construction	Concrete flat slab		S.F. Roof	14.49	4.83	
	B20 Exterior Enclosure						
2010	Exterior Walls	Face brick with common brick backup	70% of wall	S.F. Wall	33.83	6.63	
2020	Exterior Windows	Industrial, horizontal pivoted steel	30% of wall	Each	1259	6.61	14.7%
2030	Exterior Doors	Double aluminum & glass, hollow metal, overhead doors		Each	2141	.41	
	B30 Roofing						
3010	Roof Coverings	Built-up tar and gravel with flashing; perlite/EPS composite insulation		S.F. Roof	5.37	1.79	2.0%
3020	Roof Openings	Roof hatches		S.F. Roof	.30	.10	
C.	**INTERIORS**						
1010	Partitions	Gypsum board on metal studs	50 S.F. Floor/L.F. Partition	S.F. Partition	3.70	.74	
1020	Interior Doors	Single leaf fire doors	500 S.F. Floor/Door	Each	1000	2	
1030	Fittings	Toilet partitions		S.F. Floor	.61	.61	
2010	Stair Construction	Concrete		Flight	7300	.65	8.9%
3010	Wall Finishes	Paint		S.F. Surface	.75	.30	
3020	Floor Finishes	90% metallic hardener, 10% vinyl composition tile		S.F. Floor	3.52	3.52	
3030	Ceiling Finishes	Fiberglass board on exposed grid systems	10% of area	S.F. Ceiling	4.31	.43	
D.	**SERVICES**						
	D10 Conveying						
1010	Elevators & Lifts	Two hydraulic freight elevators		Each	227,700	5.06	5.4%
1020	Escalators & Moving Walks	N/A		—	—	—	
	D20 Plumbing						
2010	Plumbing Fixtures	Toilet and service fixtures, supply and drainage	1 Fixture/1345 S.F. Floor	Each	11,446	8.51	
2020	Domestic Water Distribution	Gas fired water heater		S.F. Floor	.31	.31	9.9%
2040	Rain Water Drainage	Roof drains		S.F. Roof	1.08	.36	
	D30 HVAC						
3010	Energy Supply	Oil fired hot water, unit heaters		S.F. Floor	4.07	4.07	
3020	Heat Generating Systems	N/A		—	—	—	
3030	Cooling Generating Systems	Chilled water, air cooled condenser system		S.F. Floor	10.49	10.49	15.6%
3050	Terminal & Package Units	N/A		—	—	—	
3090	Other HVAC Sys. & Equipment	N/A		—	—	—	
	D40 Fire Protection						
4010	Sprinklers	Wet pipe sprinkler system		S.F. Floor	3.26	3.26	4.0%
4020	Standpipes	Standpipe, wet, Class III		S.F. Floor	.47	.47	
	D50 Electrical						
5010	Electrical Service/Distribution	800 ampere service, panel board and feeders		S.F. Floor	.80	.80	
5020	Lighting & Branch Wiring	High intensity discharge fixtures, switches, A.C. and misc. power		S.F. Floor	9.28	9.28	12.3%
5030	Communications & Security	Addressable alarm systems and emergency lighting		S.F. Floor	1.14	1.14	
5090	Other Electrical Systems	Emergency generator, 30 kW		S.F. Floor	.20	.20	
E.	**EQUIPMENT & FURNISHINGS**						
1010	Commercial Equipment	N/A		—	—	—	
1020	Institutional Equipment	N/A		—	—	—	0.0 %
1030	Vehicular Equipment	N/A		—	—	—	
1090	Other Equipment	N/A		—	—	—	
F.	**SPECIAL CONSTRUCTION**						
1020	Integrated Construction	N/A		—	—	—	0.0 %
1040	Special Facilities	N/A		—	—	—	
G.	**BUILDING SITEWORK**	**N/A**					

	Sub-Total	93.11	100%
CONTRACTOR FEES (General Requirements: 10%, Overhead: 5%, Profit: 10%)	25%	23.26	
ARCHITECT FEES	6%	6.98	
Total Building Cost		**123.35**	

Costs per square foot of floor area

Exterior Wall	S.F. Area	4000	4500	5000	5500	6000	6500	7000	7500	8000
	L.F. Perimeter	260	280	300	310	320	336	353	370	386
Face Brick Concrete Block Back-up	Steel Joists	169.55	166.20	163.60	159.95	157.00	155.20	153.75	152.55	151.35
	Bearing Walls	164.00	160.65	158.05	154.40	151.50	149.65	148.15	146.95	145.75
Decorative Concrete Block	Steel Joists	155.00	152.30	150.25	147.40	145.15	143.65	142.55	141.55	140.60
	Bearing Walls	149.55	146.85	144.75	141.90	139.65	138.20	137.05	136.05	135.15
Limestone with Concrete Block Back-up	Steel Joists	197.15	192.60	189.05	183.90	179.65	177.15	175.15	173.45	171.85
	Bearing Walls	191.60	187.15	183.60	178.35	174.20	171.65	169.70	168.00	166.35
Perimeter Adj., Add or Deduct	Per 100 L.F.	19.50	17.35	15.55	14.25	13.05	12.00	11.15	10.40	9.75
Story Hgt. Adj., Add or Deduct	Per 1 Ft.	2.55	2.45	2.35	2.25	2.10	2.00	1.95	1.95	1.90
For Basement, add $35.95 per square foot of basement area										

The above costs were calculated using the basic specifications shown on the facing page. These costs should be adjusted where necessary for design alternatives and owner's requirements. Reported completed project costs, for this type of structure, range from $69.45 to $205.45 per S.F.

Common additives

Description	Unit	$ Cost		Description	Unit	$ Cost
Appliances				Appliances, cont.		
Cooking range, 30" free standing				Refrigerator, no frost 10-12 C.F.	Each	570 - 630
1 oven	Each	460 - 2300		14-16 C.F.	Each	610 - 945
2 oven	Each	1600 - 1850		18-20 C.F.	Each	750 - 1400
30" built-in				Lockers, Steel, single tier, 60" or 72"	Opening	200 - 350
1 oven	Each	845 - 2150		2 tier, 60" or 72" total	Opening	124 - 152
2 oven	Each	1950 - 2025		5 tier, box lockers	Opening	70 - 82.50
Counter top cook tops, 4 burner	Each	425 - 855		Locker bench, lam. maple top only	L.F.	22
Microwave oven	Each	267 - 750		Pedestals, steel pipe	Each	64
Combination range, refrig. & sink, 30" wide	Each	1625 - 5250		Sound System		
60" wide	Each	4375		Amplifier, 250 watts	Each	2400
72" wide	Each	5625		Speaker, ceiling or wall	Each	196
Combination range refrigerator, sink				Trumpet	Each	375
microwave oven & icemaker	Each	6400				
Compactor, residential, 4-1 compaction	Each	685 - 1075				
Dishwasher, built-in, 2 cycles	Each	535 - 955				
4 cycles	Each	605 - 1750				
Garbage disposer, sink type	Each	201 - 330				
Hood for range, 2 speed, vented, 30" wide	Each	279 - 1275				
42" wide	Each	475 - 2400				

Model costs calculated for a 1 story building with 14' story height and 6,000 square feet of floor area

				Unit	Unit Cost	Cost Per S.F.	% Of Sub-Total
A.	**SUBSTRUCTURE**						
1010	Standard Foundations	Poured concrete; strip and spread footings		S.F. Ground	3.03	3.03	
1020	Special Foundations	N/A		—	—	—	
1030	Slab on Grade	6" reinforced concrete with vapor barrier and granular base		S.F. Slab	6.33	6.33	12.0%
2010	Basement Excavation	Site preparation for slab and trench for foundation wall and footing		S.F. Ground	.52	.52	
2020	Basement Walls	4' foundation wall		L.F. Wall	76	4.03	
B.	**SHELL**						
	B10 Superstructure						
1010	Floor Construction	N/A		—	—	—	
1020	Roof Construction	Metal deck, open web steel joists, beams on columns		S.F. Roof	8.11	8.11	7.0 %
	B20 Exterior Enclosure						
2010	Exterior Walls	Face brick with concrete block backup	75% of wall	S.F. Wall	30.30	16.97	
2020	Exterior Windows	Aluminum insulated glass	10% of wall	Each	815	1.90	19.7%
2030	Exterior Doors	Single aluminum and glass, overhead, hollow metal	15% of wall	S.F. Door	36.52	4.09	
	B30 Roofing						
3010	Roof Coverings	Built-up tar and gravel with flashing; perlite/EPS composite insulation		S.F. Roof	6.64	6.64	5.8 %
3020	Roof Openings	Skylights, roof hatches		S.F. Roof	.14	.14	
C.	**INTERIORS**						
1010	Partitions	Concrete block	17 S.F. Floor/L.F. Partition	S.F. Partition	8.62	5.07	
1020	Interior Doors	Single leaf hollow metal	500 S.F. Floor/Door	Each	1001	2	
1030	Fittings	Toilet partitions		S.F. Floor	.61	.61	
2010	Stair Construction	N/A		—	—	—	13.2%
3010	Wall Finishes	Paint		S.F. Surface	1.86	2.19	
3020	Floor Finishes	50% vinyl tile, 50% paint		S.F. Floor	2.25	2.25	
3030	Ceiling Finishes	Fiberglass board on exposed grid, suspended	50% of area	S.F. Ceiling	6.51	3.26	
D.	**SERVICES**						
	D10 Conveying						
1010	Elevators & Lifts	N/A		—	—	—	0.0 %
1020	Escalators & Moving Walks	N/A		—	—	—	
	D20 Plumbing						
2010	Plumbing Fixtures	Kitchen, toilet and service fixtures, supply and drainage	1 Fixture/375 S.F. Floor	Each	3458	9.22	
2020	Domestic Water Distribution	Gast fired water heater		S.F. Floor	2.56	2.56	11.1%
2040	Rain Water Drainage	Roof drains		S.F. Roof	1.10	1.10	
	D30 HVAC						
3010	Energy Supply	N/A		—	—	—	
3020	Heat Generating Systems	Included in D3050		—	—	—	
3030	Cooling Generating Systems	N/A		—	—	—	18.3 %
3050	Terminal & Package Units	Rooftop multizone unit system		S.F. Floor	21.34	21.34	
3090	Other HVAC Sys. & Equipment	N/A		—	—	—	
	D40 Fire Protection						
4010	Sprinklers	Wet pipe sprinkler system		S.F. Floor	3.99	3.99	4.5%
4020	Standpipes	Standpipe, wet, Class III		S.F. Floor	1.20	1.20	
	D50 Electrical						
5010	Electrical Service/Distribution	200 ampere service, panel board and feeders		S.F. Floor	1.42	1.42	
5020	Lighting & Branch Wiring	High efficiency fluorescent fixtures, receptacles, switches, A.C. and misc. power		S.F. Floor	6.47	6.47	8.4%
5030	Communications & Security	Addressable alarm systems		S.F. Floor	1.87	1.87	
5090	Other Electrical Systems	N/A		—	—	—	
E.	**EQUIPMENT & FURNISHINGS**						
1010	Commercial Equipment	N/A		—	—	—	
1020	Institutional Equipment	N/A		—	—	—	0.0 %
1030	Vehicular Equipment	N/A		—	—	—	
1090	Other Equipment	N/A		—	—	—	
F.	**SPECIAL CONSTRUCTION**						
1020	Integrated Construction	N/A		—	—	—	0.0 %
1040	Special Facilities	N/A		—	—	—	
G.	**BUILDING SITEWORK**	**N/A**					

						Sub-Total	116.31	100%
	CONTRACTOR FEES (General Requirements: 10%, Overhead: 5%, Profit: 10%)					25%	29.06	
	ARCHITECT FEES					8%	11.63	
						Total Building Cost	**157**	

Costs per square foot of floor area

Exterior Wall	S.F. Area	6000	7000	8000	9000	10000	11000	12000	13000	14000
	L.F. Perimeter	220	240	260	280	286	303	320	336	353
Face Brick with Concrete Block Back-up	Steel Joists	175.00	169.30	164.95	161.65	157.10	154.65	152.65	150.85	149.45
	Precast Conc.	185.85	180.20	175.90	172.60	168.15	165.70	163.75	161.95	160.55
Decorative Concrete Block	Steel Joists	162.05	157.20	153.50	150.65	147.00	144.90	143.25	141.75	140.55
	Precast Conc.	172.90	168.10	164.45	161.60	158.05	155.95	154.30	152.80	151.65
Limestone with Concrete Block Back-up	Steel Joists	205.85	198.15	192.35	187.85	181.15	177.85	175.10	172.60	170.70
	Precast Conc.	216.50	208.85	203.05	198.55	192.00	188.60	185.95	183.50	181.55
Perimeter Adj., Add or Deduct	Per 100 L.F.	22.55	19.30	16.90	15.05	13.50	12.25	11.30	10.40	9.70
Story Hgt. Adj., Add or Deduct	Per 1 Ft.	2.95	2.70	2.65	2.45	2.25	2.20	2.20	2.10	2.00
For Basement, add $34.90 per square foot of basement area										

The above costs were calculated using the basic specifications shown on the facing page. These costs should be adjusted where necessary for design alternatives and owner's requirements. Reported completed project costs, for this type of structure, range from $69.25 to $205.45 per S.F.

Common additives

Description	Unit	$ Cost
Appliances		
Cooking range, 30" free standing		
1 oven	Each	460 - 2300
2 oven	Each	1600 - 1850
30" built-in		
1 oven	Each	845 - 2150
2 oven	Each	1950 - 2025
Counter top cook tops, 4 burner	Each	425 - 855
Microwave oven	Each	267 - 750
Combination range, refrig. & sink, 30" wide	Each	1625 - 5250
60" wide	Each	4375
72" wide	Each	5625
Combination range, refrigerator, sink,		
microwave oven & icemaker	Each	6400
Compactor, residential, 4-1 compaction	Each	685 - 1075
Dishwasher, built-in, 2 cycles	Each	535 - 955
4 cycles	Each	605 - 1750
Garbage diposer, sink type	Each	201 - 330
Hood for range, 2 speed, vented, 30" wide	Each	279 - 1275
42" wide	Each	475 - 2400

Description	Unit	$ Cost
Appliances, cont.		
Refrigerator, no frost 10-12 C.F.	Each	570 - 630
14-16 C.F.	Each	610 - 945
18-20 C.F.	Each	750 - 1400
Elevators, Hydraulic passenger, 2 stops		
1500# capacity	Each	60,900
2500# capacity	Each	64,300
3500# capacity	Each	67,600
Lockers, Steel, single tier, 60" or 72"	Opening	200 - 350
2 tier, 60" or 72" total	Opening	124 - 152
5 tier, box lockers	Opening	70 - 82.50
Locker bench, lam. maple top only	L.F.	22
Pedestals, steel pipe	Each	64
Sound System		
Amplifier, 250 watts	Each	2400
Speaker, ceiling or wall	Each	196
Trumpet	Each	375

Important: See the Reference Section for Location Factors

Model costs calculated for a 2 story building with 14' story height and 10,000 square feet of floor area

				Unit	Unit Cost	Cost Per S.F.	% Of Sub-Total
A.	**SUBSTRUCTURE**						
1010	Standard Foundations	Poured concrete; strip and spread footings		S.F. Ground	2.84	1.42	
1020	Special Foundations	N/A		—	—	—	
1030	Slab on Grade	6" reinforced concrete with vapor barrier and granular base		S.F. Slab	6.33	3.17	6.7%
2010	Basement Excavation	Site preparation for slab and trench for foundation wall and footing		S.F. Ground	.52	.26	
2020	Basement Walls	4' foundation wall		L.F. Wall	84	2.41	
B.	**SHELL**						
	B10 Superstructure						
1010	Floor Construction	Open web steel joists, slab form, concrete		S.F. Floor	11.20	5.60	6.8%
1020	Roof Construction	Metal deck on open web steel joists		S.F. Roof	3.58	1.79	
	B20 Exterior Enclosure						
2010	Exterior Walls	Decorative concrete block	75% of wall	S.F. Wall	17.85	10.72	
2020	Exterior Windows	Aluminum insulated glass	10% of wall	Each	703	2.45	15.2%
2030	Exterior Doors	Single aluminum and glass, steel overhead, hollow metal	15% of wall	S.F. Door	28.22	3.39	
	B30 Roofing						
3010	Roof Coverings	Built-up tar and gravel with flashing; perlite/EPS composite insulation		S.F. Roof	6.80	3.40	3.1%
3020	Roof Openings	N/A		—	—	—	
C.	**INTERIORS**						
1010	Partitions	Concrete block	10 S.F. Floor/L.F. Partition	S.F. Partition	8.62	5.07	
1020	Interior Doors	Single leaf hollow metal	500 S.F. Floor/Door	Each	1000	2	
1030	Fittings	Toilet partitions		S.F. Floor	.55	.55	
2010	Stair Construction	Concrete filled metal pan		Flight	17,125	3.43	17.6%
3010	Wall Finishes	Paint		S.F. Surface	3.19	3.75	
3020	Floor Finishes	50% vinyl tile, 50% paint		S.F. Floor	2.25	2.25	
3030	Ceiling Finishes	Fiberglass board on exposed grid, suspended	50% of area	S.F. Ceiling	4.31	2.16	
D.	**SERVICES**						
	D10 Conveying						
1010	Elevators & Lifts	One hydraulic passenger elevator		Each	78,900	7.89	7.2%
1020	Escalators & Moving Walks	N/A		—	—	—	
	D20 Plumbing						
2010	Plumbing Fixtures	Kitchen toilet and service fixtures, supply and drainage	1 Fixture/400 S.F. Floor	Each	3392	8.48	
2020	Domestic Water Distribution	Gas fired water heater		S.F. Floor	2.07	2.07	10.4%
2040	Rain Water Drainage	Roof drains		S.F. Roof	1.44	.72	
	D30 HVAC						
3010	Energy Supply	N/A		—	—	—	
3020	Heat Generating Systems	Included in D3050		—	—	—	
3030	Cooling Generating Systems	N/A		—	—	—	19.6 %
3050	Terminal & Package Units	Rooftop multizone unit system		S.F. Floor	21.34	21.34	
3090	Other HVAC Sys. & Equipment	N/A		—	—	—	
	D40 Fire Protection						
4010	Sprinklers	Wet pipe sprinkler system		S.F. Floor	3.34	3.34	4.2%
4020	Standpipes	Standpipe, wet, Class III		S.F. Floor	1.25	1.25	
	D50 Electrical						
5010	Electrical Service/Distribution	400 ampere service, panel board and feeders		S.F. Floor	1.60	1.60	
5020	Lighting & Branch Wiring	High efficiency fluorescent fixtures, receptacles, switches, A.C. and misc. power		S.F. Floor	6.67	6.67	9.2%
5030	Communications & Security	Addressable alarm systems and emergency lighting		S.F. Floor	1.45	1.45	
5090	Other Electrical Systems	Emergency generator, 15 kW		S.F. Floor	.25	.25	
E.	**EQUIPMENT & FURNISHINGS**						
1010	Commercial Equipment	N/A		—	—	—	
1020	Institutional Equipment	N/A		—	—	—	
1030	Vehicular Equipment	N/A		—	—	—	0.0 %
1090	Other Equipment	N/A		—	—	—	
F.	**SPECIAL CONSTRUCTION**						
1020	Integrated Construction	N/A		—	—	—	
1040	Special Facilities	N/A		—	—	—	0.0 %
G.	**BUILDING SITEWORK**	**N/A**					

				Sub-Total	108.88	100%
CONTRACTOR FEES (General Requirements: 10%, Overhead: 5%, Profit: 10%)				25%	27.23	
ARCHITECT FEES				8%	10.89	
			Total Building Cost		**147**	

Costs per square foot of floor area

Exterior Wall	S.F. Area	4000	5000	6000	8000	10000	12000	14000	16000	18000
	L.F. Perimeter	180	205	230	260	300	340	353	386	420
Cedar Beveled Siding	Wood Frame	173.35	163.75	157.30	147.70	142.55	139.15	135.50	133.55	132.05
Aluminum Siding	Wood Frame	172.40	162.80	156.30	146.75	141.55	138.20	134.55	132.60	131.05
Board and Batten	Wood Frame	169.20	160.95	155.45	147.00	142.55	139.60	136.30	134.60	133.25
Face Brick on Block	Bearing Wall	200.90	189.70	182.15	170.20	164.05	160.00	155.10	152.70	150.90
Stucco on Block	Bearing Wall	182.30	172.60	166.05	156.40	151.20	147.80	144.10	142.10	140.60
Decorative Block	Wood Joists	176.30	166.45	159.80	149.80	144.50	141.05	137.15	135.10	133.55
Perimeter Adj., Add or Deduct	Per 100 L.F.	15.85	12.60	10.50	7.85	6.35	5.25	4.50	3.90	3.50
Story Hgt. Adj., Add or Deduct	Per 1 Ft.	2.10	1.85	1.70	1.45	1.40	1.25	1.20	1.10	1.05
For Basement, add $22.20 per square foot of basement area										

The above costs were calculated using the basic specifications shown on the facing page. These costs should be adjusted where necessary for design alternatives and owner's requirements. Reported completed project costs, for this type of structure, range from $100.70 to $220.25 per S.F.

Common additives

Description	Unit	$ Cost	Description	Unit	$ Cost
Appliances			Appliances, cont.		
Cooking range, 30" free standing			Refrigerator, no frost 10-12 C.F.	Each	570 - 630
1 oven	Each	460 - 2300	14-16 C.F.	Each	610 - 945
2 oven	Each	1600 - 1850	18-20 C.F.	Each	750 - 1400
30" built-in			Elevators, Hydraulic passenger, 2 stops		
1 oven	Each	845 - 2150	1500# capacity	Each	60,900
2 oven	Each	1950 - 2025	2500# capacity	Each	64,300
Counter top cook tops, 4 burner	Each	425 - 855	3500# capacity	Each	67,600
Microwave oven	Each	267 - 750	Laundry Equipment		
Combination range, refrig. & sink, 30" wide	Each	1625 - 5250	Dryer, gas, 16 lb. capacity	Each	900
60" wide	Each	4375	30 lb. capacity	Each	3600
72" wide	Each	5625	Washer, 4 cycle	Each	1100
Combination range, refrigerator, sink,			Commercial	Each	1450
microwave oven & icemaker	Each	6400	Sound System		
Compactor, residential, 4-1 compaction	Each	685 - 1075	Amplifier, 250 watts	Each	2400
Dishwasher, built-in, 2 cycles	Each	535 - 955	Speaker, ceiling or wall	Each	196
4 cycles	Each	605 - 1750	Trumpet	Each	375
Garbage disposer, sink type	Each	201 - 330			
Hood for range, 2 speed, vented, 30" wide	Each	279 - 1275			
42" wide	Each	475 - 2400			

Important: See the Reference Section for Location Factors

Model costs calculated for a 2 story building with 10' story height and 10,000 square feet of floor area

Fraternity/Sorority House

				Unit	Unit Cost	Cost Per S.F.	% Of Sub-Total
A. SUBSTRUCTURE							
1010	Standard Foundations	Poured concrete; strip and spread footings		S.F. Ground	2.78	1.39	
1020	Special Foundations	N/A		—	—	—	
1030	Slab on Grade	4" reinforced concrete with vapor barrier and granular base		S.F. Slab	4.77	2.39	5.8%
2010	Basement Excavation	Site preparation for slab and trench for foundation wall and footing		S.F. Ground	.52	.26	
2020	Basement Walls	4' foundation wall		L.F. Wall	67	2.02	
B. SHELL							
	B10 Superstructure						
1010	Floor Construction	Plywood on wood joists		S.F. Floor	3.14	1.57	3.0%
1020	Roof Construction	Plywood on wood rafters (pitched)		S.F. Roof	2.89	1.62	
	B20 Exterior Enclosure						
2010	Exterior Walls	Cedar bevel siding on wood studs, insulated	80% of wall	S.F. Wall	12.85	6.17	
2020	Exterior Windows	Double hung wood	20% of wall	Each	592	2.84	10.2%
2030	Exterior Doors	Solid core wood		Each	2347	1.65	
	B30 Roofing						
3010	Roof Coverings	Asphalt shingles with flashing (pitched); rigid fiberglass insulation		S.F. Roof	4.48	2.24	2.1%
3020	Roof Openings	N/A		—	—	—	
C. INTERIORS							
1010	Partitions	Gypsum board on wood studs	25 S.F. Floor/L.F. Partition	S.F. Partition	7.22	2.31	
1020	Interior Doors	Single leaf wood	200 S.F. Floor/Door	Each	576	2.89	
1030	Fittings	N/A		—	—	—	
2010	Stair Construction	Wood		Flight	2183	.88	19.5%
3010	Wall Finishes	Paint		S.F. Surface	.73	.47	
3020	Floor Finishes	10% hardwood, 70% carpet, 20% ceramic tile		S.F. Floor	9.85	9.85	
3030	Ceiling Finishes	Gypsum board on wood furring		S.F. Ceiling	3.95	3.95	
D. SERVICES							
	D10 Conveying						
1010	Elevators & Lifts	One hydraulic passenger elevator		Each	78,900	7.89	7.5%
1020	Escalators & Moving Walks	N/A		—	—	—	
	D20 Plumbing						
2010	Plumbing Fixtures	Kitchen toilet and service fixtures, supply and drainage	1 Fixture/150 S.F. Floor	Each	1061	7.07	
2020	Domestic Water Distribution	Gas fired water heater		S.F. Floor	2.16	2.16	8.8%
2040	Rain Water Drainage	N/A		—	—	—	
	D30 HVAC						
3010	Energy Supply	Oil fired hot water, baseboard radiation		S.F. Floor	6.71	6.71	
3020	Heat Generating Systems	N/A		—	—	—	
3030	Cooling Generating Systems	N/A		—	—	—	13.5%
3050	Terminal & Package Units	Split system with air cooled condensing unit		S.F. Floor	7.41	7.41	
3090	Other HVAC Sys. & Equipment	N/A		—	—	—	
	D40 Fire Protection						
4010	Sprinklers	Wet pipe sprinkler system		S.F. Floor	3.34	3.34	3.2%
4020	Standpipes	N/A		—	—	—	
	D50 Electrical						
5010	Electrical Service/Distribution	600 ampere service, panel board and feeders		S.F. Floor	4.88	4.88	
5020	Lighting & Branch Wiring	Fluorescent fixtures, receptacles, switches, A.C. and misc. power		S.F. Floor	9.46	9.46	26.3%
5030	Communications & Security	Alarm, communication system, internet and phone wiring, and generator set		S.F. Floor	12.99	12.99	
5090	Other Electrical Systems	Emergency generator, 7.5 kW		S.F. Floor	.20	.20	
E. EQUIPMENT & FURNISHINGS							
1010	Commercial Equipment	N/A		—	—	—	
1020	Institutional Equipment	N/A		—	—	—	
1030	Vehicular Equipment	N/A		—	—	—	0.0%
1090	Other Equipment	N/A		—	—	—	
F. SPECIAL CONSTRUCTION							
1020	Integrated Construction	N/A		—	—	—	
1040	Special Facilities	N/A		—	—	—	0.0%
G. BUILDING SITEWORK	**N/A**						

			Sub-Total	104.61	**100%**
	CONTRACTOR FEES (General Requirements: 10%, Overhead: 5%, Profit: 10%)		25%	26.17	
	ARCHITECT FEES		9%	11.77	

	Total Building Cost	**142.55**

Costs per square foot of floor area

Exterior Wall	S.F. Area	4000	6000	8000	10000	12000	14000	16000	18000	20000
	L.F. Perimeter	260	320	384	424	460	484	510	540	576
Vertical Redwood Siding	Wood Frame	170.35	151.30	141.95	**135.30**	130.70	127.00	124.35	122.30	120.85
Brick Veneer	Wood Frame	183.80	162.50	152.25	144.55	139.20	134.85	131.75	129.35	127.65
Aluminum Siding	Wood Frame	169.15	150.55	141.45	135.00	130.55	127.05	124.55	122.60	121.15
Brick on Block	Wood Truss	194.30	171.70	160.85	152.45	146.60	141.85	138.40	135.75	133.95
Limestone on Block	Wood Truss	226.05	197.65	184.10	173.00	165.15	158.50	153.75	150.20	147.80
Stucco on Block	Wood Truss	177.45	157.80	148.35	141.40	136.65	132.80	130.10	127.95	126.45
Perimeter Adj., Add or Deduct	Per 100 L.F.	11.20	7.45	5.65	4.45	3.75	3.25	2.80	2.55	2.25
Story Hgt. Adj., Add or Deduct	Per 1 Ft.	1.35	1.05	1.00	0.85	0.75	0.75	0.65	0.60	0.60
For Basement, add $29.60 per square foot of basement area										

The above costs were calculated using the basic specifications shown on the facing page. These costs should be adjusted where necessary for design alternatives and owner's requirements. Reported completed project costs, for this type of structure, range from $98.60 to $303.15 per S.F.

Common additives

Description	Unit	$ Cost		Description	Unit	$ Cost
Autopsy Table, Standard	Each	10,400		Planters, Precast concrete		
Deluxe	Each	20,500		48" diam., 24" high	Each	775
Directory Boards, Plastic, glass covered				7" diam., 36" high	Each	1300
30" x 20"	Each	605		Fiberglass, 36" diam., 24" high	Each	575
36" x 48"	Each	1325		60" diam., 24" high	Each	960
Aluminum, 24" x 18"	Each	585		Smoke Detectors		
36" x 24"	Each	685		Ceiling type	Each	233
48" x 32"	Each	975		Duct type	Each	525
48" x 60"	Each	2025				
Emergency Lighting, 25 watt, battery operated						
Lead battery	Each	287				
Nickel cadmium	Each	845				
Mortuary Refrigerator, End operated						
Two capacity	Each	14,100				
Six capacity	Each	26,300				

Important: See the Reference Section for Location Factors

Model costs calculated for a 1 story building with 12' story height and 10,000 square feet of floor area

				Unit	Unit Cost	Cost Per S.F.	% Of Sub-Total
A. SUBSTRUCTURE							
1010	Standard Foundations	Poured concrete; strip and spread footings		S.F. Ground	1.66	1.66	
1020	Special Foundations	N/A		—	—	—	
1030	Slab on Grade	4" reinforced concrete with vapor barrier and granular base		S.F. Slab	4.77	4.77	10.0%
2010	Basement Excavation	Site preparation for slab and trench for foundation wall and footing		S.F. Ground	.31	.31	
2020	Basement Walls	4' foundation wall		L.F. Wall	67	3.14	
B. SHELL							
	B10 Superstructure						
1010	Floor Construction	N/A		—	—	—	4.1%
1020	Roof Construction	Plywood on wood truss		S.F. Roof	4.09	4.09	
	B20 Exterior Enclosure						
2010	Exterior Walls	1" x 4" vertical T & G redwood siding on wood studs	90% of wall	S.F. Wall	10.50	4.81	
2020	Exterior Windows	Double hung wood	10% of wall	Each	478	2.03	8.2%
2030	Exterior Doors	Wood swinging double doors, single leaf hollow metal		Each	2115	1.28	
	B30 Roofing						
3010	Roof Coverings	Single ply membrane, fully adhered; polyisocyanurate sheets		S.F. Roof	6.17	6.17	6.2%
3020	Roof Openings	N/A		—	—	—	
C. INTERIORS							
1010	Partitions	Gypsum board on wood studs with sound deadening board	15 S.F. Floor/L.F. Partition	S.F. Partition	8.57	4.57	
1020	Interior Doors	Single leaf wood	150 S.F. Floor/Door	Each	576	3.84	
1030	Fittings	N/A		—	—	—	
2010	Stair Construction	N/A		—	—	—	27.5%
3010	Wall Finishes	50% wallpaper, 25% wood paneling, 25% paint		S.F. Surface	3.38	3.60	
3020	Floor Finishes	70% carpet, 30% ceramic tile		S.F. Floor	10.94	10.94	
3030	Ceiling Finishes	Fiberglass board on exposed grid, suspended		S.F. Ceiling	4.31	4.31	
D. SERVICES							
	D10 Conveying						
1010	Elevators & Lifts	N/A		—	—	—	0.0%
1020	Escalators & Moving Walks	N/A		—	—	—	
	D20 Plumbing						
2010	Plumbing Fixtures	Toilet and service fixtures, supply and drainage	1 Fixture/770 S.F. Floor	Each	2672	3.47	
2020	Domestic Water Distribution	Electric water heater		S.F. Floor	9.63	9.63	14.1%
2040	Rain Water Drainage	Roof drain		S.F. Roof	.86	.86	
	D30 HVAC						
3010	Energy Supply	N/A		—	—	—	
3020	Heat Generating Systems	Included in D3030		—	—	—	
3030	Cooling Generating Systems	Multizone rooftop unit, gas heating, electric cooling		S.F. Floor	16.05	16.05	16.2%
3050	Terminal & Package Units	N/A		—	—	—	
3090	Other HVAC Sys. & Equipment	N/A		—	—	—	
	D40 Fire Protection						
4010	Sprinklers	Wet pipe sprinkler system		S.F. Floor	3.16	3.16	3.9%
4020	Standpipes	Standpipe, wet, Class III		S.F. Floor	.72	.72	
	D50 Electrical						
5010	Electrical Service/Distribution	400 ampere service, panel board and feeders		S.F. Floor	1.73	1.73	
5020	Lighting & Branch Wiring	High efficiency fluorescent fixtures, receptacles, switches, A.C. and misc. power		S.F. Floor	6.62	6.62	10.0%
5030	Communications & Security	Addressable alarm systems and emergency lighting		S.F. Floor	1.45	1.45	
5090	Other Electrical Systems	Emergency generator, 15 kW		S.F. Floor	.08	.08	
E. EQUIPMENT & FURNISHINGS							
1010	Commercial Equipment	N/A		—	—	—	
1020	Institutional Equipment	N/A		—	—	—	
1030	Vehicular Equipment	N/A		—	—	—	0.0%
1090	Other Equipment	N/A		—	—	—	
F. SPECIAL CONSTRUCTION							
1020	Integrated Construction	N/A		—	—	—	0.0%
1040	Special Facilities	N/A		—	—	—	
G. BUILDING SITEWORK	**N/A**						

					Sub-Total	99.29	100%
	CONTRACTOR FEES (General Requirements: 10%, Overhead: 5%, Profit: 10%)				25%	24.84	
	ARCHITECT FEES				9%	11.17	

		Total Building Cost	**135.30**

Costs per square foot of floor area

Exterior Wall	S.F. Area	12000	14000	16000	19000	21000	23000	26000	28000	30000
	L.F. Perimeter	440	474	510	556	583	607	648	670	695
E.I.F.S. on Concrete Block	Steel Frame	118.25	115.40	113.40	110.95	109.50	108.20	106.85	105.85	105.15
Tilt-up Concrete Wall	Steel Frame	115.25	112.65	110.80	108.60	107.20	106.05	104.75	103.95	103.30
Face Brick with Concrete Block Back-up	Bearing Walls	120.00	116.65	114.30	111.35	109.60	108.05	106.40	105.30	104.45
	Steel Frame	124.75	121.45	119.05	116.15	114.35	112.85	111.15	110.10	109.20
Stucco on Concrete Block	Bearing Walls	111.30	108.65	106.75	104.40	102.95	101.75	100.45	99.60	98.85
	Steel Frame	116.55	113.90	112.00	109.65	108.20	107.00	105.70	104.80	104.10
Perimeter Adj., Add or Deduct	Per 100 L.F.	6.65	5.80	5.10	4.20	3.80	3.50	3.00	2.95	2.70
Story Hgt. Adj., Add or Deduct	Per 1 Ft.	1.55	1.50	1.40	1.25	1.15	1.10	1.05	1.10	1.00
For Basement, add $33.65 per square foot of basement area										

The above costs were calculated using the basic specifications shown on the facing page. These costs should be adjusted where necessary for design alternatives and owner's requirements. Reported completed project costs, for this type of structure, range from $52.45 to $152.60 per S.F.

Common additives

Description	Unit	$ Cost
Emergency Lighting, 25 watt, battery operated		
Lead battery	Each	287
Nickel cadmium	Each	845
Smoke Detectors		
Ceiling type	Each	233
Duct type	Each	525
Sound System		
Amplifier, 250 watts	Each	2400
Speaker, ceiling or wall	Each	196
Trumpet	Each	375

Model costs calculated for a 1 story building with 14' story height and 21,000 square feet of floor area

				Unit	Unit Cost	Cost Per S.F.	% Of Sub-Total
A.	**SUBSTRUCTURE**						
1010	Standard Foundations	Poured concrete; strip and spread footings		S.F. Ground	1.30	1.30	
1020	Special Foundations	N/A		—	—	—	
1030	Slab on Grade	4" reinforced concrete with vapor barrier and granular base		S.F. Slab	7.26	7.26	13.4%
2010	Basement Excavation	Site preparation for slab and trench for foundation wall and footing		S.F. Ground	.31	.31	
2020	Basement Walls	4' foundation wall		L.F. Wall	76	2.10	
B.	**SHELL**						
	B10 Superstructure						
1010	Floor Construction	N/A		—	—	—	10.0%
1020	Roof Construction	Metal deck, open web steel joists, beams, columns		S.F. Roof	8.21	8.21	
	B20 Exterior Enclosure						
2010	Exterior Walls	E.I.F.S.	70% of wall	S.F. Wall	24.66	6.71	
2020	Exterior Windows	Window wall	30% of wall	Each	50	5.87	19.4%
2030	Exterior Doors	Double aluminum and glass, hollow metal, steel overhead		Each	3827	3.29	
	B30 Roofing						
3010	Roof Coverings	Built-up tar and gravel with flashing; perlite/EPS composite insulation		S.F. Roof	5.52	5.52	6.7%
3020	Roof Openings	N/A		—	—	—	
C.	**INTERIORS**						
1010	Partitions	Gypsum board on metal studs	28 S.F. Floor/L.F. Partition	S.F. Partition	4.04	1.73	
1020	Interior Doors	Hollow metal	280 S.F. Floor/Door	Each	1001	3.58	
1030	Fittings	N/A		—	—	—	
2010	Stair Construction	N/A		—	—	—	12.7%
3010	Wall Finishes	Paint		S.F. Surface	.75	.64	
3020	Floor Finishes	50% vinyl tile, 50% paint		S.F. Floor	2.25	2.25	
3030	Ceiling Finishes	Fiberglass board on exposed grid, suspended	50% of area	S.F. Ceiling	4.31	2.16	
D.	**SERVICES**						
	D10 Conveying						
1010	Elevators & Lifts	N/A		—	—	—	0.0%
1020	Escalators & Moving Walks	N/A		—	—	—	
	D20 Plumbing						
2010	Plumbing Fixtures	Toilet and service fixtures, supply and drainage	1 Fixture/1500 S.F. Floor	Each	2805	1.87	
2020	Domestic Water Distribution	Gas fired water heater		S.F. Floor	1.70	1.70	7.7%
2040	Rain Water Drainage	Roof drains		S.F. Roof	2.76	2.76	
	D30 HVAC						
3010	Energy Supply	N/A		—	—	—	
3020	Heat Generating Systems	N/A		—	—	—	
3030	Cooling Generating Systems	N/A		—	—	—	10.6%
3050	Terminal & Package Units	Single zone rooftop unit, gas heating, electric cooling		S.F. Floor	8.07	8.07	
3090	Other HVAC Sys. & Equipment	Underfloor garage exhaust system		S.F. Floor	.57	.57	
	D40 Fire Protection						
4010	Sprinklers	Wet pipe sprinkler system		S.F. Floor	4.11	4.11	5.5%
4020	Standpipes	Standpipe		S.F. Floor	.39	.39	
	D50 Electrical						
5010	Electrical Service/Distribution	200 ampere service, panel board and feeders		S.F. Floor	.43	.43	
5020	Lighting & Branch Wiring	T-8 fluorescent fixtures, receptacles, switches, A.C. and misc. power		S.F. Floor	6.43	6.43	12.5%
5030	Communications & Security	Addressable alarm systems, partial internet wiring and emergency lighting		S.F. Floor	3.26	3.26	
5090	Other Electrical Systems	Emergency generator, 7.5 kW		S.F. Floor	.07	.07	
E.	**EQUIPMENT & FURNISHINGS**						
1010	Commercial Equipment	N/A		—	—	—	
1020	Institutional Equipment	N/A		—	—	—	1.5%
1030	Vehicular Equipment	Hoists, compressor, fuel pump		S.F. Floor	1.24	1.24	
1090	Other Equipment	N/A		—	—	—	
F.	**SPECIAL CONSTRUCTION**						
1020	Integrated Construction	N/A		—	—	—	0.0%
1040	Special Facilities	N/A		—	—	—	
G.	**BUILDING SITEWORK**	**N/A**					

				Sub-Total	81.83	100%
	CONTRACTOR FEES (General Requirements: 10%, Overhead: 5%, Profit: 10%)			25%	20.51	
	ARCHITECT FEES			7%	7.16	

Total Building Cost   **109.50**

Costs per square foot of floor area

Exterior Wall	S.F. Area	85000	115000	145000	175000	205000	235000	265000	295000	325000
	L.F. Perimeter	529	638	723	823	923	951	1037	1057	1132
Face Brick with Concrete Block Back-up	Steel Frame	65.75	64.40	63.30	62.70	62.25	61.60	61.30	60.85	60.65
	R/Conc. Frame	54.05	52.60	51.55	50.95	50.55	49.85	49.60	49.10	48.90
Precast Concrete Precast Concrete	Steel Frame	70.20	68.35	66.95	66.20	65.65	64.70	64.30	63.65	63.45
	R/Conc. Frame	57.70	55.90	54.50	53.75	53.20	52.20	51.90	51.25	51.00
Reinforced Concrete Reinforced Concrete	Steel Frame	65.15	63.90	62.95	62.45	62.15	61.55	61.30	60.90	60.70
	R/Conc. Frame	52.35	51.10	50.20	49.70	49.35	48.75	48.50	48.10	47.95
Perimeter Adj., Add or Deduct	Per 100 L.F.	1.45	1.00	0.85	0.70	0.65	0.55	0.45	0.40	0.40
Story Hgt. Adj., Add or Deduct	Per 1 Ft.	0.50	0.35	0.35	0.35	0.40	0.35	0.30	0.30	0.30
Basement—Not Applicable										

The above costs were calculated using the basic specifications shown on the facing page. These costs should be adjusted where necessary for design alternatives and owner's requirements. Reported completed project costs, for this type of structure, range from $32.20 to $124.54 per S.F.

Common additives

Description	Unit	$ Cost
Elevators, Electric passenger, 5 stops		
2000# capacity	Each	158,200
3500# capacity	Each	166,200
5000# capacity	Each	169,700
Barrier gate w/programmable controller	Each	4075
Booth for attendant, average	Each	12,600
Fee computer	Each	15,100
Ticket spitter with time/date stamp	Each	7125
Mag strip encoding	Each	21,000
Collection station, pay on foot	Each	126,500
Parking control software	Each	25,300 - 108,500
Painting, Parking stalls	Stall	13.70
Parking Barriers		
Timber with saddles, 4″ x 4″	L.F.	6.10
Precast concrete, 6″ x 10″ x 6′	Each	80
Traffic Signs, directional, 12″ x 18″, high density	Each	84

Important: See the Reference Section for Location Factors

Model costs calculated for a 5 story building with 10' story height and 145,000 square feet of floor area

Garage, Parking

				Unit	Unit Cost	Cost Per S.F.	% Of Sub-Total
A. SUBSTRUCTURE							
1010	Standard Foundations	Poured concrete; strip and spread footings		S.F. Ground	6.30	1.26	
1020	Special Foundations	N/A		—	—	—	
1030	Slab on Grade	6" reinforced concrete with vapor barrier and granular base		S.F. Slab	6.24	1.25	10.0%
2010	Basement Excavation	Site preparation for slab and trench for foundation wall and footing		S.F. Ground	.18	.04	
2020	Basement Walls	4' foundation wall		L.F. Wall	67	1.35	
B. SHELL							
	B10 Superstructure						
1010	Floor Construction	Double tee precast concrete slab, precast concrete columns		S.F. Floor	22.45	17.96	46.2%
1020	Roof Construction	N/A		—	—	—	
	B20 Exterior Enclosure						
2010	Exterior Walls	Face brick with concrete block backup	40% of story height	S.F. Wall	30.28	3.02	
2020	Exterior Windows	N/A		—	—	—	7.8%
2030	Exterior Doors	N/A		—	—	—	
	B30 Roofing						
3010	Roof Coverings	N/A		—	—	—	0.0%
3020	Roof Openings	N/A		—	—	—	
C. INTERIORS							
1010	Partitions	Concrete block		S.F. Partition	29.90	1.15	
1020	Interior Doors	Hollow metal		Each	20,020	.14	
1030	Fittings	N/A		—	—	—	
2010	Stair Construction	Concrete		Flight	3750	.26	4.3%
3010	Wall Finishes	Paint		S.F. Surface	1.56	.12	
3020	Floor Finishes	N/A		—	—	—	
3030	Ceiling Finishes	N/A		—	—	—	
D. SERVICES							
	D10 Conveying						
1010	Elevators & Lifts	Two hydraulic passenger elevators		Each	164,575	2.27	5.8%
1020	Escalators & Moving Walks	N/A		—	—	—	
	D20 Plumbing						
2010	Plumbing Fixtures	Toilet and service fixtures, supply and drainage	1 Fixture/18,125 S.F. Floor	Each	725	.04	
2020	Domestic Water Distribution	Electric water heater		S.F. Floor	.07	.07	4.6%
2040	Rain Water Drainage	Roof drains		S.F. Roof	8.40	1.68	
	D30 HVAC						
3010	Energy Supply	N/A		—	—	—	
3020	Heat Generating Systems	N/A		—	—	—	
3030	Cooling Generating Systems	N/A		—	—	—	0.0%
3050	Terminal & Package Units	N/A		—	—	—	
3090	Other HVAC Sys. & Equipment	N/A		—	—	—	
	D40 Fire Protection						
4010	Sprinklers	Dry pipe sprinkler system		S.F. Floor	3.75	3.75	9.8%
4020	Standpipes	Standpipes and hose systems		S.F. Floor	.08	.08	
	D50 Electrical						
5010	Electrical Service/Distribution	400 ampere service, panel board and feeders		S.F. Floor	.23	.23	
5020	Lighting & Branch Wiring	T-8 fluorescent fixtures, receptacles, switches and misc. power		S.F. Floor	2.89	2.89	8.5%
5030	Communications & Security	Addressable alarm systems and emergency lighting		S.F. Floor	.11	.11	
5090	Other Electrical Systems	Emergency generator, 7.5 kW		S.F. Floor	.06	.06	
E. EQUIPMENT & FURNISHINGS							
1010	Commercial Equipment	N/A		—	—	—	
1020	Institutional Equipment	N/A		—	—	—	
1030	Vehicular Equipment	Ticket dispensers, booths, automatic gates		S.F. Floor	1.17	1.17	3.0%
1090	Other Equipment	N/A		—	—	—	
F. SPECIAL CONSTRUCTION							
1020	Integrated Construction	N/A		—	—	—	
1040	Special Facilities	N/A		—	—	—	0.0%
G. BUILDING SITEWORK	**N/A**						

		Sub-Total	38.90	100%
CONTRACTOR FEES (General Requirements: 10%, Overhead: 5%, Profit: 10%)		25%	9.73	
ARCHITECT FEES		6%	2.92	
	Total Building Cost		**51.55**	

Costs per square foot of floor area

Exterior Wall	S.F. Area	20000	30000	40000	50000	75000	100000	125000	150000	175000
	L.F. Perimeter	400	500	600	650	775	900	1000	1100	1185
Reinforced Concrete	R/Conc. Frame	92.35	86.10	82.90	79.95	75.90	73.90	72.50	71.55	70.75
Perimeter Adj., Add or Deduct	Per 100 L.F.	5.55	3.60	2.80	2.15	1.50	1.05	0.85	0.75	0.65
Story Hgt. Adj., Add or Deduct	Per 1 Ft.	2.15	1.75	1.55	1.35	1.10	0.90	0.85	0.75	0.75
Basement—Not Applicable										

The above costs were calculated using the basic specifications shown on the facing page. These costs should be adjusted where necessary for design alternatives and owner's requirements. Reported completed project costs, for this type of structure, range from $46.80 to $111.50 per S.F.

Common additives

Description	Unit	$ Cost
Elevators, Hydraulic passenger, 2 stops		
1500# capacity	Each	60,900
2500# capacity	Each	64,300
3500# capacity	Each	67,600
Barrier gate w/programmable controller	Each	4075
Booth for attendant, average	Each	12,600
Fee computer	Each	15,100
Ticket spitter with time/date stamp	Each	7125
Mag strip encoding	Each	21,000
Collection station, pay on foot	Each	126,500
Parking control software	Each	25,300 - 108,500
Painting, Parking stalls	Stall	13.70
Parking Barriers		
Timber with saddles, 4" x 4"	L.F.	6.10
Precast concrete, 6" x 10" x 6'	Each	80
Traffic Signs, directional, 12" x 18"	Each	84

Model costs calculated for a 2 story building with 10' story height and 100,000 square feet of floor area

Garage, Underground Parking

				Unit	Unit Cost	Cost Per S.F.	% Of Sub-Total
A. SUBSTRUCTURE							
1010	Standard Foundations		Poured concrete; strip and spread footings and waterproofing	S.F. Ground	8.14	4.07	
1020	Special Foundations		N/A	—	—	—	
1030	Slab on Grade		5" reinforced concrete with vapor barrier and granular base	S.F. Slab	6.33	3.17	22.4%
2010	Basement Excavation		Excavation 24' deep	S.F. Ground	10.05	5.03	
2020	Basement Walls		N/A	—	—	—	
B. SHELL							
	B10 Superstructure						
1010	Floor Construction		Cast-in-place concrete beam and slab, concrete columns	S.F. Floor	25.42	12.71	44.7%
1020	Roof Construction		Cast-in-place concrete beam and slab, concrete columns	S.F. Roof	23.52	11.76	
	B20 Exterior Enclosure						
2010	Exterior Walls		Cast-in place concrete	S.F. Wall	20.22	3.64	
2020	Exterior Windows		N/A	—	—	—	7.0%
2030	Exterior Doors		Steel overhead, hollow metal	Each	4095	.17	
	B30 Roofing						
3010	Roof Coverings		Neoprene membrane traffic deck	S.F. Roof	4.30	2.15	3.9%
3020	Roof Openings		N/A	—	—	—	
C. INTERIORS							
1010	Partitions		Concrete block	S.F. Partition	38.48	.74	
1020	Interior Doors		Hollow metal	Each	8008	.08	
1030	Fittings		N/A	—	—	—	
2010	Stair Construction		Concrete	Flight	6025	.31	2.2%
3010	Wall Finishes		Paint	S.F. Surface	2.34	.09	
3020	Floor Finishes		N/A	—	—	—	
3030	Ceiling Finishes		N/A	—	—	—	
D. SERVICES							
	D10 Conveying						
1010	Elevators & Lifts		Two hydraulic passenger elevators	Each	79,000	1.58	2.9%
1020	Escalators & Moving Walks		N/A	—	—	—	
	D20 Plumbing						
2010	Plumbing Fixtures		Drainage in parking areas, toilets, & service fixtures	1 Fixture/5000 S.F. Floor	Each	.04	.04	
2020	Domestic Water Distribution		Electric water heater		S.F. Floor	.11	.11	2.6%
2040	Rain Water Drainage		Roof drains		S.F. Roof	2.56	1.28	
	D30 HVAC						
3010	Energy Supply		N/A	—	—	—	
3020	Heat Generating Systems		N/A	—	—	—	
3030	Cooling Generating Systems		N/A	—	—	—	0.3 %
3050	Terminal & Package Units		Exhaust fans	S.F. Floor	.16	.16	
3090	Other HVAC Sys. & Equipment		N/A	—	—	—	
	D40 Fire Protection						
4010	Sprinklers		Dry pipe sprinkler system	S.F. Floor	3.68	3.68	7.0%
4020	Standpipes		Dry standpipe system, class 3	S.F. Floor	.14	.14	
	D50 Electrical						
5010	Electrical Service/Distribution		200 ampere service, panel board and feeders	S.F. Floor	.12	.12	
5020	Lighting & Branch Wiring		T-8 fluorescent fixtures, receptacles, switches and misc. power	S.F. Floor	3.11	3.11	6.3%
5030	Communications & Security		Addressable alarm systems and emergency lighting	S.F. Floor	.17	.17	
5090	Other Electrical Systems		Emergency generator, 11.5 kW	S.F. Floor	.06	.06	
E. EQUIPMENT & FURNISHINGS							
1010	Commercial Equipment		N/A	—	—	—	
1020	Institutional Equipment		N/A	—	—	—	0.7 %
1030	Vehicular Equipment		Ticket dispensers, booths, automatic gates	S.F. Floor	.37	.37	
1090	Other Equipment		N/A	—	—	—	
F. SPECIAL CONSTRUCTION							
1020	Integrated Construction		N/A	—	—	—	0.0 %
1040	Special Facilities		N/A	—	—	—	
G. BUILDING SITEWORK	**N/A**						

		Sub-Total	54.74	100%
CONTRACTOR FEES (General Requirements: 10%, Overhead: 5%, Profit: 10%)		25%	13.69	
ARCHITECT FEES		8%	5.47	
	Total Building Cost		**73.90**	

Costs per square foot of floor area

Exterior Wall	S.F. Area	2000	4000	6000	8000	10000	12000	14000	16000	18000
	L.F. Perimeter	180	260	340	420	500	580	586	600	610
Concrete Block	Wood Joists	147.25	126.15	119.05	115.55	113.40	112.00	108.35	105.85	103.85
	Steel Joists	147.50	127.00	120.10	116.65	114.65	113.25	109.65	107.20	105.15
Poured Concrete Poured Concrete	Wood Joists	157.35	133.65	125.75	121.75	119.50	117.85	113.45	110.40	107.95
	Steel Joists	158.65	134.95	127.05	123.05	120.70	119.15	114.75	111.70	109.25
Stucco	Wood Frame	151.90	127.85	119.75	115.75	113.40	111.75	107.20	104.10	101.55
Insulated Metal Panels	Steel Frame	144.40	125.75	119.50	116.35	114.55	113.20	110.15	107.95	106.20
Perimeter Adj., Add or Deduct	Per 100 L.F.	24.80	12.35	8.25	6.20	4.95	4.15	3.60	3.20	2.75
Story Hgt. Adj., Add or Deduct	Per 1 Ft.	1.90	1.35	1.15	1.10	1.10	1.00	0.90	0.85	0.65
For Basement, add $31.65 per square foot of basement area										

The above costs were calculated using the basic specifications shown on the facing page. These costs should be adjusted where necessary for design alternatives and owner's requirements. Reported completed project costs, for this type of structure, range from $67.10 to $201.60 per S.F.

Common additives

Description	Unit	$ Cost
Air Compressors		
Electric 1-1/2 H.P., standard controls	Each	1400
Dual controls	Each	1750
5 H.P. 115/230 Volt, standard controls	Each	3675
Dual controls	Each	3825
Product Dispenser		
with vapor recovery for 6 nozzles	Each	24,800
Hoists, Single post		
8000# cap., swivel arm	Each	9075
Two post, adjustable frames, 11,000# cap.	Each	12,800
24,000# cap.	Each	19,000
7500# Frame support	Each	10,700
Four post, roll on ramp	Each	9925
Lockers, Steel, single tier, 60" or 72"	Opening	200 - 350
2 tier, 60" or 72" total	Opening	124 - 152
5 tier, box lockers	Opening	70 - 82.50
Locker bench, lam. maple top only	L.F.	22
Pedestals, steel pipe	Each	64
Lube Equipment		
3 reel type, with pumps, no piping	Each	11,600
Spray Painting Booth, 26' long, complete	Each	21,000

Model costs calculated for a 1 story building with 14' story height and 10,000 square feet of floor area

				Unit	Unit Cost	Cost Per S.F.	% Of Sub-Total
A. SUBSTRUCTURE							
1010	Standard Foundations	Poured concrete; strip and spread footings		S.F. Ground	2.05	2.05	
1020	Special Foundations	N/A		—	—	—	
1030	Slab on Grade	6" reinforced concrete with vapor barrier and granular base		S.F. Slab	7.26	7.26	15.8%
2010	Basement Excavation	Site preparation for slab and trench for foundation wall and footing		S.F. Ground	.31	.31	
2020	Basement Walls	4' foundation wall		L.F. Wall	76	3.78	
B. SHELL							
B10 Superstructure							
1010	Floor Construction	N/A		—	—	—	4.7 %
1020	Roof Construction	Metal deck on open web steel joists		S.F. Roof	3.97	3.97	
B20 Exterior Enclosure							
2010	Exterior Walls	Concrete block	80% of wall	S.F. Wall	11.75	6.58	
2020	Exterior Windows	Hopper type commercial steel	5% of wall	Each	497	1.16	11.5%
2030	Exterior Doors	Steel overhead and hollow metal	15% of wall	S.F. Door	18.95	1.99	
B30 Roofing							
3010	Roof Coverings	Built-up tar and gravel; perlite/EPS composite insulation		S.F. Roof	6.30	6.30	7.4%
3020	Roof Openings	Skylight		S.F. Roof	.02	.02	
C. INTERIORS							
1010	Partitions	Concrete block	50 S.F. Floor/L.F. Partition	S.F. Partition	7.10	1.42	
1020	Interior Doors	Single leaf hollow metal	3000 S.F. Floor/Door	Each	1001	.34	
1030	Fittings	Toilet partitions		S.F. Floor	.19	.19	
2010	Stair Construction	N/A		—	—	—	7.3%
3010	Wall Finishes	Paint		S.F. Surface	6.93	2.77	
3020	Floor Finishes	90% metallic floor hardener, 10% vinyl composition tile		S.F. Floor	1.19	1.19	
3030	Ceiling Finishes	Gypsum board on wood joists in office and washrooms	10% of area	S.F. Ceiling	3.28	.32	
D. SERVICES							
D10 Conveying							
1010	Elevators & Lifts	N/A		—	—	—	0.0 %
1020	Escalators & Moving Walks	N/A		—	—	—	
D20 Plumbing							
2010	Plumbing Fixtures	Toilet and service fixtures, supply and drainage	1 Fixture/500 S.F. Floor	Each	1405	2.81	
2020	Domestic Water Distribution	Gas fired water heater		S.F. Floor	.48	.48	7.0%
2040	Rain Water Drainage	Roof drains		S.F. Roof	2.63	2.63	
D30 HVAC							
3010	Energy Supply	N/A		—	—	—	
3020	Heat Generating Systems	N/A		—	—	—	
3030	Cooling Generating Systems	N/A		—	—	—	10.7 %
3050	Terminal & Package Units	Single zone AC unit		S.F. Floor	8.26	8.26	
3090	Other HVAC Sys. & Equipment	Garage exhaust system		S.F. Floor	.80	.80	
D40 Fire Protection							
4010	Sprinklers	Sprinklers, ordinary hazard		S.F. Floor	4.11	4.11	5.8%
4020	Standpipes	Standpipe		S.F. Floor	.79	.79	
D50 Electrical							
5010	Electrical Service/Distribution	200 ampere service, panel board and feeders		S.F. Floor	.51	.51	
5020	Lighting & Branch Wiring	T-8 fluorescent fixtures, receptacles, switches, A.C. and misc. power		S.F. Floor	7.81	7.81	13.9%
5030	Communications & Security	Addressable alarm systems, partial internet wiring and emergency lighting		S.F. Floor	3.38	3.38	
5090	Other Electrical Systems	Emergency generator, 15 kW		S.F. Floor	.08	.08	
E. EQUIPMENT & FURNISHINGS							
1010	Commercial Equipment	N/A		—	—	—	
1020	Institutional Equipment	N/A		—	—	—	16.0 %
1030	Vehicular Equipment	Hoists		S.F. Floor	13.61	13.61	
1090	Other Equipment	N/A		—	—	—	
F. SPECIAL CONSTRUCTION							
1020	Integrated Construction	N/A		—	—	—	0.0 %
1040	Special Facilities	N/A		—	—	—	
G. BUILDING SITEWORK	**N/A**						

		Sub-Total	84.92	**100%**
CONTRACTOR FEES (General Requirements: 10%, Overhead: 5%, Profit: 10%)		25%	21.24	
ARCHITECT FEES		8%	8.49	

Total Building Cost	**114.65**

Costs per square foot of floor area

Exterior Wall	S.F. Area	600	800	1000	1200	1400	1600	1800	2000	2200
	L.F. Perimeter	100	120	126	140	153	160	170	180	190
Face Brick with Concrete Block Back-up	Wood Truss	258.00	241.45	222.35	213.95	207.55	200.20	195.60	192.00	189.00
	Steel Joists	249.65	233.05	214.00	205.65	199.25	191.90	187.30	183.70	180.65
Enameled Sandwich Panel	Steel Frame	232.25	217.55	201.10	193.75	188.15	181.85	177.85	174.70	172.10
Tile on Concrete Block	Steel Joists	266.45	248.40	227.10	217.90	210.85	202.70	197.55	193.55	190.25
Aluminum Siding	Wood Frame	221.85	208.95	195.00	188.75	183.85	178.55	175.15	172.55	170.25
Wood Siding	Wood Frame	224.00	210.85	196.65	190.20	185.30	179.90	176.35	173.70	171.40
Perimeter Adj., Add or Deduct	Per 100 L.F.	109.00	81.75	65.40	54.55	46.70	40.90	36.35	32.65	29.75
Story Hgt. Adj., Add or Deduct	Per 1 Ft.	7.10	6.45	5.45	5.05	4.65	4.30	4.05	3.85	3.75
Basement—Not Applicable										

The above costs were calculated using the basic specifications shown on the facing page. These costs should be adjusted where necessary for design alternatives and owner's requirements. Reported completed project costs, for this type of structure, range from $48.35 to $244.20 per S.F.

Common additives

Description	Unit	$ Cost
Air Compressors		
Electric 1-1/2 H.P., standard controls	Each	1400
Dual controls	Each	1750
5 H.P. 115/230 volt, standard controls	Each	3675
Dual controls	Each	3825
Product Dispenser		
with vapor recovery for 6 nozzles	Each	24,800
Hoists, Single post		
8000# cap. swivel arm	Each	9075
Two post, adjustable frames, 11,000# cap.	Each	12,800
24,000# cap.	Each	19,000
7500# cap.	Each	10,700
Four post, roll on ramp	Each	9925
Lockers, Steel, single tier, 60" or 72"	Opening	200 - 350
2 tier, 60" or 72" total	Opening	124 - 152
5 tier, box lockers	Each	70 - 82.50
Locker bench, lam. maple top only	L.F.	22
Pedestals, steel pipe	Each	64
Lube Equipment		
3 reel type, with pumps, no piping	Each	11,600

Important: See the Reference Section for Location Factors

Model costs calculated for a 1 story building with 12' story height and 1,400 square feet of floor area

				Unit	Unit Cost	Cost Per S.F.	% Of Sub-Total
A.	**SUBSTRUCTURE**						
1010	Standard Foundations	Poured concrete; strip and spread footings		S.F. Ground	3.89	3.89	
1020	Special Foundations	N/A		—	—	—	
1030	Slab on Grade	6" reinforced concrete with vapor barrier and granular base		S.F. Slab	7.26	7.26	12.7%
2010	Basement Excavation	Site preparation for slab and trench for foundation wall and footing		S.F. Ground	.96	.96	
2020	Basement Walls	4' foundation wall		L.F. Wall	67	7.37	
B.	**SHELL**						
	B10 Superstructure						
1010	Floor Construction	N/A		—	—	—	4.2 %
1020	Roof Construction	Plywood on wood trusses		S.F. Roof	6.46	6.46	
	B20 Exterior Enclosure						
2010	Exterior Walls	Face brick with concrete block backup	60% of wall	S.F. Wall	30.30	23.84	
2020	Exterior Windows	Store front and metal top hinged outswinging	20% of wall	Each	36.95	9.69	28.8%
2030	Exterior Doors	Steel overhead, aluminum & glass and hollow metal	20% of wall	Each	40.72	10.68	
	B30 Roofing						
3010	Roof Coverings	Asphalt shingles with flashing; perlite/EPS composite insulation		S.F. Roof	3.12	3.12	2.0%
3020	Roof Openings	N/A		—	—	—	
C.	**INTERIORS**						
1010	Partitions	Concrete block	25 S.F. Floor/L.F. Partition	S.F. Partition	7.13	2.28	
1020	Interior Doors	Single leaf hollow metal	700 S.F. Floor/Door	Each	1001	1.43	
1030	Fittings	Toilet partitions		S.F. Floor	2.58	2.58	
2010	Stair Construction	N/A		—	—	—	7.8%
3010	Wall Finishes	Paint		S.F. Surface	5.16	3.30	
3020	Floor Finishes	Vinyl composition tile	35% of floor area	S.F. Floor	2.89	1.01	
3030	Ceiling Finishes	Painted gypsum board on furring in sales area & washrooms	35% of floor area	S.F. Ceiling	3.95	1.38	
D.	**SERVICES**						
	D10 Conveying						
1010	Elevators & Lifts	N/A		—	—	—	0.0 %
1020	Escalators & Moving Walks	N/A		—	—	—	
	D20 Plumbing						
2010	Plumbing Fixtures	Toilet and service fixtures, supply and drainage	1 Fixture/235 S.F. Floor	Each	2522	10.73	
2020	Domestic Water Distribution	Gas fired water heater		S.F. Floor	3.40	3.40	9.2%
2040	Rain Water Drainage	N/A		—	—	—	
	D30 HVAC						
3010	Energy Supply	N/A		—	—	—	
3020	Heat Generating Systems	N/A		—	—	—	
3030	Cooling Generating Systems	N/A		—	—	—	15.4 %
3050	Terminal & Package Units	Single zone AC unit		S.F. Floor	18.96	18.96	
3090	Other HVAC Sys. & Equipment	Underfloor exhaust system		S.F. Floor	4.73	4.73	
	D40 Fire Protection						
4010	Sprinklers	Wet pipe sprinkler system		S.F. Floor	7.89	7.89	7.5%
4020	Standpipes	Standpipe		S.F. Floor	3.64	3.64	
	D50 Electrical						
5010	Electrical Service/Distribution	200 ampere service, panel board and feeders		S.F. Floor	4.61	4.61	
5020	Lighting & Branch Wiring	High efficiency fluorescent fixtures, receptacles, switches, A.C. and misc. power		S.F. Floor	5.82	5.82	12.4%
5030	Communications & Security	Addressable alarm systems and emergency lighting		S.F. Floor	8.09	8.09	
5090	Other Electrical Systems	Emergency generator		S.F. Floor	.61	.61	
E.	**EQUIPMENT & FURNISHINGS**						
1010	Commercial Equipment	N/A		—	—	—	
1020	Institutional Equipment	N/A		—	—	—	0.0 %
1030	Vehicular Equipment	N/A		—	—	—	
1090	Other Equipment	N/A		—	—	—	
F.	**SPECIAL CONSTRUCTION**						
1020	Integrated Construction	N/A		—	—	—	0.0 %
1040	Special Facilities	N/A		—	—	—	
G.	**BUILDING SITEWORK**	**N/A**					

					Sub-Total	153.73	**100%**
	CONTRACTOR FEES (General Requirements: 10%, Overhead: 5%, Profit: 10%)				25%	38.45	
	ARCHITECT FEES				8%	15.37	

			Total Building Cost	**207.55**

Costs per square foot of floor area

Exterior Wall	S.F. Area	12000	16000	20000	25000	30000	35000	40000	45000	50000
	L.F. Perimeter	440	520	600	700	708	780	841	910	979
Reinforced Concrete Block	Lam. Wood Arches	157.70	150.85	146.80	143.50	138.85	137.10	135.50	134.35	133.45
	Rigid Steel Frame	156.40	149.65	145.50	142.25	137.65	135.85	134.20	133.10	132.25
Face Brick with Concrete Block Back-up	Lam. Wood Arches	182.65	173.00	167.20	162.60	155.00	152.25	149.85	148.10	146.80
	Rigid Steel Frame	181.40	171.80	165.95	161.30	153.70	151.05	148.55	146.90	145.60
Metal Sandwich Panels	Lam. Wood Arches	153.20	146.90	143.10	140.05	136.00	134.35	132.90	131.85	131.10
	Rigid Steel Frame	151.95	145.65	141.85	138.85	134.80	133.10	131.70	130.65	129.80
Perimeter Adj., Add or Deduct	Per 100 L.F.	6.60	4.95	3.90	3.15	2.65	2.25	1.95	1.75	1.65
Story Hgt. Adj., Add or Deduct	Per 1 Ft.	0.90	0.85	0.65	0.70	0.65	0.55	0.50	0.55	0.50
Basement—Not Applicable										

The above costs were calculated using the basic specifications shown on the facing page. These costs should be adjusted where necessary for design alternatives and owner's requirements. Reported completed project costs, for this type of structure, range from $73.35 to $218.85 per S.F.

Common additives

Description	Unit	$ Cost
Bleachers, Telescoping, manual		
To 15 tier	Seat	118 - 164
16-20 tier	Seat	242 - 297
21-30 tier	Seat	257 - 310
For power operation, add	Seat	47 - 73.50
Gym Divider Curtain, Mesh top		
Manual roll-up	S.F.	11.85
Gym Mats		
2" naugahyde covered	S.F.	3.95
2" nylon	S.F.	6.85
1-1/2" wall pads	S.F.	9.10
1" wrestling mats	S.F.	6.05
Scoreboard		
Basketball, one side	Each	3325 - 9050
Basketball Backstop		
Wall mtd., 6' extended, fixed	Each	2150 - 2750
Swing up, wall mtd.	Each	2425 - 6800

Description	Unit	$ Cost
Lockers, Steel, single tier, 60" or 72"	Opening	200 - 350
2 tier, 60" or 72" total	Opening	124 - 152
5 tier, box lockers	Opening	70 - 82.50
Locker bench, lam. maple top only	L.F.	22
Pedestals, steel pipe	Each	64
Sound System		
Amplifier, 250 watts	Each	2400
Speaker, ceiling or wall	Each	196
Trumpet	Each	375
Emergency Lighting, 25 watt, battery operated		
Lead battery	Each	287
Nickel cadmium	Each	845

Model costs calculated for a 1 story building with 25' story height and 20,000 square feet of floor area

					Unit	Unit Cost	Cost Per S.F.	% Of Sub-Total

A. SUBSTRUCTURE

				Unit	Unit Cost	Cost Per S.F.	% Of Sub-Total
1010	Standard Foundations	Poured concrete; strip and spread footings		S.F. Ground	1.59	1.59	
1020	Special Foundations	N/A		—	—	—	
1030	Slab on Grade	4" reinforced concrete with vapor barrier and granular base		S.F. Slab	4.77	4.77	8.1%
2010	Basement Excavation	Site preparation for slab and trench for foundation wall and footing		S.F. Ground	.18	.18	
2020	Basement Walls	4' foundation wall		L.F. Wall	67	2.36	

B. SHELL

				Unit	Unit Cost	Cost Per S.F.	% Of Sub-Total
B10	**Superstructure**						
1010	Floor Construction	N/A		—	—	—	
1020	Roof Construction	Wood deck on laminated wood arches		S.F. Roof	17.95	17.95	16.4%
B20	**Exterior Enclosure**						
2010	Exterior Walls	Reinforced concrete block (end walls included)	90% of wall	S.F. Wall	12.37	8.35	
2020	Exterior Windows	Metal horizontal pivoted	10% of wall	Each	502	3.77	11.6%
2030	Exterior Doors	Aluminum and glass, hollow metal, steel overhead		Each	1884	.57	
B30	**Roofing**						
3010	Roof Coverings	EPDM, 60 mils, fully adhered; polyisocyanurate insulation		S.F. Roof	4.99	4.99	4.5%
3020	Roof Openings	N/A		—	—	—	

C. INTERIORS

				Unit	Unit Cost	Cost Per S.F.	% Of Sub-Total
1010	Partitions	Concrete block	50 S.F. Floor/L.F. Partition	S.F. Partition	8.60	1.72	
1020	Interior Doors	Single leaf hollow metal	500 S.F. Floor/Door	Each	1001	2	
1030	Fittings	Toilet partitions		S.F. Floor	.36	.36	
2010	Stair Construction	N/A		—	—	—	20.8%
3010	Wall Finishes	50% paint, 50% ceramic tile		S.F. Surface	9.15	3.66	
3020	Floor Finishes	90% hardwood, 10% ceramic tile		S.F. Floor	14.14	14.14	
3030	Ceiling Finishes	Mineral fiber tile on concealed zee bars	15% of area	S.F. Ceiling	6.51	.98	

D. SERVICES

				Unit	Unit Cost	Cost Per S.F.	% Of Sub-Total
D10	**Conveying**						
1010	Elevators & Lifts	N/A		—	—	—	0.0%
1020	Escalators & Moving Walks	N/A		—	—	—	
D20	**Plumbing**						
2010	Plumbing Fixtures	Toilet and service fixtures, supply and drainage	1 Fixture/515 S.F. Floor	Each	3198	6.21	
2020	Domestic Water Distribution	Electric water heater		S.F. Floor	4.96	4.96	10.2%
2040	Rain Water Drainage	N/A		—	—	—	
D30	**HVAC**						
3010	Energy Supply	N/A		—	—	—	
3020	Heat Generating Systems	Included in D3050		—	—	—	
3030	Cooling Generating Systems	N/A		—	—	—	9.4%
3050	Terminal & Package Units	Single zone rooftop unit, gas heating, electric cooling		S.F. Floor	10.26	10.26	
3090	Other HVAC Sys. & Equipment	N/A		—	—	—	
D40	**Fire Protection**						
4010	Sprinklers	Wet pipe sprinkler system		S.F. Floor	3.16	3.16	2.9%
4020	Standpipes	N/A		—	—	—	
D50	**Electrical**						
5010	Electrical Service/Distribution	400 ampere service, panel board and feeders		S.F. Floor	1	1	
5020	Lighting & Branch Wiring	Fluorescent fixtures, receptacles, switches, A.C. and misc. power		S.F. Floor	8.71	8.71	10.8%
5030	Communications & Security	Alarm systems, sound system and emergency lighting		S.F. Floor	1.91	1.91	
5090	Other Electrical Systems	Emergency generator, 7.5 kW		S.F. Floor	.21	.21	

E. EQUIPMENT & FURNISHINGS

				Unit	Unit Cost	Cost Per S.F.	% Of Sub-Total
1010	Commercial Equipment	N/A		—	—	—	
1020	Institutional Equipment	N/A		—	—	—	
1030	Vehicular Equipment	N/A		—	—	—	5.4%
1090	Other Equipment	Bleachers, sauna, weight room		S.F. Floor	5.91	5.91	

F. SPECIAL CONSTRUCTION

				Unit	Unit Cost	Cost Per S.F.	% Of Sub-Total
1020	Integrated Construction	N/A		—	—	—	0.0%
1040	Special Facilities	N/A		—	—	—	

G. BUILDING SITEWORK   N/A

		Sub-Total	109.72	100%
CONTRACTOR FEES (General Requirements: 10%, Overhead: 5%, Profit: 10%)		25%	27.48	
ARCHITECT FEES		7%	9.60	

	Total Building Cost	**146.80**

Costs per square foot of floor area

Exterior Wall	S.F. Area	5000	10000	15000	20000	30000	40000	50000	75000	100000
	L.F. Perimeter	300	410	500	580	710	830	930	1150	1300
Concrete Block Reinforced	Steel Frame	126.35	110.65	104.35	100.75	96.35	93.95	92.15	89.50	87.55
	Bearing Walls	130.15	114.35	107.90	104.30	99.80	97.40	95.65	92.90	90.95
Precast Concrete	Steel Frame	150.85	127.45	117.95	112.60	106.00	102.40	99.75	95.75	92.85
	Bearing Walls	154.25	130.80	121.35	116.00	109.40	105.80	103.10	99.15	96.25
Galv. Steel Siding	Steel Frame	126.35	110.65	104.35	100.75	96.30	93.95	92.15	89.45	87.55
Metal Sandwich Panel	Steel Frame	133.25	115.40	108.20	104.15	99.10	96.40	94.35	91.30	89.10
Perimeter Adj., Add or Deduct	Per 100 L.F.	16.45	8.20	5.50	4.10	2.75	2.00	1.65	1.05	0.85
Story Hgt. Adj., Add or Deduct	Per 1 Ft.	1.65	1.15	0.90	0.85	0.65	0.60	0.55	0.40	0.40
Basement—Not Applicable										

The above costs were calculated using the basic specifications shown on the facing page. These costs should be adjusted where necessary for design alternatives and owner's requirements. Reported completed project costs, for this type of structure, range from $42.10 to $192.00 per S.F.

Common additives

Description	Unit	$ Cost
Closed Circuit Surveillance, One station		
Camera and monitor	Each	1875
For additional camera stations, add	Each	1025
Emergency Lighting, 25 watt, battery operated		
Lead battery	Each	287
Nickel cadmium	Each	845
Lockers, Steel, single tier, 60" or 72"	Opening	200 - 350
2 tier, 60" or 72" total	Opening	124 - 152
5 tier, box lockers	Opening	70 - 82.50
Locker bench, lam. maple top only	L.F.	22
Pedestals, steel pipe	Each	64
Safe, Office type, 1 hour rating		
30" x 18" x 18"	Each	2400
60" x 36" x 18", double door	Each	8975
Sound System		
Amplifier, 250 watts	Each	2400
Speaker, ceiling or wall	Each	196
Trumpet	Each	375

Important: See the Reference Section for Location Factors

Model costs calculated for a 1 story building with 24' story height and 20,000 square feet of floor area

Hangar, Aircraft

						Unit	Unit Cost	Cost Per S.F.	% Of Sub-Total

A. SUBSTRUCTURE

1010	Standard Foundations	Poured concrete; strip and spread footings and		S.F. Ground	1.76	1.76	
1020	Special Foundations	N/A		—	—	—	
1030	Slab on Grade	6" reinforced concrete with vapor barrier and granular base		S.F. Slab	7.26	7.26	15.3%
2010	Basement Excavation	Site preparation for slab and trench for foundation wall and footing		S.F. Ground	.31	.31	
2020	Basement Walls	4' foundation wall		L.F. Wall	76	2.19	

B. SHELL

B10 Superstructure

1010	Floor Construction	N/A		—	—	—	
1020	Roof Construction	Metal deck, open web steel joists, beams, columns		S.F. Roof	7.98	7.98	10.6 %

B20 Exterior Enclosure

2010	Exterior Walls	Galvanized steel siding		S.F. Wall	10.14	3.53	
2020	Exterior Windows	Industrial horizontal pivoted steel	20% of wall	Each	1875	10.88	30.5%
2030	Exterior Doors	Steel overhead and sliding	30% of wall	Each	41.09	8.58	

B30 Roofing

3010	Roof Coverings	Elastomeric membrane; fiberboard insulation		S.F. Roof	3.74	3.74	
3020	Roof Openings	Roof hatches		S.F. Roof	.13	.13	5.1%

C. INTERIORS

1010	Partitions	Concrete block	200 S.F. Floor/L.F. Partition	S.F. Partition	9.20	.46	
1020	Interior Doors	Single leaf hollow metal	5000 S.F. Floor/Door	Each	1001	.20	
1030	Fittings	Toilet partitions		S.F. Floor	.27	.27	
2010	Stair Construction	N/A		—	—	—	1.4%
3010	Wall Finishes	Paint		S.F. Surface	1.10	.11	
3020	Floor Finishes	N/A		—	—	—	
3030	Ceiling Finishes	N/A		—	—	—	

D. SERVICES

D10 Conveying

1010	Elevators & Lifts	N/A		—	—	—	
1020	Escalators & Moving Walks	N/A		—	—	—	0.0 %

D20 Plumbing

2010	Plumbing Fixtures	Toilet and service fixtures, supply and drainage	1 Fixture/1000 S.F. Floor	Each	3060	3.06	
2020	Domestic Water Distribution	Gas fired water heater		S.F. Floor	.71	.71	5.0%
2040	Rain Water Drainage	N/A		—	—	—	

D30 HVAC

3010	Energy Supply	N/A		—	—	—	
3020	Heat Generating Systems	Unit heaters		Each	6.51	6.51	
3030	Cooling Generating Systems	N/A		—	—	—	9.0 %
3050	Terminal & Package Units	Exhaust fan		S.F. Floor	.26	.26	
3090	Other HVAC Sys. & Equipment	N/A		—	—	—	

D40 Fire Protection

4010	Sprinklers	Sprinklers, extra hazard		S.F. Floor	6.05	6.05	
4020	Standpipes	N/A		—	—	—	8.0%

D50 Electrical

5010	Electrical Service/Distribution	200 ampere service, panel board and feeders		S.F. Floor	.84	.84	
5020	Lighting & Branch Wiring	High intensity discharge fixtures, receptacles, switches and misc. power		S.F. Floor	9.92	9.92	
5030	Communications & Security	Alarm systems and emergency lighting		S.F. Floor	.50	.50	15.0%
5090	Other Electrical Systems	Emergency generator, 7.5 kW		S.F. Floor	.07	.07	

E. EQUIPMENT & FURNISHINGS

1010	Commercial Equipment	N/A		—	—	—	
1020	Institutional Equipment	N/A		—	—	—	
1030	Vehicular Equipment	N/A		—	—	—	0.0 %
1090	Other Equipment	N/A		—	—	—	

F. SPECIAL CONSTRUCTION

1020	Integrated Construction	N/A		—	—	—	
1040	Special Facilities	N/A		—	—	—	0.0 %

G. BUILDING SITEWORK   N/A

	Sub-Total	75.32	100%
CONTRACTOR FEES (General Requirements: 10%, Overhead: 5%, Profit: 10%)	25%	18.84	
ARCHITECT FEES	7%	6.59	
Total Building Cost		**100.75**	

147

Costs per square foot of floor area

Exterior Wall	S.F. Area	25000	40000	55000	70000	85000	100000	115000	130000	145000
	L.F. Perimeter	388	520	566	666	766	866	878	962	1045
Face Brick with Structural Facing Tile	Steel Frame	316.30	307.95	300.60	298.20	296.60	295.45	292.90	292.20	291.65
	R/Conc. Frame	324.55	316.20	308.85	306.45	304.85	303.70	301.15	300.45	299.85
Face Brick with Concrete Block Back-up	Steel Frame	308.00	301.00	295.10	293.05	291.75	290.85	288.85	288.25	287.80
	R/Conc. Frame	309.15	302.15	296.25	294.20	292.90	292.00	290.00	289.45	288.95
Precast Concrete Panels	Steel Frame	321.15	312.00	303.80	301.15	299.35	298.15	295.25	294.50	293.90
	R/Conc. Frame	329.40	320.25	312.05	309.40	307.60	306.40	303.50	302.75	302.15
Perimeter Adj., Add or Deduct	Per 100 L.F.	8.95	5.65	4.15	3.15	2.65	2.30	2.00	1.75	1.55
Story Hgt. Adj., Add or Deduct	Per 1 Ft.	2.60	2.20	1.75	1.60	1.50	1.50	1.30	1.25	1.25
For Basement, add $33.05 per square foot of basement area										

The above costs were calculated using the basic specifications shown on the facing page. These costs should be adjusted where necessary for design alternatives and owner's requirements. Reported completed project costs, for this type of structure, range from $161.85 to $408.80 per S.F.

Common additives

Description	Unit	$ Cost	Description	Unit	$ Cost
Cabinet, Base, door units, metal	L.F.	268	Nurses Call Station		
Drawer units	L.F.	525	Single bedside call station	Each	320
Tall storage cabinets, 7' high, open	L.F.	500	Ceiling speaker station	Each	147
With doors	L.F.	590	Emergency call station	Each	198
Wall, metal 12-1/2" deep, open	L.F.	200	Pillow speaker	Each	305
With doors	L.F.	360	Double bedside call station	Each	390
Closed Circuit TV (Patient monitoring)			Duty station	Each	335
One station camera & monitor	Each	1875	Standard call button	Each	190
For additional camera, add	Each	1025	Master control station for 20 stations	Each	6125
For automatic iris for low light, add	Each	2625	Sound System		
Hubbard Tank, with accessories			Amplifier, 250 watts	Each	2400
Stainless steel, 125 GPM 45 psi	Each	29,000	Speaker, ceiling or wall	Each	196
For electric hoist, add	Each	3150	Trumpet	Each	375
Mortuary Refrigerator, End operated			Station, Dietary with ice	Each	17,600
2 capacity	Each	14,100	Sterilizers		
6 capacity	Each	26,300	Single door, steam	Each	174,500
			Double door, steam	Each	224,000
			Portable, countertop, steam	Each	4175 - 6550
			Gas	Each	43,200
			Automatic washer/sterilizer	Each	59,500

Model costs calculated for a 3 story building with 12' story height and 55,000 square feet of floor area

Hospital, 2-3 Story

				Unit	Unit Cost	Cost Per S.F.	% Of Sub-Total
A.	**SUBSTRUCTURE**						
1010	Standard Foundations	Poured concrete; strip and spread footings		S.F. Ground	5.04	1.68	
1020	Special Foundations	N/A		—	—	—	
1030	Slab on Grade	4" reinforced concrete with vapor barrier and granular base		S.F. Slab	4.77	1.59	1.8%
2010	Basement Excavation	Site preparation for slab and trench for foundation wall and footing		S.F. Ground	.18	.06	
2020	Basement Walls	4' foundation wall		L.F. Wall	76	.77	
B.	**SHELL**						
	B10 Superstructure						
1010	Floor Construction	Cast-in-place concrete beam and slab		S.F. Floor	21.99	14.66	9.1%
1020	Roof Construction	Cast-in-place concrete beam and slab		S.F. Roof	17.85	5.95	
	B20 Exterior Enclosure						
2010	Exterior Walls	Face brick and structural facing tile	85% of wall	S.F. Wall	43.12	13.58	
2020	Exterior Windows	Aluminum sliding	15% of wall	Each	703	1.70	6.9%
2030	Exterior Doors	Double aluminum and glass and sliding doors		Each	2978	.32	
	B30 Roofing						
3010	Roof Coverings	Built-up tar and gravel with flashing; perlite/EPS composite insulation		S.F. Roof	5.73	1.91	0.8%
3020	Roof Openings	Roof hatches		S.F. Roof	.03	.01	
C.	**INTERIORS**						
1010	Partitions	Concrete block, gypsum board on metal studs	9 S.F. Floor/L.F. Partition	S.F. Partition	7.73	8.59	
1020	Interior Doors	Single leaf hollow metal	90 S.F. Floor/Door	Each	1012	11.24	
1030	Fittings	Hospital curtains		S.F. Floor	.89	.89	
2010	Stair Construction	Concrete filled metal pan		Flight	5625	1.33	20.1%
3010	Wall Finishes	40% vinyl wall covering, 35% ceramic tile, 25% epoxy coating		S.F. Surface	3.40	7.55	
3020	Floor Finishes	60% vinyl tile, 20% ceramic, 20% terrazzo		S.F. Floor	9.36	9.36	
3030	Ceiling Finishes	Plaster on suspended metal lath		S.F. Ceiling	6.51	6.51	
D.	**SERVICES**						
	D10 Conveying						
1010	Elevators & Lifts	Two hydraulic hospital elevators		Each	120,725	4.39	1.9%
1020	Escalators & Moving Walks	N/A		—	—	—	
	D20 Plumbing						
2010	Plumbing Fixtures	Medical, patient & specialty fixtures, supply and drainage	1 Fixture/224 S.F. Floor	Each	6129	27.36	
2020	Domestic Water Distribution	Electric water heater		S.F. Floor	11.17	11.17	17.2%
2040	Rain Water Drainage	Roof drains		S.F. Roof	1.47	.49	
	D30 HVAC						
3010	Energy Supply	Conditioned air with hot water reheat system		S.F. Floor	4.13	4.13	
3020	Heat Generating Systems	Boiler		Each	100,475	2.80	
3030	Cooling Generating Systems	Conditioned air, chillers, cooling towers		S.F. Floor	5.93	5.93	21.4%
3050	Terminal & Package Units	N/A		—	—	—	
3090	Other HVAC Sys. & Equipment	Conditioned air, ductwork, surgery air curtain		S.F. Floor	35.74	35.74	
	D40 Fire Protection						
4010	Sprinklers	Wet pipe sprinkler system		S.F. Floor	2.81	2.81	1.5%
4020	Standpipes	Standpipe		S.F. Floor	.70	.70	
	D50 Electrical						
5010	Electrical Service/Distribution	1200 ampere service, panel board and feeders		S.F. Floor	1.80	1.80	
5020	Lighting & Branch Wiring	High efficiency hospital grade light fixtures, receptacles, switches, A.C. and misc. power		S.F. Floor	16.02	16.02	
5030	Communications & Security	Addressable alarm systems, internet wiring, communications systems, emergency lighting		S.F. Floor	5.14	5.14	12.1%
5090	Other Electrical Systems	Emergency generator, 200 kW with fuel tank, Uninterruptible power supply		S.F. Floor	4.36	4.36	
E.	**EQUIPMENT & FURNISHINGS**						
1010	Commercial Equipment	N/A		—	—	—	
1020	Institutional Equipment	Medical gas system, kitchen equip, lab equip., curtain partitions		S.F. Floor	10.14	10.14	7.1%
1030	Vehicular Equipment	N/A		—	—	—	
2020	Other Equipment	Patient wall systems		S.F. Floor	6	6	
F.	**SPECIAL CONSTRUCTION**						
1020	Integrated Construction	N/A		—	—	—	
1040	Special Facilities	N/A		—	—	—	0.0%
G.	**BUILDING SITEWORK**	**N/A**					

		Sub-Total	226.68	100%
CONTRACTOR FEES (General Requirements: 10%, Overhead: 5%, Profit: 10%)		25%	56.67	
ARCHITECT FEES		9%	25.50	
	Total Building Cost		**308.85**	

Costs per square foot of floor area

Exterior Wall	S.F. Area	100000	125000	150000	175000	200000	225000	250000	275000	300000
	L.F. Perimeter	594	705	816	783	866	950	1033	1116	1200
Face Brick with Structural Facing Tile	Steel Frame	282.25	278.55	276.10	270.60	269.15	268.00	267.05	266.30	265.70
	R/Conc. Frame	288.80	285.00	282.45	276.95	275.45	274.25	273.30	272.50	271.90
Face Brick with Concrete Block Back-up	Steel Frame	277.00	273.50	271.10	266.40	265.00	264.00	263.10	262.35	261.75
	R/Conc. Frame	283.55	280.05	277.70	273.05	271.65	270.55	269.70	268.95	268.40
Precast Concrete Panels With Exposed Aggregate	Steel Frame	280.25	276.60	274.15	268.90	267.45	266.30	265.40	264.65	264.05
	R/Conc. Frame	286.90	283.20	280.70	275.55	274.00	272.90	272.00	271.20	270.60
Perimeter Adj., Add or Deduct	Per 100 L.F.	4.60	3.60	3.00	2.60	2.25	2.05	1.85	1.65	1.50
Story Hgt. Adj., Add or Deduct	Per 1 Ft.	2.05	1.95	1.85	1.60	1.45	1.50	1.45	1.35	1.40
For Basement, add $34.10 per square foot of basement area										

The above costs were calculated using the basic specifications shown on the facing page. These costs should be adjusted where necessary for design alternatives and owner's requirements. Reported completed project costs, for this type of structure, range from $165.55 to $403.70 per S.F.

Common additives

Description	Unit	$ Cost	Description	Unit	$ Cost
Cabinets, Base, door units, metal	L.F.	268	Nurses Call Station		
Drawer units	L.F.	525	Single bedside call station	Each	320
Tall storage cabinets, 7' high, open	L.F.	500	Ceiling speaker station	Each	147
With doors	L.F.	590	Emergency call station	Each	198
Wall, metal 12-1/2" deep, open	L.F.	200	Pillow speaker	Each	305
With doors	L.F.	360	Double bedside call station	Each	390
Closed Circuit TV (Patient monitoring)			Duty station	Each	335
One station camera & monitor	Each	1875	Standard call button	Each	190
For additional camera add	Each	1025	Master control station for 20 stations	Each	6125
For automatic iris for low light add	Each	2625	Sound System		
Hubbard Tank, with accessories			Amplifier, 250 watts	Each	2400
Stainless steel, 125 GPM 45 psi	Each	29,000	Speaker, ceiling or wall	Each	196
For electric hoist, add	Each	3150	Trumpet	Each	375
Mortuary Refrigerator, End operated			Station, Dietary with ice	Each	17,600
2 capacity	Each	14,100	Sterilizers		
6 capacity	Each	26,300	Single door, steam	Each	174,500
			Double door, steam	Each	224,000
			Portable, counter top, steam	Each	4175 - 6550
			Gas	Each	43,200
			Automatic washer/sterilizer	Each	59,500

Model costs calculated for a 6 story building with 12' story height and 200,000 square feet of floor area

				Unit	Unit Cost	Cost Per S.F.	% Of Sub-Total
A.	**SUBSTRUCTURE**						
1010	Standard Foundations	Poured concrete; strip and spread footings		S.F. Ground	13.20	2.20	
1020	Special Foundations	N/A		—	—	—	
1030	Slab on Grade	4" reinforced concrete with vapor barrier and granular base		S.F. Slab	7.26	1.22	1.9%
2010	Basement Excavation	Site preparation for slab and trench for foundation wall and footing		S.F. Ground	.18	.03	
2020	Basement Walls	4' foundation wall		L.F. Wall	76	.33	
B.	**SHELL**						
	B10 Superstructure						
1010	Floor Construction	Concrete slab with metal deck and beams, steel columns		S.F. Floor	18.94	15.78	8.6%
1020	Roof Construction	Metal deck, open web steel joists, beams, interior columns		S.F. Roof	7.20	1.20	
	B20 Exterior Enclosure						
2010	Exterior Walls	Face brick and structural facing tile	70% of wall	S.F. Wall	43.12	9.41	
2020	Exterior Windows	Aluminum sliding	30% of wall	Each	557	3.48	6.9%
2030	Exterior Doors	Double aluminum and glass and sliding doors		Each	5293	.74	
	B30 Roofing						
3010	Roof Coverings	Built-up tar and gravel with flashing; perlite/EPS composite insulation		S.F. Roof	6.96	1.16	0.6%
3020	Roof Openings	Roof hatches		S.F. Roof	.18	.03	
C.	**INTERIORS**						
1010	Partitions	Gypsum board on metal studs with sound deadening board	9 S.F. Floor/L.F. Partition	S.F. Partition	6.82	7.58	
1020	Interior Doors	Single leaf hollow metal	90 S.F. Floor/Door	Each	1012	11.24	
1030	Fittings	Hospital curtains		S.F. Floor	.98	.98	
2010	Stair Construction	Concrete filled metal pan		Flight	9825	1.28	23.5%
3010	Wall Finishes	40% vinyl wall covering, 35% ceramic tile, 25% epoxy coating		S.F. Surface	3.39	7.53	
3020	Floor Finishes	60% vinyl tile, 20% ceramic, 20% terrazzo		S.F. Floor	10.15	10.15	
3030	Ceiling Finishes	Plaster on suspended metal lath		S.F. Ceiling	7.59	7.60	
D.	**SERVICES**						
	D10 Conveying						
1010	Elevators & Lifts	Six geared hospital elevators		Each	214,667	6.44	3.3%
1020	Escalators & Moving Walks	N/A		—	—	—	
	D20 Plumbing						
2010	Plumbing Fixtures	Kitchen, toilet and service fixtures, supply and drainage	1 Fixture/416S.F. Floor	Each	4975	11.96	
2020	Domestic Water Distribution	Electric water heater		S.F. Floor	6.91	6.91	9.8%
2040	Rain Water Drainage	Roof drains		S.F. Floor	3.24	.54	
	D30 HVAC						
3010	Energy Supply	Oil fired hot water, wall fin radiation		S.F. Floor	3.55	3.55	
3020	Heat Generating Systems	Hot water boilers, steam boiler for services		Each	30,850	.39	
3030	Cooling Generating Systems	Chilled water units		S.F. Floor	2.75	2.75	20.2%
3050	Terminal & Package Units	N/A		—	—	—	
3090	Other HVAC Sys. & Equipment	Conditioned air with reheat, operating room air curtains		S.F. Floor	33.14	33.14	
	D40 Fire Protection						
4010	Sprinklers	Wet pipe sprinkler system		S.F. Floor	2.68	2.68	1.6%
4020	Standpipes	Standpipe		S.F. Floor	.45	.45	
	D50 Electrical						
5010	Electrical Service/Distribution	4000 ampere service, panel board and feeders		S.F. Floor	4.03	4.03	
5020	Lighting & Branch Wiring	High efficiency hospital grade light fixtures, receptacles, switches, A.C. and misc. power		S.F. Floor	18.68	18.68	
5030	Communications & Security	Addressable alarm systems, internet wiring, communications system, emergency lighting		S.F. Floor	2.39	2.39	14.9%
5090	Other Electrical Systems	Emergency generator, 800 kW with fuel tank, uninterruptible power supply		S.F. Floor	4.31	4.31	
E.	**EQUIPMENT & FURNISHINGS**						
1010	Commercial Equipment	N/A		—	—	—	
1020	Institutional Equipment	Medical gases, curtain partitions		S.F. Floor	13.41	13.41	8.8%
1030	Vehicular Equipment	N/A		—	—	—	
2020	Other Equipment	Patient wall systems		S.F. Floor	3.96	3.96	
F.	**SPECIAL CONSTRUCTION**						
1020	Integrated Construction	N/A		—	—	—	0.0%
1040	Special Facilities	N/A		—	—	—	
G.	**BUILDING SITEWORK**	**N/A**					

			Sub-Total	197.53	100%
	CONTRACTOR FEES (General Requirements: 10%, Overhead: 5%, Profit: 10%)		25%	49.40	
	ARCHITECT FEES		9%	22.22	
			Total Building Cost	**269.15**	

Costs per square foot of floor area

Exterior Wall	S.F. Area	35000	55000	75000	95000	115000	135000	155000	175000	195000
	L.F. Perimeter	314	401	497	555	639	722	754	783	850
Face Brick with Concrete Block Back-up	Steel Frame	179.90	171.15	167.45	164.00	162.45	161.35	159.50	157.95	157.40
	R/Conc. Frame	183.00	174.20	170.50	167.10	165.55	164.50	162.60	161.05	160.50
Glass and Metal Curtain Walls	Steel Frame	174.90	166.80	163.35	160.35	158.85	157.90	156.35	155.00	154.45
	R/Conc. Frame	178.20	170.15	166.70	163.70	162.20	161.25	159.65	158.30	157.80
Precast Concrete Panels	Steel Frame	191.10	179.95	175.30	170.90	168.95	167.60	165.10	163.05	162.35
	R/Conc. Frame	195.35	184.05	179.30	174.85	172.85	171.45	168.90	166.85	166.15
Perimeter Adj., Add or Deduct	Per 100 L.F.	9.30	5.90	4.35	3.40	2.90	2.45	2.10	1.85	1.70
Story Hgt. Adj., Add or Deduct	Per 1 Ft.	2.60	2.10	1.90	1.65	1.65	1.60	1.45	1.30	1.30
For Basement, add $32.45 per square foot of basement area										

The above costs were calculated using the basic specifications shown on the facing page. These costs should be adjusted where necessary for design alternatives and owner's requirements. Reported completed project costs, for this type of structure, range from $109.20 to $209.60 per S.F.

Common additives

Description	Unit	$ Cost	Description	Unit	$ Cost
Bar, Front bar	L.F.	375	Laundry Equipment		
Back bar	L.F.	300	Folders, blankets & sheets, king size	Each	67,000
Booth, Upholstered, custom, straight	L.F.	212 - 390	Ironers, 110" single roll	Each	35,600
"L" or "U" shaped	L.F.	220 - 370	Combination washer extractor 50#	Each	12,300
Closed Circuit Surveillance, One station			125#	Each	33,100
Camera and monitor	Each	1875	Sauna, Prefabricated, complete		
For additional camera stations, add	Each	1025	6' x 4'	Each	5875
Directory Boards, Plastic, glass covered			6' x 6'	Each	6975
30" x 20"	Each	605	6' x 9'	Each	8550
36" x 48"	Each	1325	8' x 8'	Each	10,100
Aluminum, 24" x 18"	Each	585	10' x 12'	Each	14,000
48" x 32"	Each	975	Smoke Detectors		
48" x 60"	Each	2025	Ceiling type	Each	233
Elevators, Electric passenger, 5 stops			Duct type	Each	525
3500# capacity	Each	166,200	Sound System		
5000# capacity	Each	169,700	Amplifier, 250 watts	Each	2400
Additional stop, add	Each	14,000	Speaker, ceiling or wall	Each	196
Emergency Lighting, 25 watt, battery operated			Trumpet	Each	375
Lead battery	Each	287	TV Antenna, Master system, 12 outlet	Outlet	320
Nickel cadmium	Each	845	30 outlet	Outlet	207
			100 outlet	Outlet	199

Model costs calculated for a 6 story building with 10' story height and 135,000 square feet of floor area

Hotel, 4-7 Story

				Unit	Unit Cost	Cost Per S.F.	% Of Sub-Total
A. SUBSTRUCTURE							
1010	Standard Foundations	Poured concrete; strip and spread footings		S.F. Ground	5.70	.95	
1020	Special Foundations	N/A		—	—	—	
1030	Slab on Grade	4" reinforced concrete with vapor barrier and granular base		S.F. Slab	4.77	.80	2.1%
2010	Basement Excavation	Site preparation for slab and trench for foundation wall and footing		S.F. Ground	.18	.03	
2020	Basement Walls	4' foundation wall		L.F. Wall	76	.80	
B. SHELL							
	B10 Superstructure						
1010	Floor Construction	Concrete slab with metal deck, beams, columns		S.F. Floor	17.27	14.39	13.0%
1020	Roof Construction	Metal deck, open web steel joists, beams, columns		S.F. Roof	8.52	1.42	
	B20 Exterior Enclosure						
2010	Exterior Walls	Face brick with concrete block backup	80% of wall	S.F. Wall	30.31	7.78	
2020	Exterior Windows	Window wall	20% of wall	Each	46.45	2.98	9.0%
2030	Exterior Doors	Glass amd metal doors and entrances		Each	3281	.24	
	B30 Roofing						
3010	Roof Coverings	Built-up tar and gravel with flashing; perlite/EPS composite insulation		S.F. Roof	5.82	.97	0.8%
3020	Roof Openings	Roof hatches		S.F. Roof	.06	.01	
C. INTERIORS							
1010	Partitions	Gypsum board and sound deadening board, steel studs	9 S.F. Floor/L.F. Partition	S.F. Partition	6.65	5.91	
1020	Interior Doors	Single leaf hollow metal	90 S.F. Floor/Door	Each	1001	11.12	
1030	Fittings	N/A		—	—	—	
2010	Stair Construction	Concrete filled metal pan		Flight	12,275	1.28	25.3%
3010	Wall Finishes	20% paint, 75% vinyl cover, 5% ceramic tile		S.F. Surface	1.87	3.32	
3020	Floor Finishes	80% carpet, 10% vinyl composition tile, 10% ceramic tile		S.F. Floor	5.08	5.08	
3030	Ceiling Finishes	Mineral fiber tile applied with adhesive		S.F. Ceiling	4.09	4.09	
D. SERVICES							
	D10 Conveying						
1010	Elevators & Lifts	Four geared passenger elevators		Each	224,438	6.65	5.5%
1020	Escalators & Moving Walks	N/A		—	—	—	
	D20 Plumbing						
2010	Plumbing Fixtures	Kitchen, toilet and service fixtures, supply and drainage	1 Fixture/155 S.F. Floor	Each	3165	20.42	
2020	Domestic Water Distribution	Oil fired water heater		S.F. Floor	.53	.53	17.4%
2040	Rain Water Drainage	Roof drains		S.F. Roof	1.32	.22	
	D30 HVAC						
3010	Energy Supply	Gas fired hot water, wall fin radiation		S.F. Floor	4.29	4.29	
3020	Heat Generating Systems	N/A		—	—	—	
3030	Cooling Generating Systems	Chilled water, fan coil units		S.F. Floor	11.35	11.35	12.8%
3050	Terminal & Package Units	N/A		—	—	—	
3090	Other HVAC Sys. & Equipment	N/A		—	—	—	
	D40 Fire Protection						
4010	Sprinklers	Sprinkler system, light hazard		S.F. Floor	2.70	2.70	2.5%
4020	Standpipes	Standpipes and hose systems		S.F. Floor	.31	.31	
	D50 Electrical						
5010	Electrical Service/Distribution	2000 ampere service, panel board and feeders		S.F. Floor	1.36	1.36	
5020	Lighting & Branch Wiring	High efficiency fluorescent fixtures, receptacles, switches, A.C. and misc. power		S.F. Floor	8.24	8.24	
5030	Communications & Security	Addressable alarm systems, internet wiring, communications systems and emergency lighting		S.F. Floor	4.04	4.04	11.6%
5090	Other Electrical Systems	Emergency generator, 250 kW		S.F. Floor	.52	.52	
E. EQUIPMENT & FURNISHINGS							
1010	Commercial Equipment	N/A		—	—	—	
1020	Institutional Equipment	N/A		—	—	—	
1030	Vehicular Equipment	N/A		—	—	—	0.0%
1090	Other Equipment	N/A		—	—	—	
F. SPECIAL CONSTRUCTION							
1020	Integrated Construction	N/A		—	—	—	
1040	Special Facilities	N/A		—	—	—	0.0%
G. BUILDING SITEWORK	**N/A**						

		Sub-Total	121.80	100%
CONTRACTOR FEES (General Requirements: 10%, Overhead: 5%, Profit: 10%)		25%	30.41	
ARCHITECT FEES		6%	9.14	
	Total Building Cost		**161.35**	

Costs per square foot of floor area

Exterior Wall	S.F. Area	140000	243000	346000	450000	552000	655000	760000	860000	965000
	L.F. Perimeter	403	587	672	800	936	1073	1213	1195	1312
Face Brick with Concrete Block Back-up	Steel Frame	166.20	159.75	154.90	152.95	151.95	151.20	150.80	149.00	148.65
	R/Conc. Frame	172.95	166.45	161.65	159.70	158.70	158.00	157.55	155.75	155.35
Face Brick Veneer On Steel Studs	Steel Frame	163.40	157.40	153.00	151.25	150.35	149.65	149.20	147.65	147.30
	R/Conc. Frame	173.00	167.05	162.65	160.90	160.00	159.35	158.90	157.30	156.95
Glass and Metal Curtain Walls	Steel Frame	198.80	184.95	177.60	174.15	172.20	170.80	169.75	167.95	167.25
	R/Conc. Frame	197.95	184.30	177.00	173.60	171.65	170.25	169.30	167.40	166.80
Perimeter Adj., Add or Deduct	Per 100 L.F.	5.50	3.15	2.20	1.75	1.40	1.20	0.95	0.85	0.75
Story Hgt. Adj., Add or Deduct	Per 1 Ft.	2.10	1.75	1.40	1.30	1.25	1.25	1.10	0.95	0.95
For Basement, add $33.70 per square foot of basement area										

The above costs were calculated using the basic specifications shown on the facing page. These costs should be adjusted where necessary for design alternatives and owner's requirements. Reported completed project costs, for this type of structure, range from $121.20 to $211.90 per S.F.

Common additives

Description	Unit	$ Cost	Description	Unit	$ Cost
Bar, Front bar	L.F.	375	Laundry Equipment		
Back bar	L.F.	300	Folders, blankets & sheets, king size	Each	67,000
Booth, Upholstered, custom, straight	L.F.	212 - 390	Ironers, 110" single roll	Each	35,600
"L" or "U" shaped	L.F.	220 - 370	Combination washer & extractor 50#	Each	12,300
Closed Circuit Surveillance, One station			125#	Each	33,100
Camera and monitor	Each	1875	Sauna, Prefabricated, complete		
For additional camera stations, add	Each	1025	6' x 4'	Each	5875
Directory Boards, Plastic, glass covered			6' x 6'	Each	6975
30" x 20"	Each	605	6' x 9'	Each	8550
36" x 48"	Each	1325	8' x 8'	Each	10,100
Aluminum, 24" x 18"	Each	585	10' x 12'	Each	14,000
48" x 32"	Each	975	Smoke Detectors		
48" x 60"	Each	2025	Ceiling type	Each	233
Elevators, Electric passenger, 10 stops			Duct type	Each	525
3500# capacity	Each	430,000	Sound System		
5000# capacity	Each	437,000	Amplifier, 250 watts	Each	2400
Additional stop, add	Each	14,000	Speaker, ceiling or wall	Each	196
Emergency Lighting, 25 watt, battery operated			Trumpet	Each	375
Lead battery	Each	287	TV Antenna, Master system, 12 outlet	Outlet	320
Nickel cadmium	Each	845	30 outlet	Outlet	207
			100 outlet	Outlet	199

Important: See the Reference Section for Location Factors

Model costs calculated for a 15 story building with 10′ story height and 450,000 square feet of floor area

					Unit	Unit Cost	Cost Per S.F.	% Of Sub-Total
A.	**SUBSTRUCTURE**							
1010	Standard Foundations	CIP concrete pile caps			S.F. Ground	5.25	.35	
1020	Special Foundations	Steel H-piles, concrete grade beams			S.F. Ground	111	7.41	
1030	Slab on Grade	4" reinforced concrete with vapor barrier and granular base			S.F. Slab	4.77	.32	6.3%
2010	Basement Excavation	Site preparation for slab, piles and grade beams			S.F. Ground	.18	.01	
2020	Basement Walls	4' foundation wall			L.F. Wall	76	.17	
B.	**SHELL**							
	B10 Superstructure							
1010	Floor Construction	Open web steel joists, slab form, concrete, columns			S.F. Floor	17.26	16.11	12.6%
1020	Roof Construction	Metal deck, open web steel joists, beams, columns			S.F. Roof	7.35	.49	
	B20 Exterior Enclosure							
2010	Exterior Walls	N/A			—	—	—	
2020	Exterior Windows	Glass and metal curtain walls	100% of wall		Each	25.40	6.78	5.3 %
2030	Exterior Doors	Glass and metal doors and entrances			Each	2965	.21	
	B30 Roofing							
3010	Roof Coverings	Built-up tar and gravel with flashing; perlite/EPS composite insulation			S.F. Roof	5.70	.38	0.3%
3020	Roof Openings	N/A			—	—	—	
C.	**INTERIORS**							
1010	Partitions	Gypsum board and sound deadening board, steel studs	9 S.F. Floor/L.F. Partition		S.F. Partition	6.45	5.73	
1020	Interior Doors	Single leaf hollow metal	90 S.F. Floor/Door		Each	1001	11.12	
1030	Fittings	N/A			—	—	—	
2010	Stair Construction	Concrete filled metal pan			Flight	12,275	2.48	23.9%
3010	Wall Finishes	20% paint, 75% vinyl cover, 5% ceramic tile			S.F. Surface	1.85	3.29	
3020	Floor Finishes	80% carpet tile, 10% vinyl composition tile, 10% ceramic tile			S.F. Floor	5.08	5.08	
3030	Ceiling Finishes	Gypsum board on resilient channel			S.F. Ceiling	3.71	3.71	
D.	**SERVICES**							
	D10 Conveying							
1010	Elevators & Lifts	One geared freight, six geared passenger elevators			Each	494,250	6.59	5.0%
1020	Escalators & Moving Walks	N/A			—	—	—	
	D20 Plumbing							
2010	Plumbing Fixtures	Kitchen, toilet and service fixtures, supply and drainage	1 Fixture/165 S.F. Floor		Each	3632	22.01	
2020	Domestic Water Distribution	Electric water heater			S.F. Floor	4.57	4.57	20.3%
2040	Rain Water Drainage	Roof drains			S.F. Roof	1.95	.13	
	D30 HVAC							
3010	Energy Supply	Gas fired hot water, wall fin radiation			S.F. Floor	2.27	2.27	
3020	Heat Generating Systems	N/A			—	—	—	
3030	Cooling Generating Systems	Chilled water, fan coil units			S.F. Floor	11.35	11.35	10.4%
3050	Terminal & Package Units	N/A			—	—	—	
3090	Other HVAC Sys. & Equipment	N/A			—	—	—	
	D40 Fire Protection							
4010	Sprinklers	Sprinkler system, light hazard			S.F. Floor	3.88	3.88	5.2%
4020	Standpipes	Standpipes, fire pump and hose systems			S.F. Floor	3.02	3.02	
	D50 Electrical							
5010	Electrical Service/Distribution	6000 ampere service, panel board and feeders			S.F. Floor	1.46	1.46	
5020	Lighting & Branch Wiring	High efficiency fluorescent fixtures, receptacles, switches, A.C. and misc. power			S.F. Floor	8.12	8.12	10.6%
5030	Communications & Security	Addressable alarm systems, internet wiring, communications systems and emergency lighting			S.F. Floor	4.02	4.02	
5090	Other Electrical Systems	Emergency generator, 500 kW			S.F. Floor	.39	.39	
E.	**EQUIPMENT & FURNISHINGS**							
1010	Commercial Equipment	N/A			—	—	—	
1020	Institutional Equipment	N/A			—	—	—	
1030	Vehicular Equipment	N/A			—	—	—	0.0 %
1090	Other Equipment	N/A			—	—	—	
F.	**SPECIAL CONSTRUCTION**							
1020	Integrated Construction	N/A			—	—	—	0.0 %
1040	Special Facilities	N/A			—	—	—	
G.	**BUILDING SITEWORK**	**N/A**						

			Sub-Total	131.45	100%
CONTRACTOR FEES (General Requirements: 10%, Overhead: 5%, Profit: 10%)			25%	32.84	
ARCHITECT FEES			6%	9.86	
			Total Building Cost	**174.15**	

Costs per square foot of floor area

Exterior Wall	S.F. Area	5500	12000	20000	30000	40000	60000	80000	100000	145000
	L.F. Perimeter	310	330	400	490	530	590	680	760	920
Face Brick with Concrete Block Back-up	Steel Frame	466.05	344.45	309.25	291.90	279.85	267.00	261.55	258.05	253.30
	R/Conc. Frame	466.50	344.85	309.65	292.25	280.25	267.35	261.95	258.40	253.65
Stucco on Concrete Block	Steel Frame	436.10	329.90	298.75	283.30	272.95	261.90	257.20	254.15	250.05
	R/Conc. Frame	436.30	330.10	298.90	283.50	273.10	262.10	257.35	254.35	250.25
Reinforced Concrete	Steel Frame	442.70	333.15	301.05	285.25	274.50	263.10	258.20	255.00	250.80
	R/Conc. Frame	442.90	333.30	301.25	285.40	274.65	263.20	258.35	255.20	251.00
Perimeter Adj., Add or Deduct	Per 100 L.F.	47.85	21.95	13.20	8.70	6.60	4.35	3.30	2.65	1.85
Story Hgt. Adj., Add or Deduct	Per 1 Ft.	11.40	5.55	4.05	3.35	2.65	1.95	1.75	1.50	1.30
For Basement, add $46.60 per square foot of basement area										

The above costs were calculated using the basic specifications shown on the facing page. These costs should be adjusted where necessary for design alternatives and owner's requirements. Reported completed project costs, for this type of structure, range from $176.35 to $468.35 per S.F.

Common additives

Description	Unit	$ Cost	Description	Unit	$ Cost
Clock System			Emergency Lighting, 25 watt, battery operated		
20 room	Each	16,000	Nickel cadmium	Each	845
50 room	Each	39,100	Flagpoles, Complete		
Closed Circuit Surveillance, One station			Aluminum, 20' high	Each	1700
Camera and monitor	Each	1875	40' High	Each	3600
For additional camera stations, add	Each	1025	70' High	Each	10,300
Elevators, Hydraulic passenger, 2 stops			Fiberglass, 23' High	Each	1825
1500# capacity	Each	60,900	39'-5" High	Each	3475
2500# capacity	Each	64,300	59' High	Each	6625
3500# capacity	Each	67,600	Laundry Equipment		
Additional stop, add	Each	8250	Folders, blankets, & sheets	Each	67,000
Emergency Generators, Complete system, gas			Ironers, 110" single roll	Each	35,600
15 KW	Each	17,100	Combination washer & extractor, 50#	Each	12,300
70 KW	Each	32,400	125#	Each	33,100
85 KW	Each	36,000	Safe, Office type, 1 hour rating		
115 KW	Each	69,500	30" x 18" x 18"	Each	2400
170 KW	Each	91,000	60" x 36" x 18", double door	Each	8975
Diesel, 50 KW	Each	31,200	Sound System		
100 KW	Each	44,600	Amplifier, 250 watts	Each	2400
150 KW	Each	54,000	Speaker, ceiling or wall	Each	196
350 KW	Each	85,000	Trumpet	Each	375

Important: See the Reference Section for Location Factors

Model costs calculated for a 3 story building with 12' story height and 40,000 square feet of floor area

				Unit	Unit Cost	Cost Per S.F.	% Of Sub-Total
A.	**SUBSTRUCTURE**						
1010	Standard Foundations	Poured concrete; strip and spread footings		S.F. Ground	1.95	.65	
1020	Special Foundations	N/A		—	—	—	
1030	Slab on Grade	4" reinforced concrete with vapor barrier and granular base		S.F. Slab	4.77	1.59	1.5%
2010	Basement Excavation	Site preparation for slab and trench for foundation wall and footing		S.F. Ground	.31	.10	
2020	Basement Walls	4' foundation wall		L.F. Wall	67	.90	
B.	**SHELL**						
	B10 Superstructure						
1010	Floor Construction	Concrete slab with metal deck and beams		S.F. Floor	26.82	17.88	11.9%
1020	Roof Construction	Concrete slab with metal deck and beams		S.F. Roof	21.24	7.08	
	B20 Exterior Enclosure						
2010	Exterior Walls	Face brick with reinforced concrete block backup	85% of wall	S.F. Wall	30.29	12.28	
2020	Exterior Windows	Bullet resisting and metal horizontal pivoted	15% of wall	Each	149	10.66	11.0%
2030	Exterior Doors	Metal		Each	6000	.15	
	B30 Roofing						
3010	Roof Coverings	Built-up tar and gravel with flashing; perlite/EPS composite insulation		S.F. Roof	5.91	1.97	1.0%
3020	Roof Openings	Roof hatches		S.F. Roof	.09	.03	
C.	**INTERIORS**						
1010	Partitions	Concrete block	45 S.F. of Floor/L.F. Partition	S.F. Partition	9.60	2.56	
1020	Interior Doors	Hollow metal fire doors	930 S.F. Floor/Door	Each	1001	1.08	
1030	Fittings	N/A		—	—	—	
2010	Stair Construction	Concrete filled metal pan		Flight	14,675	1.84	6.0%
3010	Wall Finishes	Paint		S.F. Surface	3.79	2.02	
3020	Floor Finishes	70% vinyl composition tile, 20% carpet, 10% ceramic tile		S.F. Floor	3.13	3.13	
3030	Ceiling Finishes	Mineral tile on zee runners	30% of area	S.F. Ceiling	6.51	1.95	
D.	**SERVICES**						
	D10 Conveying						
1010	Elevators & Lifts	One hydraulic passenger elevator		Each	157,600	3.94	1.9%
1020	Escalators & Moving Walks	N/A		—	—	—	
	D20 Plumbing						
2010	Plumbing Fixtures	Toilet and service fixtures, supply and drainage	1 Fixture/78 S.F. Floor	Each	3342	42.85	
2020	Domestic Water Distribution	Electric water heater		S.F. Floor	5.76	5.76	23.7%
2040	Rain Water Drainage	Roof drains		S.F. Roof	3.12	1.04	
	D30 HVAC						
3010	Energy Supply	N/A		—	—	—	
3020	Heat Generating Systems	Included in D3050		—	—	—	
3030	Cooling Generating Systems	N/A		—	—	—	6.9 %
3050	Terminal & Package Units	Rooftop multizone unit systems		S.F. Floor	14.44	14.44	
3090	Other HVAC Sys. & Equipment	N/A		—	—	—	
	D40 Fire Protection						
4010	Sprinklers	Sprinkler system, light hazard		S.F. Floor	2.74	2.74	1.6%
4020	Standpipes	Standpipes and hose systems		S.F. Floor	.52	.52	
	D50 Electrical						
5010	Electrical Service/Distribution	600 ampere service, panel board and feeders		S.F. Floor	1.21	1.21	
5020	Lighting & Branch Wiring	Fluorescent fixtures, receptacles, switches, A.C. and misc. power		S.F. Floor	9.43	9.43	7.4%
5030	Communications & Security	Addressable alarm systems, intercom, and emergency lighting		S.F. Floor	4.12	4.12	
5090	Other Electrical Systems	Emergency generator, 80 kW		S.F. Floor	.71	.71	
E.	**EQUIPMENT & FURNISHINGS**						
1010	Commercial Equipment	N/A		—	—	—	
1020	Institutional Equipment	Prefabricated cells, visitor cubicles		S.F. Floor	57	56.60	27.1 %
1030	Vehicular Equipment	N/A		—	—	—	
1090	Other Equipment	N/A		—	—	—	
F.	**SPECIAL CONSTRUCTION**						
1020	Integrated Construction	N/A		—	—	—	0.0%
1040	Special Facilities	N/A		—	—	—	
G.	**BUILDING SITEWORK**	**N/A**					

		Sub-Total	209.23	100%
CONTRACTOR FEES (General Requirements: 10%, Overhead: 5%, Profit: 10%)		25%	52.31	
ARCHITECT FEES		7%	18.31	
		Total Building Cost	279.85	

Costs per square foot of floor area

Exterior Wall	S.F. Area	1000	2000	3000	4000	5000	10000	15000	20000	25000
	L.F. Perimeter	126	179	219	253	283	400	490	568	632
Decorative Concrete Block	Steel Frame	243.75	206.95	192.45	184.40	179.10	166.85	161.90	159.05	157.05
	Bearing Walls	242.75	205.90	191.45	183.35	178.05	165.90	160.85	158.05	156.05
Face Brick with Concrete Block Back-up	Steel Frame	274.65	229.85	211.70	201.50	194.75	178.90	172.35	168.55	165.85
	Bearing Walls	271.75	226.55	208.25	197.90	191.10	175.05	168.40	164.60	161.90
Metal Sandwich Panel	Steel Frame	248.20	211.25	196.65	188.50	183.20	170.95	165.90	163.05	161.05
Precast Concrete Panel	Bearing Walls	242.65	205.85	191.35	183.25	178.00	165.80	160.85	158.00	156.05
Perimeter Adj., Add or Deduct	Per 100 L.F.	52.40	26.20	17.45	13.05	10.50	5.25	3.45	2.60	2.15
Story Hgt. Adj., Add or Deduct	Per 1 Ft.	3.55	2.55	2.00	1.70	1.60	1.20	0.90	0.80	0.70
For Basement, add $30.85 per square foot of basement area										

The above costs were calculated using the basic specifications shown on the facing page. These costs should be adjusted where necessary for design alternatives and owner's requirements. Reported completed project costs, for this type of structure, range from $98.20 to $240.00 per S.F.

Common additives

Description	Unit	$ Cost
Closed Circuit Surveillance, One station		
Camera and monitor	Each	1875
For additional camera stations, add	Each	1025
Emergency Lighting, 25 watt, battery operated		
Lead battery	Each	287
Nickel cadmium	Each	845
Laundry Equipment		
Dryers, coin operated 30 lb.	Each	3600
Double stacked	Each	7200
50 lb.	Each	3900
Dry cleaner 20 lb.	Each	42,600
30 lb.	Each	60,500
Washers, coin operated	Each	1450
Washer/extractor 20 lb.	Each	6500
30 lb.	Each	10,500
50 lb.	Each	12,300
75 lb.	Each	23,600
Smoke Detectors		
Ceiling type	Each	233
Duct type	Each	525

Important: See the Reference Section for Location Factors

Model costs calculated for a 1 story building with 12' story height and 3,000 square feet of floor area

Laundromat

				Unit	Unit Cost	Cost Per S.F.	% Of Sub-Total
A.	**SUBSTRUCTURE**						
1010	Standard Foundations	Poured concrete; strip and spread footings		S.F. Ground	3.01	3.01	
1020	Special Foundations	N/A		—	—	—	
1030	Slab on Grade	5" reinforced concrete with vapor barrier and granular base		S.F. Slab	5.17	5.17	9.5%
2010	Basement Excavation	Site preparation for slab and trench for foundation wall and footing		S.F. Ground	.52	.52	
2020	Basement Walls	4' Foundation wall		L.F. Wall	67	4.92	
B.	**SHELL**						
	B10 Superstructure						
1010	Floor Construction	N/A		—	—	—	5.7 %
1020	Roof Construction	Metal deck, open web steel joists, beams, columns		S.F. Roof	8.16	8.16	
	B20 Exterior Enclosure						
2010	Exterior Walls	Decorative concrete block	90% of wall	S.F. Wall	14.80	11.67	
2020	Exterior Windows	Store front	10% of wall	Each	38.90	3.41	11.5%
2030	Exterior Doors	Double aluminum and glass		Each	4325	1.44	
	B30 Roofing						
3010	Roof Coverings	Built-up tar and gravel with flashing; perlite/EPS composite insulation		S.F. Roof	9.17	9.17	6.4%
3020	Roof Openings	N/A		—	—	—	
C.	**INTERIORS**						
1010	Partitions	Gypsum board on metal studs	60 S.F. Floor/L.F. Partition	S.F. Partition	23.34	3.89	
1020	Interior Doors	Single leaf wood	750 S.F. Floor/Door	Each	576	.76	
1030	Fittings	N/A		—	—	—	
2010	Stair Construction	N/A		—	—	—	7.0%
3010	Wall Finishes	Paint		S.F. Surface	.75	.25	
3020	Floor Finishes	Vinyl composition tile		S.F. Floor	1.93	1.93	
3030	Ceiling Finishes	Fiberglass board on exposed grid system		S.F. Ceiling	3.28	3.28	
D.	**SERVICES**						
	D10 Conveying						
1010	Elevators & Lifts	N/A		—	—	—	0.0 %
1020	Escalators & Moving Walks	N/A		—	—	—	
	D20 Plumbing						
2010	Plumbing Fixtures	Toilet and service fixtures, supply and drainage	1 Fixture/600 S.F. Floor	Each	5292	8.82	
2020	Domestic Water Distribution	Gas fired hot water heater		S.F. Floor	35.94	35.94	31.8%
2040	Rain Water Drainage	Roof drains		S.F. Roof	1	1	
	D30 HVAC						
3010	Energy Supply	N/A		—	—	—	
3020	Heat Generating Systems	Included in D3050		—	—	—	
3030	Cooling Generating Systems	N/A		—	—	—	6.5 %
3050	Terminal & Package Units	Rooftop single zone unit systems		S.F. Floor	9.32	9.32	
3090	Other HVAC Sys. & Equipment	N/A		—	—	—	
	D40 Fire Protection						
4010	Sprinklers	Sprinkler, ordinary hazard		S.F. Floor	4.35	4.35	3.0%
4020	Standpipes	N/A		—	—	—	
	D50 Electrical						
5010	Electrical Service/Distribution	200 ampere service, panel board and feeders		S.F. Floor	3.57	3.57	
5020	Lighting & Branch Wiring	Fluorescent fixtures, receptacles, switches, A.C. and misc. power		S.F. Floor	22.82	22.82	18.7%
5030	Communications & Security	Alarm systems and emergency lighting		S.F. Floor			
5090	Other Electrical Systems	Emergency generator, 7.5 kW		S.F. Floor	.48	.48	
E.	**EQUIPMENT & FURNISHINGS**						
1010	Commercial Equipment	N/A		—	—	—	
1020	Institutional Equipment	N/A		—	—	—	0.0 %
1030	Vehicular Equipment	N/A		—	—	—	
1090	Other Equipment	N/A		—	—	—	
F.	**SPECIAL CONSTRUCTION**						
1020	Integrated Construction	N/A		—	—	—	0.0 %
1040	Special Facilities	N/A		—	—	—	
G.	**BUILDING SITEWORK**	**N/A**					

	Sub-Total	143.88	**100%**
CONTRACTOR FEES (General Requirements: 10%, Overhead: 5%, Profit: 10%)	25%	35.98	
ARCHITECT FEES	7%	12.59	
Total Building Cost		192.45	

Costs per square foot of floor area

Exterior Wall	S.F. Area	7000	10000	13000	16000	19000	22000	25000	28000	31000
	L.F. Perimeter	240	300	336	386	411	435	472	510	524
Face Brick with Concrete Block Back-up	R/Conc. Frame	177.05	166.95	158.60	154.80	150.10	146.70	144.80	143.50	141.15
	Steel Frame	174.65	164.55	156.20	152.40	147.70	144.30	142.40	141.05	138.75
Limestone with Concrete Block	R/Conc. Frame	216.10	201.20	188.10	182.30	174.75	169.20	166.30	164.20	160.40
	Steel Frame	213.70	198.70	185.65	179.90	172.35	166.80	163.95	161.80	158.00
Precast Concrete Panels	R/Conc. Frame	182.20	171.45	162.45	158.45	153.35	149.65	147.65	146.20	143.65
	Steel Frame	179.80	169.05	160.05	156.05	150.90	147.25	145.25	143.80	141.25
Perimeter Adj., Add or Deduct	Per 100 L.F.	22.35	15.70	12.05	9.80	8.25	7.10	6.30	5.60	5.05
Story Hgt. Adj., Add or Deduct	Per 1 Ft.	3.35	3.00	2.60	2.45	2.20	1.95	1.90	1.80	1.65
For Basement, add $44.15 per square foot of basement area										

The above costs were calculated using the basic specifications shown on the facing page. These costs should be adjusted where necessary for design alternatives and owner's requirements. Reported completed project costs, for this type of structure, range from $94.35 to $242.10 per S.F.

Common additives

Description	Unit	$ Cost
Carrels Hardwood	Each	605 - 1150
Closed Circuit Surveillance, One station		
Camera and monitor	Each	1875
For additional camera stations, add	Each	1025
Elevators, Hydraulic passenger, 2 stops		
1500# capacity	Each	60,900
2500# capacity	Each	64,300
3500# capacity	Each	67,600
Emergency Lighting, 25 watt, battery operated		
Lead battery	Each	287
Nickel cadmium	Each	845
Flagpoles, Complete		
Aluminum, 20' high	Each	1700
40' high	Each	3600
70' high	Each	10,300
Fiberglass, 23' high	Each	1825
39'-5" high	Each	3475
59' high	Each	6625

Description	Unit	$ Cost
Library Furnishings		
Bookshelf, 90" high, 10" shelf double face	L.F.	175
single face	L.F.	167
Charging desk, built-in with counter		
Plastic laminated top	L.F.	530
Reading table, laminated		
top 60" x 36"	Each	720

Model costs calculated for a 2 story building with 14' story height and 22,000 square feet of floor area

Library

				Unit	Unit Cost	Cost Per S.F.	% Of Sub-Total
A.	**SUBSTRUCTURE**						
1010	Standard Foundations	Poured concrete; strip and spread footings		S.F. Ground	3.54	1.77	
1020	Special Foundations	N/A		—	—	—	
1030	Slab on Grade	4" reinforced concrete with vapor barrier and granular base		S.F. Slab	4.77	2.39	5.3%
2010	Basement Excavation	Site preparation for slab and trench for foundation wall and footing		S.F. Ground	.31	.16	
2020	Basement Walls	4' foundation wall		L.F. Wall	76	1.49	
B.	**SHELL**						
	B10 Superstructure						
1010	Floor Construction	Concrete waffle slab		S.F. Floor	24.80	12.40	19.7%
1020	Roof Construction	Concrete waffle slab		S.F. Roof	17.96	8.98	
	B20 Exterior Enclosure						
2010	Exterior Walls	Face brick with concrete block backup	90% of wall	S.F. Wall	31.27	15.58	
2020	Exterior Windows	Window wall	10% of wall	Each	48.10	2.67	17.2%
2030	Exterior Doors	Double aluminum and glass, single leaf hollow metal		Each	5275	.48	
	B30 Roofing						
3010	Roof Coverings	Single ply membrane, EPDM, fully adhered; perlite/EPS composite insulation		S.F. Roof	5.02	2.51	2.3%
3020	Roof Openings	Roof hatches		S.F. Roof	.08	.04	
C.	**INTERIORS**						
1010	Partitions	Gypsum board on metal studs	30 S.F. Floor/L.F. Partition	S.F. Partition	10.58	4.23	
1020	Interior Doors	Single leaf wood	300 S.F. Floor/Door	Each	576	1.92	
1030	Fittings	N/A		—	—	—	
2010	Stair Construction	Concrete filled metal pan		Flight	8575	.78	16.5%
3010	Wall Finishes	Paint		S.F. Surface	.75	.60	
3020	Floor Finishes	50% carpet, 50% vinyl tile		S.F. Floor	3.85	3.85	
3030	Ceiling Finishes	Mineral fiber on concealed zee bars		S.F. Ceiling	6.51	6.51	
D.	**SERVICES**						
	D10 Conveying						
1010	Elevators & Lifts	One hydraulic passenger elevator		Each	82,280	3.74	3.4%
1020	Escalators & Moving Walks	N/A		—	—	—	
	D20 Plumbing						
2010	Plumbing Fixtures	Toilet and service fixtures, supply and drainage	1 Fixture/1835 S.F. Floor	Each	2771	1.51	
2020	Domestic Water Distribution	Gas fired water heater		S.F. Floor	1.02	1.02	2.7%
2040	Rain Water Drainage	Roof drains		S.F. Roof	.80	.40	
	D30 HVAC						
3010	Energy Supply	N/A		—	—	—	
3020	Heat Generating Systems	Included in D3050		—	—	—	
3030	Cooling Generating Systems	N/A		—	—	—	17.9 %
3050	Terminal & Package Units	Multizone unit, gas heating, electric cooling		S.F. Floor	19.45	19.45	
3090	Other HVAC Sys. & Equipment	N/A		—	—	—	
	D40 Fire Protection						
4010	Sprinklers	Wet pipe sprinkler system		S.F. Floor	2.86	2.86	2.6%
4020	Standpipes	N/A		—	—	—	
	D50 Electrical						
5010	Electrical Service/Distribution	400 ampere service, panel board and feeders		S.F. Floor	1.06	1.06	
5020	Lighting & Branch Wiring	Fluorescent fixtures, receptacles, switches, A.C. and misc. power		S.F. Floor	10.63	10.63	12.3%
5030	Communications & Security	Alarm systems, internet wiring, and emergency lighting		S.F. Floor	1.39	1.39	
5090	Other Electrical Systems	Emergency generator, 7.5 kW, Uninterruptible power supply		S.F. Floor	.24	.24	
E.	**EQUIPMENT & FURNISHINGS**						
1010	Commercial Equipment	N/A		—	—	—	
1020	Institutional Equipment	N/A		—	—	—	
1030	Vehicular Equipment	N/A		—	—	—	0.0 %
1090	Other Equipment	N/A		—	—	—	
F.	**SPECIAL CONSTRUCTION**						
1020	Integrated Construction	N/A		—	—	—	
1040	Special Facilities	N/A		—	—	—	0.0 %
G.	**BUILDING SITEWORK**	**N/A**					

			Sub-Total	108.66	100%
	CONTRACTOR FEES (General Requirements: 10%, Overhead: 5%, Profit: 10%)		25%	27.17	
	ARCHITECT FEES		8%	10.87	
		Total Building Cost		**146.70**	

Costs per square foot of floor area

Exterior Wall	S.F. Area	4000	5500	7000	8500	10000	11500	13000	14500	16000
	L.F. Perimeter	280	320	380	440	453	503	510	522	560
Face Brick with Concrete Block Back-up	Steel Joists	201.40	192.75	189.65	187.70	183.15	182.05	178.90	176.70	175.95
	Wood Truss	208.90	200.30	197.20	195.20	190.65	189.50	186.40	184.25	183.45
Stucco on Concrete Block	Steel Joists	189.55	182.95	180.50	178.95	175.55	174.65	172.30	170.60	170.05
	Wood Truss	197.10	190.45	188.00	186.45	183.05	182.15	179.80	178.15	177.55
Brick Veneer	Wood Truss	201.25	193.50	190.70	188.90	184.90	183.90	181.10	179.15	178.40
Wood Siding	Wood Frame	193.65	187.55	185.30	183.90	180.80	180.00	177.85	176.35	175.80
Perimeter Adj., Add or Deduct	Per 100 L.F.	16.25	11.85	9.30	7.60	6.50	5.65	5.05	4.50	4.05
Story Hgt. Adj., Add or Deduct	Per 1 Ft.	3.35	2.80	2.60	2.45	2.20	2.10	1.90	1.75	1.65
For Basement, add $28.75 per square foot of basement area										

The above costs were calculated using the basic specifications shown on the facing page. These costs should be adjusted where necessary for design alternatives and owner's requirements. Reported completed project costs, for this type of structure, range from $79.50 to $204.65 per S.F.

Common additives

Description	Unit	$ Cost	Description	Unit	$ Cost
Cabinets, Hospital, base			Directory Boards, Plastic, glass covered		
Laminated plastic	L.F.	375	30" x 20"	Each	605
Stainless steel	L.F.	600	36" x 48"	Each	1325
Counter top, laminated plastic	L.F.	72.50	Aluminum, 24" x 18"	Each	585
Stainless steel	L.F.	177	36" x 24"	Each	685
For drop-in sink, add	Each	995	48" x 32"	Each	975
Nurses station, door type			48" x 60"	Each	2025
Laminated plastic	L.F.	415	Heat Therapy Unit		
Enameled steel	L.F.	400	Humidified, 26" x 78" x 28"	Each	3725
Stainless steel	L.F.	700	Smoke Detectors		
Wall cabinets, laminated plastic	L.F.	272	Ceiling type	Each	233
Enameled steel	L.F.	320	Duct type	Each	525
Stainless steel	L.F.	565	Tables, Examining, vinyl top		
			with base cabinets	Each	1775 - 4750
			Utensil Washer, Sanitizer	Each	12,300
			X-Ray, Mobile	Each	14,700 - 83,000

Model costs calculated for a 1 story building with 10' story height and 7,000 square feet of floor area

Medical Office, 1 Story

				Unit	Unit Cost	Cost Per S.F.	% Of Sub-Total
A. SUBSTRUCTURE							
1010	Standard Foundations	Poured concrete; strip and spread footings		S.F. Ground	1.93	1.93	
1020	Special Foundations	N/A		—	—	—	
1030	Slab on Grade	4" reinforced concrete with vapor barrier and granular base		S.F. Slab	4.77	4.77	7.7%
2010	Basement Excavation	Site preparation for slab and trench for foundation wall and footing		S.F. Ground	.31	.31	
2020	Basement Walls	4' foundation wall		L.F. Wall	76	4.09	
B. SHELL							
B10 Superstructure							
1010	Floor Construction	N/A		—	—	—	4.5%
1020	Roof Construction	Plywood on wood trusses		S.F. Roof	6.46	6.46	
B20 Exterior Enclosure							
2010	Exterior Walls	Face brick with concrete block backup	70% of wall	S.F. Wall	31.26	11.88	
2020	Exterior Windows	Wood double hung	30% of wall	Each	592	5.67	13.9%
2030	Exterior Doors	Aluminum and glass doors and entrance with transoms		Each	1484	2.55	
B30 Roofing							
3010	Roof Coverings	Asphalt shingles with flashing (Pitched); rigid fiber glass insulation, gutters		S.F. Roof	4.14	4.14	2.9%
3020	Roof Openings	N/A		—	—	—	
C. INTERIORS							
1010	Partitions	Gypsum bd. & sound deadening bd. on wood studs w/insul.	6 S.F. Floor/L.F. Partition	S.F. Partition	8.29	11.05	
1020	Interior Doors	Single leaf wood	60 S.F. Floor/Door	Each	576	9.60	
1030	Fittings	N/A		—	—	—	
2010	Stair Construction	N/A		—	—	—	25.0%
3010	Wall Finishes	50% paint, 50% vinyl wall covering		S.F. Surface	1.28	3.40	
3020	Floor Finishes	50% carpet, 50% vinyl composition tile		S.F. Floor	6.60	6.60	
3030	Ceiling Finishes	Mineral fiber tile on concealed zee bars		S.F. Ceiling	5.52	5.52	
D. SERVICES							
D10 Conveying							
1010	Elevators & Lifts	N/A		—	—	—	0.0%
1020	Escalators & Moving Walks	N/A		—	—	—	
D20 Plumbing							
2010	Plumbing Fixtures	Toilet, exam room and service fixtures, supply and drainage	1 Fixture/195 S.F. Floor	Each	3569	18.30	
2020	Domestic Water Distribution	Gas fired water heater		S.F. Floor	2.87	2.87	15.2%
2040	Rain Water Drainage	Roof drains		S.F. Roof	.86	.86	
D30 HVAC							
3010	Energy Supply	N/A		—	—	—	
3020	Heat Generating Systems	Included in D3050		—	—	—	
3030	Cooling Generating Systems	N/A		—	—	—	9.5%
3050	Terminal & Package Units	Multizone unit, gas heating, electric cooling		S.F. Floor	13.75	13.75	
3090	Other HVAC Sys. & Equipment	N/A		—	—	—	
D40 Fire Protection							
4010	Sprinklers	Wet pipe sprinkler system		S.F. Floor	3.99	3.99	3.5%
4020	Standpipes	Standpipe		S.F. Floor	1.03	1.03	
D50 Electrical							
5010	Electrical Service/Distribution	200 ampere service, panel board and feeders		S.F. Floor	1.44	1.44	
5020	Lighting & Branch Wiring	High efficiency fluorescent fixtures, receptacles, switches, A.C. and misc. power		S.F. Floor	7.22	7.22	12.4%
5030	Communications & Security	Alarm systems, internet and phone wiring, intercom system, and emergency lighting		S.F. Floor	8.39	8.39	
5090	Other Electrical Systems	Emergency generator, 7.5 kW		S.F. Floor	.86	.86	
E. EQUIPMENT & FURNISHINGS							
1010	Commercial Equipment	N/A		—	—	—	
1020	Institutional Equipment	Exam room casework and countertops		S.F. Floor	8.02	8.02	5.5%
1030	Vehicular Equipment	N/A		—	—	—	
1090	Other Equipment	N/A		—	—	—	
F. SPECIAL CONSTRUCTION							
1020	Integrated Construction	N/A		—	—	—	0.0%
1040	Special Facilities	N/A		—	—	—	
G. BUILDING SITEWORK	**N/A**						

		Sub-Total	144.70	**100%**
CONTRACTOR FEES (General Requirements: 10%, Overhead: 5%, Profit: 10%)			25%	36.22
ARCHITECT FEES			9%	16.28
		Total Building Cost	**197.20**	

Costs per square foot of floor area

Exterior Wall	S.F. Area	4000	5500	7000	8500	10000	11500	13000	14500	16000
	L.F. Perimeter	180	210	240	270	286	311	336	361	386
Face Brick with Concrete Block Back-up	Steel Joists	240.90	229.70	223.30	219.15	214.65	212.25	210.35	208.90	207.65
	Wood Joists	246.60	235.40	228.95	224.90	220.35	217.95	216.05	214.60	213.30
Stucco on Concrete Block	Steel Joists	230.05	220.45	214.95	211.40	207.65	205.60	204.00	202.75	201.70
	Wood Joists	235.35	225.90	220.40	216.90	213.20	211.15	209.60	208.35	207.25
Brick Veneer	Wood Frame	240.55	230.35	224.40	220.65	216.50	214.35	212.60	211.25	210.15
Wood Siding	Wood Frame	230.25	221.55	216.50	213.35	209.95	208.10	206.65	205.50	204.60
Perimeter Adj., Add or Deduct	Per 100 L.F.	28.35	20.65	16.20	13.35	11.35	9.85	8.75	7.80	7.10
Story Hgt. Adj., Add or Deduct	Per 1 Ft.	4.20	3.55	3.15	2.95	2.65	2.50	2.40	2.30	2.25
For Basement, add $31.90 per square foot of basement area										

The above costs were calculated using the basic specifications shown on the facing page. These costs should be adjusted where necessary for design alternatives and owner's requirements. Reported completed project costs, for this type of structure, range from $81.60 to $278.75 per S.F.

Common additives

Description	Unit	$ Cost	Description	Unit	$ Cost
Cabinets, Hospital, base			Directory Boards, Plastic, glass covered		
Laminated plastic	L.F.	375	30" x 20"	Each	605
Stainless steel	L.F.	600	36" x 48"	Each	1325
Counter top, laminated plastic	L.F.	72.50	Aluminum, 24" x 18"	Each	585
Stainless steel	L.F.	177	36" x 24"	Each	685
For drop-in sink, add	Each	995	48" x 32"	Each	975
Nurses station, door type			48" x 60"	Each	2025
Laminated plastic	L.F.	415	Emergency Lighting, 25 watt, battery operated		
Enameled steel	L.F.	400	Lead battery	Each	287
Stainless steel	L.F.	700	Nickel cadmium	Each	845
Wall cabinets, laminated plastic	L.F.	272	Heat Therapy Unit		
Enameled steel	L.F.	320	Humidified, 26" x 78" x 28"	Each	3725
Stainless steel	L.F.	565	Smoke Detectors		
Elevators, Hydraulic passenger, 2 stops			Ceiling type	Each	233
1500# capacity	Each	60,900	Duct type	Each	525
2500# capacity	Each	64,300	Tables, Examining, vinyl top		
3500# capacity	Each	67,600	with base cabinets	Each	1775 - 4750
			Utensil Washer, Sanitizer	Each	12,300
			X-Ray, Mobile	Each	14,700 - 83,000

Medical Office, 2 Story

Model costs calculated for a 2 story building with 10′ story height and 7,000 square feet of floor area

				Unit	Unit Cost	Cost Per S.F.	% Of Sub-Total
A.	**SUBSTRUCTURE**						
1010	Standard Foundations	Poured concrete; strip and spread footings		S.F. Ground	3.10	1.55	
1020	Special Foundations	N/A		—	—	—	
1030	Slab on Grade	4″ reinforced concrete with vapor barrier and granular base		S.F. Slab	4.77	2.39	4.2%
2010	Basement Excavation	Site preparation for slab and trench for foundation wall and footing		S.F. Ground	.18	.09	
2020	Basement Walls	4′ foundation wall		L.F. Wall	76	2.58	
B.	**SHELL**						
	B10 Superstructure						
1010	Floor Construction	Open web steel joists, slab form, concrete, columns		S.F. Floor	11.72	5.86	5.2%
1020	Roof Construction	Metal deck, open web steel joists, beams, columns		S.F. Roof	4.74	2.37	
	B20 Exterior Enclosure						
2010	Exterior Walls	Stucco on concrete block		S.F. Wall	17.90	8.59	
2020	Exterior Windows	Outward projecting metal	30% of wall	Each	497	6.82	10.7%
2030	Exterior Doors	Aluminum and glass doors with transoms		Each	5275	1.50	
	B30 Roofing						
3010	Roof Coverings	Built-up tar and gravel with flashing; perlite/EPS composite insulation		S.F. Roof	6.96	3.48	2.3%
3020	Roof Openings	Roof hatches		S.F. Roof	.24	.12	
C.	**INTERIORS**						
1010	Partitions	Gypsum bd. & acous. insul. on metal studs	6 S.F. Floor/L.F. Partition	S.F. Partition	6.94	9.25	
1020	Interior Doors	Single leaf wood	60 S.F. Floor/Door	Each	576	9.60	
1030	Fittings	N/A		—	—	—	
2010	Stair Construction	Concrete filled metal pan		Flight	12,275	3.50	24.0%
3010	Wall Finishes	45% paint, 50% vinyl wall coating, 5% ceramic tile		S.F. Surface	1.28	3.40	
3020	Floor Finishes	50% carpet, 50% vinyl composition tile		S.F. Floor	6.60	6.60	
3030	Ceiling Finishes	Mineral fiber tile on concealed zee bars		S.F. Ceiling	5.52	5.52	
D.	**SERVICES**						
	D10 Conveying						
1010	Elevators & Lifts	One hydraulic hospital elevator		Each	102,410	14.63	9.3%
1020	Escalators & Moving Walks	N/A		—	—	—	
	D20 Plumbing						
2010	Plumbing Fixtures	Toilet, exam room and service fixtures, supply and drainage	1 Fixture/160 S.F. Floor	Each	3301	20.63	
2020	Domestic Water Distribution	Gas fired water heater		S.F. Floor	2.87	2.87	15.5%
2040	Rain Water Drainage	Roof drains		S.F. Roof	.86	.86	
	D30 HVAC						
3010	Energy Supply	N/A		—	—	—	
3020	Heat Generating Systems	Included in D3050		—	—	—	
3030	Cooling Generating Systems	N/A		—	—	—	8.7%
3050	Terminal & Package Units	Multizone unit, gas heating, electric cooling		S.F. Floor	13.75	13.75	
3090	Other HVAC Sys. & Equipment	N/A		—	—	—	
	D40 Fire Protection						
4010	Sprinklers	Wet pipe sprinkler system		S.F. Floor	3.34	3.34	3.1%
4020	Standpipes	Standpipe		S.F. Floor	1.55	1.55	
	D50 Electrical						
5010	Electrical Service/Distribution	400 ampere service, panel board and feeders		S.F. Floor	2.35	2.35	
5020	Lighting & Branch Wiring	High efficiency fluorescent fixtures, receptacles, switches, A.C. and misc. power		S.F. Floor	7.22	7.22	11.9%
5030	Communications & Security	Alarm system, internet and phone wiring, intercom system, and emergency lighting		S.F. Floor	8.39	8.39	
5090	Other Electrical Systems	Emergency generator, 7.5 kW		S.F. Floor	.86	.86	
E.	**EQUIPMENT & FURNISHINGS**						
1010	Commercial Equipment	N/A		—	—	—	
1020	Institutional Equipment	Exam room casework and contertops		S.F. Floor	8.02	8.02	5.1%
1030	Vehicular Equipment	N/A		—	—	—	
1090	Other Equipment	N/A		—	—	—	
F.	**SPECIAL CONSTRUCTION**						
1020	Integrated Construction	N/A		—	—	—	0.0%
1040	Special Facilities	N/A		—	—	—	
G.	**BUILDING SITEWORK**	**N/A**					

		Sub-Total	144.70	100%
	CONTRACTOR FEES (General Requirements: 10%, Overhead: 5%, Profit: 10%)	25%	39.45	
	ARCHITECT FEES	9%	17.75	
	Total Building Cost		**214.95**	

Costs per square foot of floor area

Exterior Wall	S.F. Area	2000	3000	4000	6000	8000	10000	12000	14000	16000
	L.F. Perimeter	240	260	280	380	480	560	580	660	740
Brick Veneer	Wood Frame	189.00	165.45	153.70	147.35	144.30	141.30	136.55	135.50	134.75
Aluminum Siding	Wood Frame	171.35	152.65	143.35	138.05	135.45	133.10	129.45	128.60	127.90
Wood Siding	Wood Frame	171.80	152.95	143.60	138.25	135.70	133.30	129.60	128.75	128.05
Wood Shingles	Wood Frame	175.30	155.45	145.65	140.05	137.35	134.85	130.95	130.05	129.35
Precast Concrete Block	Wood Truss	171.00	152.45	143.15	137.90	135.25	132.95	129.30	128.45	127.80
Brick on Concrete Block	Wood Truss	197.10	171.30	158.35	151.65	148.30	145.10	139.80	138.70	137.85
Perimeter Adj., Add or Deduct	Per 100 L.F.	27.65	18.45	13.75	9.25	6.90	5.55	4.60	3.95	3.45
Story Hgt. Adj., Add or Deduct	Per 1 Ft.	4.55	3.20	2.60	2.45	2.20	2.15	1.80	1.80	1.75
For Basement, add $21.95 per square foot of basement area										

The above costs were calculated using the basic specifications shown on the facing page. These costs should be adjusted where necessary for design alternatives and owner's requirements. Reported completed project costs, for this type of structure, range from $65.85 to $311.80 per S.F.

Common additives

Description	Unit	$ Cost
Closed Circuit Surveillance, One station		
Camera and monitor	Each	1875
For additional camera stations, add	Each	1025
Emergency Lighting, 25 watt, battery operated		
Lead battery	Each	287
Nickel cadmium	Each	845
Laundry Equipment		
Dryer, gas, 16 lb. capacity	Each	900
30 lb. capacity	Each	3600
Washer, 4 cycle	Each	1100
Commercial	Each	1450
Sauna, Prefabricated, complete		
6' x 4'	Each	5875
6' x 6'	Each	6975
6' x 9'	Each	8550
8' x 8'	Each	10,100
8' x 10'	Each	11,300
10' x 12'	Each	14,000
Smoke Detectors		
Ceiling type	Each	233
Duct type	Each	525

Description	Unit	$ Cost
Swimming Pools, Complete, gunite	S.F.	68.50 - 84.50
TV Antenna, Master system, 12 outlet	Outlet	320
30 outlet	Outlet	207
100 outlet	Outlet	199

Model costs calculated for a 1 story building with 9′ story height and 8,000 square feet of floor area

Motel, 1 Story

				Unit	Unit Cost	Cost Per S.F.	% Of Sub-Total
A.	**SUBSTRUCTURE**						
1010	Standard Foundations	Poured concrete; strip and spread footings		S.F. Ground	2.93	2.93	
1020	Special Foundations	N/A		—	—	—	
1030	Slab on Grade	4″ reinforced concrete with vapor barrier and granular base		S.F. Slab	4.77	4.77	12.9%
2010	Basement Excavation	Site preparation for slab and trench for foundation wall and footing		S.F. Ground	.31	.31	
2020	Basement Walls	4′ foundation wall		L.F. Wall	72	5.88	
B.	**SHELL**						
	B10 Superstructure						
1010	Floor Construction	N/A		—	—	—	6.0%
1020	Roof Construction	Plywood on wood trusses		S.F. Roof	6.46	6.46	
	B20 Exterior Enclosure						
2010	Exterior Walls	Face brick on wood studs with sheathing, insulation and paper	80% of wall	S.F. Wall	23.31	10.07	
2020	Exterior Windows	Wood double hung	20% of wall	Each	478	3.69	16.5%
2030	Exterior Doors	Wood solid core		Each	1484	4.08	
	B30 Roofing						
3010	Roof Coverings	Asphalt shingles with flashing (pitched); rigid fiberglass insulation		S.F. Roof	3.97	3.97	3.7%
3020	Roof Openings	N/A		—	—	—	
C.	**INTERIORS**						
1010	Partitions	Gypsum bd. and sound deadening bd. on wood studs	9 S.F. Floor/L.F. Partition	S.F. Partition	8.87	7.88	
1020	Interior Doors	Single leaf hollow core wood	300 S.F. Floor/Door	Each	498	1.66	
1030	Fittings	N/A		—	—	—	
2010	Stair Construction	N/A		—	—	—	23.8%
3010	Wall Finishes	90% paint, 10% ceramic tile		S.F. Surface	1.36	2.41	
3020	Floor Finishes	85% carpet, 15% ceramic tile		S.F. Floor	9.82	9.82	
3030	Ceiling Finishes	Painted gypsum board on furring		S.F. Ceiling	3.95	3.95	
D.	**SERVICES**						
	D10 Conveying						
1010	Elevators & Lifts	N/A		—	—	—	0.0%
1020	Escalators & Moving Walks	N/A		—	—	—	
	D20 Plumbing						
2010	Plumbing Fixtures	Toilet and service fixtures, supply and drainage	1 Fixture/90 S.F. Floor	Each	1647	18.30	
2020	Domestic Water Distribution	Gas fired water heater		S.F. Floor	1.60	1.60	19.6%
2040	Rain Water Drainage	Roof drains		S.F. Roof	1.27	1.27	
	D30 HVAC						
3010	Energy Supply	N/A		—	—	—	
3020	Heat Generating Systems	Included in D3050		—	—	—	
3030	Cooling Generating Systems	N/A		—	—	—	2.4%
3050	Terminal & Package Units	Through the wall electric heating and cooling units		S.F. Floor	2.58	2.58	
3090	Other HVAC Sys. & Equipment	N/A		—	—	—	
	D40 Fire Protection						
4010	Sprinklers	Wet pipe sprinkler system		S.F. Floor	3.99	3.99	4.5%
4020	Standpipes	Standpipe, wet, Class III		S.F. Floor	.90	.90	
	D50 Electrical						
5010	Electrical Service/Distribution	200 ampere service, panel board and feeders		S.F. Floor	1.26	1.26	
5020	Lighting & Branch Wiring	Fluorescent fixtures, receptacles, switches and misc. power		S.F. Floor	6.61	6.61	9.7%
5030	Communications & Security	Addressable alarm systems		S.F. Floor	2.57	2.57	
5090	Other Electrical Systems	N/A		—	—	—	
E.	**EQUIPMENT & FURNISHINGS**						
1010	Commercial Equipment	Laundry equipment		S.F. Floor	.91	.91	
1020	Institutional Equipment	N/A		—	—	—	
1030	Vehicular Equipment	N/A		—	—	—	0.8%
1090	Other Equipment	N/A		—	—	—	
F.	**SPECIAL CONSTRUCTION**						
1020	Integrated Construction	N/A		—	—	—	0.0%
1040	Special Facilities	N/A		—	—	—	
G.	**BUILDING SITEWORK**	**N/A**					

					Sub-Total	107.87	100%
	CONTRACTOR FEES (General Requirements: 10%, Overhead: 5%, Profit: 10%)				25%	26.99	
	ARCHITECT FEES				7%	9.44	

				Total Building Cost	**144.30**

Costs per square foot of floor area

Exterior Wall	S.F. Area	25000	37000	49000	61000	73000	81000	88000	96000	104000
	L.F. Perimeter	433	593	606	720	835	911	978	1054	1074
Decorative Concrete Block	Wood Joists	153.75	150.60	145.65	144.40	143.70	143.25	142.95	142.70	141.85
	Precast Conc.	166.90	162.80	157.35	155.85	155.00	154.45	154.10	153.75	152.90
Stucco on Concrete Block	Wood Joists	155.75	151.30	145.55	143.95	143.00	142.40	142.00	141.60	140.70
	Precast Conc.	168.05	163.60	157.85	156.25	155.25	154.70	154.30	153.95	153.00
Wood Siding	Wood Frame	150.70	147.75	143.25	142.15	141.50	141.05	140.80	140.55	139.80
Brick Veneer	Wood Frame	158.95	154.30	147.80	146.15	145.10	144.50	144.05	143.70	142.65
Perimeter Adj., Add or Deduct	Per 100 L.F.	4.50	3.00	2.25	1.85	1.50	1.35	1.30	1.10	1.10
Story Hgt. Adj., Add or Deduct	Per 1 Ft.	1.55	1.45	1.05	1.05	1.00	1.00	1.00	0.95	0.90

For Basement, add $28.80 per square foot of basement area

The above costs were calculated using the basic specifications shown on the facing page. These costs should be adjusted where necessary for design alternatives and owner's requirements. Reported completed project costs, for this type of structure, range from $60.65 to $311.80 per S.F.

Common additives

Description	Unit	$ Cost	Description	Unit	$ Cost
Closed Circuit Surveillance, One station			Sauna, Prefabricated, complete		
Camera and monitor	Each	1875	6' x 4'	Each	5875
For additional camera station, add	Each	1025	6' x 6'	Each	6975
Elevators, Hydraulic passenger, 2 stops			6' x 9'	Each	8550
1500# capacity	Each	60,900	8' x 8'	Each	10,100
2500# capacity	Each	64,300	8' x 10'	Each	11,300
3500# capacity	Each	67,600	10' x 12'	Each	14,000
Additional stop, add	Each	8250	Smoke Detectors		
Emergency Lighting, 25 watt, battery operated			Ceiling type	Each	233
Lead battery	Each	287	Duct type	Each	525
Nickel cadmium	Each	845	Swimming Pools, Complete, gunite	S.F.	68.50 - 84.50
Laundry Equipment			TV Antenna, Master system, 12 outlet	Outlet	320
Dryer, gas, 16 lb. capacity	Each	900	30 outlet	Outlet	207
30 lb. capacity	Each	3600	100 outlet	Outlet	199
Washer, 4 cycle	Each	1100			
Commercial	Each	1450			

Model costs calculated for a 3 story building with 9' story height and 49,000 square feet of floor area

				Unit	Unit Cost	Cost Per S.F.	% Of Sub-Total
A. SUBSTRUCTURE							
1010	Standard Foundations	Poured concrete; strip and spread footings		S.F. Ground	1.35	.45	
1020	Special Foundations	N/A		—	—	—	
1030	Slab on Grade	4" reinforced concrete with vapor barrier and granular base		S.F. Slab	4.77	1.59	3.3%
2010	Basement Excavation	Site preparation for slab and trench for foundation wall and footing		S.F. Ground	.18	.06	
2020	Basement Walls	4' foundation wall		L.F. Wall	76	1.87	
B. SHELL							
	B10 Superstructure						
1010	Floor Construction	Precast concrete plank		S.F. Floor	10.35	6.90	8.6%
1020	Roof Construction	Precast concrete plank		S.F. Roof	9.93	3.31	
	B20 Exterior Enclosure						
2010	Exterior Walls	Decorative concrete block	85% of wall	S.F. Wall	15.85	4.50	
2020	Exterior Windows	Aluminum sliding	15% of wall	Each	557	1.86	10.4%
2030	Exterior Doors	Aluminum and glass doors and entrance with transom		Each	1802	5.97	
	B30 Roofing						
3010	Roof Coverings	Built-up tar and gravel with flashing; perlite/EPS composite insulation		S.F. Roof	6.27	2.09	1.8%
3020	Roof Openings	Roof hatches		S.F. Roof	.12	.04	
C. INTERIORS							
1010	Partitions	Concrete block	7 S.F. Floor/L.F. Partition	S.F. Partition	17.55	20.06	
1020	Interior Doors	Wood hollow core	70 S.F. Floor/Door	Each	498	7.11	
1030	Fittings	N/A		—	—	—	
2010	Stair Construction	Concrete filled metal pan		Flight	12,275	3.01	39.4%
3010	Wall Finishes	90% paint, 10% ceramic tile		S.F. Surface	1.36	3.10	
3020	Floor Finishes	85% carpet, 5% vinyl composition tile, 10% ceramic tile		S.F. Floor	9.82	9.82	
3030	Ceiling Finishes	Textured finish		S.F. Ceiling	3.67	3.67	
D. SERVICES							
	D10 Conveying						
1010	Elevators & Lifts	Two hydraulic passenger elevators		Each	102,900	4.20	3.5%
1020	Escalators & Moving Walks	N/A		—	—	—	
	D20 Plumbing						
2010	Plumbing Fixtures	Toilet and service fixtures, supply and drainage	1 Fixture/180 S.F. Floor	Each	3778	20.99	
2020	Domestic Water Distribution	Gas fired water heater		S.F. Floor	1.05	1.05	18.9%
2040	Rain Water Drainage	Roof drains		S.F. Roof	1.14	.38	
	D30 HVAC						
3010	Energy Supply	N/A		—	—	—	
3020	Heat Generating Systems	Included in D3050		—	—	—	
3030	Cooling Generating Systems	N/A		—	—	—	2.0 %
3050	Terminal & Package Units	Through the wall electric heating and cooling units		S.F. Floor	2.33	2.33	
3090	Other HVAC Sys. & Equipment	N/A		—	—	—	
	D40 Fire Protection						
4010	Sprinklers	Sprinklers, wet, light hazard		S.F. Floor	2.74	2.74	2.5%
4020	Standpipes	Standpipe, wet, Class III		S.F. Floor	.20	.20	
	D50 Electrical						
5010	Electrical Service/Distribution	800 ampere service, panel board and feeders		S.F. Floor	1.96	1.96	
5020	Lighting & Branch Wiring	Fluorescent fixtures, receptacles, switches and misc. power		S.F. Floor	7.31	7.31	9.4%
5030	Communications & Security	Addressable alarm systems and emergency lighting		S.F. Floor	1.76	1.76	
5090	Other Electrical Systems	Emergency generator, 7.5 kW		S.F. Floor	.12	.12	
E. EQUIPMENT & FURNISHINGS							
1010	Commercial Equipment	Commercial laundry equipment		S.F. Floor	.30	.30	
1020	Institutional Equipment	N/A		—	—	—	
1030	Vehicular Equipment	N/A		—	—	—	0.3%
1090	Other Equipment	N/A		—	—	—	
F. SPECIAL CONSTRUCTION							
1020	Integrated Construction	N/A		—	—	—	
1040	Special Facilities	N/A		—	—	—	0.0 %
G. BUILDING SITEWORK	**N/A**						

			Sub-Total	118.75	100%
	CONTRACTOR FEES (General Requirements: 10%, Overhead: 5%, Profit: 10%)		25%	29.69	
	ARCHITECT FEES		6%	8.91	
			Total Building Cost	**157.35**	

Costs per square foot of floor area

Exterior Wall	S.F. Area	9000	10000	12000	13000	14000	15000	16000	18000	20000
	L.F. Perimeter	385	410	440	460	480	500	510	547	583
Decorative Concrete Block	Steel Joists	150.35	146.60	139.65	137.25	135.25	133.50	131.45	128.85	126.65
Painted Concrete Block	Steel Joists	143.30	139.95	133.70	131.50	129.70	128.05	126.25	123.90	121.90
Face Brick on Conc. Block	Steel Joists	164.55	160.25	151.85	149.10	146.65	144.60	142.05	138.95	136.35
Precast Concrete Panels	Steel Joists	173.80	169.15	159.80	156.70	154.10	151.80	148.95	145.55	142.65
Tilt-up Panels	Steel Joists	150.70	147.00	139.95	137.60	135.55	133.80	131.75	129.10	126.95
Metal Sandwich Panels	Steel Joists	146.50	142.90	136.30	134.10	132.15	130.50	128.60	126.10	124.05
Perimeter Adj., Add or Deduct	Per 100 L.F.	8.30	7.50	6.25	5.80	5.40	4.95	4.65	4.15	3.75
Story Hgt. Adj., Add or Deduct	Per 1 Ft.	1.15	1.15	1.00	0.95	0.95	0.90	0.90	0.80	0.85
Basement—Not Applicable										

The above costs were calculated using the basic specifications shown on the facing page. These costs should be adjusted where necessary for design alternatives and owner's requirements. Reported completed project costs, for this type of structure, range from $77.60 to $202.95 per S.F.

Common additives

Description	Unit	$ Cost
Emergency Lighting, 25 watt, battery operated		
Lead battery	Each	287
Nickel cadmium	Each	845
Seating		
Auditorium chair, all veneer	Each	240
Veneer back, padded seat	Each	290
Upholstered, spring seat	Each	279
Classroom, movable chair & desk	Set	65 - 120
Lecture hall, pedestal type	Each	229 - 680
Smoke Detectors		
Ceiling type	Each	233
Duct type	Each	525
Sound System		
Amplifier, 250 watts	Each	2400
Speaker, ceiling or wall	Each	196
Trumpet	Each	375

Model costs calculated for a 1 story building with 20′ story height and 12,000 square feet of floor area

Movie Theater

		Unit	Unit Cost	Cost Per S.F.	% Of Sub-Total
A. SUBSTRUCTURE					
1010	Standard Foundations	Poured concrete; strip and spread footings	S.F. Ground	1.30	1.30	
1020	Special Foundations	N/A	—	—	—	
1030	Slab on Grade	4″ reinforced concrete with vapor barrier and granular base	S.F. Slab	4.77	4.77	11.9%
2010	Basement Excavation	Site preparation for slab and trench for foundation wall and footing	S.F. Ground	3.56	3.56	
2020	Basement Walls	4′ foundation wall	L.F. Wall	76	2.77	
B. SHELL					
B10 Superstructure					
1010	Floor Construction	Open web steel joists, slab form, concrete *mezzanine 2250 S.F.*	S.F. Floor	8.98	1.68	9.7%
1020	Roof Construction	Metal deck on open web steel joists	S.F. Roof	8.47	8.47	
B20 Exterior Enclosure					
2010	Exterior Walls	Decorative concrete block *100% of wall*	S.F. Wall	17.85	13.09	
2020	Exterior Windows	N/A	—	—	—	13.6%
2030	Exterior Doors	Sliding mallfront aluminum and glass and hollow metal	Each	2308	1.15	
B30 Roofing					
3010	Roof Coverings	Built-up tar and gravel with flashing; perlite/EPS composite insulation	S.F. Roof	5.83	5.83	5.6%
3020	Roof Openings	N/A	—	—	—	
C. INTERIORS					
1010	Partitions	Concrete block *40 S.F. Floor/L.F. Partition*	S.F. Partition	8.63	3.45	
1020	Interior Doors	Single leaf hollow metal *705 S.F. Floor/Door*	Each	1001	1.42	
1030	Fittings	Toilet partitions	S.F. Floor	.75	.75	
2010	Stair Construction	Concrete filled metal pan	Flight	14,675	2.44	22.4%
3010	Wall Finishes	Paint	S.F. Surface	4.24	3.39	
3020	Floor Finishes	Carpet 50%, ceramic tile 5% *50% of area*	S.F. Floor	10.86	5.43	
3030	Ceiling Finishes	Mineral fiber tile on concealed zee runners suspended	S.F. Ceiling	6.51	6.51	
D. SERVICES					
D10 Conveying					
1010	Elevators & Lifts	N/A	—	—	—	0.0 %
1020	Escalators & Moving Walks	N/A	—	—	—	
D20 Plumbing					
2010	Plumbing Fixtures	Toilet and service fixtures, supply and drainage *1 Fixture/500 S.F. Floor*	Each	2055	4.11	
2020	Domestic Water Distribution	Gas fired water heater	S.F. Floor	.28	.28	4.6%
2040	Rain Water Drainage	Roof drains	S.F. Roof	.43	.43	
D30 HVAC					
3010	Energy Supply	N/A	—	—	—	
3020	Heat Generating Systems	Included in D3050	—	—	—	
3030	Cooling Generating Systems	N/A	—	—	—	9.0 %
3050	Terminal & Package Units	Single zone rooftop unit, gas heating, electric cooling	S.F. Floor	9.43	9.43	
3090	Other HVAC Sys. & Equipment	N/A	—	—	—	
D40 Fire Protection					
4010	Sprinklers	Wet pipe sprinkler system	S.F. Floor	3.80	3.80	3.6%
4020	Standpipes	N/A	—	—	—	
D50 Electrical					
5010	Electrical Service/Distribution	400 ampere service, panel board and feeders	S.F. Floor	1.70	1.70	
5020	Lighting & Branch Wiring	Fluorescent fixtures, receptacles, switches, A.C. and misc. power	S.F. Floor	4.66	4.66	6.8%
5030	Communications & Security	Alarm systems, sound system and emergency lighting	S.F. Floor	.54	.54	
5090	Other Electrical Systems	Emergency generator, 7.5 kW	S.F. Floor	.17	.17	
E. EQUIPMENT & FURNISHINGS					
1010	Commercial Equipment	N/A	—	—	—	
1020	Institutional Equipment	Projection equipment, screen	S.F. Floor	8.90	8.90	12.7 %
1030	Vehicular Equipment	N/A	—	—	—	
2010	Other Equipment	Movie theater seating	S.F. Floor	4.37	4.37	
F. SPECIAL CONSTRUCTION					
1020	Integrated Construction	N/A	—	—	—	0.0 %
1040	Special Facilities	N/A	—	—	—	
G. BUILDING SITEWORK N/A					

	Sub-Total		104.40	**100%**
CONTRACTOR FEES (General Requirements: 10%, Overhead: 5%, Profit: 10%)		25%	26.11	
ARCHITECT FEES		7%	9.14	
	Total Building Cost		**139.65**	

Costs per square foot of floor area

Exterior Wall	S.F. Area	10000	15000	20000	25000	30000	35000	40000	45000	50000
	L.F. Perimeter	286	370	453	457	513	568	624	680	735
Precast Concrete Panels	Bearing Walls	181.65	171.05	165.70	157.60	154.85	152.85	151.45	150.25	149.40
	Steel Frame	185.60	175.00	169.65	161.55	158.80	156.80	155.40	154.25	153.35
Face Brick with Concrete Block Back-up	Bearing Walls	172.40	163.10	158.40	151.70	149.35	147.60	146.40	145.40	144.60
	Steel Joists	177.20	167.90	163.20	156.50	154.10	152.45	151.15	150.20	149.40
Stucco on Concrete Block	Bearing Walls	162.95	154.90	150.85	145.65	143.65	142.25	141.20	140.35	139.75
	Steel Joists	167.70	159.75	155.70	150.45	148.45	147.00	146.00	145.20	144.55
Perimeter Adj., Add or Deduct	Per 100 L.F.	15.45	10.35	7.75	6.25	5.15	4.40	3.85	3.50	3.00
Story Hgt. Adj., Add or Deduct	Per 1 Ft.	3.75	3.25	2.95	2.40	2.25	2.10	2.00	2.00	1.85
For Basement, add $29.90 per square foot of basement area										

The above costs were calculated using the basic specifications shown on the facing page. These costs should be adjusted where necessary for design alternatives and owner's requirements. Reported completed project costs, for this type of structure, range from $83.30 to $208.85 per S.F.

Common additives

Description	Unit	$ Cost	Description	Unit	$ Cost
Beds, Manual	Each	830 - 2825	Kitchen Equipment, cont.		
Elevators, Hydraulic passenger, 2 stops			Ice cube maker, 50 lb. per day	Each	1875
1500# capacity	Each	60,900	Range with 1 oven	Each	2700
2500# capacity	Each	64,300	Laundry Equipment		
3500# capacity	Each	67,600	Dryer, gas, 16 lb. capacity	Each	900
Emergency Lighting, 25 watt, battery operated			30 lb. capacity	Each	3600
Lead battery	Each	287	Washer, 4 cycle	Each	1100
Nickel cadmium	Each	845	Commercial	Each	1450
Intercom System, 25 station capacity			Nurses Call System		
Master station	Each	2725	Single bedside call station	Each	320
Intercom outlets	Each	174	Pillow speaker	Each	305
Handset	Each	480	Refrigerator, Prefabricated, walk-in		
Kitchen Equipment			7'-6" high, 6' x 6'	S.F.	173
Broiler	Each	4050	10' x 10'	S.F.	136
Coffee urn, twin 6 gallon	Each	3025	12' x 14'	S.F.	120
Cooler, 6 ft. long	Each	5150	12' x 20'	S.F.	106
Dishwasher, 10-12 racks per hr.	Each	5075	TV Antenna, Master system, 12 outlet	Outlet	320
Food warmer	Each	740	30 outlet	Outlet	207
Freezer, 44 C.F., reach-in	Each	3250	100 outlet	Outlet	199
			Whirlpool Bath, Mobile, 18" x 24" x 60"	Each	4950
			X-Ray, Mobile	Each	14,700 - 83,000

Important: See the Reference Section for Location Factors

Model costs calculated for a 2 story building with 10' story height and 25,000 square feet of floor area

Nursing Home

				Unit	Unit Cost	Cost Per S.F.	% Of Sub-Total
A. SUBSTRUCTURE							
1010	Standard Foundations		Poured concrete; strip and spread footings	S.F. Ground	1.30	.65	
1020	Special Foundations		N/A	—	—	—	
1030	Slab on Grade		4" reinforced concrete with vapor barrier and granular base	S.F. Slab	4.77	2.39	4.2%
2010	Basement Excavation		Site preparation for slab and trench for foundation wall and footing	S.F. Ground	.31	.16	
2020	Basement Walls		4' foundation wall	L.F. Wall	67	1.60	
B. SHELL							
	B10 Superstructure						
1010	Floor Construction		Pre-cast double tees with concrete topping	S.F. Floor	12.52	6.26	10.2%
1020	Roof Construction		Pre-cast double tees	S.F. Roof	10.56	5.28	
	B20 Exterior Enclosure						
2010	Exterior Walls		Precast concrete panels	S.F. Wall	44.99	13.98	
2020	Exterior Windows		Wood double hung	15% of wall	Each	557	2.04	14.5%
2030	Exterior Doors		Double aluminum & glass doors, single leaf hollow metal	Each	2242	.46	
	B30 Roofing						
3010	Roof Coverings		Built-up tar and gravel with flashing; perlite/EPS composite insulation	S.F. Roof	6.28	3.14	2.8%
3020	Roof Openings		Roof hatches	S.F. Roof	.08	.04	
C. INTERIORS							
1010	Partitions	Gypsum board on metal studs	8 S.F. Floor/L.F. Partition	S.F. Partition	6.57	6.57	
1020	Interior Doors	Single leaf wood	80 S.F. Floor/Door	Each	576	7.20	
1030	Fittings	N/A		—	—	—	
2010	Stair Construction	Concrete stair		Flight	6025	1.44	22.3%
3010	Wall Finishes	50% vinyl wall coverings, 45% paint, 5% ceramic tile		S.F. Surface	1.58	3.16	
3020	Floor Finishes	95% vinyl tile, 5% ceramic tile		S.F. Floor	3.23	3.23	
3030	Ceiling Finishes	Painted gypsum board		S.F. Ceiling	3.71	3.71	
D. SERVICES							
	D10 Conveying						
1010	Elevators & Lifts		One hydraulic hospital elevator	Each	102,500	4.10	3.6%
1020	Escalators & Moving Walks		N/A	—	—	—	
	D20 Plumbing						
2010	Plumbing Fixtures	Kitchen, toilet and service fixtures, supply and drainage	1 Fixture/230 S.F. Floor	Each	3123	13.58	
2020	Domestic Water Distribution	Oil fired water heater		S.F. Floor	1.32	1.32	13.4%
2040	Rain Water Drainage	Roof drains		S.F. Roof	.66	.33	
	D30 HVAC						
3010	Energy Supply		Oil fired hot water, wall fin radiation	S.F. Floor	7.20	7.20	
3020	Heat Generating Systems		N/A	—	—	—	
3030	Cooling Generating Systems		N/A	—	—	—	12.9%
3050	Terminal & Package Units		Split systems with air cooled condensing units	S.F. Floor	7.40	7.40	
3090	Other HVAC Sys. & Equipment		N/A	—	—	—	
	D40 Fire Protection						
4010	Sprinklers		Sprinkler, light hazard	S.F. Floor	5.69	5.69	5.0%
4020	Standpipes		N/A	—	—	—	
	D50 Electrical						
5010	Electrical Service/Distribution		800 ampere service, panel board and feeders	S.F. Floor	2.07	2.07	
5020	Lighting & Branch Wiring		Incandescent fixtures, receptacles, switches, A.C. and misc. power	S.F. Floor	9.33	9.33	11.1%
5030	Communications & Security		Alarm systems and emergency lighting	S.F. Floor	.74	.74	
5090	Other Electrical Systems		Emergency generator, 15 kW	S.F. Floor	.52	.52	
E. EQUIPMENT & FURNISHINGS							
1010	Commercial Equipment		N/A	—	—	—	
1020	Institutional Equipment		N/A	—	—	—	
1030	Vehicular Equipment		N/A	—	—	—	0.0%
1090	Other Equipment		N/A	—	—	—	
F. SPECIAL CONSTRUCTION							
1020	Integrated Construction		N/A	—	—	—	
1040	Special Facilities		N/A	—	—	—	0.0%
G. BUILDING SITEWORK	**N/A**						

					Sub-Total	113.59	100%
	CONTRACTOR FEES (General Requirements: 10%, Overhead: 5%, Profit: 10%)				25%	28.39	
	ARCHITECT FEES				11%	15.62	

		Total Building Cost	**157.60**

Costs per square foot of floor area

Exterior Wall	S.F. Area	2000	3000	5000	7000	9000	12000	15000	20000	25000
	L.F. Perimeter	220	260	320	360	420	480	520	640	700
Wood Siding	Wood Truss	200.70	180.60	162.65	153.65	149.70	145.10	141.75	139.60	137.25
Brick Veneer	Wood Truss	249.50	223.75	200.70	189.10	184.00	178.00	173.60	170.85	167.70
Brick on Block	Wood Truss	192.30	165.00	140.30	127.70	122.30	115.80	110.95	108.05	104.50
	Steel Roof Deck	241.75	213.15	187.30	174.10	168.40	161.60	156.50	153.40	149.70
EIFS on Metal Studs	Steel Roof Deck	223.45	199.60	178.25	**167.55**	162.80	157.35	153.20	150.70	147.80
Tiltup Concrete Panel	Steel Roof Deck	225.65	200.95	178.90	167.80	162.90	157.20	152.95	150.40	147.35
Perimeter Adj., Add or Deduct	Per 100 L.F.	22.55	15.00	9.05	6.55	5.00	3.80	3.00	2.30	1.80
Story Hgt. Adj., Add or Deduct	Per 1 Ft.	2.70	2.10	1.55	1.30	1.15	1.00	0.85	0.80	0.70
For Basement, add $30.30 per square foot of basement area										

The above costs were calculated using the basic specifications shown on the facing page. These costs should be adjusted where necessary for design alternatives and owner's requirements. Reported completed project costs, for this type of structure, range from $63.60 to $245.90 per S.F.

Common additives

Description	Unit	$ Cost
Clock System		
20 room	Each	16,000
50 room	Each	39,100
Closed Circuit Surveillance, One station		
Camera and monitor	Each	1875
For additional camera stations, add	Each	1025
Directory Boards, Plastic, glass covered		
30" x 20"	Each	605
36" x 48"	Each	1325
Aluminum, 24" x 18"	Each	585
36" x 24"	Each	685
48" x 32"	Each	975
48" x 60"	Each	2025
Emergency Lighting, 25 watt, battery operated		
Lead battery	Each	287
Nickel cadmium	Each	845

Description	Unit	$ Cost
Smoke Detectors		
Ceiling type	Each	233
Duct type	Each	525
Sound System		
Amplifier, 250 watts	Each	2400
Speaker, ceiling or wall	Each	196
Trumpet	Each	375
TV Antenna, Master system, 12 outlet	Outlet	320
30 outlet	Outlet	207
100 outlet	Outlet	199

Model costs calculated for a 1 story building with 12' story height and 7,000 square feet of floor area

				Unit	Unit Cost	Cost Per S.F.	% Of Sub-Total
A.	**SUBSTRUCTURE**						
1010	Standard Foundations	Poured concrete; strip and spread footings		S.F. Ground	2.54	2.54	
1020	Special Foundations	N/A		—	—	—	
1030	Slab on Grade	4" reinforced concrete with vapor barrier and granular base		S.F. Slab	4.77	4.77	9.3%
2010	Basement Excavation	Site preparation for slab and trench for foundation wall and footing		S.F. Ground	.31	.31	
2020	Basement Walls	4' foundation wall		L.F. Wall	143	4.08	
B.	**SHELL**						
	B10 Superstructure						
1010	Floor Construction	N/A		—	—	—	6.5 %
1020	Roof Construction	Steel joists, girders & deck on columns		S.F. Roof	8.09	8.09	
	B20 Exterior Enclosure						
2010	Exterior Walls	E.I.F.S. on metal studs		S.F. Wall	16.61	8.20	
2020	Exterior Windows	Aluminum outward projecting	20% of wall	Each	703	3.77	11.7%
2030	Exterior Doors	Aluminum and glass, hollow metal		Each	3196	2.73	
	B30 Roofing						
3010	Roof Coverings	Single ply membrane, loose laid and ballasted		S.F. Roof	8.62	8.62	7.0%
3020	Roof Openings	Roof hatch		S.F. Roof	.16	.16	
C.	**INTERIORS**						
1010	Partitions	Gypsum board on metal studs	20 S.F. Floor/L.F. Partition	S.F. Partition	8.04	4.02	
1020	Interior Doors	Single leaf hollow metal	200 S.F. Floor/Door	Each	1001	5.01	
1030	Fittings	Toilet partitions		S.F. Floor	.50	.50	
2010	Stair Construction	N/A		—	—	—	20.1%
3010	Wall Finishes	60% vinyl wall covering, 40% paint		S.F. Surface	1.39	1.39	
3020	Floor Finishes	60% carpet, 30% vinyl composition tile, 10% ceramic tile		S.F. Floor	7.75	7.75	
3030	Ceiling Finishes	Mineral fiber tile on concealed zee bars		S.F. Ceiling	6.51	6.51	
D.	**SERVICES**						
	D10 Conveying						
1010	Elevators & Lifts	N/A		—	—	—	0.0 %
1020	Escalators & Moving Walks	N/A		—	—	—	
	D20 Plumbing						
2010	Plumbing Fixtures	Toilet and service fixtures, supply and drainage	1 Fixture/1320 S.F. Floor	Each	7009	5.31	
2020	Domestic Water Distribution	Gas fired water heater		S.F. Floor	1.16	1.16	5.7%
2040	Rain Water Drainage	Roof drains		S.F. Roof	.72	.72	
	D30 HVAC						
3010	Energy Supply	N/A		—	—	—	
3020	Heat Generating Systems	Included in D3050		—	—	—	
3030	Cooling Generating Systems	N/A		—	—	—	15.5 %
3050	Terminal & Package Units	Multizone unit gas heating, electric cooling		S.F. Floor	19.40	19.40	
3090	Other HVAC Sys. & Equipment	N/A		—	—	—	
	D40 Fire Protection						
4010	Sprinklers	Sprinkler system, light hazard		S.F. Floor	3.16	3.16	3.5%
4020	Standpipes	Standpipes and hose systems		S.F. Floor	1.24	1.24	
	D50 Electrical						
5010	Electrical Service/Distribution	400 ampere service, panel board and feeders		S.F. Floor	5.84	5.84	
5020	Lighting & Branch Wiring	High efficiency fluorescent fixtures, receptacles, switches, A.C. and misc. power		S.F. Floor	11.97	11.97	20.6%
5030	Communications & Security	Addressable alarm systems, internet and phone wiring, and emergency lighting		S.F. Floor	7.80	7.80	
5090	Other Electrical Systems	Emergency generator, 7.5 kW		S.F. Floor	.21	.21	
E.	**EQUIPMENT & FURNISHINGS**						
1010	Commercial Equipment	N/A		—	—	—	
1020	Institutional Equipment	N/A		—	—	—	0.0 %
1030	Vehicular Equipment	N/A		—	—	—	
1090	Other Equipment	N/A		—	—	—	
F.	**SPECIAL CONSTRUCTION**						
1020	Integrated Construction	N/A		—	—	—	0.0 %
1040	Special Facilities	N/A		—	—	—	
G.	**BUILDING SITEWORK**	N/A					

Sub-Total	125.26	**100%**

CONTRACTOR FEES (General Requirements: 10%, Overhead: 5%, Profit: 10%)	25%	31.33
ARCHITECT FEES	7%	10.96

Total Building Cost	**167.55**

Costs per square foot of floor area

Exterior Wall	S.F. Area	5000	8000	12000	16000	20000	35000	50000	65000	80000
	L.F. Perimeter	220	260	310	330	360	440	490	548	580
Face Brick with Concrete Block Back-up	Wood Joists	236.75	206.30	188.85	176.75	170.40	157.75	151.60	148.60	146.05
	Steel Joists	238.45	208.00	190.55	178.45	172.10	159.45	153.30	150.25	147.75
Glass and Metal Curtain Wall	Steel Frame	280.45	240.35	217.30	200.80	192.20	175.00	166.55	162.30	158.80
	R/Conc. Frame	277.70	238.05	215.25	198.95	190.45	173.40	165.00	160.80	157.35
Wood Siding	Wood Frame	190.40	168.40	155.80	147.55	143.15	134.50	130.35	128.35	126.70
Brick Veneer	Wood Frame	210.80	183.45	167.80	157.10	151.50	140.30	134.95	132.25	130.05
Perimeter Adj., Add or Deduct	Per 100 L.F.	36.35	22.70	15.20	11.35	9.05	5.20	3.70	2.75	2.30
Story Hgt. Adj., Add or Deduct	Per 1 Ft.	5.90	4.40	3.50	2.80	2.45	1.70	1.35	1.10	0.95

For Basement, add $33.80 per square foot of basement area

The above costs were calculated using the basic specifications shown on the facing page. These costs should be adjusted where necessary for design alternatives and owner's requirements. Reported completed project costs, for this type of structure, range from $66.55 to $257.85 per S.F.

Common additives

Description	Unit	$ Cost
Clock System		
20 room	Each	16,000
50 room	Each	39,100
Closed Circuit Surveillance, One station		
Camera and monitor	Each	1875
For additional camera stations, add	Each	1025
Directory Boards, Plastic, glass covered		
30" x 20"	Each	605
36" x 48"	Each	1325
Aluminum, 24" x 18"	Each	585
36" x 24"	Each	685
48" x 32"	Each	975
48" x 60"	Each	2025
Elevators, Hydraulic passenger, 2 stops		
1500# capacity	Each	60,900
2500# capacity	Each	64,300
3500# capacity	Each	67,600
Additional stop, add	Each	8250
Emergency Lighting, 25 watt, battery operated		
Lead battery	Each	287
Nickel cadmium	Each	845

Description	Unit	$ Cost
Smoke Detectors		
Ceiling type	Each	233
Duct type	Each	525
Sound System		
Amplifier, 250 watts	Each	2400
Speaker, ceiling or wall	Each	196
Trumpet	Each	375
TV Antenna, Master system, 12 outlet	Outlet	320
30 outlet	Outlet	207
100 outlet	Outlet	199

Important: See the Reference Section for Location Factors

Model costs calculated for a 3 story building with 12' story height and 20,000 square feet of floor area

				Unit	Unit Cost	Cost Per S.F.	% Of Sub-Total
A.	**SUBSTRUCTURE**						
1010	Standard Foundations	Poured concrete; strip and spread footings		S.F. Ground	6.84	2.28	
1020	Special Foundations	N/A		—	—	—	
1030	Slab on Grade	4" reinforced concrete with vapor barrier and granular base		S.F. Slab	4.77	1.59	4.3%
2010	Basement Excavation	Site preparation for slab and trench for foundation wall and footing		S.F. Ground	.18	.06	
2020	Basement Walls	4' foundation wall		L.F. Wall	71	1.60	
B.	**SHELL**						
	B10 Superstructure						
1010	Floor Construction	Open web steel joists, slab form, concrete, columns		S.F. Floor	15.72	10.48	9.6%
1020	Roof Construction	Metal deck, open web steel joists, columns		S.F. Roof	5.67	1.89	
	B20 Exterior Enclosure						
2010	Exterior Walls	Face brick with concrete block backup	80% of wall	S.F. Wall	30.30	15.71	
2020	Exterior Windows	Aluminum outward projecting	20% of wall	Each	703	3.97	16.0%
2030	Exterior Doors	Aluminum and glass, hollow metal		Each	3196	.96	
	B30 Roofing						
3010	Roof Coverings	Built-up tar and gravel with flashing; perlite/EPS composite		S.F. Roof	6.66	2.22	1.7%
3020	Roof Openings	N/A		—	—	—	
C.	**INTERIORS**						
1010	Partitions	Gypsum board on metal studs	20 S.F. Floor/L.F. Partition	S.F. Partition	9.30	3.72	
1020	Interior Doors	Single leaf hollow metal	200 S.F. Floor/Door	Each	1001	5.01	
1030	Fittings	Toilet partitions		S.F. Floor	1.05	1.05	
2010	Stair Construction	Concrete filled metal pan		Flight	12,275	4.30	22.9%
3010	Wall Finishes	60% vinyl wall covering, 40% paint		S.F. Surface	1.38	1.10	
3020	Floor Finishes	60% carpet, 30% vinyl composition tile, 10% ceramic tile		S.F. Floor	7.75	7.75	
3030	Ceiling Finishes	Mineral fiber tile on concealed zee bars		S.F. Ceiling	6.51	6.51	
D.	**SERVICES**						
	D10 Conveying						
1010	Elevators & Lifts	Two hydraulic passenger elevators		Each	116,400	11.64	9.0%
1020	Escalators & Moving Walks	N/A		—	—	—	
	D20 Plumbing						
2010	Plumbing Fixtures	Toilet and service fixtures, supply and drainage	1 Fixture/1320 S.F. Floor	Each	4171	3.16	
2020	Domestic Water Distribution	Gas fired water heater		S.F. Floor	.41	.41	3.2%
2040	Rain Water Drainage	Roof drains		S.F. Roof	1.68	.56	
	D30 HVAC						
3010	Energy Supply	N/A		—	—	—	
3020	Heat Generating Systems	Included in D3050		—	—	—	
3030	Cooling Generating Systems	N/A		—	—	—	12.2%
3050	Terminal & Package Units	Multizone unit gas heating, electric cooling		S.F. Floor	15.70	15.70	
3090	Other HVAC Sys. & Equipment	N/A		—	—	—	
	D40 Fire Protection						
4010	Sprinklers	Wet pipe sprinkler system		S.F. Floor	3.33	3.33	3.2%
4020	Standpipes	Standpipes and hose systems		S.F. Floor	.81	.81	
	D50 Electrical						
5010	Electrical Service/Distribution	1000 ampere service, panel board and feeders		S.F. Floor	4.49	4.49	
5020	Lighting & Branch Wiring	High efficiency fluorescent fixtures, receptacles, switches, A.C. and misc. power		S.F. Floor	11.48	11.48	17.8%
5030	Communications & Security	Addressable alarm systems, internet and phone wiring, and emergency lighting		S.F. Floor	6.67	6.67	
5090	Other Electrical Systems	Emergency generator, 7.5 kW, uninterruptible power supply		S.F. Floor	.23	.23	
E.	**EQUIPMENT & FURNISHINGS**						
1010	Commercial Equipment	N/A		—	—	—	
1020	Institutional Equipment	N/A		—	—	—	0.0%
1030	Vehicular Equipment	N/A		—	—	—	
1090	Other Equipment	N/A		—	—	—	
F.	**SPECIAL CONSTRUCTION**						
1020	Integrated Construction	N/A		—	—	—	0.0%
1040	Special Facilities	N/A		—	—	—	
G.	**BUILDING SITEWORK**	N/A					

	Sub-Total		128.68	**100%**
CONTRACTOR FEES (General Requirements: 10%, Overhead: 5%, Profit: 10%)		25%	32.16	
ARCHITECT FEES		7%	11.26	
	Total Building Cost		**172.10**	

Costs per square foot of floor area

Exterior Wall	S.F. Area	20000	40000	60000	80000	100000	150000	200000	250000	300000
	L.F. Perimeter	260	360	400	420	460	520	600	640	700
Precast Concrete Panel	Steel Frame	222.90	192.00	175.50	165.65	161.00	153.15	149.85	146.90	145.30
	R/Conc. Frame	223.45	192.25	175.65	165.80	161.10	153.20	149.85	146.90	145.35
Face Brick with Concrete Block Back-up	Steel Frame	205.40	179.05	165.65	157.75	153.90	147.60	144.85	142.55	141.25
	R/Conc. Frame	203.40	178.40	165.40	157.70	154.00	147.85	145.20	142.95	141.75
Limestone Panel Concrete Block Back-up	Steel Frame	251.35	211.60	189.95	177.10	171.00	160.65	156.35	152.40	150.35
	R/Conc. Frame	251.30	211.55	189.95	177.05	170.95	160.60	156.30	152.40	150.30
Perimeter Adj., Add or Deduct	Per 100 L.F.	31.10	15.55	10.35	7.80	6.25	4.15	3.10	2.45	2.10
Story Hgt. Adj., Add or Deduct	Per 1 Ft.	6.45	4.50	3.35	2.65	2.35	1.70	1.55	1.25	1.20
For Basement, add $36.55 per square foot of basement area										

The above costs were calculated using the basic specifications shown on the facing page. These costs should be adjusted where necessary for design alternatives and owner's requirements. Reported completed project costs, for this type of structure, range from $74.90 to $220.25 per S.F.

Common additives

Description	Unit	$ Cost
Clock System		
20 room	Each	16,000
50 room	Each	39,100
Closed Circuit Surveillance, One station		
Camera and monitor	Each	1875
For additional camera stations, add	Each	1025
Directory Boards, Plastic, glass covered		
30" x 20"	Each	605
36" x 48"	Each	1325
Aluminum, 24" x 18"	Each	585
36" x 24"	Each	685
48" x 32"	Each	975
48" x 60"	Each	2025
Elevators, Electric passenger, 5 stops		
2000# capacity	Each	158,200
3500# capacity	Each	166,200
5000# capacity	Each	169,700
Additional stop, add	Each	14,000
Emergency Lighting, 25 watt, battery operated		
Lead battery	Each	287
Nickel cadmium	Each	845

Description	Unit	$ Cost
Intercom System, 25 station capacity		
Master station	Each	2725
Intercom outlets	Each	174
Handset	Each	480
Smoke Detectors		
Ceiling type	Each	233
Duct type	Each	525
Sound System		
Amplifier, 250 watts	Each	2400
Speaker, ceiling or wall	Each	196
Trumpet	Each	375
TV Antenna, Master system, 12 oulet	Outlet	320
30 outlet	Outlet	207
100 outlet	Outlet	199

Important: See the Reference Section for Location Factors

Model costs calculated for a 8 story building with 12' story height and 80,000 square feet of floor area

			Unit	Unit Cost	Cost Per S.F.	% Of Sub-Total
A. SUBSTRUCTURE						
1010	Standard Foundations	Poured concrete; strip and spread footings	S.F. Ground	11.04	1.38	
1020	Special Foundations	N/A	—	—	—	
1030	Slab on Grade	4" reinforced concrete with vapor barrier and granular base	S.F. Slab	4.77	.60	2.0%
2010	Basement Excavation	Site preparation for slab and trench for foundation wall and footing	S.F. Ground	.31	.04	
2020	Basement Walls	4' foundation wall	L.F. Wall	76	.51	
B. SHELL						
B10 Superstructure						
1010	Floor Construction	Concrete slab with metal deck and beams	S.F. Floor	20.97	18.35	15.2%
1020	Roof Construction	Metal deck, open web steel joists, interior columns	S.F. Roof	5.52	.69	
B20 Exterior Enclosure						
2010	Exterior Walls	Precast concrete panels — 80% of wall	S.F. Wall	45.44	18.32	
2020	Exterior Windows	Vertical pivoted steel — 20% of wall	Each	557	3.75	17.8%
2030	Exterior Doors	Double aluminum and glass doors and entrance with transoms	Each	3851	.25	
B30 Roofing						
3010	Roof Coverings	Built-up tar and gravel with flashing; perlite/EPS composite insulation	S.F. Roof	5.76	.72	0.6%
3020	Roof Openings	N/A	—	—	—	
C. INTERIORS						
1010	Partitions	Gypsum board on metal studs — 30 S.F. Floor/L.F. Partition	S.F. Partition	8.94	2.98	
1020	Interior Doors	Single leaf hollow metal — 400 S.F. Floor/Door	Each	1001	2.50	
1030	Fittings	Toilet Partitions	S.F. Floor	.69	.69	
2010	Stair Construction	Concrete filled metal pan	Flight	12,275	2.61	19.2%
3010	Wall Finishes	60% vinyl wall covering, 40% paint	S.F. Surface	1.38	.92	
3020	Floor Finishes	60% carpet, 30% vinyl composition tile, 10% ceramic tile	S.F. Floor	7.75	7.75	
3030	Ceiling Finishes	Mineral fiber tile on concealed zee bars	S.F. Ceiling	6.51	6.51	
D. SERVICES						
D10 Conveying						
1010	Elevators & Lifts	Four geared passenger elevators	Each	294,200	14.71	11.8%
1020	Escalators & Moving Walks	N/A	—	—	—	
D20 Plumbing						
2010	Plumbing Fixtures	Toilet and service fixtures, supply and drainage — 1 Fixture/1370 S.F. Floor	Each	3069	2.24	
2020	Domestic Water Distribution	Gas fired water heater	S.F. Floor	.49	.49	2.4%
2040	Rain Water Drainage	Roof drains	S.F. Roof	2	.25	
D30 HVAC						
3010	Energy Supply	N/A	—	—	—	
3020	Heat Generating Systems	Included in D3050	—	—	—	
3030	Cooling Generating Systems	N/A	—	—	—	12.6 %
3050	Terminal & Package Units	Multizone unit gas heating, electric cooling	S.F. Floor	15.70	15.70	
3090	Other HVAC Sys. & Equipment	N/A	—	—	—	
D40 Fire Protection						
4010	Sprinklers	Wet pipe sprinkler system	S.F. Floor	2.72	2.72	2.9%
4020	Standpipes	Standpipes and hose systems	S.F. Floor	.91	.91	
D50 Electrical						
5010	Electrical Service/Distribution	1600 ampere service, panel board and feeders	S.F. Floor	1.80	1.80	
5020	Lighting & Branch Wiring	High efficiency fluorescent fixtures, receptacles, switches, A.C. and misc. power	S.F. Floor	11.38	11.38	15.6%
5030	Communications & Security	Addressable alarm systems, internet and phone wiring, emergency lighting	S.F. Floor	5.23	5.23	
5090	Other Electrical Systems	Emergency generator, 100 kW, uninterruptible power supply	S.F. Floor	1.05	1.05	
E. EQUIPMENT & FURNISHINGS						
1010	Commercial Equipment	N/A	—	—	—	
1020	Institutional Equipment	N/A	—	—	—	0.0 %
1030	Vehicular Equipment	N/A	—	—	—	
1090	Other Equipment	N/A	—	—	—	
F. SPECIAL CONSTRUCTION						
1020	Integrated Construction	N/A	—	—	—	0.0 %
1040	Special Facilities	N/A	—	—	—	
G. BUILDING SITEWORK	N/A					

		Sub-Total	125.05	**100%**
CONTRACTOR FEES (General Requirements: 10%, Overhead: 5%, Profit: 10%)		25%	31.22	
ARCHITECT FEES		6%	9.38	
	Total Building Cost		**165.65**	

Costs per square foot of floor area

Exterior Wall	S.F. Area	120000	145000	170000	200000	230000	260000	400000	600000	800000
	L.F. Perimeter	420	450	470	490	510	530	600	730	820
Double Glazed Heat Absorbing Tinted Plate Glass Panels	Steel Frame	169.40	163.75	159.10	154.95	151.85	149.45	142.50	138.65	136.25
	R/Conc. Frame	161.55	156.05	151.60	147.45	144.45	142.10	135.30	131.55	129.20
Face Brick with Concrete Block Back-up	Steel Frame	162.55	157.70	153.75	150.20	147.55	145.50	139.55	136.30	134.25
	R/Conc. Frame	179.20	174.45	170.65	167.15	164.65	162.65	156.80	153.65	151.60
Precast Concrete Panel With Exposed Aggregate	Steel Frame	168.40	162.85	158.35	154.25	151.25	148.90	142.10	138.30	135.90
	R/Conc. Frame	160.55	155.15	150.80	146.80	143.90	141.60	134.85	131.25	128.90
Perimeter Adj., Add or Deduct	Per 100 L.F.	8.45	7.05	6.05	5.15	4.45	3.95	2.55	1.65	1.25
Story Hgt. Adj., Add or Deduct	Per 1 Ft.	3.40	3.00	2.70	2.40	2.15	2.05	1.40	1.20	1.00

For Basement, add $36.55 per square foot of basement area

The above costs were calculated using the basic specifications shown on the facing page. These costs should be adjusted where necessary for design alternatives and owner's requirements. Reported completed project costs, for this type of structure, range from $93.95 to $229.25 per S.F.

Common additives

Description	Unit	$ Cost
Clock System		
20 room	Each	16,000
50 room	Each	39,100
Directory Boards, Plastic, glass covered		
30" x 20"	Each	605
36" x 48"	Each	1325
Aluminum, 24" x 18"	Each	585
36" x 24"	Each	685
48" x 32"	Each	975
48" x 60"	Each	2025
Elevators, Electric passenger, 10 stops		
3000# capacity	Each	430,000
4000# capacity	Each	433,000
5000# capacity	Each	437,000
Additional stop, add	Each	14,000
Emergency Lighting, 25 watt, battery operated		
Lead battery	Each	287
Nickel cadmium	Each	845

Description	Unit	$ Cost
Escalators, Metal		
32" wide, 10' story height	Each	139,700
20' story height	Each	168,000
48" wide, 10' story height	Each	148,200
20' story height	Each	176,000
Glass		
32" wide, 10' story height	Each	133,200
20' story height	Each	161,000
48" wide, 10' story height	Each	140,700
20' story height	Each	170,000
Smoke Detectors		
Ceiling type	Each	233
Duct type	Each	525
Sound System		
Amplifier, 250 watts	Each	2400
Speaker, ceiling or wall	Each	196
Trumpet	Each	375
TV Antenna, Master system, 12 oulet	Outlet	320
30 oulet	Outlet	207
100 outlet	Outlet	199

Important: See the Reference Section for Location Factors

Model costs calculated for a 16 story building with 10' story height and 260,000 square feet of floor area

				Unit	Unit Cost	Cost Per S.F.	% Of Sub-Total
A. SUBSTRUCTURE							
1010	Standard Foundations		CIP concrete pile caps	S.F. Ground	9.12	.57	
1020	Special Foundations		Steel H-piles, concrete grade beams	S.F. Ground	61	3.79	
1030	Slab on Grade		4" reinforced concrete with vapor barrier and granular base	S.F. Slab	4.77	.30	4.5%
2010	Basement Excavation		Site preparation for slab, piles and grade beams	S.F. Ground	.31	.02	
2020	Basement Walls		4' foundation wall	L.F. Wall	76	.36	
B. SHELL							
B10 Superstructure							
1010	Floor Construction		Concrete slab, metal deck, beams	S.F. Floor	25.39	23.80	21.5%
1020	Roof Construction		Metal deck, open web steel joists, beams, columns	S.F. Roof	6.72	.42	
B20 Exterior Enclosure							
2010	Exterior Walls		N/A	—	—	—	
2020	Exterior Windows		Double glazed heat absorbing, tinted plate glass wall panels 100% of wall	Each	42.60	13.89	12.9 %
2030	Exterior Doors		Double aluminum & glass doors	Each	5754	.61	
B30 Roofing							
3010	Roof Coverings		Single ply membrane, fully adhered; perlite/EPS composite insulation	S.F. Roof	5.12	.32	0.3%
3020	Roof Openings		N/A	—	—	—	
C. INTERIORS							
1010	Partitions		Gypsum board on metal studs 30 S.F. Floor/L.F. Partition	S.F. Partition	9.90	2.64	
1020	Interior Doors		Single leaf hollow metal 400 S.F. Floor/Door	Each	1001	2.50	
1030	Fittings		Toilet partitions	S.F. Floor	.40	.40	
2010	Stair Construction		Concrete filled metal pan	Flight	14,675	1.97	17.3%
3010	Wall Finishes		60% vinyl wall covering, 40% paint	S.F. Surface	1.39	.74	
3020	Floor Finishes		60% carpet tile, 30% vinyl composition tile, 10% ceramic tile	S.F. Floor	4.70	4.70	
3030	Ceiling Finishes		Mineral fiber tile on concealed zee bars	S.F. Ceiling	6.51	6.51	
D. SERVICES							
D10 Conveying							
1010	Elevators & Lifts		Four geared passenger elevators	Each	483,600	7.44	6.6%
1020	Escalators & Moving Walks		N/A	—	—	—	
D20 Plumbing							
2010	Plumbing Fixtures		Toilet and service fixtures, supply and drainage 1 Fixture/1345 S.F. Floor	Each	4533	3.37	
2020	Domestic Water Distribution		Oil fired water heater	S.F. Floor	.28	.28	3.4%
2040	Rain Water Drainage		Roof drains	S.F. Roof	2.56	.16	
D30 HVAC							
3010	Energy Supply		N/A	—	—	—	
3020	Heat Generating Systems		Boiler, heat exchanger and fans	Each	404,100	2.12	
3030	Cooling Generating Systems		Chilled water, fan coil units	S.F. Floor	13.87	13.87	14.2 %
3050	Terminal & Package Units		N/A	—	—	—	
3090	Other HVAC Sys. & Equipment		N/A	—	—	—	
D40 Fire Protection							
4010	Sprinklers		Sprinkler system, light hazard	S.F. Floor	2.63	2.63	2.7%
4020	Standpipes		Standpipes and hose systems	S.F. Floor	.45	.45	
D50 Electrical							
5010	Electrical Service/Distribution		2400 ampere service, panel board and feeders	S.F. Floor	1.07	1.07	
5020	Lighting & Branch Wiring		High efficiency fluorescent fixtures, receptacles, switches, A.C. and misc. power	S.F. Floor	11.26	11.26	16.8%
5030	Communications & Security		Addressable alarm systems, internet and phone wiring, emergency lighting	S.F. Floor	6.06	6.06	
5090	Other Electrical Systems		Emergency generator, 200 kW, uninterruptible power supply	S.F. Floor	.55	.55	
E. EQUIPMENT & FURNISHINGS							
1010	Commercial Equipment		N/A	—	—	—	
1020	Institutional Equipment		N/A	—	—	—	0.0 %
1030	Vehicular Equipment		N/A	—	—	—	
1090	Other Equipment		N/A	—	—	—	
F. SPECIAL CONSTRUCTION							
1020	Integrated Construction		N/A	—	—	—	0.0 %
1040	Special Facilities		N/A	—	—	—	
G. BUILDING SITEWORK	**N/A**						

		Sub-Total	112.80	100%
CONTRACTOR FEES (General Requirements: 10%, Overhead: 5%, Profit: 10%)			25%	28.19	
ARCHITECT FEES			6%	8.46	

Total Building Cost	**149.45**

Costs per square foot of floor area

Exterior Wall	S.F. Area	5000	7500	10000	12500	15000	17500	20000	22500	25000
	L.F. Perimeter	284	350	410	450	490	530	570	600	632
Brick Veneer	Wood Truss	401.60	388.95	382.15	376.50	372.75	370.10	368.10	366.10	364.65
Brick Veneer	Steel Frame	405.00	392.95	386.50	381.15	377.60	375.15	373.25	371.35	369.95
Decorative Concrete Block	Steel Frame	404.85	392.85	386.40	381.15	377.50	375.05	373.20	371.30	369.85
Decorative Concrete Block	Bearing Wall	400.10	392.30	388.15	384.95	382.75	381.20	380.10	378.95	378.10
Tiltup Concrete Panel	Steel Frame	401.55	390.20	384.10	379.10	375.70	373.40	371.65	369.85	368.55
EIFS on Concrete Block	Steel Frame	403.40	390.05	382.90	377.35	373.65	371.05	369.10	367.20	365.75
Perimeter Adj., Add or Deduct	Per 100 L.F.	18.70	12.55	9.35	7.45	6.20	5.40	4.70	4.15	3.70
Story Hgt. Adj., Add or Deduct	Per 1 Ft.	2.80	2.40	2.05	1.80	1.65	1.50	1.40	1.35	1.25
For Basement, add $28.75 per square foot of basement area										

The above costs were calculated using the basic specifications shown on the facing page. These costs should be adjusted where necessary for design alternatives and owner's requirements. Reported completed project costs, for this type of structure, range from $164.55 to $853.20 per S.F.

Common additives

Description	Unit	$ Cost
Cabinets, base, door units, metal	L.F.	268
Drawer units	L.F.	525
Tall storage cabinets, open, 7' high	L.F.	500
With glazed doors	L.F.	590
Wall cabinets, metal, 12-1/2" deep, open	L.F.	200
With doors	L.F.	360
Counter top, laminated plastic, no backsplash	L.F.	72.50
Stainless steel counter top	L.F.	177
For drop-in stainless 43" x 21" sink, add	Ea.	995
Nurse Call Station		
Single bedside call station	Ea.	320
Ceiling speaker station	Ea.	147
Emergency call station	Ea.	198
Pillow speaker	Ea.	305
Standard call button	Ea.	190
Master control station for 20 stations	Total	6125

Description	Unit	$ Cost
Directory Boards, Plastic, glass covered		
Plastic, glass covered, 30" x 20"	Ea.	605
36" x 48"	Ea.	1325
Building directory, alum., black felt panels, 1 door, 24" x 18"	Ea.	585
36" x 24"	Ea.	685
48" x 32"	Ea.	975
48" x 60"	Ea.	2025
Tables, Examining, vinyl top		
with base cabinets	Ea.	1775
Utensil washer-sanitizer	Ea.	12,300
X-ray, mobile, minimum	Ea.	14,700

Outpatient Surgery Center

Model costs calculated for a 1 story building with 15'-4" story height and 12,500 square feet of floor area

				Unit	Unit Cost	Cost Per S.F.	% Of Sub-Total
A.	**SUBSTRUCTURE**						
1010	Standard Foundations	Poured concrete; strip and spread footings		S.F. Ground	1.79	1.79	
1020	Special Foundations	N/A		—	—	—	
1030	Slab on Grade	4" reinforced concrete with vapor barrier and granular base		S.F. Slab	5.92	5.92	3.9%
2010	Basement Excavation	Site preparation for slab and trench for foundation wall and footing		S.F. Ground	.31	.31	
2020	Basement Walls	4' Foundation wall		L.F. Wall	76	2.72	
B.	**SHELL**						
	B10 Superstructure						
1010	Floor Construction	N/A		—	—	—	2.1 %
1020	Roof Construction	Steel Joists and girders on columns		S.F. Roof	5.81	5.81	
	B20 Exterior Enclosure						
2010	Exterior Walls	Brick veneer	85% of wall	S.F. Wall	19.70	9.24	
2020	Exterior Windows	Aluminum frame, fixed	15% of wall	Eacj	755	4.17	6.6%
2030	Exterior Doors	Aluminum and glass entry; steel flush doors		Each	6130	4.91	
	B30 Roofing						
3010	Roof Coverings	Single-ply membrane, loose laid and ballasted; polyisocyanurate insulation		S.F. Roof	7.35	7.35	2.7%
3020	Roof Openings	Roof hatches		S.F. Roof	.09	.09	
C.	**INTERIORS**						
1010	Partitions	Gypsum board on metal studs	7 S.F. Floor/L.F. Partition	S.F. Partition	7.02	10.03	
1020	Interior Doors	Single and double leaf fire doors	200 S.F. Floor/Door	Each	1107	5.54	
1030	Fittings	Stainless steel toilet partitions, steel lockers		S.F. Floor	1.63	1.63	
2010	Stair Construction	N/A		—	—	—	12.7%
3010	Wall Finishes	80% Paint, 10% vinyl wall covering, 10% ceramic tile		S.F. Surface	1.84	5.25	
3020	Floor Finishes	10% Carpet, 80% vinyl sheet, 10% ceramic tile		S.F. Floor	9.16	9.16	
3030	Ceiling Finishes	Mineral fiber tile and plastic coated tile on suspension system		S.F. Ceiling	3.79	3.78	
D.	**SERVICES**						
	D10 Conveying						
1010	Elevators & Lifts	N/A		—	—	—	0.0 %
1020	Escalators & Moving Walks	N/A		—	—	—	
	D20 Plumbing						
2010	Plumbing Fixtures	Medical, patient & specialty fixtures, supply and drainage	1 fixture/350 S.F. Floor	Each	2975	8.50	
2020	Domestic Water Distribution	Electric water heater		S.F. Floor	1.51	1.51	4.3%
2040	Rain Water Drainage	Roof drains		S.F. Roof	1.89	1.89	
	D30 HVAC						
3010	Energy Supply	Hot water reheat for surgery		S.F. Floor	9.65	9.65	
3020	Heat Generating Systems	Boiler		Each	50,525	5.47	
3030	Cooling Generating Systems	N/A		—	—	—	35.3%
3050	Terminal & Package Units	N/A		—	—	—	
3090	Other HVAC Sys. & Equipment	Rooftop AC, CAV, roof vents, heat exchanger, surgical air curtains		S.F. Floor	83	83.23	
	D40 Fire Protection						
4010	Sprinklers	Wet pipe sprinkler system		S.F. Floor	3.16	3.16	1.6%
4020	Standpipes	Standpipe		S.F. Floor	1.38	1.38	
	D50 Electrical						
5010	Electrical Service/Distribution	1200 ampere service, panel board and feeders		S.F. Floor	7.91	7.91	
5020	Lighting & Branch Wiring	T-8 fluorescent light fixtures, receptacles, switches, A.C. and misc. power		S.F. Floor	16.08	16.08	14.8%
5030	Communications & Security	Addressable alarm system, internet wiring, communications system, emergency lighting		S.F. Floor	9.84	9.84	
5090	Other Electrical Systems	Emergency generator, 300 kW with fuel tank		S.F. Floor	7.43	7.43	
E.	**EQUIPMENT & FURNISHINGS**						
1010	Commercial Equipment	N/A		—	—	—	
1020	Institutional Equipment	Medical gas system, cabinetry, scrub sink, sterilizer		S.F. Floor	43.54	43.54	16.0 %
1030	Vehicular Equipment	N/A		—	—	—	
1090	Other Equipment	Ice maker, washing machine		S.F. Floor	.94	.94	
F.	**SPECIAL CONSTRUCTION & DEMOLITION**						
1020	Integrated Construction	N/A		—	—	—	0.0 %
1040	Special Facilities	N/A		—	—	—	
G.	**BUILDING SITEWORK**	**N/A**					

	Sub-Total	278.23	100%
CONTRACTOR FEES (General Requirements: 10%, Overhead: 5%, Profit: 10%)	25%	69.57	
ARCHITECT FEES	9%	31.30	
Total Building Cost		**379.10**	

Costs per square foot of floor area

Exterior Wall	S.F. Area	7000	9000	11000	13000	15000	17000	19000	21000	23000
	L.F. Perimeter	240	280	303	325	354	372	397	422	447
Limestone with Concrete Block Back-up	Bearing Walls	261.75	244.30	229.35	218.85	212.30	205.65	201.30	197.85	195.00
	R/Conc. Frame	276.05	259.15	244.95	235.00	228.75	222.45	218.30	215.00	212.25
Face Brick with Concrete Block Back-up	Bearing Walls	221.80	208.05	197.20	189.60	184.70	180.00	176.80	174.30	172.20
	R/Conc. Frame	243.50	229.70	218.85	211.30	206.40	201.70	198.50	196.00	193.90
Decorative Concrete Block	Bearing Walls	210.70	197.90	188.20	181.45	177.00	172.85	170.00	167.75	165.85
	R/Conc. Frame	233.25	220.50	210.85	204.10	199.65	195.50	192.65	190.40	188.45
Perimeter Adj., Add or Deduct	Per 100 L.F.	36.10	28.20	23.00	19.45	16.85	14.90	13.40	12.00	10.95
Story Hgt. Adj., Add or Deduct	Per 1 Ft.	6.50	5.90	5.20	4.70	4.50	4.15	3.95	3.80	3.65

For Basement, add $28.75 per square foot of basement area

The above costs were calculated using the basic specifications shown on the facing page. These costs should be adjusted where necessary for design alternatives and owner's requirements. Reported completed project costs, for this type of structure, range from $107.95 to $281.60 per S.F.

Common additives

Description	Unit	$ Cost
Cells Prefabricated, 5'-6' wide,		
7'-8' high, 7'-8' deep	Each	12,500
Elevators, Hydraulic passenger, 2 stops		
1500# capacity	Each	60,900
2500# capacity	Each	64,300
3500# capacity	Each	67,600
Emergency Lighting, 25 watt, battery operated		
Lead battery	Each	287
Nickel cadmium	Each	845
Flagpoles, Complete		
Aluminum, 20' high	Each	1700
40' high	Each	3600
70' high	Each	10,300
Fiberglass, 23' high	Each	1825
39'-5" high	Each	3475
59' high	Each	6625

Description	Unit	$ Cost
Lockers, Steel, Single tier, 60" to 72"	Opening	200 - 350
2 tier, 60" or 72" total	Opening	124 - 152
5 tier, box lockers	Opening	70 - 82.50
Locker bench, lam. maple top only	L.F.	22
Pedestals, steel pipe	Each	64
Safe, Office type, 1 hour rating		
30" x 18" x 18"	Each	2400
60" x 36" x 18", double door	Each	8975
Shooting Range, Incl. bullet traps,		
target provisions, and contols,		
not incl. structural shell	Each	41,400
Smoke Detectors		
Ceiling type	Each	233
Duct type	Each	525
Sound System		
Amplifier, 250 watts	Each	2400
Speaker, ceiling or wall	Each	196
Trumpet	Each	375

Model costs calculated for a 2 story building with 12′ story height and 11,000 square feet of floor area

Police Station

				Unit	Unit Cost	Cost Per S.F.	% Of Sub-Total
A. SUBSTRUCTURE							
1010	Standard Foundations	Poured concrete; strip and spread footings		S.F. Ground	2.60	1.30	
1020	Special Foundations	N/A		—	—	—	
1030	Slab on Grade	4″ reinforced concrete with vapor barrier and granular base		S.F. Slab	4.77	2.39	3.9%
2010	Basement Excavation	Site preparation for slab and trench for foundation wall and footing		S.F. Ground	.31	.16	
2020	Basement Walls	4′ foundation wall		L.F. Wall	76	2.76	
B. SHELL							
	B10 Superstructure						
1010	Floor Construction	Open web steel joists, slab form, concrete		S.F. Floor	9.38	4.69	3.8%
1020	Roof Construction	Metal deck on open web steel joists		S.F. Roof	3.52	1.76	
	B20 Exterior Enclosure						
2010	Exterior Walls	Limestone with concrete block backup	80% of wall	S.F. Wall	67	35.17	
2020	Exterior Windows	Metal horizontal sliding	20% of wall	Each	1049	9.25	27.5%
2030	Exterior Doors	Hollow metal		Each	2488	1.82	
	B30 Roofing						
3010	Roof Coverings	Built-up tar and gravel with flashing; perlite/EPS composite insulation		S.F. Roof	6.74	3.37	2.0%
3020	Roof Openings	N/A		—	—	—	
C. INTERIORS							
1010	Partitions	Concrete block	20 S.F. Floor/L.F. Partition	S.F. Partition	8.64	4.32	
1020	Interior Doors	Single leaf kalamein fire door	200 S.F. Floor/Door	Each	1001	5.01	
1030	Fittings	Toilet partitions		S.F. Floor	1.02	1.02	
2010	Stair Construction	Concrete filled metal pan		Flight	14,675	2.67	16.0%
3010	Wall Finishes	90% paint, 10% ceramic tile		S.F. Surface	3.40	3.40	
3020	Floor Finishes	70% vinyl composition tile, 20% carpet tile, 10% ceramic tile		S.F. Floor	3.94	3.94	
3030	Ceiling Finishes	Mineral fiber tile on concealed zee bars		S.F. Ceiling	6.51	6.51	
D. SERVICES							
	D10 Conveying						
1010	Elevators & Lifts	One hydraulic passenger elevator		Each	78,870	7.17	4.3%
1020	Escalators & Moving Walks	N/A		—	—	—	
	D20 Plumbing						
2010	Plumbing Fixtures	Toilet and service fixtures, supply and drainage	1 Fixture/580 S.F. Floor	Each	4263	7.35	
2020	Domestic Water Distribution	Oil fired water heater		S.F. Floor	2.83	2.83	7.2%
2040	Rain Water Drainage	Roof drains		S.F. Roof	3.80	1.90	
	D30 HVAC						
3010	Energy Supply	N/A		—	—	—	
3020	Heat Generating Systems	N/A		—	—	—	
3030	Cooling Generating Systems	N/A		—	—	—	11.5 %
3050	Terminal & Package Units	Multizone HVAC air cooled system		S.F. Floor	19.40	19.40	
3090	Other HVAC Sys. & Equipment	N/A		—	—	—	
	D40 Fire Protection						
4010	Sprinklers	Wet pipe sprinkler system		S.F. Floor	3.34	3.34	2.5%
4020	Standpipes	Standpipe		S.F. Floor	.88	.88	
	D50 Electrical						
5010	Electrical Service/Distribution	400 ampere service, panel board and feeders		S.F. Floor	1.81	1.81	
5020	Lighting & Branch Wiring	T-8 fluorescent fixtures, receptacles, switches, A.C. and misc. power		S.F. Floor	11.63	11.63	12.2%
5030	Communications & Security	Addressable alarm systems, internet wiring, intercom and emergency lighting		S.F. Floor	6.93	6.93	
5090	Other Electrical Systems	Emergency generator, 15 kW		S.F. Floor	.22	.22	
E. EQUIPMENT & FURNISHINGS							
1010	Commercial Equipment	N/A		—	—	—	
1020	Institutional Equipment	Lockers, detention rooms, cells, gasoline dispensers		S.F. Floor	13.08	13.08	9.1 %
1030	Vehicular Equipment	Gasoline dispenser system		S.F. Floor	2.25	2.25	
1090	Other Equipment	N/A		—	—	—	
F. SPECIAL CONSTRUCTION							
1020	Integrated Construction	N/A		—	—	—	0.0 %
1040	Special Facilities	N/A		—	—	—	
G. BUILDING SITEWORK	**N/A**						

	Sub-Total		168.33	**100%**
CONTRACTOR FEES (General Requirements: 10%, Overhead: 5%, Profit: 10%)		25%	42.08	
ARCHITECT FEES		9%	18.94	
Total Building Cost			**229.35**	

Costs per square foot of floor area

Exterior Wall	S.F. Area	5000	7000	9000	11000	13000	15000	17000	19000	21000
	L.F. Perimeter	300	380	420	468	486	513	540	580	620
Face Brick with Concrete Block Back-up	Steel Frame	159.25	149.65	140.50	135.35	129.80	126.20	123.45	121.90	120.70
	Bearing Walls	157.10	147.50	138.35	133.20	127.65	124.05	121.30	119.75	118.50
Limestone with Concrete Block Back-up	Steel Frame	192.40	179.65	166.35	158.85	150.45	145.15	141.00	138.80	137.00
	Bearing Walls	189.95	177.25	163.90	156.45	148.05	142.70	138.60	136.40	134.60
Decorative Concrete Block	Steel Frame	147.80	139.35	131.70	127.25	122.70	119.75	117.45	116.15	115.05
	Bearing Walls	145.65	137.15	129.50	125.10	120.50	117.60	115.30	114.00	112.90
Perimeter Adj., Add or Deduct	Per 100 L.F.	17.10	12.25	9.60	7.80	6.60	5.80	5.05	4.60	4.05
Story Hgt. Adj., Add or Deduct	Per 1 Ft.	2.65	2.40	2.10	1.90	1.70	1.55	1.45	1.40	1.30
For Basement, add $28.00 per square foot of basement area										

The above costs were calculated using the basic specifications shown on the facing page. These costs should be adjusted where necessary for design alternatives and owner's requirements. Reported completed project costs, for this type of structure, range from $89.55 to $232.45 per S.F.

Common additives

Description	Unit	$ Cost	Description	Unit	$ Cost
Closed Circuit Surveillance, One station			Mail Boxes, Horizontal, key lock, 15" x 6" x 5"	Each	53.50
Camera and monitor	Each	1875	Double 15" x 12" x 5"	Each	98.50
For additional camera stations, add	Each	1025	Quadruple 15" x 12" x 10"	Each	126
Emergency Lighting, 25 watt, battery operated			Vertical, 6" x 5" x 15", aluminum	Each	48
Lead battery	Each	287	Bronze	Each	59
Nickel cadmium	Each	845	Steel, enameled	Each	48
Flagpoles, Complete			Scales, Dial type, 5 ton cap.		
Aluminum, 20' high	Each	1700	8' x 6' platform	Each	11,100
40' high	Each	3600	9' x 7' platform	Each	14,300
70' high	Each	10,300	Smoke Detectors		
Fiberglass, 23' high	Each	1825	Ceiling type	Each	233
39'-5" high	Each	3475	Duct type	Each	525
59' high	Each	6625			

Model costs calculated for a 1 story building with 14' story height and 13,000 square feet of floor area

Post Office

				Unit	Unit Cost	Cost Per S.F.	% Of Sub-Total
A.	**SUBSTRUCTURE**						
1010	Standard Foundations	Poured concrete; strip and spread footings		S.F. Ground	1.84	1.84	
1020	Special Foundations	N/A		—	—	—	
1030	Slab on Grade	4" reinforced concrete with vapor barrier and granular base		S.F. Slab	4.77	4.77	11.1%
2010	Basement Excavation	Site preparation for slab and trench for foundation wall and footing		S.F. Ground	.31	.31	
2020	Basement Walls	4' foundation wall		L.F. Wall	76	3.67	
B.	**SHELL**						
	B10	**Superstructure**					
1010	Floor Construction	Steel column fireproofing		L.F. Column	31.36	.09	7.4%
1020	Roof Construction	Metal deck, open web steel joists, columns		S.F. Roof	6.99	6.99	
	B20	**Exterior Enclosure**					
2010	Exterior Walls	Face brick with concrete block backup	80% of wall	S.F. Wall	30.28	12.68	
2020	Exterior Windows	Double strength window glass	20% of wall	Each	703	3.20	17.9%
2030	Exterior Doors	Double aluminum & glass, single aluminum, hollow metal, steel overhead		Each	2455	1.14	
	B30	**Roofing**					
3010	Roof Coverings	Built-up tar and gravel with flashing; perlite/EPS composite insulation		S.F. Roof	6.01	6.01	6.3%
3020	Roof Openings	N/A		—	—	—	
C.	**INTERIORS**						
1010	Partitions	Concrete block	15 S.F. Floor/L.F. Partition	S.F. Partition	8.63	6.90	
1020	Interior Doors	Single leaf hollow metal	150 S.F. Floor/Door	Each	1001	6.67	
1030	Fittings	Toilet partitions, cabinets, shelving, lockers		S.F. Floor	1.47	1.47	
2010	Stair Construction	N/A		—	—	—	23.3%
3010	Wall Finishes	Paint		S.F. Surface	2.07	3.31	
3020	Floor Finishes	50% vinyl tile, 50% paint		S.F. Floor	2.25	2.25	
3030	Ceiling Finishes	Mineral fiber tile on concealed zee bars	25% of area	S.F. Ceiling	6.51	1.63	
D.	**SERVICES**						
	D10	**Conveying**					
1010	Elevators & Lifts	N/A		—	—	—	0.0 %
1020	Escalators & Moving Walks	N/A		—	—	—	
	D20	**Plumbing**					
2010	Plumbing Fixtures	Toilet and service fixtures, supply and drainage	1 Fixture/1180 S.F. Floor	Each	2183	1.85	
2020	Domestic Water Distribution	Gas fired water heater		S.F. Floor	.37	.37	3.5%
2040	Rain Water Drainage	Roof drains		S.F. Roof	1.15	1.15	
	D30	**HVAC**					
3010	Energy Supply	N/A		—	—	—	
3020	Heat Generating Systems	Included in D3050		—	—	—	
3030	Cooling Generating Systems	N/A		—	—	—	8.3 %
3050	Terminal & Package Units	Single zone, gas heating, electric cooling		S.F. Floor	7.86	7.86	
3090	Other HVAC Sys. & Equipment	N/A		—	—	—	
	D40	**Fire Protection**					
4010	Sprinklers	Wet pipe sprinkler system		S.F. Floor	4.11	4.11	5.1%
4020	Standpipes	Standpipe		S.F. Floor	.77	.77	
	D50	**Electrical**					
5010	Electrical Service/Distribution	600 ampere service, panel board and feeders		S.F. Floor	3.06	3.06	
5020	Lighting & Branch Wiring	T-8 fluorescent fixtures, receptacles, switches, A.C. and misc. power		S.F. Floor	8.59	8.59	17.0%
5030	Communications & Security	Addressable alarm systems, internet wiring and emergency lighting		S.F. Floor	4.43	4.43	
5090	Other Electrical Systems	Emergency generator, 15 kW		S.F. Floor	.13	.13	
E.	**EQUIPMENT & FURNISHINGS**						
1010	Commercial Equipment	N/A		—	—	—	
1020	Institutional Equipment	N/A		—	—	—	0.0 %
1030	Vehicular Equipment	N/A		—	—	—	
1090	Other Equipment	N/A		—	—	—	
F.	**SPECIAL CONSTRUCTION**						
1020	Integrated Construction	N/A		—	—	—	0.0 %
1040	Special Facilities	N/A		—	—	—	
G.	**BUILDING SITEWORK**	**N/A**					

		Sub-Total	95.25	100%
	CONTRACTOR FEES (General Requirements: 10%, Overhead: 5%, Profit: 10%)	25%	23.83	
	ARCHITECT FEES	9%	10.72	
	Total Building Cost		**129.80**	

Costs per square foot of floor area

Exterior Wall	S.F. Area	5000	10000	15000	21000	25000	30000	40000	50000	60000
	L.F. Perimeter	287	400	500	600	660	700	834	900	1000
Face Brick with Concrete Block Back-up	Steel Frame	242.55	201.40	186.45	176.50	172.10	166.60	161.60	156.65	154.20
	Bearing Walls	233.15	195.70	181.95	172.75	168.60	163.35	158.70	153.95	151.60
Concrete Block	Steel Frame	210.65	178.65	167.25	159.95	156.75	153.05	149.45	146.15	144.40
Brick Veneer	Steel Frame	233.70	194.70	180.70	171.45	167.35	162.40	157.85	153.35	151.10
Galvanized Steel Siding	Steel Frame	199.50	170.70	160.55	154.05	151.25	148.05	144.90	142.10	140.60
Metal Sandwich Panel	Steel Frame	202.65	173.05	162.65	155.95	153.05	149.75	146.50	143.60	142.05
Perimeter Adj., Add or Deduct	Per 100 L.F.	29.20	14.60	9.70	6.95	5.85	4.85	3.75	2.95	2.40
Story Hgt. Adj., Add or Deduct	Per 1 Ft.	5.95	4.10	3.45	3.00	2.70	2.45	2.20	1.90	1.70
For Basement, add $26.95 per square foot of basement area										

The above costs were calculated using the basic specifications shown on the facing page. These costs should be adjusted where necessary for design alternatives and owner's requirements. Reported completed project costs, for this type of structure, range from $84.85 to $264.55 per S.F.

Common additives

Description	Unit	$ Cost
Bar, Front Bar	L.F.	375
Back Bar	L.F.	300
Booth, Upholstered, custom straight	L.F.	212 - 390
"L" or "U" shaped	L.F.	220 - 370
Bleachers, Telescoping, manual		
To 15 tier	Seat	118 - 164
21-30 tier	Seat	257 - 310
Courts		
Ceiling	Court	9275
Floor	Court	15,700
Walls	Court	26,700
Emergency Lighting, 25 watt, battery operated		
Lead battery	Each	287
Nickel cadmium	Each	845
Kitchen Equipment		
Broiler	Each	4050
Cooler, 6 ft. long, reach-in	Each	5150
Dishwasher, 10-12 racks per hr.	Each	5075
Food warmer, counter 1.2 KW	Each	740
Freezer, reach-in, 44 C.F.	Each	3250
Ice cube maker, 50 lb. per day	Each	1875

Description	Unit	$ Cost
Lockers, Steel, single tier, 60" or 72"	Opening	200 - 350
2 tier, 60" or 72" total	Opening	124 - 152
5 tier, box lockers	Opening	70 - 82.50
Locker bench, lam. maple top only	L.F.	22
Pedestals, steel pipe	Each	64
Sauna, Prefabricated, complete		
6' x 4'	Each	5875
6' x 9'	Each	8550
8' x 8'	Each	10,100
8' x 10'	Each	11,300
10' x 12'	Each	14,000
Sound System		
Amplifier, 250 watts	Each	2400
Speaker, ceiling or wall	Each	196
Trumpet	Each	375
Steam Bath, Complete, to 140 C.F.	Each	2075
To 300 C.F.	Each	2350
To 800 C.F.	Each	6425
To 2500 C.F.	Each	9600

Racquetball Court

Model costs calculated for a 2 story building with 12' story height and 30,000 square feet of floor area

				Unit	Unit Cost	Cost Per S.F.	% Of Sub-Total
A.	**SUBSTRUCTURE**						
1010	Standard Foundations	Poured concrete; strip and spread footings		S.F. Ground	3.20	1.60	
1020	Special Foundations	N/A		—	—	—	
1030	Slab on Grade	5" reinforced concrete with vapor barrier and granular base		S.F. Slab	6.33	4.11	6.2%
2010	Basement Excavation	Site preparation for slab and trench for foundation wall and footing		S.F. Ground	.31	.20	
2020	Basement Walls	4' foundation wall		L.F. Wall	76	1.77	
B.	**SHELL**						
	B10 Superstructure						
1010	Floor Construction	Open web steel joists, slab form, concrete, columns	50% of area	S.F. Floor	15.18	7.59	9.2%
1020	Roof Construction	Metal deck on open web steel joists, columns		S.F. Roof	7.54	3.77	
	B20 Exterior Enclosure						
2010	Exterior Walls	Face brick with concrete block backup	95% of wall	S.F. Wall	31.26	16.63	
2020	Exterior Windows	Storefront	5% of wall	S.F. Window	94	2.63	15.9%
2030	Exterior Doors	Aluminum and glass and hollow metal		Each	2829	.37	
	B30 Roofing						
3010	Roof Coverings	Built-up tar and gravel with flashing; perlite/EPS composite insulation		S.F. Roof	7.58	3.79	3.1%
3020	Roof Openings	Roof hatches		S.F. Roof	.12	.06	
C.	**INTERIORS**						
1010	Partitions	Concrete block, gypsum board on metal studs	25 S.F. Floor/L.F. Partition	S.F. Partition	13.05	5.22	
1020	Interior Doors	Single leaf hollow metal	810 S.F. Floor/Door	Each	1001	1.24	
1030	Fittings	Toilet partitions		S.F. Floor	.48	.48	
2010	Stair Construction	Concrete filled metal pan		Flight	12,275	1.23	12.9%
3010	Wall Finishes	Paint		S.F. Surface	1.24	.99	
3020	Floor Finishes	80% carpet tile, 20% ceramic tile	50% of floor area	S.F. Floor	5.74	2.87	
3030	Ceiling Finishes	Mineral fiber tile on concealed zee bars	60% of area	S.F. Ceiling	6.51	3.90	
D.	**SERVICES**						
	D10 Conveying						
1010	Elevators & Lifts	N/A		—	—	—	0.0 %
1020	Escalators & Moving Walks	N/A		—	—	—	
	D20 Plumbing						
2010	Plumbing Fixtures	Kitchen, bathroom and service fixtures, supply and drainage	1 Fixture/1000 S.F. Floor	Each	3290	3.29	
2020	Domestic Water Distribution	Gas fired water heater		S.F. Floor	2.33	2.33	4.9%
2040	Rain Water Drainage	Roof drains		S.F. Roof	.84	.42	
	D30 HVAC						
3010	Energy Supply	N/A		—	—	—	
3020	Heat Generating Systems	Included in D3050		—	—	—	
3030	Cooling Generating Systems	N/A		—	—	—	21.4 %
3050	Terminal & Package Units	Multizone unit, gas heating, electric cooling		S.F. Floor	26.40	26.40	
3090	Other HVAC Sys. & Equipment	N/A		—	—	—	
	D40 Fire Protection						
4010	Sprinklers	Sprinklers, light hazard		S.F. Floor	.48	.48	0.4%
4020	Standpipes	N/A		—	—	—	
	D50 Electrical						
5010	Electrical Service/Distribution	400 ampere service, panel board and feeders		S.F. Floor	.82	.82	
5020	Lighting & Branch Wiring	Fluorescent and high intensity discharge fixtures, receptacles, switches, A.C. and misc. power		S.F. Floor	12.07	12.07	10.7%
5030	Communications & Security	Alarm systems and emergency lighting		S.F. Floor	.31	.31	
5090	Other Electrical Systems	Emergency generator, 15 kW		S.F. Floor	.06	.06	
E.	**EQUIPMENT & FURNISHINGS**						
1010	Commercial Equipment	N/A		—	—	—	
1020	Institutional Equipment	N/A		—	—	—	15.2 %
1030	Vehicular Equipment	N/A		—	—	—	
1090	Other Equipment	Courts, sauna baths		S.F. Floor	18.78	18.78	
F.	**SPECIAL CONSTRUCTION**						
1020	Integrated Construction	N/A		—	—	—	0.0 %
1040	Special Facilities	N/A		—	—	—	
G.	**BUILDING SITEWORK**	**N/A**					

		Sub-Total	123.41	**100%**
CONTRACTOR FEES (General Requirements: 10%, Overhead: 5%, Profit: 10%)		25%	30.85	
ARCHITECT FEES		8%	12.34	
		Total Building Cost	**166.60**	

Costs per square foot of floor area

Exterior Wall	S.F. Area	5000	6000	7000	8000	9000	10000	11000	12000	13000
	L.F. Perimeter	286	320	353	386	397	425	454	460	486
Face Brick with Concrete Block Back-up	Steel Joists	174.10	170.05	167.00	164.70	161.00	159.40	158.15	155.60	154.65
	Wood Joists	176.10	171.80	168.60	166.20	162.35	160.65	159.35	156.65	155.65
Stucco on Concrete Block	Steel Joists	176.05	172.30	169.55	167.50	164.25	162.75	161.65	159.40	158.55
	Wood Joists	165.65	162.10	159.45	157.45	154.35	152.90	151.80	149.70	148.85
Limestone with Concrete Block Back-up	Steel Joists	214.55	208.20	203.45	199.90	193.80	191.25	189.25	185.00	183.45
	Wood Joists	204.85	198.65	194.00	190.50	184.55	182.05	180.10	175.95	174.45
Perimeter Adj., Add or Deduct	Per 100 L.F.	15.65	13.00	11.15	9.80	8.70	7.85	7.10	6.50	6.05
Story Hgt. Adj., Add or Deduct	Per 1 Ft.	2.75	2.45	2.35	2.25	2.10	2.00	1.90	1.80	1.75
For Basement, add $28.35 per square foot of basement area										

The above costs were calculated using the basic specifications shown on the facing page. These costs should be adjusted where necessary for design alternatives and owner's requirements. Reported completed project costs, for this type of structure, range from $74.10 to $229.40 per S.F.

Common additives

Description	Unit	$ Cost
Carrels Hardwood	Each	605 - 1150
Emergency Lighting, 25 watt, battery operated		
Lead battery	Each	287
Nickel cadmium	Each	845
Flagpoles, Complete		
Aluminum, 20' high	Each	1700
40' high	Each	3600
70' high	Each	10,300
Fiberglass, 23' high	Each	1825
39'-5" high	Each	3475
59' high	Each	6625
Gym Floor, Incl. sleepers and finish, maple	S.F.	13.15
Intercom System, 25 Station capacity		
Master station	Each	2725
Intercom outlets	Each	174
Handset	Each	480

Description	Unit	$ Cost
Lockers, Steel, single tier, 60" to 72"	Opening	200 - 350
2 tier, 60" to 72" total	Opening	124 - 152
5 tier, box lockers	Opening	70 - 82.50
Locker bench, lam. maple top only	L.F.	22
Pedestals, steel pipe	Each	64
Seating		
Auditorium chair, all veneer	Each	240
Veneer back, padded seat	Each	290
Upholstered, spring seat	Each	279
Classroom, movable chair & desk	Set	65 - 120
Lecture hall, pedestal type	Each	229 - 680
Smoke Detectors		
Ceiling type	Each	233
Duct type	Each	525
Sound System		
Amplifier, 250 watts	Each	2400
Speaker, ceiling or wall	Each	196
Trumpet	Each	375
Swimming Pools, Complete, gunite	S.F.	68.50 - 84.50

Important: See the Reference Section for Location Factors

Model costs calculated for a 1 story building with 12′ story height and 10,000 square feet of floor area

				Unit	Unit Cost	Cost Per S.F.	% Of Sub-Total
A.	**SUBSTRUCTURE**						
1010	Standard Foundations	Poured concrete; strip and spread footings		S.F. Ground	2.20	2.20	
1020	Special Foundations	N/A		—	—	—	
1030	Slab on Grade	4″ reinforced concrete with vapor barrier and granular base		S.F. Slab	4.77	4.77	9.0%
2010	Basement Excavation	Site preparation for slab and trench for foundation wall and footing		S.F. Ground	.31	.31	
2020	Basement Walls	4′ foundation wall		L.F. Wall	76	3.21	
B.	**SHELL**						
	B10 Superstructure						
1010	Floor Construction	N/A		—	—	—	6.3 %
1020	Roof Construction	Metal deck, open web steel joists, beams, interior columns		S.F. Roof	7.37	7.37	
	B20 Exterior Enclosure						
2010	Exterior Walls	Face brick with concrete block backup	85% of wall	S.F. Wall	30.31	13.14	
2020	Exterior Windows	Window wall	15% of wall	Each	43.60	3.34	16.1%
2030	Exterior Doors	Double aluminum and glass		Each	5813	2.33	
	B30 Roofing						
3010	Roof Coverings	Built-up tar and gravel with flashing; perlite/EPS composite insulation		S.F. Roof	6.03	6.03	5.2%
3020	Roof Openings	N/A		—	—	—	
C.	**INTERIORS**						
1010	Partitions	Concrete block	8 S.F. Floor/S.F. Partition	S.F. Partition	8.62	10.77	
1020	Interior Doors	Single leaf hollow metal	700 S.F. Floor/Door	Each	1001	1.43	
1030	Fittings	Toilet partitions		S.F. Floor	1.44	1.44	
2010	Stair Construction	N/A		—	—	—	27.2%
3010	Wall Finishes	Paint		S.F. Surface	2.09	5.23	
3020	Floor Finishes	50% vinyl tile, 50% carpet		S.F. Floor	6.40	6.40	
3030	Ceiling Finishes	Mineral fiber tile on concealed zee bars		S.F. Ceiling	6.51	6.51	
D.	**SERVICES**						
	D10 Conveying						
1010	Elevators & Lifts	N/A		—	—	—	0.0 %
1020	Escalators & Moving Walks	N/A		—	—	—	
	D20 Plumbing						
2010	Plumbing Fixtures	Toilet and service fixtures, supply and drainage	1 Fixture/455 S.F. Floor	Each	3112	6.84	
2020	Domestic Water Distribution	Gas fired water heater		S.F. Floor	1.36	1.36	7.8%
2040	Rain Water Drainage	Roof drains		S.F. Roof	.92	.92	
	D30 HVAC						
3010	Energy Supply	N/A		—	—	—	
3020	Heat Generating Systems	N/A		—	—	—	
3030	Cooling Generating Systems	N/A		—	—	—	19.3 %
3050	Terminal & Package Units	Multizone HVAC air cooled system		S.F. Floor	22.55	22.55	
3090	Other HVAC Sys. & Equipment	N/A		—	—	—	
	D40 Fire Protection						
4010	Sprinklers	Sprinkler, light hazard		S.F. Floor	3.16	3.16	3.1%
4020	Standpipes	Standpipe		S.F. Floor	.51	.51	
	D50 Electrical						
5010	Electrical Service/Distribution	200 ampere service, panel board and feeders		S.F. Floor	.98	.98	
5020	Lighting & Branch Wiring	T-8 fluorescent fixtures, receptacles, switches, A.C. and misc. power		S.F. Floor	2.72	2.72	6.1%
5030	Communications & Security	Addressable alarm systems, partial internet wiring and emergency lighting		S.F. Floor	3.38	3.38	
5090	Other Electrical Systems	Emergency generator, 15 kW		S.F. Floor	.08	.08	
E.	**EQUIPMENT & FURNISHINGS**						
1010	Commercial Equipment	N/A		—	—	—	
1020	Institutional Equipment	N/A		—	—	—	0.0 %
1030	Vehicular Equipment	N/A		—	—	—	
1090	Other Equipment	N/A		—	—	—	
F.	**SPECIAL CONSTRUCTION**						
1020	Integrated Construction	N/A		—	—	—	0.0 %
1040	Special Facilities	N/A		—	—	—	
G.	**BUILDING SITEWORK**	**N/A**					

		Sub-Total	116.98	**100%**
CONTRACTOR FEES (General Requirements: 10%, Overhead: 5%, Profit: 10%)		25%	29.26	
ARCHITECT FEES		9%	13.16	

Total Building Cost	**159.40**

Costs per square foot of floor area

Exterior Wall	S.F. Area	2000	2800	3500	4200	5000	5800	6500	7200	8000
	L.F. Perimeter	180	212	240	268	300	314	336	344	368
Wood Siding	Wood Frame	222.05	210.45	204.70	200.85	197.75	193.95	192.10	189.65	188.30
Brick Veneer	Wood Frame	234.55	221.00	214.25	209.75	206.15	201.50	199.30	196.30	194.70
Face Brick with Concrete Block Back-up	Wood Joists	241.65	227.00	219.65	214.75	210.85	205.75	203.35	200.05	198.30
	Steel Joists	234.90	218.65	210.50	205.10	200.80	195.00	192.40	188.65	186.70
Stucco on Concrete Block	Wood Joists	228.45	215.80	209.55	205.35	202.00	197.80	195.75	193.05	191.55
	Steel Joists	213.70	201.10	194.80	190.65	187.25	183.05	181.00	178.30	176.80
Perimeter Adj., Add or Deduct	Per 100 L.F.	26.10	18.65	14.95	12.40	10.45	8.95	8.05	7.25	6.50
Story Hgt. Adj., Add or Deduct	Per 1 Ft.	2.65	2.30	2.00	1.90	1.80	1.60	1.55	1.45	1.35
For Basement, add $32.80 per square foot of basement area										

The above costs were calculated using the basic specifications shown on the facing page. These costs should be adjusted where necessary for design alternatives and owner's requirements. Reported completed project costs, for this type of structure, range from $114.85 to $267.65 per S.F.

Common additives

Description	Unit	$ Cost
Bar, Front Bar	L.F.	375
Back bar	L.F.	300
Booth, Upholstered, custom straight	L.F.	212 - 390
"L" or "U" shaped	L.F.	220 - 370
Cupola, Stock unit, redwood		
30" square, 37" high, aluminum roof	Each	685
Copper roof	Each	365
Fiberglass, 5'-0" base, 63" high	Each	3500 - 4350
6'-0" base, 63" high	Each	4825 - 7025
Decorative Wood Beams, Non load bearing		
Rough sawn, 4" x 6"	L.F.	7.60
4" x 8"	L.F.	8.50
4" x 10"	L.F.	9.45
4" x 12"	L.F.	10.40
8" x 8"	L.F.	13.35
Emergency Lighting, 25 watt, battery operated		
Lead battery	Each	287
Nickel cadmium	Each	845

Description	Unit	$ Cost
Fireplace, Brick, not incl. chimney or foundation		
30" x 29" opening	Each	2875
Chimney, standard brick		
Single flue, 16" x 20"	V.L.F.	94
20" x 20"	V.L.F.	109
2 Flue, 20" x 24"	V.L.F.	137
20" x 32"	V.L.F.	166
Kitchen Equipment		
Broiler	Each	4050
Coffee urn, twin 6 gallon	Each	3025
Cooler, 6 ft. long	Each	5150
Dishwasher, 10-12 racks per hr.	Each	5075
Food warmer, counter, 1.2 KW	Each	740
Freezer, 44 C.F., reach-in	Each	3250
Ice cube maker, 50 lb. per day	Each	1875
Range with 1 oven	Each	2700
Refrigerators, Prefabricated, walk-in		
7'-6" high, 6' x 6'	S.F.	173
10' x 10'	S.F.	136
12' x 14'	S.F.	120
12' x 20'	S.F.	106

Model costs calculated for a 1 story building with 12' story height and 5,000 square feet of floor area

Restaurant

				Unit	Unit Cost	Cost Per S.F.	% Of Sub-Total
A.	**SUBSTRUCTURE**						
1010	Standard Foundations	Poured concrete; strip and spread footings		S.F. Ground	2.39	2.39	
1020	Special Foundations	N/A		—	—	—	
1030	Slab on Grade	4" reinforced concrete with vapor barrier and granular base		S.F. Slab	4.77	4.77	8.3%
2010	Basement Excavation	Site preparation for slab and trench for foundation wall and footing		S.F. Ground	.52	.52	
2020	Basement Walls	4' foundation wall		L.F. Wall	76	4.53	
B.	**SHELL**						
	B10 Superstructure						
1010	Floor Construction	Wood columns		S.F. Floor	.32	.80	5.4%
1020	Roof Construction	Plywood on wood truss (pitched)		S.F. Roof	6.45	7.22	
	B20 Exterior Enclosure						
2010	Exterior Walls	Cedar siding on wood studs with insulation	70% of wall	S.F. Wall	10.89	5.49	
2020	Exterior Windows	Storefront windows	30% of wall	Each	39.40	8.51	12.7%
2030	Exterior Doors	Aluminum and glass doors and entrance with transom		Each	4828	4.84	
	B30 Roofing						
3010	Roof Coverings	Cedar shingles with flashing (pitched); rigid fiberglass insulation		S.F. Roof	18.93	8.58	5.9%
3020	Roof Openings	Skylights		S.F. Roof	.09	.09	
C.	**INTERIORS**						
1010	Partitions	Gypsum board on wood studs	25 S.F. Floor/L.F. Partition	S.F. Partition	9.20	3.68	
1020	Interior Doors	Hollow core wood	250 S.F. Floor/Door	Each	498	1.99	
1030	Fittings	Toilet partitions		S.F. Floor	.69	.69	
2010	Stair Construction	N/A		—	—	—	15.7%
3010	Wall Finishes	75% paint, 25% ceramic tile		S.F. Surface	2.29	1.83	
3020	Floor Finishes	65% carpet, 35% quarry tile		S.F. Floor	8.51	8.51	
3030	Ceiling Finishes	Mineral fiber tile on concealed zee bars		S.F. Ceiling	6.51	6.51	
D.	**SERVICES**						
	D10 Conveying						
1010	Elevators & Lifts	N/A		—	—	—	0.0 %
1020	Escalators & Moving Walks	N/A		—	—	—	
	D20 Plumbing						
2010	Plumbing Fixtures	Kitchen, bathroom and service fixtures, supply and drainage	1 Fixture/355 S.F. Floor	Each	3397	9.57	
2020	Domestic Water Distribution	Gas fired water heater		S.F. Floor	7.14	7.14	12.3%
2040	Rain Water Drainage	Roof drains		S.F. Roof	1.55	1.55	
	D30 HVAC						
3010	Energy Supply	N/A		—	—	—	
3020	Heat Generating Systems	Included in D3050		—	—	—	
3030	Cooling Generating Systems	N/A		—	—	—	22.4 %
3050	Terminal & Package Units	Multizone unit, gas heating, electric cooling; kitchen ventilation		S.F. Floor	33.13	33.13	
3090	Other HVAC Sys. & Equipment	N/A		—	—	—	
	D40 Fire Protection						
4010	Sprinklers	Sprinklers, light hazard and ordinary hazard kitchen		S.F. Floor	6.59	6.59	5.6%
4020	Standpipes	Standpipe			1.73	1.73	
	D50 Electrical						
5010	Electrical Service/Distribution	400 ampere service, panel board and feeders		S.F. Floor	4.19	4.19	
5020	Lighting & Branch Wiring	T-8 fluorescent fixtures, receptacles, switches, A.C. and misc. power		S.F. Floor	9.29	9.29	11.6%
5030	Communications & Security	Addressable alarm systems and emergency lighting		S.F. Floor	3.39	3.39	
5090	Other Electrical Systems	Emergency generator, 15 kW		S.F. Floor	.34	.34	
E.	**EQUIPMENT & FURNISHINGS**						
1010	Commercial Equipment	N/A		—	—	—	
1020	Institutional Equipment	N/A		—	—	—	0.0 %
1030	Vehicular Equipment	N/A		—	—	—	
1090	Other Equipment	N/A		—	—	—	
F.	**SPECIAL CONSTRUCTION**						
1020	Integrated Construction	N/A		—	—	—	0.0 %
1040	Special Facilities	N/A		—	—	—	
G.	**BUILDING SITEWORK**	**N/A**					

				Sub-Total	147.87	100%
	CONTRACTOR FEES (General Requirements: 10%, Overhead: 5%, Profit: 10%)		25%	36.94		
	ARCHITECT FEES		7%	12.94		

Total Building Cost	**197.75**

Costs per square foot of floor area

Exterior Wall	S.F. Area	2000	2800	3500	4000	5000	5800	6500	7200	8000
	L.F. Perimeter	180	212	240	260	300	314	336	344	368
Face Brick with Concrete Block Back-up	Bearing Walls	217.45	207.65	202.80	200.30	196.90	192.90	191.20	188.45	187.30
	Steel Frame	220.05	210.30	205.40	203.00	199.60	195.50	193.85	191.15	189.95
Concrete Block With Stucco	Bearing Walls	206.30	198.15	194.15	192.05	189.25	185.90	184.50	182.30	181.30
	Steel Frame	209.75	201.75	197.80	195.80	193.05	189.75	188.40	186.20	185.25
Wood Siding	Wood Frame	200.70	193.65	190.20	188.40	185.95	183.05	181.85	179.90	179.05
Brick Veneer	Steel Frame	214.75	206.00	201.65	199.45	196.35	192.75	191.25	188.80	187.80
Perimeter Adj., Add or Deduct	Per 100 L.F.	34.25	24.40	19.50	17.15	13.65	11.75	10.50	9.60	8.55
Story Hgt. Adj., Add or Deduct	Per 1 Ft.	4.45	3.75	3.40	3.25	3.00	2.65	2.55	2.40	2.30
Basement—Not Applicable										

The above costs were calculated using the basic specifications shown on the facing page. These costs should be adjusted where necessary for design alternatives and owner's requirements. Reported completed project costs, for this type of structure, range from $114.60 to $223.45 per S.F.

Common additives

Description	Unit	$ Cost
Bar, Front Bar	L.F.	375
Back bar	L.F.	300
Booth, Upholstered, custom straight	L.F.	212 - 390
"L" or "U" shaped	L.F.	220 - 370
Drive-up Window	Each	7475 - 12,400
Emergency Lighting, 25 watt, battery operated		
Lead battery	Each	287
Nickel cadmium	Each	845
Kitchen Equipment		
Broiler	Each	4050
Coffee urn, twin 6 gallon	Each	3025
Cooler, 6 ft. long	Each	5150
Dishwasher, 10-12 racks per hr.	Each	5075
Food warmer, counter, 1.2 KW	Each	740
Freezer, 44 C.F., reach-in	Each	3250
Ice cube maker, 50 lb. per day	Each	1875
Range with 1 oven	Each	2700

Description	Unit	$ Cost
Refrigerators, Prefabricated, walk-in		
7'-6" High, 6' x 6'	S.F.	173
10' x 10'	S.F.	136
12' x 14'	S.F.	120
12' x 20'	S.F.	106
Serving		
Counter top (Stainless steel)	L.F.	177
Base cabinets	L.F.	375 - 600
Sound System		
Amplifier, 250 watts	Each	2400
Speaker, ceiling or wall	Each	196
Trumpet	Each	375
Storage		
Shelving	S.F.	16.40
Washing		
Stainless steel counter	L.F.	177

Model costs calculated for a 1 story building with 10' story height and 4,000 square feet of floor area

Restaurant, Fast Food

				Unit	Unit Cost	Cost Per S.F.	% Of Sub-Total
A.	**SUBSTRUCTURE**						
1010	Standard Foundations	Poured concrete; strip and spread footings		S.F. Ground	2.32	2.32	
1020	Special Foundations	N/A		—	—	—	
1030	Slab on Grade	4" reinforced concrete with vapor barrier and granular base		S.F. Slab	4.77	4.77	8.4%
2010	Basement Excavation	Site preparation for slab and trench for foundation wall and footing		S.F. Ground	.52	.52	
2020	Basement Walls	4' foundation wall		L.F. Wall	76	4.91	
B.	**SHELL**						
	B10 Superstructure						
1010	Floor Construction	N/A		—	—	—	4.0 %
1020	Roof Construction	Metal deck on open web steel joists		S.F. Roof	5.99	5.99	
	B20 Exterior Enclosure						
2010	Exterior Walls	Face brick with concrete block backup	70% of wall	S.F. Wall	30.31	13.79	
2020	Exterior Windows	Window wall	30% of wall	Each	45.35	8.85	20.1%
2030	Exterior Doors	Aluminum and glass		Each	3582	7.18	
	B30 Roofing						
3010	Roof Coverings	Built-up tar and gravel with flashing; perlite/EPS composite insulation		S.F. Roof	6.39	6.39	4.6%
3020	Roof Openings	Roof hatches		S.F. Roof	.47	.47	
C.	**INTERIORS**						
1010	Partitions	Gypsum board on metal studs	25 S.F. Floor/L.F. Partition	S.F. Partition	5.50	1.98	
1020	Interior Doors	Hollow core wood	1000 S.F. Floor/Door	Each	498	.50	
1030	Fittings	N/A		—	—	—	
2010	Stair Construction	N/A		—	—	—	17.7%
3010	Wall Finishes	Paint		S.F. Surface	2.56	1.84	
3020	Floor Finishes	Quarry tile		S.F. Floor	15.43	15.43	
3030	Ceiling Finishes	Mineral fiber tile on concealed zee bars		S.F. Ceiling	6.51	6.51	
D.	**SERVICES**						
	D10 Conveying						
1010	Elevators & Lifts	N/A		—	—	—	0.0 %
1020	Escalators & Moving Walks	N/A		—	—	—	
	D20 Plumbing						
2010	Plumbing Fixtures	Kitchen, bathroom and service fixtures, supply and drainage	1 Fixture/400 S.F. Floor	Each	3480	8.70	
2020	Domestic Water Distribution	Gas fired water heater		S.F. Floor	5.17	5.17	10.5%
2040	Rain Water Drainage	Roof drains		S.F. Roof	1.75	1.75	
	D30 HVAC						
3010	Energy Supply	N/A		—	—	—	
3020	Heat Generating Systems	Included in D3050		—	—	—	
3030	Cooling Generating Systems	N/A		—	—	—	11.8 %
3050	Terminal & Package Units	Multizone unit, gas heating, electric cooling; kitchen ventilation		S.F. Floor	17.58	17.58	
3090	Other HVAC Sys. & Equipment	N/A		—	—	—	
	D40 Fire Protection						
4010	Sprinklers	Sprinklers, light hazard		S.F. Floor	6.59	6.59	5.7%
4020	Standpipes	Standpipe		S.F. Floor	1.80	1.80	
	D50 Electrical						
5010	Electrical Service/Distribution	400 ampere service, panel board and feeders		S.F. Floor	4.33	4.33	
5020	Lighting & Branch Wiring	T-8 fluorescent fixtures, receptacles, switches, A.C. and misc. power		S.F. Floor	10.02	10.02	12.8%
5030	Communications & Security	Addressable alarm systems and emergency lighting		S.F. Floor	4.22	4.22	
5090	Other Electrical Systems	Emergency generator, 7.5 kW		S.F. Floor	.39	.39	
E.	**EQUIPMENT & FURNISHINGS**						
1010	Commercial Equipment	N/A		—	—	—	
1020	Institutional Equipment	N/A		—	—	—	4.3 %
1030	Vehicular Equipment	N/A		—	—	—	
1090	Other Equipment	Walk-in refrigerator		S.F. Floor	6.38	6.38	
F.	**SPECIAL CONSTRUCTION**						
1020	Integrated Construction	N/A		—	—	—	0.0 %
1040	Special Facilities	N/A		—	—	—	
G.	**BUILDING SITEWORK**	**N/A**					

	Sub-Total	148.38	100%
CONTRACTOR FEES (General Requirements: 10%, Overhead: 5%, Profit: 10%)	25%	37.08	
ARCHITECT FEES	8%	14.84	
	Total Building Cost	**200.30**	

Costs per square foot of floor area

Exterior Wall	S.F. Area	10000	15000	20000	25000	30000	35000	40000	45000	50000
	L.F. Perimeter	450	500	600	700	740	822	890	920	966
Face Brick with Concrete Block Back-up	Steel Frame	191.00	169.50	162.05	157.55	152.00	149.60	147.25	144.40	142.55
	Lam. Wood Truss	195.75	175.00	167.80	163.40	158.05	155.70	153.45	150.75	148.95
Concrete Block	Steel Frame	189.55	175.15	169.75	166.35	162.80	161.10	159.55	157.75	156.60
	Lam. Wood Truss	174.00	159.60	154.20	150.85	147.30	145.60	144.05	142.25	141.10
Galvanized Steel Siding	Steel Frame	185.20	170.80	165.30	162.05	158.50	156.80	155.20	153.45	152.25
Metal Sandwich Panel	Steel Joists	194.00	177.30	171.15	167.55	163.25	161.35	159.55	157.45	156.05
Perimeter Adj., Add or Deduct	Per 100 L.F.	13.05	8.75	6.60	5.25	4.30	3.75	3.30	2.90	2.60
Story Hgt. Adj., Add or Deduct	Per 1 Ft.	2.00	1.50	1.35	1.30	1.10	1.00	1.05	0.95	0.80
Basement—Not Applicable										

The above costs were calculated using the basic specifications shown on the facing page. These costs should be adjusted where necessary for design alternatives and owner's requirements. Reported completed project costs, for this type of structure, range from $67.80 to $210.95 per S.F.

Common additives

Description	Unit	$ Cost		Description	Unit	$ Cost
Bar, Front Bar	L.F.	375		Rink		
Back bar	L.F.	300		Dasher boards & top guard	Each	172,000
Booth, Upholstered, custom straight	L.F.	212 - 390		Mats, rubber	S.F.	20.50
"L" or "U" shaped	L.F.	220 - 370		Score Board	Each	12,400 - 16,700
Bleachers, Telescoping, manual						
To 15 tier	Seat	118 - 164				
16-20 tier	Seat	242 - 297				
21-30 tier	Seat	257 - 310				
For power operation, add	Seat	47 - 73.50				
Emergency Lighting, 25 watt, battery operated						
Lead battery	Each	287				
Nickel cadmium	Each	845				
Lockers, Steel, single tier, 60" or 72"	Opening	200 - 350				
2 tier, 60" or 72" total	Opening	124 - 152				
5 tier, box lockers	Opening	70 - 82.50				
Locker bench, lam. maple top only	L.F.	22				
Pedestals, steel pipe	Each	64				

Model costs calculated for a 1 story building with 24' story height and 30,000 square feet of floor area

Rink, Hockey/Indoor Soccer

			Unit	Unit Cost	Cost Per S.F.	% Of Sub-Total
A. SUBSTRUCTURE						
1010	Standard Foundations	Poured concrete; strip and spread footings	S.F. Ground	1.62	1.62	
1020	Special Foundations	N/A	—	—	—	
1030	Slab on Grade	6" reinforced concrete with vapor barrier and granular base	S.F. Slab	5.79	5.79	8.2%
2010	Basement Excavation	Site preparation for slab and trench for foundation wall and footing	S.F. Ground	.18	.18	
2020	Basement Walls	4' foundation wall	L.F. Wall	71	2.42	
B. SHELL						
	B10 Superstructure					
1010	Floor Construction	Wide flange beams and columns	S.F. Floor	28.47	19.93	26.2%
1020	Roof Construction	Metal deck on steel joist	S.F. Roof	11.93	11.93	
	B20 Exterior Enclosure					
2010	Exterior Walls	Concrete block 95% of wall	S.F. Wall	12.55	7.06	
2020	Exterior Windows	Store front 5% of wall	Each	43.60	1.28	7.5%
2030	Exterior Doors	Aluminum and glass, hollow metal, overhead	Each	2938	.80	
	B30 Roofing					
3010	Roof Coverings	Elastomeric neoprene membrane with flashing; perlite/EPS composite insulation	S.F. Roof	4.62	4.62	4.0%
3020	Roof Openings	Roof hatches	S.F. Roof	.23	.23	
C. INTERIORS						
1010	Partitions	Concrete block 140 S.F. Floor/L.F. Partition	S.F. Partition	8.63	.74	
1020	Interior Doors	Hollow metal 2500 S.F. Floor/Door	Each	1001	.40	
1030	Fittings	N/A	—	—	—	
2010	Stair Construction	N/A	—	—	—	6.1%
3010	Wall Finishes	Paint	S.F. Surface	11.08	1.90	
3020	Floor Finishes	80% rubber mat, 20% paint 50% of floor area	S.F. Floor	7.52	3.76	
3030	Ceiling Finishes	Mineral fiber tile on concealed zee bar 10% of area	S.F. Ceiling	6.51	.65	
D. SERVICES						
	D10 Conveying					
1010	Elevators & Lifts	N/A	—	—	—	0.0%
1020	Escalators & Moving Walks	N/A	—	—	—	
	D20 Plumbing					
2010	Plumbing Fixtures	Toilet and service fixtures, supply and drainage 1 Fixture/1070 S.F. Floor	Each	3510	3.28	
2020	Domestic Water Distribution	Oil fired water heater	S.F. Floor	4.59	4.59	6.5%
2040	Rain Water Drainage	N/A	—	—	—	
	D30 HVAC					
3010	Energy Supply	Oil fired hot water, unit heaters 10% of area	S.F. Floor	.84	.84	
3020	Heat Generating Systems	N/A	—	—	—	
3030	Cooling Generating Systems	N/A	—	—	—	10.8%
3050	Terminal & Package Units	Single zone, electric cooling 90% of area	S.F. Floor	12.34	12.34	
3090	Other HVAC Sys. & Equipment	N/A	—	—	—	
	D40 Fire Protection					
4010	Sprinklers	N/A	—	—	—	0.0%
4020	Standpipes	N/A	—	—	—	
	D50 Electrical					
5010	Electrical Service/Distribution	600 ampere service, panel board and feeders	S.F. Floor	1.85	1.85	
5020	Lighting & Branch Wiring	High intensity discharge and fluorescent fixtures, receptacles, switches, A.C. and misc. power	S.F. Floor	7.43	7.43	8.7%
5030	Communications & Security	Alarm systems, emergency lighting and public address	S.F. Floor	1.08	1.08	
5090	Other Electrical Systems	Emergency generator	S.F. Floor	.21	.21	
E. EQUIPMENT & FURNISHINGS						
1010	Commercial Equipment	N/A	—	—	—	
1020	Institutional Equipment	N/A	—	—	—	
1030	Vehicular Equipment	N/A	—	—	—	0.0%
1090	Other Equipment	N/A	—	—	—	
F. SPECIAL CONSTRUCTION						
1020	Integrated Construction	N/A	—	—	—	
1040	Special Facilities	Dasher boards and rink (including ice making system)	S.F. Floor	26.81	26.81	22.0%
G. BUILDING SITEWORK	**N/A**					

			Sub-Total	121.74	100%
CONTRACTOR FEES (General Requirements: 10%, Overhead: 5%, Profit: 10%)		25%	30.41	
ARCHITECT FEES		7%	10.65	
	Total Building Cost		**162.80**	

Costs per square foot of floor area

Exterior Wall	S.F. Area	25000	30000	35000	40000	45000	50000	55000	60000	65000
	L.F. Perimeter	900	1050	1200	1350	1510	1650	1800	1970	2100
Face Brick with Concrete Block Back-up	Steel Frame	170.15	168.55	167.35	166.50	166.10	165.30	164.85	164.85	164.20
	Bearing Walls	163.95	162.35	161.20	160.30	159.90	159.10	158.65	158.65	158.00
Stucco on Concrete Block	Steel Frame	163.50	162.10	161.05	160.30	159.95	159.25	158.80	158.80	158.25
	Bearing Walls	157.35	155.90	154.90	154.10	153.70	153.00	152.60	152.60	152.05
Decorative Concrete Block	Steel Frame	162.50	161.05	160.10	159.30	158.90	158.20	157.80	157.75	157.25
	Bearing Walls	160.95	159.55	158.50	157.75	157.35	156.65	156.25	156.25	155.70
Perimeter Adj., Add or Deduct	Per 100 L.F.	4.20	3.45	3.00	2.60	2.35	2.05	1.90	1.75	1.60
Story Hgt. Adj., Add or Deduct	Per 1 Ft.	1.55	1.45	1.50	1.50	1.45	1.40	1.40	1.40	1.45

For Basement, add $24.30 per square foot of basement area

The above costs were calculated using the basic specifications shown on the facing page. These costs should be adjusted where necessary for design alternatives and owner's requirements. Reported completed project costs, for this type of structure, range from $79.20 to $201.45 per S.F.

Common additives

Description	Unit	$ Cost
Bleachers, Telescoping, manual		
To 15 tier	Seat	118 - 164
16-20 tier	Seat	242 - 297
21-30 tier	Seat	257 - 310
For power operation, add	Seat	47 - 73.50
Carrels Hardwood	Each	605 - 1150
Clock System		
20 room	Each	16,000
50 room	Each	39,100
Emergency Lighting, 25 watt, battery operated		
Lead battery	Each	287
Nickel cadmium	Each	845
Flagpoles, Complete		
Aluminum, 20' high	Each	1700
40' high	Each	3600
Fiberglass, 23' high	Each	1825
39'-5" high	Each	3475
Kitchen Equipment		
Broiler	Each	4050
Cooler, 6 ft. long, reach-in	Each	5150

Description	Unit	$ Cost
Kitchen Equipment, cont.		
Dishwasher, 10-12 racks per hr.	Each	5075
Food warmer, counter, 1.2 KW	Each	740
Freezer, 44 C.F., reach-in	Each	3250
Ice cube maker, 50 lb. per day	Each	1875
Range with 1 oven	Each	2700
Lockers, Steel, single tier, 60" to 72"	Opening	200 - 350
2 tier, 60" to 72" total	Opening	124 - 152
5 tier, box lockers	Opening	70 - 82.50
Locker bench, lam. maple top only	L.F.	22
Pedestals, steel pipe	Each	64
Seating		
Auditorium chair, all veneer	Each	240
Veneer back, padded seat	Each	290
Upholstered, spring seat	Each	279
Classroom, movable chair & desk	Set	65 - 120
Lecture hall, pedestal type	Each	229 - 680
Sound System		
Amplifier, 250 watts	Each	2400
Speaker, ceiling or wall	Each	196
Trumpet	Each	375

Important: See the Reference Section for Location Factors

Model costs calculated for a 1 story building with 15′ story height and 45,000 square feet of floor area

				Unit	Unit Cost	Cost Per S.F.	% Of Sub-Total
A. SUBSTRUCTURE							
1010	Standard Foundations	Poured concrete; strip and spread footings		S.F. Ground	4.54	4.54	
1020	Special Foundations	N/A		—	—	—	
1030	Slab on Grade	4″ reinforced concrete with vapor barrier and granular base		S.F. Slab	4.77	4.77	11.5%
2010	Basement Excavation	Site preparation for slab and trench for foundation wall and footing		S.F. Ground	.18	.18	
2020	Basement Walls	4′ foundation wall		L.F. Wall	76	4.30	
B. SHELL							
	B10 Superstructure						
1010	Floor Construction	N/A		—	—	—	2.7 %
1020	Roof Construction	Metal deck on open web steel joists		S.F. Roof	3.19	3.19	
	B20 Exterior Enclosure						
2010	Exterior Walls	Face brick with concrete block backup	70% of wall	S.F. Wall	30.28	10.67	
2020	Exterior Windows	Steel outward projecting	25% of wall	Each	703	4.62	13.3%
2030	Exterior Doors	Metal and glass	5% of wall	Each	3495	.63	
	B30 Roofing						
3010	Roof Coverings	Single-ply membrane with flashing; polyisocyanurate insulation		S.F. Roof	8.02	8.02	6.7%
3020	Roof Openings	N/A		—	—	—	
C. INTERIORS							
1010	Partitions	Concrete block	20 S.F. Floor/L.F. Partition	S.F. Partition	9.18	4.59	
1020	Interior Doors	Single leaf kalamein fire doors	700 S.F. Floor/Door	Each	1001	1.43	
1030	Fittings	Toilet partitions		S.F. Floor	1.88	1.88	
2010	Stair Construction	N/A		—	—	—	21.3%
3010	Wall Finishes	75% paint, 15% glazed coating, 10% ceramic tile		S.F. Surface	4.19	4.19	
3020	Floor Finishes	65% vinyl composition tile, 25% carpet, 10% terrazzo		S.F. Floor	6.82	6.82	
3030	Ceiling Finishes	Mineral fiber tile on concealed zee bars		S.F. Ceiling	6.51	6.51	
D. SERVICES							
	D10 Conveying						
1010	Elevators & Lifts	N/A		—	—	—	0.0 %
1020	Escalators & Moving Walks	N/A		—	—	—	
	D20 Plumbing						
2010	Plumbing Fixtures	Kitchen, bathroom and service fixtures, supply and drainage	1 Fixture/625 S.F. Floor	Each	7081	11.33	
2020	Domestic Water Distribution	Gas fired water heater		S.F. Floor	.53	.53	10.8%
2040	Rain Water Drainage	Roof drains		S.F. Roof	1.08	1.08	
	D30 HVAC						
3010	Energy Supply	Oil fired hot water, wall fin radiation		S.F. Floor	9.66	9.66	
3020	Heat Generating Systems	N/A		—	—	—	
3030	Cooling Generating Systems	N/A		—	—	—	18.2%
3050	Terminal & Package Units	Split systems with air cooled condensing units		S.F. Floor	12.05	12.05	
3090	Other HVAC Sys. & Equipment	N/A		—	—	—	
	D40 Fire Protection						
4010	Sprinklers	Sprinklers, light hazard		S.F. Floor	2.78	2.78	2.6%
4020	Standpipes	Standpipe		S.F. Floor	.35	.35	
	D50 Electrical						
5010	Electrical Service/Distribution	800 ampere service, panel board and feeders		S.F. Floor	1.41	1.41	
5020	Lighting & Branch Wiring	High efficiency fluorescent fixtures, receptacles, switches, A.C. and misc. power		S.F. Floor	10.16	10.16	12.8%
5030	Communications & Security	Addressable alarm systems, internet wiring, communications systems and emergency lighting		S.F. Floor	3.63	3.63	
5090	Other Electrical Systems	Emergency generator, 15 kW		S.F. Floor	.09	.09	
E. EQUIPMENT & FURNISHINGS							
1010	Commercial Equipment	N/A		—	—	—	
1020	Institutional Equipment	Chalkboards		S.F. Floor	.13	.13	0.1 %
1030	Vehicular Equipment	N/A		—	—	—	
1090	Other Equipment	N/A		—	—	—	
F. SPECIAL CONSTRUCTION							
1020	Integrated Construction	N/A		—	—	—	0.0 %
1040	Special Facilities	N/A		—	—	—	
G. BUILDING SITEWORK	**N/A**						

		Sub-Total	119.54	100%
CONTRACTOR FEES (General Requirements: 10%, Overhead: 5%, Profit: 10%)		25%	29.90	
ARCHITECT FEES		7%	10.46	
	Total Building Cost		**159.90**	

Costs per square foot of floor area

Exterior Wall	S.F. Area	50000	70000	90000	110000	130000	150000	170000	190000	210000
	L.F. Perimeter	850	1140	1420	1700	1980	2280	2560	2840	3120
Face Brick with Concrete Block Back-up	Steel Frame	159.40	156.60	154.85	153.70	152.95	152.65	152.15	151.80	151.50
	R/Conc. Frame	166.45	162.85	160.60	159.15	158.15	157.75	157.15	156.65	156.30
Decorative Concrete Block	Steel Frame	155.05	151.70	149.65	148.35	147.45	147.00	146.45	146.05	145.75
	R/Conc. Frame	158.30	154.95	152.90	151.60	150.70	150.25	149.70	149.30	149.00
Limestone with Concrete Block Back-up	Steel Frame	179.95	175.50	172.80	170.95	169.70	169.20	168.50	167.90	167.50
	R/Conc. Frame	188.65	184.30	181.50	179.65	178.45	178.00	177.20	176.65	176.20
Perimeter Adj., Add or Deduct	Per 100 L.F.	3.85	2.75	2.10	1.80	1.45	1.30	1.15	1.00	0.95
Story Hgt. Adj., Add or Deduct	Per 1 Ft.	1.95	1.85	1.80	1.80	1.70	1.75	1.70	1.75	1.70
For Basement, add $31.75 per square foot of basement area										

The above costs were calculated using the basic specifications shown on the facing page. These costs should be adjusted where necessary for design alternatives and owner's requirements. Reported completed project costs, for this type of structure, range from $92.05 to $216.80 per S.F.

Common additives

Description	Unit	$ Cost
Bleachers, Telescoping, manual		
To 15 tier	Seat	118 - 164
16-20 tier	Seat	242 - 297
21-30 tier	Seat	257 - 310
For power operation, add	Seat	47 - 73.50
Carrels Hardwood	Each	605 - 1150
Clock System		
20 room	Each	16,000
50 room	Each	39,100
Elevators, Hydraulic passenger, 2 stops		
1500# capacity	Each	60,900
2500# capacity	Each	64,300
Emergency Lighting, 25 watt, battery operated		
Lead battery	Each	287
Nickel cadmium	Each	845
Flagpoles, Complete		
Aluminum, 20' high	Each	1700
40' high	Each	3600
Fiberglass, 23' high	Each	1825
39'-5" high	Each	3475

Description	Unit	$ Cost
Kitchen Equipment		
Broiler	Each	4050
Cooler, 6 ft. long, reach-in	Each	5150
Dishwasher, 10-12 racks per hr.	Each	5075
Food warmer, counter, 1.2 KW	Each	740
Freezer, 44 C.F., reach-in	Each	3250
Lockers, Steel, single tier, 60" or 72"	Opening	200 - 350
2 tier, 60" or 72" total	Opening	124 - 152
5 tier, box lockers	Opening	70 - 82.50
Locker bench, lam. maple top only	L.F.	22
Pedestals, steel pipe	Each	64
Seating		
Auditorium chair, all veneer	Each	240
Veneer back, padded seat	Each	290
Upholstered, spring seat	Each	279
Classroom, movable chair & desk	Set	65 - 120
Lecture hall, pedestal type	Each	229 - 680
Sound System		
Amplifier, 250 watts	Each	2400
Speaker, ceiling or wall	Each	196
Trumpet	Each	375

Model costs calculated for a 2 story building with 15' story height and 130,000 square feet of floor area

				Unit	Unit Cost	Cost Per S.F.	% Of Sub-Total
A.	**SUBSTRUCTURE**						
1010	Standard Foundations	Poured concrete; strip and spread footings		S.F. Ground	2.44	1.22	
1020	Special Foundations	N/A		—	—	—	
1030	Slab on Grade	4" reinforced concrete with vapor barrier and granular base		S.F. Slab	4.77	2.39	4.1%
2010	Basement Excavation	Site preparation for slab and trench for foundation wall and footing		S.F. Ground	.18	.09	
2020	Basement Walls	4' foundation wall		L.F. Wall	76	1.15	
B.	**SHELL**						
	B10 Superstructure						
1010	Floor Construction	Concrete slab without drop panel, concrete columns		S.F. Floor	20.22	10.11	14.5%
1020	Roof Construction	Concrete slab without drop panel		S.F. Roof	14.10	7.05	
	B20 Exterior Enclosure						
2010	Exterior Walls	Face brick with concrete block backup	75% of wall	S.F. Wall	30.32	10.39	
2020	Exterior Windows	Window wall	25% of wall	Each	67	7.62	15.6%
2030	Exterior Doors	Metal and glass		Each	2057	.45	
	B30 Roofing						
3010	Roof Coverings	Single-ply membrane and standing seam metal; polyisocyanurate insulation		S.F. Roof	11.78	5.89	5.0%
3020	Roof Openings	Roof hatches		S.F. Roof	.10	.05	
C.	**INTERIORS**						
1010	Partitions	Concrete block	25 S.F. Floor/L.F. Partition	S.F. Partition	11.54	5.54	
1020	Interior Doors	Single leaf kalamein fire doors	700 S.F. Floor/Door	Each	1001	1.43	
1030	Fittings	Toilet partitions, chalkboards		S.F. Floor	1.33	1.33	
2010	Stair Construction	Concrete filled metal pan		Flight	12,275	.57	21.1%
3010	Wall Finishes	75% paint, 15% glazed coating, 10% ceramic tile		S.F. Surface	3.19	3.06	
3020	Floor Finishes	70% vinyl composition tile, 20% carpet, 10% terrazzo		S.F. Floor	6.47	6.47	
3030	Ceiling Finishes	Mineral fiber tile on concealed zee bars		S.F. Ceiling	6.51	6.51	
D.	**SERVICES**						
	D10 Conveying						
1010	Elevators & Lifts	One hydraulic passenger elevator		Each	79,300	.61	0.5%
1020	Escalators & Moving Walks	N/A		—	—	—	
	D20 Plumbing						
2010	Plumbing Fixtures	Kitchen, bathroom and service fixtures, supply and drainage	1 Fixture/860 S.F. Floor	Each	4214	4.90	
2020	Domestic Water Distribution	Gas fired water heater		S.F. Floor	.86	.86	5.4%
2040	Rain Water Drainage	Roof drains		S.F. Roof	1.24	.62	
	D30 HVAC						
3010	Energy Supply	Oil fired hot water, wall fin radiation		S.F. Floor	4.48	4.48	
3020	Heat Generating Systems	N/A		—	—	—	
3030	Cooling Generating Systems	Chilled water, cooling tower systems		S.F. Floor	15.12	15.12	16.6%
3050	Terminal & Package Units	N/A		—	—	—	
3090	Other HVAC Sys. & Equipment	N/A		—	—	—	
	D40 Fire Protection						
4010	Sprinklers	Sprinklers, light hazard		S.F. Floor	2.42	2.42	2.3%
4020	Standpipes	Standpipe, wet, Class III		S.F. Floor	.28	.28	
	D50 Electrical						
5010	Electrical Service/Distribution	2000 ampere service, panel board and feeders		S.F. Floor	1.34	1.34	
5020	Lighting & Branch Wiring	High efficiency fluorescent fixtures, receptacles, switches, A.C. and misc. power		S.F. Floor	9.42	9.42	12.9%
5030	Communications & Security	Addressable alarm systems, internet wiring, communications systems and emergency lighting		S.F. Floor	3.92	3.92	
5090	Other Electrical Systems	Emergency generator, 250 kW		S.F. Floor	.53	.53	
E.	**EQUIPMENT & FURNISHINGS**						
1010	Commercial Equipment	N/A		—	—	—	
1020	Institutional Equipment	Laboratory casework and counters		S.F. Floor	1.64	1.64	2.1 %
1030	Vehicular Equipment	N/A		—	—	—	
1090	Other Equipment	Built-in athletic equipment , bleachers and scoreboard		S.F. Floor	.80	.80	
F.	**SPECIAL CONSTRUCTION**						
1020	Integrated Construction	N/A		—	—	—	0.0 %
1040	Special Facilities	N/A		—	—	—	
G.	**BUILDING SITEWORK**	**N/A**					

			Sub-Total	118.26	100%
	CONTRACTOR FEES (General Requirements: 10%, Overhead: 5%, Profit: 10%)		25%	29.54	
	ARCHITECT FEES		7%	10.35	
			Total Building Cost	**158.15**	

Costs per square foot of floor area

Exterior Wall	S.F. Area	50000	65000	80000	95000	110000	125000	140000	155000	170000
	L.F. Perimeter	850	1060	1280	1490	1700	1920	2140	2340	2560
Face Brick with Concrete Block Back-up	Steel Frame	164.35	162.45	161.35	160.45	159.85	159.50	159.25	158.75	158.55
	Bearing Walls	157.80	155.90	154.85	153.95	153.30	153.00	152.70	152.20	152.10
Concrete Block Stucco Face	Steel Frame	156.80	155.15	154.25	153.50	152.90	152.60	152.35	152.00	151.85
	Bearing Walls	151.00	149.30	148.40	147.65	147.10	146.80	146.55	146.20	146.00
Decorative Concrete Block	Steel Frame	158.15	156.50	155.60	154.80	154.25	153.95	153.70	153.30	153.15
	Bearing Walls	151.05	149.35	148.45	147.65	147.10	146.80	146.55	146.15	145.95
Perimeter Adj., Add or Deduct	Per 100 L.F.	3.40	2.60	2.20	1.85	1.55	1.30	1.15	1.10	1.05
Story Hgt. Adj., Add or Deduct	Per 1 Ft.	1.70	1.60	1.60	1.60	1.55	1.50	1.50	1.50	1.55
For Basement, add $33.05 per square foot of basement area										

The above costs were calculated using the basic specifications shown on the facing page. These costs should be adjusted where necessary for design alternatives and owner's requirements. Reported completed project costs, for this type of structure, range from $86.90 to $204.20 per S.F.

Common additives

Description	Unit	$ Cost
Bleachers, Telescoping, manual		
To 15 tier	Seat	118 - 164
16-20 tier	Seat	242 - 297
21-30 tier	Seat	257 - 310
For power operation, add	Seat	47 - 73.50
Carrels Hardwood	Each	605 - 1150
Clock System		
20 room	Each	16,000
50 room	Each	39,100
Elevators, Hydraulic passenger, 2 stops		
1500# capacity	Each	60,900
2500# capacity	Each	64,300
Emergency Lighting, 25 watt, battery operated		
Lead battery	Each	287
Nickel cadmium	Each	845
Flagpoles, Complete		
Aluminum, 20' high	Each	1700
40' high	Each	3600
Fiberglass, 23' high	Each	1825
39'-5" high	Each	3475

Description	Unit	$ Cost
Kitchen Equipment		
Broiler	Each	4050
Cooler, 6 ft. long, reach-in	Each	5150
Dishwasher, 10-12 racks per hr.	Each	5075
Food warmer, counter, 1.2 KW	Each	740
Freezer, 44 C.F., reach-in	Each	3250
Lockers, Steel, single tier, 60" to 72"	Opening	200 - 350
2 tier, 60" to 72" total	Opening	124 - 152
5 tier, box lockers	Opening	70 - 82.50
Locker bench, lam. maple top only	L.F.	22
Pedestals, steel pipe	Each	64
Seating		
Auditorium chair, all veneer	Each	240
Veneer back, padded seat	Each	290
Upholstered, spring seat	Each	279
Classroom, movable chair & desk	Set	65 - 120
Lecture hall, pedestal type	Each	229 - 680
Sound System		
Amplifier, 250 watts	Each	2400
Speaker, ceiling or wall	Each	196
Trumpet	Each	375

Model costs calculated for a 2 story building with 15' story height and 110,000 square feet of floor area

				Unit	Unit Cost	Cost Per S.F.	% Of Sub-Total
A. SUBSTRUCTURE							
1010	Standard Foundations	Poured concrete; strip and spread footings		S.F. Ground	2.26	1.13	
1020	Special Foundations	N/A		—	—	—	
1030	Slab on Grade	4" reinforced concrete with vapor barrier and granular base		S.F. Slab	4.77	2.39	4.0%
2010	Basement Excavation	Site preparation for slab and trench for foundation wall and footing		S.F. Ground	.18	.09	
2020	Basement Walls	4' foundation wall		L.F. Wall	76	1.16	
B. SHELL							
	B10 Superstructure						
1010	Floor Construction	Open web steel joists, slab form, concrete, columns		S.F. Floor	23.16	11.58	13.6%
1020	Roof Construction	Metal deck, open web steel joists, columns		S.F. Roof	9.44	4.72	
	B20 Exterior Enclosure						
2010	Exterior Walls	Face brick with concrete block backup	75% of wall	S.F. Wall	30.31	10.54	
2020	Exterior Windows	Window wall	25% of wall	Each	50	5.85	14.2%
2030	Exterior Doors	Double aluminum & glass		Each	2161	.56	
	B30 Roofing						
3010	Roof Coverings	Single-ply membrane and standing seam metal; polyisocyanurate insulation		S.F. Roof	11.96	5.98	5.0%
3020	Roof Openings	Roof hatches		S.F. Roof	.04	.02	
C. INTERIORS							
1010	Partitions	Concrete block	20 S.F. Floor/L.F. Partition	S.F. Partition	8.62	5.17	
1020	Interior Doors	Single leaf kalamein fire doors	750 S.F. Floor/Door	Each	1001	1.34	
1030	Fittings	Toilet partitions, chalkboards		S.F. Floor	1.39	1.39	
2010	Stair Construction	Concrete filled metal pan		Flight	12,275	.67	24.2%
3010	Wall Finishes	50% paint, 40% glazed coatings, 10% ceramic tile		S.F. Surface	3.80	4.56	
3020	Floor Finishes	50% vinyl composition tile, 30% carpet, 20% terrrazzo		S.F. Floor	9.34	9.34	
3030	Ceiling Finishes	Mineral fiberboard on concealed zee bars		S.F. Ceiling	6.51	6.51	
D. SERVICES							
	D10 Conveying						
1010	Elevators & Lifts	One hydraulic passenger elevator		Each	79,200	.72	0.6%
1020	Escalators & Moving Walks	N/A		—	—	—	
	D20 Plumbing						
2010	Plumbing Fixtures	Kitchen, toilet and service fixtures, supply and drainage	1 Fixture/1170 S.F. Floor	Each	6084	5.20	
2020	Domestic Water Distribution	Gas fired water heater		S.F. Floor	.55	.55	5.3%
2040	Rain Water Drainage	Roof drains		S.F. Roof	1.26	.63	
	D30 HVAC						
3010	Energy Supply	N/A		—	—	—	
3020	Heat Generating Systems	Included in D3050		—	—	—	
3030	Cooling Generating Systems	N/A		—	—	—	15.4 %
3050	Terminal & Package Units	Multizone unit, gas heating, electric cooling		S.F. Floor	18.40	18.40	
3090	Other HVAC Sys. & Equipment	N/A		—	—	—	
	D40 Fire Protection						
4010	Sprinklers	Sprinklers, light hazard	10% of area	S.F. Floor	2.42	2.42	2.3%
4020	Standpipes	Standpipe, wet, Class III		S.F. Floor	.34	.34	
	D50 Electrical						
5010	Electrical Service/Distribution	1600 ampere service, panel board and feeders		S.F. Floor	1.34	1.34	
5020	Lighting & Branch Wiring	High efficiency fluorescent fixtures, receptacles, switches, A.C. and misc. power		S.F. Floor	9.42	9.42	12.8%
5030	Communications & Security	Addressable alarm systems, internet wiring, communications systems and emergency lighting		S.F. Floor	4.16	4.16	
5090	Other Electrical Systems	Emergency generator, 100 kW		S.F. Floor	.41	.41	
E. EQUIPMENT & FURNISHINGS							
1010	Commercial Equipment	N/A		—	—	—	
1020	Institutional Equipment	Laboratory casework and counters		S.F. Floor	1.95	1.95	2.4 %
1030	Vehicular Equipment	N/A		—	—	—	
1090	Other Equipment	Bleachers, basketball backstops		S.F. Floor	.97	.97	
F. SPECIAL CONSTRUCTION							
1020	Integrated Construction	N/A		—	—	—	0.0 %
1040	Special Facilities	N/A		—	—	—	
G. BUILDING SITEWORK	**N/A**						

			Sub-Total	119.51	100%
CONTRACTOR FEES (General Requirements: 10%, Overhead: 5%, Profit: 10%)		25%	29.88		
ARCHITECT FEES		7%	10.46		
	Total Building Cost		**159.85**		

Costs per square foot of floor area

Exterior Wall	S.F. Area	20000	30000	40000	50000	60000	70000	80000	90000	100000
	L.F. Perimeter	440	590	740	900	1050	1200	1360	1510	1660
Face Brick with Concrete Block Back-up	Steel Frame	168.45	162.85	160.15	158.85	157.65	156.85	156.35	155.90	155.50
	Bearing Walls	162.85	157.40	154.65	153.40	152.20	151.35	151.00	150.45	150.00
Decorative Concrete Block	Steel Frame	155.15	150.95	149.00	147.95	147.05	146.45	146.10	145.75	145.45
	Bearing Walls	150.00	145.75	143.75	142.75	141.90	141.25	140.95	140.55	140.25
Steel Siding on Steel Studs	Steel Frame	150.40	146.70	144.95	144.10	143.25	142.70	142.45	142.05	141.85
Metal Sandwich Panel	Steel Frame	150.40	146.70	145.00	144.05	143.25	142.75	142.45	142.10	141.80
Perimeter Adj., Add or Deduct	Per 100 L.F.	8.70	5.85	4.30	3.45	2.90	2.45	2.20	1.90	1.70
Story Hgt. Adj., Add or Deduct	Per 1 Ft.	2.10	1.90	1.75	1.70	1.70	1.60	1.65	1.60	1.55
For Basement, add $32.05 per square foot of basement area										

The above costs were calculated using the basic specifications shown on the facing page. These costs should be adjusted where necessary for design alternatives and owner's requirements. Reported completed project costs, for this type of structure, range from $78.00 to $222.85 per S.F.

Common additives

Description	Unit	$ Cost	Description	Unit	$ Cost
Carrels Hardwood	Each	605 - 1150	Flagpoles, Complete		
Clock System			Aluminum, 20' high	Each	1700
20 room	Each	16,000	40' high	Each	3600
50 room	Each	39,100	Fiberglass, 23' high	Each	1825
Directory Boards, Plastic, glass covered			39'-5" high	Each	3475
30" x 20"	Each	605	Seating		
36" x 48"	Each	1325	Auditorium chair, all veneer	Each	240
Aluminum, 24" x 18"	Each	585	Veneer back, padded seat	Each	290
36" x 24"	Each	685	Upholstered, spring seat	Each	279
48" x 32"	Each	975	Classroom, movable chair & desk	Set	65 - 120
48" x 60"	Each	2025	Lecture hall, pedestal type	Each	229 - 680
Elevators, Hydraulic passenger, 2 stops			Shops & Workroom:		
1500# capacity	Each	60,900	Benches, metal	Each	700
2500# capacity	Each	64,300	Parts bins 6'-3" high, 3' wide, 12" deep, 72 bins	Each	655
3500# capacity	Each	67,600	Shelving, metal 1' x 3'	S.F.	11
Emergency Lighting, 25 watt, battery operated			Wide span 6' wide x 24" deep	S.F.	16.40
Lead battery	Each	287	Sound System		
Nickel cadmium	Each	845	Amplifier, 250 watts	Each	2400
			Speaker, ceiling or wall	Each	196
			Trumpet	Each	375

Model costs calculated for a 2 story building with 16' story height and 40,000 square feet of floor area

School, Vocational

				Unit	Unit Cost	Cost Per S.F.	% Of Sub-Total
A. SUBSTRUCTURE							
1010	Standard Foundations	Poured concrete; strip and spread footings		S.F. Ground	2.54	1.27	
1020	Special Foundations	N/A		—	—	—	
1030	Slab on Grade	5" reinforced concrete with vapor barrier and granular base		S.F. Slab	12.01	6.01	7.6%
2010	Basement Excavation	Site preparation for slab and trench for foundation wall and footing		S.F. Ground	.18	.09	
2020	Basement Walls	4' foundation wall		L.F. Wall	76	1.40	
B. SHELL							
	B10 Superstructure						
1010	Floor Construction	Open web steel joists, slab form, concrete, beams, columns		S.F. Floor	14.68	7.34	8.4%
1020	Roof Construction	Metal deck, open web steel joists, beams, columns		S.F. Roof	4.76	2.38	
	B20 Exterior Enclosure						
2010	Exterior Walls	Face brick with concrete block backup	85% of wall	S.F. Wall	30.31	15.25	
2020	Exterior Windows	Tubular aluminum framing with insulated glass	15% of wall	Each	50	4.49	17.5%
2030	Exterior Doors	Metal and glass doors		Each	2608	.52	
	B30 Roofing						
3010	Roof Coverings	Single-ply membrane with polyisocyanurate insulation		S.F. Roof	5.98	2.99	2.7%
3020	Roof Openings	Roof hatches		S.F. Roof	.28	.14	
C. INTERIORS							
1010	Partitions	Concrete block	20 S.F. Floor/L.F. Partition	S.F. Partition	8.62	5.17	
1020	Interior Doors	Single leaf kalamein fire doors	600 S.F. Floor/Door	Each	1001	1.67	
1030	Fittings	Toilet partitions, chalkboards		S.F. Floor	1.59	1.59	
2010	Stair Construction	Concrete filled metal pan		Flight	12,275	1.23	21.7%
3010	Wall Finishes	50% paint, 40% glazed coating, 10% ceramic tile		S.F. Surface	4.18	5.02	
3020	Floor Finishes	70% vinyl composition tile, 20% carpet, 10% terrazzo		S.F. Floor	6.47	6.47	
3030	Ceiling Finishes	Mineral fiber tile on concealed zee bars	60% of area	S.F. Ceiling	6.51	3.90	
D. SERVICES							
	D10 Conveying						
1010	Elevators & Lifts	One hydraulic passenger elevator		Each	78,800	1.97	1.7%
1020	Escalators & Moving Walks	N/A		—	—	—	
	D20 Plumbing						
2010	Plumbing Fixtures	Toilet and service fixtures, supply and drainage	1 Fixture/700 S.F. Floor	Each	3010	4.30	
2020	Domestic Water Distribution	Gas fired water heater		S.F. Floor	.89	.89	5.2%
2040	Rain Water Drainage	Roof drains		S.F. Roof	1.54	.77	
	D30 HVAC						
3010	Energy Supply	Oil fired hot water, wall fin radiation		S.F. Floor	9.13	9.13	
3020	Heat Generating Systems	N/A		—	—	—	
3030	Cooling Generating Systems	Chilled water, cooling tower systems		S.F. Floor	12.95	12.95	19.1%
3050	Terminal & Package Units	N/A		—	—	—	
3090	Other HVAC Sys. & Equipment	N/A		—	—	—	
	D40 Fire Protection						
4010	Sprinklers	Sprinklers, light hazard		S.F. Floor	2.86	2.86	3.1%
4020	Standpipes	Standpipe, wet, Class III		S.F. Floor	.77	.77	
	D50 Electrical						
5010	Electrical Service/Distribution	800 ampere service, panel board and feeders		S.F. Floor	1.58	1.58	
5020	Lighting & Branch Wiring	High efficiency fluorescent fixtures, receptacles, switches, A.C. and misc. power		S.F. Floor	10.19	10.19	12.9%
5030	Communications & Security	Addressable alarm systems, internet wiring, communications systems and emergency lighting		S.F. Floor	3.06	3.06	
5090	Other Electrical Systems	Emergency generator, 11.5 kW		S.F. Floor	.11	.11	
E. EQUIPMENT & FURNISHINGS							
1010	Commercial Equipment	N/A		—	—	—	
1020	Institutional Equipment	Stainless steel countertops		S.F. Floor	.12	.12	0.1%
1030	Vehicular Equipment	N/A		—	—	—	
1090	Other Equipment	N/A		—	—	—	
F. SPECIAL CONSTRUCTION							
1020	Integrated Construction	N/A		—	—	—	0.0%
1040	Special Facilities	N/A		—	—	—	
G. BUILDING SITEWORK	**N/A**						

		Sub-Total	115.63	100%
CONTRACTOR FEES (General Requirements: 10%, Overhead: 5%, Profit: 10%)		25%	28.90	
ARCHITECT FEES		7%	10.12	

Total Building Cost	**154.65**

Costs per square foot of floor area

Exterior Wall	S.F. Area	1000	2000	3000	4000	6000	8000	10000	12000	15000
	L.F. Perimeter	126	179	219	253	310	358	400	438	490
Wood Siding	Wood Frame	125.35	112.70	106.95	103.55	99.55	97.15	95.50	94.35	92.95
Face Brick Veneer	Wood Frame	151.30	130.40	120.90	115.35	108.65	104.65	101.95	100.00	97.75
Stucco on Concrete Block	Steel Frame	145.35	127.50	119.35	114.60	108.85	105.50	103.15	101.45	99.50
	Bearing Walls	141.70	123.85	115.70	110.95	105.20	101.85	99.55	97.80	95.90
Metal Sandwich Panel	Steel Frame	159.15	137.45	127.60	121.80	114.90	110.80	108.00	105.95	103.60
Precast Concrete	Steel Frame	196.20	163.35	148.45	139.75	129.25	123.05	118.75	115.60	112.10
Perimeter Adj., Add or Deduct	Per 100 L.F.	34.70	17.35	11.55	8.70	5.80	4.35	3.45	2.80	2.35
Story Hgt. Adj., Add or Deduct	Per 1 Ft.	2.40	1.70	1.40	1.25	1.00	0.90	0.75	0.60	0.60
For Basement, add $24.20 per square foot of basement area										

The above costs were calculated using the basic specifications shown on the facing page. These costs should be adjusted where necessary for design alternatives and owner's requirements. Reported completed project costs, for this type of structure, range from $64.60 to $223.95 per S.F.

Common additives

Description	Unit	$ Cost
Check Out Counter		
Single belt	Each	3275
Double belt	Each	4700
Emergency Lighting, 25 watt, battery operated		
Lead battery	Each	287
Nickel cadmium	Each	845
Refrigerators, Prefabricated, walk-in		
7'-6" high, 6' x 6'	S.F.	173
10' x 10'	S.F.	136
12' x 14'	S.F.	120
12' x 20'	S.F.	106
Refrigerated Food Cases		
Dairy, multi deck, 12' long	Each	11,900
Delicatessen case, single deck, 12' long	Each	8000
Multi deck, 18 S.F. shelf display	Each	7450
Freezer, self-contained chest type, 30 C.F.	Each	8575
Glass door upright, 78 C.F.	Each	11,000

Description	Unit	$ Cost
Refrigerated Food Cases, cont.		
Frozen food, chest type, 12' long	Each	8100
Glass door reach-in, 5 door	Each	15,200
Island case 12' long, single deck	Each	9125
Multi deck	Each	19,000
Meat cases, 12' long, single deck	Each	6725
Multi deck	Each	11,400
Produce, 12' long single deck	Each	8775
Multi deck	Each	9550
Safe, Office type, 1 hour rating		
30" x 18" x 18"	Each	2400
60" x 36" x 18", double door	Each	8975
Smoke Detectors		
Ceiling type	Each	233
Duct type	Each	525
Sound System		
Amplifier, 250 watts	Each	2400
Speaker, ceiling or wall	Each	196
Trumpet	Each	375

Important: See the Reference Section for Location Factors

Model costs calculated for a 1 story building with 12' story height and 4,000 square feet of floor area

Store, Convenience

			Unit	Unit Cost	Cost Per S.F.	% Of Sub-Total
A.	**SUBSTRUCTURE**					
1010	Standard Foundations	Poured concrete; strip and spread footings	S.F. Ground	1.52	1.52	
1020	Special Foundations	N/A	—	—	—	
1030	Slab on Grade	4" reinforced concrete with vapor barrier and granular base	S.F. Slab	4.77	4.77	14.3%
2010	Basement Excavation	Site preparation for slab and trench for foundation wall and footing	S.F. Ground	.52	.52	
2020	Basement Walls	4' foundation wall	L.F. Wall	67	4.27	
B.	**SHELL**					
	B10 Superstructure					
1010	Floor Construction	N/A	—	—	—	8.3 %
1020	Roof Construction	Wood truss with plywood sheathing	S.F. Roof	6.46	6.46	
	B20 Exterior Enclosure					
2010	Exterior Walls	Wood siding on wood studs, insulated 80% of wall	S.F. Wall	8.96	5.44	
2020	Exterior Windows	Storefront 20% of wall	Each	40.75	4.16	15.1%
2030	Exterior Doors	Double aluminum and glass, solid core wood	Each	2806	2.10	
	B30 Roofing					
3010	Roof Coverings	Asphalt shingles; rigid fiberglass insulation	S.F. Roof	4.23	4.23	5.5%
3020	Roof Openings	N/A	—	—	—	
C.	**INTERIORS**					
1010	Partitions	Gypsum board on wood studs 60 S.F. Floor/L.F. Partition	S.F. Partition	11.82	1.97	
1020	Interior Doors	Single leaf wood, hollow metal 1300 S.F. Floor/Door	Each	1023	.79	
1030	Fittings	N/A	—	—	—	
2010	Stair Construction	N/A	—	—	—	14.9%
3010	Wall Finishes	Paint	S.F. Surface	1.05	.35	
3020	Floor Finishes	Vinyl composition tile	S.F. Floor	2.90	2.90	
3030	Ceiling Finishes	Mineral fiber tile on wood furring	S.F. Ceiling	5.52	5.52	
D.	**SERVICES**					
	D10 Conveying					
1010	Elevators & Lifts	N/A	—	—	—	0.0 %
1020	Escalators & Moving Walks	N/A	—	—	—	
	D20 Plumbing					
2010	Plumbing Fixtures	Toilet and service fixtures, supply and drainage 1 Fixture/1000 S.F. Floor	Each	2310	2.31	
2020	Domestic Water Distribution	Gas fired water heater	S.F. Floor	1.27	1.27	4.6%
2040	Rain Water Drainage	N/A	—	—	—	
	D30 HVAC					
3010	Energy Supply	N/A	—	—	—	
3020	Heat Generating Systems	Included in D3050	—	—	—	
3030	Cooling Generating Systems	N/A	—	—	—	9.8 %
3050	Terminal & Package Units	Single zone rooftop unit, gas heating, electric cooling	S.F. Floor	7.60	7.60	
3090	Other HVAC Sys. & Equipment	N/A	—	—	—	
	D40 Fire Protection					
4010	Sprinklers	Sprinkler, ordinary hazard	S.F. Floor	3.99	3.99	5.2%
4020	Standpipes	N/A	—	—	—	
	D50 Electrical					
5010	Electrical Service/Distribution	200 ampere service, panel board and feeders	S.F. Floor	2.66	2.66	
5020	Lighting & Branch Wiring	Fluorescent fixtures, receptacles, switches, A.C. and misc. power	S.F. Floor	10.08	10.08	22.3%
5030	Communications & Security	Alarm systems and emergency lighting	S.F. Floor	4.19	4.19	
5090	Other Electrical Systems	Emergency generator, 7.5 kW	S.F. Floor	.35	.35	
E.	**EQUIPMENT & FURNISHINGS**					
1010	Commercial Equipment	N/A	—	—	—	
1020	Institutional Equipment	N/A	—	—	—	
1030	Vehicular Equipment	N/A	—	—	—	0.0 %
1090	Other Equipment	N/A	—	—	—	
F.	**SPECIAL CONSTRUCTION**					
1020	Integrated Construction	N/A	—	—	—	0.0 %
1040	Special Facilities	N/A	—	—	—	
G.	**BUILDING SITEWORK** N/A					

		Sub-Total	77.45	**100%**
CONTRACTOR FEES (General Requirements: 10%, Overhead: 5%, Profit: 10%)		25%	19.32	
ARCHITECT FEES		7%	6.78	
		Total Building Cost	**103.55**	

Costs per square foot of floor area

Exterior Wall	S.F. Area	50000	65000	80000	95000	110000	125000	140000	155000	170000
	L.F. Perimeter	920	1065	1167	1303	1333	1433	1533	1633	1733
Face Brick with Concrete Block Back-up	R/Conc. Frame	118.40	116.00	113.90	112.90	111.30	110.55	109.90	109.45	109.10
	Steel Frame	100.00	97.50	95.50	94.45	92.80	92.10	91.50	91.00	90.70
Decorative Concrete Block	R/Conc. Frame	112.70	110.85	109.40	108.65	107.50	107.00	106.55	106.20	105.90
	Steel Joists	95.30	93.35	91.75	90.95	89.75	89.15	88.70	88.35	88.00
Precast Concrete Panels	R/Conc. Frame	116.50	114.35	112.50	111.60	110.20	109.50	109.00	108.60	108.25
	Steel Joists	97.60	95.40	93.60	92.70	91.30	90.60	90.05	89.65	89.35
Perimeter Adj., Add or Deduct	Per 100 L.F.	1.80	1.35	1.15	0.90	0.80	0.65	0.65	0.60	0.55
Story Hgt. Adj., Add or Deduct	Per 1 Ft.	0.90	0.75	0.80	0.70	0.55	0.55	0.55	0.50	0.50
For Basement, add $23.80 per square foot of basement area										

The above costs were calculated using the basic specifications shown on the facing page. These costs should be adjusted where necessary for design alternatives and owner's requirements. Reported completed project costs, for this type of structure, range from $55.75 to $138.70 per S.F.

Common additives

Description	Unit	$ Cost
Closed Circuit Surveillance, One station		
Camera and monitor	Each	1875
For additional camera stations, add	Each	1025
Directory Boards, Plastic, glass covered		
30" x 20"	Each	605
36" x 48"	Each	1325
Aluminum, 24" x 18"	Each	585
36" x 24"	Each	685
48" x 32"	Each	975
48" x 60"	Each	2025
Emergency Lighting, 25 watt, battery operated		
Lead battery	Each	287
Nickel cadmium	Each	845
Safe, Office type, 1 hour rating		
30" x 18" x 18"	Each	2400
60" x 36" x 18", double door	Each	8975
Sound System		
Amplifier, 250 watts	Each	2400
Speaker, ceiling or wall	Each	196
Trumpet	Each	375

Model costs calculated for a 1 story building with 14' story height and 110,000 square feet of floor area

Store, Department, 1 Story

				Unit	Unit Cost	Cost Per S.F.	% Of Sub-Total
A. SUBSTRUCTURE							
1010	Standard Foundations	Poured concrete; strip and spread footings		S.F. Ground	.83	.83	
1020	Special Foundations	N/A		—	—	—	
1030	Slab on Grade	4" reinforced concrete with vapor barrier and granular base		S.F. Slab	4.77	4.77	8.1%
2010	Basement Excavation	Site preparation for slab and trench for foundation wall and footing		S.F. Ground	.28	.28	
2020	Basement Walls	4' foundation wall		L.F. Wall	76	.92	
B. SHELL							
	B10 Superstructure						
1010	Floor Construction	Cast-in-place concrete columns		L.F. Column	66	.83	27.1%
1020	Roof Construction	Precast concrete beam and plank, concrete columns		S.F. Roof	21.95	21.95	
	B20 Exterior Enclosure						
2010	Exterior Walls	Face brick with concrete block backup	90% of wall	S.F. Wall	31.24	4.77	
2020	Exterior Windows	Storefront	10% of wall	Each	61	1.04	7.4%
2030	Exterior Doors	Sliding electric operated entrance, hollow metal		Each	6984	.38	
	B30 Roofing						
3010	Roof Coverings	Built-up tar and gravel with flashing; perlite/EPS composite insulation		S.F. Roof	4.98	4.98	6.0%
3020	Roof Openings	Roof hatches		S.F. Roof	.05	.05	
C. INTERIORS							
1010	Partitions	Gypsum board on metal studs	60 S.F. Floor/L.F. Partition	S.F. Partition	9.24	1.54	
1020	Interior Doors	Single leaf hollow metal	600 S.F. Floor/Door	Each	1001	1.67	
1030	Fittings	N/A		—	—	—	
2010	Stair Construction	N/A		—	—	—	24.0%
3010	Wall Finishes	Paint		S.F. Surface	.75	.25	
3020	Floor Finishes	50% ceramic tile, 50% carpet tile		S.F. Floor	12.61	12.61	
3030	Ceiling Finishes	Mineral fiber board on exposed grid system, suspended		S.F. Ceiling	4.09	4.09	
D. SERVICES							
	D10 Conveying						
1010	Elevators & Lifts	N/A		—	—	—	0.0 %
1020	Escalators & Moving Walks	N/A		—	—	—	
	D20 Plumbing						
2010	Plumbing Fixtures	Toilet and service fixtures, supply and drainage	1 Fixture/4075 S.F. Floor	Each	5420	1.33	
2020	Domestic Water Distribution	Gas fired water heater		S.F. Floor	.32	.32	2.8%
2040	Rain Water Drainage	Roof drains		S.F. Roof	.66	.66	
	D30 HVAC						
3010	Energy Supply	N/A		—	—	—	
3020	Heat Generating Systems	Included in D3050		—	—	—	
3030	Cooling Generating Systems	N/A		—	—	—	8.6 %
3050	Terminal & Package Units	Single zone unit gas heating, electric cooling		S.F. Floor	7.23	7.23	
3090	Other HVAC Sys. & Equipment	N/A		—	—	—	
	D40 Fire Protection						
4010	Sprinklers	Sprinklers, light hazard		S.F. Floor	2.78	2.78	3.6%
4020	Standpipes	Standpipe		S.F. Floor	.23	.23	
	D50 Electrical						
5010	Electrical Service/Distribution	1200 ampere service, panel board and feeders		S.F. Floor	.90	.90	
5020	Lighting & Branch Wiring	High efficiency fluorescent fixtures, receptacles, switches, A.C. and misc. power		S.F. Floor	8.28	8.28	12.4%
5030	Communications & Security	Addressable alarm systems, internet wiring and emergency lighting		S.F. Floor	1.23	1.23	
5090	Other Electrical Systems	Emergency generator, 7.5 kW		S.F. Floor	.04	.04	
E. EQUIPMENT & FURNISHINGS							
1010	Commercial Equipment	N/A		—	—	—	
1020	Institutional Equipment	N/A		—	—	—	0.0 %
1030	Vehicular Equipment	N/A		—	—	—	
1090	Other Equipment	N/A		—	—	—	
F. SPECIAL CONSTRUCTION							
1020	Integrated Construction	N/A		—	—	—	0.0 %
1040	Special Facilities	N/A		—	—	—	
G. BUILDING SITEWORK	**N/A**						

			Sub-Total	83.96	100%
CONTRACTOR FEES (General Requirements: 10%, Overhead: 5%, Profit: 10%)			25%	21.04	
ARCHITECT FEES			6%	6.30	
		Total Building Cost		**111.30**	

Costs per square foot of floor area

Exterior Wall	S.F. Area	50000	65000	80000	95000	110000	125000	140000	155000	170000
	L.F. Perimeter	533	593	670	715	778	840	871	923	976
Face Brick with Concrete Block Back-up	Steel Frame	142.90	137.20	134.30	131.40	129.70	128.40	126.80	125.85	125.15
	R/Conc. Frame	144.40	138.75	135.75	132.90	131.20	129.90	128.30	127.40	126.65
Face Brick on Steel Studs	Steel Frame	138.00	133.05	130.45	127.95	126.45	125.30	123.90	123.15	122.50
	R/Conc. Frame	141.65	136.75	134.10	131.60	130.10	128.95	127.60	126.80	126.20
Precast Concrete Panels Exposed Aggregate	Steel Frame	146.75	140.55	137.30	134.15	132.30	130.90	129.10	128.15	127.30
	R/Conc. Frame	149.85	143.70	140.45	137.30	135.40	134.00	132.20	131.20	130.40
Perimeter Adj., Add or Deduct	Per 100 L.F.	4.85	3.75	3.00	2.55	2.20	1.95	1.75	1.60	1.40
Story Hgt. Adj., Add or Deduct	Per 1 Ft.	1.45	1.30	1.10	1.05	0.95	0.90	0.80	0.85	0.75

For Basement, add $40.35 per square foot of basement area

The above costs were calculated using the basic specifications shown on the facing page. These costs should be adjusted where necessary for design alternatives and owner's requirements. Reported completed project costs, for this type of structure, range from $57.60 to $154.15 per S.F.

Common additives

Description	Unit	$ Cost
Closed Circuit Surveillance, One station		
Camera and monitor	Each	1875
For additional camera stations, add	Each	1025
Directory Boards, Plastic, glass covered		
30" x 20"	Each	605
36" x 48"	Each	1325
Aluminum, 24" x 18"	Each	585
36" x 24"	Each	685
48" x 32"	Each	975
48" x 60"	Each	2025
Elevators, Hydraulic passenger, 2 stops		
1500# capacity	Each	60,900
2500# capacity	Each	64,300
3500# capacity	Each	67,600
Additional stop, add	Each	8250
Emergency Lighting, 25 watt, battery operated		
Lead battery	Each	287
Nickel cadmium	Each	845

Description	Unit	$ Cost
Escalators, Metal		
32" wide, 10' story height	Each	139,700
20' story height	Each	168,000
48" wide, 10' story height	Each	148,200
20' story height	Each	176,000
Glass		
32" wide, 10' story height	Each	133,200
20' story height	Each	161,000
48" wide, 10' story height	Each	140,700
20' story height	Each	170,000
Safe, Office type, 1 hour rating		
30" x 18" x 18"	Each	2400
60" x 36" x 18", double door	Each	8975
Sound System		
Amplifier, 250 watts	Each	2400
Speaker, ceiling or wall	Each	196
Trumpet	Each	375

Model costs calculated for a 3 story building with 16' story height and 95,000 square feet of floor area

Store, Department, 3 Story

				Unit	Unit Cost	Cost Per S.F.	% Of Sub-Total
A. SUBSTRUCTURE							
1010	Standard Foundations	Poured concrete; strip and spread footings		S.F. Ground	3.33	1.11	
1020	Special Foundations	N/A		—	—	—	
1030	Slab on Grade	4" reinforced concrete with vapor barrier and granular base		S.F. Slab	4.77	1.59	3.4%
2010	Basement Excavation	Site preparation for slab and trench for foundation wall and footing		S.F. Ground	.18	.06	
2020	Basement Walls	4' foundation wall		L.F. Wall	71	.61	
B. SHELL							
	B10 Superstructure						
1010	Floor Construction	Concrete slab with metal deck and beams, steel columns		S.F. Floor	25.40	16.93	19.8%
1020	Roof Construction	Metal deck, open web steel joists, beams, columns		S.F. Roof	8.22	2.74	
	B20 Exterior Enclosure						
2010	Exterior Walls	Face brick with concrete block backup	90% of wall	S.F. Wall	31.25	10.16	
2020	Exterior Windows	Storefront	10% of wall	Each	43.40	1.58	13.5%
2030	Exterior Doors	Revolving and sliding panel, mall-front		Each	10,915	1.60	
	B30 Roofing						
3010	Roof Coverings	Built-up tar and gravel with flashing; perlite/EPS composite insulation		S.F. Roof	5.43	1.81	1.9%
3020	Roof Openings	Roof hatches		S.F. Roof	.21	.07	
C. INTERIORS							
1010	Partitions	Gypsum board on metal studs	60 S.F. Floor/L.F. Partition	S.F. Partition	5.52	.92	
1020	Interior Doors	Single leaf hollow metal	600 S.F. Floor/Door	Each	1001	1.67	
1030	Fittings	N/A		—	—	—	
2010	Stair Construction	Concrete filled metal pan		Flight	17,125	1.81	25.6%
3010	Wall Finishes	70% paint, 20% vinyl wall covering, 10% ceramic tile		S.F. Surface	4.38	1.46	
3020	Floor Finishes	50% carpet tile, 40% marble tile, 10% terrazzo		S.F. Floor	13.02	13.02	
3030	Ceiling Finishes	Mineral fiber tile on concealed zee bars		S.F. Ceiling	6.51	6.51	
D. SERVICES							
	D10 Conveying						
1010	Elevators & Lifts	One hydraulic passenger, one hydraulic freight		Each	322,050	3.39	9.6%
1020	Escalators & Moving Walks	Four escalators		Each	146,063	6.15	
	D20 Plumbing						
2010	Plumbing Fixtures	Toilet and service fixtures, supply and drainage	1 Fixture/2570 S.F. Floor	Each	3675	1.43	
2020	Domestic Water Distribution	Gas fired water heater		S.F. Floor	.38	.38	2.3%
2040	Rain Water Drainage	Roof drains		S.F. Roof	1.32	.44	
	D30 HVAC						
3010	Energy Supply	N/A		—	—	—	
3020	Heat Generating Systems	Included in D3050		—	—	—	
3030	Cooling Generating Systems	N/A		—	—	—	7.3%
3050	Terminal & Package Units	Multizone rooftop unit, gas heating, electric cooling		S.F. Floor	7.23	7.23	
3090	Other HVAC Sys. & Equipment	N/A		—	—	—	
	D40 Fire Protection						
4010	Sprinklers	Sprinklers, light hazard		S.F. Floor	2.31	2.31	3.7%
4020	Standpipes	Standpipes		S.F. Floor	1.37	1.37	
	D50 Electrical						
5010	Electrical Service/Distribution	1200 ampere service, panel board and feeders		S.F. Floor	1.40	1.40	
5020	Lighting & Branch Wiring	High efficiency fluorescent fixtures, receptacles, switches, A.C. and misc. power		S.F. Floor	8.68	8.68	12.9%
5030	Communications & Security	Addressable alarm systems, internet wiring and emergency lighting		S.F. Floor	2.41	2.41	
5090	Other Electrical Systems	Emergency generator, 50 kW		S.F. Floor	.33	.33	
E. EQUIPMENT & FURNISHINGS							
1010	Commercial Equipment	N/A		—	—	—	
1020	Institutional Equipment	N/A		—	—	—	
1030	Vehicular Equipment	N/A		—	—	—	0.0%
1090	Other Equipment	N/A		—	—	—	
F. SPECIAL CONSTRUCTION							
1020	Integrated Construction	N/A		—	—	—	0.0%
1040	Special Facilities	N/A		—	—	—	
G. BUILDING SITEWORK	**N/A**						

				Sub-Total	99.17	100%
CONTRACTOR FEES (General Requirements: 10%, Overhead: 5%, Profit: 10%)				25%	24.79	
ARCHITECT FEES				6%	7.44	

	Total Building Cost	**131.40**

Costs per square foot of floor area

Exterior Wall	S.F. Area	4000	6000	8000	10000	12000	15000	18000	20000	22000
	L.F. Perimeter	260	340	360	410	440	490	540	565	594
Split Face Concrete Block	Steel Joists	134.85	123.25	112.95	108.65	104.70	101.00	98.50	96.95	95.90
Stucco on Concrete Block	Steel Joists	133.15	121.75	111.75	107.60	103.80	100.20	97.75	96.30	95.25
Painted Concrete Block	Steel Joists	127.10	116.00	106.55	102.45	98.80	95.30	93.05	91.65	90.65
Face Brick on Concrete Block	Steel Joists	152.00	138.20	124.90	119.50	114.40	109.65	106.50	104.50	103.10
Painted Reinforced Concrete	Steel Joists	142.55	129.95	118.30	113.55	109.10	104.90	102.15	100.40	99.20
Tilt-up Concrete Panels	Steel Joists	132.50	121.15	111.35	107.20	103.45	99.85	97.50	96.00	95.00
Perimeter Adj., Add or Deduct	Per 100 L.F.	14.95	9.95	7.50	5.95	4.95	3.95	3.40	3.00	2.70
Story Hgt. Adj., Add or Deduct	Per 1 Ft.	1.85	1.60	1.25	1.15	1.05	0.95	0.85	0.90	0.75
For Basement, add $33.50 per square foot of basement area										

The above costs were calculated using the basic specifications shown on the facing page. These costs should be adjusted where necessary for design alternatives and owner's requirements. Reported completed project costs, for this type of structure, range from $56.20 to $195.45 per S.F.

Common additives

Description	Unit	$ Cost
Emergency Lighting, 25 watt, battery operated		
Lead battery	Each	287
Nickel cadmium	Each	845
Safe, Office type, 1 hour rating		
30" x 18" x 18"	Each	2400
60" x 36" x 18", double door	Each	8975
Smoke Detectors		
Ceiling type	Each	233
Duct type	Each	525
Sound System		
Amplifier, 250 watts	Each	2400
Speaker, ceiling or wall	Each	196
Trumpet	Each	375

Important: See the Reference Section for Location Factors

				Unit	Unit Cost	Cost Per S.F.	% Of Sub-Total
A. SUBSTRUCTURE							
1010	Standard Foundations	Poured concrete; strip and spread footings		S.F. Ground	1.72	1.72	
1020	Special Foundations	N/A		—	—	—	
1030	Slab on Grade	4" reinforced concrete with vapor barrier and granular base		S.F. Slab	4.77	4.77	12.2%
2010	Basement Excavation	Site preparation for slab and trench for foundation wall and footing		S.F. Ground	.31	.31	
2020	Basement Walls	4' foundation wall		L.F. Wall	76	3.40	
B. SHELL							
	B10 Superstructure						
1010	Floor Construction	N/A		—	—	—	6.8 %
1020	Roof Construction	Metal deck, open web steel joists, beams, interior columns		S.F. Roof	5.69	5.69	
	B20 Exterior Enclosure						
2010	Exterior Walls	Decorative concrete block	90% of wall	S.F. Wall	15.84	8.98	
2020	Exterior Windows	Storefront windows	10% of wall	Each	43.40	2.73	14.6%
2030	Exterior Doors	Sliding entrance door and hollow metal service doors		Each	2077	.53	
	B30 Roofing						
3010	Roof Coverings	Built-up tar and gravel with flashing; perlite/EPS composite insulation		S.F. Roof	6.11	6.11	7.4%
3020	Roof Openings	Roof hatches		S.F. Roof	.11	.11	
C. INTERIORS							
1010	Partitions	Gypsum board on metal studs	60 S.F. Floor/L.F. Partition	S.F. Partition	5.52	.92	
1020	Interior Doors	Single leaf hollow metal	600 S.F. Floor/Door	Each	1001	1.67	
1030	Fittings	N/A		—	—	—	
2010	Stair Construction	N/A		—	—	—	16.6%
3010	Wall Finishes	Paint		S.F. Surface	5.67	1.89	
3020	Floor Finishes	Vinyl tile		S.F. Floor	2.90	2.90	
3030	Ceiling Finishes	Mineral fiber tile on concealed zee bars		S.F. Ceiling	6.51	6.51	
D. SERVICES							
	D10 Conveying						
1010	Elevators & Lifts	N/A		—	—	—	0.0 %
1020	Escalators & Moving Walks	N/A		—	—	—	
	D20 Plumbing						
2010	Plumbing Fixtures	Toilet and service fixtures, supply and drainage	1 Fixture/890 S.F. Floor	Each	2634	2.96	
2020	Domestic Water Distribution	Gas fired water heater		S.F. Floor	4.45	4.45	10.5%
2040	Rain Water Drainage	Roof drains		S.F. Roof	1.38	1.38	
	D30 HVAC						
3010	Energy Supply	N/A		—	—	—	
3020	Heat Generating Systems	Included in D3050		—	—	—	
3030	Cooling Generating Systems	N/A		—	—	—	8.6 %
3050	Terminal & Package Units	Single zone unit, gas heating, electric cooling		S.F. Floor	7.23	7.23	
3090	Other HVAC Sys. & Equipment	N/A		—	—	—	
	D40 Fire Protection						
4010	Sprinklers	Wet pipe sprinkler system		S.F. Floor	4.11	4.11	6.0%
4020	Standpipes	Standpipe, wet, Class III		S.F. Floor	.90	.90	
	D50 Electrical						
5010	Electrical Service/Distribution	400 ampere service, panel board and feeders		S.F. Floor	2.50	2.50	
5020	Lighting & Branch Wiring	High efficiency fluorescent fixtures, receptacles, switches, A.C. and misc. power		S.F. Floor	10.09	10.09	17.2%
5030	Communications & Security	Addressable alarm systems and emergency lighting		S.F. Floor	1.59	1.59	
5090	Other Electrical Systems	Emergency generator, 15 kW		S.F. Floor	.21	.21	
E. EQUIPMENT & FURNISHINGS							
1010	Commercial Equipment	N/A		—	—	—	
1020	Institutional Equipment	N/A		—	—	—	0.0 %
1030	Vehicular Equipment	N/A		—	—	—	
1090	Other Equipment	N/A		—	—	—	
F. SPECIAL CONSTRUCTION							
1020	Integrated Construction	N/A		—	—	—	0.0 %
1040	Special Facilities	N/A		—	—	—	
G. BUILDING SITEWORK	**N/A**						

		Sub-Total	83.66	100%
CONTRACTOR FEES (General Requirements: 10%, Overhead: 5%, Profit: 10%)		25%	20.92	
ARCHITECT FEES		8%	8.37	
		Total Building Cost	**112.95**	

Costs per square foot of floor area

Exterior Wall	S.F. Area	12000	16000	20000	26000	32000	38000	44000	52000	60000
	L.F. Perimeter	450	510	570	650	716	780	840	920	1020
Face Brick with Concrete Block Back-up	Steel Frame	134.40	123.85	117.55	111.25	106.90	103.80	101.40	99.10	97.90
	Bearing Walls	132.05	121.45	115.15	108.90	104.45	101.40	99.00	96.75	95.45
Stucco on Concrete Block	Steel Frame	124.35	115.30	109.90	104.55	100.85	98.30	96.30	94.35	93.30
	Bearing Walls	124.50	115.45	110.05	104.70	101.00	98.45	96.45	94.50	93.45
Precast Concrete Panels	Steel Frame	141.55	129.90	122.95	116.05	111.15	107.65	105.05	102.45	101.10
Metal Sandwich Panels	Steel Frame	116.70	108.80	104.05	99.45	96.30	94.10	92.40	90.75	89.85
Perimeter Adj., Add or Deduct	Per 100 L.F.	9.35	7.00	5.60	4.35	3.50	2.95	2.60	2.20	1.85
Story Hgt. Adj., Add or Deduct	Per 1 Ft.	1.95	1.65	1.45	1.30	1.15	1.10	1.00	0.95	0.80
For Basement, add $24.35 per square foot of basement area										

The above costs were calculated using the basic specifications shown on the facing page. These costs should be adjusted where necessary for design alternatives and owner's requirements. Reported completed project costs, for this type of structure, range from $65.65 to $188.05 per S.F.

Common additives

Description	Unit	$ Cost	Description	Unit	$ Cost
Check Out Counter			Refrigerated Food Cases, cont.		
Single belt	Each	3275	Frozen food, chest type, 12' long	Each	8100
Double belt	Each	4700	Glass door reach in, 5 door	Each	15,200
Scanner, registers, guns & memory 2 lanes	Each	16,700	Island case 12' long, single deck	Each	9125
10 lanes	Each	158,500	Multi deck	Each	19,000
Power take away	Each	5525	Meat case 12' long, single deck	Each	6725
Emergency Lighting, 25 watt, battery operated			Multi deck	Each	11,400
Lead battery	Each	287	Produce, 12' long single deck	Each	8775
Nickel cadmium	Each	845	Multi deck	Each	9550
Refrigerators, Prefabricated, walk-in			Safe, Office type, 1 hour rating		
7'-6" High, 6' x 6'	S.F.	173	30" x 18" x 18"	Each	2400
10' x 10'	S.F.	136	60" x 36" x 18", double door	Each	8975
12' x 14'	S.F.	120	Smoke Detectors		
12' x 20'	S.F.	106	Ceiling type	Each	233
Refrigerated Food Cases			Duct type	Each	525
Dairy, multi deck, 12' long	Each	11,900	Sound System		
Delicatessen case, single deck, 12' long	Each	8000	Amplifier, 250 watts	Each	2400
Multi deck, 18 S.F. shelf display	Each	7450	Speaker, ceiling or wall	Each	196
Freezer, self-contained chest type, 30 C.F.	Each	8575	Trumpet	Each	375
Glass door upright, 78 C.F.	Each	11,000			

Important: See the Reference Section for Location Factors

Model costs calculated for a 1 story building with 18' story height and 44,000 square feet of floor area

Supermarket

				Unit	Unit Cost	Cost Per S.F.	% Of Sub-Total
A.	**SUBSTRUCTURE**						
1010	Standard Foundations	Poured concrete; strip and spread footings		S.F. Ground	1.02	1.02	
1020	Special Foundations	N/A		—	—	—	
1030	Slab on Grade	4" reinforced concrete with vapor barrier and granular base		S.F. Slab	4.77	4.77	10.0%
2010	Basement Excavation	Site preparation for slab and trench for foundation wall and footing		S.F. Ground	.18	.18	
2020	Basement Walls	4' foundation wall		L.F. Wall	76	1.44	
B.	**SHELL**						
	B10 Superstructure						
1010	Floor Construction	N/A		—	—	—	8.4 %
1020	Roof Construction	Metal deck, open web steel joists, beams, interior columns		S.F. Roof	6.20	6.20	
	B20 Exterior Enclosure						
2010	Exterior Walls	Face brick with concrete block backup	85% of wall	S.F. Wall	30.30	8.85	
2020	Exterior Windows	Storefront windows	15% of wall	Each	69	3.57	18.4%
2030	Exterior Doors	Sliding entrance doors with electrical operator, hollow metal		Each	4948	1.24	
	B30 Roofing						
3010	Roof Coverings	Built-up tar and gravel with flashing; perlite/EPS composite insulation		S.F. Roof	5.22	5.22	7.2%
3020	Roof Openings	Roof hatches		S.F. Roof	.10	.10	
C.	**INTERIORS**						
1010	Partitions	50% concrete block, 50% gypsum board on metal studs	50 S.F. Floor/L.F. Partition	S.F. Partition	7.04	1.69	
1020	Interior Doors	Single leaf hollow metal	2000 S.F. Floor/Door	Each	1001	.50	
1030	Fittings	N/A		—	—	—	
2010	Stair Construction	N/A		—	—	—	18.7%
3010	Wall Finishes	Paint		S.F. Surface	4.63	2.22	
3020	Floor Finishes	Vinyl composition tile		S.F. Floor	2.90	2.90	
3030	Ceiling Finishes	Mineral fiber tile on concealed zee bars		S.F. Ceiling	6.51	6.51	
D.	**SERVICES**						
	D10 Conveying						
1010	Elevators & Lifts	N/A		—	—	—	0.0 %
1020	Escalators & Moving Walks	N/A		—	—	—	
	D20 Plumbing						
2010	Plumbing Fixtures	Toilet and service fixtures, supply and drainage	1 Fixture/1820 S.F. Floor	Each	1765	.97	
2020	Domestic Water Distribution	Gas fired water heater		S.F. Floor	.35	.35	2.9%
2040	Rain Water Drainage	Roof drains		S.F. Roof	.81	.81	
	D30 HVAC						
3010	Energy Supply	N/A		—	—	—	
3020	Heat Generating Systems	Included in D3050		—	—	—	
3030	Cooling Generating Systems	N/A		—	—	—	6.4 %
3050	Terminal & Package Units	Single zone rooftop unit, gas heating, electric cooling		S.F. Floor	4.75	4.75	
3090	Other HVAC Sys. & Equipment	N/A		—	—	—	
	D40 Fire Protection						
4010	Sprinklers	Sprinklers, light hazard		S.F. Floor	3.16	3.16	4.9%
4020	Standpipes	Standpipe		S.F. Floor	.48	.48	
	D50 Electrical						
5010	Electrical Service/Distribution	1600 ampere service, panel board and feeders		S.F. Floor	3.32	3.32	
5020	Lighting & Branch Wiring	High efficiency fluorescent fixtures, receptacles, switches, A.C. and misc. power		S.F. Floor	11.63	11.63	23.1%
5030	Communications & Security	Addressable alarm systems, partial internet wiring and emergency lighting		S.F. Floor	2.12	2.12	
5090	Other Electrical Systems	Emergency generator, 15 kW		S.F. Floor	.04	.04	
E.	**EQUIPMENT & FURNISHINGS**						
1010	Commercial Equipment	N/A		—	—	—	
1020	Institutional Equipment	N/A		—	—	—	0.0 %
1030	Vehicular Equipment	N/A		—	—	—	
1090	Other Equipment	N/A		—	—	—	
F.	**SPECIAL CONSTRUCTION**						
1020	Integrated Construction	N/A		—	—	—	0.0 %
1040	Special Facilities	N/A		—	—	—	
G.	**BUILDING SITEWORK**	**N/A**					

		Sub-Total	74.04	**100%**
	CONTRACTOR FEES (General Requirements: 10%, Overhead: 5%, Profit: 10%)		25%	18.48
	ARCHITECT FEES		7%	6.48
		Total Building Cost	99	

Costs per square foot of floor area

Exterior Wall	S.F. Area	10000	16000	20000	22000	24000	26000	28000	30000	32000
	L.F. Perimeter	420	510	600	640	660	684	706	740	737
Face Brick with Concrete Block Back-up	Wood Truss	242.60	221.70	217.10	215.05	212.05	209.75	207.60	206.50	203.50
	Precast Conc.	279.55	254.90	249.60	247.25	243.65	240.85	238.35	236.95	233.40
Metal Sandwich Panel	Wood Truss	238.65	222.75	219.10	217.55	215.30	213.60	212.05	211.15	209.00
Precast Concrete Panel	Precast Conc.	251.05	233.05	229.05	227.35	224.75	222.80	221.05	220.00	217.60
Painted Concrete Block	Wood Frame	238.95	223.35	219.75	218.20	216.05	214.35	212.80	211.90	209.80
	Precast Conc.	257.40	238.05	233.75	231.90	229.10	226.95	225.00	223.95	221.25
Perimeter Adj., Add or Deduct	Per 100 L.F.	16.35	10.20	8.15	7.55	6.80	6.25	5.85	5.40	5.15
Story Hgt. Adj., Add or Deduct	Per 1 Ft.	2.05	1.55	1.45	1.50	1.30	1.30	1.30	1.20	1.15
For Basement, add $36.10 per square foot of basement area										

The above costs were calculated using the basic specifications shown on the facing page. These costs should be adjusted where necessary for design alternatives and owner's requirements. Reported completed project costs, for this type of structure, range from $103.70 to $304.90 per S.F.

Common additives

Description	Unit	$ Cost
Bleachers, Telescoping, manual		
To 15 tier	Seat	118 - 164
16-20 tier	Seat	242 - 297
21-30 tier	Seat	257 - 310
For power operation, add	Seat	47 - 73.50
Emergency Lighting, 25 watt, battery operated		
Lead battery	Each	287
Nickel cadmium	Each	845
Lockers, Steel, single tier, 60" or 72"	Opening	200 - 350
2 tier, 60" or 72" total	Opening	124 - 152
5 tier, box lockers	Opening	70 - 82.50
Locker bench, lam. maple top only	L.F.	22
Pedestal, steel pipe	Each	64
Pool Equipment		
Diving stand, 3 meter	Each	14,100
1 meter	Each	7175
Diving board, 16' aluminum	Each	4075
Fiberglass	Each	3175
Lifeguard chair, fixed	Each	3650
Portable	Each	3000
Lights, underwater, 12 volt, 300 watt	Each	875

Description	Unit	$ Cost
Sauna, Prefabricated, complete		
6' x 4'	Each	5875
6' x 6'	Each	6975
6' x 9'	Each	8550
8' x 8'	Each	10,100
8' x 10'	Each	11,300
10' x 12'	Each	14,000
Sound System		
Amplifier, 250 watts	Each	2400
Speaker, ceiling or wall	Each	196
Trumpet	Each	375
Steam Bath, Complete, to 140 C.F.	Each	2075
To 300 C.F.	Each	2350
To 800 C.F.	Each	6425
To 2500 C.F.	Each	9600

Important: See the Reference Section for Location Factors

Model costs calculated for a 1 story building with 24' story height and 20,000 square feet of floor area

Swimming Pool, Enclosed

				Unit	Unit Cost	Cost Per S.F.	% Of Sub-Total
A.	**SUBSTRUCTURE**						
1010	Standard Foundations	Poured concrete; strip and spread footings		S.F. Ground	1.48	1.48	
1020	Special Foundations	N/A		—	—	—	
1030	Slab on Grade	4" reinforced concrete with vapor barrier and granular base		S.F. Slab	4.77	4.77	15.8%
2010	Basement Excavation	Site preparation for slab and trench for foundation wall and footing		S.F. Ground	10.05	10.05	
2020	Basement Walls	4' foundation wall		L.F. Wall	76	9.08	
B.	**SHELL**						
	B10 Superstructure						
1010	Floor Construction	N/A		—	—	—	9.8 %
1020	Roof Construction	Wood deck on laminated wood truss		S.F. Roof	15.72	15.72	
	B20 Exterior Enclosure						
2010	Exterior Walls	Face brick with concrete block backup	80% of wall	S.F. Wall	31.25	18	
2020	Exterior Windows	Outward projecting steel	20% of wall	Each	43.40	6.25	15.5%
2030	Exterior Doors	Double aluminum and glass, hollow metal		Each	2426	.62	
	B30 Roofing						
3010	Roof Coverings	Asphalt shingles with flashing; perlite/EPS composite insulation		S.F. Roof	2.21	2.21	1.4%
3020	Roof Openings	N/A		—	—	—	
C.	**INTERIORS**						
1010	Partitions	Concrete block	100 S.F. Floor/L.F. Partition	S.F. Partition	8.70	.87	
1020	Interior Doors	Single leaf hollow metal	1000 S.F. Floor/Door	Each	1001	1	
1030	Fittings	Toilet partitions		S.F. Floor	.17	.17	
2010	Stair Construction	N/A		—	—	—	15.5%
3010	Wall Finishes	Acrylic glazed coating		S.F. Surface	10.35	2.07	
3020	Floor Finishes	70% terrazzo, 30% ceramic tile		S.F. Floor	20.06	20.06	
3030	Ceiling Finishes	Mineral fiber tile on concealed zee bars	20% of area	S.F. Ceiling	4.09	.82	
D.	**SERVICES**						
	D10 Conveying						
1010	Elevators & Lifts	N/A		—	—	—	0.0 %
1020	Escalators & Moving Walks	N/A		—	—	—	
	D20 Plumbing						
2010	Plumbing Fixtures	Toilet and service fixtures, supply and drainage	1 Fixture/210 S.F. Floor	Each	2900	13.81	
2020	Domestic Water Distribution	Gas fired water heater		S.F. Floor	3.30	3.30	10.6%
2040	Rain Water Drainage	N/A		—	—	—	
	D30 HVAC						
3010	Energy Supply	Terminal unit heaters	10% of area	S.F. Floor	.84	.84	
3020	Heat Generating Systems	N/A		—	—	—	
3030	Cooling Generating Systems	N/A		—	—	—	8.2%
3050	Terminal & Package Units	Single zone unit gas heating, electric cooling		S.F. Floor	12.34	12.34	
3090	Other HVAC Sys. & Equipment	N/A		—	—	—	
	D40 Fire Protection						
4010	Sprinklers	Sprinklers, light hazard		S.F. Floor	1.59	1.59	1.0%
4020	Standpipes	N/A		—	—	—	
	D50 Electrical						
5010	Electrical Service/Distribution	400 ampere service, panel board and feeders		S.F. Floor	1.04	1.04	
5020	Lighting & Branch Wiring	Fluorescent fixtures, receptacles, switches, A.C. and misc. power		S.F. Floor	8.89	8.89	6.4%
5030	Communications & Security	Alarm systems and emergency lighting		S.F. Floor	.35	.35	
5090	Other Electrical Systems	Emergency generator, 15 kW		S.F. Floor	.08	.08	
E.	**EQUIPMENT & FURNISHINGS**						
1010	Commercial Equipment	N/A		—	—	—	
1020	Institutional Equipment	N/A		—	—	—	15.8 %
1030	Vehicular Equipment	N/A		—	—	—	
1090	Other Equipment	Swimming pool		S.F. Floor	25.40	25.40	
F.	**SPECIAL CONSTRUCTION**						
1020	Integrated Construction	N/A		—	—	—	0.0 %
1040	Special Facilities	N/A		—	—	—	
G.	**BUILDING SITEWORK**	**N/A**					

			Sub-Total	160.81	**100%**
CONTRACTOR FEES (General Requirements: 10%, Overhead: 5%, Profit: 10%)		25%	40.21		
ARCHITECT FEES		8%	16.08		

Total Building Cost **217.10**

Costs per square foot of floor area

Exterior Wall	S.F. Area	2000	3000	4000	5000	6000	7000	8000	9000	10000
	L.F. Perimeter	180	220	260	286	320	353	368	397	425
Face Brick with Concrete Block Back-up	Steel Frame	199.55	177.95	167.15	158.10	153.35	149.75	145.05	142.75	140.85
	Bearing Walls	195.55	173.95	163.20	154.15	149.40	145.75	141.10	138.75	136.85
Limestone with Concrete Block Back-up	Steel Frame	242.95	213.30	198.55	185.70	179.05	174.00	167.20	164.00	161.30
	Bearing Walls	239.00	209.30	194.55	181.75	175.05	170.05	163.25	160.05	157.35
Decorative Concrete Block	Steel Frame	184.60	165.75	156.40	148.60	144.45	141.40	137.40	135.45	133.75
	Bearing Walls	180.65	161.80	152.45	144.65	140.50	137.40	133.45	131.50	129.80
Perimeter Adj., Add or Deduct	Per 100 L.F.	45.70	30.45	22.85	18.25	15.20	13.00	11.40	10.15	9.10
Story Hgt. Adj., Add or Deduct	Per 1 Ft.	5.35	4.30	3.85	3.45	3.15	2.95	2.70	2.60	2.50
For Basement, add $35.00 per square foot of basement area										

The above costs were calculated using the basic specifications shown on the facing page. These costs should be adjusted where necessary for design alternatives and owner's requirements. Reported completed project costs, for this type of structure, range from $91.65 to $330.75 per S.F.

Common additives

Description	Unit	$ Cost
Emergency Lighting, 25 watt, battery operated		
Lead battery	Each	287
Nickel cadmium	Each	845
Smoke Detectors		
Ceiling type	Each	233
Duct type	Each	525
Emergency Generators, complete system, gas		
15 kw	Each	17,100
85 kw	Each	36,000
170 kw	Each	91,000
Diesel, 50 kw	Each	31,200
150 kw	Each	54,000
350 kw	Each	85,000

Important: See the Reference Section for Location Factors

Model costs calculated for a 1 story building with 12' story height and 5,000 square feet of floor area

				Unit	Unit Cost	Cost Per S.F.	% Of Sub-Total
A.	**SUBSTRUCTURE**						
1010	Standard Foundations	Poured concrete; strip and spread footings		S.F. Ground	2.51	2.51	
1020	Special Foundations	N/A		—	—	—	
1030	Slab on Grade	4" reinforced concrete with vapor barrier and granular base		S.F. Slab	4.77	4.77	10.6%
2010	Basement Excavation	Site preparation for slab and trench for foundation wall and footing		S.F. Ground	.52	.52	
2020	Basement Walls	4' foundation wall		L.F. Wall	76	4.32	
B.	**SHELL**						
	B10 Superstructure						
1010	Floor Construction	N/A		—	—	—	7.2 %
1020	Roof Construction	Metal deck, open web steel joists, beams, columns		S.F. Roof	8.16	8.16	
	B20 Exterior Enclosure						
2010	Exterior Walls	Face brick with concrete block backup	80% of wall	S.F. Wall	30.30	16.64	
2020	Exterior Windows	Outward projecting steel	20% of wall	Each	1615	11.08	25.6%
2030	Exterior Doors	Single aluminum glass with transom		Each	3557	1.43	
	B30 Roofing						
3010	Roof Coverings	Built-up tar and gravel with flashing; perlite/EPS composite insulation		S.F. Roof	6.54	6.54	5.7%
3020	Roof Openings						
C.	**INTERIORS**						
1010	Partitions	Double layer gypsum board on metal studs	15 S.F. Floor/L.F. Partition	S.F. Partition	5.51	3.67	
1020	Interior Doors	Single leaf hollow metal	150 S.F. Floor/Door	Each	1001	6.67	
1030	Fittings	Toilet partitions		S.F. Floor	1.44	1.44	
2010	Stair Construction	N/A		—	—	—	24.8%
3010	Wall Finishes	Paint		S.F. Surface	1.94	2.58	
3020	Floor Finishes	90% carpet, 10% terrazzo		S.F. Floor	9.55	9.55	
3030	Ceiling Finishes	Fiberglass board on exposed grid system, suspended		S.F. Ceiling	4.31	4.31	
D.	**SERVICES**						
	D10 Conveying						
1010	Elevators & Lifts	N/A		—	—	—	0.0 %
1020	Escalators & Moving Walks	N/A		—	—	—	
	D20 Plumbing						
2010	Plumbing Fixtures	Kitchen, toilet and service fixtures, supply and drainage	1 Fixture/715 S.F. Floor	Each	2831	3.96	
2020	Domestic Water Distribution	Gas fired water heater		S.F. Floor	1.37	1.37	6.8%
2040	Rain Water Drainage	Roof drains		S.F. Roof	2.47	2.47	
	D30 HVAC						
3010	Energy Supply	N/A		—	—	—	
3020	Heat Generating Systems	Included in D3050		—	—	—	
3030	Cooling Generating Systems	N/A		—	—	—	7.4 %
3050	Terminal & Package Units	Single zone unit, gas heating, electric cooling		S.F. Floor	8.47	8.47	
3090	Other HVAC Sys. & Equipment	N/A		—	—	—	
	D40 Fire Protection						
4010	Sprinklers	Wet pipe sprinkler system		S.F. Floor	4.35	4.35	3.8%
4020	Standpipes	N/A		—	—	—	
	D50 Electrical						
5010	Electrical Service/Distribution	200 ampere service, panel board and feeders		S.F. Floor	1.70	1.70	
5020	Lighting & Branch Wiring	Fluorescent fixtures, receptacles, switches, A.C. and misc. power		S.F. Floor	5.85	5.85	8.0%
5030	Communications & Security	Alarm systems and emergency lighting		S.F. Floor	1.41	1.41	
5090	Other Electrical Systems	Emergency generator, 15 kW		S.F. Floor	.18	.18	
E.	**EQUIPMENT & FURNISHINGS**						
1010	Commercial Equipment	N/A		—	—	—	
1020	Institutional Equipment	N/A		—	—	—	0.0 %
1030	Vehicular Equipment	N/A		—	—	—	
1090	Other Equipment	N/A		—	—	—	
F.	**SPECIAL CONSTRUCTION**						
1020	Integrated Construction	N/A		—	—	—	0.0 %
1040	Special Facilities	N/A		—	—	—	
G.	**BUILDING SITEWORK**	**N/A**					

	Sub-Total	113.95	**100%**
CONTRACTOR FEES (General Requirements: 10%, Overhead: 5%, Profit: 10%)	25%	28.48	
ARCHITECT FEES	11%	15.67	
Total Building Cost		**158.10**	

Costs per square foot of floor area

Exterior Wall	S.F. Area	5000	6500	8000	9500	11000	14000	17500	21000	24000
	L.F. Perimeter	300	360	386	396	435	510	550	620	680
Face Brick with Concrete Block Back-up	Steel Joists	147.80	142.20	135.50	129.70	127.45	124.25	119.75	117.90	116.80
	Wood Joists	149.65	143.90	137.05	131.10	128.75	125.50	120.90	119.00	117.85
Stone with Concrete Block Back-up	Steel Joists	153.75	147.65	140.30	133.85	131.35	127.90	122.90	120.85	119.60
	Wood Joists	155.60	149.40	141.85	135.20	132.70	129.10	124.05	121.90	120.65
Brick Veneer	Wood Frame	140.45	135.25	129.25	123.95	121.90	119.00	115.00	113.30	112.25
E.I.F.S.	Wood Frame	129.05	124.45	119.45	115.05	113.30	110.85	107.50	106.05	105.20
Perimeter Adj., Add or Deduct	Per 100 L.F.	14.95	11.50	9.40	7.90	6.75	5.30	4.30	3.55	3.10
Story Hgt. Adj., Add or Deduct	Per 1 Ft.	2.70	2.55	2.20	1.90	1.80	1.65	1.45	1.40	1.30
For Basement, add $ 28.80 per square foot of basement area										

The above costs were calculated using the basic specifications shown on the facing page. These costs should be adjusted where necessary for design alternatives and owner's requirements. Reported completed project costs, for this type of structure, range from $ 86.65 to $ 227.90 per S.F.

Common additives

Description	Unit	$ Cost
Directory Boards, Plastic, glass covered		
30" x 20"	Each	605
36" x 48"	Each	1325
Aluminum, 24" x 18"	Each	585
36" x 24"	Each	685
48" x 32"	Each	975
48" x 60"	Each	2025
Emergency Lighting, 25 watt, battery operated		
Lead battery	Each	287
Nickel cadmium	Each	845
Flagpoles, Complete		
Aluminum, 20' high	Each	1700
40' high	Each	3600
70' high	Each	10,300
Fiberglass, 23' high	Each	1825
39'-5" high	Each	3475
59' high	Each	6625
Safe, Office type, 1 hour rating		
30" x 18" x 18"	Each	2400
60" x 36" x 18", double door	Each	8975

Description	Unit	$ Cost
Smoke Detectors		
Ceiling type	Each	233
Duct type	Each	525
Vault Front, Door & frame		
1 Hour test, 32" x 78"	Opening	5375
2 Hour test, 32" door	Opening	5825
40" door	Opening	6775
4 Hour test, 32" door	Opening	6300
40" door	Opening	7525
Time lock movement; two movement	Each	1875

Important: See the Reference Section for Location Factors

Model costs calculated for a 1 story building with 12' story height and 11,000 square feet of floor area

				Unit	Unit Cost	Cost Per S.F.	% Of Sub-Total
A. SUBSTRUCTURE							
1010	Standard Foundations	Poured concrete; strip and spread footings		S.F. Ground	1.55	1.55	
1020	Special Foundations	N/A		—	—	—	
1030	Slab on Grade	4" reinforced concrete with vapor barrier and granular base		S.F. Slab	4.77	4.77	10.3%
2010	Basement Excavation	Site preparation for slab and trench for foundation wall and footing		S.F. Ground	.31	.31	
2020	Basement Walls	4' foundation wall		L.F. Wall	76	2.99	
B. SHELL							
	B10 Superstructure						
1010	Floor Construction	N/A		—	—	—	6.4 %
1020	Roof Construction	Metal deck, open web steel joists, beams, interior columns		S.F. Roof	5.99	5.99	
	B20 Exterior Enclosure						
2010	Exterior Walls	Face brick with concrete block backup	70% of wall	S.F. Wall	30.31	10.07	
2020	Exterior Windows	Metal outward projecting	30% of wall	Each	703	4.36	16.4%
2030	Exterior Doors	Metal and glass with transom		Each	2604	.95	
	B30 Roofing						
3010	Roof Coverings	Built-up tar and gravel with flashing; perlite/EPS composite insulation		S.F. Roof	6.10	6.10	6.5%
3020	Roof Openings	Roof hatches		S.F. Roof			
C. INTERIORS							
1010	Partitions	Gypsum board on metal studs	20 S.F. Floor/L.F. Partition	S.F. Partition	8.22	4.11	
1020	Interior Doors	Wood solid core	200 S.F. Floor/Door	Each	576	2.89	
1030	Fittings	Toilet partitions		S.F. Floor	.48	.48	
2010	Stair Construction	N/A		—	—	—	24.4%
3010	Wall Finishes	90% paint, 10% ceramic tile		S.F. Surface	1.35	1.35	
3020	Floor Finishes	70% carpet tile, 15% terrazzo, 15% vinyl composition tile		S.F. Floor	7.50	7.50	
3030	Ceiling Finishes	Mineral fiber tile on concealed zee bars		S.F. Ceiling	6.51	6.51	
D. SERVICES							
	D10 Conveying						
1010	Elevators & Lifts	N/A		—	—	—	0.0 %
1020	Escalators & Moving Walks	N/A		—	—	—	
	D20 Plumbing						
2010	Plumbing Fixtures	Kitchen, toilet and service fixtures, supply and drainage	1 Fixture/500 S.F. Floor	Each	2040	4.08	
2020	Domestic Water Distribution	Gas fired water heater		S.F. Floor	1.27	1.27	7.1%
2040	Rain Water Drainage	Roof drains		S.F. Roof	1.30	1.30	
	D30 HVAC						
3010	Energy Supply	N/A		—	—	—	
3020	Heat Generating Systems	Included in D3050		—	—	—	
3030	Cooling Generating Systems	N/A		—	—	—	8.4 %
3050	Terminal & Package Units	Multizone unit, gas heating, electric cooling		S.F. Floor	7.86	7.86	
3090	Other HVAC Sys. & Equipment	N/A		—	—	—	
	D40 Fire Protection						
4010	Sprinklers	Wet pipe sprinkler system		S.F. Floor	3.16	3.16	4.1%
4020	Standpipes	Standpipe, wet, Class III		S.F. Floor	.66	.66	
	D50 Electrical						
5010	Electrical Service/Distribution	400 ampere service, panel board and feeders		S.F. Floor	1.81	1.81	
5020	Lighting & Branch Wiring	High efficiency fluorescent fixtures, receptacles, switches, A.C. and misc. power		S.F. Floor	9.97	9.97	16.3%
5030	Communications & Security	Addressable alarm systems, internet wiring, and emergency lighting		S.F. Floor	3.35	3.35	
5090	Other Electrical Systems	Emergency generator, 15 kW		S.F. Floor	.15	.15	
E. EQUIPMENT & FURNISHINGS							
1010	Commercial Equipment	N/A		—	—	—	
1020	Institutional Equipment	N/A		—	—	—	0.0 %
1030	Vehicular Equipment	N/A		—	—	—	
1090	Other Equipment	N/A		—	—	—	
F. SPECIAL CONSTRUCTION							
1020	Integrated Construction	N/A		—	—	—	0.0 %
1040	Special Facilities	N/A		—	—	—	
G. BUILDING SITEWORK	**N/A**						

		Sub-Total	93.54	100%
CONTRACTOR FEES (General Requirements: 10%, Overhead: 5%, Profit: 10%)			25%	23.39
ARCHITECT FEES			9%	10.52

Total Building Cost	**127.45**

Costs per square foot of floor area

Exterior Wall	S.F. Area	8000	10000	12000	15000	18000	24000	28000	35000	40000
	L.F. Perimeter	206	233	260	300	320	360	393	451	493
Face Brick with Concrete Block Back-up	Steel Frame	196.50	186.35	179.55	172.65	166.15	157.85	154.75	151.00	149.20
	R/Conc. Frame	205.00	194.85	188.00	181.10	174.60	166.30	163.25	159.50	157.65
Stone with Concrete Block Back-up	Steel Frame	192.35	183.10	176.90	170.60	164.45	156.65	153.75	150.25	148.60
	R/Conc. Frame	212.65	201.80	194.45	187.05	179.85	170.80	167.40	163.30	161.35
Limestone with Concrete Block Back-up	Steel Frame	219.05	205.90	197.10	188.15	178.85	167.25	162.95	157.80	155.25
	R/Conc. Frame	227.55	214.40	205.55	196.65	187.35	175.70	171.40	166.25	163.70
Perimeter Adj., Add or Deduct	Per 100 L.F.	22.90	18.30	15.30	12.25	10.20	7.65	6.60	5.20	4.60
Story Hgt. Adj., Add or Deduct	Per 1 Ft.	3.45	3.20	3.00	2.70	2.40	2.05	1.95	1.80	1.65

For Basement, add $28.05 per square foot of basement area

The above costs were calculated using the basic specifications shown on the facing page. These costs should be adjusted where necessary for design alternatives and owner's requirements. Reported completed project costs, for this type of structure, range from $77.65 to $227.90 per S.F.

Common additives

Description	Unit	$ Cost
Directory Boards, Plastic, glass covered		
30" x 20"	Each	605
36" x 48"	Each	1325
Aluminum, 24" x 18"	Each	585
36" x 24"	Each	685
48" x 32"	Each	975
48" x 60"	Each	2025
Elevators, Hydraulic passenger, 2 stops		
1500# capacity	Each	60,900
2500# capacity	Each	64,300
3500# capacity	Each	67,600
Additional stop, add	Each	8250
Emergency Lighting, 25 watt, battery operated		
Lead battery	Each	287
Nickel cadmium	Each	845

Description	Unit	$ Cost
Flagpoles, Complete		
Aluminum, 20' high	Each	1700
40' high	Each	3600
70' high	Each	10,300
Fiberglass, 23' high	Each	1825
39'-5" high	Each	3475
59' high	Each	6625
Safe, Office type, 1 hour rating		
30" x 18" x 18"	Each	2400
60" x 36" x 18", double door	Each	8975
Smoke Detectors		
Ceiling type	Each	233
Duct type	Each	525
Vault Front, Door & frame		
1 Hour test, 32" x 78"	Opening	5375
2 Hour test, 32" door	Opening	5825
40" door	Opening	6775
4 Hour test, 32" door	Opening	6300
40" door	Opening	7525
Time lock movement; two movement	Each	1875

Important: See the Reference Section for Location Factors

Model costs calculated for a 3 story building with 12' story height and 18,000 square feet of floor area

				Unit	Unit Cost	Cost Per S.F.	% Of Sub-Total
A.	**SUBSTRUCTURE**						
1010	Standard Foundations	Poured concrete; strip and spread footings		S.F. Ground	3.72	1.24	
1020	Special Foundations	N/A		—	—	—	
1030	Slab on Grade	4" reinforced concrete with vapor barrier and granular base		S.F. Slab	4.77	1.59	3.6%
2010	Basement Excavation	Site preparation for slab and trench for foundation wall and footing		S.F. Ground	.18	.06	
2020	Basement Walls	4' foundation wall		L.F. Wall	71	1.49	
B.	**SHELL**						
	B10	**Superstructure**					
1010	Floor Construction	Open web steel joists, slab form, concrete, wide flange steel columns		S.F. Floor	22.32	14.88	14.1%
1020	Roof Construction	Metal deck, open web steel joists, beams, interior columns		S.F. Roof	6.57	2.19	
	B20	**Exterior Enclosure**					
2010	Exterior Walls	Stone with concrete block backup	70% of wall	S.F. Wall	38.95	17.45	
2020	Exterior Windows	Metal outward projecting	10% of wall	Each	1040	2.40	17.0%
2030	Exterior Doors	Metal and glass with transoms		Each	2426	.68	
	B30	**Roofing**					
3010	Roof Coverings	Built-up tar and gravel with flashing; perlite/EPS composite insulation		S.F. Roof	6.06	2.02	1.7%
3020	Roof Openings	N/A		—	—	—	
C.	**INTERIORS**						
1010	Partitions	Gypsum board on metal studs	20 S.F. Floor/L.F. Partition	S.F. Partition	9.18	4.59	
1020	Interior Doors	Wood solid core	200 S.F. Floor/Door	Each	576	2.89	
1030	Fittings	Toilet partitions		S.F. Floor	.30	.30	
2010	Stair Construction	Concrete filled metal pan		Flight	14,675	4.90	23.2%
3010	Wall Finishes	90% paint, 10% ceramic tile		S.F. Surface	1.35	1.35	
3020	Floor Finishes	70% carpet tile, 15% terrazzo, 15% vinyl composition tile		S.F. Floor	7.50	7.50	
3030	Ceiling Finishes	Mineral fiber tile on concealed zee bars		S.F. Ceiling	6.51	6.51	
D.	**SERVICES**						
	D10	**Conveying**					
1010	Elevators & Lifts	Two hydraulic elevators		Each	116,370	12.93	10.7%
1020	Escalators & Moving Walks	N/A		—	—	—	
	D20	**Plumbing**					
2010	Plumbing Fixtures	Toilet and service fixtures, supply and drainage	1 Fixture/1385 S.F. Floor	Each	5498	3.97	
2020	Domestic Water Distribution	Gas fired water heater		S.F. Floor	.40	.40	4.0%
2040	Rain Water Drainage	Roof drains		S.F. Roof	1.53	.51	
	D30	**HVAC**					
3010	Energy Supply	N/A		—	—	—	
3020	Heat Generating Systems	Included in D3050		—	—	—	
3030	Cooling Generating Systems	N/A		—	—	—	6.5 %
3050	Terminal & Package Units	Multizone unit, gas heating, electric cooling		S.F. Floor	7.86	7.86	
3090	Other HVAC Sys. & Equipment	N/A		—	—	—	
	D40	**Fire Protection**					
4010	Sprinklers	Sprinklers, light hazard		S.F. Floor	2.97	2.97	4.8%
4020	Standpipes	Standpipe, wet, Class III		S.F. Floor	2.87	2.87	
	D50	**Electrical**					
5010	Electrical Service/Distribution	400 ampere service, panel board and feeders		S.F. Floor	1.57	1.57	
5020	Lighting & Branch Wiring	High efficiency fluorescent fixtures, receptacles, switches, A.C. and misc. power		S.F. Floor	11.93	11.93	14.2%
5030	Communications & Security	Addressable alarm systems, internet wiring, and emergency lighting		S.F. Floor	3.48	3.48	
5090	Other Electrical Systems	Emergency generator, 15 kW		S.F. Floor	.18	.18	
E.	**EQUIPMENT & FURNISHINGS**						
1010	Commercial Equipment	N/A		—	—	—	
1020	Institutional Equipment	N/A		—	—	—	
1030	Vehicular Equipment	N/A		—	—	—	0.0 %
1090	Other Equipment	N/A		—	—	—	
F.	**SPECIAL CONSTRUCTION**						
1020	Integrated Construction	N/A		—	—	—	
1040	Special Facilities	N/A		—	—	—	0.0 %
G.	**BUILDING SITEWORK**	**N/A**					

		Sub-Total	120.71	**100%**
CONTRACTOR FEES (General Requirements: 10%, Overhead: 5%, Profit: 10%)		25%	30.16	
ARCHITECT FEES		9%	13.58	

	Total Building Cost	**164.45**

Costs per square foot of floor area

Exterior Wall	S.F. Area	10000	15000	20000	25000	30000	35000	40000	50000	60000
	L.F. Perimeter	410	500	600	640	700	766	833	966	1000
Tiltup Concrete Panels	Steel Frame	107.55	98.15	93.80	89.50	87.05	85.45	84.30	82.60	80.25
Brick with Block Back-up	Bearing Walls	125.40	111.95	105.75	99.10	95.55	93.15	91.40	88.90	85.20
Concrete Block Concrete Block	Steel Frame	107.30	100.05	96.80	93.20	91.30	90.00	89.10	87.75	85.80
	Bearing Walls	104.30	96.95	93.55	89.95	88.00	86.70	85.75	84.45	82.40
Galvanized Steel Siding	Steel Frame	114.30	106.80	103.40	99.70	97.70	96.35	95.40	94.00	91.90
Metal Sandwich Panels	Steel Frame	115.65	106.90	102.95	98.50	96.05	94.50	93.35	91.65	89.20
Perimeter Adj., Add or Deduct	Per 100 L.F.	7.40	4.95	3.75	2.95	2.45	2.10	1.85	1.45	1.20
Story Hgt. Adj., Add or Deduct	Per 1 Ft.	0.90	0.75	0.70	0.50	0.50	0.45	0.45	0.35	0.35
For Basement, add $26.95 per square foot of basement area										

The above costs were calculated using the basic specifications shown on the facing page. These costs should be adjusted where necessary for design alternatives and owner's requirements. Reported completed project costs, for this type of structure, range from $34.50 to $137.75 per S.F.

Common additives

Description	Unit	$ Cost	Description	Unit	$ Cost
Dock Leveler, 10 ton cap.			Sound System		
6' x 8'	Each	6375	Amplifier, 250 watts	Each	2400
7' x 8'	Each	6975	Speaker, ceiling or wall	Each	196
Emergency Lighting, 25 watt, battery operated			Trumpet	Each	375
Lead battery	Each	287	Yard Lighting, 20' aluminum pole	Each	2900
Nickel cadmium	Each	845	with 400 watt		
Fence, Chain link, 6' high			high pressure sodium		
9 ga. wire	L.F.	22.50	fixture.		
6 ga. wire	L.F.	31.50			
Gate	Each	335			
Flagpoles, Complete					
Aluminum, 20' high	Each	1700			
40' high	Each	3600			
70' high	Each	10,300			
Fiberglass, 23' high	Each	1825			
39'-5" high	Each	3475			
59' high	Each	6625			
Paving, Bituminous					
Wearing course plus base course	S.Y.	8.75			
Sidewalks, Concrete 4" thick	S.F.	4.09			

Important: See the Reference Section for Location Factors

Model costs calculated for a 1 story building with 24' story height and 30,000 square feet of floor area

Warehouse

				Unit	Unit Cost	Cost Per S.F.	% Of Sub-Total
A.	**SUBSTRUCTURE**						
1010	Standard Foundations	Poured concrete; strip and spread footings		S.F. Ground	1.31	1.31	
1020	Special Foundations	N/A		—	—	—	
1030	Slab on Grade	5" reinforced concrete with vapor barrier and granular base		S.F. Slab	12.01	12.01	24.0%
2010	Basement Excavation	Site preparation for slab and trench for foundation wall and footing		S.F. Ground	.18	.18	
2020	Basement Walls	4' foundation wall		L.F. Wall	76	2.29	
B.	**SHELL**						
	B10 Superstructure						
1010	Floor Construction	Mezzanine: open web steel joists, slab form, concrete beams, columns	10% of area	S.F. Floor	17.30	1.73	11.3%
1020	Roof Construction	Metal deck, open web steel joists, beams, columns		S.F. Roof	5.68	5.68	
	B20 Exterior Enclosure						
2010	Exterior Walls	Concrete block	95% of wall	S.F. Wall	11.77	6.26	
2020	Exterior Windows	N/A		—	—	—	11.1%
2030	Exterior Doors	Steel overhead, hollow metal	5% of wall	Each	2870	1.05	
	B30 Roofing						
3010	Roof Coverings	Built-up tar and gravel with flashing; perlite/EPS composite insulation		S.F. Roof	5.38	5.38	8.6%
3020	Roof Openings	Roof hatches and skylight		S.F. Roof	.30	.30	
C.	**INTERIORS**						
1010	Partitions	Concrete block (office and washrooms)	100 S.F. Floor/L.F. Partition	S.F. Partition	8.63	.69	
1020	Interior Doors	Single leaf hollow metal	5000 S.F. Floor/Door	Each	1001	.20	
1030	Fittings	N/A					
2010	Stair Construction	Steel gate with rails		Flight	11,650	.77	9.5%
3010	Wall Finishes	Paint		S.F. Surface	12.56	2.01	
3020	Floor Finishes	90% hardener, 10% vinyl composition tile		S.F. Floor	1.90	1.90	
3030	Ceiling Finishes	Suspended mineral tile on zee channels in office area	10% of area	S.F. Ceiling	6.51	.65	
D.	**SERVICES**						
	D10 Conveying						
1010	Elevators & Lifts	N/A		—	—	—	0.0 %
1020	Escalators & Moving Walks	N/A		—	—	—	
	D20 Plumbing						
2010	Plumbing Fixtures	Toilet and service fixtures, supply and drainage	1 Fixture/2500 S.F. Floor	Each	3175	1.27	
2020	Domestic Water Distribution	Gas fired water heater		S.F. Floor	.24	.24	4.6%
2040	Rain Water Drainage	Roof drains		S.F. Roof	1.52	1.52	
	D30 HVAC						
3010	Energy Supply	N/A		—	—	—	
3020	Heat Generating Systems	Ventilation with heat		Each	89,200	3.21	
3030	Cooling Generating Systems	N/A		—	—	—	6.2 %
3050	Terminal & Package Units	Single zone unit gas, heating, electric cooling	10% of area	S.F. Floor	.89	.89	
3090	Other HVAC Sys. & Equipment	N/A		—	—	—	
	D40 Fire Protection						
4010	Sprinklers	Sprinklers, ordinary hazard		S.F. Floor	3.19	3.19	7.6%
4020	Standpipes	Standpipe		S.F. Floor	1.84	1.84	
	D50 Electrical						
5010	Electrical Service/Distribution	200 ampere service, panel board and feeders		S.F. Floor	.47	.47	
5020	Lighting & Branch Wiring	High efficiency fluorescent fixtures, receptacles, switches, A.C. and misc. power		S.F. Floor	5.79	5.79	13.4%
5030	Communications & Security	Addressable alarm systems		S.F. Floor	2.56	2.56	
5090	Other Electrical Systems	N/A		—	—	—	
E.	**EQUIPMENT & FURNISHINGS**						
1010	Commercial Equipment	N/A		—	—	—	
1020	Institutional Equipment	N/A		—	—	—	3.7 %
1030	Vehicular Equipment	Dock boards, dock levelers		S.F. Floor	2.41	2.41	
1090	Other Equipment	N/A		—	—	—	
F.	**SPECIAL CONSTRUCTION**						
1020	Integrated Construction	N/A		—	—	—	0.0 %
1040	Special Facilities	N/A		—	—	—	
G.	**BUILDING SITEWORK**	**N/A**					

			Sub-Total	65.80	100%
CONTRACTOR FEES (General Requirements: 10%, Overhead: 5%, Profit: 10%)		25%	16.44	
ARCHITECT FEES		7%	5.76	
Total Building Cost			**88**	

Costs per square foot of floor area

Exterior Wall	S.F. Area	2000	3000	5000	8000	12000	20000	30000	50000	100000
	L.F. Perimeter	180	220	300	420	580	900	1300	2100	4100
Concrete Block	Steel Frame	171.80	149.90	132.40	122.50	117.05	112.65	110.50	108.75	107.40
	R/Conc. Frame	154.00	140.45	129.55	123.40	120.00	117.25	115.90	114.80	114.00
Metal Sandwich Panel	Steel Frame	148.35	133.50	121.60	114.95	111.20	108.20	106.75	105.55	104.70
Tilt-up Concrete Panel	R/Conc. Frame	164.05	149.90	138.60	132.20	128.70	125.80	124.45	123.25	122.45
Precast Concrete Panel	Steel Frame	153.55	137.60	124.85	117.65	113.70	110.45	108.85	107.60	106.65
	R/Conc. Frame	178.10	161.20	147.60	139.95	135.75	132.35	130.65	129.30	128.30
Perimeter Adj., Add or Deduct	Per 100 L.F.	26.80	17.85	10.70	6.70	4.50	2.70	1.75	1.05	0.50
Story Hgt. Adj., Add or Deduct	Per 1 Ft.	2.80	2.25	1.85	1.65	1.55	1.35	1.25	1.20	1.25
Basement—Not Applicable										

The above costs were calculated using the basic specifications shown on the facing page. These costs should be adjusted where necessary for design alternatives and owner's requirements. Reported completed project costs, for this type of structure, range from $33.05 to $189.50 per S.F.

Common additives

Description	Unit	$ Cost
Dock Leveler, 10 ton cap.		
6' x 8'	Each	6375
7' x 8'	Each	6975
Emergency Lighting, 25 watt, battery operated		
Lead battery	Each	287
Nickel cadmium	Each	845
Fence, Chain link, 6' high		
9 ga. wire	L.F.	22.50
6 ga. wire	L.F.	31.50
Gate	Each	335
Flagpoles, Complete		
Aluminum, 20' high	Each	1700
40' high	Each	3600
70' high	Each	10,300
Fiberglass, 23' high	Each	1825
39'-5" high	Each	3475
59' high	Each	6625
Paving, Bituminous		
Wearing course plus base course	S.Y.	8.75
Sidewalks, Concrete 4" thick	S.F.	4.09

Description	Unit	$ Cost
Sound System		
Amplifier, 250 watts	Each	2400
Speaker, ceiling or wall	Each	196
Trumpet	Each	375
Yard Lighting,	Each	2900
20' aluminum pole with 400 watt high pressure sodium fixture		

Model costs calculated for a 1 story building with 12' story height and 20,000 square feet of floor area

Warehouse, Mini

					Unit	Unit Cost	Cost Per S.F.	% Of Sub-Total
A.	**SUBSTRUCTURE**							
1010	Standard Foundations	Poured concrete; strip and spread footings			S.F. Ground	1.75	1.75	
1020	Special Foundations	N/A			—	—	—	
1030	Slab on Grade	4" reinforced concrete with vapor barrier and granular base			S.F. Slab	9.80	9.80	17.7%
2010	Basement Excavation	Site preparation for slab and trench for foundation wall and footing			S.F. Ground	.31	.31	
2020	Basement Walls	4' foundation wall			L.F. Wall	67	3.03	
B.	**SHELL**							
	B10 Superstructure							
1010	Floor Construction	N/A			—	—	—	8.8 %
1020	Roof Construction	Metal deck, open web steel joists, beams, columns			S.F. Roof	7.37	7.37	
	B20 Exterior Enclosure							
2010	Exterior Walls	Concrete block		70% of wall	S.F. Wall	14.21	5.37	
2020	Exterior Windows	Aluminum projecting		5% of wall	Each	781	7.03	18.8%
2030	Exterior Doors	Steel overhead, hollow metal		25% of wall	Each	1540	3.39	
	B30 Roofing							
3010	Roof Coverings	Built-up tar and gravel with flashing; perlite/EPS composite insulation			S.F. Roof	4.81	4.81	5.7%
3020	Roof Openings	N/A			—	—	—	
C.	**INTERIORS**							
1010	Partitions	Concrete block, gypsum board on metal studs	10.65 S.F. Floor/L.F. Partition		S.F. Partition	8.45	14.49	
1020	Interior Doors	Single leaf hollow metal	300 S.F. Floor/Door		Each	1533	5.11	
1030	Fittings	N/A			—	—	—	
2010	Stair Construction	N/A			—	—	—	23.3%
3010	Wall Finishes	N/A			—	—	—	
3020	Floor Finishes	N/A			—	—	—	
3030	Ceiling Finishes	N/A			—	—	—	
D.	**SERVICES**							
	D10 Conveying							
1010	Elevators & Lifts	N/A			—	—	—	0.0 %
1020	Escalators & Moving Walks	N/A			—	—	—	
	D20 Plumbing							
2010	Plumbing Fixtures	Toilet and service fixtures, supply and drainage	1 Fixture/5000 S.F. Floor		Each	2350	.47	
2020	Domestic Water Distribution	Gas fired water heater			S.F. Floor	.22	.22	2.1%
2040	Rain Water Drainage	Roof drains			S.F. Roof	1.09	1.09	
	D30 HVAC							
3010	Energy Supply	N/A			—	—	—	
3020	Heat Generating Systems	Oil fired hot water, unit heaters			Each	8.24	8.24	
3030	Cooling Generating Systems	N/A			—	—	—	9.8 %
3050	Terminal & Package Units	N/A			—	—	—	
3090	Other HVAC Sys. & Equipment	N/A			—	—	—	
	D40 Fire Protection							
4010	Sprinklers	Wet pipe sprinkler system			S.F. Floor	4.11	4.11	4.9%
4020	Standpipes	N/A			—	—	—	
	D50 Electrical							
5010	Electrical Service/Distribution	200 ampere service, panel board and feeders			S.F. Floor	1.59	1.59	
5020	Lighting & Branch Wiring	Fluorescent fixtures, receptacles, switches and misc. power			S.F. Floor	5.50	5.50	9.0%
5030	Communications & Security	Alarm systems			S.F. Floor	.42	.42	
5090	Other Electrical Systems	Emergency generator, 7.5 kW			S.F. Floor	.11	.11	
E.	**EQUIPMENT & FURNISHINGS**							
1010	Commercial Equipment	N/A			—	—	—	
1020	Institutional Equipment	N/A			—	—	—	0.0 %
1030	Vehicular Equipment	N/A			—	—	—	
1090	Other Equipment	N/A			—	—	—	
F.	**SPECIAL CONSTRUCTION**							
1020	Integrated Construction	N/A			—	—	—	0.0 %
1040	Special Facilities	N/A			—	—	—	
G.	**BUILDING SITEWORK**	**N/A**						

		Sub-Total	84.21	**100%**
CONTRACTOR FEES (General Requirements: 10%, Overhead: 5%, Profit: 10%)			25%	21.07	
ARCHITECT FEES			7%	7.37	
		Total Building Cost	**112.65**	

Deterioration occurs within the structure itself and is determined by the observation of both materials and equipment.

Curable deterioration can be remedied either by maintenance, repair or replacement, within prudent economic limits.

Incurable deterioration that has progressed to the point of actually affecting the structural integrity of the structure, making repair or replacement not economically feasible.

Actual Versus Observed Age

The observed age of a structure refers to the age that the structure appears to be. Periodic maintenance, remodeling, and renovation all tend to reduce the amount of deterioration that has taken place, thereby decreasing the observed age. Actual age, on the other hand, relates solely to the year that the structure was built.

The Depreciation Table shown here relates to the observed age.

Obsolescence arises from conditions either occurring within the structure (functional) or caused by factors outside the limits of the structure (economic).

Functional obsolescence is any inadequacy caused by outmoded design, dated construction materials, or oversized or undersized areas, all of which cause excessive operation costs.

Incurable is so costly as not to economically justify the capital expenditure required to correct the deficiency.

Economic obsolescence is caused by factors outside the limits of the structure. The prime causes of economic obsolescence are:

- zoning and environmental laws
- government legislation
- negative neighborhood influences
- business climate
- proximity to transportation facilities

Depreciation Table
Commercial/Industrial/Institutional

Building Material			
Observed Age (Years)	Frame	Masonry On Wood	Masonry On Masonry Or Steel
1	1%	0%	0%
2	2	1	0
3	3	2	1
4	4	3	2
5	6	5	3
10	20	15	8
15	25	20	15
20	30	25	20
25	35	30	25
30	40	35	30
35	45	40	35
40	50	45	40
45	55	50	45
50	60	55	50
55	65	60	55
60	70	65	60

Assemblies Section

Table of Contents

Table of Contents

Introduction to the Assemblies Section

This section of the manual contains the technical description and installed cost of all components used in the commercial/industrial/institutional section. In most cases, there is an illustration of the component to aid in identification.

All Costs Include:

Materials purchased in lot sizes typical of normal construction projects with discounts appropriate to established contractors.

Installation work performed by union labor working under standard conditions at a normal pace.

Installing Contractor's Overhead & Profit

Standard installing contractor's overhead & profit are included in the assemblies costs.

These assemblies costs contain no provisions for General Contractor's overhead and profit or Architect's fees, nor do they include premiums for labor or materials, or savings which may be realized under certain economic situations.

The assemblies tables are arranged by physical size (dimensions) of the component wherever possible.

How to Use the Assemblies Cost Tables

The following is a detailed explanation of a sample Assemblies Cost Table. Most Assembly Tables are separated into two parts:
1) an illustration of the system to be estimated with descriptive notes; and 2) the costs for similar systems with dimensional and/or size variations. Next to each bold number below is the item being described with the appropriate component of the sample entry following in parenthesis. In most cases, if the work is to be subcontracted, the general contractor will need to add an additional markup (RSMeans suggests using 10%) to the "Total" figures.

System/Line Numbers (A1030 120 2240)

Each Assemblies Cost Line has been assigned a unique identification number based on the UNIFORMAT II classification sstem.

UNIFORMAT II Major Group

A1030 **120** **2240**

UNIFORMAT II Level 3

Means Major Classification

Means Individual Line Number

A10 Foundations

A1030 Slab on Grade **1**

2

Fibre Expansion Joint
Wire Mesh Reinforcing
Concrete Slab
Vapor Barrier
Compacted Gravel

Reinforced Slab on Grade

A Slab on Grade system includes fine grading; 6″ of compacted gravel; vapor barrier; 3500 p.s.i. concrete; bituminous fiber expansion joint; all necessary edge forms 4 uses; steel trowel finish; and sprayed on membrane curing compound. Wire mesh reinforcing used in all reinforced slabs.

Non-industrial slabs are for foot traffic only with negligible abrasion. Light industrial slabs are for pneumatic wheels and light abrasion. Industrial slabs are for solid rubber wheels and moderate abrasion. Heavy industrial slabs are for steel wheels and severe abrasion.

4

A1030 120	Plain & Reinforced			COST PER S.F.	
			MAT.	INST.	TOTAL
2220	Slab on grade, 4" thick, non industrial, non reinforced		1.84	2.45	4.29
2240	Reinforced		1.97	2.80	4.77
2260	Light industrial, non reinforced		2.44	3	5.44
2280	Reinforced	**3**	2.57	3.35	5.92
2300	Industrial, non reinforced		3.07	6.25	9.32
2320	Reinforced		3.20	6.60	9.80
3340	5" thick, non industrial, non reinforced		2.17	2.52	4.69
3360	Reinforced		2.30	2.87	5.17
3380	Light industrial, non reinforced		2.78	3.07	5.85
3400	Reinforced		2.91	3.42	6.33
3420	Heavy industrial, non reinforced		4.04	7.45	11.49
3440	Reinforced		4.16	7.85	12.01
4460	6" thick, non industrial, non reinforced		2.61	2.46	5.07
4480	Reinforced		2.86	2.93	5.79
4500	Light industrial, non reinforced		3.23	3.01	6.24
4520	Reinforced		3.63	3.63	7.26
4540	Heavy industrial, non reinforced		4.50	7.55	12.05
4560	Reinforced		4.75	8.05	12.80
5580	7" thick, non industrial, non reinforced		2.95	2.54	5.49
5600	Reinforced		3.27	3.04	6.31

5 **6** **7**

Illustration

At the top of most assembly pages is an illustration, a brief description, and the design criteria used to develop the cost.

System Description

The components of a typical system are listed in the description to show what has been included in the development of the total system price. The rest of the table contains prices for other similar systems with dimensional and/or size variations.

Unit of Measure for Each System (S.F.)

Costs shown in the three right hand columns have been adjusted by the component quantity and unit of measure for the entire system. In this example, "Cost per S.F." is the unit of measure for this system or "assembly."

Materials (1.97)

This column contains the Materials Cost of each component. These cost figures are bare costs plus 10% for profit.

Installation (2.80)

Installation includes labor and equipment plus the installing contractor's overhead and profit. Equipment costs are the bare rental costs plus 10% for profit. The labor overhead and profit is defined on the inside back cover of this book.

Total (4.77)

The figure in this column is the sum of the material and installation costs.

Material Cost	+	Installation Cost	=	Total
$1.97	+	$2.80	=	$4.77

Example

Assemblies costs can be used to calculate the cost for each component of a building and, accumulated with appropriate markups, produce a cost for the complete structure. Components from a model building with similar characteristics in conjunction with Assemblies components may be used to calculate costs for a complete structure.

Example:

Outline Specifications

General	Building size—60' x 100', 8 suspended floors, 12' floor to floor, 4' high parapet above roof, full basement 11'-8" floor to floor, bay size 25' x 30', ceiling heights - 9' in office area and 8' in core area.
A Substructure	Concrete, spread and strip footings, concrete walls waterproofed, 4" slab on grade.
B Shell	
B10 Superstructure	Columns, wide flange; 3 hr. fire rated; Floors, composite steel frame & deck with concrete slab; Roof, steel beams, open web joists and deck.
B20 Exterior Enclosure	Walls; North, East & West, brick & lightweight concrete block with 2" cavity insulation, 25% window; South, 8" lightweight concrete block insulated, 10% window. Doors, aluminum & glass. Windows, aluminum, 3'-0" x 5'-4", insulating glass.
B30 Roofing	Tar & gravel, 2" rigid insulation, R 12.5.
C Interiors	Core—6" lightweight concrete block, full height with plaster on exposed faces to ceiling height.
	Corridors—1st & 2nd floor; 3-5/8" steel studs with F.R. gypsum board, full height.
	Exterior Wall—plaster, ceiling height.
	Fittings—Toilet accessories, directory boards.
	Doors—hollow metal.
	Wall Finishes—lobby, mahogany paneling on furring, remainder plaster & gypsum board, paint.
	Floor Finishes—1st floor lobby, corridors & toilet rooms, terrazzo; remainder concrete, tenant developed. 2nd thru 8th, toilet rooms, ceramic tile; office & corridor, carpet.
	Ceiling Finishes—24" x 48" fiberglass board on Tee grid.
D Services	
D10 Conveying	2-2500 lb. capacity, 200 F.P.M. geared elevators, 9 stops.
D20 Plumbing	Fixtures, see sketch.
	Roof Drains—2-4" C.I. pipe.
D30 HVAC	Heating—fin tube radiation, forced hot water.
	Air Conditioning—chilled water with water cooled condenser.
D40 Fire Protection	Fire Protection—4" standpipe, 9 hose cabinets.
D50 Electrical	Lighting, 1st thru 8th, 15 fluorescent fixtures/1000 S.F., 3 Watts/S.F.
	Basement, 10 fluorescent fixtures/1000 S.F., 2 Watts/S.F.
	Receptacles, 1st thru 8th, 16.5/1000 S.F., 2 Watts/S.F.
	Basement, 10 receptacles/1000 S.F., 1.2 Watts/S.F.
	Air Conditioning, 4 Watts/S.F.
	Miscellaneous Connections, 1.2 Watts/S.F.
	Elevator Power, 2-10 H.P., 230 volt motors.
	Wall Switches, 2/1000 S.F.
	Service, panel board & feeder, 2000 Amp.
	Fire Detection System, pull stations, signals, smoke and heat detectors.
	Emergency Lighting Generator, 30 KW.
E Equipment & Furnishings	NA.
F Special Construction	NA.
G Building Sitework	NA.

Front Elevation

Basement Plan

Ground Floor Plan

Typical Floor Plan

Example

If spread footing & column sizes are unknown, develop approximate loads as follows. Enter tables with these loads to determine costs.

Superimposed Load Ranges

Apartments & Residential Structures	65	to	75 psf
Assembly Areas & Retail Stores	110	to	125 psf
Commercial & Manufacturing	150	to	250 psf
Offices	75	to	100 psf

Approximate loads/S.F. for roof & floors.
Roof. Assume 40 psf superimposed load.
Steel joists, beams & deck.
Table B1020 112—Line 3900

B1020 112		**Steel Joists, Beams, & Deck on Columns**						
	BAY SIZE (FT.)	SUPERIMPOSED LOAD (P.S.F.)	DEPTH (IN.)	TOTAL LOAD (P.S.F.)	COLUMN ADD	COST PER S.F.		
						MAT.	INST.	TOTAL
3500	25x30	20	22	40		6.65	1.49	8.14
3600					columns	1.36	.27	1.63
3900		40	25	60		8	1.76	9.76
4000					columns	1.63	.32	1.95

B1010 256		**Composite Beams, Deck & Slab**						
	BAY SIZE (FT.)	SUPERIMPOSED LOAD (P.S.F.)	SLAB THICKNESS (IN.)	TOTAL DEPTH (FT.-IN.)	TOTAL LOAD (P.S.F.)	COST PER S.F.		
						MAT.	INST.	TOTAL
3400	25x30	40	5-1/2	1 - 11-1/2	83	14	5.20	19.20
3600		75	5-1/2	1 - 11-1/2	119	15.10	5.25	20.35
3900		125	5-1/2	1 - 11-1/2	170	17.60	5.95	23.55
4000		200	6-1/4	2 - 6-1/4	252	22	6.80	28.80

Floors–Total load, 119 psf.
Interior foundation load.
Roof

[(25' x 30' x 60 psf) + 8 floors x (25' x 30' x 119 psf)] x 1/1000 lb./Kip =	759 Kips
Approximate Footing Loads, Interior footing =	759 Kips
Exterior footing (1/2 bay) 759 k x .6 =	455 Kips
Corner footing (1/4 bay) 759 k x .45 =	342 Kips
[Factors to convert Interior load to Exterior & Corner loads]	
Approximate average Column load 759 k/2 =	379 Kips

PRELIMINARY ESTIMATE

PROJECT **Office Building**	TOTAL SITE AREA
BUILDING TYPE	OWNER
LOCATION	ARCHITECT
DATE OF CONSTRUCTION	ESTIMATED CONSTRUCTION PERIOD

BRIEF DESCRIPTION **Building size: 60' x 100' - 8 structural floors - 12' floor to floor**

4' high parapet above roof, full basement 11' - 8" floor to floor, bay size

25' x 30', ceiling heights - 9' in office area + 8' in core area

TYPE OF PLAN		TYPE OF CONSTRUCTION
QUALITY		BUILDING CAPACITY

Floor			**Wall Area**					
Below Grade Levels			Foundation Walls	L.F.		Ht.		S.F.
Area		S.F.	Frost Walls	L.F.		Ht.		S.F.
Area		S.F.	Exterior Closure		Total			S.F.
Total Area		S.F.	Comment					
Ground Floor			Fenestration		%			S.F.
Area		S.F.			%			S.F.
Area		S.F.	Exterior Wall		%			S.F.
Total Area		S.F.			%			S.F.
Supported Levels			**Site Work**					
Area			Parking		S.F. (For			Cars)
Area		S.F.	Access Roads		L.F. (X			Ft. Wide)
Area		S.F.	Sidewalk		L.F. (X			Ft. Wide)
Area		S.F.	Landscaping		S.F. (		% Unbuilt Site)	
Area		S.F.	**Building Codes**					
Total Area		S.F.	City			Country		
Miscellaneous			National			Other		
Area		S.F.	**Loading**					
Area		S.F.	Roof		psf	Ground Floor		psf
Area		S.F.	Supported Floors		psf	Corridor		psf
Area		S.F.	Balcony		psf	Partition, allow		psf
Total Area		S.F.	Miscellaneous					psf
Net Finished Area		S.F.	Live Load Reduction					
Net Floor Area		S.F.	Wind					
Gross Floor Area	**54,000**	S.F.	Earthquake			Zone		
Roof			Comment					
Total Area		S.F.	Soil Type					
Comments			Bearing Capacity					K.S.F.
			Frost Depth					Ft.
Volume			**Frame**					
Depth of Floor System			Type		Bay Spacing			
Minimum		In.	Foundation					
Maximum		In.	Special					
Foundation Wall Height		Ft.	Substructure					
Floor to Floor Height		Ft.	Comment					
Floor to Ceiling Height		Ft.	Superstructure, Vertical					
Subgrade Volume		C.F.	Fireproofing		☐ Columns			Hrs.
Above Grade Volume		C.F.	☐ Girders		Hrs.	☐ Beams		Hrs.
Total Building Volume	**648,000**	C.F.	☐ Floor		Hrs.	☐ None		

NUMBER			QTY.	UNIT	TOTAL COST UNIT	TOTAL	COST PER S.F.
A	**SUBSTRUCTURE**						
A1010 210 7900	Corner Footings 8'-6" SQ. x 27"		4	Ea.	1,735.00	6,940	
8010	Exterior	9' -6" SQ. x 30"	8		2,300.00	18,400	
8300	Interior	12" SQ.	3		4,275.00	12,825	
A1010 110 2700	Strip	2' Wide x 1' Thick					
	320 L.F. [(4 x 8.5) + 8 x 9.5)] =		320	L.F.	35.55	11,376	
A2020 110 7260	Foundation Wall 12' High, 1' Thick		320		234.00	74,880	
A1010 320 2800	Foundation Waterproofing		320		20.70	6,624	
A1030 120 2240	4", non-industrial, reinf. slab on grade		6000	S.F.	4.77	28,620	
A2010 110 3440	Building Excavation + Backfill		6000	S.F.	6.57	39,420	
3500	(Interpolated ; 12' Between						
4620	8' and 16' ; 6,000 Between						
4680	4,000 and 10,000 S.F.						
		Total				199,085	3.69
B	**SHELL**						
B10	**Superstructure**						
B1010 208 5800	Columns: Load 379K Use 400K						
	Exterior 12 x 96' + Interior 3 x 108'		1476	V.L.F.	126.15	186,197	
B1010 720 3650	Column Fireproofing - Interior 4 sides						
3700	Exterior 1/2 (Interpolated for 12")		900	V.L.F	31.69	28,521	
B1010 256 3600	Floors: Composite steel + lt. wt. conc.		48000	S.F.	16.15	775,200	
B1020 112 3900	Roof: Open web joists, beams + decks		6000	S.F.	6.57	39,420	
		Total				1,029,338	19.06
B20	**Exterior Enclosure**						
B2010 134 1200	4" Brick + 6" Block - Insulated						
	75% NE + W Walls 220' x 100' x .75		16500	S.F.	28.85	476,025	
B2010 110 3410	8" Block - Insulated 90%						
	100' x 100' x .9		9000	S.F.	11.60	104,400	
B2030 110 6950	Double Aluminum + Glass Door		1	Pr.	5,275.00	5,275	
6300	Single		2	Ea.	2,585.00	5,170	
B2020 106 8800	Aluminum windows - Insul. glass						
	(5,500 S.F. + 1,000 S.F.)/(3 x 5.33)		406	Ea.	867.00	352,002	
B2010 105 8480	Precast concrete coping		320	L.F.	33.95	10,864	
		Total				953,736	17.66
B30	**Roofing**						
B3010 320 4020	Insulation: Perlite/Polyisocyanurate		6000	S.F.	1.74	10,440	
B3010 105 1400	Roof: 4 Ply T & G composite		6000		2.67	16,020	
B3010 430 0300	Flashing: Aluminum - fabric backed		640		2.98	1,907	
B3020 210 0500	Roof hatch		1	Ea.	1,158.00	1,158	
		Total				29,525	0.55

ASSEMBLY NUMBER		QTY.	UNIT	TOTAL COST		COST PER S.F.
				UNIT	TOTAL	
C	INTERIORS					
C1010 104 5500	Core partitions: 6" Lt. Wt. concrete					
	block 288 L.F. x 11.5' x 8 Floors	26,449	S.F.	8.62	227,990	
C1010 144 0920	Plaster: [(196 L.F. x 8') + (144 L.F. x 9')] 8					
	+ Ext. Wall [(320 x 9 x 8 Floors) - 6500]	39,452	S.F.	2.90	114,411	
C1010 126 5400	Corridor partitions 1st + 2nd floors					
	steel studs + F.R. gypsum board	2,530	S.F.	3.69	9,336	
C1020 102 2600	Interior doors	66	Ea.	1,001.00	66,066	
	Wall finishes - Lobby: Mahogany w/furring					
C1010 128 0652	Furring: 150 L.F. x 9' + Ht.	1,350	S.F.	142.00	191,700	
C1030 710 0120	Towel Dispenser	16	Ea.	70.50	1,128	
C1030 710 0140	Grab Bar	32	Ea.	55.00	1,760	
C1030 710 0160	Mirror	31	Ea.	233.50	7,239	
C1030 710 0180	Toilet Tissue Dispenser	31	Ea.	36.25	1,124	
C1030 510 0140	Directory Board	8	Ea.	287.00	2,296	
C2010 110 0780	Stairs: Steel w/conc. fill	18	Flt.	17,125.00	308,250	
C3010 230 1662	Mahogany paneling	1,350	S.F.	5.32	7,182	
C3010 230 0080	Paint, primer + 2 coats finish, on plaster or gyp. board					
	39,452 + [(220' x 9' x 2) - 1350] + 72	42,134	S.F.	1.05	44,241	
	Floor Finishes					
C3020 410 1100	1st floor lobby + terrazzo	2,175	S.F.	18.11	39,389	
	Remainder: Concrete - Tenant finished					
C3020 410 0060	2nd - 8th: Carpet - 4725 S.F. x 7	33,075	S.F.	4.14	136,931	
C3020 410 1720	Ceramic Tile - 300 S.F. x 7	2,100	S.F.	9.47	19,887	
C3030 240 2780	Suspended Ceiling Tiles- 5200 S.F. x 8	41,600	S.F.	2.05	85,280	
C3030 240 3260	Suspended Ceiling Grid- 5200 S.F. x 8	41,600	S.F.	1.39	57,824	
C1030 110 0700	Toilet partitions	31	Ea.	756.00	23,436	
0760	Handicapped addition	16	Ea.	315.00	5,040	
	Total				1,350,509	25.01
D	SERVICES					
D10	Conveying					
	2 Elevators - 2500# 200'/min. Geared					
D1010 140 1600	5 Floors $162,200					
1800	15 Floors $ 414,500					
	Diff. $252,300/10 = 25,230/Flr.					
	$162,200 + (4 x 25,230) = $263,120/Ea.	2	Ea.	263,120.00	526,240	
	Total				526,240	9.75
D20	Plumbing					
D2010 310 1600	Plumbing - Lavatories	31	Ea.	1,140.00	35,340	
D2010 440 4340	-Service Sink	8	Ea.	3,050.00	24,400	
D2010 210 2000	-Urinals	8	Ea.	1,310.00	10,480	

ASSEMBLY NUMBER		QTY.	UNIT	TOTAL COST UNIT	TOTAL	COST PER S.F.
D2010 110 2080	-Water Closets	31	Ea.	2,175.00	67,425	
	Water Control, Waste Vent Piping	45%			61,940	
D2040 210 4200	- Roof Drains, 4" C.I.	2	Ea.	1,550.00	3,100	
4240	- Pipe, 9 Flr. x 12' x 12' Ea.	216	L.F.	34.45	7,441	
	Total				210,126	3.89
D30	HVAC					
D3010 520 2000	10,000 S.F. @ $9.66/S.F. Interpolate					
2040	100,000 S.F. @ $3.90/S.F.	48,000	S.F.	7.23	347,040	
	Cooling - Chilled Water, Air Cool, Cond.					
D3030 115 4000	40,000 S.F. @11.85/S.F. Interpolate	48,000	S.F.	12.15	583,200	
4040	60,000 S.F. @12.60/S.F.					
	Total				930,240	17.23
D40	Fire Protection					
D4020 310 0560	Wet Stand Pipe: 4" x 10' 1st floor	12/10	Ea.	7,100.00	8,520	
0580	Additional Floors: (12'/10 x 8)	9.6	Ea.	2,005.00	19,248	
D4020 410 8400	Cabinet Assembly	9	Ea.	1,005.00	9,045	
	Total				36,813	0.68
D50	Electrical					
D5020 210 0280	Office Lighting, 15/1000 S.F. - 3 Watts/S.F.	48,000	S.F.	7.51	360,480	
0240	Basement Lighting, 10/1000 S.F. - 2 Watts/S.F.	6,000	S.F.	4.98	29,880	
D5020 110 0640	Office Receptacles 16.5/1000 S.F. - 2 Watts/S.F.	48,000	S.F.	4.42	212,160	
0560	Basement Receptacles 10/1000 S.F. - 1.2 Watts/S.F.	6,000	S.F.	3.36	20,160	
D5020 140 0280	Central A.C. - 4 Watts/S.F.	48,000	S.F.	0.58	27,840	
D5020 135 0320	Misc. Connections - 1.2 Watts/S.F.	48,000	S.F.	0.31	14,880	
D5020 145 0680	Elevator Motor Power - 10 H.P.	2	Ea.	3,025.00	6,050	
D5020 130 0280	Wall Switches - 2/1000 S.F.	54,000	S.F.	0.44	23,760	
D5010 120 0560	2000 Amp Service	1	Ea.	42,000.00	42,000	
D5010 240 0400	Switchgear	1	Ea.	55,100.00	55,100	
D5010 230 0560	Feeder	50	L.F.	564.00	28,200	
D5030 910 0400	Fire Detection System - 50 Detectors	1	Ea.	33,000.00	33,000	
D5090 210 0320	Emergency Generator - 30 kW	30	kW	578.00	17,340	
	Total				870,850	16.13
E	EQUIPMENT & FURNISHINGS					
G	BUILDING SITEWORK					
	Miscellaneous					

Preliminary Estimate Cost Summary

PROJECT: Office Building	TOTAL AREA	54,000 S.F.	SHEET NO.
LOCATION	TOTAL VOLUME	648,000 C.F.	ESTIMATE NO.
ARCHITECT	COST PER S.F.		DATE
OWNER	COST PER C.F.		NO OF STORIES
QUANTITIES BY	EXTENSIONS BY		CHECKED BY

DIV	DESCRIPTION	SUBTOTAL COST	COST/S.F.	PERCENTAGE
A	SUBSTRUCTURE	199,085	$ 3.69	
B10	SHELL: SUPERSTRUCTURE	1,029,338	$ 19.06	
B20	SHELL: EXTERIOR ENCLOSURE	953,736	$ 17.66	
B30	SHELL: ROOFING	29,525	$ 0.55	
C	INTERIORS	1,350,509	$ 25.01	
D10	SERVICES: CONVEYING	526,240	$ 9.75	
D20	SERVICES: PLUMBING	210,126	$ 3.89	
D30	SERVICES: HVAC	930,240	$ 17.23	
D40	SERVICES: FIRE PROTECTION	36,813	$ 0.68	
D50	SERVICES: ELECTRICAL	870,850	$ 16.13	
E	EQUIPMENT & FURNISHINGS			
F	SPECIAL CONSTRUCTION & DEMO.			
G	BUILDING SITEWORK			

BUILDING SUBTOTAL	$ 6,136,462		$ 6,136,462	

Sales Tax   % x Subtotal /2   N/A $ -

General Conditions 10 % x Subtotal $ 613,646

     Subtotal "A" $ 6,750,108

Overhead   5 % x Subtotal "A" $ 337,505

     Subtotal "B" $ 7,087,614

Profit   10% x Subtotal "B" $ 708,761

     Subtotal "C" $ 7,796,375

Location Factor   % x Subtotal "C"   N/A   Localized Cost

Architects Fee 6.5% x Localized Cost = $ 506,764
Contingency   x Localized Cost =   N/A $ -
     Project Total Cost $ 8,303,140

Square Foot Cost $8,303,140/ 54,000 S.F. =   S.F. Cost $ 153.76
Cubic Foot Cost $8,303,140/ 648,000 C.F. =   C.F. Cost $ 12.81

241

A1010 Standard Foundations

The Strip Footing System includes: excavation; hand trim; all forms needed for footing placement; forms for 2" x 6" keyway (four uses); dowels; and 3,000 p.s.i. concrete.

The footing size required varies for different soils. Soil bearing capacities are listed for 3 KSF and 6 KSF. Depths of the system range from 8" to 24". Widths range from 16" to 96". Smaller strip footings may not require reinforcement.

A1010 110	Strip Footings	COST PER L.F.		
		MAT.	INST.	TOTAL
2100	Strip footing, load 2.6 KLF, soil capacity 3 KSF, 16" wide x 8" deep plain	7.35	12.75	20.10
2300	Load 3.9 KLF, soil capacity, 3 KSF, 24" wide x 8" deep, plain	9.20	13.95	23.15
2500	Load 5.1KLF, soil capacity 3 KSF, 24" wide x 12" deep, reinf.	14.55	21	35.55
2700	Load 11.1KLF, soil capacity 6 KSF, 24" wide x 12" deep, reinf.	14.55	21	35.55
2900	Load 6.8 KLF, soil capacity 3 KSF, 32" wide x 12" deep, reinf.	17.75	23	40.75
3100	Load 14.8 KLF, soil capacity 6 KSF, 32" wide x 12" deep, reinf.	17.75	23	40.75
3300	Load 9.3 KLF, soil capacity 3 KSF, 40" wide x 12" deep, reinf.	21	25	46
3500	Load 18.4 KLF, soil capacity 6 KSF, 40" wide x 12" deep, reinf.	21	25	46
4500	Load 10KLF, soil capacity 3 KSF, 48" wide x 16" deep, reinf.	30	31.50	61.50
4700	Load 22KLF, soil capacity 6 KSF, 48" wide, 16" deep, reinf.	30.50	32	62.50
5700	Load 15KLF, soil capacity 3 KSF, 72" wide x 20" deep, reinf.	52	45	97
5900	Load 33KLF, soil capacity 6 KSF, 72" wide x 20" deep, reinf.	54.50	48	102.50

A10 Foundations

A1010 Standard Foundations

Waterproofing

Spread Footings

Strip Footing

A1010 210	Spread Footings	COST EACH		
		MAT.	INST.	TOTAL
7090	Spread footings, 3000 psi concrete, chute delivered			
7100	Load 25K, soil capacity 3 KSF, 3'-0" sq. x 12" deep	50.50	110	160.50
7150	Load 50K, soil capacity 3 KSF, 4'-6" sq. x 12" deep	110	189	299
7200	Load 50K, soil capacity 6 KSF, 3'-0" sq. x 12" deep	50.50	110	160.50
7250	Load 75K, soil capacity 3 KSF, 5'-6" sq. x 13" deep	176	267	443
7300	Load 75K, soil capacity 6 KSF, 4'-0" sq. x 12" deep	88.50	162	250.50
7350	Load 100K, soil capacity 3 KSF, 6'-0" sq. x 14" deep	223	320	543
7410	Load 100K, soil capacity 6 KSF, 4'-6" sq. x 15" deep	136	223	359
7450	Load 125K, soil capacity 3 KSF, 7'-0" sq. x 17" deep	355	460	815
7500	Load 125K, soil capacity 6 KSF, 5'-0" sq. x 16" deep	176	267	443
7550	Load 150K, soil capacity 3 KSF 7'-6" sq. x 18" deep	430	540	970
7610	Load 150K, soil capacity 6 KSF, 5'-6" sq. x 18" deep	236	340	576
7650	Load 200K, soil capacity 3 KSF, 8'-6" sq. x 20" deep	615	710	1,325
7700	Load 200K, soil capacity 6 KSF, 6'-0" sq. x 20" deep	310	415	725
7750	Load 300K, soil capacity 3 KSF, 10'-6" sq. x 25" deep	1,150	1,175	2,325
7810	Load 300K, soil capacity 6 KSF, 7'-6" sq. x 25" deep	590	700	1,290
7850	Load 400K, soil capacity 3 KSF, 12'-6" sq. x 28" deep	1,825	1,700	3,525
7900	Load 400K, soil capacity 6 KSF, 8'-6" sq. x 27" deep	825	910	1,735
8010	Load 500K, soil capacity 6 KSF, 9'-6" sq. x 30" deep	1,125	1,175	2,300
8100	Load 600K, soil capacity 6 KSF, 10'-6" sq. x 33" deep	1,525	1,525	3,050
8200	Load 700K, soil capacity 6 KSF, 11'-6" sq. x 36" deep	1,975	1,875	3,850
8300	Load 800K, soil capacity 6 KSF, 12'-0" sq. x 37" deep	2,225	2,050	4,275
8400	Load 900K, soil capacity 6 KSF, 13'-0" sq. x 39" deep	2,725	2,425	5,150
8500	Load 1000K, soil capacity 6 KSF, 13'-6" sq. x 41" deep	3,100	2,700	5,800

A1010 320	Foundation Dampproofing	COST PER L.F.		
		MAT.	INST.	TOTAL
1000	Foundation dampproofing, bituminous, 1 coat, 4' high	.88	4.07	4.95
1400	8' high	1.76	8.15	9.91
1800	12' high	2.64	12.60	15.24
2000	2 coats, 4' high	1.76	4.99	6.75
2400	8' high	3.52	10	13.52
2800	12' high	5.30	15.40	20.70
3000	Asphalt with fibers, 1/16" thick, 4' high	1.88	4.99	6.87
3400	8' high	3.76	10	13.76

A1010 Standard Foundations

A1010 320	Foundation Dampproofing	COST PER L.F.		
		MAT.	INST.	TOTAL
3800	12' high	5.65	15.40	21.05
4000	1/8" thick, 4' high	3.32	5.95	9.27
4400	8' high	6.65	11.90	18.55
4800	12' high	9.95	18.25	28.20
5000	Asphalt coated board and mastic, 1/4" thick, 4' high	3.48	5.45	8.93
5400	8' high	6.95	10.85	17.80
5800	12' high	10.45	16.70	27.15
6000	1/2" thick, 4' high	5.20	7.55	12.75
6400	8' high	10.45	15.05	25.50
6800	12' high	15.65	23	38.65
7000	Cementitious coating, on walls, 1/8" thick coating, 4' high	7.20	7.95	15.15
7400	8' high	14.40	15.95	30.35
7800	12' high	21.50	24	45.50
8000	Cementitious/metallic slurry, 2 coat, 1/4" thick, 2' high	26.50	187	213.50
8400	4' high	53	375	428
8800	6' high	79.50	560	639.50

A1030 Slab on Grade

Fibre Expansion Joint

Wire Mesh Reinforcing

Concrete Slab

Vapor Barrier

Compacted Gravel

Reinforced Slab on Grade

A Slab on Grade system includes fine grading; 6″ of compacted gravel; vapor barrier; 3500 p.s.i. concrete; bituminous fiber expansion joint; all necessary edge forms 4 uses; steel trowel finish; and sprayed on membrane curing compound. Wire mesh reinforcing used in all reinforced slabs.

Non-industrial slabs are for foot traffic only with negligible abrasion. Light industrial slabs are for pneumatic wheels and light abrasion. Industrial slabs are for solid rubber wheels and moderate abrasion. Heavy industrial slabs are for steel wheels and severe abrasion.

A1030 120	Plain & Reinforced	COST PER S.F.		
		MAT.	INST.	TOTAL
2220	Slab on grade, 4″ thick, non industrial, non reinforced	1.84	2.45	4.29
2240	Reinforced	1.97	2.80	4.77
2260	Light industrial, non reinforced	2.44	3	5.44
2280	Reinforced	2.57	3.35	5.92
2300	Industrial, non reinforced	3.07	6.25	9.32
2320	Reinforced	3.20	6.60	9.80
3340	5″ thick, non industrial, non reinforced	2.17	2.52	4.69
3360	Reinforced	2.30	2.87	5.17
3380	Light industrial, non reinforced	2.78	3.07	5.85
3400	Reinforced	2.91	3.42	6.33
3420	Heavy industrial, non reinforced	4.04	7.45	11.49
3440	Reinforced	4.16	7.85	12.01
4460	6″ thick, non industrial, non reinforced	2.61	2.46	5.07
4480	Reinforced	2.86	2.93	5.79
4500	Light industrial, non reinforced	3.23	3.01	6.24
4520	Reinforced	3.63	3.63	7.26
4540	Heavy industrial, non reinforced	4.50	7.55	12.05
4560	Reinforced	4.75	8.05	12.80
5580	7″ thick, non industrial, non reinforced	2.95	2.54	5.49
5600	Reinforced	3.27	3.04	6.31
5620	Light industrial, non reinforced	3.58	3.09	6.67
5640	Reinforced	3.90	3.59	7.49
5660	Heavy industrial, non reinforced	4.85	7.45	12.30
5680	Reinforced	5.10	7.95	13.05
6700	8″ thick, non industrial, non reinforced	3.28	2.58	5.86
6720	Reinforced	3.55	3	6.55
6740	Light industrial, non reinforced	3.92	3.13	7.05
6760	Reinforced	4.19	3.55	7.74
6780	Heavy industrial, non reinforced	5.20	7.55	12.75
6800	Reinforced	5.55	8	13.55

A2010 Basement Excavation

Sand & Gravel 1:1

Clay 1/2:1

Line of Excavation

In general, the following items are accounted for in the table below.

Costs:
1) Excavation for building or other structure to depth and extent indicated.
2) Backfill compacted in place
3) Haul of excavated waste
4) Replacement of unsuitable backfill material with bank run gravel.

A2010 110	Building Excavation & Backfill	COST PER S.F.		
		MAT.	INST.	TOTAL
2220	Excav & fill, 1000 S.F. 4' sand, gravel, or common earth, on site storage		.96	.96
2240	Off site storage		1.46	1.46
2260	Clay excavation, bank run gravel borrow for backfill	2.40	2.24	4.64
2280	8' deep, sand, gravel, or common earth, on site storage		5.80	5.80
2300	Off site storage		12.90	12.90
2320	Clay excavation, bank run gravel borrow for backfill	9.75	10.45	20.20
2340	16' deep, sand, gravel, or common earth, on site storage		15.75	15.75
2350	Off site storage		31.50	31.50
2360	Clay excavation, bank run gravel borrow for backfill	27	27	54
3380	4000 S.F.,4' deep, sand, gravel, or common earth, on site storage		.52	.52
3400	Off site storage		1.03	1.03
3420	Clay excavation, bank run gravel borrow for backfill	1.22	1.12	2.34
3440	8' deep, sand, gravel, or common earth, on site storage		4.10	4.10
3460	Off site storage		7.25	7.25
3480	Clay excavation, bank run gravel borrow for backfill	4.44	6.25	10.69
3500	16' deep, sand, gravel, or common earth, on site storage		9.80	9.80
3520	Off site storage		19.70	19.70
3540	Clay, excavation, bank run gravel borrow for backfill	11.90	14.85	26.75
4560	10,000 S.F., 4' deep, sand gravel, or common earth, on site storage		.31	.31
4580	Off site storage		.62	.62
4600	Clay excavation, bank run gravel borrow for backfill	.74	.69	1.43
4620	8' deep, sand, gravel, or common earth, on site storage		3.56	3.56
4640	Off site storage		5.45	5.45
4660	Clay excavation, bank run gravel borrow for backfill	2.69	4.90	7.59
4680	16' deep, sand, gravel, or common earth, on site storage		8.05	8.05
4700	Off site storage		13.80	13.80
4720	Clay excavation, bank run gravel borrow for backfill	7.15	11.15	18.30
5740	30,000 S.F., 4' deep, sand, gravel, or common earth, on site storage		.18	.18
5760	Off site storage		.36	.36
5780	Clay excavation, bank run gravel borrow for backfill	.43	.41	.84
5860	16' deep, sand, gravel, or common earth, on site storage		6.90	6.90
5880	Off site storage		10.05	10.05
5900	Clay excavation, bank run gravel borrow for backfill	3.96	8.65	12.61
6910	100,000 S.F., 4' deep, sand, gravel, or common earth, on site storage		.09	.09
6940	8' deep, sand, gravel, or common earth, on site storage		3	3
6970	16' deep, sand, gravel, or common earth, on site storage		6.25	6.25
6980	Off site storage		7.90	7.90
6990	Clay excavation, bank run gravel borrow for backfill	2.14	7.20	9.34

A20 Basement Construction

A2020 Basement Walls

The Foundation Bearing Wall System includes: forms up to 16' high (four uses); 3,000 p.s.i. concrete placed and vibrated; and form removal with breaking form ties and patching walls. The wall systems list walls from 6" to 16" thick and are designed with minimum reinforcement.

Excavation and backfill are not included.

Please see the reference section for further design and cost information.

A2020 110			Walls, Cast in Place					
	WALL HEIGHT (FT.)	PLACING METHOD	CONCRETE (C.Y. per L.F.)	REINFORCING (LBS. per L.F.)	WALL THICKNESS (IN.)	COST PER L.F.		
						MAT.	INST.	TOTAL
1500	4'	direct chute	.074	3.3	6	16.10	47	63.10
1520			.099	4.8	8	19.40	48	67.40
1540			.123	6.0	10	22.50	49	71.50
1561			.148	7.2	12	25.50	50	75.50
1580			.173	8.1	14	28.50	51	79.50
1600			.197	9.44	16	32	52	84
3000	6'	direct chute	.111	4.95	6	24	70	94
3020			.149	7.20	8	29	72.50	101.50
3040			.184	9.00	10	33.50	73.50	107
3061			.222	10.8	12	38.50	75.50	114
5000	8'	direct chute	.148	6.6	6	32	93.50	125.50
5020			.199	9.6	8	39	96.50	135.50
5040			.250	12	10	45.50	98	143.50
5061			.296	14.39	12	51.50	100	151.50
6020	10'	direct chute	.248	12	8	48.50	120	168.50
6040			.307	14.99	10	56	123	179
6061			.370	17.99	12	64	125	189
7220	12'	pumped	.298	14.39	8	58.50	149	207.50
7240			.369	17.99	10	67.50	152	219.50
7262			.444	21.59	12	77	157	234
9220	16'	pumped	.397	19.19	8	77.50	199	276.50
9240			.492	23.99	10	90	203	293
9260			.593	28.79	12	103	209	312

B1010 Floor Construction

General: It is desirable for purposes of consistency and simplicity to maintain constant column sizes throughout building height. To do this, concrete strength may be varied (higher strength concrete at lower stories and lower strength concrete at upper stories), as well as varying the amount of reinforcing.

The table provides probably minimum column sizes with related costs and weight per lineal foot of story height.

B1010 203		C.I.P. Column, Square Tied						
	LOAD (KIPS)	STORY HEIGHT (FT.)	COLUMN SIZE (IN.)	COLUMN WEIGHT (P.L.F.)	CONCRETE STRENGTH (PSI)	COST PER V.L.F.		
						MAT.	INST.	TOTAL
0640	100	10	10	96	4000	9.60	43.50	53.10
0680		12	10	97	4000	9.75	44	53.75
0720		14	12	142	4000	12.70	53	65.70
0840	200	10	12	140	4000	13.85	55	68.85
0860		12	12	142	4000	14.10	55.50	69.60
0900		14	14	196	4000	16.40	62	78.40
0920	300	10	14	192	4000	17.35	64	81.35
0960		12	14	194	4000	17.65	64.50	82.15
0980		14	16	253	4000	20.50	70.50	91
1020	400	10	16	248	4000	21.50	72.50	94
1060		12	16	251	4000	22	73.50	95.50
1080		14	16	253	4000	22.50	74	96.50
1200	500	10	18	315	4000	25.50	81	106.50
1250		12	20	394	4000	30.50	92	122.50
1300		14	20	397	4000	31.50	93	124.50
1350	600	10	20	388	4000	32	94	126
1400		12	20	394	4000	32.50	95	127.50
1600		14	20	397	4000	33	96.50	129.50
3400	900	10	24	560	4000	43.50	117	160.50
3800		12	24	567	4000	44	118	162
4000		14	24	571	4000	45	120	165
7300	300	10	14	192	6000	18.25	63	81.25
7500		12	14	194	6000	18.55	64	82.55
7600		14	14	196	6000	18.90	64.50	83.40
8000	500	10	16	248	6000	23.50	72.50	96
8050		12	16	251	6000	23.50	73.50	97
8100		14	16	253	6000	24	74	98
8200	600	10	18	315	6000	27	81.50	108.50
8300		12	18	319	6000	27.50	82.50	110
8400		14	18	321	6000	28	83.50	111.50
8800	800	10	20	388	6000	33	92	125
8900		12	20	394	6000	34	93	127
9000		14	20	397	6000	34.50	94	128.50

B10 Superstructure

B1010 Floor Construction

B1010 203 — C.I.P. Column, Square Tied

	LOAD (KIPS)	STORY HEIGHT (FT.)	COLUMN SIZE (IN.)	COLUMN WEIGHT (P.L.F.)	CONCRETE STRENGTH (PSI)	COST PER V.L.F.		
						MAT.	INST.	TOTAL
9100	900	10	20	388	6000	35.50	96.50	132
9300		12	20	394	6000	36.50	97.50	134
9600		14	20	397	6000	37	98.50	135.50

B1010 204 — C.I.P. Column, Square Tied-Minimum Reinforcing

	LOAD (KIPS)	STORY HEIGHT (FT.)	COLUMN SIZE (IN.)	COLUMN WEIGHT (P.L.F.)	CONCRETE STRENGTH (PSI)	COST PER V.L.F.		
						MAT.	INST.	TOTAL
9913	150	10-14	12	135	4000	11.60	51	62.60
9918	300	10-14	16	240	4000	18.50	67	85.50
9924	500	10-14	20	375	4000	27.50	86.50	114
9930	700	10-14	24	540	4000	38	107	145
9936	1000	10-14	28	740	4000	50	131	181
9942	1400	10-14	32	965	4000	61.50	148	209.50
9948	1800	10-14	36	1220	4000	76.50	173	249.50

B1010 Floor Construction

Concentric Load

Eccentric Load

General: Data presented here is for plant produced members transported 50 miles to 100 miles to the site and erected.

Design and pricing assumptions:
Normal wt. concrete, f'c = 5 KSI

Main reinforcement, fy = 60 KSI
Ties, fy = 40 KSI

Minimum design eccentricity, 0.1t.

Concrete encased structural steel haunches are assumed where practical; otherwise galvanized rebar haunches are assumed.

Base plates are integral with columns.

Foundation anchor bolts, nuts and washers are included in price.

B1010 206 — Tied, Concentric Loaded Precast Concrete Columns

	LOAD (KIPS)	STORY HEIGHT (FT.)	COLUMN SIZE (IN.)	COLUMN WEIGHT (P.L.F.)	LOAD LEVELS	COST PER V.L.F.		
						MAT.	INST.	TOTAL
0560	100	10	12x12	164	2	122	11.75	133.75
0570		12	12x12	162	2	118	9.80	127.80
0580		14	12x12	161	2	117	9.80	126.80
0590	150	10	12x12	166	3	117	9.80	126.80
0600		12	12x12	169	3	117	8.85	125.85
0610		14	12x12	162	3	117	8.85	125.85
0620	200	10	12x12	168	4	117	10.80	127.80
0630		12	12x12	170	4	117	9.80	126.80
0640		14	12x12	220	4	117	9.80	126.80

B1010 207 — Tied, Eccentric Loaded Precast Concrete Columns

	LOAD (KIPS)	STORY HEIGHT (FT.)	COLUMN SIZE (IN.)	COLUMN WEIGHT (P.L.F.)	LOAD LEVELS	COST PER V.L.F.		
						MAT.	INST.	TOTAL
1130	100	10	12x12	161	2	115	11.75	126.75
1140		12	12x12	159	2	115	9.80	124.80
1150		14	12x12	159	2	115	9.80	124.80
1390	600	10	18x18	385	4	195	10.80	205.80
1400		12	18x18	380	4	194	9.80	203.80
1410		14	18x18	375	4	193	9.80	202.80
1480	800	10	20x20	490	4	195	10.80	205.80
1490		12	20x20	480	4	194	9.80	203.80
1500		14	20x20	475	4	195	9.80	204.80

B10 Superstructure

B1010 Floor Construction

I (A) Wide Flange

○ (B) Pipe

⊙ (C) Pipe, Concrete Filled

▢ (G) Rectangular Tube, Concrete Filled

B1010 208					Steel Columns			
	LOAD (KIPS)	UNSUPPORTED HEIGHT (FT.)	WEIGHT (P.L.F.)	SIZE (IN.)	TYPE	COST PER V.L.F.		
						MAT.	INST.	TOTAL
1000	25	10	13	4	A	19.05	10.15	29.20
1020			7.58	3	B	11.10	10.15	21.25
1040			15	3-1/2	C	13.55	10.15	23.70
1120			20	4x3	G	15.65	10.15	25.80
1200		16	16	5	A	21.50	7.60	29.10
1220			10.79	4	B	14.60	7.60	22.20
1240			36	5-1/2	C	20.50	7.60	28.10
1320			64	8x6	G	31.50	7.60	39.10
1600	50	10	16	5	A	23.50	10.15	33.65
1620			14.62	5	B	21.50	10.15	31.65
1640			24	4-1/2	C	16.20	10.15	26.35
1720			28	6x3	G	21	10.15	31.15
1800		16	24	8	A	32.50	7.60	40.10
1840			36	5-1/2	C	20.50	7.60	28.10
1920			64	8x6	G	31.50	7.60	39.10
2000		20	28	8	A	36	7.60	43.60
2040			49	6-5/8	C	25	7.60	32.60
2120			64	8x6	G	30	7.60	37.60
2200	75	10	20	6	A	29.50	10.15	39.65
2240			36	4-1/2	C	40.50	10.15	50.65
2320			35	6x4	G	23.50	10.15	33.65
2400		16	31	8	A	42	7.60	49.60
2440			49	6-5/8	C	26.50	7.60	34.10
2520			64	8x6	G	31.50	7.60	39.10
2600		20	31	8	A	40	7.60	47.60
2640			81	8-5/8	C	38	7.60	45.60
2720			64	8x6	G	30	7.60	37.60

B10 Superstructure

B1010 Floor Construction

B1010 208				Steel Columns				

	LOAD (KIPS)	UNSUPPORTED HEIGHT (FT.)	WEIGHT (P.L.F.)	SIZE (IN.)	TYPE	COST PER V.L.F.		
						MAT.	INST.	TOTAL
2800	100	10	24	8	A	35	10.15	45.15
2840			35	4-1/2	C	40.50	10.15	50.65
2920			46	8x4	G	28.50	10.15	38.65
3000		16	31	8	A	42	7.60	49.60
3040			56	6-5/8	C	39.50	7.60	47.10
3120			64	8x6	G	31.50	7.60	39.10
3200		20	40	8	A	51.50	7.60	59.10
3240			81	8-5/8	C	38	7.60	45.60
3320			70	8x6	G	43	7.60	50.60
3400	125	10	31	8	A	45.50	10.15	55.65
3440			81	8	C	43.50	10.15	53.65
3520			64	8x6	G	34	10.15	44.15
3600		16	40	8	A	54	7.60	61.60
3640			81	8	C	40	7.60	47.60
3720			64	8x6	G	31.50	7.60	39.10
3800		20	48	8	A	61.50	7.60	69.10
3840			81	8	C	38	7.60	45.60
3920			60	8x6	G	43	7.60	50.60
4000	150	10	35	8	A	51	10.15	61.15
4040			81	8-5/8	C	43.50	10.15	53.65
4120			64	8x6	G	34	10.15	44.15
4200		16	45	10	A	61	7.60	68.60
4240			81	8-5/8	C	40	7.60	47.60
4320			70	8x6	G	45.50	7.60	53.10
4400		20	49	10	A	63	7.60	70.60
4440			123	10-3/4	C	54.50	7.60	62.10
4520			86	10x6	G	42.50	7.60	50.10
4600	200	10	45	10	A	66	10.15	76.15
4640			81	8-5/8	C	43.50	10.15	53.65
4720			70	8x6	G	49	10.15	59.15
4800		16	49	10	A	66.50	7.60	74.10
4840			123	10-3/4	C	57.50	7.60	65.10
4920			85	10x6	G	52.50	7.60	60.10
5200	300	10	61	14	A	89.50	10.15	99.65
5240			169	12-3/4	C	76	10.15	86.15
5320			86	10x6	G	73	10.15	83.15
5400		16	72	12	A	97.50	7.60	105.10
5440			169	12-3/4	C	70.50	7.60	78.10
5600		20	79	12	A	101	7.60	108.60
5640			169	12-3/4	C	67	7.60	74.60
5800	400	10	79	12	A	116	10.15	126.15
5840			178	12-3/4	C	99	10.15	109.15
6000		16	87	12	A	118	7.60	125.60
6040			178	12-3/4	C	92	7.60	99.60
6400	500	10	99	14	A	145	10.15	155.15
6600		16	109	14	A	148	7.60	155.60
6800		20	120	12	A	154	7.60	161.60
7000	600	10	120	12	A	176	10.15	186.15
7200		16	132	14	A	179	7.60	186.60
7400		20	132	14	A	169	7.60	176.60
7600	700	10	136	12	A	199	10.15	209.15
7800		16	145	14	A	197	7.60	204.60
8000		20	145	14	A	186	7.60	193.60

B10 Superstructure

B1010 Floor Construction

B1010 208				Steel Columns				
	LOAD (KIPS)	UNSUPPORTED HEIGHT (FT.)	WEIGHT (P.L.F.)	SIZE (IN.)	TYPE	COST PER V.L.F.		
						MAT.	INST.	TOTAL
8200	800	10	145	14	A	212	10.15	222.15
8300		16	159	14	A	215	7.60	222.60
8400		20	176	14	A	226	7.60	233.60
8800	900	10	159	14	A	233	10.15	243.15
8900		16	176	14	A	239	7.60	246.60
9000		20	193	14	A	248	7.60	255.60
9100	1000	10	176	14	A	258	10.15	268.15
9200		16	193	14	A	262	7.60	269.60
9300		20	211	14	A	271	7.60	278.60

B1010 Floor Construction

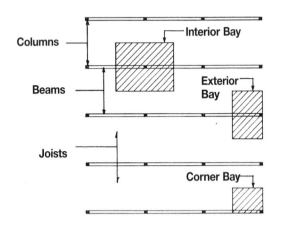

Description: Table below lists costs per S.F. of bay size for wood columns of various sizes and unsupported heights and the maximum allowable total load per S.F. per bay size.

B1010 210	Wood Columns						
	NOMINAL COLUMN SIZE (IN.)	BAY SIZE (FT.)	UNSUPPORTED HEIGHT (FT.)	MATERIAL (BF per M.S.F.)	TOTAL LOAD (P.S.F.)	COST PER S.F.		
						MAT.	INST.	TOTAL
1000	4 x 4	10 x 8	8	133	100	.19	.23	.42
1050			10	167	60	.23	.28	.51
1200		10 x 10	8	106	80	.15	.18	.33
1250			10	133	50	.19	.23	.42
1400		10 x 15	8	71	50	.10	.12	.22
1450			10	88	30	.12	.15	.27
1600		15 x 15	8	47	30	.07	.08	.15
1650			10	59	15	.08	.10	.18
2000	6 x 6	10 x 15	8	160	230	.23	.25	.48
2050			10	200	210	.29	.32	.61
2200		15 x 15	8	107	150	.16	.17	.33
2250			10	133	140	.19	.21	.40
2400		15 x 20	8	80	110	.12	.13	.25
2450			10	100	100	.15	.16	.31
2600		20 x 20	8	60	80	.09	.09	.18
2650			10	75	70	.11	.12	.23
2800		20 x 25	8	48	60	.07	.08	.15
2850			10	60	50	.09	.09	.18
3400	8 x 8	20 x 20	8	107	160	.16	.16	.32
3450			10	133	160	.20	.20	.40
3600		20 x 25	8	85	130	.12	.14	.26
3650			10	107	130	.15	.18	.33
3800		25 x 25	8	68	100	.11	.09	.20
3850			10	85	100	.14	.12	.26
4200	10 x 10	20 x 25	8	133	210	.21	.18	.39
4250			10	167	210	.27	.23	.50
4400		25 x 25	8	107	160	.17	.15	.32
4450			10	133	160	.21	.18	.39
4700	12 x 12	20 x 25	8	192	310	.32	.24	.56
4750			10	240	310	.40	.31	.71
4900		25 x 25	8	154	240	.26	.20	.46
4950			10	192	240	.32	.24	.56

B10 Superstructure

B1010 Floor Construction

"T" Shaped Precast Beams

"L" Shaped Precast Beams

B1010 214		"T" Shaped Precast Beams						
	SPAN (FT.)	SUPERIMPOSED LOAD (K.L.F.)	SIZE W X D (IN.)	BEAM WEIGHT (P.L.F.)	TOTAL LOAD (K.L.F.)	COST PER L.F.		
						MAT.	INST.	TOTAL
2300	15	2.8	12x16	260	3.06	121	19.35	140.35
2500		8.37	12x28	515	8.89	175	21	196
8900	45	3.34	12x60	1165	4.51	345	12.45	357.45
9900		6.5	24x60	1915	8.42	465	17.40	482.40

B1010 215		"L" Shaped Precast Beams						
	SPAN (FT.)	SUPERIMPOSED LOAD (K.L.F.)	SIZE W X D (IN.)	BEAM WEIGHT (P.L.F.)	TOTAL LOAD (K.L.F.)	COST PER L.F.		
						MAT.	INST.	TOTAL
2250	15	2.58	12x16	230	2.81	93	19.35	112.35
2400		5.92	12x24	370	6.29	126	19.35	145.35
4000	25	2.64	12x28	435	3.08	142	12.55	154.55
4450		6.44	18x36	790	7.23	203	14.95	217.95
5300	30	2.80	12x36	565	3.37	170	11.35	181.35
6400		8.66	24x44	1245	9.90	279	16.35	295.35

B10 Superstructure

B1010 Floor Construction

Cast in Place Floor Slab, One Way
General: Solid concrete slabs of uniform depth reinforced for flexure in one direction and for temperature and shrinkage in the other direction.

Cast in Place Beam & Slab, One Way
General: Solid concrete one way slab cast monolithically with reinforced concrete beams and girders.

B1010 217	Cast in Place Slabs, One Way							
	SLAB DESIGN & SPAN (FT.)	SUPERIMPOSED LOAD (P.S.F.)	THICKNESS (IN.)	TOTAL LOAD (P.S.F.)		COST PER S.F.		
						MAT.	INST.	TOTAL
2500	Single 8	40	4	90		4.14	7.45	11.59
2600		75	4	125		4.16	8.10	12.26
2700		125	4-1/2	181		4.32	8.15	12.47
2800		200	5	262		4.73	8.45	13.18
3000	Single 10	40	4	90		4.29	8.05	12.34
3100		75	4	125		4.29	8.05	12.34
3200		125	5	188		4.71	8.20	12.91
3300		200	7-1/2	293		6.05	9.25	15.30
3500	Single 15	40	5-1/2	90		5.05	8.15	13.20
3600		75	6-1/2	156		5.55	8.60	14.15
3700		125	7-1/2	219		6	8.80	14.80
3800		200	8-1/2	306		6.45	9	15.45
4000	Single 20	40	7-1/2	115		5.95	8.60	14.55
4100		75	9	200		6.50	8.60	15.10
4200		125	10	250		7.10	8.95	16.05
4300		200	10	324		7.55	9.35	16.90

B1010 219	Cast in Place Beam & Slab, One Way							
	BAY SIZE (FT.)	SUPERIMPOSED LOAD (P.S.F.)	MINIMUM COL. SIZE (IN.)	SLAB THICKNESS (IN.)	TOTAL LOAD (P.S.F.)	COST PER S.F.		
						MAT.	INST.	TOTAL
5000	20x25	40	12	5-1/2	121	4.92	10.25	15.17
5200		125	16	5-1/2	215	5.80	11.75	17.55
5500	25x25	40	12	6	129	5.20	10.10	15.30
5700		125	18	6	227	6.55	12.40	18.95
7500	30x35	40	16	8	158	6.45	11.40	17.85
7600		75	18	8	196	6.85	11.80	18.65
7700		125	22	8	254	7.65	13.10	20.75
7800		200	26	8	332	8.25	13.55	21.80
8000	35x35	40	16	9	169	7.20	11.75	18.95
8200		75	20	9	213	7.75	12.85	20.60
8400		125	24	9	272	8.45	13.35	21.80
8600		200	26	9	355	9.25	14.25	23.50

B10 Superstructure

B1010 Floor Construction

General: Solid concrete two way slab cast monolithically with reinforced concrete support beams and girders.

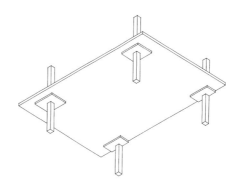

General: Flat Slab: Solid uniform depth concrete two way slabs with drop panels at columns and no column capitals.

B1010 220		Cast in Place Beam & Slab, Two Way						
	BAY SIZE (FT.)	SUPERIMPOSED LOAD (P.S.F.)	MINIMUM COL. SIZE (IN.)	SLAB THICKNESS (IN.)	TOTAL LOAD (P.S.F.)	COST PER S.F.		
						MAT.	INST.	TOTAL
4000	20 x 25	40	12	7	141	5.65	10.35	16
4300		75	14	7	181	6.45	11.35	17.80
4500		125	16	7	236	6.55	11.65	18.20
5100	25 x 25	40	12	7-1/2	149	5.95	10.45	16.40
5200		75	16	7-1/2	185	6.40	11.25	17.65
5300		125	18	7-1/2	250	7	12.20	19.20
7600	30 x 35	40	16	10	188	7.80	11.90	19.70
7700		75	18	10	225	8.25	12.35	20.60
8000		125	22	10	282	9.10	13.30	22.40
8500	35 x 35	40	16	10-1/2	193	8.30	12.15	20.45
8600		75	20	10-1/2	233	8.70	12.60	21.30
9000		125	24	10-1/2	287	9.65	13.50	23.15

B1010 222		Cast in Place Flat Slab with Drop Panels						
	BAY SIZE (FT.)	SUPERIMPOSED LOAD (P.S.F.)	MINIMUM COL. SIZE (IN.)	SLAB & DROP (IN.)	TOTAL LOAD (P.S.F.)	COST PER S.F.		
						MAT.	INST.	TOTAL
1960	20 x 20	40	12	7 - 3	132	4.99	8.25	13.24
1980		75	16	7 - 4	168	5.30	8.45	13.75
2000		125	18	7 - 6	221	5.85	8.75	14.60
3200	25 x 25	40	12	8-1/2 - 5-1/2	154	5.85	8.65	14.50
4000		125	20	8-1/2 - 8-1/2	243	6.55	9.30	15.85
4400		200	24	9 - 8-1/2	329	6.90	9.55	16.45
5000	25 x 30	40	14	9-1/2 - 7	168	6.35	9	15.35
5200		75	18	9-1/2 - 7	203	6.75	9.35	16.10
5600		125	22	9-1/2 - 8	256	7.05	9.55	16.60
6400	30 x 30	40	14	10-1/2 - 7-1/2	182	6.90	9.20	16.10
6600		75	18	10-1/2 - 7-1/2	217	7.30	9.60	16.90
6800		125	22	10-1/2 - 9	269	7.60	9.85	17.45
7400	30 x 35	40	16	11-1/2 - 9	196	7.50	9.60	17.10
7900		75	20	11-1/2 - 9	231	7.95	10	17.95
8000		125	24	11-1/2 - 11	284	8.25	10.25	18.50
9000	35 x 35	40	16	12 - 9	202	7.70	9.65	17.35
9400		75	20	12 - 11	240	8.25	10.15	18.40
9600		125	24	12 - 11	290	8.50	10.35	18.85

B10 Superstructure

B1010 Floor Construction

General: Flat Plates: Solid uniform depth concrete two way slab without drops or interior beams. Primary design limit is shear at columns.

Multispan Joist Slab

General: Combination of thin concrete slab and monolithic ribs at uniform spacing to reduce dead weight and increase rigidity.

B1010 223		Cast in Place Flat Plate						
	BAY SIZE (FT.)	SUPERIMPOSED LOAD (P.S.F.)	MINIMUM COL. SIZE (IN.)	SLAB THICKNESS (IN.)	TOTAL LOAD (P.S.F.)	COST PER S.F.		
						MAT.	INST.	TOTAL
3000	15 x 20	40	14	7	127	4.73	7.95	12.68
3400		75	16	7-1/2	169	5.05	8.10	13.15
3600		125	22	8-1/2	231	5.55	8.35	13.90
3800		175	24	8-1/2	281	5.55	8.35	13.90
4200	20 x 20	40	16	7	127	4.74	7.95	12.69
4400		75	20	7-1/2	175	5.10	8.10	13.20
4600		125	24	8-1/2	231	5.55	8.30	13.85
5000		175	24	8-1/2	281	5.60	8.35	13.95
5600	20 x 25	40	18	8-1/2	146	5.50	8.30	13.80
6000		75	20	9	188	5.70	8.40	14.10
6400		125	26	9-1/2	244	6.10	8.70	14.80
6600		175	30	10	300	6.35	8.80	15.15
7000	25 x 25	40	20	9	152	5.70	8.40	14.10
7400		75	24	9-1/2	194	6	8.65	14.65
7600		125	30	10	250	6.40	8.85	15.25

B1010 226		Cast in Place Multispan Joist Slab						
	BAY SIZE (FT.)	SUPERIMPOSED LOAD (P.S.F.)	MINIMUM COL. SIZE (IN.)	RIB DEPTH (IN.)	TOTAL LOAD (P.S.F.)	COST PER S.F.		
						MAT.	INST.	TOTAL
2000	15 x 15	40	12	8	115	6.20	9.70	15.90
2100		75	12	8	150	6.25	9.75	16
2200		125	12	8	200	6.35	9.85	16.20
2300		200	14	8	275	6.50	10.25	16.75
2600	15 x 20	40	12	8	115	6.30	9.70	16
2800		75	12	8	150	6.40	10.25	16.65
3000		125	14	8	200	6.60	10.45	17.05
3300		200	16	8	275	6.90	10.65	17.55
3600	20 x 20	40	12	10	120	6.40	9.60	16
3900		75	14	10	155	6.70	10.15	16.85
4000		125	16	10	205	6.75	10.35	17.10
4100		200	18	10	280	7	10.80	17.80
6200	30 x 30	40	14	14	131	7.10	10.05	17.15
6400		75	18	14	166	7.30	10.40	17.70
6600		125	20	14	216	7.70	10.95	18.65
6700		200	24	16	297	8.20	11.35	19.55

B10 Superstructure

B1010 Floor Construction

Waffle Slab

B1010 227	Cast in Place Waffle Slab							
	BAY SIZE (FT.)	SUPERIMPOSED LOAD (P.S.F.)	MINIMUM COL. SIZE (IN.)	RIB DEPTH (IN.)	TOTAL LOAD (P.S.F.)	COST PER S.F.		
						MAT.	INST.	TOTAL
3900	20 x 20	40	12	8	144	7.75	9.65	17.40
4000		75	12	8	179	7.85	9.80	17.65
4100		125	16	8	229	8	9.90	17.90
4200		200	18	8	304	8.40	10.25	18.65
4400	20 x 25	40	12	8	146	7.90	9.75	17.65
4500		75	14	8	181	8.05	9.90	17.95
4600		125	16	8	231	8.20	10.05	18.25
4700		200	18	8	306	8.55	10.35	18.90
4900	25 x 25	40	12	10	150	8.10	9.85	17.95
5000		75	16	10	185	8.30	10.05	18.35
5300		125	18	10	235	8.50	10.25	18.75
5500		200	20	10	310	8.75	10.45	19.20
5700	25 x 30	40	14	10	154	8.25	9.90	18.15
5800		75	16	10	189	8.45	10.10	18.55
5900		125	18	10	239	8.65	10.30	18.95
6000		200	20	12	329	9.45	10.80	20.25
6400	30 x 30	40	14	12	169	8.80	10.15	18.95
6500		75	18	12	204	8.95	10.30	19.25
6600		125	20	12	254	9.10	10.45	19.55
6700		200	24	12	329	9.85	11.05	20.90
6900	30 x 35	40	16	12	169	8.95	10.25	19.20
7000		75	18	12	204	8.95	10.25	19.20
7100		125	22	12	254	9.35	10.60	19.95
7200		200	26	14	334	10.30	11.25	21.55
7400	35 x 35	40	16	14	174	9.45	10.50	19.95
7500		75	20	14	209	9.65	10.65	20.30
7600		125	24	14	259	9.85	10.85	20.70
7700		200	26	16	346	10.45	11.45	21.90
8000	35 x 40	40	18	14	176	9.70	10.70	20.40
8300		75	22	14	211	9.95	10.95	20.90
8500		125	26	16	271	10.45	11.20	21.65
8750		200	30	20	372	11.40	11.90	23.30
9200	40 x 40	40	18	14	176	9.95	10.95	20.90
9400		75	24	14	211	10.30	11.25	21.55
9500		125	26	16	271	10.60	11.35	21.95
9700	40 x 45	40	20	16	186	10.35	11.10	21.45
9800		75	24	16	221	10.70	11.40	22.10
9900		125	28	16	271	10.85	11.55	22.40

B1010   Floor Construction

Precast Plank with No Topping

Precast Plank with 2″ Concrete Topping

B1010 229		Precast Plank with No Topping						
	SPAN (FT.)	SUPERIMPOSED LOAD (P.S.F.)	TOTAL DEPTH (IN.)	DEAD LOAD (P.S.F.)	TOTAL LOAD (P.S.F.)	COST PER S.F.		
						MAT.	INST.	TOTAL
0720	10	40	4	50	90	6.10	3.27	9.37
0750		75	6	50	125	7.15	2.80	9.95
0770		100	6	50	150	7.15	2.80	9.95
0800	15	40	6	50	90	7.15	2.80	9.95
0820		75	6	50	125	7.15	2.80	9.95
0850		100	6	50	150	7.15	2.80	9.95
0950	25	40	6	50	90	7.15	2.80	9.95
0970		75	8	55	130	7.90	2.45	10.35
1000		100	8	55	155	7.90	2.45	10.35
1200	30	40	8	55	95	7.90	2.45	10.35
1300		75	8	55	130	7.90	2.45	10.35
1400		100	10	70	170	8.25	2.18	10.43
1500	40	40	10	70	110	8.25	2.18	10.43
1600		75	12	70	145	8.80	1.96	10.76
1700	45	40	12	70	110	8.80	1.96	10.76

B1010 230		Precast Plank with 2″ Concrete Topping						
	SPAN (FT.)	SUPERIMPOSED LOAD (P.S.F.)	TOTAL DEPTH (IN.)	DEAD LOAD (P.S.F.)	TOTAL LOAD (P.S.F.)	COST PER S.F.		
						MAT.	INST.	TOTAL
2000	10	40	6	75	115	7	5.40	12.40
2100		75	8	75	150	8.05	4.94	12.99
2200		100	8	75	175	8.05	4.94	12.99
2500	15	40	8	75	115	8.05	4.94	12.99
2600		75	8	75	150	8.05	4.94	12.99
2700		100	8	75	175	8.05	4.94	12.99
3100	25	40	8	75	115	8.05	4.94	12.99
3200		75	8	75	150	8.05	4.94	12.99
3300		100	10	80	180	8.80	4.59	13.39
3400	30	40	10	80	120	8.80	4.59	13.39
3500		75	10	80	155	8.80	4.59	13.39
3600		100	10	80	180	8.80	4.59	13.39
4000	40	40	12	95	135	9.15	4.32	13.47
4500		75	14	95	170	9.70	4.10	13.80
5000	45	40	14	95	135	9.70	4.10	13.80

B1010 Floor Construction

Precast Double "T" Beams with No Topping

Most widely used for moderate span floors and moderate and long span roofs. At shorter spans, they tend to be competitive with hollow core slabs. They are also used as wall panels.

Precast Double "T" Beams with 2″ Topping

B1010 234	Precast Double "T" Beams with No Topping							
	SPAN (FT.)	SUPERIMPOSED LOAD (P.S.F.)	DBL. "T" SIZE D (IN.) W (FT.)	CONCRETE "T" TYPE	TOTAL LOAD (P.S.F.)	COST PER S.F.		
						MAT.	INST.	TOTAL
4300	50	30	20x8	Lt. Wt.	66	8.60	1.48	10.08
4400		40	20x8	Lt. Wt.	76	9.20	1.36	10.56
4500		50	20x8	Lt. Wt.	86	9.30	1.64	10.94
4600		75	20x8	Lt. Wt.	111	9.40	1.87	11.27
5600	70	30	32x10	Lt. Wt.	78	10.05	1.02	11.07
5750		40	32x10	Lt. Wt.	88	10.15	1.23	11.38
5900		50	32x10	Lt. Wt.	98	10.20	1.42	11.62
6000		75	32x10	Lt. Wt.	123	10.30	1.72	12.02
6100		100	32x10	Lt. Wt.	148	10.55	2.33	12.88
6200	80	30	32x10	Lt. Wt.	78	10.20	1.42	11.62
6300		40	32x10	Lt. Wt.	88	10.30	1.72	12.02
6400		50	32x10	Lt. Wt.	98	10.40	2.02	12.42

B1010 235	Precast Double "T" Beams With 2″ Topping							
	SPAN (FT.)	SUPERIMPOSED LOAD (P.S.F.)	DBL. "T" SIZE D (IN.) W (FT.)	CONCRETE "T" TYPE	TOTAL LOAD (P.S.F.)	COST PER S.F.		
						MAT.	INST.	TOTAL
7100	40	30	18x8	Reg. Wt.	120	7.80	2.95	10.75
7200		40	20x8	Reg. Wt.	130	7.85	2.82	10.67
7300		50	20x8	Reg. Wt.	140	7.95	3.09	11.04
7400		75	20x8	Reg. Wt.	165	8	3.20	11.20
7500		100	20x8	Reg. Wt.	190	8.15	3.59	11.74
7550	50	30	24x8	Reg. Wt.	120	8.40	2.95	11.35
7600		40	24x8	Reg. Wt.	130	8.45	3.05	11.50
7750		50	24x8	Reg. Wt.	140	8.45	3.07	11.52
7800		75	24x8	Reg. Wt.	165	8.60	3.46	12.06
7900		100	32x10	Reg. Wt.	189	9.70	2.82	12.52

B1010 Floor Construction

Precast Beam and Plank with No Topping

Precast Beam and Plank with 2″ Topping

B1010 236		Precast Beam & Plank with No Topping						
	BAY SIZE (FT.)	SUPERIMPOSED LOAD (P.S.F.)	PLANK THICKNESS (IN.)	TOTAL DEPTH (IN.)	TOTAL LOAD (P.S.F.)	COST PER S.F.		
						MAT.	INST.	TOTAL
6200	25x25	40	8	28	118	18.40	3.26	21.66
6400		75	8	36	158	19	3.26	22.26
6500		100	8	36	183	19	3.26	22.26
7000	25x30	40	8	36	110	17.50	3.15	20.65
7200		75	10	36	159	18.45	2.90	21.35
7400		100	12	36	188	19.50	2.64	22.14
7600	30x30	40	8	36	121	18.10	3.06	21.16
8000		75	10	44	140	18.85	3.06	21.91
8250		100	12	52	206	21.50	2.59	24.09
8500	30x35	40	12	44	135	18.40	2.51	20.91
8750		75	12	52	176	19.55	2.50	22.05
9000	35x35	40	12	52	141	19.55	2.44	21.99
9250		75	12	60	181	20.50	2.45	22.95
9500	35x40	40	12	52	137	19.55	2.44	21.99

B1010 238		Precast Beam & Plank with 2″ Topping						
	BAY SIZE (FT.)	SUPERIMPOSED LOAD (P.S.F.)	PLANK THICKNESS (IN.)	TOTAL DEPTH (IN.)	TOTAL LOAD (P.S.F.)	COST PER S.F.		
						MAT.	INST.	TOTAL
4300	20x20	40	6	22	135	18.25	5.70	23.95
4400		75	6	24	173	19.20	5.80	25
4500		100	6	28	200	19.80	5.75	25.55
4600	20x25	40	6	26	134	17.65	5.60	23.25
5000		75	8	30	177	18.60	5.20	23.80
5200		100	8	30	202	18.60	5.20	23.80
5400	25x25	40	6	38	143	19.60	5.35	24.95
5600		75	8	38	183	16.20	5.05	21.25
6000		100	8	46	216	17.35	4.66	22.01
6200	25x30	40	8	38	144	18.25	4.83	23.08
6400		75	10	46	200	19.85	4.55	24.40
6600		100	10	46	225	20	4.56	24.56
7000	30x30	40	8	46	150	20	4.75	24.75
7200		75	10	54	181	21.50	4.76	26.26
7600		100	10	54	231	21.50	4.76	26.26
7800	30x35	40	10	54	166	20	4.38	24.38
8000		75	12	54	200	20	4.14	24.14

B10 Superstructure

B1010 Floor Construction

With Topping

General: Beams and double tees priced here are for plant produced prestressed members transported to the site and erected.

The 2″ structural topping is applied after the beams and double tees are in place and is reinforced with W.W.F.

Note: Deduct from prices 20% for Southern States. Add to prices 10% for Western States.

B1010 239		**Precast Double "T" & 2″ Topping on Precast Beams**						
	BAY SIZE (FT.)	SUPERIMPOSED LOAD (P.S.F.)	DEPTH (IN.)		TOTAL LOAD (P.S.F.)	COST PER S.F.		
						MAT.	INST.	TOTAL
3000	25x30	40	38		130	18.35	4.01	22.36
3100		75	38		168	18.35	4.01	22.36
3300		100	46		196	19.10	3.99	23.09
3600	30x30	40	46		150	19.65	3.89	23.54
3750		75	46		174	19.65	3.89	23.54
4000		100	54		203	20.50	3.88	24.38
4100	30x40	40	46		136	16.55	3.53	20.08
4300		75	54		173	17.15	3.50	20.65
4400		100	62		204	17.95	3.52	21.47
4600	30x50	40	54		138	15.90	3.25	19.15
4800		75	54		181	16.55	3.28	19.83
5000		100	54		219	18.10	3	21.10
5200	30x60	40	62		151	16.85	2.95	19.80
5400		75	62		192	17.55	2.98	20.53
5600		100	62		215	17.70	2.99	20.69
5800	35x40	40	54		139	17.60	3.46	21.06
6000		75	62		179	18.80	3.30	22.10
6250		100	62		212	19.35	3.42	22.77
6500	35x50	40	62		142	16.90	3.20	20.10
6750		75	62		186	17.45	3.31	20.76
7300		100	62		231	20	3.02	23.02
7600	35x60	40	54		154	18.95	2.79	21.74
7750		75	54		179	19.35	2.81	22.16
8000		100	62		224	20.50	2.85	23.35
8250	40x40	40	62		145	20	3.46	23.46
8400		75	62		187	20.50	3.47	23.97
8750		100	62		223	22	3.55	25.55
9000	40x50	40	62		151	19.05	3.23	22.28
9300		75	62		193	19.05	2.43	21.48
9800	40x60	40	62		164	23.50	2.35	25.85

B10 Superstructure

B1010 Floor Construction

General: The following table is based upon structural wide flange (WF) beam and girder framing. Non-composite action is assumed between WF framing and decking. Deck costs not included.

The deck spans the short direction. The steel beams and girders are fireproofed with sprayed fiber fireproofing.

No columns included in price.

B1010 241		W Shape Beams & Girders						
	BAY SIZE (FT.) BEAM X GIRD	**SUPERIMPOSED LOAD (P.S.F.)**	**STEEL FRAMING DEPTH (IN.)**	**FIREPROOFING (S.F. PER S.F.)**	**TOTAL LOAD (P.S.F.)**	**COST PER S.F.**		
						MAT.	**INST.**	**TOTAL**
1350	15x20	40	12	.535	50	4.25	1.81	6.06
1400		40	16	.65	90	5.55	2.34	7.89
1450		75	18	.694	125	7.30	2.95	10.25
1500		125	24	.796	175	10.05	4.03	14.08
2000	20x20	40	12	.55	50	4.75	2	6.75
2050		40	14	.579	90	6.50	2.60	9.10
2100		75	16	.672	125	7.75	3.09	10.84
2150		125	16	.714	175	9.25	3.69	12.94
2900	20x25	40	16	.53	50	5.25	2.14	7.39
2950		40	18	.621	96	8.25	3.20	11.45
3000		75	18	.651	131	9.50	3.65	13.15
3050		125	24	.77	200	12.50	4.85	17.35
5200	25x25	40	18	.486	50	5.70	2.26	7.96
5300		40	18	.592	96	8.45	3.27	11.72
5400		75	21	.668	131	10.25	3.91	14.16
5450		125	24	.738	191	13.45	5.15	18.60
6300	25x30	40	21	.568	50	6.50	2.59	9.09
6350		40	21	.694	90	8.75	3.44	12.19
6400		75	24	.776	125	11.25	4.34	15.59
6450		125	30	.904	175	13.55	5.30	18.85
7700	30x30	40	21	.52	50	6.95	2.70	9.65
7750		40	24	.629	103	10.70	4.03	14.73
7800		75	30	.715	138	12.70	4.77	17.47
7850		125	36	.822	206	16.70	6.30	23
8250	30x35	40	21	.508	50	7.90	3.02	10.92
8300		40	24	.651	109	11.70	4.39	16.09
8350		75	33	.732	150	14.20	5.25	19.45
8400		125	36	.802	225	17.70	6.65	24.35
9550	35x35	40	27	.560	50	8.20	3.15	11.35
9600		40	36	.706	109	14.90	5.50	20.40
9650		75	36	.750	150	15.45	5.70	21.15
9820		125	36	.797	225	20.50	7.65	28.15

B1010 Floor Construction

Description: Table below lists costs for light gage CEE or PUNCHED DOUBLE joists to suit the span and loading with the minimum thickness subfloor required by the joist spacing.

B1010 244		Light Gauge Steel Floor Systems						
	SPAN (FT.)	SUPERIMPOSED LOAD (P.S.F.)	FRAMING DEPTH (IN.)	FRAMING SPAC. (IN.)	TOTAL LOAD (P.S.F.)	COST PER S.F.		
						MAT.	INST.	TOTAL
1500	15	40	8	16	54	2.35	2.01	4.36
1550			8	24	54	2.40	2.05	4.45
1600		65	10	16	80	3.12	2.46	5.58
1650			10	24	80	3.01	2.40	5.41
1700		75	10	16	90	3.12	2.46	5.58
1750			10	24	90	3.01	2.40	5.41
1800		100	10	16	116	4.12	3.13	7.25
1850			10	24	116	3.01	2.40	5.41
1900		125	10	16	141	4.12	3.13	7.25
1950			10	24	141	3.01	2.40	5.41
2500	20	40	8	16	55	2.72	2.27	4.99
2550			8	24	55	2.40	2.05	4.45
2600		65	8	16	80	3.18	2.60	5.78
2650			10	24	80	3.02	2.42	5.44
2700		75	10	16	90	2.93	2.34	5.27
2750			12	24	90	3.43	2.11	5.54
2800		100	10	16	115	2.93	2.34	5.27
2850			10	24	116	3.96	2.70	6.66
2900		125	12	16	142	4.73	2.67	7.40
2950			12	24	141	4.32	2.89	7.21
3500	25	40	10	16	55	3.12	2.46	5.58
3550			10	24	55	3.01	2.40	5.41
3600		65	10	16	81	4.12	3.13	7.25
3650			12	24	81	3.82	2.29	6.11
3700		75	12	16	92	4.72	2.67	7.39
3750			12	24	91	4.32	2.89	7.21
3800		100	12	16	117	5.35	2.95	8.30
3850		125	12	16	143	6.10	3.88	9.98
4500	30	40	12	16	57	4.72	2.67	7.39
4550			12	24	56	3.81	2.28	6.09
4600		65	12	16	82	5.30	2.94	8.24
4650		75	12	16	92	5.30	2.94	8.24

B1010 Floor Construction

Description: The table below lists costs per S.F. for a floor system on bearing walls using open web steel joists, galvanized steel slab form and 2-1/2" concrete slab reinforced with welded wire fabric. Costs of the bearing walls are not included.

B1010 246		Deck & Joists on Bearing Walls						
	SPAN (FT.)	SUPERIMPOSED LOAD (P.S.F.)	JOIST SPACING FT. - IN.	DEPTH (IN.)	TOTAL LOAD (P.S.F.)	COST PER S.F.		
						MAT.	INST.	TOTAL
1050	20	40	2-0	14-1/2	83	4.04	3.60	7.64
1070		65	2-0	16-1/2	109	4.35	3.81	8.16
1100		75	2-0	16-1/2	119	4.35	3.81	8.16
1120		100	2-0	18-1/2	145	4.37	3.82	8.19
1150		125	1-9	18-1/2	170	5.20	4.54	9.74
1170	25	40	2-0	18-1/2	84	4.74	4.23	8.97
1200		65	2-0	20-1/2	109	4.89	4.33	9.22
1220		75	2-0	20-1/2	119	4.77	4.08	8.85
1250		100	2-0	22-1/2	145	4.84	4.12	8.96
1270		125	1-9	22-1/2	170	5.45	4.51	9.96
1300	30	40	2-0	22-1/2	84	4.84	3.72	8.56
1320		65	2-0	24-1/2	110	5.20	3.89	9.09
1350		75	2-0	26-1/2	121	5.40	3.97	9.37
1370		100	2-0	26-1/2	146	5.70	4.11	9.81
1400		125	2-0	24-1/2	172	6.15	5.30	11.45
1420	35	40	2-0	26-1/2	85	5.85	5.05	10.90
1450		65	2-0	28-1/2	111	6	5.20	11.20
1470		75	2-0	28-1/2	121	6	5.20	11.20
1500		100	1-11	28-1/2	147	6.45	5.55	12
1520		125	1-8	28-1/2	172	7.05	5.95	13
1550	5/8" gyp. fireproof.							
1560	On metal furring, add					.70	2.66	3.36

B10 Superstructure

B1010 Floor Construction

The table below lists costs for a floor system on exterior bearing walls and interior columns and beams using open web steel joists, galvanized steel slab form, 2-1/2″ concrete slab reinforced with welded wire fabric.

Columns are sized to accommodate one floor plus roof loading but costs for columns are from suspended floor to slab on grade only. Costs of the bearing walls are not included.

Steel Joists and Concrete Slab on Walls and Beams

B1010 248		Steel Joists on Beam & Wall						
	BAY SIZE (FT.)	SUPERIMPOSED LOAD (P.S.F.)	DEPTH (IN.)	TOTAL LOAD (P.S.F.)	COLUMN ADD	COST PER S.F.		
						MAT.	INST.	TOTAL
1720	25x25	40	23	84		6.25	4.52	10.77
1730					columns	.35	.12	.47
1750	25x25	65	29	110		6.55	4.67	11.22
1760					columns	.35	.12	.47
1770	25x25	75	26	120		6.85	4.58	11.43
1780					columns	.40	.13	.53
1800	25x25	100	29	145		7.65	5	12.65
1810					columns	.40	.13	.53
1820	25x25	125	29	170		8.05	5.20	13.25
1830					columns	.45	.14	.59
2020	30x30	40	29	84		6.70	4.14	10.84
2030					columns	.31	.10	.41
2050	30x30	65	29	110		7.60	4.47	12.07
2060					columns	.31	.10	.41
2070	30x30	75	32	120		7.75	4.55	12.30
2080					columns	.36	.12	.48
2100	30x30	100	35	145		8.60	4.87	13.47
2110					columns	.42	.14	.56
2120	30x30	125	35	172		9.30	6.20	15.50
2130					columns	.50	.17	.67
2170	30x35	40	29	85		7.45	4.47	11.92
2180					columns	.30	.10	.40
2200	30x35	65	29	111		8.25	5.75	14
2210					columns	.34	.12	.46
2220	30x35	75	32	121		8.25	5.75	14
2230					columns	.35	.12	.47
2250	30x35	100	35	148		9	5.10	14.10
2260					columns	.43	.14	.57
2270	30x35	125	38	173		10	5.50	15.50
2280					columns	.44	.14	.58
2320	35x35	40	32	85		7.70	4.55	12.25
2330					columns	.30	.10	.40
2350	35x35	65	35	111		8.80	6	14.80
2360					columns	.37	.12	.49
2370	35x35	75	35	121		9	6.05	15.05
2380					columns	.37	.12	.49
2400	35x35	100	38	148		9.35	6.25	15.60
2410					columns	.45	.15	.60

B1010 Floor Construction

B1010 248	Steel Joists on Beam & Wall							

	BAY SIZE (FT.)	SUPERIMPOSED LOAD (P.S.F.)	DEPTH (IN.)	TOTAL LOAD (P.S.F.)	COLUMN ADD	COST PER S.F.		
						MAT.	INST.	TOTAL
2460 2475	5/8 gyp. fireproof. On metal furring, add					.70	2.66	3.36

B10 Superstructure

B1010 Floor Construction

The table below lists costs per S.F. for a floor system on steel columns and beams using open web steel joists, galvanized steel slab form, 2-1/2″ concrete slab reinforced with welded wire fabric.

Columns are sized to accommodate one floor plus roof loading but costs for columns are from floor to grade only.

Steel Joists and Concrete Slab on Steel Columns and Beams

B1010 250		Steel Joists, Beams & Slab on Columns						
	BAY SIZE (FT.)	SUPERIMPOSED LOAD (P.S.F.)	DEPTH (IN.)	TOTAL LOAD (P.S.F.)	COLUMN ADD	COST PER S.F.		
						MAT.	INST.	TOTAL
2350	15x20	40	17	83		6.15	4.16	10.31
2400					column	1.14	.38	1.52
2450	15x20	65	19	108		6.95	4.41	11.36
2500					column	1.14	.38	1.52
2550	15x20	75	19	119		7.15	4.56	11.71
2600					column	1.25	.41	1.66
2650	15x20	100	19	144		7.60	4.75	12.35
2700					column	1.25	.41	1.66
2750	15x20	125	19	170		8.45	5.35	13.80
2800					column	1.66	.55	2.21
2850	20x20	40	19	83		6.85	4.35	11.20
2900					column	.94	.31	1.25
2950	20x20	65	23	109		7.50	4.65	12.15
3000					column	1.25	.41	1.66
3100	20x20	75	26	119		8	4.80	12.80
3200					column	1.25	.41	1.66
3400	20x20	100	23	144		8.35	4.92	13.27
3450					column	1.25	.41	1.66
3500	20x20	125	23	170		9.20	5.30	14.50
3600					column	1.50	.49	1.99
3700	20x25	40	44	83		7.65	4.98	12.63
3800					column	1	.33	1.33
3900	20x25	65	26	110		8.40	5.25	13.65
4000					column	1	.33	1.33
4100	20x25	75	26	120		8.30	5	13.30
4200					column	1.20	.40	1.60
4300	20x25	100	26	145		8.80	5.25	14.05
4400					column	1.20	.40	1.60
4500	20x25	125	29	170		9.70	5.70	15.40
4600					column	1.40	.46	1.86
4700	25x25	40	23	84		8.35	5.20	13.55
4800					column	.96	.32	1.28
4900	25x25	65	29	110		8.85	5.40	14.25
5000					column	.96	.32	1.28
5100	25x25	75	26	120		9.30	5.40	14.70
5200					column	1.12	.37	1.49

B1010 Floor Construction

B1010 250	Steel Joists, Beams & Slab on Columns

	BAY SIZE (FT.)	SUPERIMPOSED LOAD (P.S.F.)	DEPTH (IN.)	TOTAL LOAD (P.S.F.)	COLUMN ADD	COST PER S.F.		
						MAT.	INST.	TOTAL
5300	25x25	100	29	145		10.25	5.85	16.10
5400					column	1.12	.37	1.49
5500	25x25	125	32	170		10.90	6.10	17
5600					column	1.24	.41	1.65
5700	25x30	40	29	84		8.60	5.40	14
5800					column	.93	.31	1.24
5900	25x30	65	29	110		8.80	5.60	14.40
6000					column	.93	.31	1.24
6050	25x30	75	29	120		9.65	5.15	14.80
6100					column	1.03	.34	1.37
6150	25x30	100	29	145		10.45	5.45	15.90
6200					column	1.03	.34	1.37
6250	25x30	125	32	170		11.15	6.65	17.80
6300					column	1.19	.39	1.58
6350	30x30	40	29	84		9.05	4.90	13.95
6400					column	.86	.28	1.14
6500	30x30	65	29	110		10.30	5.35	15.65
6600					column	.86	.28	1.14
6700	30x30	75	32	120		10.50	5.45	15.95
6800					column	.99	.32	1.31
6900	30x30	100	35	145		11.65	5.85	17.50
7000					column	1.15	.39	1.54
7100	30x30	125	35	172		12.70	7.25	19.95
7200					column	1.28	.43	1.71
7300	30x35	40	29	85		10.10	5.35	15.45
7400					column	.74	.24	.98
7500	30x35	65	29	111		11.15	6.70	17.85
7600					column	.95	.31	1.26
7700	30x35	75	32	121		11.15	6.70	17.85
7800					column	.97	.32	1.29
7900	30x35	100	35	148		12.15	6.10	18.25
8000					column	1.19	.39	1.58
8100	30x35	125	38	173		13.35	6.60	19.95
8200					column	1.21	.40	1.61
8300	35x35	40	32	85		10.45	5.45	15.90
8400					column	.85	.28	1.13
8500	35x35	65	35	111		11.80	6.95	18.75
8600					column	1.02	.34	1.36
9300	35x35	75	38	121		12.15	7.05	19.20
9400					column	1.02	.34	1.36
9500	35x35	100	38	148		13.10	7.50	20.60
9600					column	1.26	.41	1.67
9750	35x35	125	41	173		14.25	6.85	21.10
9800					column	1.28	.43	1.71
9810	5/8 gyp. fireproof.							
9815	On metal furring, add					.70	2.66	3.36

B1010 Floor Construction

Composite Beam & Cast in Place Slab

General: Composite construction of wide flange beams and concrete slabs is most efficiently used when loads are heavy and spans are moderately long. It is stiffer with less deflection than non-composite construction of similar depth and spans.

B1010 252	Composite Beam & Cast in Place Slab						
	BAY SIZE (FT.)	SUPERIMPOSED LOAD (P.S.F.)	SLAB THICKNESS (IN.)	TOTAL DEPTH (FT.-IN.)	TOTAL LOAD (P.S.F.)	COST PER S.F.	
						MAT.	INST.	TOTAL
3800	20x25	40	4	1 - 8	94	9.30	9.95	19.25
3900		75	4	1 - 8	130	10.70	10.55	21.25
4000		125	4	1 - 10	181	12.35	11.30	23.65
4100		200	5	2 - 2	272	15.95	12.85	28.80
4200	25x25	40	4-1/2	1 - 8-1/2	99	9.65	10.05	19.70
4300		75	4-1/2	1 - 10-1/2	136	11.55	10.90	22.45
4400		125	5-1/2	2 - 0-1/2	200	13.55	11.80	25.35
4500		200	5-1/2	2 - 2-1/2	278	16.70	13.10	29.80
4600	25x30	40	4-1/2	1 - 8-1/2	100	10.75	10.50	21.25
4700		75	4-1/2	1 - 10-1/2	136	12.50	11.25	23.75
4800		125	5-1/2	2 - 2-1/2	202	15.20	12.35	27.55
4900		200	5-1/2	2 - 5-1/2	279	18	13.55	31.55
5000	30x30	40	4	1 - 8	95	10.75	10.50	21.25
5200		75	4	2 - 1	131	12.40	11.15	23.55
5400		125	4	2 - 4	183	15.15	12.25	27.40
5600		200	5	2 - 10	274	19.20	13.95	33.15
5800	30x35	40	4	2 - 1	95	11.35	10.70	22.05
6000		75	4	2 - 4	139	13.60	11.60	25.20
6250		125	4	2 - 4	185	17.45	13.10	30.55
6500		200	5	2 - 11	276	21	14.65	35.65
8000	35x35	40	4	2 - 1	96	11.90	10.90	22.80
8250		75	4	2 - 4	133	14.10	11.80	25.90
8500		125	4	2 - 10	185	16.80	12.90	29.70
8750		200	5	3 - 5	276	22	14.85	36.85
9000	35x40	40	4	2 - 4	97	13.20	11.45	24.65
9250		75	4	2 - 4	134	15.35	12.15	27.50
9500		125	4	2 - 7	186	18.55	13.50	32.05
9750		200	5	3 - 5	278	24	15.55	39.55

B10 Superstructure

B1010   Floor Construction

The table below lists costs per S.F. for floors using steel beams and girders, composite steel deck, concrete slab reinforced with W.W.F. and sprayed fiber fireproofing (non- asbestos) on the steel beams and girders and on the steel deck.

Wide Flange, Composite Deck & Slab

B1010 254		W Shape, Composite Deck, & Slab						
	BAY SIZE (FT.) BEAM X GIRD	SUPERIMPOSED LOAD (P.S.F.)	SLAB THICKNESS (IN.)	TOTAL DEPTH (FT.-IN.)	TOTAL LOAD (P.S.F.)	COST PER S.F.		
						MAT.	INST.	TOTAL
0540	15x20	40	5	1-7	89	9.55	6.10	15.65
0560		75	5	1-9	125	10.55	6.50	17.05
0580		125	5	1-11	176	12.40	7.10	19.50
0600		200	5	2-2	254	15.20	8.15	23.35
0700	20x20	40	5	1-9	90	10.45	6.40	16.85
0720		75	5	1-9	126	11.85	6.90	18.75
0740		125	5	1-11	177	13.55	7.55	21.10
0760		200	5	2-5	255	16.25	8.45	24.70
0780	20x25	40	5	1-9	90	10.45	6.35	16.80
0800		75	5	1-9	127	13.20	7.35	20.55
0820		125	5	1-11	180	15.70	8.05	23.75
0840		200	5	2-5	256	17.90	9.05	26.95
0940	25x25	40	5	1-11	91	11.95	6.85	18.80
0960		75	5	2-5	178	13.85	7.55	21.40
0980		125	5	2-5	181	16.90	8.40	25.30
1000		200	5-1/2	2-8-1/2	263	19.65	9.70	29.35
1400	25x30	40	5	2-5	91	12.05	6.95	19
1500		75	5	2-5	128	14.55	7.80	22.35
1600		125	5	2-8	180	16.80	8.65	25.45
1700		200	5	2-11	259	21	10.05	31.05
2200	30x30	40	5	2-2	92	13.40	7.40	20.80
2300		75	5	2-5	129	15.75	8.25	24
2400		125	5	2-11	182	18.75	9.30	28.05
2500		200	5	3-2	263	25.50	11.70	37.20
2600	30x35	40	5	2-5	94	14.65	7.90	22.55
2700		75	5	2-11	131	17.15	8.75	25.90
2800		125	5	3-2	183	20	9.80	29.80
2900		200	5-1/2	3-5-1/2	268	24.50	11.35	35.85
3800	35x35	40	5	2-8	94	14.90	7.90	22.80
3900		75	5	2-11	131	17.40	8.80	26.20
4000		125	5	3-5	184	21	10.05	31.05
4100		200	5-1/2	3-5-1/2	270	27	11.90	38.90

B1010 Floor Construction

The table below lists costs per S.F. for floors using composite steel beams with welded shear studs, composite steel deck and light weight concrete slab reinforced with W.W.F.

Composite Beam, Deck & Slab

B1010 256			**Composite Beams, Deck & Slab**					
	BAY SIZE (FT.)	SUPERIMPOSED LOAD (P.S.F.)	SLAB THICKNESS (IN.)	TOTAL DEPTH (FT.-IN.)	TOTAL LOAD (P.S.F.)	COST PER S.F.		
						MAT.	INST.	TOTAL
2400	20x25	40	5-1/2	1 - 5-1/2	80	10	5.70	15.70
2500		75	5-1/2	1 - 9-1/2	115	10.40	5.70	16.10
2750		125	5-1/2	1 - 9-1/2	167	12.80	6.75	19.55
2900		200	6-1/4	1 - 11-1/2	251	14.50	7.25	21.75
3000	25x25	40	5-1/2	1 - 9-1/2	82	9.65	5.45	15.10
3100		75	5-1/2	1 - 11-1/2	118	10.80	5.55	16.35
3200		125	5-1/2	2 - 2-1/2	169	11.30	6	17.30
3300		200	6-1/4	2 - 6-1/4	252	15.35	7	22.35
3400	25x30	40	5-1/2	1 - 11-1/2	83	9.85	5.40	15.25
3600		75	5-1/2	1 - 11-1/2	119	10.65	5.50	16.15
3900		125	5-1/2	1 - 11-1/2	170	12.50	6.20	18.70
4000		200	6-1/4	2 - 6-1/4	252	15.40	7	22.40
4200	30x30	40	5-1/2	1 - 11-1/2	81	10.10	5.55	15.65
4400		75	5-1/2	2 - 2-1/2	116	10.90	5.80	16.70
4500		125	5-1/2	2 - 5-1/2	168	13.30	6.50	19.80
4700		200	6-1/4	2 - 9-1/4	252	16.10	7.55	23.65
4900	30x35	40	5-1/2	2 - 2-1/2	82	10.60	5.75	16.35
5100		75	5-1/2	2 - 5-1/2	117	11.65	5.90	17.55
5300		125	5-1/2	2 - 5-1/2	169	13.75	6.65	20.40
5500		200	6-1/4	2 - 9-1/4	254	16.25	7.60	23.85
5750	35x35	40	5-1/2	2 - 5-1/2	84	11.10	5.80	16.90
6000		75	5-1/2	2 - 5-1/2	121	12.75	6.20	18.95
7000		125	5-1/2	2 - 8-1/2	170	15.15	7.10	22.25
7200		200	5-1/2	2 - 11-1/2	254	17.25	7.85	25.10
7400	35x40	40	5-1/2	2 - 5-1/2	85	12.35	6.25	18.60
7600		75	5-1/2	2 - 5-1/2	121	13.45	6.45	19.90
8000		125	5-1/2	2 - 5-1/2	171	15.55	7.25	22.80
9000		200	5-1/2	2 - 11-1/2	255	18.80	8.20	27

B1010 Floor Construction

Description: Table B1010 258 lists S.F. costs for steel deck and concrete slabs for various spans.

Description: Table B1010 261 lists the S.F. costs for wood joists and a minimum thickness plywood subfloor.

Description: Table B1010 264 lists the S.F. costs, total load, and member sizes, for various bay sizes and loading conditions.

Metal Deck/Concrete Fill

Wood Joist

Wood Beam and Joist

B1010 258	Metal Deck/Concrete Fill							
	SUPERIMPOSED LOAD (P.S.F.)	DECK SPAN (FT.)	DECK GAGE DEPTH	SLAB THICKNESS (IN.)	TOTAL LOAD (P.S.F.)	COST PER S.F.		
						MAT.	INST.	TOTAL
0900	125	6	22 1-1/2	4	164	2.24	2.30	4.54
0920		7	20 1-1/2	4	164	2.45	2.44	4.89
0950		8	20 1-1/2	4	165	2.45	2.44	4.89
0970		9	18 1-1/2	4	165	2.82	2.44	5.26
1000		10	18 2	4	165	2.77	2.57	5.34
1020		11	18 3	5	169	3.27	2.80	6.07

B1010 261	Wood Joist	COST PER S.F.		
		MAT.	INST.	TOTAL
2902	Wood joist, 2" x 8",12" O.C.	1.15	1.73	2.88
2950	16" O.C.	.99	1.48	2.47
3000	24" O.C.	1.01	1.38	2.39
3300	2"x10", 12" O.C.	1.35	1.98	3.33
3350	16" O.C.	1.14	1.66	2.80
3400	24" O.C.	1.10	1.49	2.59
3700	2"x12", 12" O.C.	1.77	2	3.77
3750	16" O.C.	1.45	1.68	3.13
3800	24" O.C.	1.31	1.50	2.81
4100	2"x14", 12" O.C.	2.89	2.18	5.07
4150	16" O.C.	2.29	1.82	4.11
4200	24" O.C.	1.88	1.60	3.48
7101	Note: Subfloor cost is included in these prices.			

B1010 264	Wood Beam & Joist						
	BAY SIZE (FT.)	SUPERIMPOSED LOAD (P.S.F.)	GIRDER BEAM (IN.)	JOISTS (IN.)	TOTAL LOAD (P.S.F.)	COST PER S.F.		
						MAT.	INST.	TOTAL
2000	15x15	40	8 x 12 4 x 12	2 x 6 @ 16	53	8.35	4.03	12.38
2050		75	8 x 16 4 x 16	2 x 8 @ 16	90	10.90	4.39	15.29
2100		125	12 x 16 6 x 16	2 x 8 @ 12	144	17.15	5.65	22.80
2150		200	14 x 22 12 x 16	2 x 10 @ 12	227	34.50	8.65	43.15
3000	20x20	40	10 x 14 10 x 12	2 x 8 @ 16	63	10.20	4.04	14.24
3050		75	12 x 16 8 x 16	2 x 10 @ 16	102	14.60	4.55	19.15
3100		125	14 x 22 12 x 16	2 x 10 @ 12	163	28.50	7.25	35.75

B10 Superstructure

B1010 Floor Construction

Steel Column Fireproofing

"T" Shaped
Precast Beams

"L" Shaped
Precast Beams

B1010 720	Steel Column Fireproofing							
	ENCASEMENT SYSTEM	COLUMN SIZE (IN.)	THICKNESS (IN.)	FIRE RATING (HRS.)	WEIGHT (P.L.F.)	COST PER V.L.F.		
						MAT.	INST.	TOTAL
3000	Concrete	8	1	1	110	5.60	33	38.60
3300		14	1	1	258	9.46	47.67	57.17
3400			2	3	325	11.35	53.82	65.22
3450	Gypsum board	8	1/2	2	8	2.96	19.85	22.81
3550	1 layer	14	1/2	2	18	3.20	21	24.20
3600	Gypsum board	8	1	3	14	4.08	25.50	29.58
3650	1/2" fire rated	10	1	3	17	4.36	27	31.36
3700	2 layers	14	1	3	22	4.52	27.50	32.02
3750	Gypsum board	8	1-1/2	3	23	5.40	32	37.40
3800	1/2" fire rated	10	1-1/2	3	27	6.05	35.50	41.55
3850	3 layers	14	1-1/2	3	35	6.70	38.50	45.20
3900	Sprayed fiber	8	1-1/2	2	6.3	3.93	6.40	10.33
3950	Direct application		2	3	8.3	5.40	8.80	14.20
4050		10	1-1/2	2	7.9	4.75	7.70	12.45
4200		14	1-1/2	2	10.8	5.90	9.55	15.45

B1020 Roof Construction

The table below lists prices per S.F. for roof rafters and sheathing by nominal size and spacing. Sheathing is 5/16″ CDX for 12″ and 16″ spacing and 3/8″ CDX for 24″ spacing.

Factors for Converting Inclined to Horizontal

Roof Slope	Approx. Angle	Factor	Roof Slope	Approx. Angle	Factor
Flat	0°	1.000	12 in 12	45.0°	1.414
1 in 12	4.8°	1.003	13 in 12	47.3°	1.474
2 in 12	9.5°	1.014	14 in 12	49.4°	1.537
3 in 12	14.0°	1.031	15 in 12	51.3°	1.601
4 in 12	18.4°	1.054	16 in 12	53.1°	1.667
5 in 12	22.6°	1.083	17 in 12	54.8°	1.734
6 in 12	26.6°	1.118	18 in 12	56.3°	1.803
7 in 12	30.3°	1.158	19 in 12	57.7°	1.873
8 in 12	33.7°	1.202	20 in 12	59.0°	1.943
9 in 12	36.9°	1.250	21 in 12	60.3°	2.015
10 in 12	39.8°	1.302	22 in 12	61.4°	2.088
11 in 12	42.5°	1.357	23 in 12	62.4°	2.162

B1020 102	Wood/Flat or Pitched	COST PER S.F.		
		MAT.	INST.	TOTAL
2500	Flat rafter, 2″x4″, 12″ O.C.	.91	1.56	2.47
2550	16″ O.C.	.84	1.34	2.18
2600	24″ O.C.	.57	1.15	1.72
2900	2″x6″, 12″ O.C.	1.07	1.57	2.64
2950	16″ O.C.	.95	1.35	2.30
3000	24″ O.C.	.65	1.15	1.80
3300	2″x8″, 12″ O.C.	1.28	1.69	2.97
3350	16″ O.C.	1.12	1.44	2.56
3400	24″ O.C.	.76	1.22	1.98
3700	2″x10″, 12″ O.C.	1.48	1.94	3.42
3750	16″ O.C.	1.27	1.62	2.89
3800	24″ O.C.	.85	1.33	2.18
4100	2″x12″, 12″ O.C.	1.90	1.96	3.86
4150	16″ O.C.	1.58	1.64	3.22
4200	24″ O.C.	1.06	1.34	2.40
4500	2″x14″, 12″ O.C.	3.02	2.86	5.88
4550	16″ O.C.	2.42	2.32	4.74
4600	24″ O.C.	1.63	1.80	3.43
4900	3″x6″, 12″ O.C.	2.30	1.67	3.97
4950	16″ O.C.	1.88	1.42	3.30
5000	24″ O.C.	1.27	1.20	2.47
5300	3″x8″, 12″ O.C.	3.14	1.86	5
5350	16″ O.C.	2.51	1.56	4.07
5400	24″ O.C.	1.69	1.29	2.98
5700	3″x10″, 12″ O.C.	3.98	2.11	6.09
5750	16″ O.C.	3.15	1.76	4.91
5800	24″ O.C.	2.11	1.42	3.53
6100	3″x12″, 12″ O.C.	4.90	2.55	7.45
6150	16″ O.C.	3.83	2.08	5.91
6200	24″ O.C.	2.57	1.64	4.21
7001	Wood truss, 4 in 12 slope, 24″ O.C., 24′ to 29′ span	3.88	2.38	6.26
7100	30′ to 43′ span	4.08	2.38	6.46
7200	44′ to 60′ span	4.34	2.38	6.72

B1020   Roof Construction

The table below lists the cost per S.F. for a roof system with steel columns, beams and deck using open web steel joists and 1-1/2″ galvanized metal deck. Perimeter of system is supported on bearing walls.

Fireproofing is not included. Costs/S.F. are based on a building 4 bays long and 4 bays wide.

Column costs are additive. Costs for the bearing walls are not included.

Steel Joists, Beams and Deck on Bearing Walls

B1020 108		Steel Joists, Beams & Deck on Columns & Walls						
	BAY SIZE (FT.)	SUPERIMPOSED LOAD (P.S.F.)	DEPTH (IN.)	TOTAL LOAD (P.S.F.)	COLUMN ADD	COST PER S.F.		
						MAT.	INST.	TOTAL
3000	25x25	20	18	40		2.97	1.34	4.31
3100					columns	.32	.09	.41
3200		30	22	50		3.36	1.58	4.94
3300					columns	.43	.11	.54
3400		40	20	60		3.57	1.57	5.14
3500					columns	.43	.11	.54
3600	25x30	20	22	40		3.11	1.29	4.40
3700					columns	.36	.09	.45
3800		30	20	50		3.39	1.65	5.04
3900					columns	.36	.09	.45
4000		40	25	60		3.67	1.48	5.15
4100					columns	.43	.11	.54
4200	30x30	20	25	42		3.46	1.38	4.84
4300					columns	.30	.08	.38
4400		30	22	52		3.80	1.51	5.31
4500					columns	.36	.09	.45
4600		40	28	62		3.98	1.56	5.54
4700					columns	.36	.09	.45
4800	30x35	20	22	42		3.57	1.45	5.02
4900					columns	.31	.08	.39
5000		30	28	52		3.78	1.51	5.29
5100					columns	.31	.08	.39
5200		40	25	62		4.12	1.63	5.75
5300					columns	.36	.09	.45
5400	35x35	20	28	42		3.56	1.46	5.02
5500					columns	.26	.07	.33

B1020 Roof Construction

Description: The table below lists the cost per S.F. for a roof system with steel columns, beams, and deck, using open web steel joists and 1-1/2″ galvanized metal deck.

Column costs are additive. Fireproofing is not included.

Steel Joists, Beams and Deck on Columns

B1020 112		Steel Joists, Beams, & Deck on Columns						
	BAY SIZE (FT.)	SUPERIMPOSED LOAD (P.S.F.)	DEPTH (IN.)	TOTAL LOAD (P.S.F.)	COLUMN ADD	COST PER S.F.		
						MAT.	INST.	TOTAL
1100	15x20	20	16	40		2.93	1.33	4.26
1200					columns	1.87	.50	2.37
1500		40	18	60		3.28	1.47	4.75
1600					columns	1.87	.50	2.37
2300	20x25	20	18	40		3.37	1.48	4.85
2400					columns	1.12	.29	1.41
2700		40	20	60		3.80	1.67	5.47
2800					columns	1.50	.40	1.90
2900	25x25	20	18	40		4.07	1.67	5.74
3000					columns	.90	.24	1.14
3100		30	22	50		4.39	1.89	6.28
3200					columns	1.20	.32	1.52
3300		40	20	60		4.74	1.90	6.64
3400					columns	1.20	.32	1.52
3500	25x30	20	22	40		3.94	1.51	5.45
3600					columns	1	.27	1.27
3900		40	25	60		4.77	1.80	6.57
4000					columns	1.20	.32	1.52
4100	30x30	20	25	42		4.51	1.68	6.19
4200					columns	.83	.22	1.05
4300		30	22	52		5	1.84	6.84
4400					columns	1	.27	1.27
4500		40	28	62		5.30	1.92	7.22
4600					columns	1	.27	1.27
5300	35x35	20	28	42		4.91	1.83	6.74
5400					columns	.73	.19	.92
5500		30	25	52		5.60	2.02	7.62
5600					columns	.86	.23	1.09
5700		40	28	62		6.05	2.16	8.21
5800					columns	.95	.25	1.20

JOCWorks™ Software

JOC Knowledgeable Staff
Facility managers gain knowledge of the JOC process, how to interact with JOC contractors, an understanding of the master bidding document (i.e., Unit Price Book) and the development of JOC project specifications and contracts.

RSMeans Unit Price Book (UPB)
A line item analysis of JOC tasks and the components of those tasks. When both the JOC client and contractor agree to this document, negotiations are reduced to project-specific peculiarities.

JOC Training
Designed for working facilities professionals, training increases understanding of how to use JOCWorks™, as well as the Unit Price Book.

JOCWorks™ – a leading project management and document management software program with specialized features for renovation and repair projects.

RSMeans Business Solutions

B10 Superstructure

B1020 Roof Construction

Description: The table below lists the cost per S.F. for a roof system using open web steel joists and 1-1/2" galvanized metal deck. The system is assumed supported on bearing walls or other suitable support. Costs for the supports are not included.

B1020 116		Steel Joists & Deck on Bearing Walls						
	BAY SIZE (FT.)	SUPERIMPOSED LOAD (P.S.F.)	DEPTH (IN.)	TOTAL LOAD (P.S.F.)		COST PER S.F.		
						MAT.	INST.	TOTAL
1100	20	20	13-1/2	40		1.69	.97	2.66
1200		30	15-1/2	50		1.73	1	2.73
1300		40	15-1/2	60		1.85	1.07	2.92
1400	25	20	17-1/2	40		1.85	1.07	2.92
1500		30	17-1/2	50		1.99	1.16	3.15
1600		40	19-1/2	60		2.02	1.17	3.19
1700	30	20	19-1/2	40		2.02	1.17	3.19
1800		30	21-1/2	50		2.06	1.20	3.26
1900		40	23-1/2	60		2.21	1.29	3.50
2000	35	20	23-1/2	40		2.19	1.09	3.28
2100		30	25-1/2	50		2.26	1.12	3.38
2200		40	25-1/2	60		2.40	1.18	3.58
2300	40	20	25-1/2	41		2.43	1.20	3.63
2400		30	25-1/2	51		2.58	1.26	3.84
2500		40	25-1/2	61		2.67	1.30	3.97
2600	45	20	27-1/2	41		2.69	1.76	4.45
2700		30	31-1/2	51		2.84	1.87	4.71
2800		40	31-1/2	61		2.99	1.97	4.96
2900	50	20	29-1/2	42		3	1.97	4.97
3000		30	31-1/2	52		3.26	2.17	5.43
3100		40	31-1/2	62		3.45	2.31	5.76
3200	60	20	37-1/2	42		3.84	2.01	5.85
3300		30	37-1/2	52		4.26	2.22	6.48
3400		40	37-1/2	62		4.26	2.22	6.48
3500	70	20	41-1/2	42		4.25	2.22	6.47
3600		30	41-1/2	52		4.53	2.37	6.90
3700		40	41-1/2	64		5.50	2.86	8.36
3800	80	20	45-1/2	44		6.10	3.27	9.37
3900		30	45-1/2	54		6.10	3.27	9.37
4000		40	45-1/2	64		6.75	3.62	10.37
4400	100	20	57-1/2	44		5.25	2.65	7.90
4500		30	57-1/2	54		6.25	3.14	9.39
4600		40	57-1/2	65		6.90	3.46	10.36
4700	125	20	69-1/2	44		7.35	3.69	11.04
4800		30	69-1/2	56		8.60	4.30	12.90
4900		40	69-1/2	67		9.80	4.90	14.70

B10 Superstructure

B1020 Roof Construction

Description: The table below lists costs for a roof system supported on exterior bearing walls and interior columns. Costs include bracing, joist girders, open web steel joists and 1-1/2″ galvanized metal deck.

Column costs are additive. Fireproofing is not included.

Costs/S.F. are based on a building 4 bays long and 4 bays wide. Costs for bearing walls are not included.

B1020 120		Steel Joists, Joist Girders & Deck on Columns & Walls						
	BAY SIZE (FT.) GIRD X JOISTS	SUPERIMPOSED LOAD (P.S.F.)	DEPTH (IN.)	TOTAL LOAD (P.S.F.)	COLUMN ADD	COST PER S.F.		
						MAT.	INST.	TOTAL
2350	30x35	20	32-1/2	40		2.46	1.39	3.85
2400					columns	.31	.08	.39
2550		40	36-1/2	60		2.76	1.53	4.29
2600					columns	.36	.09	.45
3000	35x35	20	36-1/2	40		2.73	1.70	4.43
3050					columns	.26	.07	.33
3200		40	36-1/2	60		3.06	1.87	4.93
3250					columns	.34	.09	.43
3300	35x40	20	36-1/2	40		2.73	1.56	4.29
3350					columns	.27	.07	.34
3500		40	36-1/2	60		3.07	1.72	4.79
3550					columns	.30	.08	.38
3900	40x40	20	40-1/2	41		3.03	1.98	5.01
3950					columns	.26	.07	.33
4100		40	40-1/2	61		3.37	2.17	5.54
4150					columns	.26	.07	.33
5100	45x50	20	52-1/2	41		3.39	2.28	5.67
5150					columns	.19	.05	.24
5300		40	52-1/2	61		4.09	2.74	6.83
5350					columns	.27	.07	.34
5400	50x45	20	56-1/2	41		3.28	2.20	5.48
5450					columns	.19	.05	.24
5600		40	56-1/2	61		4.19	2.81	7
5650					columns	.25	.07	.32
5700	50x50	20	56-1/2	42		3.67	2.49	6.16
5750					columns	.19	.05	.24
5900		40	59	64		4.98	2.79	7.77
5950					columns	.26	.07	.33
6300	60x50	20	62-1/2	43		3.71	2.52	6.23
6350					columns	.30	.08	.38
6500		40	71	65		5.10	2.87	7.97
6550					columns	.40	.10	.50

B10 Superstructure

B1020 Roof Construction

Description: The table below lists the cost per S.F. for a roof system supported on columns. Costs include joist girders, open web steel joists and 1-1/2″ galvanized metal deck. Column costs are additive.

Fireproofing is not included. Costs/S.F. are based on a building 4 bays long and 4 bays wide.

Costs for wind columns are not included.

B1020 124	Steel Joists & Joist Girders on Columns							
	BAY SIZE (FT.) GIRD X JOISTS	SUPERIMPOSED LOAD (P.S.F.)	DEPTH (IN.)	TOTAL LOAD (P.S.F.)	COLUMN ADD	COST PER S.F.		
						MAT.	INST.	TOTAL
2000	30x30	20	17-1/2	40		2.52	1.61	4.13
2050					columns	1	.26	1.26
2100		30	17-1/2	50		2.72	1.73	4.45
2150					columns	1	.26	1.26
2200		40	21-1/2	60		2.79	1.76	4.55
2250					columns	1	.26	1.26
2300	30x35	20	32-1/2	40		2.64	1.53	4.17
2350					columns	.86	.22	1.08
2500		40	36-1/2	60		2.97	1.71	4.68
2550					columns	1	.26	1.26
3200	35x40	20	36-1/2	40		3.31	2.34	5.65
3250					columns	.75	.20	.95
3400		40	36-1/2	60		3.71	2.60	6.31
3450					columns	.83	.22	1.05
3800	40x40	20	40-1/2	41		3.35	2.34	5.69
3850					columns	.73	.19	.92
4000		40	40-1/2	61		3.74	2.59	6.33
4050					columns	.73	.19	.92
4100	40x45	20	40-1/2	41		3.61	2.65	6.26
4150					columns	.65	.17	.82
4200		30	40-1/2	51		3.92	2.84	6.76
4250					columns	.65	.17	.82
4300		40	40-1/2	61		4.33	3.12	7.45
4350					columns	.73	.19	.92
5000	45x50	20	52-1/2	41		3.78	2.71	6.49
5050					columns	.52	.13	.65
5200		40	52-1/2	61		4.58	3.26	7.84
5250					columns	.67	.17	.84
5300	50x45	20	56-1/2	41		3.70	2.64	6.34
5350					columns	.52	.13	.65
5500		40	56-1/2	61		4.71	3.33	8.04
5550					columns	.67	.17	.84
5600	50x50	20	56-1/2	42		4.11	2.93	7.04
5650					columns	.53	.14	.67
5800		40	59	64		4.39	2.65	7.04
5850					columns	.74	.20	.94

B2010 Exterior Walls

Cast in Place Concrete Wall

Flat Precast Concrete Wall

Table B2010 101 describes a concrete wall system for exterior closure. There are several types of wall finishes priced from plain finish to a finish with 3/4″ rustication strip.

In Table B2010 103, precast concrete wall panels are either solid or insulated. Prices below are based on delivery within 50 miles of a plant. Small jobs can double the prices below. For large, highly repetitive jobs, deduct up to 15%.

B2010 101	Cast In Place Concrete	COST PER S.F.		
		MAT.	INST.	TOTAL
2100	Conc wall reinforced, 8' high, 6" thick, plain finish, 3000 PSI	4.16	14.70	18.86
2700	Aged wood liner, 3000 PSI	5.50	16.70	22.20
3000	Sand blast light 1 side, 3000 PSI	5	16.75	21.75
3700	3/4" bevel rustication strip, 3000 PSI	4.31	15.55	19.86
4000	8" thick, plain finish, 3000 PSI	4.95	15.10	20.05
4100	4000 PSI	5.10	15.10	20.20
4300	Rub concrete 1 side, 3000 PSI	4.95	17.75	22.70
4550	8" thick, aged wood liner, 3000 PSI	6.30	17.10	23.40
4750	Sand blast light 1 side, 3000 PSI	5.80	17.10	22.90
5300	3/4" bevel rustication strip, 3000 PSI	5.10	15.95	21.05
5600	10" thick, plain finish, 3000 PSI	5.70	15.45	21.15
5900	Rub concrete 1 side, 3000 PSI	5.70	18.10	23.80
6200	Aged wood liner, 3000 PSI	7.05	17.45	24.50
6500	Sand blast light 1 side, 3000 PSI	6.55	17.50	24.05
7100	3/4" bevel rustication strip, 3000 PSI	5.85	16.30	22.15
7400	12" thick, plain finish, 3000 PSI	6.60	15.90	22.50
7700	Rub concrete 1 side, 3000 PSI	6.60	18.55	25.15
8000	Aged wood liner, 3000 PSI	7.95	17.90	25.85
8300	Sand blast light 1 side, 3000 PSI	7.45	17.90	25.35
8900	3/4" bevel rustication strip, 3000 PSI	6.75	16.80	23.55

B2010 102		Flat Precast Concrete				COST PER S.F.		
	THICKNESS (IN.)	PANEL SIZE (FT.)	FINISHES	RIGID INSULATION (IN)	TYPE	MAT.	INST.	TOTAL
3000	4	5x18	smooth gray	none	low rise	12.10	7.50	19.60
3050		6x18				10.10	6.25	16.35
3100		8x20				18.35	3.03	21.38
3150		12x20				17.35	2.86	20.21
3200	6	5x18	smooth gray	2	low rise	13.25	8	21.25
3250		6x18				11.30	6.80	18.10
3300		8x20				19.80	3.64	23.44
3350		12x20				17.95	3.32	21.27
3400	8	5x18	smooth gray	2	low rise	25.50	4.58	30.08
3450		6x18				24.50	4.36	28.86
3500		8x20				22	4.01	26.01
3550		12x20				20	3.69	23.69

B20 Exterior Enclosure

B2010 Exterior Walls

B2010 102 — Flat Precast Concrete

	THICKNESS (IN.)	PANEL SIZE (FT.)	FINISHES	RIGID INSULATION (IN)	TYPE	COST PER S.F.		
						MAT.	INST.	TOTAL
3600	4	4x8	white face	none	low rise	43.50	4.33	47.83
3650		8x8				32.50	3.24	35.74
3700		10x10				28.50	2.84	31.34
3750		20x10				26	2.58	28.58
3800	5	4x8	white face	none	low rise	44.50	4.41	48.91
3850		8x8				33.50	3.34	36.84
3900		10x10				29.50	2.95	32.45
3950		20x20				27	2.70	29.70
4000	6	4x8	white face	none	low rise	46	4.58	50.58
4050		8x8				35	3.48	38.48
4100		10x10				31	3.06	34.06
4150		20x10				28.50	2.81	31.31
4200	6	4x8	white face	2	low rise	47.50	5.15	52.65
4250		8x8				36	4.06	40.06
4300		10x10				32	3.64	35.64
4350		20x10				28.50	2.81	31.31
4400	7	4x8	white face	none	low rise	47	4.69	51.69
4450		8x8				36	3.60	39.60
4500		10x10				32.50	3.23	35.73
4550		20x10				29.50	2.95	32.45
4600	7	4x8	white face	2	low rise	48.50	5.25	53.75
4650		8x8				37.50	4.18	41.68
4700		10x10				33.50	3.81	37.31
4750		20x10				31	3.53	34.53
4800	8	4x8	white face	none	low rise	48	4.79	52.79
4850		8x8				37	3.69	40.69
4900		10x10				33.50	3.32	36.82
4950		20x10				31	3.06	34.06
5000	8	4x8	white face	2	low rise	49.50	5.35	54.85
5050		8x8				38.50	4.27	42.77
5100		10x10				34.50	3.90	38.40
5150		20x10				32	3.64	35.64

B2010 103 — Fluted Window or Mullion Precast Concrete

	THICKNESS (IN.)	PANEL SIZE (FT.)	FINISHES	RIGID INSULATION (IN)	TYPE	COST PER S.F.		
						MAT.	INST.	TOTAL
5200	4	4x8	smooth gray	none	high rise	27.50	17.20	44.70
5250		8x8				19.80	12.25	32.05
5300		10x10				35	5.75	40.75
5350		20x10				30.50	5.05	35.55
5400	5	4x8	smooth gray	none	high rise	28	17.45	45.45
5450		8x8				20.50	12.70	33.20
5500		10x10				36.50	6	42.50
5550		20x10				32.50	5.30	37.80
5600	6	4x8	smooth gray	none	high rise	29	17.95	46.95
5650		8x8				21	13.15	34.15
5700		10x10				37.50	6.20	43.70
5750		20x10				33.50	5.50	39
5800	6	4x8	smooth gray	2	high rise	30	18.50	48.50
5850		8x8				22.50	13.70	36.20
5900		10x10				39	6.80	45.80
5950		20x10				34.50	6.10	40.60

B2010 Exterior Walls

B2010 103	Fluted Window or Mullion Precast Concrete

	THICKNESS (IN.)	PANEL SIZE (FT.)	FINISHES	RIGID INSULATION (IN)	TYPE	COST PER S.F.		
						MAT.	INST.	TOTAL
6000	7	4x8	smooth gray	none	high rise	29.50	18.30	47.80
6050		8x8				22	13.50	35.50
6100		10x10				39	6.45	45.45
6150		20x10				34.50	5.65	40.15
6200	7	4x8	smooth gray	2	high rise	31	18.90	49.90
6250		8x8				23	14.10	37.10
6300		10x10				40.50	7.05	47.55
6350		20x10				35.50	6.25	41.75
6400	8	4x8	smooth gray	none	high rise	30	18.55	48.55
6450		8x8				22.50	13.85	36.35
6500		10x10				40.50	6.65	47.15
6550		20x10				36	5.95	41.95
6600	8	4x8	smooth gray	2	high rise	31	19.15	50.15
6650		8x8				23.50	14.45	37.95
6700		10x10				41.50	7.25	48.75
6750		20x10				37.50	6.55	44.05

B2010 104	Ribbed Precast Concrete

	THICKNESS (IN.)	PANEL SIZE (FT.)	FINISHES	RIGID INSULATION(IN.)	TYPE	COST PER S.F.		
						MAT.	INST.	TOTAL
6800	4	4x8	aggregate	none	high rise	32.50	17.15	49.65
6850		8x8				24	12.50	36.50
6900		10x10				41	4.53	45.53
6950		20x10				36.50	4	40.50
7000	5	4x8	aggregate	none	high rise	33.50	17.45	50.95
7050		8x8				24.50	12.85	37.35
7100		10x10				42.50	4.70	47.20
7150		20x10				38	4.20	42.20
7200	6	4x8	aggregate	none	high rise	34	17.80	51.80
7250		8x8				25.50	13.25	38.75
7300		10x10				44	4.85	48.85
7350		20x10				39.50	4.35	43.85
7400	6	4x8	aggregate	2	high rise	35	18.40	53.40
7450		8x8				26.50	13.80	40.30
7500		10x10				45.50	5.45	50.95
7550		20x10				40.50	4.93	45.43
7600	7	4x8	aggregate	none	high rise	34.50	18.20	52.70
7650		8x8				26	13.55	39.55
7700		10x10				45.50	5	50.50
7750		20x10				40.50	4.47	44.97
7800	7	4x8	aggregate	2	high rise	36	18.75	54.75
7850		8x8				27	14.15	41.15
7900		10x10				46.50	5.60	52.10
7950		20x10				42	5.05	47.05
8000	8	4x8	aggregate	none	high rise	35.50	18.50	54
8050		8x8				26.50	14	40.50
8100		10x10				47	5.15	52.15
8150		20x10				42.50	4.67	47.17
8200	8	4x8	aggregate	2	high rise	36.50	19.10	55.60
8250		8x8				28	14.60	42.60
8300		10x10				48.50	5.75	54.25
8350		20x10				43.50	5.25	48.75

B20 Exterior Enclosure

B2010 Exterior Walls

B2010 105						**Precast Concrete Specialties**		

	TYPE	SIZE				COST PER L.F.		
						MAT.	INST.	TOTAL
8400	Coping, precast	6" wide				19.30	12.15	31.45
8450	Stock units	10" wide				19.30	13	32.30
8460		12" wide				22.50	14	36.50
8480		14" wide				18.80	15.15	33.95
8500	Window sills	6" wide				13.50	13	26.50
8550	Precast	10" wide				17.60	15.15	32.75
8600		14" wide				17.60	18.20	35.80
8610								

B2010 Exterior Walls

The advantage of tilt up construction is in the low cost of forms and placing of concrete and reinforcing. Tilt up has been used for several types of buildings, including warehouses, stores, offices, and schools. The panels are cast in forms on the ground, or floor slab. Most jobs use 5-1/2″ thick solid reinforced concrete panels.

Tilt Up Concrete Panel

B2010 106	Tilt-Up Concrete Panel	COST PER S.F.		
		MAT.	INST.	TOTAL
3200	Tilt up conc panels, broom finish, 5-1/2″ thick, 3000 PSI	4.85	5.60	10.45
3250	5000 PSI	5	5.55	10.55
3300	6″ thick, 3000 PSI	5.40	5.80	11.20
3350	5000 PSI	5.55	5.70	11.25
3400	7-1/2″ thick, 3000 PSI	7.15	6	13.15
3450	5000 PSI	7.35	5.90	13.25
3500	8″ thick, 3000 PSI	7.70	6.15	13.85
3550	5000 PSI	7.90	6.05	13.95
3700	Steel trowel finish, 5-1/2″ thick, 3000 PSI	4.88	5.75	10.63
3750	5000 PSI	5	5.60	10.60
3800	6″ thick, 3000 PSI	5.40	5.85	11.25
3850	5000 PSI	5.55	5.75	11.30
3900	7-1/2″ thick, 3000 PSI	7.15	6.10	13.25
3950	5000 PSI	7.35	5.95	13.30
4000	8″ thick, 3000 PSI	7.70	6.20	13.90
4050	5000 PSI	7.90	6.10	14
4200	Exp. aggregate finish, 5-1/2″ thick, 3000 PSI	5.20	5.80	11
4250	5000 PSI	5.35	5.70	11.05
4300	6″ thick, 3000 PSI	5.75	5.95	11.70
4350	5000 PSI	5.90	5.85	11.75
4400	7-1/2″ thick, 3000 PSI	7.45	6.15	13.60
4450	5000 PSI	7.70	6.05	13.75
4500	8″ thick, 3000 PSI	8	6.30	14.30
4550	5000 PSI	8.25	6.20	14.45
4600	Exposed aggregate & vert. rustication 5-1/2″ thick, 3000 PSI	7.80	7.10	14.90
4650	5000 PSI	7.90	7	14.90
4700	6″ thick, 3000 PSI	8.30	7.25	15.55
4750	5000 PSI	8.45	7.15	15.60
4800	7-1/2″ thick, 3000 PSI	10.05	7.45	17.50
4850	5000 PSI	10.25	7.35	17.60
4900	8″ thick, 3000 PSI	10.60	7.60	18.20
4950	5000 PSI	10.85	7.50	18.35
5000	Vertical rib & light sandblast, 5-1/2″ thick, 3000 PSI	7.75	9.35	17.10
5100	6″ thick, 3000 PSI	8.30	9.50	17.80
5200	7-1/2″ thick, 3000 PSI	10	9.70	19.70
5300	8″ thick, 3000 PSI	10.55	9.85	20.40
6000	Broom finish w/2″ polystyrene insulation, 6″ thick, 3000 PSI	4.90	6.95	11.85
6100	Broom finish 2″ fiberplank insulation, 6″ thick, 3000 PSI	5.60	6.90	12.50
6200	Exposed aggregate w/2″ polystyrene insulation, 6″ thick, 3000 PSI	5.15	6.95	12.10
6300	Exposed aggregate 2″ fiberplank insulation, 6″ thick, 3000 PSI	5.85	6.90	12.75

B20 Exterior Enclosure

B2010 Exterior Walls

Exterior concrete block walls are defined in the following terms; structural reinforcement, weight, percent solid, size, strength and insulation. Within each of these categories, two to four variations are shown. No costs are included for brick shelf or relieving angles.

B2010 109		Concrete Block Wall - Regular Weight				COST PER S.F.		
	TYPE	SIZE (IN.)	STRENGTH (P.S.I.)	CORE FILL		MAT.	INST.	TOTAL
1200	Hollow	4x8x16	2,000	none		1.76	5.60	7.36
1250			4,500	none		2.22	5.60	7.82
1400		8x8x16	2,000	perlite		3.27	6.80	10.07
1410				styrofoam		4.76	6.40	11.16
1440				none		2.77	6.40	9.17
1450			4,500	perlite		3.84	6.80	10.64
1460				styrofoam		5.35	6.40	11.75
1490				none		3.34	6.40	9.74
2000	75% solid	4x8x16	2,000	none		2.12	5.65	7.77
2100		6x8x16	2,000	perlite		3.14	6.25	9.39
2500	Solid	4x8x16	2,000	none		2.15	5.80	7.95
2700		8x8x16	2,000	none		3.74	6.65	10.39

B2010 110		Concrete Block Wall - Lightweight				COST PER S.F.		
	TYPE	SIZE (IN.)	WEIGHT (P.C.F.)	CORE FILL		MAT.	INST.	TOTAL
3100	Hollow	8x4x16	105	perlite		3.04	6.40	9.44
3110				styrofoam		4.53	6	10.53
3200		4x8x16	105	none		1.96	5.50	7.46
3250			85	none		2.37	5.35	7.72
3300		6x8x16	105	perlite		3.11	6.15	9.26
3310				styrofoam		4.76	5.85	10.61
3340				none		2.77	5.85	8.62
3400		8x8x16	105	perlite		3.84	6.65	10.49
3410				styrofoam		5.35	6.25	11.60
3440				none		3.34	6.25	9.59
3450			85	perlite		4.22	6.50	10.72
4000	75% solid	4x8x16	105	none		2.84	5.55	8.39
4050			85	none		2.92	5.40	8.32
4100		6x8x16	105	perlite		4.23	6.10	10.33
4500	Solid	4x8x16	105	none		2.74	5.75	8.49
4700		8x8x16	105	none		3.89	6.55	10.44

B2010 Exterior Walls

| B2010 111 | | | | Reinforced Concrete Block Wall - Regular Weight | | | | |

	TYPE	SIZE (IN.)	STRENGTH (P.S.I.)	VERT. REINF & GROUT SPACING		COST PER S.F.		
						MAT.	INST.	TOTAL
5200	Hollow	4x8x16	2,000	#4 @ 48"		1.89	6.20	8.09
5300		6x8x16	2,000	#4 @ 48"		2.68	6.65	9.33
5330				#5 @ 32"		2.87	6.85	9.72
5340				#5 @ 16"		3.27	7.75	11.02
5350			4,500	#4 @ 28"		2.96	6.65	9.61
5390				#5 @ 16"		3.55	7.75	11.30
5400		8x8x16	2,000	#4 @ 48"		3.08	7.05	10.13
5430				#5 @ 32"		3.26	7.45	10.71
5440				#5 @ 16"		3.75	8.50	12.25
5450		8x8x16	4,500	#4 @ 48"		3.62	7.15	10.77
5490				#5 @ 16"		4.32	8.50	12.82
5500		12x8x16	2,000	#4 @ 48"		4.35	9.10	13.45
5540				#5 @ 16"		5.30	10.50	15.80
6100	75% solid	6x8x16	2,000	#4 @ 48"		3.09	6.60	9.69
6140				#5 @ 16"		3.46	7.55	11.01
6150			4,500	#4 @ 48"		3.46	6.60	10.06
6190				#5 @ 16"		3.83	7.55	11.38
6200		8x8x16	2,000	#4 @ 48"		3.47	7.15	10.62
6230				#5 @ 32"		3.61	7.40	11.01
6240				#5 @ 16"		3.87	8.25	12.12
6250			4,500	#4 @ 48"		4.22	7.15	11.37
6280				#5 @ 32"		4.36	7.40	11.76
6290				#5 @ 16"		4.62	8.25	12.87
6500	Solid-double	2-4x8x16	2,000	#4 @ 48" E.W.		5.15	13.10	18.25
6530	Wythe			#5 @ 16" E.W.		5.75	13.90	19.65
6550			4,500	#4 @ 48" E.W.		6.95	13	19.95
6580				#5 @ 16" E.W.		7.55	13.80	21.35

| B2010 112 | | | | Reinforced Concrete Block Wall - Lightweight | | | | |

	TYPE	SIZE (IN.)	WEIGHT (P.C.F.)	VERT REINF. & GROUT SPACING		COST PER S.F.		
						MAT.	INST.	TOTAL
7100	Hollow	8x4x16	105	#4 @ 48"		2.82	6.75	9.57
7140				#5 @ 16"		3.52	8.10	11.62
7150			85	#4 @ 48"		3.97	6.60	10.57
7190				#5 @ 16"		4.67	7.95	12.62
7400		8x8x16	105	#4 @ 48"		3.62	7	10.62
7440				#5 @ 16"		4.32	8.35	12.67
7450		8x8x16	85	#4 @ 48"		4	6.85	10.85
7490				#5 @ 16"		4.70	8.20	12.90
7800		8x8x24	105	#4 @ 48"		2.78	7.55	10.33
7840				#5 @ 16"		3.48	8.90	12.38
7850			85	#4 @ 48"		6.30	6.40	12.70
7890				#5 @ 16"		7	7.75	14.75
8100	75% solid	6x8x16	105	#4 @ 48"		4.18	6.45	10.63
8130				#5 @ 32"		4.31	6.65	10.96
8150			85	#4 @ 48"		4.23	6.30	10.53
8180				#5 @ 32"		4.36	6.50	10.86
8200		8x8x16	105	#4 @ 48"		5	6.95	11.95
8230				#5 @ 32"		5.15	7.20	12.35
8250			85	#4 @ 48"		4.71	6.80	11.51
8280				#5 @ 32"		4.85	7.05	11.90

B2010 Exterior Walls

B2010 112	Reinforced Concrete Block Wall - Lightweight

	TYPE	SIZE (IN.)	WEIGHT (P.C.F.)	VERT REINF. & GROUT SPACING		COST PER S.F.		
						MAT.	INST.	TOTAL
8500	Solid-double	2-4x8x16	105	#4 @ 48"		6.35	13	19.35
8530	Wythe		105	#5 @ 16"		6.90	13.80	20.70
8600		2-6x8x16	105	#4 @ 48"		6.95	13.85	20.80
8630				#5 @ 16"		7.55	14.65	22.20
8650			85	#4 @ 48"		10.45	13.25	23.70

B2010 Exterior Walls

Exterior split ribbed block walls are defined in the following terms; structural reinforcement, weight, percent solid, size, number of ribs and insulation. Within each of these categories two to four variations are shown. No costs are included for brick shelf or relieving angles. Costs include control joints every 20′ and horizontal reinforcing.

B2010 113	Split Ribbed Block Wall - Regular Weight							
	TYPE	SIZE (IN.)	RIBS	CORE FILL		COST PER S.F.		
						MAT.	INST.	TOTAL
1220	Hollow	4x8x16	4	none		3.50	6.95	10.45
1250			8	none		3.80	6.95	10.75
1280			16	none		4.10	7.05	11.15
1430		8x8x16	8	perlite		5.50	8.20	13.70
1440				styrofoam		7	7.80	14.80
1450				none		4.99	7.80	12.79
1530		12x8x16	8	perlite		6.75	10.90	17.65
1540				styrofoam		8.40	10.15	18.55
1550				none		5.95	10.15	16.10
2120	75% solid	4x8x16	4	none		4.39	7.05	11.44
2150			8	none		4.78	7.05	11.83
2180			16	none		5.15	7.15	12.30
2520	Solid	4x8x16	4	none		4.99	7.15	12.14
2550			8	none		5.90	7.15	13.05
2580			16	none		5.90	7.25	13.15

B2010 115	Reinforced Split Ribbed Block Wall - Regular Weight							
	TYPE	SIZE (IN.)	RIBS	VERT. REINF. & GROUT SPACING		COST PER S.F.		
						MAT.	INST.	TOTAL
5200	Hollow	4x8x16	4	#4 @ 48″		3.63	7.55	11.18
5230			8	#4 @ 48″		3.93	7.55	11.48
5260			16	#4 @ 48″		4.23	7.65	11.88
5430		8x8x16	8	#4 @ 48″		5.25	8.55	13.80
5440				#5 @ 32″		5.50	8.85	14.35
5450				#5 @ 16″		5.95	9.90	15.85
5530		12x8x16	8	#4 @ 48″		6.35	10.95	17.30
5540				#5 @ 32″		6.60	11.25	17.85
5550				#5 @ 16″		7.30	12.35	19.65
6230	75% solid	6x8x16	8	#4 @ 48″		5.65	7.95	13.60
6240				#5 @ 32″		5.75	8.15	13.90
6250				#5 @ 16″		6	8.90	14.90
6330		8x8x16	8	#4 @ 48″		6.35	8.60	14.95
6340				#5 @ 32″		6.50	8.85	15.35
6350				#5 @ 16″		6.75	9.70	16.45

B2010 Exterior Walls

Exterior split face block walls are defined in the following terms; structural reinforcement, weight, percent solid, size, scores and insulation. Within each of these categories two to four variations are shown. No costs are included for brick shelf or relieving angles. Costs include control joints every 20' and horizontal reinforcing.

B2010 117				Split Face Block Wall - Regular Weight				
	TYPE	SIZE (IN.)	SCORES	CORE FILL		COST PER S.F.		
						MAT.	INST.	TOTAL
1200	Hollow	8x4x16	0	perlite		5.35	9.05	14.40
1210				styrofoam		6.80	8.65	15.45
1240				none		4.83	8.65	13.48
1250			1	perlite		5.65	9.05	14.70
1260				styrofoam		7.15	8.65	15.80
1290				none		5.15	8.65	13.80
1300		12x4x16	0	perlite		6.85	10.30	17.15
1310				styrofoam		8.50	9.55	18.05
1340				none		6.05	9.55	15.60
1350			1	perlite		7.25	10.30	17.55
1360				styrofoam		8.90	9.55	18.45
1390				none		6.45	9.55	16
1400		4x8x16	0	none		3.16	6.85	10.01
1450			1	none		3.42	6.95	10.37
1500		6x8x16	0	perlite		3.99	7.85	11.84
1510				styrofoam		5.65	7.55	13.20
1540				none		3.65	7.55	11.20
1550			1	perlite		4.29	8	12.29
1560				styrofoam		5.95	7.70	13.65
1590				none		3.95	7.70	11.65
1600		8x8x16	0	perlite		4.66	8.45	13.11
1610				styrofoam		6.15	8.05	14.20
1640				none		4.16	8.05	12.21
1650		8x8x16	1	perlite		4.99	8.60	13.59
1660				styrofoam		6.50	8.20	14.70
1690				none		4.49	8.20	12.69
1700		12x8x16	0	perlite		5.90	11.10	17
1710				styrofoam		7.55	10.35	17.90
1740				none		5.10	10.35	15.45
1750			1	perlite		6.30	11.30	17.60
1760				styrofoam		7.95	10.55	18.50
1790				none		5.50	10.55	16.05

B2010 Exterior Walls

| B2010 117 | Split Face Block Wall - Regular Weight | | | | | | | |

	TYPE	SIZE (IN.)	SCORES	CORE FILL		COST PER S.F.		
						MAT.	INST.	TOTAL
1800	75% solid	8x4x16	0	perlite		6.05	9	15.05
1840				none		5.80	8.80	14.60
1852		8x4x16	1	perlite		6.50	9	15.50
1890				none		6.25	8.80	15.05
2000		4x8x16	0	none		3.94	6.95	10.89
2050			1	none		4.28	7.05	11.33
2400	Solid	8x4x16	0	none		6.45	8.95	15.40
2450			1	none		6.95	8.95	15.90
2900		12x8x16	0	none		7.10	10.75	17.85
2950			1	none		7.65	10.95	18.60

| B2010 119 | Reinforced Split Face Block Wall - Regular Weight | | | | | | | |

	TYPE	SIZE (IN.)	SCORES	VERT. REINF. & GROUT SPACING		COST PER S.F.		
						MAT.	INST.	TOTAL
5200	Hollow	8x4x16	0	#4 @ 48"		5.10	9.40	14.50
5210				#5 @ 32"		5.30	9.70	15
5240				#5 @ 16"		5.80	10.75	16.55
5250			1	#4 @ 48"		5.45	9.40	14.85
5260				#5 @ 32"		5.65	9.70	15.35
5290				#5 @ 16"		6.15	10.75	16.90
5700		12x8x16	0	#4 @ 48"		5.50	11.15	16.65
5710				#5 @ 32"		5.75	11.45	17.20
5740				#5 @ 16"		6.45	12.55	19
5750			1	#4 @ 48"		5.90	11.35	17.25
5760				#5 @ 32"		6.15	11.65	17.80
5790				#5 @ 16"		6.85	12.75	19.60
6000	75% solid	8x4x16	0	#4 @ 48"		5.95	9.45	15.40
6010				#5 @ 32"		6.10	9.70	15.80
6040				#5 @ 16"		6.35	10.55	16.90
6050			1	#4 @ 48"		6.40	9.45	15.85
6060				#5 @ 32"		6.55	9.70	16.25
6090				#5 @ 16"		6.80	10.55	17.35
6100		12x4x16	0	#4 @ 48"		7.45	10.40	17.85
6110				#5 @ 32"		7.65	10.65	18.30
6140				#5 @ 16"		8	11.30	19.30
6150			1	#4 @ 48"		7.95	10.40	18.35
6160				#5 @ 32"		8.15	10.65	18.80
6190				#5 @ 16"		8.55	11.50	20.05
6700	Solid-double Wythe	2-4x8x16	0	#4 @ 48" E.W.		9.80	15.60	25.40
6710				#5 @ 32" E.W.		10	15.75	25.75
6740				#5 @ 16" E.W.		10.40	16.40	26.80
6750			1	#4 @ 48" E.W.		10.55	15.80	26.35
6760				#5 @ 32" E.W.		10.80	15.95	26.75
6790				#5 @ 16" E.W.		11.15	16.60	27.75
6800		2-6x8x16	0	#4 @ 48" E.W.		11.15	17.15	28.30
6810				#5 @ 32" E.W.		11.35	17.25	28.60
6840				#5 @ 16" E.W.		11.70	17.95	29.65
6850			1	#4 @ 48" E.W.		11.95	17.45	29.40
6860				#5 @ 32" E.W.		12.20	17.55	29.75
6890				#5 @ 16" E.W.		12.55	18.25	30.80

B2010 Exterior Walls

Exterior solid brick walls are defined in the following terms; structural reinforcement, type of brick, thickness, and bond. Nine different types of face bricks are presented, with single wythes shown in four different bonds. Shelf angles are included in system components. These walls do not include ties and as such are not tied to a backup wall.

B2010 125			Solid Brick Walls - Single Wythe					
	TYPE	THICKNESS (IN.)	BOND			MAT.	INST.	TOTAL
1000	Common	4	running			4.67	12.60	17.27
1010			common			5.65	14.40	20.05
1050			Flemish			6.25	17.50	23.75
1100			English			7.05	18.50	25.55
1150	Standard	4	running			6.15	13.05	19.20
1160			common			7.30	15.05	22.35
1200			Flemish			8.15	18	26.15
1250			English			9.15	19.10	28.25
1300	Glazed	4	running			19.10	13.55	32.65
1310			common			23	15.75	38.75
1350			Flemish			25.50	19.10	44.60
1400			English			28.50	20.50	49
1600	Economy	4	running			5	10	15
1610			common			5.90	11.45	17.35
1650			Flemish			6.50	13.55	20.05
1700			English			16.25	14.40	30.65
1900	Fire	4-1/2	running			12.20	11.45	23.65
1910			common			14.55	13.05	27.60
1950			Flemish			16.10	15.75	31.85
2000			English			18.15	16.60	34.75
2050	King	3-1/2	running			4.02	10.25	14.27
2060			common			4.70	11.80	16.50
2100			Flemish			5.70	14.10	19.80
2150			English			6.30	14.75	21.05
2200	Roman	4	running			7.15	11.80	18.95
2210			common			8.55	13.55	22.10
2250			Flemish			9.45	16.15	25.60
2300			English			10.70	17.50	28.20
2350	Norman	4	running			6.20	9.75	15.95
2360			common			7.35	11.10	18.45
2400			Flemish			16.85	13.30	30.15
2450			English			9.15	14.10	23.25

B2010 Exterior Walls

B2010 125 — Solid Brick Walls - Single Wythe

	TYPE	THICKNESS (IN.)	BOND			COST PER S.F.		
						MAT.	INST.	TOTAL
2500	Norwegian	4	running			5.05	8.70	13.75
2510			common			5.95	9.85	15.80
2550			Flemish			6.60	11.80	18.40
2600			English			7.40	12.40	19.80

B2010 126 — Solid Brick Walls - Double Wythe

	TYPE	THICKNESS (IN.)	COLLAR JOINT THICKNESS (IN.)			COST PER S.F.		
						MAT.	INST.	TOTAL
4100	Common	8	3/4			9	21	30
4150	Standard	8	1/2			11.85	22	33.85
4200	Glazed	8	1/2			37.50	22.50	60
4250	Engineer	8	1/2			9.85	19.30	29.15
4300	Economy	8	3/4			9.55	16.55	26.10
4350	Double	8	3/4			12.10	13.40	25.50
4400	Fire	8	3/4			23.50	19.30	42.80
4450	King	7	3/4			7.55	16.95	24.50
4500	Roman	8	1			13.80	19.90	33.70
4550	Norman	8	3/4			11.85	16.20	28.05
4600	Norwegian	8	3/4			9.60	14.35	23.95
4650	Utility	8	3/4			8.90	12.40	21.30
4700	Triple	8	3/4			6.90	11.15	18.05
4750	SCR	12	3/4			13.30	16.55	29.85
4800	Norwegian	12	3/4			11.35	14.65	26
4850	Jumbo	12	3/4			12.45	12.80	25.25

B2010 127 — Reinforced Brick Walls

	TYPE	THICKNESS (IN.)	WYTHE	REINF. & SPACING		COST PER S.F.		
						MAT.	INST.	TOTAL
7200	Common	4"	1	#4 @ 48"vert		4.84	13.10	17.94
7220				#5 @ 32"vert		4.96	13.25	18.21
7230				#5 @ 16"vert		5.20	13.80	19
7700	King	9"	2	#4 @ 48" E.W.		7.95	17.75	25.70
7720				#5 @ 32" E.W.		8.15	17.90	26.05
7730				#5 @ 16" E.W.		8.55	18.40	26.95
7800	Common	10"	2	#4 @ 48" E.W.		9.40	22	31.40
7820				#5 @ 32" E.W.		9.60	22	31.60
7830				#5 @ 16" E.W.		10	22.50	32.50
7900	Standard	10"	2	#4 @ 48" E.W.		12.25	22.50	34.75
7920				#5 @ 32" E.W.		12.45	22.50	34.95
7930				#5 @ 16" E.W.		12.85	23.50	36.35
8100	Economy	10"	2	#4 @ 48" E.W.		9.95	17.35	27.30
8120				#5 @ 32" E.W.		10.15	17.50	27.65
8130				#5 @ 16" E.W.		10.55	18	28.55
8200	Double	10"	2	#4 @ 48" E.W.		12.50	14.20	26.70
8220				#5 @ 32" E.W.		12.70	14.35	27.05
8230				#5 @ 16" E.W.		13.10	14.85	27.95
8300	Fire	10"	2	#4 @ 48" E.W.		24	20	44
8320				#5 @ 32" E.W.		24.50	20	44.50
8330				#5 @ 16" E.W.		24.50	20.50	45
8400	Roman	10"	2	#4 @ 48" E.W.		14.20	20.50	34.70
8420				#5 @ 32" E.W.		14.40	20.50	34.90
8430				#5 @ 16" E.W.		14.80	21.50	36.30

B20 Exterior Enclosure

B2010 Exterior Walls

B2010 127			Reinforced Brick Walls				

	TYPE	THICKNESS (IN.)	WYTHE	REINF. & SPACING		COST PER S.F.		
						MAT.	INST.	TOTAL
8500	Norman	10″	2	#4 @ 48″ E.W.		12.25	17	29.25
8520				#5 @ 32″ E.W.		12.45	17.15	29.60
8530				#5 @ 16″ E.W.		12.85	17.65	30.50
8600	Norwegian	10″	2	#4 @ 48″ E.W.		10	15.15	25.15
8620				#5 @ 32″ E.W.		10.20	15.30	25.50
8630				#5 @ 16″ E.W.		10.60	15.80	26.40
8800	Triple	10″	2	#4 @ 48″ E.W.		7.30	11.95	19.25
8820				#5 @ 32″ E.W.		7.50	12.10	19.60
8830				#5 @ 16″ E.W.		7.90	12.60	20.50

B2010 Exterior Walls

Stone Veneer

The table below lists costs per S.F. for stone veneer walls on various backup using different stone.

B2010 128	Stone Veneer	COST PER S.F.		
		MAT.	INST.	TOTAL
2000	Ashlar veneer, 4", 2"x4" stud backup, 16" O.C., 8' high, low priced stone	12.05	19.85	31.90
2100	Metal stud backup, 8' high, 16" O.C.	12.95	20	32.95
2150	24" O.C.	12.60	19.65	32.25
2200	Conc. block backup, 4" thick	13.45	23.50	36.95
2300	6" thick	14.15	24	38.15
2350	8" thick	14.45	24.50	38.95
2400	10" thick	15.15	25.50	40.65
2500	12" thick	15.60	27.50	43.10
3100	High priced stone, wood stud backup, 10' high, 16" O.C.	17.60	22.50	40.10
3200	Metal stud backup, 10' high, 16" O.C.	18.45	23	41.45
3250	24" O.C.	18.10	22.50	40.60
3300	Conc. block backup, 10' high, 4" thick	18.95	26.50	45.45
3350	6" thick	19.65	26.50	46.15
3400	8" thick	19.95	27	46.95
3450	10" thick	20.50	28.50	49
3500	12" thick	21	30	51
4000	Indiana limestone 2" thick, sawn finish, wood stud backup, 10' high, 16" O.C.	40.50	13	53.50
4100	Metal stud backup, 10' high, 16" O.C.	41	13.25	54.25
4150	24" O.C.	41	12.80	53.80
4200	Conc. block backup, 4" thick	41.50	16.75	58.25
4250	6" thick	42.50	17	59.50
4300	8" thick	42.50	17.50	60
4350	10" thick	43.50	18.75	62.25
4400	12" thick	44	20.50	64.50
4450	2" thick, smooth finish, wood stud backup, 8' high, 16" O.C.	40.50	13	53.50
4550	Metal stud backup, 8' high, 16" O.C.	41	13.10	54.10
4600	24" O.C.	41	12.65	53.65
4650	Conc. block backup, 4" thick	41.50	16.55	58.05
4700	6" thick	42.50	16.80	59.30
4750	8" thick	42.50	17.30	59.80
4800	10" thick	43.50	18.75	62.25
4850	12" thick	44	20.50	64.50
5350	4" thick, smooth finish, wood stud backup, 8' high, 16" O.C.	47	13	60
5450	Metal stud backup, 8' high, 16" O.C.	47.50	13.25	60.75
5500	24" O.C.	47.50	12.80	60.30
5550	Conc. block backup, 4" thick	48	16.75	64.75
5600	6" thick	49	17	66
5650	8" thick	49	17.50	66.50

B2010 Exterior Walls

B2010 128	Stone Veneer	COST PER S.F.		
		MAT.	INST.	TOTAL
5700	10" thick	50	18.75	68.75
5750	12" thick	50.50	20.50	71
6000	Granite, gray or pink, 2" thick, wood stud backup, 8' high, 16" O.C.	24	24.50	48.50
6100	Metal studs, 8' high, 16" O.C.	24.50	25	49.50
6150	24" O.C.	24.50	24.50	49
6200	Conc. block backup, 4" thick	25	28.50	53.50
6250	6" thick	26	29	55
6300	8" thick	26	29.50	55.50
6350	10" thick	27	30.50	57.50
6400	12" thick	27.50	32.50	60
6900	4" thick, wood stud backup, 8' high, 16" O.C.	38.50	27.50	66
7000	Metal studs, 8' high, 16" O.C.	39	28	67
7050	24" O.C.	39	27.50	66.50
7100	Conc. block backup, 4" thick	39.50	31.50	71
7150	6" thick	40.50	32	72.50
7200	8" thick	40.50	32.50	73
7250	10" thick	41.50	33.50	75
7300	12" thick	42	35.50	77.50

B2010 Exterior Walls

Exterior brick veneer/stud backup walls are defined in the following terms; type of brick and studs, stud spacing and bond. All systems include a brick shelf, ties to the backup and all necessary dampproofing and insulation.

B2010 129				Brick Veneer/Wood Stud Backup				
	FACE BRICK	STUD BACKUP	STUD SPACING (IN.)	BOND		COST PER S.F.		
						MAT.	INST.	TOTAL
1100	Standard	2x4-wood	16	running		7.60	15.45	23.05
1120				common		8.75	17.45	26.20
1140				Flemish		9.60	20.50	30.10
1160				English		10.60	21.50	32.10
1400		2x6-wood	16	running		7.75	15.55	23.30
1420				common		8.90	17.55	26.45
1440				Flemish		9.75	20.50	30.25
1460				English		10.75	21.50	32.25
1700	Glazed	2x4-wood	16	running		20.50	15.95	36.45
1720				common		24.50	18.15	42.65
1740				Flemish		27	21.50	48.50
1760				English		30	23	53
2300	Engineer	2x4-wood	16	running		6.60	13.85	20.45
2320				common		7.55	15.45	23
2340				Flemish		8.20	18.15	26.35
2360				English		9.05	19	28.05
2900	Roman	2x4-wood	16	running		8.60	14.20	22.80
2920				common		10	15.95	25.95
2940				Flemish		10.90	18.55	29.45
2960				English		12.15	19.90	32.05
4100	Norwegian	2x4-wood	16	running		6.50	11.10	17.60
4120				common		7.40	12.25	19.65
4140				Flemish		8.05	14.20	22.25
4160				English		8.85	14.80	23.65

B2010 130				Brick Veneer/Metal Stud Backup				
	FACE BRICK	STUD BACKUP	STUD SPACING (IN.)	BOND		COST PER S.F.		
						MAT.	INST.	TOTAL
5100	Standard	25ga.x6"NLB	24	running		7.50	15.10	22.60
5120				common		8.65	17.10	25.75
5140				Flemish		9.10	19.40	28.50
5160				English		10.50	21	31.50
5200		20ga.x3-5/8"NLB	16	running		7.50	15.75	23.25
5220				common		8.65	17.75	26.40
5240				Flemish		9.50	20.50	30
5260				English		10.50	22	32.50

B2010 Exterior Walls

B2010 130			Brick Veneer/Metal Stud Backup					

	FACE BRICK	STUD BACKUP	STUD SPACING (IN.)	BOND		COST PER S.F.		
						MAT.	INST.	TOTAL
5400	Standard	16ga.x3-5/8"LB	16	running		8.25	15.95	24.20
5420				common		9.40	17.95	27.35
5440				Flemish		10.25	21	31.25
5460				English		11.25	22	33.25
5700	Glazed	25ga.x6"NLB	24	running		20.50	15.60	36.10
5720				common		24.50	17.80	42.30
5740				Flemish		27	21	48
5760				English		30	22.50	52.50
5800		20ga.x3-5/8"NLB	24	running		20.50	15.75	36.25
5820				common		24.50	17.95	42.45
5840				Flemish		27	21.50	48.50
5860				English		30	22.50	52.50
6000		16ga.x3-5/8"LB	16	running		21	16.45	37.45
6020				common		25	18.65	43.65
6040				Flemish		27.50	22	49.50
6060				English		30.50	23.50	54
6300	Engineer	25ga.x6"NLB	24	running		6.50	13.50	20
6320				common		7.45	15.10	22.55
6340				Flemish		8.10	17.80	25.90
6360				English		8.95	18.65	27.60
6400		20ga.x3-5/8"NLB	16	running		6.55	14.15	20.70
6420				common		7.45	15.75	23.20
6440				Flemish		8.10	18.45	26.55
6460				English		8.95	19.30	28.25
6900	Roman	25ga.x6"NLB	24	running		8.50	13.85	22.35
6920				common		9.90	15.55	25.45
6940				Flemish		10.80	18.20	29
6960				English		12.05	19.55	31.60
7000		20ga.x3-5/8"NLB	16	running		8.50	14.50	23
7020				common		9.90	16.25	26.15
7040				Flemish		10.80	18.85	29.65
7060				English		12.05	20	32.05
7500	Norman	25ga.x6"NLB	24	running		7.55	11.80	19.35
7520				common		8.70	13.15	21.85
7540				Flemish		18.20	15.35	33.55
7560				English		10.50	16.15	26.65
7600		20ga.x3-5/8"NLB	24	running		7.45	11.95	19.40
7620				common		8.60	13.30	21.90
7640				Flemish		18.10	15.50	33.60
7660				English		10.40	16.30	26.70
8100	Norwegian	25ga.x6"NLB	24	running		6.40	10.75	17.15
8120				common		7.30	11.90	19.20
8140				Flemish		7.95	13.85	21.80
8160				English		8.75	14.45	23.20
8400		16ga.x3-5/8"LB	16	running		7.20	11.60	18.80
8420				common		8.10	12.75	20.85
8440				Flemish		8.70	14.70	23.40
8460				English		9.50	15.30	24.80

Brick Face Composite Wall

B2010 132	Brick Face Composite Wall - Double Wythe						

	FACE BRICK	BACKUP MASONRY	BACKUP THICKNESS (IN.)	BACKUP CORE FILL		COST PER S.F.		
						MAT.	INST.	TOTAL
1000	Standard	common brick	4	none		10.25	24	34.25
1040		SCR brick	6	none		12.50	21.50	34
1080		conc. block	4	none		8.30	19.20	27.50
1120			6	perlite		9.30	19.90	29.20
1160				styrofoam		10.95	19.60	30.55
1200			8	perlite		9.80	20.50	30.30
1240				styrofoam		11.25	20	31.25
1520		glazed block	4	none		16.05	20.50	36.55
1560			6	perlite		17.05	21	38.05
1600				styrofoam		18.70	21	39.70
1640			8	perlite		17.35	21.50	38.85
1680				styrofoam		18.80	21.50	40.30
1720		clay tile	4	none		14.35	18.45	32.80
1760			6	none		16.25	18.95	35.20
1800			8	none		18.90	19.60	38.50
1840		glazed tile	4	none		18.55	24.50	43.05
1880								
2000	Glazed	common brick	4	none		23	24.50	47.50
2040		SCR brick	6	none		25.50	22	47.50
2080		conc. block	4	none		21	19.70	40.70
2200			8	perlite		22.50	21	43.50
2240				styrofoam		24	20.50	44.50
2280		L.W. block	4	none		21.50	19.60	41.10
2400			8	perlite		23.50	20.50	44
2520		glazed block	4	none		29	21	50
2560			6	perlite		30	21.50	51.50
2600				styrofoam		31.50	21.50	53
2640			8	perlite		30.50	22	52.50
2680				styrofoam		32	22	54

B20 Exterior Enclosure

B2010 Exterior Walls

| B2010 132 | Brick Face Composite Wall - Double Wythe |

	FACE BRICK	BACKUP MASONRY	BACKUP THICKNESS (IN.)	BACKUP CORE FILL		COST PER S.F.		
						MAT.	INST.	TOTAL
2720	Glazed	clay tile	4	none		27.50	18.95	46.45
2760			6	none		29	19.45	48.45
2800			8	none		32	20	52
2840		glazed tile	4	none		31.50	25	56.50
3000	Engineer	common brick	4	none		9.30	22.50	31.80
3040		SCR brick	6	none		11.50	19.70	31.20
3080		conc. block	4	none		7.30	17.60	24.90
3200			8	perlite		8.80	18.80	27.60
3280		L.W. block	4	none		5.80	12.50	18.30
3322		L.W. block	6	perlite		8.65	18.15	26.80
3520		glazed block	4	none		15.05	18.95	34
3560			6	perlite		16.10	19.55	35.65
3600				styrofoam		17.75	19.25	37
3640			8	perlite		16.35	20	36.35
3680				styrofoam		17.85	19.70	37.55
3720		clay tile	4	none		13.35	16.85	30.20
3800			8	none		17.90	18	35.90
3840		glazed tile	4	none		17.55	22.50	40.05
4000	Roman	common brick	4	none		11.25	22.50	33.75
4040		SCR brick	6	none		13.50	20	33.50
4080		conc. block	4	none		9.30	17.95	27.25
4120			6	perlite		10.30	18.65	28.95
4200			8	perlite		10.80	19.15	29.95
5000	Norman	common brick	4	none		10.30	20.50	30.80
5040		SCR brick	6	none		12.55	18	30.55
5280		L.W. block	4	none		8.55	15.80	24.35
5320			6	perlite		9.70	16.55	26.25
5720		clay tile	4	none		14.40	15.15	29.55
5760			6	none		16.30	15.65	31.95
5840		glazed tile	4	none		18.60	21	39.60
6000	Norwegian	common brick	4	none		9.15	19.55	28.70
6040		SCR brick	6	none		11.40	16.95	28.35
6080		conc. block	4	none		7.20	14.85	22.05
6120			6	perlite		8.20	15.55	23.75
6160				styrofoam		9.85	15.25	25.10
6200			8	perlite		8.70	16.05	24.75
6520		glazed block	4	none		14.95	16.20	31.15
6560			6	perlite		16	16.80	32.80
7000	Utility	common brick	4	none		8.80	18.50	27.30
7040		SCR brick	6	none		11.05	15.90	26.95
7080		conc. block	4	none		6.85	13.80	20.65
7120			6	perlite		7.85	14.50	22.35
7160				styrofoam		9.50	14.20	23.70
7200			8	perlite		8.35	15	23.35
7240				styrofoam		9.80	14.60	24.40
7280		L.W. block	4	none		7.05	13.70	20.75
7320			6	perlite		8.20	14.35	22.55
7520		glazed block	4	none		14.60	15.15	29.75
7560			6	perlite		15.60	15.75	31.35
7720		clay tile	4	none		12.90	13.05	25.95
7760			6	none		14.80	13.55	28.35
7840		glazed tile	4	none		17.10	18.95	36.05

B20 Exterior Enclosure

B2010 Exterior Walls

B2010 133	Brick Face Composite Wall - Triple Wythe

	FACE BRICK	MIDDLE WYTHE	INSIDE MASONRY	TOTAL THICKNESS (IN.)		COST PER S.F.		
						MAT.	INST.	TOTAL
8000	Standard	common brick	standard brick	12		15.80	34	49.80
8100		4" conc. brick	standard brick	12		15.20	35	50.20
8120		4" conc. brick	common brick	12		13.70	34.50	48.20
8200	Glazed	common brick	standard brick	12		29	34.50	63.50
8300		4" conc. brick	standard brick	12		28	35.50	63.50
8320		4" conc. brick	glazed brick	12		26.50	35	61.50

B2010 Exterior Walls

Brick Face Cavity Wall

B2010 134				Brick Face Cavity Wall				
	FACE BRICK	BACKUP MASONRY	TOTAL THICKNESS (IN.)	CAVITY INSULATION		COST PER S.F.		
						MAT.	INST.	TOTAL
1000	Standard	4″ common brick	10	polystyrene		10.30	24	34.30
1020				none		9.90	23.50	33.40
1040		6″ SCR brick	12	polystyrene		12.55	21.50	34.05
1060				none		12.15	20.50	32.65
1080		4″ conc. block	10	polystyrene		8.35	19.25	27.60
1100				none		7.95	18.65	26.60
1120		6″ conc. block	12	polystyrene		9.05	19.65	28.70
1140				none		8.65	19.05	27.70
1160		4″ L.W. block	10	polystyrene		8.55	19.15	27.70
1180				none		8.15	18.55	26.70
1200		6″ L.W. block	12	polystyrene		9.35	19.50	28.85
1220				none		8.95	18.90	27.85
1240		4″ glazed block	10	polystyrene		16.10	20.50	36.60
1260				none		15.70	20	35.70
1280		6″ glazed block	12	polystyrene		16.10	20.50	36.60
1300				none		15.70	20	35.70
1320		4″ clay tile	10	polystyrene		14.40	18.50	32.90
1340				none		14	17.90	31.90
1360		4″ glazed tile	10	polystyrene		18.60	24.50	43.10
1380				none		18.20	24	42.20
1500	Glazed	4″ common brick	10	polystyrene		23.50	24.50	48
1520				none		23	24	47
1580		4″ conc. block	10	polystyrene		21.50	19.75	41.25
1600				none		21	19.15	40.15
1660		4″ L.W. block	10	polystyrene		21.50	19.65	41.15
1680				none		21	19.05	40.05
1740		4″ glazed block	10	polystyrene		29	21	50
1760				none		28.50	20.50	49
1820		4″ clay tile	10	polystyrene		27.50	19	46.50
1840				none		27	18.40	45.40
1860		4″ glazed tile	10	polystyrene		29	19.50	48.50
1880				none		29	18.90	47.90
2000	Engineer	4″ common brick	10	polystyrene		9.35	22.50	31.85
2020				none		8.95	21.50	30.45
2080		4″ conc. block	10	polystyrene		7.40	17.65	25.05
2100				none		7	17.05	24.05

B20 Exterior Enclosure

B2010 Exterior Walls

B2010 134						COST PER S.F.		

Brick Face Cavity Wall

	FACE BRICK	BACKUP MASONRY	TOTAL THICKNESS (IN.)	CAVITY INSULATION		MAT.	INST.	TOTAL
2162	Engineer	4" L.W. block	10	polystyrene		7.60	17.55	25.15
2180				none		7.20	16.95	24.15
2240		4" glazed block	10	polystyrene		15.10	19	34.10
2260				none		14.70	18.40	33.10
2320		4" clay tile	10	polystyrene		13.40	16.90	30.30
2340				none		13	16.30	29.30
2360		4" glazed tile	10	polystyrene		17.60	23	40.60
2380				none		17.20	22	39.20
2500	Roman	4" common brick	10	polystyrene		11.30	22.50	33.80
2520				none		10.90	22	32.90
2580		4" conc. block	10	polystyrene		9.35	18	27.35
2600				none		8.95	17.40	26.35
2660		4" L.W. block	10	polystyrene		9.55	17.90	27.45
2680				none		9.15	17.30	26.45
2740		4" glazed block	10	polystyrene		17.10	19.35	36.45
2760				none		16.70	18.75	35.45
2820		4" clay tile	10	polystyrene		15.40	17.25	32.65
2840				none		15	16.65	31.65
2860		4" glazed tile	10	polystyrene		19.60	23	42.60
2880				none		19.20	22.50	41.70
3000	Norman	4" common brick	10	polystyrene		10.35	20.50	30.85
3020				none		9.95	20	29.95
3080		4" conc. block	10	polystyrene		8.40	15.95	24.35
3100				none		8	15.35	23.35
3160		4" L.W. block	10	polystyrene		8.60	15.85	24.45
3180				none		8.20	15.25	23.45
3240		4" glazed block	10	polystyrene		16.15	17.30	33.45
3260				none		15.75	16.70	32.45
3320		4" clay tile	10	polystyrene		14.45	15.20	29.65
3340				none		14.05	14.60	28.65
3360		4" glazed tile	10	polystyrene		18.65	21	39.65
3380				none		18.25	20.50	38.75
3500	Norwegian	4" common brick	10	polystyrene		9.20	19.60	28.80
3520				none		8.80	19	27.80
3580		4" conc. block	10	polystyrene		7.25	14.90	22.15
3600				none		6.85	14.30	21.15
3660		4" L.W. block	10	polystyrene		7.45	14.80	22.25
3680				none		7.05	14.20	21.25
3740		4" glazed block	10	polystyrene		15	16.25	31.25
3760				none		14.60	15.65	30.25
3820		4" clay tile	10	polystyrene		13.30	14.15	27.45
3840				none		12.90	13.55	26.45
3860		4" glazed tile	10	polystyrene		17.50	20	37.50
3880				none		17.10	19.45	36.55
4000	Utility	4" common brick	10	polystyrene		8.85	18.55	27.40
4020				none		8.45	17.95	26.40
4080		4" conc. block	10	polystyrene		6.90	13.85	20.75
4100				none		6.50	13.25	19.75
4160		4" L.W. block	10	polystyrene		7.10	13.75	20.85
4180				none		6.70	13.15	19.85
4240		4" glazed block	10	polystyrene		14.65	15.20	29.85
4260				none		14.25	14.60	28.85

B2010 Exterior Walls

B2010 134 — Brick Face Cavity Wall

	FACE BRICK	BACKUP MASONRY	TOTAL THICKNESS (IN.)	CAVITY INSULATION		COST PER S.F. MAT.	INST.	TOTAL
4320	Utility	4" clay tile	10	polystyrene		12.95	13.10	26.05
4340				none		12.55	12.50	25.05
4360		4" glazed tile	10	polystyrene		17.15	19	36.15
4380				none		16.75	18.40	35.15

B2010 135 — Brick Face Cavity Wall - Insulated Backup

	FACE BRICK	BACKUP MASONRY	TOTAL THICKNESS (IN.)	BACKUP CORE FILL		COST PER S.F. MAT.	INST.	TOTAL
5100	Standard	6" conc. block	10	perlite		8.95	19.35	28.30
5120				styrofoam		10.60	19.05	29.65
5180		6" L.W. block	10	perlite		9.30	19.20	28.50
5200				styrofoam		10.95	18.90	29.85
5260		6" glazed block	10	perlite		16.75	20.50	37.25
5280				styrofoam		18.40	20.50	38.90
5340		6" clay tile	10	none		15.90	18.40	34.30
5360		8" clay tile	12	none		18.55	19.05	37.60
5600	Glazed	6" conc. block	10	perlite		22	19.85	41.85
5620				styrofoam		23.50	19.55	43.05
5680		6" L.W. block	10	perlite		22	19.70	41.70
5700				styrofoam		24	19.40	43.40
5760		6" glazed block	10	perlite		29.50	21	50.50
5780				styrofoam		31.50	21	52.50
5840		6" clay tile	10	none		29	18.90	47.90
5860		8" clay tile	8	none		31.50	19.55	51.05
6100	Engineer	6" conc. block	10	perlite		8	17.75	25.75
6120				styrofoam		9.65	17.45	27.10
6180		6" L.W. block	10	perlite		8.30	17.60	25.90
6200				styrofoam		9.95	17.30	27.25
6260		6" glazed block	10	perlite		15.75	19	34.75
6280				styrofoam		17.40	18.70	36.10
6340		6" clay tile	10	none		14.90	16.80	31.70
6360		8" clay tile	12	none		17.60	17.45	35.05
6600	Roman	6" conc. block	10	perlite		9.95	18.10	28.05
6620				styrofoam		11.60	17.80	29.40
6680		6" L.W. block	10	perlite		10.30	17.95	28.25
6700				styrofoam		11.95	17.65	29.60
6760		6" glazed block	10	perlite		17.75	19.35	37.10
6780				styrofoam		19.40	19.05	38.45
6840		6" clay tile	10	none		16.90	17.15	34.05
6860		8" clay tile	12	none		19.55	17.80	37.35
7100	Norman	6" conc. block	10	perlite		9	16.05	25.05
7120				styrofoam		10.65	15.75	26.40
7180		6" L.W. block	10	perlite		9.35	15.90	25.25
7200				styrofoam		11	15.60	26.60
7260		6" glazed block	10	perlite		16.80	17.30	34.10
7280				styrofoam		18.45	17	35.45
7340		6" clay tile	10	none		15.95	15.10	31.05
7360		8" clay tile	12	none		18.60	15.75	34.35
7600	Norwegian	6" conc. block	10	perlite		7.90	15	22.90
7620				styrofoam		9.55	14.70	24.25
7680		6" L.W. block	10	perlite		8.20	14.85	23.05
7700				styrofoam		9.85	14.55	24.40

B20 Exterior Enclosure

B2010 Exterior Walls

| B2010 135 | | | | | | Brick Face Cavity Wall - Insulated Backup | | | |

	FACE BRICK	BACKUP MASONRY	TOTAL THICKNESS (IN.)	BACKUP CORE FILL		COST PER S.F.		
						MAT.	INST.	TOTAL
7760	Norwegian	6" glazed block	10	perlite		15.65	16.25	31.90
7780				styrofoam		17.30	15.95	33.25
7840		6" clay tile	10	none		14.80	14.05	28.85
7860		8" clay tile	12	none		17.45	14.70	32.15
8100	Utility	6" conc. block	10	perlite		7.50	13.95	21.45
8120				styrofoam		9.15	13.65	22.80
8180		6" L.W. block	10	perlite		7.85	13.80	21.65
8200				styrofoam		9.50	13.50	23
8260		6" glazed block	10	perlite		15.30	15.20	30.50
8280				styrofoam		16.95	14.90	31.85
8340		6" clay tile	10	none		14.45	13	27.45
8360		8" clay tile	12	none		17.10	13.65	30.75

B2010 Exterior Walls

Glass Block

B2010 140	Glass Block	COST PER S.F.		
		MAT.	INST.	TOTAL
2300	Glass block 4″ thick, 6″x6″ plain, under 1,000 S.F.	28	22.50	50.50
2400	1,000 to 5,000 S.F.	27.50	19.50	47
2500	Over 5,000 S.F.	26.50	18.30	44.80
2600	Solar reflective, under 1,000 S.F.	39.50	30.50	70
2700	1,000 to 5,000 S.F.	38.50	26.50	65
2800	Over 5,000 S.F.	37	24.50	61.50
3500	8″x8″ plain, under 1,000 S.F.	17.20	16.80	34
3600	1,000 to 5,000 S.F.	16.85	14.55	31.40
3700	Over 5,000 S.F.	16.35	13.10	29.45
3800	Solar reflective, under 1,000 S.F.	24	22.50	46.50
3900	1,000 to 5,000 S.F.	23.50	19.45	42.95
4000	Over 5,000 S.F.	23	17.45	40.45
5000	12″x12″ plain, under 1,000 S.F.	22	15.55	37.55
5100	1,000 to 5,000 S.F.	21.50	13.10	34.60
5200	Over 5,000 S.F.	21	12	33
5300	Solar reflective, under 1,000 S.F.	31	21	52
5400	1,000 to 5,000 S.F.	30	17.45	47.45
5600	Over 5,000 S.F.	29.50	15.90	45.40
5800	3″ thinline, 6″x6″ plain, under 1,000 S.F.	19.20	22.50	41.70
5900	Over 5,000 S.F.	19.20	22.50	41.70
6000	Solar reflective, under 1,000 S.F.	27	30.50	57.50
6100	Over 5,000 S.F.	26	24.50	50.50
6200	8″x8″ plain, under 1,000 S.F.	10.80	16.80	27.60
6300	Over 5,000 S.F.	10.45	13.10	23.55
6400	Solar reflective, under 1,000 S.F.	15.10	22.50	37.60
6500	Over 5,000 S.F.	14.60	17.45	32.05

B2010 Exterior Walls

Metal Siding

The table below lists costs for metal siding of various descriptions, not including the steel frame, or the structural steel, of a building. Costs are per S.F. including all accessories and insulation.

B2010 146	Metal Siding Panel	COST PER S.F.		
		MAT.	INST.	TOTAL
1400	Metal siding aluminum panel, corrugated, .024" thick, natural	2.91	3.68	6.59
1450	Painted	3.08	3.68	6.76
1500	.032" thick, natural	3.25	3.68	6.93
1550	Painted	3.82	3.68	7.50
1600	Ribbed 4" pitch, .032" thick, natural	3.18	3.68	6.86
1650	Painted	3.86	3.68	7.54
1700	.040" thick, natural	3.63	3.68	7.31
1750	Painted	4.17	3.68	7.85
1800	.050" thick, natural	4.11	3.68	7.79
1850	Painted	4.66	3.68	8.34
1900	8" pitch panel, .032" thick, natural	3.03	3.53	6.56
1950	Painted	3.69	3.55	7.24
2000	.040" thick, natural	3.49	3.55	7.04
2050	Painted	3.99	3.58	7.57
2100	.050" thick, natural	3.96	3.57	7.53
2150	Painted	4.54	3.60	8.14
3000	Steel, corrugated or ribbed, 29 Ga. .0135" thick, galvanized	2.07	3.38	5.45
3050	Colored	2.94	3.42	6.36
3100	26 Ga. .0179" thick, galvanized	2.14	3.39	5.53
3150	Colored	3.03	3.43	6.46
3200	24 Ga. .0239" thick, galvanized	2.75	3.41	6.16
3250	Colored	3.42	3.45	6.87
3300	22 Ga. .0299" thick, galvanized	2.77	3.43	6.20
3350	Colored	4.16	3.47	7.63
3400	20 Ga. .0359" thick, galvanized	2.77	3.43	6.20
3450	Colored	4.43	3.67	8.10
4100	Sandwich panels, factory fab., 1" polystyrene, steel core, 26 Ga., galv.	5.80	5.30	11.10
4200	Colored, 1 side	7.05	5.30	12.35
4300	2 sides	9.10	5.30	14.40
4400	2" polystyrene, steel core, 26 Ga., galvanized	6.85	5.30	12.15
4500	Colored, 1 side	8.10	5.30	13.40
4600	2 sides	10.15	5.30	15.45
4700	22 Ga., baked enamel exterior	13.45	5.55	19
4800	Polyvinyl chloride exterior	14.15	5.55	19.70
5100	Textured aluminum, 4' x 8' x 5/16" plywood backing, single face	4.04	3.17	7.21
5200	Double face	5.50	3.17	8.67

B2010 Exterior Walls

Exterior Wall Siding

The table below lists costs per S.F. for exterior walls with wood siding. A variety of systems are presented using both wood and metal studs at 16″ and 24″ O.C.

B2010 148	Panel, Shingle & Lap Siding	MAT.	INST.	TOTAL
		COST PER S.F.		
1400	Wood siding w/2″x4″studs, 16″O.C., insul. wall, 5/8″text 1-11 fir plywood	3.14	5.25	8.39
1450	5/8″ text 1-11 cedar plywood	4.30	5.05	9.35
1500	1″ x 4″ vert T.&G. redwood	5.05	6.05	11.10
1600	1″ x 8″ vert T.&G. redwood	5.20	5.30	10.50
1650	1″ x 5″ rabbetted cedar bev. siding	6.85	5.45	12.30
1700	1″ x 6″ cedar drop siding	6.95	5.50	12.45
1750	1″ x 12″ rough sawn cedar	2.99	5.30	8.29
1800	1″ x 12″ sawn cedar, 1″ x 4″ battens	5.95	4.95	10.90
1850	1″ x 10″ redwood shiplap siding	5.30	5.10	10.40
1900	18″ no. 1 red cedar shingles, 5-1/2″ exposed	4.63	6.90	11.53
1950	6″ exposed	4.41	6.70	11.11
2000	6-1/2″ exposed	4.20	6.45	10.65
2100	7″ exposed	3.98	6.25	10.23
2150	7-1/2″ exposed	3.77	6	9.77
3000	8″ wide aluminum siding	3.32	4.70	8.02
3150	8″ plain vinyl siding	2.48	4.78	7.26
3250	8″ insulated vinyl siding	2.78	5.30	8.08
3300				
3400	2″ x 6″ studs, 16″ O.C., insul. wall, w/ 5/8″ text 1-11 fir plywood	3.56	5.40	8.96
3500	5/8″ text 1-11 cedar plywood	4.72	5.40	10.12
3600	1″ x 4″ vert T.&G. redwood	5.45	6.20	11.65
3700	1″ x 8″ vert T.&G. redwood	5.60	5.45	11.05
3800	1″ x 5″ rabbetted cedar bev siding	7.25	5.60	12.85
3900	1″ x 6″ cedar drop siding	7.35	5.65	13
4000	1″ x 12″ rough sawn cedar	3.41	5.45	8.86
4200	1″ x 12″ sawn cedar, 1″ x 4″ battens	6.40	5.10	11.50
4500	1″ x 10″ redwood shiplap siding	5.70	5.25	10.95
4550	18″ no. 1 red cedar shingles, 5-1/2″ exposed	5.05	7.05	12.10
4600	6″ exposed	4.83	6.85	11.68
4650	6-1/2″ exposed	4.62	6.60	11.22
4700	7″ exposed	4.40	6.40	10.80
4750	7-1/2″ exposed	4.19	6.15	10.34
4800	8″ wide aluminum siding	3.74	4.85	8.59
4850	8″ plain vinyl siding	2.90	4.93	7.83
4900	8″ insulated vinyl siding	3.20	5.45	8.65
4910				
5000	2″ x 6″ studs, 24″ O.C., insul. wall, 5/8″ text 1-11, fir plywood	3.45	5.10	8.55
5050	5/8″ text 1-11 cedar plywood	4.61	5.10	9.71
5100	1″ x 4″ vert T.&G. redwood	5.35	5.90	11.25
5150	1″ x 8″ vert T.&G. redwood	5.50	5.10	10.60

B2010 Exterior Walls

B2010 148	Panel, Shingle & Lap Siding	COST PER S.F.		
		MAT.	INST.	TOTAL
5200	1" x 5" rabbetted cedar bev siding	7.15	5.30	12.45
5250	1" x 6" cedar drop siding	7.25	5.35	12.60
5300	1" x 12" rough sawn cedar	3.30	5.15	8.45
5400	1" x 12" sawn cedar, 1" x 4" battens	6.25	4.78	11.03
5450	1" x 10" redwood shiplap siding	5.60	4.93	10.53
5500	18" no. 1 red cedar shingles, 5-1/2" exposed	4.94	6.75	11.69
5550	6" exposed	4.72	6.50	11.22
5650	7" exposed	4.29	6.05	10.34
5700	7-1/2" exposed	4.08	5.85	9.93
5750	8" wide aluminum siding	3.63	4.53	8.16
5800	8" plain vinyl siding	2.79	4.61	7.40
5850	8" insulated vinyl siding	3.09	5.10	8.19
5900	3-5/8" metal studs, 16 Ga., 16" OC insul.wall, 5/8"text 1-11 fir plywood	4.02	5.55	9.57
5950	5/8" text 1-11 cedar plywood	5.20	5.55	10.75
6000	1" x 4" vert T.&G. redwood	5.90	6.35	12.25
6050	1" x 8" vert T.&G. redwood	6.05	5.55	11.60
6100	1" x 5" rabbetted cedar bev siding	7.70	5.75	13.45
6150	1" x 6" cedar drop siding	7.80	5.80	13.60
6200	1" x 12" rough sawn cedar	3.87	5.60	9.47
6250	1" x 12" sawn cedar, 1" x 4" battens	6.85	5.25	12.10
6300	1" x 10" redwood shiplap siding	6.15	5.40	11.55
6350	18" no. 1 red cedar shingles, 5-1/2" exposed	5.50	7.20	12.70
6500	6" exposed	5.30	6.95	12.25
6550	6-1/2" exposed	5.10	6.75	11.85
6600	7" exposed	4.86	6.50	11.36
6650	7-1/2" exposed	4.86	6.35	11.21
6700	8" wide aluminum siding	4.20	4.98	9.18
6750	8" plain vinyl siding	3.36	4.88	8.24
6800	8" insulated vinyl siding	3.66	5.40	9.06
7000	3-5/8" metal studs, 16 Ga. 24" OC insul wall, 5/8" text 1-11 fir plywood	3.69	4.91	8.60
7050	5/8" text 1-11 cedar plywood	4.85	4.91	9.76
7100	1" x 4" vert T.&G. redwood	5.60	5.90	11.50
7150	1" x 8" vert T.&G. redwood	5.75	5.10	10.85
7200	1" x 5" rabbetted cedar bev siding	7.40	5.30	12.70
7250	1" x 6" cedar drop siding	7.50	5.35	12.85
7300	1" x 12" rough sawn cedar	3.54	5.15	8.69
7350	1" x 12" sawn cedar 1" x 4" battens	6.50	4.79	11.29
7400	1" x 10" redwood shiplap siding	5.85	4.94	10.79
7450	18" no. 1 red cedar shingles, 5-1/2" exposed	5.40	7	12.40
7500	6" exposed	5.20	6.75	11.95
7550	6-1/2" exposed	4.96	6.55	11.51
7600	7" exposed	4.75	6.30	11.05
7650	7-1/2" exposed	4.32	5.85	10.17
7700	8" wide aluminum siding	3.87	4.54	8.41
7750	8" plain vinyl siding	3.03	4.62	7.65
7800	8" insul. vinyl siding	3.33	5.15	8.48

B2010 Exterior Walls

The table below lists costs for some typical stucco walls including all the components as demonstrated in the component block below. Prices are presented for backup walls using wood studs, metal studs and CMU.

Exterior Stucco Wall

B2010 151	Stucco Wall	COST PER S.F.		
		MAT.	INST.	TOTAL
2100	Cement stucco, 7/8" th., plywood sheathing, stud wall, 2" x 4", 16" O.C.	2.39	8.45	10.84
2200	24" O.C.	2.32	8.20	10.52
2300	2" x 6", 16" O.C.	2.81	8.60	11.41
2400	24" O.C.	2.70	8.30	11
2500	No sheathing, metal lath on stud wall, 2" x 4", 16" O.C.	1.77	7.35	9.12
2600	24" O.C.	1.70	7.10	8.80
2700	2" x 6", 16" O.C.	2.19	7.50	9.69
2800	24" O.C.	2.08	7.15	9.23
2900	1/2" gypsum sheathing, 3-5/8" metal studs, 16" O.C.	2.88	7.65	10.53
2950	24" O.C.	2.55	7.20	9.75
3000	Cement stucco, 5/8" th., 2 coats on std. CMU block, 8"x 16", 8" thick	3.17	10.40	13.57
3100	10" thick	4.20	10.65	14.85
3200	12" thick	4.35	12.25	16.60
3300	Std. light Wt. block 8" x 16", 8" thick	3.70	10.20	13.90
3400	10" thick	4.38	10.45	14.83
3500	12" thick	5.30	12	17.30
3600	3 coat stucco, self furring metal lath 3.4 Lb/SY, on 8" x 16", 8" thick	3.21	10.95	14.16
3700	10" thick	4.24	11.20	15.44
3800	12" thick	4.39	12.80	17.19
3900	Lt. Wt. block, 8" thick	3.74	10.75	14.49
4000	10" thick	4.42	11	15.42
4100	12" thick	5.35	12.55	17.90

B2010 152	E.I.F.S.	COST PER S.F.		
		MAT.	INST.	TOTAL
5100	E.I.F.S., plywood sheathing, stud wall, 2" x 4", 16" O.C., 1" EPS	3.99	8.55	12.54
5110	2" EPS	4.39	8.55	12.94
5120	3" EPS	4.78	8.55	13.33
5130	4" EPS	5.20	8.55	13.75
5140	2" x 6", 16" O.C., 1" EPS	4.41	8.70	13.11
5150	2" EPS	4.81	8.70	13.51
5160	3" EPS	5.20	8.70	13.90
5170	4" EPS	5.60	8.70	14.30
5180	Cement board sheathing, 3-5/8" metal studs, 16" O.C., 1" EPS	5	10.50	15.50
5190	2" EPS	5.40	10.50	15.90
5200	3" EPS	5.80	10.50	16.30
5210	4" EPS	6.95	10.50	17.45
5220	6" metal studs, 16" O.C., 1" EPS	5.65	10.55	16.20
5230	2" EPS	6.05	10.55	16.60

B2010 Exterior Walls

B2010 152	E.I.F.S.	COST PER S.F.		
		MAT.	INST.	TOTAL
5240	3" EPS	6.45	10.55	17
5250	4" EPS	7.55	10.55	18.10
5260	CMU block, 8" x 8" x 16", 1" EPS	5.25	12.20	17.45
5270	2" EPS	5.65	12.20	17.85
5280	3" EPS	6.05	12.20	18.25
5290	4" EPS	6.45	12.20	18.65
5300	8" x 10" x 16", 1" EPS	6.30	12.45	18.75
5310	2" EPS	6.70	12.45	19.15
5320	3" EPS	7.10	12.45	19.55
5330	4" EPS	7.50	12.45	19.95
5340	8" x 12" x 16", 1" EPS	6.45	14.05	20.50
5350	2" EPS	6.85	14.05	20.90
5360	3" EPS	7.25	14.05	21.30
5370	4" EPS	7.65	14.05	21.70

B2010 Exterior Walls

Description: The table below lists costs, $/S.F., for channel girts with sag rods and connector angles top and bottom for various column spacings, building heights and wind loads. Additive costs are shown for wind columns.

B2010 154		Metal Siding Support						
	BLDG. HEIGHT (FT.)	WIND LOAD (P.S.F.)	COL. SPACING (FT.)		INTERMEDIATE COLUMNS	COST PER S.F.		
						MAT.	INST.	TOTAL
3000	18	20	20			2	3.71	5.71
3100					wind cols.	1.04	.27	1.31
3200		20	25			2.19	3.76	5.95
3300					wind cols.	.83	.22	1.05
3400		20	30			2.42	3.82	6.24
3500					wind cols.	.69	.18	.87
3600		20	35			2.66	3.89	6.55
3700					wind cols.	.59	.15	.74
3800		30	20			2.22	3.77	5.99
3900					wind cols.	1.04	.27	1.31
4000		30	25			2.42	3.82	6.24
4100					wind cols.	.83	.22	1.05
4200		30	30			2.67	3.89	6.56
4300					wind cols.	.94	.24	1.18
4600	30	20	20			1.94	2.61	4.55
4700					wind cols.	1.46	.38	1.84
4800		20	25			2.18	2.68	4.86
4900					wind cols.	1.17	.31	1.48
5000		20	30			2.46	2.75	5.21
5100					wind cols.	1.15	.31	1.46
5200		20	35			2.75	2.83	5.58
5300					wind cols.	1.14	.31	1.45
5400		30	20			2.21	2.69	4.90
5500					wind cols.	1.73	.46	2.19
5600		30	25			2.45	2.75	5.20
5700					wind cols.	1.65	.44	2.09
5800		30	30			2.77	2.83	5.60
5900					wind cols.	1.55	.41	1.96

B2020 Exterior Windows

The table below lists window systems by material, type and size. Prices between sizes listed can be interpolated with reasonable accuracy. Prices include frame, hardware, and casing as illustrated in the component block below.

B2020 102				Wood Windows				
	MATERIAL	**TYPE**	**GLAZING**	**SIZE**	**DETAIL**	**COST PER UNIT**		
						MAT.	**INST.**	**TOTAL**
3000	Wood	double hung	std. glass	2'-8" x 4'-6"		237	224	461
3050				3'-0" x 5'-6"		315	257	572
3100			insul. glass	2'-8" x 4'-6"		254	224	478
3150				3'-0" x 5'-6"		335	257	592
3200		sliding	std. glass	3'-4" x 2'-7"		315	186	501
3250				4'-4" x 3'-3"		325	204	529
3300				5'-4" x 6'-0"		415	246	661
3350			insul. glass	3'-4" x 2'-7"		385	218	603
3400				4'-4" x 3'-3"		400	237	637
3450				5'-4" x 6'-0"		505	280	785
3500		awning	std. glass	2'-10" x 1'-9"		223	112	335
3600				4'-4" x 2'-8"		355	109	464
3700			insul. glass	2'-10" x 1'-9"		272	129	401
3800				4'-4" x 2'-8"		435	119	554
3900		casement	std. glass	1'-10" x 3'-2"	1 lite	320	147	467
3950				4'-2" x 4'-2"	2 lite	575	196	771
4000				5'-11" x 5'-2"	3 lite	865	252	1,117
4050				7'-11" x 6'-3"	4 lite	1,225	305	1,530
4100				9'-11" x 6'-3"	5 lite	1,600	355	1,955
4150			insul. glass	1'-10" x 3'-2"	1 lite	320	147	467
4200				4'-2" x 4'-2"	2 lite	585	196	781
4250				5'-11" x 5'-2"	3 lite	915	252	1,167
4300				7'-11" x 6'-3"	4 lite	1,300	305	1,605
4350				9'-11" x 6'-3"	5 lite	1,675	355	2,030
4400		picture	std. glass	4'-6" x 4'-6"		455	253	708
4450				5'-8" x 4'-6"		510	281	791
4500			insul. glass	4'-6" x 4'-6"		555	294	849
4550				5'-8" x 4'-6"		625	325	950
4600		fixed bay	std. glass	8' x 5'		1,600	460	2,060
4650				9'-9" x 5'-4"		1,125	650	1,775
4700			insul. glass	8' x 5'		2,150	460	2,610
4750				9'-9" x 5'-4"		1,225	650	1,875
4800		casement bay	std. glass	8' x 5'		1,600	530	2,130
4850			insul. glass	8' x 5'		1,800	530	2,330
4900		vert. bay	std. glass	8' x 5'		1,800	530	2,330
4950			insul. glass	8' x 5'		1,875	530	2,405

B20 Exterior Enclosure

B2020 Exterior Windows

B2020 104				Steel Windows				

	MATERIAL	TYPE	GLAZING	SIZE	DETAIL	COST PER UNIT		
						MAT.	INST.	TOTAL
5000	Steel	double hung	1/4" tempered	2'-8" x 4'-6"		840	176	1,016
5050				3'-4" x 5'-6"		1,275	268	1,543
5100			insul. glass	2'-8" x 4'-6"		860	202	1,062
5150				3'-4" x 5'-6"		1,300	305	1,605
5202		horiz. pivoted	std. glass	2' x 2'		255	58.50	313.50
5250				3' x 3'		575	132	707
5300				4' x 4'		1,025	234	1,259
5350				6' x 4'		1,525	350	1,875
5400			insul. glass	2' x 2'		263	67	330
5450				3' x 3'		590	151	741
5500				4' x 4'		1,050	269	1,319
5550				6' x 4'		1,575	405	1,980
5600		picture window	std. glass	3' x 3'		360	132	492
5650				6' x 4'		955	350	1,305
5700			insul. glass	3' x 3'		375	151	526
5750				6' x 4'		1,000	405	1,405
5800		industrial security	std. glass	2'-9" x 4'-1"		760	164	924
5850				4'-1" x 5'-5"		1,500	325	1,825
5900			insul. glass	2'-9" x 4'-1"		780	188	968
5950				4'-1" x 5'-5"		1,550	370	1,920
6000		comm. projected	std. glass	3'-9" x 5'-5"		1,250	298	1,548
6050				6'-9" x 4'-1"		1,700	405	2,105
6100			insul. glass	3'-9" x 5'-5"		1,275	340	1,615
6150				6'-9" x 4'-1"		1,750	465	2,215
6200		casement	std. glass	4'-2" x 4'-2"	2 lite	985	254	1,239
6250			insul. glass	4'-2" x 4'-2"		1,025	291	1,316
6300			std. glass	5'-11" x 5'-2"	3 lite	1,850	450	2,300
6350			insul. glass	5'-11" x 5'-2"		1,925	515	2,440

B2020 106				Aluminum Windows				

	MATERIAL	TYPE	GLAZING	SIZE	DETAIL	COST PER UNIT		
						MAT.	INST.	TOTAL
6400	Aluminum	awning	std. glass	3'-1" x 3'-2"		370	132	502
6450				4'-5" x 5'-3"		420	165	585
6500			insul. glass	3'-1" x 3'-2"		445	158	603
6550				4'-5" x 5'-3"		505	198	703
6600		sliding	std. glass	3' x 2'		230	132	362
6650				5' x 3'		350	147	497
6700				8' x 4'		370	220	590
6750				9' x 5'		560	330	890
6800			insul. glass	3' x 2'		246	132	378
6850				5' x 3'		410	147	557
6900				8' x 4'		595	220	815
6950				9' x 5'		900	330	1,230
7000		single hung	std. glass	2' x 3'		218	132	350
7050				2'-8" x 6'-8"		385	165	550
7100				3'-4" x 5'-0"		315	147	462
7150			insul. glass	2' x 3'		264	132	396
7200				2'-8" x 6'-8"		495	165	660
7250				3'-4" x 5'		350	147	497
7300		double hung	std. glass	2' x 3'		290	88	378
7350				2'-8" x 6'-8"		860	261	1,121

319

B2020 Exterior Windows

B2020 106		Aluminum Windows						

	MATERIAL	TYPE	GLAZING	SIZE	DETAIL	COST PER UNIT		
						MAT.	INST.	TOTAL
7400	Aluminum	double hung		3'-4" x 5'		805	244	1,049
7450			insul. glass	2' x 3'		300	101	401
7500				2'-8" x 6'-8"		895	299	1,194
7550				3'-4" x 5'-0"		835	280	1,115
7600		casement	std. glass	3'-1" x 3'-2"		249	143	392
7650				4'-5" x 5'-3"		605	345	950
7700			insul. glass	3'-1" x 3'-2"		267	164	431
7750				4'-5" x 5'-3"		645	395	1,040
7800		hinged swing	std. glass	3' x 4'		530	176	706
7850				4' x 5'		885	293	1,178
7900			insul. glass	3' x 4'		555	202	757
7950				4' x 5'		925	335	1,260
8200		picture unit	std. glass	2'-0" x 3'-0"		160	88	248
8250				2'-8" x 6'-8"		475	261	736
8300				3'-4" x 5'-0"		445	244	689
8350			insul. glass	2'-0" x 3'-0"		171	101	272
8400				2'-8" x 6'-8"		505	299	804
8450				3'-4" x 5'-0"		475	280	755
8500		awning type	std. glass	3'-0" x 3'-0"	2 lite	445	94	539
8550				3'-0" x 4'-0"	3 lite	520	132	652
8600				3'-0" x 5'-4"	4 lite	625	132	757
8650				4'-0" x 5'-4"	4 lite	685	147	832
8700			insul. glass	3'-0" x 3'-0"	2 lite	475	94	569
8750				3'-0" x 4'-0"	3 lite	595	132	727
8800				3'-0" x 5'-4"	4 lite	735	132	867
8850				4'-0" x 5'-4"	4 lite	825	147	972
9051								
9052								

B2020 210	Tubular Aluminum Framing	COST/S.F. OPNG.		
		MAT.	INST.	TOTAL
1100	Alum flush tube frame, for 1/4" glass, 1-3/4"x4", 5'x6'opng, no inter horiz	12.60	10.40	23
1150	One intermediate horizontal	17.20	12.45	29.65
1200	Two intermediate horizontals	22	14.55	36.55
1250	5' x 20' opening, three intermediate horizontals	11.60	8.85	20.45
1400	1-3/4" x 4-1/2", 5' x 6' opening, no intermediate horizontals	14.55	10.40	24.95
1450	One intermediate horizontal	19.30	12.45	31.75
1500	Two intermediate horizontals	24	14.55	38.55
1550	5' x 20' opening, three intermediate horizontals	13.05	8.85	21.90
1700	For insulating glass, 2"x4-1/2", 5'x6' opening, no intermediate horizontals	15.05	11	26.05
1750	One intermediate horizontal	19.50	13.15	32.65
1800	Two intermediate horizontals	24	15.35	39.35
1850	5' x 20' opening, three intermediate horizontals	13.10	9.35	22.45
2000	Thermal break frame, 2-1/4"x4-1/2", 5'x6'opng, no intermediate horizontals	15.65	11.15	26.80
2050	One intermediate horizontal	21	13.70	34.70
2100	Two intermediate horizontals	26	16.30	42.30
2150	5' x 20' opening, three intermediate horizontals	14.20	9.80	24

B2020 Exterior Windows

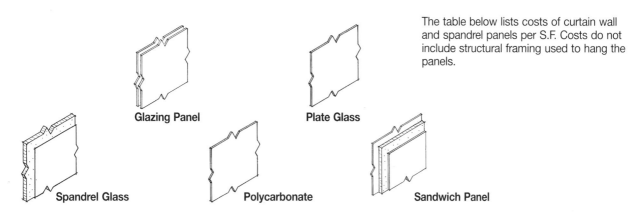

The table below lists costs of curtain wall and spandrel panels per S.F. Costs do not include structural framing used to hang the panels.

Glazing Panel

Plate Glass

Spandrel Glass

Polycarbonate

Sandwich Panel

B2020 220	Curtain Wall Panels	COST PER S.F.		
		MAT.	INST.	TOTAL
1000	Glazing panel, insulating, 1/2" thick, 2 lites 1/8" float, clear	10.20	10.20	20.40
1100	Tinted	14.20	10.20	24.40
1200	5/8" thick units, 2 lites 3/16" float, clear	14.25	10.75	25
1300	Tinted	14.65	10.75	25.40
1400	1" thick units, 2 lites, 1/4" float, clear	17.05	12.90	29.95
1500	Tinted	24.50	12.90	37.40
1600	Heat reflective film inside	28.50	11.40	39.90
1700	Light and heat reflective glass, tinted	32.50	11.40	43.90
2000	Plate glass, 1/4" thick, clear	5.90	8.05	13.95
2050	Tempered	8.35	8.05	16.40
2100	Tinted	8.15	8.05	16.20
2200	3/8" thick, clear	10.25	12.90	23.15
2250	Tempered	16.95	12.90	29.85
2300	Tinted	16.30	12.90	29.20
2400	1/2" thick, clear	20.50	17.60	38.10
2450	Tempered	25	17.60	42.60
2500	Tinted	28.50	17.60	46.10
2600	3/4" thick, clear	36.50	27.50	64
2650	Tempered	43	27.50	70.50
3000	Spandrel glass, panels, 1/4" plate glass insul w/fiberglass, 1" thick	17.40	8.05	25.45
3100	2" thick	20.50	8.05	28.55
3200	Galvanized steel backing, add	6.05		6.05
3300	3/8" plate glass, 1" thick	29	8.05	37.05
3400	2" thick	32	8.05	40.05
4000	Polycarbonate, masked, clear or colored, 1/8" thick	10.65	5.70	16.35
4100	3/16" thick	12.20	5.85	18.05
4200	1/4" thick	13.75	6.25	20
4300	3/8" thick	24.50	6.45	30.95
5000	Facing panel, textured al, 4' x 8' x 5/16" plywood backing, sgl face	4.04	3.17	7.21
5100	Double face	5.50	3.17	8.67
5200	4' x 10' x 5/16" plywood backing, single face	4.32	3.17	7.49
5300	Double face	5.80	3.17	8.97
5400	4' x 12' x 5/16" plywood backing, single face	3.66	3.17	6.83
5500	Sandwich panel, 22 Ga. galv., both sides 2" insulation, enamel exterior	13.45	5.55	19
5600	Polyvinylidene fluoride exterior finish	14.15	5.55	19.70
5700	26 Ga., galv. both sides, 1" insulation, colored 1 side	7.05	5.30	12.35
5800	Colored 2 sides	9.10	5.30	14.40

B2030 Exterior Doors

Costs are listed for exterior door systems by material, type and size. Prices between sizes listed can be interpolated with reasonable accuracy. Prices are per opening for a complete door system including frame.

B2030 110				Glazed Doors, Steel or Aluminum				
	MATERIAL	TYPE	DOORS	SPECIFICATION	OPENING	COST PER OPNG.		
						MAT.	INST.	TOTAL
5600	St. Stl. & glass	revolving	stock unit	manual oper.	6'-0" x 7'-0"	41,600	8,750	50,350
5650				auto Cntrls.	6'-10" x 7'-0"	57,500	9,325	66,825
5700	Bronze		stock unit	manual oper.	6'-10" x 7'-0"	48,500	17,600	66,100
5750				auto Cntrls.	6'-10" x 7'-0"	64,500	18,200	82,700
5800	St. Stl. & glass	balanced	standard	economy	3'-0" x 7'-0"	9,025	1,475	10,500
5850				premium	3'-0" x 7'-0"	15,600	1,925	17,525
6300	Alum. & glass	w/o transom	narrow stile	w/panic Hrdwre.	3'-0" x 7'-0"	1,625	960	2,585
6350				dbl. door, Hrdwre.	6'-0" x 7'-0"	2,725	1,600	4,325
6400			wide stile	hdwre.	3'-0" x 7'-0"	1,850	940	2,790
6450				dbl. door, Hdwre.	6'-0" x 7'-0"	3,575	1,875	5,450
6500			full vision	hdwre.	3'-0" x 7'-0"	2,400	1,500	3,900
6550				dbl. door, Hdwre.	6'-0" x 7'-0"	3,250	2,150	5,400
6600			non-standard	hdwre.	3'-0" x 7'-0"	2,025	940	2,965
6650				dbl. door, Hdwre.	6'-0" x 7'-0"	4,050	1,875	5,925
6700			bronze fin.	hdwre.	3'-0" x 7'-0"	1,500	940	2,440
6750				dbl. door, Hrdwre.	6'-0" x 7'-0"	2,975	1,875	4,850
6800			black fin.	hdwre.	3'-0" x 7'-0"	2,150	940	3,090
6850				dbl. door, Hdwre.	6'-0" x 7'-0"	4,300	1,875	6,175
6900		w/transom	narrow stile	hdwre.	3'-0" x 10'-0"	2,150	1,100	3,250
6950				dbl. door, Hdwre.	6'-0" x 10'-0"	3,375	1,900	5,275
7000			wide stile	hdwre.	3'-0" x 10'-0"	2,450	1,300	3,750
7050				dbl. door, Hdwre.	6'-0" x 10'-0"	3,725	2,275	6,000
7100			full vision	hdwre.	3'-0" x 10'-0"	2,725	1,450	4,175
7150				dbl. door, Hdwre.	6'-0" x 10'-0"	4,050	2,500	6,550

B2030 Exterior Doors

B2030 110	Glazed Doors, Steel or Aluminum

	MATERIAL	TYPE	DOORS	SPECIFICATION	OPENING	COST PER OPNG.		
						MAT.	INST.	TOTAL
7200	Alum. & glass	w/transom	non-standard	hdwre.	3'-0" x 10'-0"	2,100	1,025	3,125
7250				dbl. door, Hdwre.	6'-0" x 10'-0"	4,200	2,025	6,225
7300			bronze fin.	hdwre.	3'-0" x 10'-0"	1,575	1,025	2,600
7350				dbl. door, Hdwre.	6'-0" x 10'-0"	3,125	2,025	5,150
7400			black fin.	hdwre.	3'-0" x 10'-0"	2,225	1,025	3,250
7450				dbl. door, Hdwre.	6'-0" x 10'-0"	4,450	2,025	6,475
7500		revolving	stock design	minimum	6'-10" x 7'-0"	20,900	3,525	24,425
7550				average	6'-0" x 7'-0"	25,100	4,400	29,500
7600				maximum	6'-10" x 7'-0"	43,200	5,875	49,075
7650				min., automatic	6'-10" x 7'-0"	36,700	4,100	40,800
7700				avg., automatic	6'-10" x 7'-0"	40,900	4,975	45,875
7750				max., automatic	6'-10" x 7'-0"	59,000	6,450	65,450
7800		balanced	standard	economy	3'-0" x 7'-0"	6,475	1,475	7,950
7850				premium	3'-0" x 7'-0"	8,025	1,875	9,900
7900		mall front	sliding panels	alum. fin.	16'-0" x 9'-0"	3,400	745	4,145
7950					24'-0" x 9'-0"	4,850	1,375	6,225
8000				bronze fin.	16'-0" x 9'-0"	3,950	870	4,820
8050					24'-0" x 9'-0"	5,650	1,600	7,250
8100			fixed panels	alum. fin.	48'-0" x 9'-0"	9,025	1,075	10,100
8150				bronze fin.	48'-0" x 9'-0"	10,500	1,250	11,750
8200		sliding entrance	5' x 7' door	electric oper.	12'-0" x 7'-6"	8,575	1,375	9,950
8250		sliding patio	temp. glass	economy	6'-0" x 7'-0"	1,750	256	2,006
8300			temp. glass	economy	12'-0" x 7'-0"	4,175	340	4,515
8350				premium	6'-0" x 7'-0"	2,625	385	3,010
8400					12'-0" x 7'-0"	6,275	510	6,785

B2030 Exterior Doors

Exterior Door System

The table below lists exterior door systems by material, type and size. Prices between sizes listed can be interpolated with reasonable accuracy. Prices are per opening for a complete door system including frame and required hardware.

Wood doors in this table are designed with wood frames and metal doors with hollow metal frames. Depending upon quality the total material and installation cost of a wood door frame is about the same as a hollow metal frame.

B2030 210				Wood Doors			
	MATERIAL	TYPE	DOORS	SPECIFICATION	OPENING	COST PER OPNG.		
						MAT.	INST.	TOTAL
2350	Birch	solid core	single door	hinged	2'-6" x 6'-8"	1,175	277	1,452
2400					2'-6" x 7'-0"	1,175	280	1,455
2450					2'-8" x 7'-0"	1,200	280	1,480
2500					3'-0" x 7'-0"	1,200	284	1,484
2550			double door	hinged	2'-6" x 6'-8"	2,175	505	2,680
2600					2'-6" x 7'-0"	2,175	510	2,685
2650					2'-8" x 7'-0"	2,175	510	2,685
2700					3'-0" x 7'-0"	2,225	520	2,745
2750	Wood	combination	storm & screen	hinged	3'-0" x 6'-8"	325	68.50	393.50
2800					3'-0" x 7'-0"	360	75.50	435.50
2850		overhead	panels, H.D.	manual oper.	8'-0" x 8'-0"	850	510	1,360
2900					10'-0" x 10'-0"	1,250	570	1,820
2950					12'-0" x 12'-0"	1,825	685	2,510
3000					14'-0" x 14'-0"	2,800	790	3,590
3050					20'-0" x 16'-0"	6,575	1,575	8,150
3100				electric oper.	8'-0" x 8'-0"	1,825	765	2,590
3150					10'-0" x 10'-0"	2,225	825	3,050
3200					12'-0" x 12'-0"	2,800	940	3,740
3250					14'-0" x 14'-0"	3,775	1,050	4,825
3300					20'-0" x 16'-0"	7,700	2,075	9,775

B2030 220				Steel Doors			
	MATERIAL	TYPE	DOORS	SPECIFICATION	OPENING	COST PER OPNG.		
						MAT.	INST.	TOTAL
3350	Steel 18 Ga.	hollow metal	1 door w/frame	no label	2'-6" x 7'-0"	1,425	289	1,714
3400					2'-8" x 7'-0"	1,425	289	1,714
3450					3'-0" x 7'-0"	1,425	289	1,714
3500					3'-6" x 7'-0"	1,525	305	1,830
3550			1 door w/frame	no label	4'-0" x 8'-0"	1,675	305	1,980
3600			2 doors w/frame	no label	5'-0" x 7'-0"	2,725	535	3,260
3650					5'-4" x 7'-0"	2,725	535	3,260
3700					6'-0" x 7'-0"	2,725	535	3,260
3750					7'-0" x 7'-0"	2,925	565	3,490
3800					8'-0" x 8'-0"	3,200	565	3,765
3850			1 door w/frame	"A" label	2'-6" x 7'-0"	1,625	350	1,975

B2030 Exterior Doors

B2030 220 — Steel Doors

	MATERIAL	TYPE	DOORS	SPECIFICATION	OPENING	COST PER OPNG. MAT.	INST.	TOTAL
3900	Steel 18 Ga.	hollow metal	1 door w/frame	"A" label	2'-8" x 7'-0"	1,600	350	1,950
3950					3'-0" x 7'-0"	1,600	350	1,950
4000					3'-6" x 7'-0"	1,750	360	2,110
4050					4'-0" x 8'-0"	1,950	375	2,325
4100			2 doors w/frame	"A" label	5'-0" x 7'-0"	3,075	645	3,720
4150					5'-4" x 7'-0"	3,125	655	3,780
4200					6'-0" x 7'-0"	3,125	655	3,780
4250					7'-0" x 7'-0"	3,400	665	4,065
4300					8'-0" x 8'-0"	3,400	670	4,070
4350	Steel 24 Ga.	overhead	sectional	manual oper.	8'-0" x 8'-0"	840	510	1,350
4400					10'-0" x 10'-0"	1,075	570	1,645
4450					12'-0" x 12'-0"	1,325	685	2,010
4500					20'-0" x 14'-0"	3,775	1,475	5,250
4550				electric oper.	8'-0" x 8'-0"	1,825	765	2,590
4600					10'-0" x 10'-0"	2,050	825	2,875
4650					12'-0" x 12'-0"	2,300	940	3,240
4700					20'-0" x 14'-0"	4,900	1,975	6,875
4750	Steel	overhead	rolling	manual oper.	8'-0" x 8'-0"	1,175	825	2,000
4800					10'-0" x 10'-0"	1,950	940	2,890
4850					12'-0" x 12'-0"	1,975	1,100	3,075
4900					14'-0" x 14'-0"	3,200	1,650	4,850
4950					20'-0" x 12'-0"	2,250	1,475	3,725
5000					20'-0" x 16'-0"	3,575	2,200	5,775
5050				electric oper.	8'-0" x 8'-0"	2,300	1,100	3,400
5100					10'-0" x 10'-0"	3,075	1,200	4,275
5150					12'-0" x 12'-0"	3,100	1,375	4,475
5200					14'-0" x 14'-0"	4,325	1,925	6,250
5250					20'-0" x 12'-0"	3,425	1,750	5,175
5300					20'-0" x 16'-0"	4,750	2,475	7,225
5350				fire rated	10'-0" x 10'-0"	2,100	1,200	3,300
5400			rolling grille	manual oper.	10'-0" x 10'-0"	2,700	1,325	4,025
5450					15'-0" x 8'-0"	3,125	1,650	4,775
5500		vertical lift	1 door w/frame	motor operator	16'-0" x 16'-0"	23,400	5,200	28,600
5550					32'-0" x 24'-0"	50,500	3,450	53,950

B2030 230 — Aluminum Doors

	MATERIAL	TYPE	DOORS	SPECIFICATION	OPENING	COST PER OPNG. MAT.	INST.	TOTAL
6000	Aluminum	combination	storm & screen	hinged	3'-0" x 6'-8"	320	73	393
6050					3'-0" x 7'-0"	350	80.50	430.50
6100		overhead	rolling grille	manual oper.	12'-0" x 12'-0"	4,325	2,325	6,650
6150				motor oper.	12'-0" x 12'-0"	5,700	2,575	8,275
6200	Alum. & Fbrgls.	overhead	heavy duty	manual oper.	12'-0" x 12'-0"	2,850	685	3,535
6250				electric oper.	12'-0" x 12'-0"	3,825	940	4,765

B3010 Roof Coverings

Multiple ply roofing is the most popular covering for minimum pitch roofs.

Built Up Ply

B3010 105	Built-Up	COST PER S.F.		
		MAT.	INST.	TOTAL
1200	Asphalt flood coat w/gravel; not incl. insul, flash., nailers			
1300				
1400	Asphalt base sheets & 3 plies #15 asphalt felt, mopped	1.03	1.64	2.67
1500	On nailable deck	1.08	1.72	2.80
1600	4 plies #15 asphalt felt, mopped	1.43	1.80	3.23
1700	On nailable deck	1.27	1.90	3.17
1800	Coated glass base sheet, 2 plies glass (type IV), mopped	1.01	1.64	2.65
1900	For 3 plies	1.21	1.80	3.01
2000	On nailable deck	1.13	1.90	3.03
2300	4 plies glass fiber felt (type IV), mopped	1.48	1.80	3.28
2400	On nailable deck	1.33	1.90	3.23
2500	Organic base sheet & 3 plies #15 organic felt, mopped	1.11	1.82	2.93
2600	On nailable deck	1.06	1.90	2.96
2700	4 plies #15 organic felt, mopped	1.33	1.64	2.97
2750				
2800	Asphalt flood coat, smooth surface, not incl. insul, flash., nailers			
2900	Asphalt base sheet & 3 plies #15 asphalt felt, mopped	1.10	1.50	2.60
3000	On nailable deck	1.02	1.57	2.59
3100	Coated glass fiber base sheet & 2 plies glass fiber felt, mopped	.95	1.44	2.39
3200	On nailable deck	.90	1.50	2.40
3300	For 3 plies, mopped	1.15	1.57	2.72
3400	On nailable deck	1.07	1.64	2.71
3700	4 plies glass fiber felt (type IV), mopped	1.35	1.57	2.92
3800	On nailable deck	1.27	1.64	2.91
3900	Organic base sheet & 3 plies #15 organic felt, mopped	1.08	1.50	2.58
4000	On nailable decks	1	1.57	2.57
4100	4 plies #15 organic felt, mopped	1.27	1.64	2.91
4200	Coal tar pitch with gravel surfacing			
4300	4 plies #15 tarred felt, mopped	2.01	1.72	3.73
4400	3 plies glass fiber felt (type IV), mopped	1.65	1.90	3.55
4500	Coated glass fiber base sheets 2 plies glass fiber felt, mopped	1.63	1.90	3.53
4600	On nailable decks	1.43	2	3.43
4800	3 plies glass fiber felt (type IV), mopped	2.26	1.72	3.98
4900	On nailable decks	2.07	1.80	3.87

B3010 Roof Coverings

Full Adhered Single Ply Membrane

Preformed Metal

Flat Seam

Batten Seam

Standing Seam

B3010 120	Single Ply Membrane	COST PER S.F.		
		MAT.	INST.	TOTAL
1000	CSPE (Chlorosulfonated polyethylene), 35 mils, fully adhered	2.04	.90	2.94
2000	EPDM (Ethylene propylene diene monomer), 45 mils, fully adhered	1.14	.90	2.04
4000	Modified bit., SBS modified, granule surface cap sheet, mopped, 150 mils	.64	1.80	2.44
4500	APP modified, granule surface cap sheet, torched, 180 mils	.65	1.17	1.82
6000	Reinforced PVC, 48 mils, loose laid and ballasted with stone	1.07	.46	1.53
6200	Fully adhered with adhesive	1.30	.90	2.20

B3010 130	Preformed Metal Roofing	COST PER S.F.		
		MAT.	INST.	TOTAL
0200	Corrugated roofing, aluminum, mill finish, .0175" thick, .272 P.S.F.	1.06	1.67	2.73
0250	.0215" thick, .334 P.S.F.	1.40	1.67	3.07

B3010 135	Formed Metal	COST PER S.F.		
		MAT.	INST.	TOTAL
1000	Batten seam, formed copper roofing, 3" min slope, 16 oz., 1.2 P.S.F.	12.15	5.50	17.65
1100	18 oz., 1.35 P.S.F.	13.15	6.05	19.20
2000	Zinc copper alloy, 3" min slope, .020" thick, .88 P.S.F.	12.15	5.05	17.20
3000	Flat seam, copper, 1/4" min. slope, 16 oz., 1.2 P.S.F.	12.15	5.05	17.20
3100	18 oz., 1.35 P.S.F.	13.15	5.25	18.40
5000	Standing seam, copper, 2-1/2" min. slope, 16 oz., 1.25 P.S.F.	12.15	4.68	16.83
5100	18 oz., 1.40 P.S.F.	13.15	5.05	18.20
6000	Zinc copper alloy, 2-1/2" min. slope, .020" thick, .87 P.S.F.	12.15	5.05	17.20
6100	.032" thick, 1.39 P.S.F.	20	5.50	25.50

B3010 Roof Coverings

Shingle and Tile

Shingles and tiles are practical in applications where the roof slope is more than 3-1/2″ per foot of rise. Lines 1100 through 6000 list the various materials and the weight per square foot.

Roof Edges

The table below lists the costs for various types of roof perimeter edge treatments.

Roof edge systems include the cost per L.F. for a 2″ x 8″ treated wood nailer fastened at 4′-0″ O.C. and a diagonally cut 4″ x 6″ treated wood cant.

Roof edge and base flashing are assumed to be made from the same material.

B3010 140	Shingle & Tile	COST PER S.F.		
		MAT.	INST.	TOTAL
1095	Asphalt roofing			
1100	Strip shingles, 4″ slope, inorganic class A 210-235 lb/Sq.	.87	1	1.87
1150	Organic, class C, 235-240 lb./sq.	.76	1.08	1.84
1200	Premium laminated multi-layered, class A, 260-300 lb./Sq.	1.16	1.49	2.65
1545	Metal roofing			
1550	Alum., shingles, colors, 3″ min slope, .019″ thick, 0.4 PSF	2.46	1.10	3.56
1850	Steel, colors, 3″ min slope, 26 gauge, 1.0 PSF	3.30	2.23	5.53
2795	Slate roofing			
2800	4″ min. slope, shingles, 3/16″ thick, 8.0 PSF	5.40	2.78	8.18
3495	Wood roofing			
3500	4″ min slope, cedar shingles, 16″ x 5″, 5″ exposure 1.6 PSF	3.53	2.21	5.74
4000	Shakes, 18″, 8-1/2″ exposure, 2.8 PSF	1.74	2.72	4.46
5095	Tile roofing			
5100	Aluminum, mission, 3″ min slope, .019″ thick, 0.65 PSF	9.10	2.13	11.23
6000	Clay, Americana, 3″ minimum slope, 8 PSF	7.45	2.95	10.40

B3010 Roof Coverings

B3010 320	Roof Deck Rigid Insulation	COST PER S.F.		
		MAT.	INST.	TOTAL
0100	Fiberboard low density, 1/2" thick, R1.39			
0150	1" thick R2.78	.46	.59	1.05
0300	1 1/2" thick R4.17	.70	.59	1.29
0350	2" thick R5.56	.94	.59	1.53
0370	Fiberboard high density, 1/2" thick R1.3	.24	.47	.71
0380	1" thick R2.5	.48	.59	1.07
0390	1 1/2" thick R3.8	.73	.59	1.32
0410	Fiberglass, 3/4" thick R2.78	.62	.47	1.09
0450	15/16" thick R3.70	.81	.47	1.28
0550	1-5/16" thick R5.26	1.39	.47	1.86
0650	2 7/16" thick R10	1.71	.59	2.30
1510	Polyisocyanurate 2#/CF density, 1" thick	.56	.34	.90
1550	1 1/2" thick	.68	.38	1.06
1600	2" thick	.95	.43	1.38
1650	2 1/2" thick	1.11	.45	1.56
1700	3" thick	1.36	.47	1.83
1750	3 1/2" thick	2.15	.47	2.62
1800	Tapered for drainage	2.15	.34	2.49
1810	Expanded polystyrene, 1#/CF density, 3/4" thick R2.89	.30	.31	.61
1820	2" thick R7.69	.79	.38	1.17
1830	Extruded polystyrene, 15 PSI compressive strength, 1" thick, R5	.54	.31	.85
1835	2" thick R10	.68	.38	1.06
1840	3" thick R15	1.40	.47	1.87
2550	40 PSI compressive strength, 1" thick R5	.56	.31	.87
2600	2" thick R10	1.06	.38	1.44
2650	3" thick R15	1.54	.47	2.01
2700	4" thick R20	2.07	.47	2.54
2750	Tapered for drainage	.78	.34	1.12
2810	60 PSI compressive strength, 1" thick R5	.65	.32	.97
2850	2" thick R10	1.16	.39	1.55
2900	Tapered for drainage	.94	.34	1.28
3050	Composites with 2" EPS			
3060	1" Fiberboard	1.36	.50	1.86
3070	7/16" oriented strand board	1.03	.59	1.62
3080	1/2" plywood	1.25	.59	1.84
3090	1" perlite	1.30	.59	1.89
4000	Composites with 1-1/2" polyisocyanurate			
4010	1" fiberboard	1.25	.59	1.84
4020	1" perlite	1.19	.55	1.74
4030	7/16" oriented strand board	.92	.59	1.51

B3010 420			Roof Edges			COST PER L.F.		
	EDGE TYPE	DESCRIPTION	SPECIFICATION	FACE HEIGHT		MAT.	INST.	TOTAL
1000	Aluminum	mill finish	.050" thick	4"		12.75	9.85	22.60
1100				6"		13.25	10.15	23.40
1300		duranodic	.050" thick	4"		12.50	9.85	22.35
1400				6"		13.65	10.15	23.80
1600		painted	.050" thick	4"		13.50	9.85	23.35
1700				6"		14.90	10.15	25.05
2000	Copper	plain	16 oz.	4"		8.65	9.85	18.50
2100				6"		10	10.15	20.15
2300			20 oz.	4"		9.50	11.20	20.70
2400				6"		10.65	10.80	21.45
2700	Sheet Metal	galvanized	20 Ga.	4"		9.80	11.60	21.40
2800				6"		11.70	11.60	23.30

B30 Roofing

B3010 Roof Coverings

B3010 420 — Roof Edges

	EDGE TYPE	DESCRIPTION	SPECIFICATION	FACE HEIGHT		COST PER L.F.		
						MAT.	INST.	TOTAL
3000	sheet metal	galvanized	24 Ga.	4"		8.65	9.85	18.50
3100				6"		10	9.85	19.85

B3010 430 — Flashing

	MATERIAL	BACKING	SIDES	SPECIFICATION	QUANTITY	COST PER S.F.		
						MAT.	INST.	TOTAL
0040	Aluminum	none		.019"		1.12	3.25	4.37
0050				.032"		1.63	3.25	4.88
0300		fabric	2	.004"		1.55	1.43	2.98
0400		mastic		.004"		1.55	1.43	2.98
0700	Copper	none		16 oz.	<500 lbs.	8.30	4.09	12.39
0800				24 oz.	<500 lbs.	12.45	4.48	16.93
2000	Copper lead	fabric	1	2 oz.		2.66	1.43	4.09
3500	PVC black	none		.010"		.25	1.65	1.90
3700				.030"		.46	1.65	2.11
4200	Neoprene			1/16"		2.23	1.65	3.88
4500	Stainless steel	none		.015"	<500 lbs.	5.75	4.09	9.84
4600	Copper clad				>2000 lbs.	5.55	3.04	8.59
5000	Plain			32 ga.		4.17	3.04	7.21

B30 Roofing

B3010 Roof Coverings

B3010 610 Gutters

	SECTION	MATERIAL	THICKNESS	SIZE	FINISH	COST PER L.F.		
						MAT.	INST.	TOTAL
0050	Box	aluminum	.027″	5″	enameled	2.68	4.96	7.64
0100					mill	3.03	4.96	7.99
0200			.032″	5″	enameled	3.19	4.96	8.15
0500		copper	16 Oz.	4″	lead coated	12.50	4.96	17.46
0600					mill	7.50	4.96	12.46
1000		steel galv.	28 Ga.	5″	enameled	1.66	4.96	6.62
1200			26 Ga.	5″	mill	1.75	4.96	6.71
1800		vinyl		4″	colors	1.19	4.66	5.85
1900				5″	colors	1.52	4.66	6.18
2300		hemlock or fir		4″x5″	treated	10.40	5.10	15.50
3000	Half round	copper	16 Oz.	4″	lead coated	10.40	4.96	15.36
3100					mill	7.65	4.96	12.61
3600		steel galv.	28 Ga.	5″	enameled	1.66	4.96	6.62
4102		stainless steel		5″	mill	6.95	4.96	11.91
5000		vinyl		4″	white	1.19	4.66	5.85

B3010 620 Downspouts

	MATERIALS	SECTION	SIZE	FINISH	THICKNESS	COST PER V.L.F.		
						MAT.	INST.	TOTAL
0100	Aluminum	rectangular	2″x3″	embossed mill	.020″	1.14	3.13	4.27
0150				enameled	.020″	1.75	3.13	4.88
0250			3″x4″	enameled	.024″	2.38	4.25	6.63
0300		round corrugated	3″	enameled	.020″	1.56	3.13	4.69
0350			4″	enameled	.025″	2.63	4.25	6.88
0500	Copper	rectangular corr.	2″x3″	mill	16 Oz.	8.10	3.13	11.23
0600		smooth		mill	16 Oz.	7.70	3.13	10.83
0700		rectangular corr.	3″x4″	mill	16 Oz.	9.45	4.10	13.55
1300	Steel	rectangular corr.	2″x3″	galvanized	28 Ga.	1.41	3.13	4.54
1350				epoxy coated	24 Ga.	2.20	3.13	5.33
1400		smooth		galvanized	28 Ga.	1.41	3.13	4.54
1450		rectangular corr.	3″x4″	galvanized	28 Ga.	2.38	4.10	6.48
1500				epoxy coated	24 Ga.	2.75	4.10	6.85
1550		smooth		galvanized	28 Ga.	2.90	4.10	7
1652		round corrugated	3″	galvanized	28 Ga.	2.46	3.13	5.59
1700			4″	galvanized	28 Ga.	3	4.10	7.10
1750			5″	galvanized	28 Ga.	3.88	4.58	8.46
2552	S.S. tubing sch.5	rectangular	3″x4″	mill		69	4.10	73.10

B3010 630 Gravel Stop

	MATERIALS	SECTION	SIZE	FINISH	THICKNESS	COST PER L.F.		
						MAT.	INST.	TOTAL
5100	Aluminum	extruded	4″	mill	.050″	6.75	4.10	10.85
5200			4″	duranodic	.050″	6.50	4.10	10.60
5300			8″	mill	.050″	9.05	4.76	13.81
5400			8″	duranodic	.050″	8.80	4.76	13.56
6000			12″-2 pc.	duranodic	.050″	11.05	5.95	17
6100	Stainless	formed	6″	mill	24 Ga.	12.60	4.41	17.01

B3020 Roof Openings

Roof Hatch

Smoke Hatch

Skylight

B3020 110	Skylights	COST PER S.F.		
		MAT.	INST.	TOTAL
5100	Skylights, plastic domes, insul curbs, nom. size to 10 S.F., single glaze	27	12.55	39.55
5200	Double glazing	31	15.45	46.45
5300	10 S.F. to 20 S.F., single glazing	26	5.10	31.10
5400	Double glazing	26.50	6.35	32.85
5500	20 S.F. to 30 S.F., single glazing	21.50	4.32	25.82
5600	Double glazing	24.50	5.10	29.60
5700	30 S.F. to 65 S.F., single glazing	18.70	3.29	21.99
5800	Double glazing	24.50	4.32	28.82
6000	Sandwich panels fiberglass, 1-9/16" thick, 2 S.F. to 10 S.F.	17.80	10.05	27.85
6100	10 S.F. to 18 S.F.	16.05	7.55	23.60
6200	2-3/4" thick, 25 S.F. to 40 S.F.	25.50	6.80	32.30
6300	40 S.F. to 70 S.F.	21	6.10	27.10

B3020 210	Hatches	COST PER OPNG.		
		MAT.	INST.	TOTAL
0200	Roof hatches with curb and 1" fiberglass insulation, 2'-6"x3'-0", aluminum	1,075	201	1,276
0300	Galvanized steel 165 lbs.	655	201	856
0400	Primed steel 164 lbs.	550	201	751
0500	2'-6"x4'-6" aluminum curb and cover, 150 lbs.	935	223	1,158
0600	Galvanized steel 220 lbs.	660	223	883
0650	Primed steel 218 lbs.	910	223	1,133
0800	2'x6"x8'-0" aluminum curb and cover, 260 lbs.	1,825	305	2,130
0900	Galvanized steel, 360 lbs.	1,750	305	2,055
0950	Primed steel 358 lbs.	1,325	305	1,630
1200	For plexiglass panels, add to the above.	440		440
2100	Smoke hatches, unlabeled not incl. hand winch operator, 2'-6"x3', galv	795	243	1,038
2200	Plain steel, 160 lbs.	665	243	908
2400	2'-6"x8'-0",galvanized steel, 360 lbs.	1,925	335	2,260
2500	Plain steel, 350 lbs.	1,450	335	1,785
3000	4'-0"x8'-0", double leaf low profile, aluminum cover, 359 lb.	2,850	251	3,101
3100	Galvanized steel 475 lbs.	2,425	251	2,676
3200	High profile, aluminum cover, galvanized curb, 361 lbs.	2,750	251	3,001

The Concrete Block Partition Systems are defined by weight and type of block, thickness, type of finish and number of sides finished. System components include joint reinforcing on alternate courses and vertical control joints.

C1010 102			Concrete Block Partitions - Regular Weight					
	TYPE	THICKNESS (IN.)	TYPE FINISH	SIDES FINISHED		COST PER S.F.		
						MAT.	INST.	TOTAL
1000	Hollow	4	none	0		1.76	5.60	7.36
1010			gyp. plaster 2 coat	1		2.14	8.15	10.29
1020				2		2.52	10.70	13.22
1200			portland - 3 coat	1		2.03	8.50	10.53
1400			5/8″ drywall	1		2.26	7.35	9.61
1500		6	none	0		2.45	6	8.45
1510			gyp. plaster 2 coat	1		2.83	8.55	11.38
1520				2		3.21	11.10	14.31
1700			portland - 3 coat	1		2.72	8.90	11.62
1900			5/8″ drywall	1		2.95	7.75	10.70
1910				2		3.45	9.50	12.95
2000		8	none	0		2.77	6.40	9.17
2010			gyp. plaster 2 coat	1		3.15	8.95	12.10
2020			gyp. plaster 2 coat	2		3.53	11.55	15.08
2200			portland - 3 coat	1		3.04	9.35	12.39
2400			5/8″ drywall	1		3.27	8.15	11.42
2410				2		3.77	9.90	13.67
2500		10	none	0		3.48	6.70	10.18
2510			gyp. plaster 2 coat	1		3.86	9.25	13.11
2520				2		4.24	11.85	16.09
2700			portland - 3 coat	1		3.75	9.65	13.40
2900			5/8″ drywall	1		3.98	8.45	12.43
2910				2		4.48	10.20	14.68
3000	Solid	2	none	0		1.62	5.55	7.17
3010			gyp. plaster	1		2	8.10	10.10
3020				2		2.38	10.65	13.03
3200			portland - 3 coat	1		1.89	8.45	10.34
3400			5/8″ drywall	1		2.12	7.30	9.42
3410				2		2.62	9.05	11.67
3500		4	none	0		2.15	5.80	7.95
3510			gyp. plaster	1		2.62	8.35	10.97
3520				2		2.91	10.90	13.81
3700			portland - 3 coat	1		2.42	8.70	11.12
3900			5/8″ drywall	1		2.65	7.55	10.20
3910				2		3.15	9.30	12.45

C1010 Partitions

C1010 102	Concrete Block Partitions - Regular Weight

	TYPE	THICKNESS (IN.)	TYPE FINISH	SIDES FINISHED		COST PER S.F.		
						MAT.	INST.	TOTAL
4002	Solid	6	none	0		2.52	6.25	8.77
4010			gyp. plaster	1		2.90	8.80	11.70
4020				2		3.28	11.35	14.63
4200			portland - 3 coat	1		2.79	9.15	11.94
4400			5/8" drywall	1		3.02	8	11.02
4410				2		3.52	9.75	13.27

C1010 104	Concrete Block Partitions - Lightweight

	TYPE	THICKNESS (IN.)	TYPE FINISH	SIDES FINISHED		COST PER S.F.		
						MAT.	INST.	TOTAL
5000	Hollow	4	none	0		1.96	5.50	7.46
5010			gyp. plaster	1		2.34	8.05	10.39
5020				2		2.72	10.60	13.32
5200			portland - 3 coat	1		2.23	8.40	10.63
5400			5/8" drywall	1		2.46	7.25	9.71
5410				2		2.96	9	11.96
5500		6	none	0		2.77	5.85	8.62
5520			gyp. plaster	2		3.53	10.95	14.48
5700			portland - 3 coat	1		3.04	8.75	11.79
5900			5/8" drywall	1		3.27	7.60	10.87
5910				2		3.77	9.35	13.12
6000		8	none	0		3.34	6.25	9.59
6010			gyp. plaster	1		3.72	8.80	12.52
6020				2		4.10	11.40	15.50
6200			portland - 3 coat	1		3.61	9.20	12.81
6400			5/8" drywall	1		3.84	8	11.84
6410				2		4.34	9.75	14.09
6500		10	none	0		4.01	6.55	10.56
6510			gyp. plaster	1		4.39	9.10	13.49
6520				2		4.77	11.70	16.47
6700			portland - 3 coat	1		4.28	9.50	13.78
6900			5/8" drywall	1		4.51	8.30	12.81
6910				2		5	10.05	15.05
7000	Solid	4	none	0		2.74	5.75	8.49
7010			gyp. plaster	1		3.12	8.30	11.42
7020				2		3.50	10.85	14.35
7200			portland - 3 coat	1		3.01	8.65	11.66
7400			5/8" drywall	1		3.24	7.50	10.74
7410				2		3.74	9.25	12.99
7500		6	none	0		3.06	6.15	9.21
7510			gyp. plaster	1		3.52	8.75	12.27
7520				2		3.82	11.25	15.07
7700			portland - 3 coat	1		3.33	9.05	12.38
7900			5/8" drywall	1		3.56	7.90	11.46
7910				2		4.06	9.65	13.71
8000		8	none	0		3.89	6.55	10.44
8010			gyp. plaster	1		4.27	9.10	13.37
8020				2		4.65	11.70	16.35
8200			portland - 3 coat	1		4.16	9.50	13.66
8400			5/8" drywall	1		4.39	8.30	12.69
8410				2		4.89	10.05	14.94

C1010 Partitions

Single Wythe

Structural facing tile
8W series
8″ x 16″

Double Wythe

Single Wythe

Structural facing tile
6T series
5-1/3″ x 12″

Double Wythe

C1010 120	Tile Partitions	COST PER S.F.		
		MAT.	INST.	TOTAL
1000	8W series 8″x16″, 4″ thick wall, reinf every 2 courses, glazed 1 side	12.20	6.75	18.95
1100	Glazed 2 sides	14.80	7.15	21.95
1200	Glazed 2 sides, using 2 wythes of 2″ thick tile	19.30	12.90	32.20
1300	6″ thick wall, horizontal reinf every 2 courses, glazed 1 side	22.50	7.05	29.55
1400	Glazed 2 sides, each face different color, 2″ and 4″ tile	22	13.20	35.20
1500	8″ thick wall, glazed 2 sides using 2 wythes of 4″ thick tile	24.50	13.50	38
1600	10″ thick wall, glazed 2 sides using 1 wythe of 4″ tile & 1 wythe of 6″ tile	34.50	13.80	48.30
1700	Glazed 2 sides cavity wall, using 2 wythes of 4″ thick tile	24.50	13.50	38
1800	12″ thick wall, glazed 2 sides using 2 wythes of 6″ thick tile	45	14.10	59.10
1900	Glazed 2 sides cavity wall, using 2 wythes of 4″ thick tile	24.50	13.50	38
2100	6T series 5-1/3″x12″ tile, 4″ thick, non load bearing glazed one side,	11.85	10.55	22.40
2200	Glazed two sides	15.25	11.90	27.15
2300	Glazed two sides, using two wythes of 2″ thick tile	17.50	20.50	38
2400	6″ thick, glazed one side	17.30	11.05	28.35
2500	Glazed two sides	20.50	12.55	33.05
2600	Glazed two sides using 2″ thick tile and 4″ thick tile	20.50	21	41.50
2700	8″ thick, glazed one side	23	12.90	35.90
2800	Glazed two sides using two wythes of 4″ thick tile	23.50	21	44.50
2900	Glazed two sides using 6″ thick tile and 2″ thick tile	26	21.50	47.50
3000	10″ thick cavity wall, glazed two sides using two wythes of 4″ tile	23.50	21	44.50
3100	12″ thick, glazed two sides using 4″ thick tile and 8″ thick tile	35	23.50	58.50
3200	2″ thick facing tile, glazed one side, on 6″ concrete block	11.95	12.25	24.20
3300	On 8″ concrete block	12.25	12.65	24.90
3400	On 10″ concrete block	12.95	12.90	25.85

C10 Interior Construction

C1010 Partitions

Gypsum board, single layer each side on wood studs.

Gypsum board, sound deadening board each side, with 1-1/2" insulation on wood studs.

Gypsum board, two layers each side on wood studs.

Gypsum board, single layer each side on metal studs.

Gypsum board, sound deadening board each side on metal studs.

Gypsum board two layers one side, single layer opposite side, with 3-1/2" insulation on metal studs.

C1010 124 — Drywall Partitions/Wood Stud Framing

	FACE LAYER	BASE LAYER	FRAMING	OPPOSITE FACE	INSULATION	COST PER S.F.		
						MAT.	INST.	TOTAL
1200	5/8" FR drywall	none	2 x 4, @ 16" O.C.	same	0	.99	3.07	4.06
1250				5/8" reg. drywall	0	1	3.07	4.07
1300				nothing	0	.64	2.05	2.69
1400		1/4" SD gypsum	2 x 4 @ 16" O.C.	same	1-1/2" fiberglass	2.41	4.72	7.13
1450				5/8" FR drywall	1-1/2" fiberglass	2.05	4.15	6.20
1500				nothing	1-1/2" fiberglass	1.70	3.13	4.83
1600		resil. channels	2 x 4 @ 16", O.C.	same	1-1/2" fiberglass	2.03	6	8.03
1650				5/8" FR drywall	1-1/2" fiberglass	1.86	4.79	6.65
1700				nothing	1-1/2" fiberglass	1.51	3.77	5.28
1800		5/8" FR drywall	2 x 4 @ 24" O.C.	same	0	1.52	3.88	5.40
1850				5/8" FR drywall	0	1.22	3.37	4.59
1900				nothing	0	.87	2.35	3.22
2200		5/8" FR drywall	2 rows-2 x 4	same	2" fiberglass	2.63	5.60	8.23
2250			16"O.C.	5/8" FR drywall	2" fiberglass	2.33	5.10	7.43
2300				nothing	2" fiberglass	1.98	4.09	6.07
2400	5/8" WR drywall	none	2 x 4, @ 16" O.C.	same	0	1.33	3.07	4.40
2450				5/8" FR drywall	0	1.16	3.07	4.23
2500				nothing	0	.81	2.05	2.86
2600		5/8" FR drywall	2 x 4, @ 24" O.C.	same	0	1.86	3.88	5.74
2650				5/8" FR drywall	0	1.39	3.37	4.76
2700				nothing	0	1.04	2.35	3.39
2800	5/8 VF drywall	none	2 x 4, @ 16" O.C.	same	0	1.85	3.31	5.16
2850				5/8" FR drywall	0	1.42	3.19	4.61
2900				nothing	0	1.07	2.17	3.24

C10 Interior Construction

C1010 Partitions

C1010 124		Drywall Partitions/Wood Stud Framing						

	FACE LAYER	BASE LAYER	FRAMING	OPPOSITE FACE	INSULATION	COST PER S.F.		
						MAT.	INST.	TOTAL
3000	5/8 VF drywall	5/8" FR drywall	2 x 4 , 24" O.C.	same	0	2.38	4.12	6.50
3050				5/8" FR drywall	0	1.65	3.49	5.14
3100				nothing	0	1.30	2.47	3.77
3200	1/2" reg drywall	3/8" reg drywall	2 x 4, @ 16" O.C.	same	0	1.43	4.09	5.52
3252				5/8"FR drywall	0	1.21	3.58	4.79
3300				nothing	0	.86	2.56	3.42

C1010 126		Drywall Partitions/Metal Stud Framing						

	FACE LAYER	BASE LAYER	FRAMING	OPPOSITE FACE	INSULATION	COST PER S.F.		
						MAT.	INST.	TOTAL
5200	5/8" FR drywall	none	1-5/8" @ 24" O.C.	same	0	.89	2.71	3.60
5250				5/8" reg. drywall	0	.90	2.71	3.61
5300				nothing	0	.54	1.69	2.23
5400			3-5/8" @ 24" O.C.	same	0	.96	2.73	3.69
5450				5/8" reg. drywall	0	.97	2.73	3.70
5500				nothing	0	.61	1.71	2.32
5600		1/4" SD gypsum	1-5/8" @ 24" O.C.	same	0	1.61	3.85	5.46
5650				5/8" FR drywall	0	1.25	3.28	4.53
5700				nothing	0	.90	2.26	3.16
5800			2-1/2" @ 24" O.C.	same	0	1.65	3.86	5.51
5850				5/8" FR drywall	0	1.29	3.29	4.58
5900				nothing	0	.94	2.27	3.21
6000		5/8" FR drywall	2-1/2" @ 16" O.C.	same	0	1.71	4.37	6.08
6050				5/8" FR drywall	0	1.41	3.86	5.27
6100				nothing	0	1.06	2.84	3.90
6200			3-5/8" @ 24" O.C.	same	0	1.56	3.75	5.31
6250				5/8"FR drywall	3-1/2" fiberglass	1.85	3.62	5.47
6300				nothing	0	.91	2.22	3.13
6400	5/8" WR drywall	none	1-5/8" @ 24" O.C.	same	0	1.23	2.71	3.94
6450				5/8" FR drywall	0	1.06	2.71	3.77
6500				nothing	0	.71	1.69	2.40
6600			3-5/8" @ 24" O.C.	same	0	1.30	2.73	4.03
6650				5/8" FR drywall	0	1.13	2.73	3.86
6700				nothing	0	.78	1.71	2.49
6800		5/8" FR drywall	2-1/2" @ 16" O.C.	same	0	2.05	4.37	6.42
6850				5/8" FR drywall	0	1.58	3.86	5.44
6900				nothing	0	1.23	2.84	4.07
7000			3-5/8" @ 24" O.C.	same	0	1.90	3.75	5.65
7050				5/8"FR drywall	3-1/2" fiberglass	2.02	3.62	5.64
7100				nothing	0	1.08	2.22	3.30
7200	5/8" VF drywall	none	1-5/8" @ 24" O.C.	same	0	1.75	2.95	4.70
7250				5/8" FR drywall	0	1.32	2.83	4.15
7300				nothing	0	.97	1.81	2.78
7400			3-5/8" @ 24" O.C.	same	0	1.82	2.97	4.79
7450				5/8" FR drywall	0	1.39	2.85	4.24
7500				nothing	0	1.04	1.83	2.87
7600		5/8" FR drywall	2-1/2" @ 16" O.C.	same	0	2.57	4.61	7.18
7650				5/8" FR drywall	0	1.84	3.98	5.82
7700				nothing	0	1.49	2.96	4.45
7800			3-5/8" @ 24" O.C.	same	0	2.42	3.99	6.41
7850				5/8"FR drywall	3-1/2" fiberglass	2.28	3.74	6.02
7900				nothing	0	1.34	2.34	3.68

338

C1010 Partitions

C1010 128	Drywall Components	COST PER S.F.		
		MAT.	INST.	TOTAL
0060	Metal studs, 24" O.C. including track, load bearing, 20 gage, 2-1/2"	.62	.96	1.58
0080	3-5/8"	.74	.98	1.72
0100	4"	.67	1	1.67
0120	6"	.98	1.02	2
0140	Metal studs, 24" O.C. including track, load bearing, 18 gage, 2-1/2"	.62	.96	1.58
0160	3-5/8"	.74	.98	1.72
0180	4"	.67	1	1.67
0200	6"	.98	1.02	2
0220	16 gage, 2-1/2"	.72	1.09	1.81
0240	3-5/8"	.86	1.12	1.98
0260	4"	.91	1.14	2.05
0280	6"	1.14	1.17	2.31
0300	Non load bearing, 25 gage, 1-5/8"	.19	.67	.86
0340	3-5/8"	.26	.69	.95
0360	4"	.27	.69	.96
0380	6"	.40	.71	1.11
0400	20 gage, 2-1/2"	.30	.85	1.15
0420	3-5/8"	.33	.87	1.20
0440	4"	.39	.87	1.26
0460	6"	.46	.88	1.34
0540	Wood studs including blocking, shoe and double top plate, 2"x4", 12" O.C.	.36	1.28	1.64
0560	16" O.C.	.29	1.03	1.32
0580	24" O.C.	.22	.82	1.04
0600	2"x6", 12" O.C.	.54	1.47	2.01
0620	16" O.C.	.44	1.14	1.58
0640	24" O.C.	.33	.89	1.22
0642	Furring one side only, steel channels, 3/4", 12" O.C.	.34	1.85	2.19
0644	16" O.C.	.31	1.64	1.95
0646	24" O.C.	.20	1.24	1.44
0647	1-1/2" , 12" O.C.	.46	2.07	2.53
0648	16" O.C.	.41	1.81	2.22
0649	24" O.C.	.27	1.43	1.70
0650	Wood strips, 1" x 3", on wood, 12" O.C.	.38	.93	1.31
0651	16" O.C.	.29	.70	.99
0652	On masonry, 12" O.C.	.38	1.04	1.42
0653	16" O.C.	.29	.78	1.07
0654	On concrete, 12" O.C.	.38	1.97	2.35
0655	16" O.C.	.29	1.48	1.77
0665	Gypsum board, one face only, exterior sheathing, 1/2"	.42	.91	1.33
0680	Interior, fire resistant, 1/2"	.31	.51	.82
0700	5/8"	.30	.51	.81
0720	Sound deadening board 1/4"	.36	.57	.93
0740	Standard drywall 3/8"	.26	.51	.77
0760	1/2"	.26	.51	.77
0780	5/8"	.31	.51	.82
0800	Tongue & groove coreboard 1"	.61	2.13	2.74
0820	Water resistant, 1/2"	.39	.51	.90
0840	5/8"	.47	.51	.98
0860	Add for the following:, foil backing	.15		.15
0880	Fiberglass insulation, 3-1/2"	.59	.38	.97
0900	6"	.85	.38	1.23
0920	Rigid insulation 1"	.54	.51	1.05
0940	Resilient furring @ 16" O.C.	.22	1.61	1.83
0960	Taping and finishing	.05	.51	.56
0980	Texture spray	.06	.54	.60
1000	Thin coat plaster	.09	.64	.73
1040	2"x4" staggered studs 2"x6" plates & blocking	.41	1.13	1.54

Gypsum plaster and gypsum lath on metal studs.

Gypsum plaster and diamond metal lath on metal studs.

Gypsum plaster and gypsum lath on wood studs.

Gypsum plaster and diamond metal lath on wood studs.

C1010 140		Plaster Partitions/Metal Stud Framing							
	TYPE	FRAMING	LATH	OPPOSITE FACE		COST PER S.F.			
						MAT.	INST.	TOTAL	
1000	2 coat gypsum	2-1/2" @ 16"O.C.	3/8" gypsum	same		2.70	7.35	10.05	
1010				nothing		1.57	4.29	5.86	
1100		3-1/4" @ 24"O.C.	1/2" gypsum	same		2.63	7.10	9.73	
1110				nothing		1.50	3.98	5.48	
1500	2 coat vermiculite	2-1/2" @ 16"O.C.	3/8" gypsum	same		2.58	8.05	10.63	
1510				nothing		1.51	4.64	6.15	
1600		3-1/4" @ 24"O.C.	1/2" gypsum	same		2.51	7.80	10.31	
1610				nothing		1.44	4.33	5.77	
2000	3 coat gypsum	2-1/2" @ 16"O.C.	3/8" gypsum	same		2.58	8.35	10.93	
2010				nothing		1.51	4.76	6.27	
2020			3.4lb. diamond	same		2.50	8.35	10.85	
2030				nothing		1.47	4.76	6.23	
2040			2.75lb. ribbed	same		2.25	8.35	10.60	
2050				nothing		1.34	4.76	6.10	
2100		3-1/4" @ 24"O.C.	1/2" gypsum	same		2.51	8.05	10.56	
2110				nothing		1.44	4.45	5.89	
2120			3.4lb. ribbed	same		2.69	8.05	10.74	
2130				nothing		1.53	4.45	5.98	
3500	3 coat gypsum W/med. Keenes	2-1/2" @ 16"O.C.	3/8" gypsum	same		3.18	10.70	13.88	
3510				nothing		1.81	5.95	7.76	
3520			3.4lb. diamond	same		3.10	10.70	13.80	
3530				nothing		1.77	5.95	7.72	
3540			2.75lb. ribbed	same		2.85	10.70	13.55	
3550				nothing		1.64	5.95	7.59	
3600		3-1/4" @ 24"O.C.	1/2" gypsum	same		3.11	10.45	13.56	
3610				nothing		1.74	5.60	7.34	
3620			3.4lb. ribbed	same		3.29	10.45	13.74	
3630				nothing		1.83	5.60	7.43	

C10 Interior Construction

C1010 Partitions

| C1010 140 | Plaster Partitions/Metal Stud Framing | | | | | | | |

	TYPE	FRAMING	LATH	OPPOSITE FACE		COST PER S.F.		
						MAT.	INST.	TOTAL
4000	3 coat gypsum	2-1/2" @ 16"O.C.	3/8" gypsum	same		3.18	11.85	15.03
4010	W/hard Keenes			nothing		1.81	6.50	8.31
4022			3.4 lb. diamond	same		3.10	11.85	14.95
4032				nothing		1.77	6.50	8.27
4040			2.75lb. ribbed	same		2.85	11.85	14.70
4050				nothing		1.64	6.50	8.14
4100		3-1/4" @ 24"O.C.	1/2" gypsum	same		3.11	11.60	14.71
4110				nothing		1.74	6.20	7.94
4120			3.4lb. ribbed	same		3.29	11.60	14.89
4130				nothing		1.83	6.20	8.03

| C1010 142 | Plaster Partitions/Wood Stud Framing | | | | | | | |

	TYPE	FRAMING	LATH	OPPOSITE FACE		COST PER S.F.		
						MAT.	INST.	TOTAL
5000	2 coat gypsum	2"x4" @ 16"O.C.	3/8" gypsum	same		2.68	7.15	9.83
5010				nothing		1.55	4.20	5.75
5100		2"x4" @ 24"O.C.	1/2" gypsum	same		2.59	7.05	9.64
5110				nothing		1.46	4.02	5.48
5500	2 coat vermiculite	2"x4" @ 16"O.C.	3/8" gypsum	same		2.56	7.90	10.46
5510				nothing		1.49	4.55	6.04
5600		2"x4" @ 24"O.C.	1/2" gypsum	same		2.47	7.75	10.22
5610				nothing		1.40	4.37	5.77
6000	3 coat gypsum	2"x4" @ 16"O.C.	3/8" gypsum	same		2.56	8.15	10.71
6010				nothing		1.49	4.67	6.16
6020			3.4lb. diamond	same		2.48	8.25	10.73
6030				nothing		1.45	4.70	6.15
6040			2.75lb. ribbed	same		2.20	8.30	10.50
6050				nothing		1.31	4.73	6.04
6100		2"x4" @ 24"O.C.	1/2" gypsum	same		2.47	8.05	10.52
6110				nothing		1.40	4.49	5.89
6120			3.4lb. ribbed	same		2.16	8.10	10.26
6130				nothing		1.25	4.53	5.78
7500	3 coat gypsum	2"x4" @ 16"O.C.	3/8" gypsum	same		3.16	10.55	13.71
7510	W/med Keenes			nothing		1.79	5.85	7.64
7520			3.4lb. diamond	same		3.08	10.60	13.68
7530				nothing		1.75	5.85	7.60
7540			2.75lb. ribbed	same		2.80	10.65	13.45
7550				nothing		1.61	5.90	7.51
7600		2"x4" @ 24"O.C.	1/2" gypsum	same		3.07	10.40	13.47
7610				nothing		1.70	5.65	7.35
7620			3.4lb. ribbed	same		3.25	10.55	13.80
7630				nothing		1.79	5.75	7.54
8000	3 coat gypsum	2"x4" @ 16"O.C.	3/8" gypsum	same		3.16	11.65	14.81
8010	W/hard Keenes			nothing		1.79	6.45	8.24
8020			3.4lb. diamond	same		3.08	11.75	14.83
8030				nothing		1.75	6.45	8.20
8040			2.75lb. ribbed	same		2.80	11.80	14.60
8050				nothing		1.61	6.50	8.11
8100		2"x4" @ 24"O.C.	1/2" gypsum	same		3.07	11.55	14.62
8110				nothing		1.70	6.25	7.95
8120			3.4lb. ribbed	same		3.25	11.70	14.95
8130				nothing		1.79	6.35	8.14

C10 Interior Construction

C1010 Partitions

C1010 144	Plaster Partition Components	COST PER S.F.		
		MAT.	INST.	TOTAL
0060	Metal studs, 16" O.C., including track, non load bearing, 25 gage, 1-5/8"	.31	1.05	1.36
0080	2-1/2"	.31	1.05	1.36
0100	3-1/4"	.36	1.07	1.43
0120	3-5/8"	.36	1.07	1.43
0140	4"	.36	1.08	1.44
0160	6"	.54	1.09	1.63
0180	Load bearing, 20 gage, 2-1/2"	.84	1.33	2.17
0200	3-5/8"	1	1.35	2.35
0220	4"	1.05	1.39	2.44
0240	6"	1.33	1.41	2.74
0260	16 gage 2-1/2"	.99	1.51	2.50
0280	3-5/8"	1.19	1.56	2.75
0300	4"	1.25	1.58	2.83
0320	6"	1.57	1.60	3.17
0340	Wood studs, including blocking, shoe and double plate, 2"x4", 12" O.C.	.36	1.28	1.64
0360	16" O.C.	.29	1.03	1.32
0380	24" O.C.	.22	.82	1.04
0400	2"x6", 12" O.C.	.54	1.47	2.01
0420	16" O.C.	.44	1.14	1.58
0440	24" O.C.	.33	.89	1.22
0460	Furring one face only, steel channels, 3/4", 12" O.C.	.34	1.85	2.19
0480	16" O.C.	.31	1.64	1.95
0500	24" O.C.	.20	1.24	1.44
0520	1-1/2", 12" O.C.	.46	2.07	2.53
0540	16" O.C.	.41	1.81	2.22
0560	24"O.C.	.27	1.43	1.70
0580	Wood strips 1"x3", on wood., 12" O.C.	.38	.93	1.31
0600	16"O.C.	.29	.70	.99
0620	On masonry, 12" O.C.	.38	1.04	1.42
0640	16" O.C.	.29	.78	1.07
0660	On concrete, 12" O.C.	.38	1.97	2.35
0680	16" O.C.	.29	1.48	1.77
0700	Gypsum lath. plain or perforated, nailed to studs, 3/8" thick	.62	.57	1.19
0720	1/2" thick	.60	.60	1.20
0740	Clipped to studs, 3/8" thick	.62	.64	1.26
0760	1/2" thick	.60	.69	1.29
0780	Metal lath, diamond painted, nailed to wood studs, 2.5 lb.	.55	.57	1.12
0800	3.4 lb.	.58	.60	1.18
0820	Screwed to steel studs, 2.5 lb.	.55	.60	1.15
0840	3.4 lb.	.49	.64	1.13
0860	Rib painted, wired to steel, 2.75 lb	.45	.64	1.09
0880	3.4 lb	.69	.69	1.38
0900	4.0 lb	.70	.74	1.44
0910				
0920	Gypsum plaster, 2 coats	.46	2.44	2.90
0940	3 coats	.65	2.91	3.56
0960	Perlite or vermiculite plaster, 2 coats	.46	2.79	3.25
0980	3 coats	.77	3.45	4.22
1000	Stucco, 3 coats, 1" thick, on wood framing	.68	4.81	5.49
1020	On masonry	.28	3.80	4.08
1100	Metal base galvanized and painted 2-1/2" high	.78	1.81	2.59

C10 Interior Construction

C1010 Partitions

Folding Accordion

Folding Leaf

Movable and Borrow Lites

C1010 205	Partitions	COST PER S.F.		
		MAT.	INST.	TOTAL
0360	Folding accordion, vinyl covered, acoustical, 3 lb. S.F., 17 ft max. hgt	28	10.25	38.25
0380	5 lb. per S.F. 27 ft max height	39	10.80	49.80
0400	5.5 lb. per S.F., 17 ft. max height	46	11.40	57.40
0420	Commercial, 1.75 lb per S.F., 8 ft. max height	24.50	4.55	29.05
0440	2.0 Lb per S.F., 17 ft. max height	25	6.85	31.85
0460	Industrial, 4.0 lb. per S.F. 27 ft. max height	36	13.65	49.65
0480	Vinyl clad wood or steel, electric operation 6 psf	53	6.40	59.40
0500	Wood, non acoustic, birch or mahogany	28.50	3.42	31.92
0560	Folding leaf, aluminum framed acoustical 12 ft.high.,5.5 lb per S.F., min.	40.50	17.10	57.60
0580	Maximum	49	34	83
0600	6.5 lb. per S.F., minimum	43	17.10	60.10
0620	Maximum	53	34	87
0640	Steel acoustical, 7.5 per S.F., vinyl faced, minimum	60.50	17.10	77.60
0660	Maximum	73.50	34	107.50
0680	Wood acoustic type, vinyl faced to 18' high 6 psf, minimum	56.50	17.10	73.60
0700	Average	67.50	23	90.50
0720	Maximum	87.50	34	121.50
0740	Formica or hardwood faced, minimum	58.50	17.10	75.60
0760	Maximum	62.50	34	96.50
0780	Wood, low acoustical type to 12 ft. high 4.5 psf	42.50	20.50	63
0840	Demountable, trackless wall, cork finish, semi acous, 1-5/8" th, min	42.50	3.15	45.65
0860	Maximum	40	5.40	45.40
0880	Acoustic, 2" thick, minimum	34	3.36	37.36
0900	Maximum	59	4.55	63.55
0920	In-plant modular office system, w/prehung steel door			
0940	3" thick honeycomb core panels			
0960	12' x 12', 2 wall	12.10	.65	12.75
0970	4 wall	10.20	.86	11.06
0980	16' x 16', 2 wall	10.65	.45	11.10
0990	4 wall	7.20	.46	7.66
1000	Gypsum, demountable, 3" to 3-3/4" thick x 9' high, vinyl clad	7.05	2.39	9.44
1020	Fabric clad	17.55	2.61	20.16
1040	1.75 system, vinyl clad hardboard, paper honeycomb core panel			
1060	1-3/4" to 2-1/2" thick x 9' high	11.90	2.39	14.29
1080	Unitized gypsum panel system, 2" to 2-1/2" thick x 9' high			
1100	Vinyl clad gypsum	15.20	2.39	17.59
1120	Fabric clad gypsum	25	2.61	27.61
1140	Movable steel walls, modular system			
1160	Unitized panels, 48" wide x 9' high			
1180	Baked enamel, pre-finished	17.20	1.90	19.10
1200	Fabric clad	25	2.03	27.03

C1020 Interior Doors

Single Leaf   **Sliding Entrance**   **Rolling Overhead**

C1020 102	Special Doors	COST PER OPNG.		
		MAT.	**INST.**	**TOTAL**
2500	Single leaf, wood, 3'-0"x7'-0"x1 3/8", birch, solid core	385	191	576
2510	Hollow core	315	183	498
2530	Hollow core, lauan	300	183	483
2540	Louvered pine	445	183	628
2550	Paneled pine	445	183	628
2600	Hollow metal, comm. quality, flush, 3'-0"x7'-0"x1-3/8"	800	201	1,001
2650	3'-0"x10'-0" openings with panel	1,300	260	1,560
2700	Metal fire, comm. quality, 3'-0"x7'-0"x1-3/8"	845	209	1,054
2800	Kalamein fire, comm. quality, 3'-0"x7'-0"x1-3/4"	615	385	1,000
3200	Double leaf, wood, hollow core, 2 - 3'-0"x7'-0"x1-3/8"	555	350	905
3300	Hollow metal, comm. quality, B label, 2'-3'-0"x7'-0"x1-3/8"	1,625	440	2,065
3400	6'-0"x10'-0" opening, with panel	2,550	545	3,095
3500	Double swing door system, 12'-0"x7'-0", mill finish	8,250	2,875	11,125
3700	Black finish	8,350	2,975	11,325
3800	Sliding entrance door and system mill finish	9,750	2,425	12,175
3900	Bronze finish	10,700	2,625	13,325
4000	Black finish	11,100	2,725	13,825
4100	Sliding panel mall front, 16'x9' opening, mill finish	3,400	745	4,145
4200	Bronze finish	4,425	970	5,395
4300	Black finish	5,450	1,200	6,650
4400	24'x9' opening mill finish	4,850	1,375	6,225
4500	Bronze finish	6,300	1,800	8,100
4600	Black finish	7,750	2,200	9,950
4700	48'x9' opening mill finish	9,025	1,075	10,100
4800	Bronze finish	11,700	1,400	13,100
4900	Black finish	14,400	1,725	16,125
5000	Rolling overhead steel door, manual, 8' x 8' high	1,175	825	2,000
5100	10' x 10' high	1,950	940	2,890
5200	20' x 10' high	3,250	1,325	4,575
5300	12' x 12' high	1,975	1,100	3,075
5400	Motor operated, 8' x 8' high	2,300	1,100	3,400
5500	10' x 10' high	3,075	1,200	4,275
5600	20' x 10' high	4,375	1,600	5,975
5700	12' x 12' high	3,100	1,375	4,475
5800	Roll up grille, aluminum, manual, 10' x 10' high, mill finish	3,000	1,600	4,600
5900	Bronze anodized	4,800	1,600	6,400
6000	Motor operated, 10' x 10' high, mill finish	4,375	1,875	6,250
6100	Bronze anodized	6,175	1,875	8,050
6200	Steel, manual, 10' x 10' high	2,700	1,325	4,025
6300	15' x 8' high	3,125	1,650	4,775
6400	Motor operated, 10' x 10' high	4,075	1,600	5,675
6500	15' x 8' high	4,500	1,925	6,425

C1030 Fittings

Toilet Units

Entrance Screens

Urinal Screens

C1030 110	Toilet Partitions	COST PER UNIT		
		MAT.	INST.	TOTAL
0380	Toilet partitions, cubicles, ceiling hung, marble	1,600	480	2,080
0400	Painted metal	490	256	746
0420	Plastic laminate	615	256	871
0460	Stainless steel	1,550	256	1,806
0480	Handicap addition	445		445
0520	Floor and ceiling anchored, marble	2,000	385	2,385
0540	Painted metal	590	205	795
0560	Plastic laminate	840	205	1,045
0600	Stainless steel	1,675	205	1,880
0620	Handicap addition	315		315
0660	Floor mounted marble	1,175	320	1,495
0680	Painted metal	590	146	736
0700	Plastic laminate	610	146	756
0740	Stainless steel	1,625	146	1,771
0760	Handicap addition	315		315
0780	Juvenile deduction	45.50		45.50
0820	Floor mounted with handrail marble	1,125	320	1,445
0840	Painted metal	515	171	686
0860	Plastic laminate	880	171	1,051
0900	Stainless steel	1,300	171	1,471
0920	Handicap addition	365		365
0960	Wall hung, painted metal	685	146	831
1020	Stainless steel	1,825	146	1,971
1040	Handicap addition	365		365
1080	Entrance screens, floor mounted, 54" high, marble	745	107	852
1100	Painted metal	256	68.50	324.50
1140	Stainless steel	985	68.50	1,053.50
1300	Urinal screens, floor mounted, 24" wide, laminated plastic	242	128	370
1320	Marble	630	152	782
1340	Painted metal	230	128	358
1380	Stainless steel	630	128	758
1428	Wall mounted wedge type, painted metal	148	102	250
1460	Stainless steel	620	102	722
1500	Partitions, shower stall, single wall, painted steel, 2'-8" x 2'-8"	1,025	238	1,263
1510	Fiberglass, 2'-8" x 2'-8"	800	264	1,064
1520	Double wall, enameled steel, 2'-8" x 2'-8"	935	238	1,173
1530	Stainless steel, 2'-8" x 2'-8"	2,200	238	2,438
1560	Tub enclosure, sliding panels, tempered glass, aluminum frame	415	297	712
1570	Chrome/brass frame-deluxe	1,225	395	1,620

C1030 Fittings

C1030 310	Storage Specialties, EACH	COST EACH		
		MAT.	INST.	TOTAL
0200	Lockers, steel, single tier, 5' to 6' high, per opening, min.	157	42.50	199.50
0210	Maximum	300	49.50	349.50
0600	Shelving, metal industrial, braced, 3' wide, 1' deep	22	11.30	33.30
0610	2' deep	31	12	43

C1030 510	Identifying/Visual Aid Specialties, EACH	COST EACH		
		MAT.	INST.	TOTAL
0100	Control boards, magnetic, porcelain finish, framed, 24" x 18"	219	128	347
0110	96" x 48"	1,175	205	1,380
0120	Directory boards, outdoor, black plastic, 36" x 24"	790	510	1,300
0130	36" x 36"	915	685	1,600
0140	Indoor, economy, open faced, 18" x 24"	141	146	287
0500	Signs, interior electric exit sign, wall mounted, 6"	122	73	195
0510	Street, reflective alum., dbl. face, 4 way, w/bracket	190	34	224
0520	Letters, cast aluminum, 1/2" deep, 4" high	24.50	28.50	53
0530	1" deep, 10" high	54.50	28.50	83
0540	Plaques, cast aluminum, 20" x 30"	1,150	256	1,406
0550	Cast bronze, 36" x 48"	4,650	510	5,160

C1030 520	Identifying/Visual Aid Specialties, S.F.	COST PER S.F.		
		MAT.	INST.	TOTAL
0100	Bulletin board, cork sheets, no frame, 1/4" thick	3.56	3.53	7.09
0120	Aluminum frame, 1/4" thick, 3' x 5'	8.30	4.29	12.59
0200	Chalkboards, wall hung, alum, frame & chalktrough	12.10	2.26	14.36
0210	Wood frame & chalktrough	11.95	2.45	14.40
0220	Sliding board, one board with back panel	51	2.19	53.19
0230	Two boards with back panel	86	2.19	88.19
0240	Liquid chalk type, alum. frame & chalktrough	11.45	2.26	13.71
0250	Wood frame & chalktrough	26	2.26	28.26

C1030 710	Bath and Toilet Accessories, EACH	COST EACH		
		MAT.	INST.	TOTAL
0100	Specialties, bathroom accessories, st. steel, curtain rod, 5' long, 1" diam	38	39.50	77.50
0120	Dispenser, towel, surface mounted	38.50	32	70.50
0140	Grab bar, 1-1/4" diam., 12" long	33.50	21.50	55
0160	Mirror, framed with shelf, 18" x 24"	208	25.50	233.50
0170	72" x 24"	291	85.50	376.50
0180	Toilet tissue dispenser, surface mounted, single roll	19.15	17.10	36.25
0200	Towel bar, 18" long	43.50	22.50	66
0300	Medicine cabinets, sliding mirror doors, 20" x 16" x 4-3/4", unlighted	121	73	194
0310	24" x 19" x 8-1/2", lighted	175	102	277

C1030 730	Bath and Toilet Accessories, L.F.	COST PER L.F.		
		MAT.	INST.	TOTAL
0100	Partitions, hospital curtain, ceiling hung, polyester oxford cloth	19.55	5	24.55
0110	Designer oxford cloth	36.50	6.30	42.80

C1030 830	Fabricated Cabinets & Counters, L.F.	COST PER L.F.		
		MAT.	INST.	TOTAL
0110	Household, base, hardwood, one top drawer & one door below x 12" wide	252	41.50	293.50
0120	Four drawer x 24" wide	335	46	381
0130	Wall, hardwood, 30" high with one door x 12" wide	204	46.50	250.50
0140	Two doors x 48" wide	475	55.50	530.50
0150	Counter top-laminated plastic, stock, economy	10	17.10	27.10
0160	Custom-square edge, 7/8" thick	14.45	38.50	52.95

C10 Interior Construction

C1030 Fittings

C1030 830	Fabricated Cabinets & Counters, L.F.	COST PER L.F.		
		MAT.	INST.	TOTAL
0170	School, counter, wood, 32" high	236	51	287
0180	Metal, 84" high	425	68.50	493.50

C1030 910	Other Fittings, EACH	COST EACH		
		MAT.	INST.	TOTAL
0500	Mail boxes, horizontal, rear loaded, aluminum, 5" x 6" x 15" deep	38.50	15.05	53.55
0510	Front loaded, aluminum, 10" x 12" x 15" deep	119	25.50	144.50
0520	Vertical, front loaded, aluminum, 15" x 5" x 6" deep	33	15.05	48.05
0530	Bronze, duranodic finish	44	15.05	59.05
0540	Letter slot, post office	116	64	180
0550	Mail counter, window, post office, with grille	605	256	861
0700	Turnstiles, one way, 4' arm, 46" diam., manual	885	205	1,090
0710	Electric	1,450	855	2,305
0720	3 arm, 5'-5" diam. & 7' high, manual	4,375	1,025	5,400
0730	Electric	4,950	1,700	6,650

C2010   Stair Construction

The table below lists the cost per flight for 4'-0" wide stairs. Side walls are not included. Railings are included in the prices.

C2010 110	Stairs	COST PER FLIGHT		
		MAT.	INST.	TOTAL
0470	Stairs, C.I.P. concrete, w/o landing, 12 risers, w/o nosing	960	2,175	3,135
0480	With nosing	1,675	2,475	4,150
0550	W/landing, 12 risers, w/o nosing	1,100	2,650	3,750
0560	With nosing	1,825	2,950	4,775
0570	16 risers, w/o nosing	1,350	3,325	4,675
0580	With nosing	2,300	3,725	6,025
0590	20 risers, w/o nosing	1,600	4,025	5,625
0600	With nosing	2,775	4,525	7,300
0610	24 risers, w/o nosing	1,850	4,675	6,525
0620	With nosing	3,275	5,300	8,575
0630	Steel, grate type w/nosing & rails, 12 risers, w/o landing	4,750	1,150	5,900
0640	With landing	6,200	1,575	7,775
0660	16 risers, with landing	7,775	1,925	9,700
0680	20 risers, with landing	9,350	2,300	11,650
0700	24 risers, with landing	10,900	2,675	13,575
0701				
0710	Concrete fill metal pan & picket rail, 12 risers, w/o landing	6,175	1,150	7,325
0720	With landing	8,125	1,700	9,825
0740	16 risers, with landing	10,200	2,075	12,275
0760	20 risers, with landing	12,200	2,475	14,675
0780	24 risers, with landing	14,300	2,825	17,125
0790	Cast iron tread & pipe rail, 12 risers, w/o landing	6,175	1,150	7,325
0800	With landing	8,125	1,700	9,825
1120	Wood, prefab box type, oak treads, wood rails 3'-6" wide, 14 risers	2,100	425	2,525
1150	Prefab basement type, oak treads, wood rails 3'-0" wide, 14 risers	1,050	105	1,155

C30 Interior Finishes

C3010 Wall Finishes

C3010 230	Paint & Covering	COST PER S.F.		
		MAT.	INST.	TOTAL
0060	Painting, interior on plaster and drywall, brushwork, primer & 1 coat	.11	.67	.78
0080	Primer & 2 coats	.17	.88	1.05
0100	Primer & 3 coats	.23	1.08	1.31
0120	Walls & ceilings, roller work, primer & 1 coat	.11	.44	.55
0140	Primer & 2 coats	.17	.57	.74
0160	Woodwork incl. puttying, brushwork, primer & 1 coat	.11	.96	1.07
0180	Primer & 2 coats	.17	1.27	1.44
0200	Primer & 3 coats	.23	1.73	1.96
0260	Cabinets and casework, enamel, primer & 1 coat	.11	1.08	1.19
0280	Primer & 2 coats	.17	1.33	1.50
0300	Masonry or concrete, latex, brushwork, primer & 1 coat	.22	.90	1.12
0320	Primer & 2 coats	.30	1.29	1.59
0340	Addition for block filler	.19	1.11	1.30
0380	Fireproof paints, intumescent, 1/8" thick 3/4 hour	2.14	.88	3.02
0400	3/16" thick 1 hour	4.78	1.33	6.11
0420	7/16" thick 2 hour	6.15	3.09	9.24
0440	1-1/16" thick 3 hour	10.05	6.20	16.25
0500	Gratings, primer & 1 coat	.29	1.35	1.64
0600	Pipes over 12" diameter	.58	4.33	4.91
0700	Structural steel, brushwork, light framing 300-500 S.F./Ton	.09	1.67	1.76
0720	Heavy framing 50-100 S.F./Ton	.09	.84	.93
0740	Spraywork, light framing 300-500 S.F./Ton	.10	.37	.47
0760	Heavy framing 50-100 S.F./Ton	.10	.41	.51
0800	Varnish, interior wood trim, no sanding sealer & 1 coat	.08	1.08	1.16
0820	Hardwood floor, no sanding 2 coats	.17	.23	.40
0840	Wall coatings, acrylic glazed coatings, minimum	.31	.82	1.13
0860	Maximum	.66	1.42	2.08
0880	Epoxy coatings, minimum	.41	.82	1.23
0900	Maximum	1.23	2.55	3.78
0940	Exposed epoxy aggregate, troweled on, 1/16" to 1/4" aggregate, minimum	.62	1.84	2.46
0960	Maximum	1.32	3.33	4.65
0980	1/2" to 5/8" aggregate, minimum	1.22	3.33	4.55
1000	Maximum	2.09	5.40	7.49
1020	1" aggregate, minimum	2.12	4.81	6.93
1040	Maximum	3.25	7.85	11.10
1060	Sprayed on, minimum	.56	1.47	2.03
1080	Maximum	1.06	2.98	4.04
1100	High build epoxy 50 mil, minimum	.68	1.11	1.79
1120	Maximum	1.17	4.56	5.73
1140	Laminated epoxy with fiberglass minimum	.74	1.47	2.21
1160	Maximum	1.31	2.98	4.29
1180	Sprayed perlite or vermiculite 1/16" thick, minimum	.26	.15	.41
1200	Maximum	.75	.68	1.43
1260	Wall coatings, vinyl plastic, minimum	.33	.59	.92
1280	Maximum	.83	1.80	2.63
1300	Urethane on smooth surface, 2 coats, minimum	.26	.38	.64
1320	Maximum	.58	.65	1.23
1340	3 coats, minimum	.35	.52	.87
1360	Maximum	.79	.92	1.71
1380	Ceramic-like glazed coating, cementitious, minimum	.48	.98	1.46
1400	Maximum	.81	1.25	2.06
1420	Resin base, minimum	.33	.68	1.01
1440	Maximum	.54	1.31	1.85
1460	Wall coverings, aluminum foil	1.12	1.58	2.70
1500	Vinyl backing	6	1.81	7.81
1520	Cork tiles, 12"x12", light or dark, 3/16" thick	4.92	1.81	6.73
1540	5/16" thick	4.18	1.85	6.03
1560	Basketweave, 1/4" thick	6.45	1.81	8.26

C3010 Wall Finishes

C3010 230	Paint & Covering	COST PER S.F.		
		MAT.	INST.	TOTAL
1580	Natural, non-directional, 1/2" thick	8.65	1.81	10.46
1600	12"x36", granular, 3/16" thick	1.39	1.13	2.52
1620	1" thick	1.79	1.18	2.97
1640	12"x12", polyurethane coated, 3/16" thick	4.35	1.81	6.16
1660	5/16" thick	6.20	1.85	8.05
1661	Paneling, prefinished plywood, birch	1.49	2.44	3.93
1662	Mahogany, African	2.76	2.56	5.32
1663	Philippine (lauan)	1.19	2.05	3.24
1664	Oak or cherry	3.54	2.56	6.10
1665	Rosewood	5.05	3.20	8.25
1666	Teak	3.54	2.56	6.10
1667	Chestnut	5.25	2.73	7.98
1668	Pecan	2.27	2.56	4.83
1669	Walnut	5.75	2.56	8.31
1670	Wood board, knotty pine, finished	1.89	4.24	6.13
1671	Rough sawn cedar	2.35	4.24	6.59
1672	Redwood	5.75	4.24	9.99
1673	Aromatic cedar	3.96	4.55	8.51
1680	Cork wallpaper, paper backed, natural	2.19	.91	3.10
1700	Color	3.06	.91	3.97
1720	Gypsum based, fabric backed, minimum	.90	.54	1.44
1740	Average	1.35	.60	1.95
1760	Maximum	1.51	.68	2.19
1780	Vinyl wall covering, fabric back, light weight	.76	.68	1.44
1800	Medium weight	.90	.91	1.81
1820	Heavy weight	1.80	1	2.80
1840	Wall paper, double roll, solid pattern, avg. workmanship	.41	.68	1.09
1860	Basic pattern, avg. workmanship	.78	.81	1.59
1880	Basic pattern, quality workmanship	2.72	1	3.72
1900	Grass cloths with lining paper, minimum	.86	1.09	1.95
1920	Maximum	2.74	1.24	3.98
1940	Ceramic tile, thin set, 4-1/4" x 4-1/4"	2.58	4.34	6.92
1942				
1960	12" x 12"	3.82	2.75	6.57

C3010 235	Paint Trim	COST PER L.F.		
		MAT.	INST.	TOTAL
2040	Painting, wood trim, to 6" wide, enamel, primer & 1 coat	.11	.54	.65
2060	Primer & 2 coats	.17	.69	.86
2080	Misc. metal brushwork, ladders	.59	5.40	5.99
2100	Pipes, to 4" dia.	.07	1.14	1.21
2120	6" to 8" dia.	.14	2.28	2.42
2140	10" to 12" dia.	.46	3.41	3.87
2160	Railings, 2" pipe	.21	2.71	2.92
2180	Handrail, single	.13	1.08	1.21
2185	Caulking & Sealants, Polyureathane, In place, 1 or 2 component, 1/2" X 1/4"	.36	1.75	2.11

C3020 Floor Finishes

C3020 410	Tile & Covering	COST PER S.F.		
		MAT.	INST.	TOTAL
0060	Carpet tile, nylon, fusion bonded, 18″ x 18″ or 24″ x 24″, 24 oz.	3.50	.64	4.14
0080	35 oz.	4.16	.64	4.80
0100	42 oz.	5.55	.64	6.19
0140	Carpet, tufted, nylon, roll goods, 12′ wide, 26 oz.	5.45	.73	6.18
0160	36 oz.	8.25	.73	8.98
0180	Woven, wool, 36 oz.	14.10	.73	14.83
0200	42 oz.	14.45	.73	15.18
0220	Padding, add to above, minimum	.56	.34	.90
0240	Maximum	.95	.34	1.29
0260	Composition flooring, acrylic, 1/4″ thick	1.60	5	6.60
0280	3/8″ thick	2.04	5.75	7.79
0300	Epoxy, minimum	2.74	3.85	6.59
0320	Maximum	3.32	5.30	8.62
0340	Epoxy terrazzo, minimum	5.80	5.60	11.40
0360	Maximum	8.50	7.50	16
0380	Mastic, hot laid, 1-1/2″ thick, minimum	4.02	3.77	7.79
0400	Maximum	5.15	5	10.15
0420	Neoprene 1/4″ thick, minimum	3.95	4.76	8.71
0440	Maximum	5.45	6.05	11.50
0460	Polyacrylate with ground granite 1/4″, minimum	3.28	3.53	6.81
0480	Maximum	5.65	5.40	11.05
0500	Polyester with colored quart 2 chips 1/16″, minimum	2.94	2.44	5.38
0520	Maximum	4.64	3.85	8.49
0540	Polyurethane with vinyl chips, minimum	7.70	2.44	10.14
0560	Maximum	11.10	3.02	14.12
0600	Concrete topping, granolithic concrete, 1/2″ thick	.33	4.15	4.48
0620	1″ thick	.67	4.25	4.92
0640	2″ thick	1.33	4.89	6.22
0660	Heavy duty 3/4″ thick, minimum	.45	6.40	6.85
0680	Maximum	.79	7.65	8.44
0700	For colors, add to above, minimum	.47	1.48	1.95
0720	Maximum	.78	1.64	2.42
0740	Exposed aggregate finish, minimum	.25	.74	.99
0760	Maximum	.77	1	1.77
0780	Abrasives, .25 P.S.F. add to above, minimum	.46	.55	1.01
0800	Maximum	.74	.55	1.29
0820	Dust on coloring, add, minimum	.47	.36	.83
0840	Maximum	.78	.74	1.52
0860	Floor coloring using 0.6 psf powdered color, 1/2″ integral, minimum	5.60	4.15	9.75
0880	Maximum	5.95	4.15	10.10
0900	Dustproofing, add, minimum	.19	.24	.43
0920	Maximum	.68	.36	1.04
0930	Paint	.30	1.29	1.59
0940	Hardeners, metallic add, minimum	.56	.55	1.11
0960	Maximum	1.67	.81	2.48
0980	Non-metallic, minimum	.25	.55	.80
1000	Maximum	.75	.81	1.56
1020	Integral topping and finish, 1:1:2 mix, 3/16″ thick	.11	2.44	2.55
1040	1/2″ thick	.30	2.58	2.88
1060	3/4″ thick	.45	2.87	3.32
1080	1″ thick	.60	3.26	3.86
1100	Terrazzo, minimum	3.16	14.95	18.11
1120	Maximum	5.90	18.70	24.60
1280	Resilient, asphalt tile, 1/8″ thick on concrete, minimum	1.31	1.15	2.46
1300	Maximum	1.43	1.15	2.58
1320	On wood, add for felt underlay	.22		.22
1340	Cork tile, minimum	6.60	1.46	8.06
1360	Maximum	14.15	1.46	15.61

C3020 Floor Finishes

C3020 410	Tile & Covering	COST PER S.F.		
		MAT.	INST.	TOTAL
1380	Polyethylene, in rolls, minimum	3.52	1.67	5.19
1400	Maximum	6.70	1.67	8.37
1420	Polyurethane, thermoset, minimum	5.30	4.60	9.90
1440	Maximum	6.05	9.20	15.25
1460	Rubber, sheet goods, minimum	7.10	3.83	10.93
1480	Maximum	11.35	5.10	16.45
1500	Tile, minimum	7.05	1.15	8.20
1520	Maximum	13.05	1.67	14.72
1580	Vinyl, composition tile, minimum	1.01	.92	1.93
1600	Maximum	1.98	.92	2.90
1620	Vinyl tile, minimum	3.30	.92	4.22
1640	Maximum	5.45	.92	6.37
1660	Sheet goods, minimum	3.85	1.84	5.69
1680	Maximum	6.70	2.30	9
1720	Tile, ceramic natural clay	4.96	4.51	9.47
1730	Marble, synthetic 12"x12"x5/8"	6.65	13.75	20.40
1740	Porcelain type, minimum	5.20	4.51	9.71
1760	Maximum	7.15	5.30	12.45
1800	Quarry tile, mud set, minimum	7.50	5.90	13.40
1820	Maximum	9.95	7.50	17.45
1840	Thin set, deduct		1.18	1.18
1860	Terrazzo precast, minimum	4.71	6.15	10.86
1880	Maximum	9.85	6.15	16
1900	Non-slip, minimum	20	15.15	35.15
1920	Maximum	21.50	20	41.50
2020	Wood block, end grain factory type, natural finish, 2" thick	5.10	4.10	9.20
2040	Fir, vertical grain, 1"x4", no finish, minimum	2.84	2.01	4.85
2060	Maximum	3.01	2.01	5.02
2080	Prefinished white oak, prime grade, 2-1/4" wide	5.25	3.01	8.26
2100	3-1/4" wide	6.75	2.77	9.52
2120	Maple strip, sanded and finished, minimum	4.51	4.39	8.90
2140	Maximum	5.45	4.39	9.84
2160	Oak strip, sanded and finished, minimum	3.50	4.39	7.89
2180	Maximum	5.05	4.39	9.44
2200	Parquetry, sanded and finished, minimum	5.70	4.58	10.28
2220	Maximum	6.30	6.50	12.80
2260	Add for sleepers on concrete, treated, 24" O.C., 1"x2"	2.18	2.63	4.81
2280	1"x3"	2.30	2.05	4.35
2300	2"x4"	.57	1.03	1.60
2340	Underlayment, plywood, 3/8" thick	.85	.68	1.53
2350	1/2" thick	.88	.71	1.59
2360	5/8" thick	1.29	.73	2.02
2370	3/4" thick	1.35	.79	2.14
2380	Particle board, 3/8" thick	.40	.68	1.08
2390	1/2" thick	.43	.71	1.14
2400	5/8" thick	.55	.73	1.28
2410	3/4" thick	.72	.79	1.51
2420	Hardboard, 4' x 4', .215" thick	.63	.68	1.31

C3030 Ceiling Finishes

2 Coats of Plaster on Gypsum
Lath on Wood Furring

Fiberglass Board on
Exposed Suspended Grid System

Plaster and Metal Lath
on Metal Furring

C3030 105				Plaster Ceilings				
	TYPE	LATH	FURRING	SUPPORT		COST PER S.F.		
						MAT.	INST.	TOTAL
2400	2 coat gypsum	3/8″ gypsum	1″x3″ wood, 16″ O.C.	wood		1.48	5.10	6.58
2500	Painted			masonry		1.48	5.25	6.73
2600				concrete		1.48	5.85	7.33
2700	3 coat gypsum	3.4# metal	1″x3″ wood, 16″ O.C.	wood		1.63	5.50	7.13
2800	Painted			masonry		1.63	5.60	7.23
2900				concrete		1.63	6.20	7.83
3000	2 coat perlite	3/8″ gypsum	1″x3″ wood, 16″ O.C.	wood		1.48	5.30	6.78
3100	Painted			masonry		1.48	5.40	6.88
3200				concrete		1.48	6.05	7.53
3300	3 coat perlite	3.4# metal	1″x3″ wood, 16″ O.C.	wood		1.44	5.45	6.89
3400	Painted			masonry		1.44	5.55	6.99
3500				concrete		1.44	6.15	7.59
3600	2 coat gypsum	3/8″ gypsum	3/4″ CRC, 12″ O.C.	1-1/2″ CRC, 48″O.C.		1.53	6.10	7.63
3700	Painted		3/4″ CRC, 16″ O.C.	1-1/2″ CRC, 48″O.C.		1.50	5.55	7.05
3800			3/4″ CRC, 24″ O.C.	1-1/2″ CRC, 48″O.C.		1.39	5.05	6.44
3900	2 coat perlite	3/8″ gypsum	3/4″ CRC, 12″ O.C.	1-1/2″ CRC, 48″O.C		1.53	6.50	8.03
4000	Painted		3/4″ CRC, 16″ O.C.	1-1/2″ CRC, 48″O.C.		1.50	5.95	7.45
4100			3/4″ CRC, 24″ O.C.	1-1/2″ CRC, 48″O.C.		1.39	5.45	6.84
4200	3 coat gypsum	3.4# metal	3/4″ CRC, 12″ O.C.	1-1/2″ CRC, 36″ O.C.		1.93	8	9.93
4300	Painted		3/4″ CRC, 16″ O.C.	1-1/2″ CRC, 36″ O.C.		1.90	7.40	9.30
4400			3/4″ CRC, 24″ O.C.	1-1/2″ CRC, 36″ O.C.		1.79	6.95	8.74
4500	3 coat perlite	3.4# metal	3/4″ CRC, 12″ O.C.	1-1/2″ CRC,36″ O.C.		2.05	8.75	10.80
4600	Painted		3/4″ CRC, 16″ O.C.	1-1/2″ CRC, 36″ O.C.		2.02	8.20	10.22
4700			3/4″ CRC, 24″ O.C.	1-1/2″ CRC, 36″ O.C.		1.91	7.70	9.61

C3030 110				Drywall Ceilings				
	TYPE	FINISH	FURRING	SUPPORT		COST PER S.F.		
						MAT.	INST.	TOTAL
4800	1/2″ F.R. drywall	painted and textured	1″x3″ wood, 16″ O.C.	wood		.81	3.14	3.95
4900				masonry		.81	3.24	4.05
5000				concrete		.81	3.87	4.68
5100	5/8″ F.R. drywall	painted and textured	1″x3″ wood, 16″ O.C.	wood		.80	3.14	3.94
5200				masonry		.80	3.24	4.04
5300				concrete		.80	3.87	4.67
5400	1/2″ F.R. drywall	painted and textured	7/8″resil. channels	24″ O.C.		.66	3.05	3.71

C30 Interior Finishes

C3030 Ceiling Finishes

C3030 110 — Drywall Ceilings

	TYPE	FINISH	FURRING	SUPPORT		COST PER S.F. MAT.	COST PER S.F. INST.	COST PER S.F. TOTAL
5500	1/2" F.R. drywall	painted and textured	1"x2" wood	stud clips		.86	2.93	3.79
5602		painted	1-5/8" metal studs	24" O.C.		.71	2.71	3.42
5700	5/8" F.R. drywall	painted and textured	1-5/8"metal studs	24" O.C.		.70	2.71	3.41
5702								

C3030 140 — Plaster Ceiling Components

		COST PER S.F. MAT.	COST PER S.F. INST.	COST PER S.F. TOTAL
0060	Plaster, gypsum incl. finish, 2 coats			
0080	3 coats	.65	3.26	3.91
0100	Perlite, incl. finish, 2 coats	.46	3.21	3.67
0120	3 coats	.77	4.04	4.81
0140	Thin coat on drywall	.09	.64	.73
0200	Lath, gypsum, 3/8" thick	.62	.79	1.41
0220	1/2" thick	.60	.82	1.42
0240	5/8" thick	.30	.96	1.26
0260	Metal, diamond, 2.5 lb.	.55	.64	1.19
0280	3.4 lb.	.58	.69	1.27
0300	Flat rib, 2.75 lb.	.45	.64	1.09
0320	3.4 lb.	.69	.69	1.38
0440	Furring, steel channels, 3/4" galvanized , 12" O.C.	.34	2.07	2.41
0460	16" O.C.	.31	1.50	1.81
0480	24" O.C.	.20	1.04	1.24
0500	1-1/2" galvanized , 12" O.C.	.46	2.29	2.75
0520	16" O.C.	.41	1.67	2.08
0540	24" O.C.	.27	1.11	1.38
0560	Wood strips, 1"x3", on wood, 12" O.C.	.38	1.46	1.84
0580	16" O.C.	.29	1.10	1.39
0600	24" O.C.	.19	.73	.92
0620	On masonry, 12" O.C.	.38	1.60	1.98
0640	16" O.C.	.29	1.20	1.49
0660	24" O.C.	.19	.80	.99
0680	On concrete, 12" O.C.	.38	2.44	2.82
0700	16" O.C.	.29	1.83	2.12
0720	24" O.C.	.19	1.22	1.41
0940	Paint on plaster or drywall, roller work, primer + 1 coat	.11	.44	.55
0960	Primer + 2 coats	.17	.57	.74

C3030 210 — Acoustical Ceilings

	TYPE	TILE	GRID	SUPPORT		COST PER S.F. MAT.	COST PER S.F. INST.	COST PER S.F. TOTAL
5800	5/8" fiberglass board	24" x 48"	tee	suspended		1.74	1.54	3.28
5900		24" x 24"	tee	suspended		1.92	1.69	3.61
6000	3/4" fiberglass board	24" x 48"	tee	suspended		2.74	1.57	4.31
6100		24" x 24"	tee	suspended		2.92	1.72	4.64
6500	5/8" mineral fiber	12" x 12"	1"x3" wood, 12" O.C.	wood		2.35	3.17	5.52
6600				masonry		2.35	3.31	5.66
6700				concrete		2.35	4.15	6.50
6800	3/4" mineral fiber	12" x 12"	1"x3" wood, 12" O.C.	wood		2.48	3.17	5.65
6900				masonry		2.48	3.17	5.65
7000				concrete		2.48	3.17	5.65
7100	3/4"mineral fiber on	12" x 12"	25 ga. channels	runners		2.84	3.91	6.75
7102	5/8" F.R. drywall							

C30 Interior Finishes

C3030 Ceiling Finishes

C3030 210			Acoustical Ceilings					

	TYPE	TILE	GRID	SUPPORT		COST PER S.F.		
						MAT.	INST.	TOTAL
7200 7201 7202	5/8" plastic coated Mineral fiber	12" x 12"	25 ga. channels	adhesive backed		2.38	1.71	4.09
7300 7301 7302	3/4" plastic coated Mineral fiber	12" x 12"		adhesive backed		2.51	1.71	4.22
7400 7401 7402	3/4" mineral fiber	12" x 12"	conceal 2" bar & channels	suspended		2.64	3.87	6.51

C3030 240	Acoustical Ceiling Components	COST PER S.F.		
		MAT.	INST.	TOTAL
2480	Ceiling boards, eggcrate, acrylic, 1/2" x 1/2" x 1/2" cubes	2	1.02	3.02
2500	Polystyrene, 3/8" x 3/8" x 1/2" cubes	1.68	1	2.68
2520	1/2" x 1/2" x 1/2" cubes	1.93	1.02	2.95
2540	Fiberglass boards, plain, 5/8" thick	.77	.82	1.59
2560	3/4" thick	1.77	.85	2.62
2580	Grass cloth faced, 3/4" thick	3.16	1.02	4.18
2600	1" thick	2.40	1.06	3.46
2620	Luminous panels, prismatic, acrylic	2.44	1.28	3.72
2640	Polystyrene	1.25	1.28	2.53
2660	Flat or ribbed, acrylic	4.25	1.28	5.53
2680	Polystyrene	2.92	1.28	4.20
2700	Drop pan, white, acrylic	6.25	1.28	7.53
2720	Polystyrene	5.20	1.28	6.48
2740	Mineral fiber boards, 5/8" thick, standard	.86	.76	1.62
2760	Plastic faced	1.35	1.28	2.63
2780	2 hour rating	1.29	.76	2.05
2800	Perforated aluminum sheets, .024 thick, corrugated painted	2.48	1.05	3.53
2820	Plain	4.14	1.02	5.16
3080	Mineral fiber, plastic coated, 12" x 12" or 12" x 24", 5/8" thick	1.97	1.71	3.68
3100	3/4" thick	2.10	1.71	3.81
3120	Fire rated, 3/4" thick, plain faced	1.54	1.71	3.25
3140	Mylar faced	1.30	1.71	3.01
3160	Add for ceiling primer	.14		.14
3180	Add for ceiling cement	.41		.41
3240	Suspension system, furring, 1" x 3" wood 12" O.C.	.38	1.46	1.84
3260	T bar suspension system, 2' x 4' grid	.75	.64	1.39
3280	2' x 2' grid	.93	.79	1.72
3300	Concealed Z bar suspension system 12" module	.76	.99	1.75
3320	Add to above for 1-1/2" carrier channels 4' O.C.	.12	1.09	1.21
3340	Add to above for carrier channels for recessed lighting	.21	1.11	1.32

D10 Conveying

D1010 Elevators and Lifts

Hydraulic

Traction Geared

Traction Gearless

D1010 110	Hydraulic	COST EACH		
		MAT.	INST.	TOTAL
1300	Pass. elev., 1500 lb., 2 Floors, 100 FPM	45,800	15,100	60,900
1400	5 Floors, 100 FPM	96,000	40,000	136,000
1600	2000 lb., 2 Floors, 100 FPM	48,000	15,100	63,100
1700	5 floors, 100 FPM	98,000	40,000	138,000
1900	2500 lb., 2 Floors, 100 FPM	49,200	15,100	64,300
2000	5 floors, 100 FPM	99,500	40,000	139,500
2200	3000 lb., 2 Floors, 100 FPM	50,500	15,100	65,600
2300	5 floors, 100 FPM	100,500	40,000	140,500
2500	3500 lb., 2 Floors, 100 FPM	52,500	15,100	67,600
2600	5 floors, 100 FPM	102,500	40,000	142,500
2800	4000 lb., 2 Floors, 100 FPM	54,000	15,100	69,100
2900	5 floors, 100 FPM	104,000	40,000	144,000
3100	4500 lb., 2 Floors, 100 FPM	55,500	15,100	70,600
3200	5 floors, 100 FPM	105,500	40,000	145,500
4000	Hospital elevators, 3500 lb., 2 Floors, 100 FPM	80,000	15,100	95,100
4100	5 floors, 100 FPM	140,000	40,000	180,000
4300	4000 lb., 2 Floors, 100 FPM	80,000	15,100	95,100
4400	5 floors, 100 FPM	140,000	40,000	180,000
4600	4500 lb., 2 Floors, 100 FPM	87,500	15,100	102,600
4800	5 floors, 100 FPM	147,500	40,000	187,500
4900	5000 lb., 2 Floors, 100 FPM	89,500	15,100	104,600
5000	5 floors, 100 FPM	149,500	40,000	189,500
6700	Freight elevators (Class "B"), 3000 lb., 2 Floors, 50 FPM	127,500	19,200	146,700
6800	5 floors, 100 FPM	252,000	50,000	302,000
7000	4000 lb., 2 Floors, 50 FPM	136,500	19,200	155,700
7100	5 floors, 100 FPM	261,000	50,000	311,000
7500	10,000 lb., 2 Floors, 50 FPM	175,000	19,200	194,200
7600	5 floors, 100 FPM	299,500	50,000	349,500
8100	20,000 lb., 2 Floors, 50 FPM	243,000	19,200	262,200
8200	5 Floors, 100 FPM	367,500	50,000	417,500

D1010 140	Traction Geared Elevators	COST EACH		
		MAT.	INST.	TOTAL
1300	Passenger, 2000 Lb., 5 floors, 200 FPM	120,500	37,700	158,200
1500	15 floors, 350 FPM	296,500	114,000	410,500
1600	2500 Lb., 5 floors, 200 FPM	124,500	37,700	162,200
1800	15 floors, 350 FPM	300,500	114,000	414,500
1900	3000 Lb., 5 floors, 200 FPM	125,000	37,700	162,700
2100	15 floors, 350 FPM	301,000	114,000	415,000
2200	3500 Lb., 5 floors, 200 FPM	128,500	37,700	166,200
2400	15 floors, 350 FPM	304,500	114,000	418,500
2500	4000 Lb., 5 floors, 200 FPM	128,000	37,700	165,700
2700	15 floors, 350 FPM	304,000	114,000	418,000

D10 Conveying

D1010 Elevators and Lifts

D1010 140	Traction Geared Elevators	COST EACH		
		MAT.	INST.	TOTAL
2800	4500 Lb., 5 floors, 200 FPM	130,000	37,700	167,700
3000	15 floors, 350 FPM	306,000	114,000	420,000
3100	5000 Lb., 5 floors, 200 FPM	132,000	37,700	169,700
3300	15 floors, 350 FPM	308,000	114,000	422,000
4000	Hospital, 3500 Lb., 5 floors, 200 FPM	124,000	37,700	161,700
4200	15 floors, 350 FPM	297,000	114,000	411,000
4300	4000 Lb., 5 floors, 200 FPM	124,000	37,700	161,700
4500	15 floors, 350 FPM	297,000	114,000	411,000
4600	4500 Lb., 5 floors, 200 FPM	130,000	37,700	167,700
4800	15 floors, 350 FPM	303,000	114,000	417,000
4900	5000 Lb., 5 floors, 200 FPM	131,000	37,700	168,700
5100	15 floors, 350 FPM	304,000	114,000	418,000
6000	Freight, 4000 Lb., 5 floors, 50 FPM class 'B'	139,500	39,000	178,500
6200	15 floors, 200 FPM class 'B'	351,000	135,500	486,500
6300	8000 Lb., 5 floors, 50 FPM class 'B'	171,500	39,000	210,500
6500	15 floors, 200 FPM class 'B'	382,500	135,500	518,000
7000	10,000 Lb., 5 floors, 50 FPM class 'B'	189,000	39,000	228,000
7200	15 floors, 200 FPM class 'B'	542,500	135,500	678,000
8000	20,000 Lb., 5 floors, 50 FPM class 'B'	219,000	39,000	258,000
8200	15 floors, 200 FPM class 'B'	572,000	135,500	707,500

D1010 150	Traction Gearless Elevators	COST EACH		
		MAT.	INST.	TOTAL
1700	Passenger, 2500 Lb., 10 floors, 200 FPM	331,000	98,500	429,500
1900	30 floors, 600 FPM	694,500	251,000	945,500
2000	3000 Lb., 10 floors, 200 FPM	331,500	98,500	430,000
2200	30 floors, 600 FPM	695,500	251,000	946,500
2300	3500 Lb., 10 floors, 200 FPM	335,000	98,500	433,500
2500	30 floors, 600 FPM	699,000	251,000	950,000
2700	50 floors, 800 FPM	1,036,500	403,500	1,440,000
2800	4000 lb., 10 floors, 200 FPM	334,500	98,500	433,000
3000	30 floors, 600 FPM	698,500	251,000	949,500
3200	50 floors, 800 FPM	1,036,000	403,500	1,439,500
3300	4500 lb., 10 floors, 200 FPM	336,500	98,500	435,000
3500	30 floors, 600 FPM	700,500	251,000	951,500
3700	50 floors, 800 FPM	1,038,000	403,500	1,441,500
3800	5000 lb., 10 floors, 200 FPM	338,500	98,500	437,000
4000	30 floors, 600 FPM	702,000	251,000	953,000
4200	50 floors, 800 FPM	1,040,000	403,500	1,443,500
6000	Hospital, 3500 Lb., 10 floors, 200 FPM	329,000	98,500	427,500
6200	30 floors, 600 FPM	688,000	251,000	939,000
6400	4000 Lb., 10 floors, 200 FPM	329,000	98,500	427,500
6600	30 floors, 600 FPM	688,000	251,000	939,000
6800	4500 Lb., 10 floors, 200 FPM	335,000	98,500	433,500
7000	30 floors, 600 FPM	694,000	251,000	945,000
7200	5000 Lb., 10 floors, 200 FPM	336,000	98,500	434,500
7400	30 floors, 600 FPM	695,000	251,000	946,000

D1090 Other Conveying Systems

D1090 510	Chutes	COST EACH		
		MAT.	INST.	TOTAL
0100	Chutes, linen or refuse incl. sprinklers, galv. steel, 18" diam., per floor	1,425	340	1,765
0110	Aluminized steel, 36" diam., per floor	1,900	425	2,325
0120	Mail, 8-3/4" x 3-1/2", aluminum & glass, per floor	715	238	953
0130	Bronze or stainless, per floor	1,100	264	1,364

D2010 Plumbing Fixtures

Minimum Plumbing Fixture Requirements

Type of Building or Occupancy (2)	Water Closets (14) (Fixtures per Person)		Urinals (5,10) (Fixtures per Person)		Lavatories (Fixtures per Person)		Bathtubs or Showers (Fixtures per Person)	Drinking Fountains (Fixtures per Person) (3, 13)
	Male	Female	Male	Female	Male	Female		
Assembly Places- Theatres, Auditoriums, Convention Halls, etc.-for permanent employee use	1: 1 - 15 2: 16 - 35 3: 36 - 55 Over 55, add 1 fixture for each additional 40 persons	1: 1 - 15 2: 16 - 35 3: 36 - 55	0: 1 - 9 1: 10 - 50 Add one fixture for each additional 50 males		1 per 40	1 per 40		
Assembly Places- Theatres, Auditoriums, Convention Halls, etc. - for public use	1: 1 - 100 2: 101 - 200 3: 201 - 400 Over 400, add 1 fixture for each additional 500 males and 1 for each additional 125 females	3: 1 - 50 4: 51 - 100 8: 101 - 200 11: 201 - 400	1: 1 - 100 2: 101 - 200 3: 201 - 400 4: 401 - 600 Over 600, add 1 fixture for each additional 300 males		1: 1 - 200 2: 201 - 400 3: 401 - 750 Over 750, add 1 fixture for each additional 500 persons	1: 1 - 200 2: 201 - 400 3: 401 - 750		1: 1 - 150 2: 151 - 400 3: 401 - 750 Over 750, add one fixture for each additional 500 persons
Dormitories (9) School or Labor	1 per 10 Add 1 fixture for each additional 25 males (over 10) and 1 for each additional 20 females (over 8)	1 per 8	1 per 25 Over 150, add 1 fixture for each additional 50 males		1 per 12 Over 12 add 1 fixture for each additional 20 males and 1 for each 15 additional females	1 per 12	1 per 8 For females add 1 bathtub per 30. Over 150, add 1 per 20	1 per 150 (12)
Dormitories- for Staff Use	1: 1 - 15 2: 16 - 35 3: 36 - 55 Over 55, add 1 fixture for each additional 40 persons	1: 1 - 15 3: 16 - 35 4: 36 - 55	1 per 50		1 per 40	1 per 40	1 per 8	
Dwellings: Single Dwelling Multiple Dwelling or Apartment House	1 per dwelling 1 per dwelling or apartment unit				1 per dwelling 1 per dwelling or apartment unit		1 per dwelling 1 per dwelling or apartment unit	
Hospital Waiting rooms	1 per room				1 per room			1 per 150 (12)
Hospitals- for employee use	1: 1 - 15 2: 16 - 35 3: 36 - 55 Over 55, add 1 fixture for each additional 40 persons	1: 1 - 15 3: 16 - 35 4: 36 - 55	0: 1 - 9 1: 10 - 50 Add 1 fixture for each additional 50 males		1 per 40	1 per 40		
Hospitals: Individual Room Ward Room	1 per room 1 per 8 patients				1 per room 1 per 10 patients		1 per room 1 per 20 patients	1 per 150 (12)
Industrial (6) Warehouses Workshops, Foundries and similar establishments- for employee use	1: 1 -10 2: 11 - 25 3: 26 - 50 4: 51 - 75 5: 76 - 100 Over 100, add 1 fixture for each additional 30 persons	1: 1 -10 2: 11 - 25 3: 26 - 50 4: 51 - 75 5: 76 - 100			Up to 100, per 10 persons Over 100, 1 per 15 persons (7, 8)		1 shower for each 15 persons exposed to excessive heat or to skin contamination with poisonous, infectious or irritating material	1 per 150 (12)
Institutional - Other than Hospitals or Penal Institutions (on each occupied floor)	1 per 25	1 per 20	0: 1 - 9 1: 10 - 50 Add 1 fixture for each additional 50 males		1 per 10	1 per 10	1 per 8	1 per 150 (12)
Institutional - Other than Hospitals or Penal Institutions (on each occupied floor)- for employee use	1: 1 - 15 2: 16 - 35 3: 36 - 55 Over 55, add 1 fixture for each additional 40 persons	1: 1 - 15 3: 16 - 35 4: 36 - 55	0: 1 - 9 1: 10 - 50 Add 1 fixture for each additional 50 males		1 per 40	1 per 40	1 per 8	1 per 150 (12)
Office or Public Buildings	1: 1 - 100 2: 101 - 200 3: 201 - 400 Over 400, add 1 fixture for each additional 500 males and 1 for each additional 150 females	3: 1 - 50 4: 51 - 100 8: 101 - 200 11: 201 - 400	1: 1 - 100 2: 101 - 200 3: 201 - 400 4: 401 - 600 Over 600, add 1 fixture for each additional 300 males		1: 1 - 200 2: 201 - 400 3: 401 - 750 Over 750, add 1 fixture for each additional 500 persons	1: 1 - 200 2: 201 - 400 3: 401 - 750		1 per 150 (12)
Office or Public Buildings - for employee use	1: 1 - 15 2: 16 - 35 3: 36 - 55 Over 55, add 1 fixture for each additional 40 persons	1: 1 - 15 3: 16 - 35 4: 36 - 55	0: 1 - 9 1: 10 - 50 Add 1 fixture for each additional 50 males		1 per 40	1 per 40		

D2010   Plumbing Fixtures

Minimum Plumbing Fixture Requirements

Type of Building or Occupancy	Water Closets (14) (Fixtures per Person)		Urinals (5, 10) (Fixtures per Person)		Lavatories (Fixtures per Person)		Bathtubs or Showers (Fixtures per Person)	Drinking Fountains (Fixtures per Person) (3, 13)
	Male	Female	Male	Female	Male	Female		
Penal Institutions - for employee use	1: 1 - 15 2: 16 - 35 3: 36 - 55 Over 55, add 1 fixture for each additional 40 persons	1: 1 - 15 3: 16 - 35 4: 36 - 55	0: 1 - 9 1: 10 - 50 Add 1 fixture for each additional 50 males		1 per 40	1 per 40		1 per 150 (12)
Penal Institutions - for prison use Cell	1 per cell				1 per cell			1 per cellblock floor
Exercise room	1 per exercise room		1 per exercise room		1 per exercise room			1 per exercise room
Restaurants, Pubs and Lounges (11)	1: 1 - 50 2: 51 - 150 3: 151 - 300 Over 300, add 1 fixture for each additional 200 persons	1: 1 - 50 2: 51 - 150 4: 151 - 300	1: 1 - 150 Over 150, add 1 fixture for each additional 150 males		1: 1 - 150 2: 151 - 200 3: 201 - 400 Over 400, add 1 fixture for each additional 400 persons	1: 1 - 150 2: 151 - 200 3: 201 - 400		
Schools - for staff use All Schools	1: 1 - 15 2: 16 - 35 3: 36 - 55 Over 55, add 1 fixture for each additional 40 persons	1: 1 - 15 3: 16 - 35 4: 36 - 55	1 per 50		1 per 40	1 per 40		
Schools - for student use: Nursery	1: 1 - 20 2: 21 - 50 Over 50, add 1 fixture for each additional 50 persons	1: 1 - 20 2: 21 - 50			1: 1 - 25 2: 26 - 50 Over 50, add 1 fixture for each additional 50 persons	1: 1 - 25 2: 26 - 50		1 per 150 (12)
Elementary	1 per 30	1 per 25	1 per 75		1 per 35	1 per 35		1 per 150 (12)
Secondary	1 per 40	1 per 30	1 per 35		1 per 40	1 per 40		1 per 150 (12)
Others (Colleges, Universities, Adult Centers, etc.)	1 per 40	1 per 30	1 per 35		1 per 40	1 per 40		1 per 150 (12)
Worship Places Educational and Activities Unit	1 per 150	1 per 75	1 per 150		1 per 2 water closets			1 per 150 (12)
Worship Places Principal Assembly Place	1 per 150	1 per 75	1 per 150		1 per 2 water closets			1 per 150 (12)

Notes:
1. The figures shown are based upon one (1) fixture being the minimum required for the number of persons indicated or any fraction thereof.
2. Building categories not shown on this table shall be considered separately by the Administrative Authority.
3. Drinking fountains shall not be installed in toilet rooms.
4. Laundry trays. One (1) laundry tray or one (1) automatic washer standpipe for each dwelling unit or one (1) laundry trays or one (1) automatic washer standpipes, or combination thereof, for each twelve (12) apartments. Kitchen sinks, one (1) for each dwelling or apartment unit.
5. For each urinal added in excess of the minimum required, one water closet may be deducted. The number of water closets shall not be reduced to less than two-thirds (2/3) of the minimum requirement.
6. As required by ANSI Z4.1-1968, Sanitation in Places of Employment.
7. Where there is exposure to skin contamination with poisonous, infectious, or irritating materials, provide one (1) lavatory for each five (5) persons.
8. Twenty-four (24) lineal inches of wash sink or eighteen (18) inches of a circular basin, when provided with water outlets for such space shall be considered equivalent to one (1) lavatory.
9. Laundry trays, one (1) for each fifty (50) persons. Service sinks, one (1) for each hundred (100) persons.
10. General. In applying this schedule of facilities, consideration shall be given to the accessibility of the fixtures. Conformity purely on a numerical basis may not result in an installation suited to the need of the individual establishment. For example, schools should be provided with toilet facilities on each floor having classrooms.
    a. Surrounding materials, wall and floor space to a point two (2) feet in front of urinal lip and four (4) feet above the floor, and at least two (2) feet to each side of the urinal shall be lined with non-absorbent materials.
    b. Trough urinals shall be prohibited.
11. A restaurant is defined as a business which sells food to be consumed on the premises.
    a. The number of occupants for a drive-in restaurant shall be considered as equal to the number of parking stalls.
    b. Employee toilet facilities shall not to be included in the above restaurant requirements. Hand washing facilities shall be available in the kitchen for employees.
12. Where food is consumed indoors, water stations may be substituted for drinking fountains. Offices, or public buildings for use by more than six (6) persons shall have one (1) drinking fountain for the first one hundred fifty (150) persons and one additional fountain for each three hundred (300) persons thereafter.
13. There shall be a minimum of one (1) drinking fountain per occupied floor in schools, theaters, auditoriums, dormitories, offices of public building.
14. The total number of water closets for females shall be at least equal to the total number of water closets and urinals required for males.

D20 Plumbing

D2010 Plumbing Fixtures

One Piece Wall Hung Water Closet

Wall Hung Urinal

Side By Side Water Closet Group

Floor Mount Water Closet

Stall Type Urinal

Back to Back Water Closet Group

D2010 110	Water Closet Systems	MAT.	INST.	TOTAL
		COST EACH		
1800	Water closet, vitreous china, elongated			
1840	Tank type, wall hung			
1880	Close coupled two piece	1,650	660	2,310
1920	Floor mount, one piece	930	705	1,635
1960	One piece low profile	970	705	1,675
2000	Two piece close coupled	625	705	1,330
2040	Bowl only with flush valve			
2080	Wall hung	1,425	750	2,175
2120	Floor mount	750	715	1,465
2122				
2160	Floor mount, ADA compliant with 18" high bowl	770	735	1,505

D2010 120	Water Closets, Group	MAT.	INST.	TOTAL
		COST EACH		
1760	Water closets, battery mount, wall hung, side by side, first closet	1,500	775	2,275
1800	Each additional water closet, add	1,425	730	2,155
3000	Back to back, first pair of closets	2,250	1,025	3,275
3100	Each additional pair of closets, back to back	2,200	1,000	3,200

D2010 210	Urinal Systems	MAT.	INST.	TOTAL
		COST EACH		
2000	Urinal, vitreous china, wall hung	565	745	1,310
2040	Stall type	1,200	890	2,090

D2010 Plumbing Fixtures

Systems are complete with trim and rough-in (supply, waste and vent) to connect to supply branches and waste mains.

Vanity Top

Wall Hung

Counter Top Single Bowl

Counter Top Double Bowl

D2010 310	Lavatory Systems	COST EACH		
		MAT.	INST.	TOTAL
1560	Lavatory w/trim, vanity top, PE on CI, 20" x 18", Vanity top by others.	565	665	1,230
1600	19" x 16" oval	475	665	1,140
1640	18" round	550	665	1,215
1680	Cultured marble, 19" x 17"	485	665	1,150
1720	25" x 19"	520	665	1,185
1760	Stainless, self-rimming, 25" x 22"	660	665	1,325
1800	17" x 22"	650	665	1,315
1840	Steel enameled, 20" x 17"	480	680	1,160
1880	19" round	475	680	1,155
1920	Vitreous china, 20" x 16"	585	695	1,280
1960	19" x 16"	585	695	1,280
2000	22" x 13"	595	695	1,290
2040	Wall hung, PE on CI, 18" x 15"	845	730	1,575
2080	19" x 17"	845	730	1,575
2120	20" x 18"	815	730	1,545
2160	Vitreous china, 18" x 15"	700	755	1,455
2200	19" x 17"	645	755	1,400
2240	24" x 20"	915	755	1,670
2300	20" x 27", handicap	915	755	1,670

D2010 410	Kitchen Sink Systems	COST EACH		
		MAT.	INST.	TOTAL
1720	Kitchen sink w/trim, countertop, PE on CI, 24"x21", single bowl	580	725	1,305
1760	30" x 21" single bowl	630	725	1,355
1800	32" x 21" double bowl	745	785	1,530
1840	42" x 21" double bowl	925	795	1,720
1880	Stainless steel, 19" x 18" single bowl	910	725	1,635
1920	25" x 22" single bowl	980	725	1,705
1960	33" x 22" double bowl	1,300	785	2,085
2000	43" x 22" double bowl	1,475	795	2,270
2040	44" x 22" triple bowl	1,725	830	2,555
2080	44" x 24" corner double bowl	1,125	795	1,920
2120	Steel, enameled, 24" x 21" single bowl	500	725	1,225
2160	32" x 21" double bowl	550	785	1,335
2240	Raised deck, PE on CI, 32" x 21", dual level, double bowl	690	990	1,680
2280	42" x 21" dual level, triple bowl	1,400	1,075	2,475

D2010 Plumbing Fixtures

Laboratory Sink

Service Sink

Systems are complete with trim and rough-in (supply, waste and vent) to connect to supply branches and waste mains.

Single Compartment Sink

Double Compartment Sink

D2010 420	Laundry Sink Systems	COST EACH		
		MAT.	INST.	TOTAL
1740	Laundry sink w/trim, PE on CI, black iron frame			
1760	24" x 20", single compartment	810	715	1,525
1840	48" x 21" double compartment	1,225	775	2,000
1920	Molded stone, on wall, 22" x 21" single compartment	460	715	1,175
1960	45"x 21" double compartment	605	775	1,380
2040	Plastic, on wall or legs, 18" x 23" single compartment	455	700	1,155
2080	20" x 24" single compartment	455	700	1,155
2120	36" x 23" double compartment	505	755	1,260
2160	40" x 24" double compartment	595	755	1,350

D2010 430	Laboratory Sink Systems	COST EACH		
		MAT.	INST.	TOTAL
1580	Laboratory sink w/trim, polyethylene, single bowl,			
1600	Double drainboard, 54" x 24" O.D.	1,325	930	2,255
1640	Single drainboard, 47" x 24"O.D.	1,525	930	2,455
1680	70" x 24" O.D.	1,400	930	2,330
1760	Flanged, 14-1/2" x 14-1/2" O.D.	425	835	1,260
1800	18-1/2" x 18-1/2" O.D.	515	835	1,350
1840	23-1/2" x 20-1/2" O.D.	545	835	1,380
1920	Polypropylene, cup sink, oval, 7" x 4" O.D.	305	740	1,045
1960	10" x 4-1/2" O.D.	315	740	1,055
1961				

D2010 Plumbing Fixtures

D2010 440	Service Sink Systems	COST EACH		
		MAT.	INST.	TOTAL
4260	Service sink w/trim, PE on CI, corner floor, 28" x 28", w/rim guard	1,575	935	2,510
4300	Wall hung w/rim guard, 22" x 18"	1,900	1,100	3,000
4340	24" x 20"	1,950	1,100	3,050
4380	Vitreous china, wall hung 22" x 20"	1,750	1,100	2,850
4383				

D2010 Plumbing Fixtures

Recessed Bathtub

Corner Bathtub

Systems are complete with trim and rough-in (supply, waste and vent) to connect to supply branches and waste mains.

Circular Wash Fountain

Semi-Circular Wash Fountain

D2010 510	Bathtub Systems	COST EACH		
		MAT.	INST.	TOTAL
2000	Bathtub, recessed, P.E. on Cl., 48″ x 42″	2,225	825	3,050
2040	72″ x 36″	2,850	915	3,765
2080	Mat bottom, 5′ long	1,450	795	2,245
2120	5′-6″ long	2,050	825	2,875
2160	Corner, 48″ x 42″	2,775	795	3,570
2200	Formed steel, enameled, 4′-6″ long	825	735	1,560

D2010 610	Group Wash Fountain Systems	COST EACH		
		MAT.	INST.	TOTAL
1740	Group wash fountain, precast terrazzo			
1760	Circular, 36″ diameter	5,050	1,200	6,250
1800	54″ diameter	6,175	1,325	7,500
1840	Semi-circular, 36″ diameter	4,700	1,200	5,900
1880	54″ diameter	5,575	1,325	6,900
1960	Stainless steel, circular, 36″ diameter	4,975	1,125	6,100
2000	54″ diameter	6,275	1,250	7,525
2040	Semi-circular, 36″ diameter	4,375	1,125	5,500
2080	54″ diameter	5,475	1,250	6,725
2160	Thermoplastic, circular, 36″ diameter	3,825	905	4,730
2200	54″ diameter	4,375	1,050	5,425
2240	Semi-circular, 36″ diameter	3,575	905	4,480
2280	54″ diameter	4,250	1,050	5,300

D2010 Plumbing Fixtures

Square Shower Stall

Corner Angle Shower Stall

Wall Mounted, Low Back

Wall Mounted, No Back

Systems are complete with trim and rough-in (supply, waste and vent) to connect to supply branches and waste mains.

D2010 710	Shower Systems	COST EACH		
		MAT.	INST.	TOTAL
1560	Shower, stall, baked enamel, molded stone receptor, 30" square	1,225	765	1,990
1600	32" square	1,225	775	2,000
1640	Terrazzo receptor, 32" square	1,675	775	2,450
1680	36" square	1,825	780	2,605
1720	36" corner angle	2,075	780	2,855
1800	Fiberglass one piece, three walls, 32" square	830	750	1,580
1840	36" square	845	750	1,595
1880	Polypropylene, molded stone receptor, 30" square	925	1,100	2,025
1920	32" square	940	1,100	2,040
1960	Built-in head, arm, bypass, stops and handles	107	288	395

D2010 810	Drinking Fountain Systems	COST EACH		
		MAT.	INST.	TOTAL
1740	Drinking fountain, one bubbler, wall mounted			
1760	Non recessed			
1800	Bronze, no back	1,300	440	1,740
1840	Cast iron, enameled, low back	850	440	1,290
1880	Fiberglass, 12" back	2,125	440	2,565
1920	Stainless steel, no back	1,200	440	1,640
1960	Semi-recessed, poly marble	1,225	440	1,665
2040	Stainless steel	1,000	440	1,440
2080	Vitreous china	915	440	1,355
2120	Full recessed, poly marble	1,875	440	2,315
2200	Stainless steel	1,575	440	2,015
2240	Floor mounted, pedestal type, aluminum	2,500	590	3,090
2320	Bronze	2,175	590	2,765
2360	Stainless steel	1,975	590	2,565

D2010 Plumbing Fixtures

Wall Hung Water Cooler

Floor Mounted Water Cooler

Systems are complete with trim and rough-in (supply, waste and vent) to connect to supply branches and waste mains.

D2010 820	Water Cooler Systems	COST EACH		
		MAT.	INST.	TOTAL
1840	Water cooler, electric, wall hung, 8.2 GPH	925	565	1,490
1880	Dual height, 14.3 GPH	1,375	580	1,955
1920	Wheelchair type, 7.5 G.P.H.	2,000	565	2,565
1960	Semi recessed, 8.1 G.P.H.	870	565	1,435
2000	Full recessed, 8 G.P.H.	2,025	600	2,625
2040	Floor mounted, 14.3 G.P.H.	995	490	1,485
2080	Dual height, 14.3 G.P.H.	1,350	590	1,940
2120	Refrigerated compartment type, 1.5 G.P.H.	1,600	490	2,090

D2010 Plumbing Fixtures

Two Fixture   Three Fixture

*Common wall is with adjacent bathroom.

D2010 920	Two Fixture Bathroom, Two Wall Plumbing	COST EACH		
		MAT.	INST.	TOTAL
1180	Bathroom, lavatory & water closet, 2 wall plumbing, stand alone	1,750	1,825	3,575
1200	Share common plumbing wall*	1,550	1,575	3,125

D2010 922	Two Fixture Bathroom, One Wall Plumbing	COST EACH		
		MAT.	INST.	TOTAL
2220	Bathroom, lavatory & water closet, one wall plumbing, stand alone	1,625	1,650	3,275
2240	Share common plumbing wall*	1,400	1,400	2,800

D2010 924	Three Fixture Bathroom, One Wall Plumbing	COST EACH		
		MAT.	INST.	TOTAL
1150	Bathroom, three fixture, one wall plumbing			
1160	Lavatory, water closet & bathtub			
1170	Stand alone	2,775	2,150	4,925
1180	Share common plumbing wall *	2,400	1,550	3,950

D2010 926	Three Fixture Bathroom, Two Wall Plumbing	COST EACH		
		MAT.	INST.	TOTAL
2130	Bathroom, three fixture, two wall plumbing			
2140	Lavatory, water closet & bathtub			
2160	Stand alone	2,775	2,175	4,950
2180	Long plumbing wall common *	2,500	1,725	4,225
3610	Lavatory, bathtub & water closet			
3620	Stand alone	3,100	2,475	5,575
3640	Long plumbing wall common *	2,850	2,225	5,075
4660	Water closet, corner bathtub & lavatory			
4680	Stand alone	4,125	2,200	6,325
4700	Long plumbing wall common *	3,775	1,650	5,425
6100	Water closet, stall shower & lavatory			
6120	Stand alone	3,075	2,450	5,525
6140	Long plumbing wall common *	2,900	2,275	5,175
7060	Lavatory, corner stall shower & water closet			
7080	Stand alone	3,675	2,175	5,850
7100	Short plumbing wall common *	3,125	1,450	4,575

D2010 Plumbing Fixtures

Four Fixture   Five Fixture

*Common wall is with adjacent bathroom.

D2010 928	Four Fixture Bathroom, Two Wall Plumbing		COST EACH		
			MAT.	INST.	TOTAL
1140	Bathroom, four fixture, two wall plumbing				
1150		Bathtub, water closet, stall shower & lavatory			
1160		Stand alone	3,825	2,375	6,200
1180		Long plumbing wall common *	3,475	1,825	5,300
2260		Bathtub, lavatory, corner stall shower & water closet			
2280		Stand alone	4,675	2,375	7,050
2320		Long plumbing wall common *	4,325	1,825	6,150
3620		Bathtub, stall shower, lavatory & water closet			
3640		Stand alone	4,200	2,975	7,175
3660		Long plumbing wall (opp. door) common *	3,875	2,425	6,300
4680	Bathroom, four fixture, three wall plumbing				
4700		Bathtub, stall shower, lavatory & water closet			
4720		Stand alone	5,325	3,275	8,600
4761		Long plumbing wall (opposite door) common*	5,100	3,025	8,125

D2010 932	Five Fixture Bathroom, Two Wall Plumbing		COST EACH		
			MAT.	INST.	TOTAL
1320	Bathroom, five fixture, two wall plumbing				
1340		Bathtub, water closet, stall shower & two lavatories			
1360		Stand alone	5,025	3,775	8,800
1400		One short plumbing wall common *	4,675	3,250	7,925
2360	Bathroom, five fixture, three wall plumbing				
2380		Water closet, bathtub, two lavatories & stall shower			
2400		Stand alone	5,875	3,800	9,675
2440		One short plumbing wall common *	5,525	3,250	8,775
4080	Bathroom, five fixture, one wall plumbing				
4100		Bathtub, two lavatories, corner stall shower & water closet			
4120		Stand alone	5,625	3,400	9,025
4160		Share common wall *	4,825	2,200	7,025

D2020 Domestic Water Distribution

Installation includes piping and fittings
within 10' of heater. Gas heaters require
vent piping (not included with these units).

Gas Fired

Oil Fired

D2020 220	Gas Fired Water Heaters - Residential Systems	COST EACH		
		MAT.	INST.	TOTAL
2200	Gas fired water heater, residential, 100° F rise			
2260	30 gallon tank, 32 GPH	1,500	1,300	2,800
2300	40 gallon tank, 32 GPH	1,700	1,450	3,150
2340	50 gallon tank, 63 GPH	1,725	1,450	3,175
2380	75 gallon tank, 63 GPH	2,300	1,625	3,925
2420	100 gallon tank, 63 GPH	2,625	1,700	4,325
2422				

D2020 230	Oil Fired Water Heaters - Residential Systems	COST EACH		
		MAT.	INST.	TOTAL
2200	Oil fired water heater, residential, 100° F rise			
2220	30 gallon tank, 103 GPH	1,225	1,200	2,425
2260	50 gallon tank, 145 GPH	1,825	1,350	3,175
2300	70 gallon tank, 164 GPH	2,825	1,525	4,350
2340	85 gallon tank, 181 GPH	6,150	1,550	7,700

D2020 Domestic Water Distribution

Systems below include piping and fittings within 10' of heater. Electric water heaters do not require venting. Gas fired heaters require vent piping (not included in these prices).

Electric

Gas Fired

D2020 240	Electric Water Heaters - Commercial Systems	COST EACH		
		MAT.	INST.	TOTAL
1800	Electric water heater, commercial, 100°F rise			
1820	50 gallon tank, 9 KW 37 GPH	5,025	1,075	6,100
1860	80 gal, 12 KW 49 GPH	6,500	1,350	7,850
1900	36 KW 147 GPH	8,850	1,450	10,300
1940	120 gal, 36 KW 147 GPH	9,550	1,550	11,100
1980	150 gal, 120 KW 490 GPH	29,200	1,675	30,875
2020	200 gal, 120 KW 490 GPH	30,900	1,725	32,625
2060	250 gal, 150 KW 615 GPH	34,500	2,000	36,500
2100	300 gal, 180 KW 738 GPH	38,700	2,125	40,825
2140	350 gal, 30 KW 123 GPH	28,400	2,275	30,675
2180	180 KW 738 GPH	37,900	2,275	40,175
2220	500 gal, 30 KW 123 GPH	36,400	2,675	39,075
2260	240 KW 984 GPH	54,000	2,675	56,675
2300	700 gal, 30 KW 123 GPH	45,300	3,075	48,375
2340	300 KW 1230 GPH	64,000	3,075	67,075
2380	1000 gal, 60 KW 245 GPH	53,000	4,275	57,275
2420	480 KW 1970 GPH	86,000	4,275	90,275
2460	1500 gal, 60 KW 245 GPH	77,500	5,275	82,775
2500	480 KW 1970 GPH	122,000	5,275	127,275

D2020 250	Gas Fired Water Heaters - Commercial Systems	COST EACH		
		MAT.	INST.	TOTAL
1760	Gas fired water heater, commercial, 100°F rise			
1780	75.5 MBH input, 63 GPH	2,675	1,675	4,350
1820	95 MBH input, 86 GPH	6,000	1,675	7,675
1860	100 MBH input, 91 GPH	6,400	1,750	8,150
1900	115 MBH input, 110 GPH	7,225	1,800	9,025
1980	155 MBH input, 150 GPH	9,100	2,025	11,125
2020	175 MBH input, 168 GPH	9,250	2,150	11,400
2060	200 MBH input, 192 GPH	9,725	2,450	12,175
2100	240 MBH input, 230 GPH	10,100	2,650	12,750
2140	300 MBH input, 278 GPH	11,200	3,025	14,225
2180	390 MBH input, 374 GPH	13,100	3,050	16,150
2220	500 MBH input, 480 GPH	17,700	3,300	21,000
2260	600 MBH input, 576 GPH	18,700	3,575	22,275

D2020 Domestic Water Distribution

Oil Fired

Oil fired water heater systems include piping and fittings within 10' of heater. Oil fired heaters require vent piping (not included in these systems).

D2020 260	Oil Fired Water Heaters - Commercial Systems	COST EACH		
		MAT.	INST.	TOTAL
1800	Oil fired water heater, commercial, 100°F rise			
1820	140 gal., 140 MBH input, 134 GPH	16,900	1,425	18,325
1900	140 gal., 255 MBH input, 247 GPH	18,300	1,775	20,075
1940	140 gal., 270 MBH input, 259 GPH	22,400	2,025	24,425
1980	140 gal., 400 MBH input, 384 GPH	23,300	2,350	25,650
2060	140 gal., 720 MBH input, 691 GPH	24,700	2,450	27,150
2100	221 gal., 300 MBH input, 288 GPH	32,900	2,700	35,600
2140	221 gal., 600 MBH input, 576 GPH	36,500	2,725	39,225
2180	221 gal., 800 MBH input, 768 GPH	37,000	2,825	39,825
2220	201 gal., 1000 MBH input, 960 GPH	37,800	2,825	40,625
2260	201 gal., 1250 MBH input, 1200 GPH	38,600	2,900	41,500
2300	201 gal., 1500 MBH input, 1441 GPH	41,800	2,975	44,775
2340	411 gal., 600 MBH input, 576 GPH	42,200	3,050	45,250
2380	411 gal., 800 MBH input, 768 GPH	42,500	3,150	45,650
2420	411 gal., 1000 MBH input, 960 GPH	46,000	3,625	49,625
2460	411 gal., 1250 MBH input, 1200 GPH	47,000	3,725	50,725
2500	397 gal., 1500 MBH input, 1441 GPH	47,400	3,850	51,250
2540	397 gal., 1750 MBH input, 1681 GPH	51,000	3,975	54,975

D2040 Rain Water Drainage

Design Assumptions: Vertical conductor size is based on a maximum rate of rainfall of 4″ per hour. To convert roof area to other rates multiply "Max. S.F. Roof Area" shown by four and divide the result by desired local rate. The answer is the local roof area that may be handled by the indicated pipe diameter.

Basic cost is for roof drain, 10′ of vertical leader and 10′ of horizontal, plus connection to the main.

Pipe Dia.	Max. S.F. Roof Area	Gallons per Min.
2″	544	23
3″	1610	67
4″	3460	144
5″	6280	261
6″	10,200	424
8″	22,000	913

D2040 210	Roof Drain Systems	COST EACH		
		MAT.	INST.	TOTAL
1880	Roof drain, DWV PVC, 2″ diam., piping, 10′ high	229	625	854
1920	For each additional foot add	2.65	19.05	21.70
1960	3″ diam., 10′ high	294	725	1,019
2000	For each additional foot add	4.96	21	25.96
2040	4″ diam., 10′ high	350	825	1,175
2080	For each additional foot add	6.75	23.50	30.25
2120	5″ diam., 10′ high	1,075	945	2,020
2160	For each additional foot add	15.35	26	41.35
2200	6″ diam., 10′ high	1,050	1,050	2,100
2240	For each additional foot add	12.20	29	41.20
2280	8″ diam., 10′ high	2,300	1,775	4,075
2320	For each additional foot add	28.50	36.50	65
3940	C.I., soil, single hub, service wt., 2″ diam. piping, 10′ high	455	685	1,140
3980	For each additional foot add	7.70	17.85	25.55
4120	3″ diam., 10′ high	640	740	1,380
4160	For each additional foot add	10.80	18.75	29.55
4200	4″ diam., 10′ high	740	810	1,550
4240	For each additional foot add	13.95	20.50	34.45
4280	5″ diam., 10′ high	1,150	895	2,045
4320	For each additional foot add	19.40	23	42.40
4360	6″ diam., 10′ high	1,300	955	2,255
4400	For each additional foot add	23.50	24	47.50
4440	8″ diam., 10′ high	2,575	1,950	4,525
4480	For each additional foot add	37	40.50	77.50
6040	Steel galv. sch 40 threaded, 2″ diam. piping, 10′ high	755	665	1,420
6080	For each additional foot add	19	17.55	36.55
6120	3″ diam., 10′ high	1,525	950	2,475
6160	For each additional foot add	40.50	26	66.50
6200	4″ diam., 10′ high	2,150	1,225	3,375
6240	For each additional foot add	57.50	31	88.50
6280	5″ diam., 10′ high	2,100	1,025	3,125
6320	For each additional foot add	56.50	30.50	87
6360	6″ diam, 10′ high	2,400	1,300	3,700
6400	For each additional foot add	66	41.50	107.50
6440	8″ diam., 10′ high	3,750	1,900	5,650
6480	For each additional foot add	78	47	125

D2090 Other Plumbing Systems

D2090 810	Piping - Installed - Unit Costs	COST PER L.F.		
		MAT.	INST.	TOTAL
0840	Cast iron, soil, B & S, service weight, 2" diameter	7.70	17.85	25.55
0860	3" diameter	10.80	18.75	29.55
0880	4" diameter	13.95	20.50	34.45
0900	5" diameter	19.40	23	42.40
0920	6" diameter	23.50	24	47.50
0940	8" diameter	37	40.50	77.50
0960	10" diameter	61	44	105
0980	12" diameter	87.50	49.50	137
1040	No hub, 1-1/2" diameter	7.75	15.85	23.60
1060	2" diameter	8.10	16.75	24.85
1080	3" diameter	11.20	17.55	28.75
1100	4" diameter	14.55	19.35	33.90
1120	5" diameter	20.50	21	41.50
1140	6" diameter	25	22	47
1160	8" diameter	45	34.50	79.50
1180	10" diameter	75	39	114
1220	Copper tubing, hard temper, solder, type K, 1/2" diameter	3.70	8	11.70
1260	3/4" diameter	6.60	8.45	15.05
1280	1" diameter	8.75	9.45	18.20
1300	1-1/4" diameter	11	11.15	22.15
1320	1-1/2" diameter	14.30	12.50	26.80
1340	2" diameter	22	15.60	37.60
1360	2-1/2" diameter	33.50	18.75	52.25
1380	3" diameter	47	21	68
1400	4" diameter	80	29.50	109.50
1480	5" diameter	180	35	215
1500	6" diameter	263	46	309
1520	8" diameter	440	51.50	491.50
1560	Type L, 1/2" diameter	2.74	7.70	10.44
1600	3/4" diameter	4.21	8.20	12.41
1620	1" diameter	6.55	9.20	15.75
1640	1-1/4" diameter	9.20	10.75	19.95
1660	1-1/2" diameter	11.80	12	23.80
1680	2" diameter	18.40	14.85	33.25
1700	2-1/2" diameter	28.50	18.10	46.60
1720	3" diameter	39	20	59
1740	4" diameter	66.50	29	95.50
1760	5" diameter	143	33	176
1780	6" diameter	200	43.50	243.50
1800	8" diameter	335	48.50	383.50
1840	Type M, 1/2" diameter	2.05	7.45	9.50
1880	3/4" diameter	3.22	8	11.22
1900	1" diameter	5.20	8.90	14.10
1920	1-1/4" diameter	7.65	10.40	18.05
1940	1-1/2" diameter	10.35	11.55	21.90
1960	2" diameter	16.25	14.20	30.45
1980	2-1/2" diameter	25	17.55	42.55
2000	3" diameter	33.50	19.35	52.85
2020	4" diameter	61	28	89
2060	6" diameter	250	41.50	291.50
2080	8" diameter	335	46	381
2120	Type DWV, 1-1/4" diameter	10.80	10.40	21.20
2160	1-1/2" diameter	10.20	11.55	21.75
2180	2" diameter	13.80	14.20	28
2200	3" diameter	26	19.35	45.35
2220	4" diameter	46	28	74
2240	5" diameter	129	31	160
2260	6" diameter	186	41.50	227.50

D2090 Other Plumbing Systems

D2090 810	Piping - Installed - Unit Costs	COST PER L.F.		
		MAT.	INST.	TOTAL
2280	8" diameter	450	46	496
2800	Plastic,PVC, DWV, schedule 40, 1-1/4" diameter	2.19	14.85	17.04
2820	1-1/2" diameter	2.20	17.35	19.55
2830	2" diameter	2.65	19.05	21.70
2840	3" diameter	4.96	21	25.96
2850	4" diameter	6.75	23.50	30.25
2890	6" diameter	12.20	29	41.20
3010	Pressure pipe 200 PSI, 1/2" diameter	1.70	11.55	13.25
3030	3/4" diameter	1.88	12.25	14.13
3040	1" diameter	2.06	13.55	15.61
3050	1-1/4" diameter	2.62	14.85	17.47
3060	1-1/2" diameter	2.83	17.35	20.18
3070	2" diameter	3.41	19.05	22.46
3080	2-1/2" diameter	6.15	20	26.15
3090	3" diameter	6.95	21	27.95
3100	4" diameter	11.35	23.50	34.85
3110	6" diameter	20.50	29	49.50
3120	8" diameter	31.50	36.50	68
4000	Steel, schedule 40, black, threaded, 1/2" diameter	3.97	9.90	13.87
4020	3/4" diameter	4.71	10.25	14.96
4030	1" diameter	6.90	11.75	18.65
4040	1-1/4" diameter	8.75	12.65	21.40
4050	1-1/2" diameter	10.40	14.05	24.45
4060	2" diameter	13.85	17.55	31.40
4070	2-1/2" diameter	21.50	22.50	44
4080	3" diameter	27.50	26	53.50
4090	4" diameter	40.50	31	71.50
4100	Grooved, 5" diameter	25	30.50	55.50
4110	6" diameter	31.50	41.50	73
4120	8" diameter	48.50	47	95.50
4130	10" diameter	71.50	56.50	128
4140	12" diameter	91	64.50	155.50
4200	Galvanized, threaded, 1/2" diameter	5.65	9.90	15.55
4220	3/4" diameter	5.50	10.25	15.75
4230	1" diameter	9.30	11.75	21.05
4240	1-1/4" diameter	12	12.65	24.65
4250	1-1/2" diameter	14.20	14.05	28.25
4260	2" diameter	19	17.55	36.55
4270	2-1/2" diameter	31.50	22.50	54
4280	3" diameter	40.50	26	66.50
4290	4" diameter	57.50	31	88.50
4300	Grooved, 5" diameter	56.50	30.50	87
4310	6" diameter	66	41.50	107.50
4320	8" diameter	78	47	125
4330	10" diameter	115	56.50	171.50
4340	12" diameter	150	64.50	214.50
5010	Flanged, black, 1" diameter	9.95	16.95	26.90
5020	1-1/4" diameter	11.25	18.55	29.80
5030	1-1/2" diameter	11.85	20.50	32.35
5040	2" diameter	14.10	26.50	40.60
5050	2-1/2" diameter	17.75	33	50.75
5060	3" diameter	20.50	37	57.50
5070	4" diameter	27.50	45.50	73
5080	5" diameter	40	56.50	96.50
5090	6" diameter	46.50	72.50	119
5100	8" diameter	72.50	95.50	168
5110	10" diameter	118	113	231
5120	12" diameter	158	130	288

D2090 Other Plumbing Systems

D2090 810	Piping - Installed - Unit Costs	COST PER L.F.		
		MAT.	INST.	TOTAL
5720	Grooved joints, black, 3/4" diameter	5.80	8.80	14.60
5730	1" diameter	3.72	9.90	13.62
5740	1-1/4" diameter	4.87	10.75	15.62
5750	1-1/2" diameter	5.60	12.25	17.85
5760	2" diameter	6.90	15.60	22.50
5770	2-1/2" diameter	10.10	19.70	29.80
5900	3" diameter	12.70	22.50	35.20
5910	4" diameter	17.95	25	42.95
5920	5" diameter	25	30.50	55.50
5930	6" diameter	31.50	41.50	73
5940	8" diameter	48.50	47	95.50
5950	10" diameter	71.50	56.50	128
5960	12" diameter	91	64.50	155.50

Reed Construction Data
The leader in construction information and BIM solutions

Reed Construction Data, Inc. is a leading provider of construction information and building information modeling (BIM) solutions. The company's portfolio of information products and services is designed specifically to help construction industry professionals advance their business with timely and accurate project, product, and cost data. Reed Construction Data is a division of Reed Business Information, a member of the Reed Elsevier PLC group of companies.

Cost Information

RSMeans, the undisputed market leader in construction costs, provides current cost and estimating information through its innovative MeansCostworks.com® web-based solution. In addition, RSMeans publishes annual cost books, estimating software, and a rich library of reference books. RSMeans also conducts a series of professional seminars and provides construction cost consulting for owners, manufacturers, designers, and contractors to sharpen personal skills and maximize the effective use of cost estimating and management tools.

Project Data

Reed Construction Data assembles one of the largest databases of public and private project data for use by contractors, distributors, and building product manufacturers in the U.S. and Canadian markets. In addition, Reed Construction Data is the North American construction community's premier resource for project leads and bid documents. Reed Bulletin and Reed CONNECT™ provide project leads and project data through all stages of construction for many of the country's largest public sector, commercial, industrial, and multi-family residential projects.

Research and Analytics

Reed Construction Data's forecasting tools cover most aspects of the construction business in the U.S. and Canada. With a vast network of resources, Reed Construction Data is uniquely qualified to give you the information you need to keep your business profitable.

SmartBIM Solutions

Reed Construction Data has emerged as the leader in the field of building information modeling (BIM) with products and services that have helped to advance the evolution of BIM. Through an in-depth SmartBIM Object Creation Program, BPMs can rely on Reed to create high-quality, real world objects embedded with superior cost data (RSMeans). In addition, Reed has made it easy for architects to manage these manufacturer-specific objects as well as generic objects in Revit with the SmartBIM Library.

SmartBuilding Index

The leading industry source for product research, product documentation, BIM objects, design ideas, and source locations. Search, select, and specify available building products with our online directories of manufacturer profiles, MANU-SPEC, SPEC-DATA, guide specs, manufacturer catalogs, building codes, historical project data, and BIM objects.

SmartBuilding Studio

A vibrant online resource library which brings together a comprehensive catalog of commercial interior finishes and products under a single standard for high-definition imagery, product data, and searchable attributes.

Associated Construction Publications (ACP)

Reed Construction Data's regional construction magazines cover the nation through a network of 14 regional magazines. Serving the construction market for more than 100 years, our magazines are a trusted source of news and information in the local and national construction communities.

For more information, please visit our website at www.reedconstructiondata.com

The leader in construction information and BIM solutions.

Reed Construction Data®

D30 HVAC

D3010 Energy Supply

Forced Hot Water Heating System

Fin Tube Radiation

Terminal Unit Heater

D3010 510	Apartment Building Heating - Fin Tube Radiation	COST PER S.F.		
		MAT.	INST.	TOTAL
1740	Heating systems, fin tube radiation, forced hot water			
1760	1,000 S.F. area, 10,000 C.F. volume	6.18	5.95	12.13
1800	10,000 S.F. area, 100,000 C.F. volume	2.51	3.59	6.10
1840	20,000 S.F. area, 200,000 C.F. volume	2.83	4.03	6.86
1880	30,000 S.F. area, 300,000 C.F. volume	2.73	3.92	6.65

D3010 520	Commercial Building Heating - Fin Tube Radiation	COST PER S.F.		
		MAT.	INST.	TOTAL
1940	Heating systems, fin tube radiation, forced hot water			
1960	1,000 S.F. bldg, one floor	15.70	13.50	29.20
2000	10,000 S.F., 100,000 C.F., total two floors	4.56	5.10	9.66
2040	100,000 S.F., 1,000,000 C.F., total three floors	1.63	2.27	3.90
2080	1,000,000 S.F., 10,000,000 C.F., total five floors	.86	1.20	2.06

D3010 530	Commercial Bldg. Heating - Terminal Unit Heaters	COST PER S.F.		
		MAT.	INST.	TOTAL
1860	Heating systems, terminal unit heaters, forced hot water			
1880	1,000 S.F. bldg., one floor	15.20	12.30	27.50
1920	10,000 S.F. bldg., 100,000 C.F. total two floors	4.08	4.25	8.33
1960	100,000 S.F. bldg., 1,000,000 C.F. total three floors	1.63	2.07	3.70
2000	1,000,000 S.F. bldg., 10,000,000 C.F. total five floors	1.10	1.35	2.45

D3020   Heat Generating Systems

**Small Electric Boiler Systems
Considerations:**

1. Terminal units are fin tube baseboard radiation rated at 720 BTU/hr with 200° water temperature or 820 BTU/hr steam.
2. Primary use being for residential or smaller supplementary areas, the floor levels are based on 7-1/2" ceiling heights.
3. All distribution piping is copper for boilers through 205 MBH. All piping for larger systems is steel pipe.

**Large Electric Boiler System
Considerations:**

1. Terminal units are all unit heaters of the same size. Quantities are varied to accommodate total requirements.
2. All air is circulated through the heaters a minimum of three times per hour.
3. As the capacities are adequate for commercial use, gross output rating by floor levels are based on 10' ceiling height.
4. All distribution piping is black steel pipe.

D3020 102	Small Heating Systems, Hydronic, Electric Boilers	COST PER S.F.		
		MAT.	INST.	TOTAL
1100	Small heating systems, hydronic, electric boilers			
1120	Steam, 1 floor, 1480 S.F., 61 M.B.H.	6.40	7.64	14.04
1160	3,000 S.F., 123 M.B.H.	4.90	6.70	11.60
1200	5,000 S.F., 205 M.B.H.	4.28	6.20	10.48
1240	2 floors, 12,400 S.F., 512 M.B.H.	3.64	6.15	9.79
1280	3 floors, 24,800 S.F., 1023 M.B.H.	3.79	6.05	9.84
1320	34,750 S.F., 1,433 M.B.H.	3.44	5.90	9.34
1360	Hot water, 1 floor, 1,000 S.F., 41 M.B.H.	10	4.26	14.26
1400	2,500 S.F., 103 M.B.H.	7.45	7.65	15.10
1440	2 floors, 4,850 S.F., 205 M.B.H.	7.35	9.20	16.55
1480	3 floors, 9,700 S.F., 410 M.B.H.	7.70	9.50	17.20

D3020 104	Large Heating Systems, Hydronic, Electric Boilers	COST PER S.F.		
		MAT.	INST.	TOTAL
1230	Large heating systems, hydronic, electric boilers			
1240	9,280 S.F., 135 K.W., 461 M.B.H., 1 floor	3.52	2.88	6.40
1280	14,900 S.F., 240 K.W., 820 M.B.H., 2 floors	4.28	4.68	8.96
1320	18,600 S.F., 296 K.W., 1,010 M.B.H., 3 floors	4.22	5.10	9.32
1360	26,100 S.F., 420 K.W., 1,432 M.B.H., 4 floors	4.19	4.99	9.18
1400	39,100 S.F., 666 K.W., 2,273 M.B.H., 4 floors	3.59	4.16	7.75
1440	57,700 S.F., 900 K.W., 3,071 M.B.H., 5 floors	3.42	4.08	7.50
1480	111,700 S.F., 1,800 K.W., 6,148 M.B.H., 6 floors	3.14	3.52	6.66
1520	149,000 S.F., 2,400 K.W., 8,191 M.B.H., 8 floors	3.11	3.49	6.60
1560	223,300 S.F., 3,600 K.W., 12,283 M.B.H., 14 floors	3.30	3.98	7.28

D3020 Heat Generating Systems

Boiler Selection: The maximum allowable working pressures are limited by ASME "Code for Heating Boilers" to 15 PSI for steam and 160 PSI for hot water heating boilers, with a maximum temperature limitation of 250° F. Hot water boilers are generally rated for a working pressure of 30 PSI. High pressure boilers are governed by the ASME "Code for Power Boilers" which is used almost universally for boilers operating over 15 PSIG. High pressure boilers used for a combination of heating/process loads are usually designed for 150 PSIG.

Boiler ratings are usually indicated as either Gross or Net Output. The Gross Load is equal to the Net Load plus a piping and pickup allowance. When this allowance cannot be determined, divide the gross output rating by 1.25 for a value equal to or greater than the next heat loss requirement of the building.

Table below lists installed cost per boiler and includes insulating jacket, standard controls, burner and safety controls. Costs do not include piping or boiler base pad. Outputs are Gross.

D3020 106	Boilers, Hot Water & Steam	COST EACH		
		MAT.	INST.	TOTAL
0600	Boiler, electric, steel, hot water, 12 K.W., 41 M.B.H.	4,125	1,300	5,425
0620	30 K.W., 103 M.B.H.	4,925	1,425	6,350
0640	60 K.W., 205 M.B.H.	6,025	1,550	7,575
0660	120 K.W., 410 M.B.H.	6,750	1,900	8,650
0680	210 K.W., 716 M.B.H.	7,900	2,850	10,750
0700	510 K.W., 1,739 M.B.H.	18,900	5,275	24,175
0720	720 K.W., 2,452 M.B.H.	22,900	5,950	28,850
0740	1,200 K.W., 4,095 M.B.H.	29,600	6,850	36,450
0760	2,100 K.W., 7,167 M.B.H.	56,500	8,600	65,100
0780	3,600 K.W., 12,283 M.B.H.	87,000	14,500	101,500
0820	Steam, 6 K.W., 20.5 M.B.H.	3,850	1,425	5,275
0840	24 K.W., 81.8 M.B.H.	4,775	1,550	6,325
0860	60 K.W., 205 M.B.H.	6,625	1,700	8,325
0880	150 K.W., 512 M.B.H.	9,575	2,625	12,200
0900	510 K.W., 1,740 M.B.H.	24,400	6,450	30,850
0920	1,080 K.W., 3,685 M.B.H.	34,100	9,300	43,400
0940	2,340 K.W., 7,984 M.B.H.	70,000	14,500	84,500
0980	Gas, cast iron, hot water, 80 M.B.H.	1,975	1,625	3,600
1000	100 M.B.H.	2,525	1,750	4,275
1020	163 M.B.H.	3,100	2,375	5,475
1040	280 M.B.H.	4,550	2,625	7,175
1060	544 M.B.H.	9,025	4,675	13,700
1080	1,088 M.B.H.	13,600	5,925	19,525
1100	2,000 M.B.H.	19,700	9,250	28,950
1120	2,856 M.B.H.	23,700	11,900	35,600
1140	4,720 M.B.H.	72,000	16,400	88,400
1160	6,970 M.B.H.	103,500	26,600	130,100
1180	For steam systems under 2,856 M.B.H., add 8%			
1520	Oil, cast iron, hot water, 109 M.B.H.	2,175	1,975	4,150
1540	173 M.B.H.	2,750	2,375	5,125
1560	236 M.B.H.	3,525	2,800	6,325
1580	1,084 M.B.H.	10,300	6,300	16,600
1600	1,600 M.B.H.	13,300	9,050	22,350
1620	2,480 M.B.H.	20,400	11,600	32,000
1640	3,550 M.B.H.	26,300	13,900	40,200
1660	Steam systems same price as hot water			

D3020 Heat Generating Systems

Unit Heater

Fossil Fuel Boiler System Considerations:

1. Terminal units are horizontal unit heaters. Quantities are varied to accommodate total heat loss per building.
2. Unit heater selection was determined by their capacity to circulate the building volume a minimum of three times per hour in addition to the BTU output.
3. Systems shown are forced hot water. Steam boilers cost slightly more than hot water boilers. However, this is compensated for by the smaller size or fewer terminal units required with steam.
4. Floor levels are based on 10' story heights.
5. MBH requirements are gross boiler output.

D3020 108	Heating Systems, Unit Heaters	COST PER S.F.		
		MAT.	INST.	TOTAL
1260	Heating systems, hydronic, fossil fuel, terminal unit heaters,			
1280	Cast iron boiler, gas, 80 M.B.H., 1,070 S.F. bldg.	10.47	9.18	19.65
1320	163 M.B.H., 2,140 S.F. bldg.	7	6.25	13.25
1360	544 M.B.H., 7,250 S.F. bldg.	4.52	4.43	8.95
1400	1,088 M.B.H., 14,500 S.F. bldg.	3.80	4.20	8
1440	3,264 M.B.H., 43,500 S.F. bldg.	2.95	3.14	6.09
1480	5,032 M.B.H., 67,100 S.F. bldg.	3.58	3.26	6.84
1520	Oil, 109 M.B.H., 1,420 S.F. bldg.	10.40	8.20	18.60
1560	235 M.B.H., 3,150 S.F. bldg.	6.30	5.85	12.15
1600	940 M.B.H., 12,500 S.F. bldg.	4.42	3.82	8.24
1640	1,600 M.B.H., 21,300 S.F. bldg.	3.98	3.66	7.64
1680	2,480 M.B.H., 33,100 S.F. bldg.	3.98	3.29	7.27
1720	3,350 M.B.H., 44,500 S.F. bldg.	3.65	3.38	7.03
1760	Coal, 148 M.B.H., 1,975 S.F. bldg.	9.15	5.30	14.45
1800	300 M.B.H., 4,000 S.F. bldg.	6.90	4.20	11.10
1840	2,360 M.B.H., 31,500 S.F. bldg.	4.42	3.36	7.78

D3020 Heat Generating Systems

Fin Tube Radiator

Fossil Fuel Boiler System Considerations:

1. Terminal units are commercial steel fin tube radiation. Quantities are varied to accommodate total heat loss per building.
2. Systems shown are forced hot water. Steam boilers cost slightly more than hot water boilers. However, this is compensated for by the smaller size or fewer terminal units required with steam.
3. Floor levels are based on 10' story heights.
4. MBH requirements are gross boiler output.

D3020 110	Heating System, Fin Tube Radiation	COST PER S.F.		
		MAT.	INST.	TOTAL
3230	Heating systems, hydronic, fossil fuel, fin tube radiation			
3240	Cast iron boiler, gas, 80 MBH, 1,070 S.F. bldg.	12.99	13.74	26.73
3280	169 M.B.H., 2,140 S.F. bldg.	8.20	8.70	16.90
3320	544 M.B.H., 7,250 S.F. bldg.	6.30	7.50	13.80
3360	1,088 M.B.H., 14,500 S.F. bldg.	5.70	7.35	13.05
3400	3,264 M.B.H., 43,500 S.F. bldg.	4.97	6.45	11.42
3440	5,032 M.B.H., 67,100 S.F. bldg.	5.60	6.55	12.15
3480	Oil, 109 M.B.H., 1,420 S.F. bldg.	14.60	15	29.60
3520	235 M.B.H., 3,150 S.F. bldg.	8.10	8.85	16.95
3560	940 M.B.H., 12,500 S.F. bldg.	6.35	7	13.35
3600	1,600 M.B.H., 21,300 S.F. bldg.	6	6.95	12.95
3640	2,480 M.B.H., 33,100 S.F. bldg.	6	6.60	12.60
3680	3,350 M.B.H., 44,500 S.F. bldg.	5.65	6.65	12.30
3720	Coal, 148 M.B.H., 1,975 S.F. bldg.	10.85	8.30	19.15
3760	300 M.B.H., 4,000 S.F. bldg.	8.60	7.15	15.75
3800	2,360 M.B.H., 31,500 S.F. bldg.	6.40	6.60	13
4080	Steel boiler, oil, 97 M.B.H., 1,300 S.F. bldg.	10.90	12.35	23.25
4120	315 M.B.H., 4,550 S.F. bldg.	6.05	6.25	12.30
4160	525 M.B.H., 7,000 S.F. bldg.	7.65	7.25	14.90
4200	1,050 M.B.H., 14,000 S.F. bldg.	6.60	7.25	13.85
4240	2,310 M.B.H., 30,800 S.F. bldg.	6.15	6.65	12.80
4280	3,150 M.B.H., 42,000 S.F. bldg.	5.95	6.75	12.70

D3030 Cooling Generating Systems

Chilled Water Supply & Return Piping

Air Cooled Water Chiller Unit

Roof

Insulate

Return

Fan Coil Unit

Supply

Finish Ceiling

*Cooling requirements would lead to a choice of multiple chillers.

D3030 110	Chilled Water, Air Cooled Condenser Systems	COST PER S.F.		
		MAT.	INST.	TOTAL
1180	Packaged chiller, air cooled, with fan coil unit			
1200	Apartment corridors, 3,000 S.F., 5.50 ton	6.60	7.60	14.20
1240	6,000 S.F., 11.00 ton	5.50	6.05	11.55
1280	10,000 S.F., 18.33 ton	4.83	4.61	9.44
1320	20,000 S.F., 36.66 ton	3.56	3.34	6.90
1360	40,000 S.F., 73.33 ton	4.16	3.35	7.51
1440	Banks and libraries, 3,000 S.F., 12.50 ton	10.40	8.95	19.35
1480	6,000 S.F., 25.00 ton	9.55	7.25	16.80
1520	10,000 S.F., 41.66 ton	7.05	5.05	12.10
1560	20,000 S.F., 83.33 ton	7.50	4.56	12.06
1600	40,000 S.F., 167 ton*			
1680	Bars and taverns, 3,000 S.F., 33.25 ton	17.40	10.60	28
1720	6,000 S.F., 66.50 ton	16.45	8.10	24.55
1760	10,000 S.F., 110.83 ton	13.25	2.63	15.88
1800	20,000 S.F., 220 ton*			
1840	40,000 S.F., 440 ton*			
1920	Bowling alleys, 3,000 S.F., 17.00 ton	13.25	10.05	23.30
1960	6,000 S.F., 34.00 ton	9.55	7.05	16.60
2000	10,000 S.F., 56.66 ton	8.75	5.05	13.80
2040	20,000 S.F., 113.33 ton	8.45	4.63	13.08
2080	40,000 S.F., 227 ton*			
2160	Department stores, 3,000 S.F., 8.75 ton	9.30	8.45	17.75
2200	6,000 S.F., 17.50 ton	7.25	6.55	13.80
2240	10,000 S.F., 29.17 ton	5.60	4.72	10.32
2280	20,000 S.F., 58.33 ton	4.93	3.47	8.40
2320	40,000 S.F., 116.66 ton	5.40	3.47	8.87
2400	Drug stores, 3,000 S.F., 20.00 ton	15.05	10.35	25.40
2440	6,000 S.F., 40.00 ton	11.40	7.70	19.10
2480	10,000 S.F., 66.66 ton	11.65	6.55	18.20
2520	20,000 S.F., 133.33 ton	10.45	4.89	15.34
2560	40,000 S.F., 267 ton*			
2640	Factories, 2,000 S.F., 10.00 ton	9.05	8.60	17.65
2680	6,000 S.F., 20.00 ton	8.25	6.90	15.15
2720	10,000 S.F., 33.33 ton	6.05	4.85	10.90
2760	20,000 S.F., 66.66 ton	6.45	4.34	10.79
2800	40,000 S.F., 133.33 ton	5.95	3.58	9.53
2880	Food supermarkets, 3,000 S.F., 8.50 ton	9.15	8.40	17.55
2920	6,000 S.F., 17.00 ton	7.10	6.50	13.60
2960	10,000 S.F., 28.33 ton	5.30	4.60	9.90
3000	20,000 S.F., 56.66 ton	4.74	3.40	8.14

D30 HVAC

D3030 Cooling Generating Systems

D3030 110	Chilled Water, Air Cooled Condenser Systems	COST PER S.F.		
		MAT.	INST.	TOTAL
3040	40,000 S.F., 113.33 ton	5.10	3.48	8.58
3120	Medical centers, 3,000 S.F., 7.00 ton	8.20	8.20	16.40
3160	6,000 S.F., 14.00 ton	6.30	6.30	12.60
3200	10,000 S.F., 23.33 ton	5.70	4.80	10.50
3240	20,000 S.F., 46.66 ton	4.28	3.53	7.81
3280	40,000 S.F., 93.33 ton	4.62	3.41	8.03
3360	Offices, 3,000 S.F., 9.50 ton	8.40	8.35	16.75
3400	6,000 S.F., 19.00 ton	7.95	6.80	14.75
3440	10,000 S.F., 31.66 ton	5.90	4.79	10.69
3480	20,000 S.F., 63.33 ton	6.45	4.41	10.86
3520	40,000 S.F., 126.66 ton	5.85	3.59	9.44
3600	Restaurants, 3,000 S.F., 15.00 ton	11.70	9.30	21
3640	6,000 S.F., 30.00 ton	9.10	7.05	16.15
3680	10,000 S.F., 50.00 ton	8.35	5.15	13.50
3720	20,000 S.F., 100.00 ton	8.60	4.85	13.45
3760	40,000 S.F., 200 ton*			
3840	Schools and colleges, 3,000 S.F., 11.50 ton	9.85	8.80	18.65
3880	6,000 S.F., 23.00 ton	9.05	7.10	16.15
3920	10,000 S.F., 38.33 ton	6.65	4.98	11.63
3960	20,000 S.F., 76.66 ton	7.30	4.60	11.90
4000	40,000 S.F., 153 ton*			

D3030 Cooling Generating Systems

*Cooling requirements would lead to a choice of multiple chillers.

D3030 115	Chilled Water, Cooling Tower Systems	COST PER S.F.		
		MAT.	INST.	TOTAL
1300	Packaged chiller, water cooled, with fan coil unit			
1320	Apartment corridors, 4,000 S.F., 7.33 ton	5.98	7.43	13.41
1360	6,000 S.F., 11.00 ton	4.95	6.40	11.35
1400	10,000 S.F., 18.33 ton	4.94	4.74	9.68
1440	20,000 S.F., 26.66 ton	3.86	3.54	7.40
1480	40,000 S.F., 73.33 ton	3.77	3.62	7.39
1520	60,000 S.F., 110.00 ton	3.94	3.71	7.65
1600	Banks and libraries, 4,000 S.F., 16.66 ton	10.40	8.15	18.55
1640	6,000 S.F., 25.00 ton	8.85	7.20	16.05
1680	10,000 S.F., 41.66 ton	7.85	5.35	13.20
1720	20,000 S.F., 83.33 ton	6.55	5.10	11.65
1760	40,000 S.F., 166.66 ton	7.25	6.25	13.50
1800	60,000 S.F., 250.00 ton	7.40	6.55	13.95
1880	Bars and taverns, 4,000 S.F., 44.33 ton	18.75	10.30	29.05
1920	6,000 S.F., 66.50 ton	18.70	10.60	29.30
1960	10,000 S.F., 110.83 ton	15.10	8.45	23.55
2000	20,000 S.F., 221.66 ton	18.50	8.55	27.05
2040	40,000 S.F., 440 ton*			
2080	60,000 S.F., 660 ton*			
2160	Bowling alleys, 4,000 S.F., 22.66 ton	12.45	8.95	21.40
2200	6,000 S.F., 34.00 ton	10.70	7.90	18.60
2240	10,000 S.F., 56.66 ton	9.40	5.80	15.20
2280	20,000 S.F., 113.33 ton	8.35	5.40	13.75
2320	40,000 S.F., 226.66 ton	10.30	6.25	16.55
2360	60,000 S.F., 340 ton			
2440	Department stores, 4,000 S.F., 11.66 ton	6.75	8.05	14.80
2480	6,000 S.F., 17.50 ton	7.65	6.65	14.30
2520	10,000 S.F., 29.17 ton	5.80	4.95	10.75
2560	20,000 S.F., 58.33 ton	4.90	3.68	8.58
2600	40,000 S.F., 116.66 ton	4.90	3.81	8.71
2640	60,000 S.F., 175.00 ton	6.60	6	12.60
2720	Drug stores, 4,000 S.F., 26.66 ton	13.20	9.20	22.40
2760	6,000 S.F., 40.00 ton	12.40	7.90	20.30
2800	10,000 S.F., 66.66 ton	11.85	7.15	19
2840	20,000 S.F., 133.33 ton	9.70	5.65	15.35
2880	40,000 S.F., 266.67 ton	10.20	7.05	17.25
2920	60,000 S.F., 400 ton*			
3000	Factories, 4,000 S.F., 13.33 ton	8.95	7.75	16.70
3040	6,000 S.F., 20.00 ton	7.85	6.70	14.55
3080	10,000 S.F., 33.33 ton	6.35	5.10	11.45
3120	20,000 S.F., 66.66 ton	6.45	4.59	11.04
3160	40,000 S.F., 133.33 ton	5.55	3.94	9.49

D3030 Cooling Generating Systems

D3030 115	Chilled Water, Cooling Tower Systems	COST PER S.F.		
		MAT.	INST.	TOTAL
3200	60,000 S.F., 200.00 ton	6.60	6.30	12.90
3280	Food supermarkets, 4,000 S.F., 11.33 ton	6.65	8	14.65
3320	6,000 S.F., 17.00 ton	7	6.50	13.50
3360	10,000 S.F., 28.33 ton	5.70	4.92	10.62
3400	20,000 S.F., 56.66 ton	4.97	3.70	8.67
3440	40,000 S.F., 113.33 ton	4.87	3.82	8.69
3480	60,000 S.F., 170.00 ton	6.55	6	12.55
3560	Medical centers, 4.000 S.F., 9.33 ton	5.70	7.35	13.05
3600	6,000 S.F., 14.00 ton	6.65	6.35	13
3640	10,000 S.F., 23.33 ton	5.30	4.77	10.07
3680	20,000 S.F., 46.66 ton	4.44	3.61	8.05
3720	40,000 S.F., 93.33 ton	4.37	3.74	8.11
3760	60,000 S.F., 140.00 ton	5.30	6.05	11.35
3840	Offices, 4,000 S.F., 12.66 ton	8.65	7.70	16.35
3880	6,000 S.F., 19.00 ton	7.85	6.85	14.70
3920	10,000 S.F., 31.66 ton	6.35	5.15	11.50
3960	20,000 S.F., 63.33 ton	6.35	4.60	10.95
4000	40,000 S.F., 126.66 ton	5.90	5.95	11.85
4040	60,000 S.F., 190.00 ton	6.35	6.25	12.60
4120	Restaurants, 4,000 S.F., 20.00 ton	11.10	8.30	19.40
4160	6,000 S.F., 30.00 ton	9.60	7.40	17
4200	10,000 S.F., 50.00 ton	8.65	5.55	14.20
4240	20,000 S.F., 100.00 ton	7.80	5.35	13.15
4280	40,000 S.F., 200.00 ton	7.75	6.05	13.80
4320	60,000 S.F., 300.00 ton	8.30	6.80	15.10
4400	Schools and colleges, 4,000 S.F., 15.33 ton	9.85	8	17.85
4440	6,000 S.F., 23.00 ton	8.40	7.05	15.45
4480	10,000 S.F., 38.33 ton	7.40	5.25	12.65
4520	20,000 S.F., 76.66 ton	6.30	5.05	11.35
4560	40,000 S.F., 153.33 ton	6.85	6.10	12.95
4600	60,000 S.F., 230.00 ton	6.80	6.35	13.15

D3050 Terminal & Package Units

Roof Top Unit

Roof

Return Duct

Fan Coil Unit

Insulated Supply Duct

Return Grille

Finish Ceiling

Supply Diff.

*Above normal capacity

Rooftop, Single Zone System

D3050 150	Rooftop Single Zone Unit Systems	COST PER S.F.		
		MAT.	INST.	TOTAL
1260	Rooftop, single zone, air conditioner			
1280	Apartment corridors, 500 S.F., .92 ton	4	3.55	7.55
1320	1,000 S.F., 1.83 ton	3.96	3.52	7.48
1360	1500 S.F., 2.75 ton	2.59	2.91	5.50
1400	3,000 S.F., 5.50 ton	2.42	2.79	5.21
1440	5,000 S.F., 9.17 ton	2.59	2.54	5.13
1480	10,000 S.F., 18.33 ton	2.67	2.39	5.06
1560	Banks or libraries, 500 S.F., 2.08 ton	9	8	17
1600	1,000 S.F., 4.17 ton	5.90	6.60	12.50
1640	1,500 S.F., 6.25 ton	5.50	6.35	11.85
1680	3,000 S.F., 12.50 ton	5.85	5.75	11.60
1720	5,000 S.F., 20.80 ton	6.05	5.40	11.45
1760	10,000 S.F., 41.67 ton	4.86	5.40	10.26
1840	Bars and taverns, 500 S.F. 5.54 ton	13.25	10.65	23.90
1880	1,000 S.F., 11.08 ton	14.25	9.10	23.35
1920	1,500 S.F., 16.62 ton	13.55	8.50	22.05
1960	3,000 S.F., 33.25 ton	11.75	8.10	19.85
2000	5,000 S.F., 55.42 ton	11	8.10	19.10
2040	10,000 S.F., 110.83 ton*			
2080	Bowling alleys, 500 S.F., 2.83 ton	8	9	17
2120	1,000 S.F., 5.67 ton	7.45	8.65	16.10
2160	1,500 S.F., 8.50 ton	8	7.85	15.85
2200	3,000 S.F., 17.00 ton	7.65	7.55	15.20
2240	5,000 S.F., 28.33 ton	6.70	7.35	14.05
2280	10,000 S.F., 56.67 ton	6.35	7.35	13.70
2360	Department stores, 500 S.F., 1.46 ton	6.30	5.60	11.90
2400	1,000 S.F., 2.92 ton	4.12	4.64	8.76
2480	3,000 S.F., 8.75 ton	4.11	4.04	8.15
2520	5,000 S.F., 14.58 ton	3.94	3.87	7.81
2560	10,000 S.F., 29.17 ton	3.46	3.77	7.23
2640	Drug stores, 500 S.F., 3.33 ton	9.40	10.55	19.95
2680	1,000 S.F., 6.67 ton	8.80	10.15	18.95
2720	1,500 S.F., 10.00 ton	9.40	9.25	18.65
2760	3,000 S.F., 20.00 ton	9.70	8.70	18.40
2800	5,000 S.F., 33.33 ton	7.90	8.65	16.55
2840	10,000 S.F., 66.67 ton	7.45	8.60	16.05
2920	Factories, 500 S.F., 1.67 ton	7.20	6.40	13.60
3000	1,500 S.F., 5.00 ton	4.40	5.10	9.50
3040	3,000 S.F., 10.00 ton	4.70	4.62	9.32
3080	5,000 S.F., 16.67 ton	4.50	4.43	8.93

D3050 Terminal & Package Units

D3050 150	Rooftop Single Zone Unit Systems	COST PER S.F.		
		MAT.	INST.	TOTAL
3120	10,000 S.F., 33.33 ton	3.95	4.31	8.26
3200	Food supermarkets, 500 S.F., 1.42 ton	6.15	5.45	11.60
3240	1,000 S.F., 2.83 ton	3.96	4.48	8.44
3280	1,500 S.F., 4.25 ton	3.74	4.32	8.06
3320	3,000 S.F., 8.50 ton	4	3.93	7.93
3360	5,000 S.F., 14.17 ton	3.83	3.77	7.60
3400	10,000 S.F., 28.33 ton	3.36	3.67	7.03
3480	Medical centers, 500 S.F., 1.17 ton	5.05	4.50	9.55
3520	1,000 S.F., 2.33 ton	5.05	4.49	9.54
3560	1,500 S.F., 3.50 ton	3.29	3.71	7
3640	5,000 S.F., 11.67 ton	3.29	3.24	6.53
3680	10,000 S.F., 23.33 ton	3.39	3.03	6.42
3760	Offices, 500 S.F., 1.58 ton	6.85	6.10	12.95
3800	1,000 S.F., 3.17 ton	4.47	5.05	9.52
3840	1,500 S.F., 4.75 ton	4.18	4.83	9.01
3880	3,000 S.F., 9.50 ton	4.46	4.38	8.84
3920	5,000 S.F., 15.83 ton	4.27	4.20	8.47
3960	10,000 S.F., 31.67 ton	3.76	4.10	7.86
4000	Restaurants, 500 S.F., 2.50 ton	10.80	9.65	20.45
4040	1,000 S.F., 5.00 ton	6.60	7.65	14.25
4080	1,500 S.F., 7.50 ton	7.05	6.95	14
4120	3,000 S.F., 15.00 ton	6.75	6.65	13.40
4160	5,000 S.F., 25.00 ton	7.30	6.50	13.80
4200	10,000 S.F., 50.00 ton	5.60	6.45	12.05
4240	Schools and colleges, 500 S.F., 1.92 ton	8.30	7.40	15.70
4280	1,000 S.F., 3.83 ton	5.40	6.10	11.50
4360	3,000 S.F., 11.50 ton	5.40	5.30	10.70
4400	5,000 S.F., 19.17 ton	5.55	4.99	10.54

D3050 Terminal & Package Units

Roof
Rooftop Unit
Return Ducts
Insulated Supply Ducts
Finish Ceiling
Return Grille
Supply Diffusers

*Note A: Small single zone unit recommended.

*Note B: A combination of multizone units recommended.

D3050 155	Rooftop Multizone Unit Systems	COST PER S.F.		
		MAT.	INST.	TOTAL
1240	Rooftop, multizone, air conditioner			
1260	Apartment corridors, 1,500 S.F., 2.75 ton. See Note A.			
1280	3,000 S.F., 5.50 ton	9.77	4.50	14.27
1320	10,000 S.F., 18.30 ton	6.80	4.34	11.14
1360	15,000 S.F., 27.50 ton	6.60	4.32	10.92
1400	20,000 S.F., 36.70 ton	6.90	4.34	11.24
1440	25,000 S.F., 45.80 ton	6.45	4.36	10.81
1520	Banks or libraries, 1,500 S.F., 6.25 ton	22	10.25	32.25
1560	3,000 S.F., 12.50 ton	18.70	10.05	28.75
1600	10,000 S.F., 41.67 ton	14.65	9.90	24.55
1640	15,000 S.F., 62.50 ton	10.80	9.85	20.65
1680	20,000 S.F., 83.33 ton	10.80	9.85	20.65
1720	25,000 S.F., 104.00 ton	9.65	9.80	19.45
1800	Bars and taverns, 1,500 S.F., 16.62 ton	48	14.80	62.80
1840	3,000 S.F., 33.24 ton	40	14.25	54.25
1880	10,000 S.F., 110.83 ton	24	14.20	38.20
1920	15,000 S.F., 165 ton, See Note B			
1960	20,000 S.F., 220 ton, See Note B			
2000	25,000 S.F., 275 ton, See Note B			
2080	Bowling alleys, 1,500 S.F., 8.50 ton	30	13.95	43.95
2120	3,000 S.F., 17.00 ton	25.50	13.65	39.15
2160	10,000 S.F., 56.70 ton	19.90	13.50	33.40
2200	15,000 S.F., 85.00 ton	14.70	13.40	28.10
2240	20,000 S.F., 113.00 ton	13.10	13.30	26.40
2280	25,000 S.F., 140.00 ton see Note B			
2360	Department stores, 1,500 S.F., 4.37 ton, See Note A.			
2400	3,000 S.F., 8.75 ton	15.55	7.20	22.75
2440	10,000 S.F., 29.17 ton	10.95	6.90	17.85
2520	20,000 S.F., 58.33 ton	7.55	6.90	14.45
2560	25,000 S.F., 72.92 ton	7.55	6.90	14.45
2640	Drug stores, 1,500 S.F., 10.00 ton	35.50	16.40	51.90
2680	3,000 S.F., 20.00 ton	24.50	15.80	40.30
2720	10,000 S.F., 66.66 ton	17.30	15.75	33.05
2760	15,000 S.F., 100.00 ton	15.45	15.70	31.15
2800	20,000 S.F., 135 ton, See Note B.			
2840	25,000 S.F., 165 ton, See Note B.			
2920	Factories, 1,500 S.F., 5 ton, See Note A			
3000	10,000 S.F., 33.33 ton	12.50	7.90	20.40

D30 HVAC

D3050 Terminal & Package Units

D3050 155	Rooftop Multizone Unit Systems	COST PER S.F.		
		MAT.	INST.	TOTAL
3040	15,000 S.F., 50.00 ton	11.70	7.95	19.65
3080	20,000 S.F., 66.66 ton	8.65	7.85	16.50
3120	25,000 S.F., 83.33 ton	8.65	7.85	16.50
3200	Food supermarkets, 1,500 S.F., 4.25 ton, See Note A.			
3240	3,000 S.F., 8.50 ton	15.10	6.95	22.05
3280	10,000 S.F., 28.33 ton	10.25	6.70	16.95
3320	15,000 S.F., 42.50 ton	9.95	6.75	16.70
3360	20,000 S.F., 56.67 ton	7.35	6.70	14.05
3400	25,000 S.F., 70.83 ton	7.35	6.70	14.05
3480	Medical centers, 1,500 S.F., 3.5 ton, See Note A.			
3520	3,000 S.F., 7.00 ton	12.45	5.75	18.20
3560	10,000 S.F., 23.33 ton	8.25	5.50	13.75
3600	15,000 S.F., 35.00 ton	8.75	5.50	14.25
3640	20,000 S.F., 46.66 ton	8.20	5.55	13.75
3680	25,000 S.F., 58.33 ton	6.05	5.50	11.55
3760	Offices, 1,500 S.F., 4.75 ton, See Note A.			
3800	3,000 S.F., 9.50 ton	16.85	7.80	24.65
3840	10,000 S.F., 31.66 ton	11.90	7.50	19.40
3920	20,000 S.F., 63.33 ton	8.20	7.50	15.70
3960	25,000 S.F., 79.16 ton	8.20	7.50	15.70
4000	Restaurants, 1,500 S.F., 7.50 ton	26.50	12.30	38.80
4040	3,000 S.F., 15.00 ton	22.50	12.05	34.55
4080	10,000 S.F., 50.00 ton	17.55	11.90	29.45
4120	15,000 S.F., 75.00 ton	13	11.80	24.80
4160	20,000 S.F., 100.00 ton	11.60	11.80	23.40
4200	25,000 S.F., 125 ton, See Note B.			
4240	Schools and colleges, 1,500 S.F., 5.75 ton	20.50	9.45	29.95
4320	10,000 S.F., 38.33 ton	13.45	9.10	22.55
4360	15,000 S.F., 57.50 ton	9.95	9.05	19
4400	20,000 S.F.,76.66 ton	9.95	9.05	19
4440	25,000 S.F., 95.83 ton	9.35	9.05	18.40
2280	25,000 S.F., 140.00 ton see Note B.			
4200	25,000 S.F., 125 ton, See Note B.			

D3050 Terminal & Package Units

Self-Contained Water Cooled System

D3050 160	Self-contained, Water Cooled Unit Systems	COST PER S.F.		
		MAT.	INST.	TOTAL
1280	Self-contained, water cooled unit	2.99	2.86	5.85
1300	Apartment corridors, 500 S.F., .92 ton	3	2.85	5.85
1320	1,000 S.F., 1.83 ton	2.99	2.85	5.84
1360	3,000 S.F., 5.50 ton	2.49	2.39	4.88
1400	5,000 S.F., 9.17 ton	2.43	2.21	4.64
1440	10,000 S.F., 18.33 ton	3.45	1.98	5.43
1520	Banks or libraries, 500 S.F., 2.08 ton	6.55	2.94	9.49
1560	1,000 S.F., 4.17 ton	5.65	5.40	11.05
1600	3,000 S.F., 12.50 ton	5.55	5	10.55
1640	5,000 S.F., 20.80 ton	7.80	4.49	12.29
1680	10,000 S.F., 41.66 ton	6.80	4.49	11.29
1760	Bars and taverns, 500 S.F., 5.54 ton	14.35	4.96	19.31
1800	1,000 S.F., 11.08 ton	14.30	8.60	22.90
1840	3,000 S.F., 33.25 ton	18.45	6.80	25.25
1880	5,000 S.F., 55.42 ton	17.30	7.10	24.40
1920	10,000 S.F., 110.00 ton	16.95	7	23.95
2000	Bowling alleys, 500 S.F., 2.83 ton	8.95	4	12.95
2040	1,000 S.F., 5.66 ton	7.70	7.35	15.05
2080	3,000 S.F., 17.00 ton	10.65	6.10	16.75
2120	5,000 S.F., 28.33 ton	9.65	5.90	15.55
2160	10,000 S.F., 56.66 ton	9	6	15
2200	Department stores, 500 S.F., 1.46 ton	4.61	2.06	6.67
2240	1,000 S.F., 2.92 ton	3.96	3.79	7.75
2280	3,000 S.F., 8.75 ton	3.87	3.51	7.38
2320	5,000 S.F., 14.58 ton	5.50	3.15	8.65
2360	10,000 S.F., 29.17 ton	4.96	3.03	7.99
2440	Drug stores, 500 S.F., 3.33 ton	10.55	4.72	15.27
2480	1,000 S.F., 6.66 ton	9.05	8.70	17.75
2520	3,000 S.F., 20.00 ton	12.55	7.20	19.75
2560	5,000 S.F., 33.33 ton	13.15	7.10	20.25
2600	10,000 S.F., 66.66 ton	10.60	7.15	17.75
2680	Factories, 500 S.F., 1.66 ton	5.25	2.35	7.60
2720	1,000 S.F. 3.37 ton	4.56	4.39	8.95
2760	3,000 S.F., 10.00 ton	4.43	4	8.43
2800	5,000 S.F., 16.66 ton	6.25	3.59	9.84
2840	10,000 S.F., 33.33 ton	5.65	3.45	9.10

D3050 Terminal & Package Units

D3050 160	Self-contained, Water Cooled Unit Systems	COST PER S.F.		
		MAT.	INST.	TOTAL
2920	Food supermarkets, 500 S.F., 1.42 ton	4.47	2	6.47
2960	1,000 S.F., 2.83 ton	4.62	4.41	9.03
3000	3,000 S.F., 8.50 ton	3.84	3.68	7.52
3040	5,000 S.F., 14.17 ton	3.75	3.41	7.16
3080	10,000 S.F., 28.33 ton	4.82	2.94	7.76
3160	Medical centers, 500 S.F., 1.17 ton	3.68	1.65	5.33
3200	1,000 S.F., 2.33 ton	3.82	3.64	7.46
3240	3,000 S.F., 7.00 ton	3.17	3.04	6.21
3280	5,000 S.F., 11.66 ton	3.10	2.81	5.91
3320	10,000 S.F., 23.33 ton	4.39	2.52	6.91
3400	Offices, 500 S.F., 1.58 ton	5	2.24	7.24
3440	1,000 S.F., 3.17 ton	5.15	4.93	10.08
3480	3,000 S.F., 9.50 ton	4.20	3.81	8.01
3520	5,000 S.F., 15.83 ton	5.95	3.42	9.37
3560	10,000 S.F., 31.67 ton	5.40	3.29	8.69
3640	Restaurants, 500 S.F., 2.50 ton	7.90	3.54	11.44
3680	1,000 S.F., 5.00 ton	6.80	6.50	13.30
3720	3,000 S.F., 15.00 ton	6.65	6	12.65
3760	5,000 S.F., 25.00 ton	9.40	5.40	14.80
3800	10,000 S.F., 50.00 ton	5.75	4.95	10.70
3880	Schools and colleges, 500 S.F., 1.92 ton	6.05	2.71	8.76
3920	1,000 S.F., 3.83 ton	5.20	4.99	10.19
3960	3,000 S.F., 11.50 ton	5.10	4.62	9.72
4000	5,000 S.F., 19.17 ton	7.20	4.14	11.34
4040	10,000 S.F., 38.33 ton	6.25	4.13	10.38

D3050 Terminal & Package Units

Self-Contained Air Cooled System

D3050 165	Self-contained, Air Cooled Unit Systems	COST PER S.F.		
		MAT.	INST.	TOTAL
1300	Self-contained, air cooled unit			
1320	Apartment corridors, 500 S.F., .92 ton	3.95	3.65	7.60
1360	1,000 S.F., 1.83 ton	3.88	3.61	7.49
1400	3,000 S.F., 5.50 ton	3.26	3.38	6.64
1440	5,000 S.F., 9.17 ton	2.89	3.22	6.11
1480	10,000 S.F., 18.33 ton	2.74	2.96	5.70
1560	Banks or libraries, 500 S.F., 2.08 ton	8.55	4.67	13.22
1600	1,000 S.F., 4.17 ton	7.40	7.65	15.05
1640	3,000 S.F., 12.50 ton	6.60	7.30	13.90
1680	5,000 S.F., 20.80 ton	6.30	6.70	13
1720	10,000 S.F., 41.66 ton	7.05	6.55	13.60
1800	Bars and taverns, 500 S.F., 5.54 ton	17.45	10.20	27.65
1840	1,000 S.F., 11.08 ton	17.15	14.70	31.85
1880	3,000 S.F., 33.25 ton	17.65	12.70	30.35
1920	5,000 S.F., 55.42 ton	18.65	12.75	31.40
1960	10,000 S.F., 110.00 ton	18.80	12.65	31.45
2040	Bowling alleys, 500 S.F., 2.83 ton	11.70	6.40	18.10
2080	1,000 S.F., 5.66 ton	10.10	10.45	20.55
2120	3,000 S.F., 17.00 ton	8.50	9.10	17.60
2160	5,000 S.F., 28.33 ton	9.20	8.90	18.10
2200	10,000 S.F., 56.66 ton	9.70	8.90	18.60
2240	Department stores, 500 S.F., 1.46 ton	6	3.28	9.28
2280	1,000 S.F., 2.92 ton	5.20	5.35	10.55
2320	3,000 S.F., 8.75 ton	4.60	5.10	9.70
2360	5,000 S.F., 14.58 ton	4.60	5.10	9.70
2400	10,000 S.F., 29.17 ton	4.72	4.59	9.31
2480	Drug stores, 500 S.F., 3.33 ton	13.75	7.50	21.25
2520	1,000 S.F., 6.66 ton	11.85	12.25	24.10
2560	3,000 S.F., 20.00 ton	10.05	10.75	20.80
2600	5,000 S.F., 33.33 ton	10.80	10.50	21.30
2640	10,000 S.F., 66.66 ton	11.45	10.50	21.95
2720	Factories, 500 S.F., 1.66 ton	7	3.80	10.80
2760	1,000 S.F., 3.33 ton	5.95	6.15	12.10
2800	3,000 S.F., 10.00 ton	5.25	5.85	11.10
2840	5,000 S.F., 16.66 ton	4.96	5.35	10.31
2880	10,000 S.F., 33.33 ton	5.40	5.25	10.65
2960	Food supermarkets, 500 S.F., 1.42 ton	5.85	3.20	9.05
3000	1,000 S.F., 2.83 ton	5.95	5.60	11.55
3040	3,000 S.F., 8.50 ton	5.05	5.20	10.25
3080	5,000 S.F., 14.17 ton	4.46	4.97	9.43

D30 HVAC

D3050 Terminal & Package Units

D3050 165	Self-contained, Air Cooled Unit Systems	COST PER S.F.		
		MAT.	INST.	TOTAL
3120	10,000 S.F., 28.33 ton	4.59	4.46	9.05
3200	Medical centers, 500 S.F., 1.17 ton	4.84	2.64	7.48
3240	1,000 S.F., 2.33 ton	4.97	4.62	9.59
3280	3,000 S.F., 7.00 ton	4.16	4.29	8.45
3320	5,000 S.F., 16.66 ton	3.69	4.09	7.78
3360	10,000 S.F., 23.33 ton	3.50	3.75	7.25
3440	Offices, 500 S.F., 1.58 ton	6.55	3.57	10.12
3480	1,000 S.F., 3.16 ton	6.75	6.25	13
3520	3,000 S.F., 9.50 ton	5	5.55	10.55
3560	5,000 S.F., 15.83 ton	4.77	5.10	9.87
3600	10,000 S.F., 31.66 ton	5.15	4.98	10.13
3680	Restaurants, 500 S.F., 2.50 ton	10.30	5.60	15.90
3720	1,000 S.F., 5.00 ton	8.90	9.20	18.10
3760	3,000 S.F., 15.00 ton	7.90	8.75	16.65
3800	5,000 S.F., 25.00 ton	8.05	7.85	15.90
3840	10,000 S.F., 50.00 ton	8.60	7.80	16.40
3920	Schools and colleges, 500 S.F., 1.92 ton	7.90	4.33	12.23
3960	1,000 S.F., 3.83 ton	6.85	7.05	13.90
4000	3,000 S.F., 11.50 ton	6.05	6.75	12.80
4040	5,000 S.F., 19.17 ton	5.75	6.15	11.90
4080	10,000 S.F., 38.33 ton	6.50	6.05	12.55

D3050   Terminal & Package Units

*Cooling requirements would lead to more
 than one system.

D3050 170	Split Systems With Air Cooled Condensing Units	COST PER S.F.		
		MAT.	INST.	TOTAL
1260	Split system, air cooled condensing unit			
1280	Apartment corridors, 1,000 S.F., 1.83 ton	2.45	2.28	4.73
1320	2,000 S.F., 3.66 ton	1.75	2.30	4.05
1360	5,000 S.F., 9.17 ton	1.89	2.86	4.75
1400	10,000 S.F., 18.33 ton	2.15	3.14	5.29
1440	20,000 S.F., 36.66 ton	2.34	3.19	5.53
1520	Banks and libraries, 1,000 S.F., 4.17 ton	3.98	5.25	9.23
1560	2,000 S.F., 8.33 ton	4.29	6.55	10.84
1600	5,000 S.F., 20.80 ton	4.89	7.10	11.99
1640	10,000 S.F., 41.66 ton	5.30	7.25	12.55
1680	20,000 S.F., 83.32 ton	6.20	7.55	13.75
1760	Bars and taverns, 1,000 S.F., 11.08 ton	10.30	10.30	20.60
1800	2,000 S.F., 22.16 ton	13.95	12.35	26.30
1840	5,000 S.F., 55.42 ton	12.15	11.55	23.70
1880	10,000 S.F., 110.84 ton	14.15	12.05	26.20
1920	20,000 S.F., 220 ton*			
2000	Bowling alleys, 1,000 S.F., 5.66 ton	5.45	9.75	15.20
2040	2,000 S.F., 11.33 ton	5.85	8.85	14.70
2080	5,000 S.F., 28.33 ton	6.65	9.70	16.35
2120	10,000 S.F., 56.66 ton	7.25	9.85	17.10
2160	20,000 S.F., 113.32 ton	9.10	10.70	19.80
2320	Department stores, 1,000 S.F., 2.92 ton	2.80	3.60	6.40
2360	2,000 S.F., 5.83 ton	2.81	5.05	7.86
2400	5,000 S.F., 14.58 ton	3	4.58	7.58
2440	10,000 S.F., 29.17 ton	3.43	4.98	8.41
2480	20,000 S.F., 58.33 ton	3.72	5.05	8.77
2560	Drug stores, 1,000 S.F., 6.66 ton	6.40	11.50	17.90
2600	2,000 S.F., 13.32 ton	6.85	10.45	17.30
2640	5,000 S.F., 33.33 ton	8.50	11.60	20.10
2680	10,000 S.F., 66.66 ton	8.85	12.10	20.95
2720	20,000 S.F., 133.32 ton*			
2800	Factories, 1,000 S.F., 3.33 ton	3.20	4.09	7.29
2840	2,000 S.F., 6.66 ton	3.21	5.75	8.96
2880	5,000 S.F., 16.66 ton	3.91	5.70	9.61
2920	10,000 S.F., 33.33 ton	4.25	5.80	10.05
2960	20,000 S.F., 66.66 ton	4.41	6.05	10.46
3040	Food supermarkets, 1,000 S.F., 2.83 ton	2.71	3.49	6.20
3080	2,000 S.F., 5.66 ton	2.73	4.88	7.61
3120	5,000 S.F., 14.66 ton	2.92	4.44	7.36
3160	10,000 S.F., 28.33 ton	3.34	4.85	8.19
3200	20,000 S.F., 56.66 ton	3.63	4.93	8.56
3280	Medical centers, 1,000 S.F., 2.33 ton	2.44	2.82	5.26

D30 HVAC

D3050 Terminal & Package Units

D3050 170	Split Systems With Air Cooled Condensing Units	COST PER S.F.		
		MAT.	INST.	TOTAL
3320	2,000 S.F., 4.66 ton	2.25	4.01	6.26
3360	5,000 S.F., 11.66 ton	2.40	3.66	6.06
3400	10,000 S.F., 23.33 ton	2.74	4	6.74
3440	20,000 S.F., 46.66 ton	2.98	4.07	7.05
3520	Offices, 1,000 S.F., 3.17 ton	3.04	3.90	6.94
3560	2,000 S.F., 6.33 ton	3.05	5.45	8.50
3600	5,000 S.F., 15.83 ton	3.26	4.95	8.21
3640	10,000 S.F., 31.66 ton	3.72	5.40	9.12
3680	20,000 S.F., 63.32 ton	4.20	5.75	9.95
3760	Restaurants, 1,000 S.F., 5.00 ton	4.82	8.65	13.47
3800	2,000 S.F., 10.00 ton	5.15	7.85	13
3840	5,000 S.F., 25.00 ton	5.90	8.55	14.45
3880	10,000 S.F., 50.00 ton	6.40	8.70	15.10
3920	20,000 S.F., 100.00 ton	8.05	9.45	17.50
4000	Schools and colleges, 1,000 S.F., 3.83 ton	3.66	4.82	8.48
4040	2,000 S.F., 7.66 ton	3.95	6	9.95
4080	5,000 S.F., 19.17 ton	4.51	6.55	11.06
4120	10,000 S.F., 38.33 ton	4.90	6.65	11.55

D3090 Other HVAC Systems/Equip

Cast Iron Garage Exhaust System

Dual Exhaust System

D3090 320	Garage Exhaust Systems	COST PER BAY		
		MAT.	INST.	TOTAL
1040	Garage, single exhaust, 3" outlet, cars & light trucks, one bay	3,575	1,700	5,275
1060	Additional bays up to seven bays	900	460	1,360
1500	4" outlet, trucks, one bay	3,600	1,700	5,300
1520	Additional bays up to six bays	935	460	1,395
1600	5" outlet, diesel trucks, one bay	3,950	1,700	5,650
1650	Additional single bays up to six	1,375	545	1,920
1700	Two adjoining bays	3,950	1,700	5,650
2000	Dual exhaust, 3" outlets, pair of adjoining bays	4,475	2,150	6,625
2100	Additional pairs of adjoining bays	1,450	545	1,995

R10520-120 System Classification

System Classification
Rules for installation of sprinkler systems vary depending on the classification of occupancy falling into one of three categories as follows:

Light Hazard Occupancy
The protection area allotted per sprinkler should not exceed 225 S.F., with the maximum distance between lines and sprinklers on lines being 15'. The sprinklers do not need to be staggered. Branch lines should not exceed eight sprinklers on either side of a cross main. Each large area requiring more than 100 sprinklers and without a sub-dividing partition should be supplied by feed mains or risers sized for ordinary hazard occupancy.
Maximum system area = 52,000 S.F.

Included in this group are:
Churches	Nursing Homes
Clubs	Offices
Educational	Residential
Hospitals	Restaurants
Institutional	Theaters and Auditoriums
Libraries	(except stages and prosceniums)
(except large stack rooms)	Unused Attics
Museums	

Ordinary Hazard Occupancy
The protection area allotted per sprinkler shall not exceed 130 S.F. of noncombustible ceiling and 130 S.F. of combustible ceiling. The maximum allowable distance between sprinkler lines and sprinklers on line is 15'. Sprinklers shall be staggered if the distance between heads exceeds 12'. Branch lines should not exceed eight sprinklers on either side of a cross main.
Maximum system area = 52,000 S.F.

Included in this group are:
Group 1	Group 2
Automotive Parking and Showrooms	Cereal Mills
Bakeries	Chemical Plants—Ordinary
Beverage manufacturing	Confectionery Products
Canneries	Distilleries
Dairy Products Manufacturing/Processing	Dry Cleaners
Electronic Plans	Feed Mills
Glass and Glass Products Manufacturing	Horse Stables
Laundries	Leather Goods Manufacturing
Restaurant Service Areas	Libraries—Large Stack Room Areas
	Machine Shops
	Metal Working
	Mercantile
	Paper and Pulp Mills
	Paper Process Plants
	Piers and Wharves
	Post Offices
	Printing and Publishing
	Repair Garages
	Stages
	Textile Manufacturing
	Tire Manufacturing
	Tobacco Products Manufacturing
	Wood Machining
	Wood Product Assembly

Extra Hazard Occupancy
The protection area allotted per sprinkler shall not exceed 100 S.F. of noncombustible ceiling and 100 S.F. of combustible ceiling. The maximum allowable distance between lines and between sprinklers on lines is 12'. Sprinklers on alternate lines shall be staggered if the distance between sprinklers on lines exceeds 8'. Branch lines should not exceed six sprinklers on either side of a cross main.
Maximum system area:
   Design by pipe schedule = 25,000 S.F.
   Design by hydraulic calculation = 40,000 S.F.

Included in this group are:
Group 1	Group 2
Aircraft hangars	Asphalt Saturating
Combustible Hydraulic Fluid Use Area	Flammable Liquids Spraying
Die Casting	Flow Coating
Metal Extruding	Manufactured/Modular Home
Plywood/Particle Board Manufacturing	Building Assemblies (where
Printing (inkls with flash points < 100 degrees F	finished enclosure is present and has combustible interiors)
Rubber Reclaiming, Compounding, Drying, Milling, Vulcanizing	Open Oil Quenching
Saw Mills	Plastics Processing
Textile Picking, Opening, Blending, Garnetting, Carding, Combing of Cotton, Synthetics, Wood Shoddy, or Burlap	Solvent Cleaning
Upholstering with Plastic Foams	Varnish and Paint Dipping

R10520-110   Sprinkler Systems (Automatic)

Sprinkler systems may be classified by type as follows:

1. **Wet Pipe System.** A system employing automatic sprinklers attached to a piping system containing water and connected to a water supply so that water discharges immediately from sprinklers opened by a fire.

2. **Dry Pipe System.** A system employing automatic sprinklers attached to a piping system containing air under pressure, the release of which as from the opening of sprinklers permits the water pressure to open a valve known as a "dry pipe valve". The water then flows into the piping system and out the opened sprinklers.

3. **Pre-Action System.** A system employing automatic sprinklers attached to a piping system containing air that may or may not be under pressure, with a supplemental heat responsive system of generally more sensitive characteristics than the automatic sprinklers themselves, installed in the same areas as the sprinklers; actuation of the heat responsive system, as from a fire, opens a valve which permits water to flow into the sprinkler piping system and to be discharged from any sprinklers which may be open.

4. **Deluge System.** A system employing open sprinklers attached to a piping system connected to a water supply through a valve which is opened by the operation of a heat responsive system installed in the same areas as the sprinklers. When this valve opens, water flows into the piping system and discharges from all sprinklers attached thereto.

5. **Combined Dry Pipe and Pre-Action Sprinkler System.** A system employing automatic sprinklers attached to a piping system containing air under pressure with a supplemental heat responsive system of generally more sensitive characteristics than the automatic sprinklers themselves, installed in the same areas as the sprinklers; operation of the heat responsive system, as from a fire, actuates tripping devices which open dry pipe valves simultaneously and without loss of air pressure in the system. Operation of the heat responsive system also opens

approved air exhaust valves at the end of the feed main which facilitate the filling of the system with water which usually precedes the opening of sprinklers. The heat responsive system also serves as an automatic fire alarm system.

6. **Limited Water Supply System.** A system employing automatic sprinklers and conforming to these standards but supplied by a pressure tank of limited capacity.

7. **Chemical Systems.** Systems using halon, carbon dioxide, dry chemical or high expansion foam as selected for special requirements. Agent may extinguish flames by chemically inhibiting flame propagation, suffocate flames by excluding oxygen, interrupting chemical action of oxygen uniting with fuel or sealing and cooling the combustion center.

8. **Firecycle System.** Firecycle is a fixed fire protection sprinkler system utilizing water as its extinguishing agent. It is a time delayed, recycling preaction type which automatically shuts the water off when heat is reduced below the detector operating temperature and turns the water back on when that temperature is exceeded. The system senses a fire condition through a closed circuit electrical detector system which controls water flow to the fire automatically. Batteries supply up to 90 hour emergency power supply for system operation. The piping system dry (until water is required) and is monitored with pressurized air. Should any leak in the system piping occur, an alarm will sound, but water will not enter the system until heat is sensed by a firecycle detector.

Area coverage sprinkler systems may be laid out and fed from the supply in any one of several patterns as shown below. It is desirable, if possible, to utilize a central feed and achieve a shorter flow path from the riser to the furthest sprinkler. This permits use of the smallest sizes of pipe possible with resulting savings.

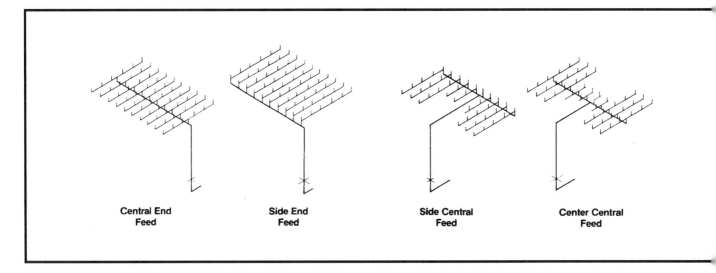

| Central End Feed | Side End Feed | Side Central Feed | Center Central Feed |

D4010 Sprinklers

Dry Pipe System: A system employing automatic sprinklers attached to a piping system containing air under pressure, the release of which as from the opening of sprinklers permits the water pressure to open a valve known as a "dry pipe valve". The water then flows into the piping system and out the opened sprinklers.

All areas are assumed to be open.

D4010 310	Dry Pipe Sprinkler Systems	COST PER S.F.		
		MAT.	INST.	TOTAL
0520	Dry pipe sprinkler systems, steel, black, sch. 40 pipe			
0530	Light hazard, one floor, 500 S.F.	8	5.45	13.45
0560	1000 S.F.	4.55	3.20	7.75
0580	2000 S.F.	4.24	3.25	7.49
0600	5000 S.F.	2.38	2.18	4.56
0620	10,000 S.F.	1.76	1.84	3.60
0640	50,000 S.F.	1.53	1.61	3.14
0660	Each additional floor, 500 S.F.	2.14	2.68	4.82
0680	1000 S.F.	1.66	2.20	3.86
0700	2000 S.F.	1.59	2.04	3.63
0720	5000 S.F.	1.38	1.74	3.12
0740	10,000 S.F.	1.30	1.62	2.92
0760	50,000 S.F.	1.17	1.43	2.60
1000	Ordinary hazard, one floor, 500 S.F.	8.15	5.55	13.70
1020	1000 S.F.	4.66	3.23	7.89
1040	2000 S.F.	4.37	3.37	7.74
1060	5000 S.F.	2.69	2.34	5.03
1080	10,000 S.F.	2.31	2.40	4.71
1100	50,000 S.F.	1.99	2.27	4.26
1140	Each additional floor, 500 S.F.	2.29	2.75	5.04
1160	1000 S.F.	1.96	2.47	4.43
1180	2000 S.F.	1.87	2.26	4.13
1200	5000 S.F.	1.75	1.94	3.69
1220	10,000 S.F.	1.63	1.89	3.52
1240	50,000 S.F.	1.43	1.63	3.06
1500	Extra hazard, one floor, 500 S.F.	10.85	6.90	17.75
1520	1000 S.F.	6.75	5	11.75
1540	2000 S.F.	4.96	4.33	9.29
1560	5000 S.F.	3.23	3.25	6.48
1580	10,000 S.F.	3.13	3.14	6.27
1600	50,000 S.F.	3.33	3.04	6.37
1660	Each additional floor, 500 S.F.	2.81	3.43	6.24
1680	1000 S.F.	2.74	3.26	6
1700	2000 S.F.	2.59	3.25	5.84
1720	5000 S.F.	2.27	2.82	5.09
1740	10,000 S.F.	2.42	2.60	5.02
1760	50,000 S.F.	2.44	2.50	4.94
2020	Grooved steel, black, sch. 40 pipe, light hazard, one floor, 2000 S.F.	3.78	2.74	6.52

D4010 Sprinklers

D4010 310	Dry Pipe Sprinkler Systems	COST PER S.F.		
		MAT.	INST.	TOTAL
2060	10,000 S.F.	1.43	1.60	3.03
2100	Each additional floor, 2000 S.F.	1.28	1.63	2.91
2150	10,000 S.F.	.97	1.38	2.35
2200	Ordinary hazard, one floor, 2000 S.F.	3.95	2.92	6.87
2250	10,000 S.F.	1.76	2.03	3.79
2300	Each additional floor, 2000 S.F.	1.45	1.81	3.26
2350	10,000 S.F.	1.31	1.82	3.13
2400	Extra hazard, one floor, 2000 S.F.	4.49	3.69	8.18
2450	10,000 S.F.	2.47	2.63	5.10
2500	Each additional floor, 2000 S.F.	2.13	2.67	4.80
2550	10,000 S.F.	1.90	2.31	4.21
3050	Grooved steel black sch. 10 pipe, light hazard, one floor, 2000 S.F.	3.72	2.73	6.45
3100	10,000 S.F.	1.39	1.58	2.97
3150	Each additional floor, 2000 S.F.	1.22	1.62	2.84
3200	10,000 S.F.	.93	1.36	2.29
3250	Ordinary hazard, one floor, 2000 S.F.	3.90	2.90	6.80
3300	10,000 S.F.	1.70	2.01	3.71
3350	Each additional floor, 2000 S.F.	1.40	1.79	3.19
3400	10,000 S.F.	1.25	1.80	3.05
3450	Extra hazard, one floor, 2000 S.F.	4.44	3.69	8.13
3500	10,000 S.F.	2.35	2.59	4.94
3550	Each additional floor, 2000 S.F.	2.08	2.67	4.75
3600	10,000 S.F.	1.83	2.28	4.11
4050	Copper tubing, type M, light hazard, one floor, 2000 S.F.	4.38	2.73	7.11
4100	10,000 S.F.	2	1.58	3.58
4150	Each additional floor, 2000 S.F.	1.90	1.65	3.55
4200	10,000 S.F.	1.54	1.37	2.91
4250	Ordinary hazard, one floor, 2000 S.F.	4.69	3.05	7.74
4300	10,000 S.F.	2.44	1.88	4.32
4350	Each additional floor, 2000 S.F.	2.51	1.91	4.42
4400	10,000 S.F.	1.92	1.63	3.55
4450	Extra hazard, one floor, 2000 S.F.	5.45	3.73	9.18
4500	10,000 S.F.	5.15	2.81	7.96
4550	Each additional floor, 2000 S.F.	3.11	2.71	5.82
4600	10,000 S.F.	3.70	2.47	6.17
5050	Copper tubing, type M, T-drill system, light hazard, one floor			
5060	2000 S.F.	4.33	2.56	6.89
5100	10,000 S.F.	1.82	1.32	3.14
5150	Each additional floor, 2000 S.F.	1.85	1.48	3.33
5200	10,000 S.F.	1.36	1.11	2.47
5250	Ordinary hazard, one floor, 2000 S.F.	4.41	2.61	7.02
5300	10,000 S.F.	2.29	1.67	3.96
5350	Each additional floor, 2000 S.F.	1.91	1.50	3.41
5400	10,000 S.F.	1.75	1.39	3.14
5450	Extra hazard, one floor, 2000 S.F.	4.90	3.09	7.99
5500	10,000 S.F.	4	2.08	6.08
5550	Each additional floor, 2000 S.F.	2.54	2.07	4.61
5600	10,000 S.F.	2.57	1.74	4.31

D4010 Sprinklers

Preaction System: A system employing automatic sprinklers attached to a piping system containing air that may or may not be under pressure, with a supplemental heat responsive system of generally more sensitive characteristics than the automatic sprinklers themselves, installed in the same areas as the sprinklers. Actuation of the heat responsive system, as from a fire, opens a valve which permits water to flow into the sprinkler piping system and to be discharged from those sprinklers which were opened by heat from the fire.

All areas are assumed to be opened.

D4010 350	Preaction Sprinkler Systems	COST PER S.F.		
		MAT.	INST.	TOTAL
0520	Preaction sprinkler systems, steel, black, sch. 40 pipe			
0530	Light hazard, one floor, 500 S.F.	7.80	4.39	12.19
0560	1000 S.F.	4.60	3.28	7.88
0580	2000 S.F.	4.13	3.24	7.37
0600	5000 S.F.	2.29	2.17	4.46
0620	10,000 S.F.	1.68	1.84	3.52
0640	50,000 S.F.	1.46	1.60	3.06
0660	Each additional floor, 500 S.F.	2.15	2.39	4.54
0680	1000 S.F.	1.68	2.20	3.88
0700	2000 S.F.	1.61	2.04	3.65
0720	5000 S.F.	1.29	1.73	3.02
0740	10,000 S.F.	1.22	1.62	2.84
0760	50,000 S.F.	1.26	1.47	2.73
1000	Ordinary hazard, one floor, 500 S.F.	8.25	4.75	13
1020	1000 S.F.	4.55	3.22	7.77
1040	2000 S.F.	4.47	3.36	7.83
1060	5000 S.F.	2.50	2.32	4.82
1080	10,000 S.F.	2.08	2.39	4.47
1100	50,000 S.F.	1.75	2.25	4
1140	Each additional floor, 500 S.F.	2.59	2.75	5.34
1160	1000 S.F.	1.66	2.22	3.88
1180	2000 S.F.	1.56	2.23	3.79
1200	5000 S.F.	1.66	2.09	3.75
1220	10,000 S.F.	1.63	2.18	3.81
1240	50,000 S.F.	1.40	1.93	3.33
1500	Extra hazard, one floor, 500 S.F.	10.65	6.10	16.75
1520	1000 S.F.	6.25	4.58	10.83
1540	2000 S.F.	4.58	4.30	8.88

D4010 Sprinklers

D4010 350	Preaction Sprinkler Systems	COST PER S.F.		
		MAT.	INST.	TOTAL
1560	5000 S.F.	3.22	3.51	6.73
1580	10,000 S.F.	2.91	3.48	6.39
1600	50,000 S.F.	3.03	3.36	6.39
1660	Each additional floor, 500 S.F.	2.84	3.42	6.26
1680	1000 S.F.	2.36	3.23	5.59
1700	2000 S.F.	2.21	3.22	5.43
1720	5000 S.F.	1.94	2.84	4.78
1740	10,000 S.F.	1.96	2.63	4.59
1760	50,000 S.F.	1.88	2.45	4.33
2020	Grooved steel, black, sch. 40 pipe, light hazard, one floor, 2000 S.F.	3.80	2.74	6.54
2060	10,000 S.F.	1.35	1.60	2.95
2100	Each additional floor of 2000 S.F.	1.30	1.63	2.93
2150	10,000 S.F.	.89	1.38	2.27
2200	Ordinary hazard, one floor, 2000 S.F.	3.84	2.90	6.74
2250	10,000 S.F.	1.53	2.02	3.55
2300	Each additional floor, 2000 S.F.	1.34	1.79	3.13
2350	10,000 S.F.	1.08	1.81	2.89
2400	Extra hazard, one floor, 2000 S.F.	4.11	3.66	7.77
2450	10,000 S.F.	2	2.61	4.61
2500	Each additional floor, 2000 S.F.	1.75	2.64	4.39
2550	10,000 S.F.	1.40	2.28	3.68
3050	Grooved steel, black, sch. 10 pipe light hazard, one floor, 2000 S.F.	3.74	2.73	6.47
3100	10,000 S.F.	1.31	1.58	2.89
3150	Each additional floor, 2000 S.F.	1.24	1.62	2.86
3200	10,000 S.F.	.85	1.36	2.21
3250	Ordinary hazard, one floor, 2000 S.F.	3.73	2.71	6.44
3300	10,000 S.F.	1.26	1.98	3.24
3350	Each additional floor, 2000 S.F.	1.29	1.77	3.06
3400	10,000 S.F.	1.02	1.79	2.81
3450	Extra hazard, one floor, 2000 S.F.	4.06	3.66	7.72
3500	10,000 S.F.	1.85	2.56	4.41
3550	Each additional floor, 2000 S.F.	1.70	2.64	4.34
3600	10,000 S.F.	1.33	2.25	3.58
4050	Copper tubing, type M, light hazard, one floor, 2000 S.F.	4.41	2.73	7.14
4100	10,000 S.F.	1.93	1.58	3.51
4150	Each additional floor, 2000 S.F.	1.93	1.65	3.58
4200	10,000 S.F.	1.25	1.35	2.60
4250	Ordinary hazard, one floor, 2000 S.F.	4.58	3.03	7.61
4300	10,000 S.F.	2.21	1.87	4.08
4350	Each additional floor, 2000 S.F.	1.94	1.68	3.62
4400	10,000 S.F.	1.54	1.49	3.03
4450	Extra hazard, one floor, 2000 S.F.	5.10	3.70	8.80
4500	10,000 S.F.	4.60	2.77	7.37
4550	Each additional floor, 2000 S.F.	2.73	2.68	5.41
4600	10,000 S.F.	3.20	2.44	5.64
5050	Copper tubing, type M, T-drill system, light hazard, one floor			
5060	2000 S.F.	4.35	2.56	6.91
5100	10,000 S.F.	1.74	1.32	3.06
5150	Each additional floor, 2000 S.F.	1.87	1.48	3.35
5200	10,000 S.F.	1.28	1.11	2.39
5250	Ordinary hazard, one floor, 2000 S.F.	4.30	2.59	6.89
5300	10,000 S.F.	2.06	1.66	3.72
5350	Each additional floor, 2000 S.F.	1.81	1.49	3.30
5400	10,000 S.F.	1.61	1.45	3.06
5450	Extra hazard, one floor, 2000 S.F.	4.52	3.06	7.58
5500	10,000 S.F.	3.47	2.04	5.51
5550	Each additional floor, 2000 S.F.	2.16	2.04	4.20
5600	10,000 S.F.	2.07	1.71	3.78

Deluge System: A system employing open sprinklers attached to a piping system connected to a water supply through a valve which is opened by the operation of a heat responsive system installed in the same areas as the sprinklers. When this valve opens, water flows into the piping system and discharges from all sprinklers attached thereto.

D4010 370	Deluge Sprinkler Systems	COST PER S.F.		
		MAT.	INST.	TOTAL
0520	Deluge sprinkler systems, steel, black, sch. 40 pipe			
0530	Light hazard, one floor, 500 S.F.	21.50	4.44	25.94
0560	1000 S.F.	11.45	3.16	14.61
0580	2000 S.F.	7.85	3.25	11.10
0600	5000 S.F.	3.77	2.18	5.95
0620	10,000 S.F.	2.43	1.84	4.27
0640	50,000 S.F.	1.62	1.60	3.22
0660	Each additional floor, 500 S.F.	2.15	2.39	4.54
0680	1000 S.F.	1.68	2.20	3.88
0700	2000 S.F.	1.61	2.04	3.65
0720	5000 S.F.	1.29	1.73	3.02
0740	10,000 S.F.	1.22	1.62	2.84
0760	50,000 S.F.	1.26	1.47	2.73
1000	Ordinary hazard, one floor, 500 S.F.	22.50	5.05	27.55
1020	1000 S.F.	11.45	3.25	14.70
1040	2000 S.F.	8.15	3.37	11.52
1060	5000 S.F.	3.98	2.33	6.31
1080	10,000 S.F.	2.83	2.39	5.22
1100	50,000 S.F.	1.95	2.28	4.23
1140	Each additional floor, 500 S.F.	2.59	2.75	5.34
1160	1000 S.F.	1.66	2.22	3.88
1180	2000 S.F.	1.56	2.23	3.79
1200	5000 S.F.	1.56	1.92	3.48
1220	10,000 S.F.	1.50	1.93	3.43
1240	50,000 S.F.	1.29	1.76	3.05
1500	Extra hazard, one floor, 500 S.F.	24.50	6.15	30.65
1520	1000 S.F.	13.50	4.75	18.25
1540	2000 S.F.	8.30	4.31	12.61
1560	5000 S.F.	4.32	3.24	7.56
1580	10,000 S.F.	3.45	3.18	6.63
1600	50,000 S.F.	3.43	3.09	6.52
1660	Each additional floor, 500 S.F.	2.84	3.42	6.26

D4010 Sprinklers

D4010 370	Deluge Sprinkler Systems	COST PER S.F.		
		MAT.	INST.	TOTAL
1680	1000 S.F.	2.36	3.23	5.59
1700	2000 S.F.	2.21	3.22	5.43
1720	5000 S.F.	1.94	2.84	4.78
1740	10,000 S.F.	2.04	2.73	4.77
1760	50,000 S.F.	2.01	2.62	4.63
2000	Grooved steel, black, sch. 40 pipe, light hazard, one floor			
2020	2000 S.F.	7.50	2.75	10.25
2060	10,000 S.F.	2.12	1.62	3.74
2100	Each additional floor, 2,000 S.F.	1.30	1.63	2.93
2150	10,000 S.F.	.89	1.38	2.27
2200	Ordinary hazard, one floor, 2000 S.F.	3.84	2.90	6.74
2250	10,000 S.F.	2.28	2.02	4.30
2300	Each additional floor, 2000 S.F.	1.34	1.79	3.13
2350	10,000 S.F.	1.08	1.81	2.89
2400	Extra hazard, one floor, 2000 S.F.	7.80	3.67	11.47
2450	10,000 S.F.	2.77	2.62	5.39
2500	Each additional floor, 2000 S.F.	1.75	2.64	4.39
2550	10,000 S.F.	1.40	2.28	3.68
3000	Grooved steel, black, sch. 10 pipe, light hazard, one floor			
3050	2000 S.F.	7	2.62	9.62
3100	10,000 S.F.	2.06	1.58	3.64
3150	Each additional floor, 2000 S.F.	1.24	1.62	2.86
3200	10,000 S.F.	.85	1.36	2.21
3250	Ordinary hazard, one floor, 2000 S.F.	7.50	2.89	10.39
3300	10,000 S.F.	2.01	1.98	3.99
3350	Each additional floor, 2000 S.F.	1.29	1.77	3.06
3400	10,000 S.F.	1.02	1.79	2.81
3450	Extra hazard, one floor, 2000 S.F.	7.75	3.67	11.42
3500	10,000 S.F.	2.60	2.56	5.16
3550	Each additional floor, 2000 S.F.	1.70	2.64	4.34
3600	10,000 S.F.	1.33	2.25	3.58
4000	Copper tubing, type M, light hazard, one floor			
4050	2000 S.F.	8.10	2.74	10.84
4100	10,000 S.F.	2.67	1.58	4.25
4150	Each additional floor, 2000 S.F.	1.91	1.65	3.56
4200	10,000 S.F.	1.25	1.35	2.60
4250	Ordinary hazard, one floor, 2000 S.F.	8.30	3.04	11.34
4300	10,000 S.F.	2.96	1.87	4.83
4350	Each additional floor, 2000 S.F.	1.94	1.68	3.62
4400	10,000 S.F.	1.54	1.49	3.03
4450	Extra hazard, one floor, 2000 S.F.	8.80	3.71	12.51
4500	10,000 S.F.	5.40	2.79	8.19
4550	Each additional floor, 2000 S.F.	2.73	2.68	5.41
4600	10,000 S.F.	3.20	2.44	5.64
5000	Copper tubing, type M, T-drill system, light hazard, one floor			
5050	2000 S.F.	8.05	2.57	10.62
5100	10,000 S.F.	2.49	1.32	3.81
5150	Each additional floor, 2000 S.F.	1.74	1.50	3.24
5200	10,000 S.F.	1.28	1.11	2.39
5250	Ordinary hazard, one floor, 2000 S.F.	8	2.60	10.60
5300	10,000 S.F.	2.81	1.66	4.47
5350	Each additional floor, 2000 S.F.	1.80	1.48	3.28
5400	10,000 S.F.	1.61	1.45	3.06
5450	Extra hazard, one floor, 2000 S.F.	8.20	3.07	11.27
5500	10,000 S.F.	4.22	2.04	6.26
5550	Each additional floor, 2000 S.F.	2.16	2.04	4.20
5600	10,000 S.F.	2.07	1.71	3.78

D4010 Sprinklers

Firecycle is a fixed fire protection sprinkler system utilizing water as its extinguishing agent. It is a time delayed, recycling, preaction type which automatically shuts the water off when heat is reduced below the detector operating temperature and turns the water back on when that temperature is exceeded.

The system senses a fire condition through a closed circuit electrical detector system which controls water flow to the fire automatically. Batteries supply up to 90 hours emergency power supply for system operation. The piping system is dry (until water is required) and is monitored with pressurized air. Shouldany leak in the system piping occur, an alarm will sound, but water will not enter the system until heat is sensed by a Firecycle detector.

D4010 390	Firecycle Sprinkler Systems	COST PER S.F.		
		MAT.	INST.	TOTAL
0520	Firecycle sprinkler systems, steel black sch. 40 pipe			
0530	Light hazard, one floor, 500 S.F.	30.50	8.60	39.10
0560	1000 S.F.	15.85	5.40	21.25
0580	2000 S.F.	9.70	3.98	13.68
0600	5000 S.F.	4.52	2.46	6.98
0620	10,000 S.F.	2.85	1.98	4.83
0640	50,000 S.F.	1.73	1.63	3.36
0660	Each additional floor of 500 S.F.	2.19	2.41	4.60
0680	1000 S.F.	1.70	2.21	3.91
0700	2000 S.F.	1.42	2.03	3.45
0720	5000 S.F.	1.30	1.73	3.03
0740	10,000 S.F.	1.28	1.62	2.90
0760	50,000 S.F.	1.30	1.47	2.77
1000	Ordinary hazard, one floor, 500 S.F.	31	8.95	39.95
1020	1000 S.F.	15.80	5.30	21.10
1040	2000 S.F.	9.80	4.09	13.89
1060	5000 S.F.	4.73	2.61	7.34
1080	10,000 S.F.	3.25	2.53	5.78
1100	50,000 S.F.	2.22	2.56	4.78
1140	Each additional floor, 500 S.F.	2.63	2.77	5.40
1160	1000 S.F.	1.68	2.23	3.91
1180	2000 S.F.	1.66	2.05	3.71
1200	5000 S.F.	1.57	1.92	3.49
1220	10,000 S.F.	1.46	1.88	3.34
1240	50,000 S.F.	1.30	1.68	2.98
1500	Extra hazard, one floor, 500 S.F.	33	10.30	43.30
1520	1000 S.F.	17.25	6.65	23.90
1540	2000 S.F.	10.15	5.05	15.20
1560	5000 S.F.	5.05	3.52	8.57
1580	10,000 S.F.	4.03	3.60	7.63

D4010 Sprinklers

D4010 390	Firecycle Sprinkler Systems	COST PER S.F.		
		MAT.	INST.	TOTAL
1600	50,000 S.F.	3.57	3.99	7.56
1660	Each additional floor, 500 S.F.	2.88	3.44	6.32
1680	1000 S.F.	2.38	3.24	5.62
1700	2000 S.F.	2.23	3.23	5.46
1720	5000 S.F.	1.95	2.84	4.79
1740	10,000 S.F.	2.02	2.63	4.65
1760	50,000 S.F.	2	2.54	4.54
2020	Grooved steel, black, sch. 40 pipe, light hazard, one floor			
2030	2000 S.F.	9.35	3.48	12.83
2060	10,000 S.F.	2.75	2.40	5.15
2100	Each additional floor, 2000 S.F.	1.32	1.64	2.96
2150	10,000 S.F.	.95	1.38	2.33
2200	Ordinary hazard, one floor, 2000 S.F.	9.40	3.64	13.04
2250	10,000 S.F.	2.90	2.29	5.19
2300	Each additional floor, 2000 S.F.	1.36	1.80	3.16
2350	10,000 S.F.	1.14	1.81	2.95
2400	Extra hazard, one floor, 2000 S.F.	9.65	4.40	14.05
2450	10,000 S.F.	3.12	2.73	5.85
2500	Each additional floor, 2000 S.F.	1.77	2.65	4.42
2550	10,000 S.F.	1.46	2.28	3.74
3050	Grooved steel, black, sch. 10 pipe light hazard, one floor,			
3060	2000 S.F.	9.30	3.47	12.77
3100	10,000 S.F.	2.48	1.72	4.20
3150	Each additional floor, 2000 S.F.	1.26	1.63	2.89
3200	10,000 S.F.	.91	1.36	2.27
3250	Ordinary hazard, one floor, 2000 S.F.	9.35	3.62	12.97
3300	10,000 S.F.	2.64	2.14	4.78
3350	Each additional floor, 2000 S.F.	1.31	1.78	3.09
3400	10,000 S.F.	1.08	1.79	2.87
3450	Extra hazard, one floor, 2000 S.F.	9.60	4.40	14
3500	10,000 S.F.	3	2.69	5.69
3550	Each additional floor, 2000 S.F.	1.72	2.65	4.37
3600	10,000 S.F.	1.39	2.25	3.64
4060	Copper tubing, type M, light hazard, one floor, 2000 S.F.	9.95	3.47	13.42
4100	10,000 S.F.	3.09	1.72	4.81
4150	Each additional floor, 2000 S.F.	1.94	1.66	3.60
4200	10,000 S.F.	1.52	1.37	2.89
4250	Ordinary hazard, one floor, 2000 S.F.	10.15	3.77	13.92
4300	10,000 S.F.	3.38	2.01	5.39
4350	Each additional floor, 2000 S.F.	1.96	1.69	3.65
4400	10,000 S.F.	1.58	1.47	3.05
4450	Extra hazard, one floor, 2000 S.F.	10.65	4.44	15.09
4500	10,000 S.F.	5.85	2.94	8.79
4550	Each additional floor, 2000 S.F.	2.75	2.69	5.44
4600	10,000 S.F.	3.26	2.44	5.70
5060	Copper tubing, type M, T-drill system, light hazard, one floor 2000 S.F.	9.90	3.30	13.20
5100	10,000 S.F.	2.91	1.46	4.37
5150	Each additional floor, 2000 S.F.	2.03	1.58	3.61
5200	10,000 S.F.	1.34	1.11	2.45
5250	Ordinary hazard, one floor, 2000 S.F.	9.85	3.33	13.18
5300	10,000 S.F.	3.23	1.80	5.03
5350	Each additional floor, 2000 S.F.	1.82	1.49	3.31
5400	10,000 S.F.	1.67	1.45	3.12
5450	Extra hazard, one floor, 2000 S.F.	10.10	3.80	13.90
5500	10,000 S.F.	4.62	2.17	6.79
5550	Each additional floor, 2000 S.F.	2.18	2.05	4.23
5600	10,000 S.F.	2.13	1.71	3.84

D4010 Sprinklers

Wet Pipe System. A system employing automatic sprinklers attached to a piping system containing water and connected to a water supply so that water discharges immediately from sprinklers opened by heat from a fire.

All areas are assumed to be open.

D4010 410	Wet Pipe Sprinkler Systems	COST PER S.F.		
		MAT.	INST.	TOTAL
0520	Wet pipe sprinkler systems, steel, black, sch. 40 pipe			
0530	Light hazard, one floor, 500 S.F.	2.49	2.83	5.32
0560	1000 S.F.	3.61	2.91	6.52
0580	2000 S.F.	3.70	2.94	6.64
0600	5000 S.F.	1.95	2.04	3.99
0620	10,000 S.F.	1.39	1.77	3.16
0640	50,000 S.F.	1.21	1.57	2.78
0660	Each additional floor, 500 S.F.	1.35	2.40	3.75
0680	1000 S.F.	1.29	2.24	3.53
0700	2000 S.F.	1.20	2.02	3.22
0720	5000 S.F.	.96	1.71	2.67
0740	10,000 S.F.	.93	1.60	2.53
0760	50,000 S.F.	.80	1.24	2.04
1000	Ordinary hazard, one floor, 500 S.F.	2.76	3.02	5.78
1020	1000 S.F.	3.56	2.85	6.41
1040	2000 S.F.	3.83	3.07	6.90
1060	5000 S.F.	2.16	2.19	4.35
1080	10,000 S.F.	1.79	2.32	4.11
1100	50,000 S.F.	1.47	2.19	3.66
1140	Each additional floor, 500 S.F.	1.77	2.71	4.48
1160	1000 S.F.	1.25	2.20	3.45
1180	2000 S.F.	1.35	2.22	3.57
1200	5000 S.F.	1.35	2.10	3.45
1220	10,000 S.F.	1.34	2.16	3.50
1240	50,000 S.F.	1.14	1.92	3.06
1500	Extra hazard, one floor, 500 S.F.	6.65	4.69	11.34
1520	1000 S.F.	4.53	4.07	8.60
1540	2000 S.F.	4.17	4.14	8.31
1560	5000 S.F.	3.15	3.58	6.73
1580	10,000 S.F.	2.61	3.44	6.05
1600	50,000 S.F.	2.81	3.32	6.13
1660	Each additional floor, 500 S.F.	2.02	3.38	5.40
1680	1000 S.F.	1.95	3.21	5.16
1700	2000 S.F.	1.80	3.20	5
1720	5000 S.F.	1.61	2.82	4.43
1740	10,000 S.F.	1.67	2.61	4.28
1760	50,000 S.F.	1.66	2.49	4.15
2020	Grooved steel, black sch. 40 pipe, light hazard, one floor, 2000 S.F.	3.37	2.46	5.83

D40 Fire Protection

D4010 Sprinklers

D4010 410	Wet Pipe Sprinkler Systems	MAT.	INST.	TOTAL
2060	10,000 S.F.	1.30	1.59	2.89
2100	Each additional floor, 2000 S.F.	.89	1.61	2.50
2150	10,000 S.F.	.60	1.36	1.96
2200	Ordinary hazard, one floor, 2000 S.F.	3.41	2.62	6.03
2250	10,000 S.F.	1.24	1.95	3.19
2300	Each additional floor, 2000 S.F.	.93	1.77	2.70
2350	10,000 S.F.	.79	1.79	2.58
2400	Extra hazard, one floor, 2000 S.F.	3.68	3.38	7.06
2450	10,000 S.F.	1.68	2.53	4.21
2500	Each additional floor, 2000 S.F.	1.34	2.62	3.96
2550	10,000 S.F.	1.11	2.26	3.37
3050	Grooved steel black sch. 10 pipe, light hazard, one floor, 2000 S.F.	3.31	2.45	5.76
3100	10,000 S.F.	1.02	1.51	2.53
3150	Each additional floor, 2000 S.F.	.83	1.60	2.43
3200	10,000 S.F.	.56	1.34	1.90
3250	Ordinary hazard, one floor, 2000 S.F.	3.36	2.60	5.96
3300	10,000 S.F.	1.18	1.93	3.11
3350	Each additional floor, 2000 S.F.	.88	1.75	2.63
3400	10,000 S.F.	.73	1.77	2.50
3450	Extra hazard, one floor, 2000 S.F.	3.63	3.38	7.01
3500	10,000 S.F.	1.56	2.49	4.05
3550	Each additional floor, 2000 S.F.	1.29	2.62	3.91
3600	10,000 S.F.	1.04	2.23	3.27
4050	Copper tubing, type M, light hazard, one floor, 2000 S.F.	3.97	2.45	6.42
4100	10,000 S.F.	1.63	1.51	3.14
4150	Each additional floor, 2000 S.F.	1.51	1.63	3.14
4200	10,000 S.F.	1.17	1.35	2.52
4250	Ordinary hazard, one floor, 2000 S.F.	4.15	2.75	6.90
4300	10,000 S.F.	1.92	1.80	3.72
4350	Each additional floor, 2000 S.F.	1.73	1.80	3.53
4400	10,000 S.F.	1.40	1.60	3
4450	Extra hazard, one floor, 2000 S.F.	4.66	3.42	8.08
4500	10,000 S.F.	4.31	2.70	7.01
4550	Each additional floor, 2000 S.F.	2.32	2.66	4.98
4600	10,000 S.F.	2.91	2.42	5.33
5050	Copper tubing, type M, T-drill system, light hazard, one floor			
5060	2000 S.F.	3.92	2.28	6.20
5100	10,000 S.F.	1.45	1.25	2.70
5150	Each additional floor, 2000 S.F.	1.46	1.46	2.92
5200	10,000 S.F.	.99	1.09	2.08
5250	Ordinary hazard, one floor, 2000 S.F.	3.87	2.31	6.18
5300	10,000 S.F.	1.77	1.59	3.36
5350	Each additional floor, 2000 S.F.	1.39	1.46	2.85
5400	10,000 S.F.	1.32	1.43	2.75
5450	Extra hazard, one floor, 2000 S.F.	4.09	2.78	6.87
5500	10,000 S.F.	3.18	1.97	5.15
5550	Each additional floor, 2000 S.F.	1.82	2.06	3.88
5600	10,000 S.F.	1.78	1.69	3.47

D4020 Standpipes

Roof — Roof connections with hose gate valves (for combustible roof)

Hose connections on each floor (size based on class of service)

Check Valve — Siamese inlet connections (for fire department use)

D4020 310	Wet Standpipe Risers, Class I	COST PER FLOOR		
		MAT.	INST.	TOTAL
0550	Wet standpipe risers, Class I, steel black sch. 40, 10' height			
0560	4" diameter pipe, one floor	4,025	3,075	7,100
0580	Additional floors	1,050	955	2,005
0600	6" diameter pipe, one floor	6,225	5,400	11,625
0620	Additional floors	1,525	1,525	3,050
0640	8" diameter pipe, one floor	9,225	6,550	15,775
0660	Additional floors	2,250	1,850	4,100

D4020 310	Wet Standpipe Risers, Class II	COST PER FLOOR		
		MAT.	INST.	TOTAL
1030	Wet standpipe risers, Class II, steel black sch. 40, 10' height			
1040	2" diameter pipe, one floor	1,575	1,125	2,700
1060	Additional floors	525	430	955
1080	2-1/2" diameter pipe, one floor	2,300	1,625	3,925
1100	Additional floors	625	500	1,125

D4020 310	Wet Standpipe Risers, Class III	COST PER FLOOR		
		MAT.	INST.	TOTAL
1530	Wet standpipe risers, Class III, steel black sch. 40, 10' height			
1540	4" diameter pipe, one floor	4,125	3,075	7,200
1560	Additional floors	910	800	1,710
1580	6" diameter pipe, one floor	6,325	5,400	11,725
1600	Additional floors	1,575	1,525	3,100
1620	8" diameter pipe, one floor	9,325	6,550	15,875
1640	Additional floors	2,300	1,850	4,150

D4020 Standpipes

Roof — Roof connections with hose gate valves (for combustible roof)

Hose connections on each floor (size based on class of service)

Check valve — Siamese inlet connections (for fire department use)

D4020 330	Dry Standpipe Risers, Class I	COST PER FLOOR		
		MAT.	INST.	TOTAL
0530	Dry standpipe riser, Class I, steel black sch. 40, 10' height			
0540	4" diameter pipe, one floor	2,650	2,475	5,125
0560	Additional floors	985	905	1,890
0580	6" diameter pipe, one floor	4,575	4,275	8,850
0600	Additional floors	1,475	1,475	2,950
0620	8" diameter pipe, one floor	6,825	5,225	12,050
0640	Additional floors	2,175	1,775	3,950

D4020 330	Dry Standpipe Risers, Class II	COST PER FLOOR		
		MAT.	INST.	TOTAL
1030	Dry standpipe risers, Class II, steel black sch. 40, 10' height			
1040	2" diameter pipe, one floor	1,350	1,150	2,500
1060	Additional floor	460	380	840
1080	2-1/2" diameter pipe, one floor	1,875	1,350	3,225
1100	Additional floors	560	450	1,010

D4020 330	Dry Standpipe Risers, Class III	COST PER FLOOR		
		MAT.	INST.	TOTAL
1530	Dry standpipe risers, Class III, steel black sch. 40, 10' height			
1540	4" diameter pipe, one floor	2,675	2,425	5,100
1560	Additional floors	860	815	1,675
1580	6" diameter pipe, one floor	4,625	4,275	8,900
1600	Additional floors	1,525	1,475	3,000
1620	8" diameter pipe, one floor	6,875	5,225	12,100
1640	Additional floor	2,225	1,775	4,000

D40 Fire Protection

D4020 Standpipes

D4020 410	Fire Hose Equipment	COST EACH		
		MAT.	INST.	TOTAL
0100	Adapters, reducing, 1 piece, FxM, hexagon, cast brass, 2-1/2" x 1-1/2"	50		50
0200	Pin lug, 1-1/2" x 1"	43.50		43.50
0250	3" x 2-1/2"	113		113
0300	For polished chrome, add 75% mat.			
0400	Cabinets, D.S. glass in door, recessed, steel box, not equipped			
0500	Single extinguisher, steel door & frame	102	136	238
0550	Stainless steel door & frame	214	136	350
0600	Valve, 2-1/2" angle, steel door & frame	123	90.50	213.50
0650	Aluminum door & frame	149	90.50	239.50
0700	Stainless steel door & frame	200	90.50	290.50
0750	Hose rack assy, 2-1/2" x 1-1/2" valve & 100' hose, steel door & frame	229	181	410
0800	Aluminum door & frame	335	181	516
0850	Stainless steel door & frame	450	181	631
0900	Hose rack assy,& extinguisher,2-1/2"x1-1/2" valve & hose,steel door & frame	252	217	469
0950	Aluminum	430	217	647
1000	Stainless steel	515	217	732
1550	Compressor, air, dry pipe system, automatic, 200 gal., 3/4 H.P.	850	465	1,315
1600	520 gal., 1 H.P.	865	465	1,330
1650	Alarm, electric pressure switch (circuit closer)	265	23	288
2500	Couplings, hose, rocker lug, cast brass, 1-1/2"	48		48
2550	2-1/2"	63		63
3000	Escutcheon plate, for angle valves, polished brass, 1-1/2"	12.15		12.15
3050	2-1/2"	29		29
3500	Fire pump, electric, w/controller, fittings, relief valve			
3550	4" pump, 30 H.P., 500 G.P.M.	16,500	3,375	19,875
3600	5" pump, 40 H.P., 1000 G.P.M.	17,200	3,825	21,025
3650	5" pump, 100 H.P., 1000 G.P.M.	19,600	4,250	23,850
3700	For jockey pump system, add	2,200	545	2,745
5000	Hose, per linear foot, synthetic jacket, lined,			
5100	300 lb. test, 1-1/2" diameter	2.63	.42	3.05
5150	2-1/2" diameter	4.38	.49	4.87
5200	500 lb. test, 1-1/2" diameter	2.72	.42	3.14
5250	2-1/2" diameter	4.72	.49	5.21
5500	Nozzle, plain stream, polished brass, 1-1/2" x 10"	43.50		43.50
5550	2-1/2" x 15" x 13/16" or 1-1/2"	78.50		78.50
5600	Heavy duty combination adjustable fog and straight stream w/handle 1-1/2"	405		405
5650	2-1/2" direct connection	505		505
6000	Rack, for 1-1/2" diameter hose 100 ft. long, steel	54	54.50	108.50
6050	Brass	66.50	54.50	121
6500	Reel, steel, for 50 ft. long 1-1/2" diameter hose	118	77.50	195.50
6550	For 75 ft. long 2-1/2" diameter hose	192	77.50	269.50
7050	Siamese, w/plugs & chains, polished brass, sidewalk, 4" x 2-1/2" x 2-1/2"	520	435	955
7100	6" x 2-1/2" x 2-1/2"	710	545	1,255
7200	Wall type, flush, 4" x 2-1/2" x 2-1/2"	470	217	687
7250	6" x 2-1/2" x 2-1/2"	655	236	891
7300	Projecting, 4" x 2-1/2" x 2-1/2"	430	217	647
7350	6" x 2-1/2" x 2-1/2"	725	236	961
7400	For chrome plate, add 15% mat.			
8000	Valves, angle, wheel handle, 300 Lb., rough brass, 1-1/2"	82	50.50	132.50
8050	2-1/2"	66	86	152
8100	Combination pressure restricting, 1-1/2"	77	50.50	127.50
8150	2-1/2"	153	86	239
8200	Pressure restricting, adjustable, satin brass, 1-1/2"	109	50.50	159.50
8250	2-1/2"	152	86	238
8300	Hydrolator, vent and drain, rough brass, 1-1/2"	64.50	50.50	115
8350	2-1/2"	64.50	50.50	115
8400	Cabinet assy, incls. adapter, rack, hose, and nozzle	675	330	1,005

D4090 Other Fire Protection Systems

General: Automatic fire protection (suppression) systems other than water sprinklers may be desired for special environments, high risk areas, isolated locations or unusual hazards. Some typical applications would include:

1. Paint dip tanks
2. Securities vaults
3. Electronic data processing
4. Tape and data storage
5. Transformer rooms
6. Spray booths
7. Petroleum storage
8. High rack storage

Piping and wiring costs are highly variable and are not included.

D4090 910	FM200 Fire Suppression	COST EACH		
		MAT.	INST.	TOTAL
0020	Detectors with brackets			
0040	Fixed temperature heat detector	54	73	127
0060	Rate of temperature rise detector	54	73	127
0080	Ion detector (smoke) detector	139	94	233
0200	Extinguisher agent			
0240	200 lb FM200, container	7,175	281	7,456
0280	75 lb carbon dioxide cylinder	1,175	187	1,362
0320	Dispersion nozzle			
0340	FM200 1-1/2" dispersion nozzle	62	44.50	106.50
0380	Carbon dioxide 3" x 5" dispersion nozzle	62	34.50	96.50
0420	Control station			
0440	Single zone control station with batteries	1,600	585	2,185
0470	Multizone (4) control station with batteries	3,025	1,175	4,200
0500	Electric mechanical release	154	300	454
0550	Manual pull station	55.50	104	159.50
0640	Battery standby power 10" x 10" x 17"	435	146	581
0740	Bell signalling device	77.50	73	150.50
0900	Standard low-rise sprinkler accessory package, 3 story	2,850	1,625	4,475
1000	Standard mid-rise sprinkler accessory package, 8 story	6,225	3,100	9,325
1100	Standard high-rise sprinkler accessory package, 16 story	13,000	6,050	19,050

D4090 920	FM200 Systems	COST PER C.F.		
		MAT.	INST.	TOTAL
0820	Average FM200 system, minimum			1.63
0840	Maximum			3.23

D5010   Electrical Service/Distribution

Feeder Installation

Electric Service

Switchgear

D5010 120	Electric Service, 3 Phase - 4 Wire	COST EACH		
		MAT.	INST.	TOTAL
0200	Service installation, includes breakers, metering, 20' conduit & wire			
0220	3 phase, 4 wire, 120/208 volts, 60 A	995	900	1,895
0240	100 A	1,200	1,075	2,275
0280	200 A	1,825	1,675	3,500
0320	400 A	4,300	3,050	7,350
0360	600 A	8,100	4,150	12,250
0400	800 A	10,400	4,975	15,375
0440	1000 A	12,900	5,750	18,650
0480	1200 A	16,000	5,850	21,850
0520	1600 A	29,300	8,400	37,700
0560	2000 A	32,400	9,600	42,000
0570	Add 25% for 277/480 volt			

D5010 230	Feeder Installation	COST PER L.F.		
		MAT.	INST.	TOTAL
0200	Feeder installation 600 V, including RGS conduit and XHHW wire, 60 A	7.05	10.90	17.95
0240	100 A	10.60	14.40	25
0280	200 A	23.50	22.50	46
0320	400 A	47	44.50	91.50
0360	600 A	104	73	177
0400	800 A	149	86.50	235.50
0440	1000 A	169	112	281
0480	1200 A	218	114	332
0520	1600 A	297	173	470
0560	2000 A	340	224	564
1200	Branch installation 600 V, including EMT conduit and THW wire, 15 A	1.24	5.55	6.79
1240	20 A	1.24	5.55	6.79
1280	30 A	1.90	6.80	8.70
1320	50 A	3.18	7.85	11.03
1360	65 A	3.88	8.35	12.23
1400	85 A	6.25	9.90	16.15
1440	100 A	7.85	10.50	18.35
1480	130 A	10.30	11.80	22.10
1520	150 A	12.70	13.50	26.20
1560	200 A	16.70	15.20	31.90

D50 Electrical

D5010 Electrical Service/Distribution

D5010 240	Switchgear	COST EACH		
		MAT.	INST.	TOTAL
0200	Switchgear inst., incl. swbd., panels & circ bkr, 400 A, 120/208volt	4,425	3,675	8,100
0240	600 A	13,800	4,975	18,775
0280	800 A	17,500	7,075	24,575
0320	1200 A	21,000	10,900	31,900
0360	1600 A	28,300	15,200	43,500
0400	2000 A	35,700	19,400	55,100
0410	Add 20% for 277/480 volt			

D5020 Lighting and Branch Wiring

Duplex Wall Receptacle

Undercarpet Receptacle System

D5020 110	Receptacle (by Wattage)	COST PER S.F.		
		MAT.	INST.	TOTAL
0190	Receptacles include plate, box, conduit, wire & transformer when required			
0200	2.5 per 1000 S.F., .3 watts per S.F.	.41	1.31	1.72
0240	With transformer	.48	1.37	1.85
0280	4 per 1000 S.F., .5 watts per S.F.	.47	1.53	2
0320	With transformer	.57	1.63	2.20
0360	5 per 1000 S.F., .6 watts per S.F.	.55	1.80	2.35
0400	With transformer	.68	1.93	2.61
0440	8 per 1000 S.F., .9 watts per S.F.	.58	1.99	2.57
0480	With transformer	.76	2.17	2.93
0520	10 per 1000 S.F., 1.2 watts per S.F.	.61	2.16	2.77
0560	With transformer	.91	2.45	3.36
0600	16.5 per 1000 S.F., 2.0 watts per S.F.	.70	2.70	3.40
0640	With transformer	1.22	3.20	4.42
0680	20 per 1000 S.F.,2.4 watts per S.F.	.74	2.93	3.67
0720	With transformer	1.34	3.51	4.85

D5020 115	Receptacles, Floor	COST PER S.F.		
		MAT.	INST.	TOTAL
0200	Receptacle systems, underfloor duct, 5' on center, low density	6.55	2.87	9.42
0240	High density	7.05	3.69	10.74
0280	7' on center, low density	5.15	2.46	7.61
0320	High density	5.70	3.28	8.98
0400	Poke thru fittings, low density	1.23	1.39	2.62
0440	High density	2.44	2.75	5.19
0520	Telepoles, using Romex, low density	1.16	.86	2.02
0560	High density	2.33	1.73	4.06
0600	Using EMT, low density	1.24	1.15	2.39
0640	High density	2.46	2.28	4.74
0720	Conduit system with floor boxes, low density	1.12	.98	2.10
0760	High density	2.26	1.98	4.24
0840	Undercarpet power system, 3 conductor with 5 conductor feeder, low density	1.48	.34	1.82
0880	High density	2.87	.73	3.60

D5020 Lighting and Branch Wiring

Description: Table D5020 130 includes the cost for switch, plate, box, conduit in slab or EMT exposed and copper wire. Add 20% for exposed conduit.

No power required for switches.

Federal energy guidelines recommend the maximum lighting area controlled per switch shall not exceed 1000 S.F. and that areas over 500 S.F. shall be so controlled that total illumination can be reduced by at least 50%.

D5020 130	Wall Switch by Sq. Ft.	COST PER S.F.		
		MAT.	INST.	TOTAL
0200	Wall switches, 1.0 per 1000 S.F.	.07	.21	.28
0240	1.2 per 1000 S.F.	.07	.22	.29
0280	2.0 per 1000 S.F.	.10	.34	.44
0320	2.5 per 1000 S.F.	.12	.42	.54
0360	5.0 per 1000 S.F.	.26	.91	1.17
0400	10.0 per 1000 S.F.	.54	1.83	2.37

D5020 135	Miscellaneous Power	COST PER S.F.		
		MAT.	INST.	TOTAL
0200	Miscellaneous power, to .5 watts	.03	.11	.14
0240	.8 watts	.05	.14	.19
0280	1 watt	.07	.20	.27
0320	1.2 watts	.08	.23	.31
0360	1.5 watts	.09	.27	.36
0400	1.8 watts	.11	.32	.43
0440	2 watts	.12	.38	.50
0480	2.5 watts	.16	.46	.62
0520	3 watts	.19	.54	.73

D5020 140	Central A. C. Power (by Wattage)	COST PER S.F.		
		MAT.	INST.	TOTAL
0200	Central air conditioning power, 1 watt	.07	.23	.30
0220	2 watts	.09	.26	.35
0240	3 watts	.12	.29	.41
0280	4 watts	.18	.40	.58
0320	6 watts	.41	.55	.96
0360	8 watts	.40	.57	.97
0400	10 watts	.55	.66	1.21

D50 Electrical

D5020 Lighting and Branch Wiring

Motor Installation

D5020 145	Motor Installation	COST EACH		
		MAT.	INST.	TOTAL
0200	Motor installation, single phase, 115V, 1/3 HP motor size	675	890	1,565
0240	1 HP motor size	700	890	1,590
0280	2 HP motor size	745	945	1,690
0320	3 HP motor size	835	960	1,795
0360	230V, 1 HP motor size	675	900	1,575
0400	2 HP motor size	710	900	1,610
0440	3 HP motor size	790	965	1,755
0520	Three phase, 200V, 1-1/2 HP motor size	795	990	1,785
0560	3 HP motor size	845	1,075	1,920
0600	5 HP motor size	905	1,200	2,105
0640	7-1/2 HP motor size	935	1,225	2,160
0680	10 HP motor size	1,500	1,525	3,025
0720	15 HP motor size	1,925	1,700	3,625
0760	20 HP motor size	2,450	1,950	4,400
0800	25 HP motor size	2,500	1,975	4,475
0840	30 HP motor size	3,900	2,325	6,225
0880	40 HP motor size	4,975	2,750	7,725
0920	50 HP motor size	8,350	3,200	11,550
0960	60 HP motor size	8,750	3,375	12,125
1000	75 HP motor size	11,400	3,875	15,275
1040	100 HP motor size	27,600	4,550	32,150
1080	125 HP motor size	28,600	5,000	33,600
1120	150 HP motor size	33,800	5,850	39,650
1160	200 HP motor size	33,600	7,175	40,775
1240	230V, 1-1/2 HP motor size	755	975	1,730
1280	3 HP motor size	805	1,075	1,880
1320	5 HP motor size	865	1,175	2,040
1360	7-1/2 HP motor size	865	1,175	2,040
1400	10 HP motor size	1,325	1,450	2,775
1440	15 HP motor size	1,550	1,575	3,125
1480	20 HP motor size	2,275	1,900	4,175
1520	25 HP motor size	2,450	1,950	4,400
1560	30 HP motor size	2,500	1,975	4,475
1600	40 HP motor size	4,625	2,675	7,300
1640	50 HP motor size	4,875	2,825	7,700
1680	60 HP motor size	8,350	3,225	11,575
1720	75 HP motor size	10,400	3,650	14,050
1760	100 HP motor size	11,800	4,050	15,850
1800	125 HP motor size	28,000	4,700	32,700
1840	150 HP motor size	30,500	5,325	35,825
1880	200 HP motor size	32,200	5,925	38,125
1960	460V, 2 HP motor size	930	985	1,915
2000	5 HP motor size	980	1,075	2,055
2040	10 HP motor size	1,000	1,175	2,175

D5020 Lighting and Branch Wiring

D5020 145	Motor Installation	COST EACH		
		MAT.	INST.	TOTAL
2080	15 HP motor size	1,300	1,350	2,650
2120	20 HP motor size	1,375	1,450	2,825
2160	25 HP motor size	1,525	1,525	3,050
2200	30 HP motor size	1,900	1,650	3,550
2240	40 HP motor size	2,450	1,775	4,225
2280	50 HP motor size	2,750	1,975	4,725
2320	60 HP motor size	3,975	2,300	6,275
2360	75 HP motor size	4,675	2,550	7,225
2400	100 HP motor size	5,175	2,825	8,000
2440	125 HP motor size	8,650	3,225	11,875
2480	150 HP motor size	10,600	3,600	14,200
2520	200 HP motor size	12,500	4,075	16,575
2600	575V, 2 HP motor size	930	985	1,915
2640	5 HP motor size	980	1,075	2,055
2680	10 HP motor size	1,000	1,175	2,175
2720	20 HP motor size	1,300	1,350	2,650
2760	25 HP motor size	1,375	1,450	2,825
2800	30 HP motor size	1,900	1,650	3,550
2840	50 HP motor size	2,050	1,700	3,750
2880	60 HP motor size	3,950	2,275	6,225
2920	75 HP motor size	3,975	2,300	6,275
2960	100 HP motor size	4,675	2,550	7,225
3000	125 HP motor size	8,125	3,175	11,300
3040	150 HP motor size	8,650	3,225	11,875
3080	200 HP motor size	11,100	3,650	14,750

D5020 155	Motor Feeder	COST PER L.F.		
		MAT.	INST.	TOTAL
0200	Motor feeder systems, single phase, feed up to 115V 1HP or 230V 2 HP	2.35	6.90	9.25
0240	115V 2HP, 230V 3HP	2.49	7	9.49
0280	115V 3HP	2.82	7.30	10.12
0360	Three phase, feed to 200V 3HP, 230V 5HP, 460V 10HP, 575V 10HP	2.47	7.45	9.92
0440	200V 5HP, 230V 7.5HP, 460V 15HP, 575V 20HP	2.69	7.60	10.29
0520	200V 10HP, 230V 10HP, 460V 30HP, 575V 30HP	3.19	8.05	11.24
0600	200V 15HP, 230V 15HP, 460V 40HP, 575V 50HP	4.28	9.20	13.48
0680	200V 20HP, 230V 25HP, 460V 50HP, 575V 60HP	6.30	11.65	17.95
0760	200V 25HP, 230V 30HP, 460V 60HP, 575V 75HP	6.95	11.85	18.80
0840	200V 30HP	10.50	12.25	22.75
0920	230V 40HP, 460V 75HP, 575V 100HP	10.25	13.35	23.60
1000	200V 40HP	15.80	14.25	30.05
1080	230V 50HP, 460V 100HP, 575V 125HP	14.20	15.75	29.95
1160	200V 50HP, 230V 60HP, 460V 125HP, 575V 150HP	23	16.70	39.70
1240	200V 60HP, 460V 150HP	21	19.60	40.60
1320	230V 75HP, 575V 200HP	30	20.50	50.50
1400	200V 75HP	31	21	52
1480	230V 100HP, 460V 200HP	36.50	24.50	61
1560	200V 100HP	50	30.50	80.50
1640	230V 125HP	56.50	33	89.50
1720	200V 125HP, 230V 150HP	66	38	104
1800	200V 150HP	70	41	111
1880	200V 200HP	99.50	61	160.50
1960	230V 200HP	89.50	45.50	135

D5020 Lighting and Branch Wiring

Fluorescent Fixture

Incandescent Fixture

D5020 210	Fluorescent Fixtures (by Wattage)	COST PER S.F.		
		MAT.	INST.	TOTAL
0190	Fluorescent fixtures recess mounted in ceiling			
0195	T-12, standard 40 watt lamps			
0200	1 watt per S.F., 20 FC, 5 fixtures @40 watts per 1000 S.F.	.81	1.72	2.53
0240	2 watt per S.F., 40 FC, 10 fixtures @40 watt per 1000 S.F.	1.61	3.37	4.98
0280	3 watt per S.F., 60 FC, 15 fixtures @40 watt per 1000 S.F	2.41	5.10	7.51
0320	4 watt per S.F., 80 FC, 20 fixtures @40 watt per 1000 S.F.	3.22	6.75	9.97
0400	5 watt per S.F., 100 FC, 25 fixtures @40 watt per 1000 S.F.	4.03	8.45	12.48
0450	T-8, energy saver 32 watt lamps			
0500	0.8 watt per S.F., 20 FC, 5 fixtures @32 watt per 1000 S.F.	.88	1.72	2.60
0520	1.6 watt per S.F., 40 FC, 10 fixtures @32 watt per 1000 S.F.	1.76	3.37	5.13
0540	2.4 watt per S.F., 60 FC, 15 fixtures @ 32 watt per 1000 S.F	2.63	5.10	7.73
0560	3.2 watt per S.F., 80 FC, 20 fixtures @32 watt per 1000 S.F.	3.51	6.75	10.26
0580	4 watt per S.F., 100 FC, 25 fixtures @32 watt per 1000 S.F.	4.39	8.45	12.84

D5020 216	Incandescent Fixture (by Wattage)	COST PER S.F.		
		MAT.	INST.	TOTAL
0190	Incandescent fixture recess mounted, type A			
0200	1 watt per S.F., 8 FC, 6 fixtures per 1000 S.F.	1.02	1.37	2.39
0240	2 watt per S.F., 16 FC, 12 fixtures per 1000 S.F.	2.04	2.75	4.79
0280	3 watt per S.F., 24 FC, 18 fixtures, per 1000 S.F.	3.04	4.09	7.13
0320	4 watt per S.F., 32 FC, 24 fixtures per 1000 S.F.	4.06	5.45	9.51
0400	5 watt per S.F., 40 FC, 30 fixtures per 1000 S.F.	5.10	6.85	11.95

D5020 Lighting and Branch Wiring

High Bay Fixture

Low Bay Fixture

D5020 226	H.I.D. Fixture, High Bay, 16′ (by Wattage)	COST PER S.F.		
		MAT.	INST.	TOTAL
0190	High intensity discharge fixture, 16′ above work plane			
0240	1 watt/S.F., type E, 42 FC, 1 fixture/1000 S.F.	1.06	1.45	2.51
0280	Type G, 52 FC, 1 fixture/1000 S.F.	1.06	1.45	2.51
0320	Type C, 54 FC, 2 fixture/1000 S.F.	1.25	1.61	2.86
0440	2 watt/S.F., type E, 84 FC, 2 fixture/1000 S.F.	2.15	2.93	5.08
0480	Type G, 105 FC, 2 fixture/1000 S.F.	2.15	2.93	5.08
0520	Type C, 108 FC, 4 fixture/1000 S.F.	2.48	3.22	5.70
0640	3 watt/S.F., type E, 126 FC, 3 fixture/1000 S.F.	3.20	4.38	7.58
0680	Type G, 157 FC, 3 fixture/1000 S.F.	3.20	4.38	7.58
0720	Type C, 162 FC, 6 fixture/1000 S.F.	3.73	4.82	8.55
0840	4 watt/S.F., type E, 168 FC, 4 fixture/1000 S.F.	4.29	5.85	10.14
0880	Type G, 210 FC, 4 fixture/1000 S.F.	4.29	5.85	10.14
0920	Type C, 243 FC, 9 fixture/1000 S.F.	5.40	6.75	12.15
1040	5 watt/S.F., type E, 210 FC, 5 fixture/1000 S.F.	5.35	7.30	12.65
1080	Type G, 262 FC, 5 fixture/1000 S.F.	5.35	7.30	12.65
1120	Type C, 297 FC, 11 fixture/1000 S.F.	6.65	8.35	15

D5020 234	H.I.D. Fixture, High Bay, 30′ (by Wattage)	COST PER S.F.		
		MAT.	INST.	TOTAL
0190	High intensity discharge fixture, 30′ above work plane			
0240	1 watt/S.F., type E, 37 FC, 1 fixture/1000 S.F.	1.20	1.83	3.03
0280	Type G, 45 FC., 1 fixture/1000 S.F.	1.20	1.83	3.03
0320	Type F, 50 FC, 1 fixture/1000 S.F.	1.01	1.43	2.44
0440	2 watt/S.F., type E, 74 FC, 2 fixtures/1000 S.F.	2.40	3.64	6.04
0480	Type G, 92 FC, 2 fixtures/1000 S.F.	2.40	3.64	6.04
0520	Type F, 100 FC, 2 fixtures/1000 S.F.	2.04	2.91	4.95
0640	3 watt/S.F., type E, 110 FC, 3 fixtures/1000 S.F.	3.61	5.55	9.16
0680	Type G, 138 FC, 3 fixtures/1000 S.F.	3.61	5.55	9.16
0720	Type F, 150 FC, 3 fixtures/1000 S.F.	3.04	4.34	7.38
0840	4 watt/S.F., type E, 148 FC, 4 fixtures/1000 S.F.	4.81	7.35	12.16
0880	Type G, 185 FC, 4 fixtures/1000 S.F.	4.81	7.35	12.16
0920	Type F, 200 FC, 4 fixtures/1000 S.F.	4.07	5.80	9.87
1040	5 watt/S.F., type E, 185 FC, 5 fixtures/1000 S.F.	6	9.20	15.20
1080	Type G, 230 FC, 5 fixtures/1000 S.F.	6	9.20	15.20
1120	Type F, 250 FC, 5 fixtures/1000 S.F.	5.10	7.25	12.35

D5020 238	H.I.D. Fixture, Low Bay, 8′-10′ (by Wattage)	COST PER S.F.		
		MAT.	INST.	TOTAL
0190	High intensity discharge fixture, 8′-10′ above work plane			
0240	1 watt/S.F., type J, 30 FC, 4 fixtures/1000 S.F.	2.42	2.99	5.41
0280	Type K, 29 FC, 5 fixtures/1000 S.F.	2.48	2.68	5.16
0400	2 watt/S.F., type J, 52 FC, 7 fixtures/1000 S.F.	4.32	5.45	9.77

D50 Electrical

D5020 Lighting and Branch Wiring

D5020 238	H.I.D. Fixture, Low Bay, 8'-10' (by Wattage)	COST PER S.F.		
		MAT.	INST.	TOTAL
0440	Type K, 63 FC, 11 fixtures/1000 S.F.	5.35	5.60	10.95
0560	3 watt/S.F., type J, 81 FC, 11 fixtures/1000 S.F.	6.70	8.25	14.95
0600	Type K, 92 FC, 16 fixtures/1000 S.F.	7.85	8.30	16.15
0720	4 watt/S.F., type J, 103 FC, 14 fixtures/1000 S.F.	8.65	10.85	19.50
0760	Type K, 127 FC, 22 fixtures/1000 S.F.	10.70	11.25	21.95
0880	5 watt/S.F., type J, 133 FC, 18 fixtures/1000 S.F.	11	13.70	24.70
0920	Type K, 155 FC, 27 fixtures/1000 S.F.	13.20	13.90	27.10

D5020 242	H.I.D. Fixture, Low Bay, 16' (by Wattage)	COST PER S.F.		
		MAT.	INST.	TOTAL
0190	High intensity discharge fixture, mounted 16' above work plane			
0240	1 watt/S.F., type J, 28 FC, 4 fixt./1000 S.F.	2.56	3.39	5.95
0280	Type K, 27 FC, 5 fixt./1000 S.F.	2.87	3.79	6.66
0400	2 watt/S.F., type J, 48 FC, 7 fixt/1000 S.F.	4.70	6.55	11.25
0440	Type K, 58 FC, 11 fixt/1000 S.F.	6.10	7.75	13.85
0560	3 watt/S.F., type J, 75 FC, 11 fixt/1000 S.F.	7.25	9.90	17.15
0600	Type K, 85 FC, 16 fixt/1000 S.F.	8.95	11.50	20.45
0720	4 watt/S.F., type J, 95 FC, 14 fixt/1000 S.F.	9.40	13.10	22.50
0760	Type K, 117 FC, 22 fixt/1000 S.F.	12.20	15.45	27.65
0880	5 watt/S.F., type J, 122 FC, 18 fixt/1000 S.F.	11.95	16.45	28.40
0920	Type K, 143 FC, 27 fixt/1000 S.F.	15.10	19.25	34.35

D5030 Communications and Security

D5030 910	Communication & Alarm Systems	COST EACH		
		MAT.	INST.	TOTAL
0200	Communication & alarm systems, includes outlets, boxes, conduit & wire			
0210	Sound system, 6 outlets	6,375	7,400	13,775
0220	12 outlets	8,900	11,800	20,700
0240	30 outlets	15,400	22,300	37,700
0280	100 outlets	46,500	75,000	121,500
0320	Fire detection systems, non-addressable, 12 detectors	3,600	6,125	9,725
0360	25 detectors	6,375	10,400	16,775
0400	50 detectors	12,500	20,500	33,000
0440	100 detectors	23,400	37,400	60,800
0450	Addressable type, 12 detectors	4,925	6,175	11,100
0452	25 detectors	8,950	10,500	19,450
0454	50 detectors	17,300	20,500	37,800
0456	100 detectors	33,400	37,700	71,100
0458	Fire alarm control panel, 8 zone	835	1,175	2,010
0459	12 zone	2,525	1,750	4,275
0460	Fire alarm command center, addressable without voice	4,750	1,025	5,775
0462	Addressable with voice	10,000	1,600	11,600
0480	Intercom systems, 6 stations	3,575	5,050	8,625
0560	25 stations	12,500	19,400	31,900
0640	100 stations	47,500	71,000	118,500
0680	Master clock systems, 6 rooms	5,275	8,250	13,525
0720	12 rooms	8,300	14,100	22,400
0760	20 rooms	11,500	19,900	31,400
0800	30 rooms	19,000	36,700	55,700
0840	50 rooms	30,800	62,000	92,800
0920	Master TV antenna systems, 6 outlets	3,175	5,225	8,400
0960	12 outlets	5,925	9,725	15,650
1000	30 outlets	12,000	22,500	34,500
1040	100 outlets	40,000	73,500	113,500

D5090 Other Electrical Systems

Generator System

A: Engine
B: Battery
C: Charger
D: Transfer Switch
E: Muffler

D5090 210	Generators (by kW)	COST PER kW		
		MAT.	INST.	TOTAL
0190	Generator sets, include battery, charger, muffler & transfer switch			
0200	Gas/gasoline operated, 3 phase, 4 wire, 277/480 volt, 7.5 kW	1,175	255	1,430
0240	11.5 kW	1,075	195	1,270
0280	20 kW	730	126	856
0320	35 kW	495	83	578
0360	80 kW	355	50	405
0400	100 kW	310	48	358
0440	125 kW	510	45.50	555.50
0480	185 kW	455	34.50	489.50
0560	Diesel engine with fuel tank, 30 kW	740	96.50	836.50
0600	50 kW	550	75.50	625.50
0640	75 kW	475	60.50	535.50
0680	100 kW	395	51.50	446.50
0760	150 kW	320	41	361
0840	200 kW	268	33	301
0880	250 kW	252	27.50	279.50
0960	350 kW	220	22.50	242.50
1040	500 kW	240	17.65	257.65

E10 Equipment

E1010 Commercial Equipment

E1010 110	Security/Vault, EACH	COST EACH		
		MAT.	INST.	TOTAL
0100	Bank equipment, drive up window, drawer & mike, no glazing, economy	6,150	1,325	7,475
0110	Deluxe	9,775	2,650	12,425
0120	Night depository, economy	7,975	1,325	9,300
0130	Deluxe	11,300	2,650	13,950
0140	Pneumatic tube systems, 2 station, standard	27,100	4,400	31,500
0150	Teller, automated, 24 hour, single unit	47,000	4,400	51,400
0160	Teller window, bullet proof glazing, 44" x 60"	3,600	805	4,405
0170	Pass through, painted steel, 72" x 40"	3,900	1,650	5,550
0300	Safe, office type, 1 hr. rating, 34" x 20" x 20"	2,400		2,400
0310	4 hr. rating, 62" x 33" x 20"	9,375		9,375
0320	Data storage, 4 hr. rating, 63" x 44" x 16"	14,400		14,400
0330	Jewelers, 63" x 44" x 16"	24,700		24,700
0340	Money, "B" label, 9" x 14" x 14"	580		580
0350	Tool and torch resistive, 24" x 24" x 20"	10,100	204	10,304
0500	Security gates-scissors type, painted steel, single, 6' high, 5-1/2' wide	235	330	565
0510	Double gate, 7-1/2' high, 14' wide	665	660	1,325

E1010 510	Mercantile Equipment, EACH	COST EACH		
		MAT.	INST.	TOTAL
0015	Barber equipment, chair, hydraulic, economy	460	21.50	481.50
0020	Deluxe	3,275	32	3,307
0100	Checkout counter, single belt	3,200	81.50	3,281.50
0110	Double belt, power take-away	4,600	91	4,691
0200	Display cases, freestanding, glass and aluminum, 3'-6" x 3' x 1'-0" deep	1,350	128	1,478
0220	Wall mounted, glass and aluminum, 3' x 4' x 1'-4" deep	2,350	205	2,555
0320	Frozen food, chest type, 12 ft. long	7,750	340	8,090

E1010 610	Laundry/Dry Cleaning, EACH	COST EACH		
		MAT.	INST.	TOTAL
0100	Laundry equipment, dryers, gas fired, residential, 16 lb. capacity	690	208	898
0110	Commercial, 30 lb. capacity, single	3,400	208	3,608
0120	Dry cleaners, electric, 20 lb. capacity	36,600	6,025	42,625
0130	30 lb. capacity	52,500	8,050	60,550
0140	Ironers, commercial, 120" with canopy, 8 roll	178,500	17,200	195,700
0150	Institutional, 110", single roll	32,700	2,925	35,625
0160	Washers, residential, 4 cycle	880	208	1,088
0170	Commercial, coin operated, deluxe	3,325	208	3,533

E10 Equipment

E1020 Institutional Equipment

E1020 110	**Ecclesiastical Equipment, EACH**	COST EACH		
		MAT.	INST.	TOTAL
0090	Church equipment, altar, wood, custom, plain	2,100	365	2,465
0100	Granite, custom, deluxe	33,400	4,800	38,200
0110	Baptistry, fiberglass, economy	4,200	1,325	5,525
0120	Bells & carillons, keyboard operation	18,000	8,100	26,100
0130	Confessional, wood, single, economy	2,675	855	3,530
0140	Double, deluxe	16,700	2,550	19,250
0150	Steeples, translucent fiberglas, 30" square, 15' high	6,000	1,650	7,650

E1020 130	**Ecclesiastical Equipment, L.F.**	COST PER L.F.		
		MAT.	INST.	TOTAL
0100	Arch. equip., church equip. pews, bench type, hardwood, economy	81.50	25.50	107
0110	Deluxe	161	34	195

E1020 210	**Library Equipment, EACH**	COST EACH		
		MAT.	INST.	TOTAL
0110	Library equipment, carrels, metal, economy	276	102	378
0120	Hardwood, deluxe	1,025	128	1,153

E1020 230	**Library Equipment, L.F.**	COST PER L.F.		
		MAT.	INST.	TOTAL
0100	Library equipment, book shelf, metal, single face, 90" high x 10" shelf	124	42.50	166.50
0110	Double face, 90" high x 10" shelf	270	92.50	362.50
0120	Charging desk, built-in, with counter, plastic laminate	455	73	528

E1020 310	**Theater and Stage Equipment, EACH**	COST EACH		
		MAT.	INST.	TOTAL
0200	Movie equipment, changeover, economy	510		510
0210	Film transport, incl. platters and autowind, economy	5,525		5,525
0220	Lamphouses, incl. rectifiers, xenon, 1000W	7,300	291	7,591
0230	4000W	10,400	390	10,790
0240	Projector mechanisms, 35 mm, economy	12,200		12,200
0250	Deluxe	16,700		16,700
0260	Sound systems, incl. amplifier, single, economy	3,625	650	4,275
0270	Dual, Dolby/super sound	18,200	1,450	19,650
0280	Projection screens, wall hung, manual operation, 50 S.F., economy	325	103	428
0290	Electric operation, 100 S.F., deluxe	2,175	510	2,685

E1020 320	**Theater and Stage Equipment, S.F.**	COST PER S.F.		
		MAT.	INST.	TOTAL
0090	Movie equipment, projection screens, rigid in wall, acrylic, 1/4" thick	46	4.97	50.97
0100	1/2" thick	53.50	7.45	60.95
0110	Stage equipment, curtains, velour, medium weight	8.70	1.71	10.41
0120	Silica based yarn, fireproof	16.70	20.50	37.20
0130	Stages, portable with steps, folding legs, 8" high	30.50		30.50
0140	Telescoping platforms, aluminum, deluxe	44.50	26.50	71

E1020 330	**Theater and Stage Equipment, L.F.**	COST PER L.F.		
		MAT.	INST.	TOTAL
0100	Stage equipment, curtain track, heavy duty	67	57	124
0110	Lights, border, quartz, colored	190	29	219

E1020 610	**Detention Equipment, EACH**	COST EACH		
		MAT.	INST.	TOTAL
0110	Detention equipment, cell front rolling door, 7/8" bars, 5' x 7' high	5,525	1,400	6,925
0120	Cells, prefab., including front, 5' x 7' x 7' deep	10,600	1,875	12,475

E1020 Institutional Equipment

E1020 610	Detention Equipment, EACH	COST EACH		
		MAT.	INST.	TOTAL
0130	Doors and frames, 3' x 7', single plate	4,625	705	5,330
0140	Double plate	5,700	705	6,405
0150	Toilet apparatus, incl wash basin	3,700	890	4,590
0160	Visitor cubicle, vision panel, no intercom	3,225	1,400	4,625

E1020 710	Laboratory Equipment, EACH	COST EACH		
		MAT.	INST.	TOTAL
0110	Laboratory equipment, glassware washer, distilled water, economy	6,225	670	6,895
0120	Deluxe	13,000	1,200	14,200
0140	Radio isotope	18,600		18,600

E1020 720	Laboratory Equipment, S.F.	COST PER S.F.		
		MAT.	INST.	TOTAL
0100	Arch. equip., lab equip., counter tops, acid proof, economy	36	12.50	48.50
0110	Stainless steel	108	12.50	120.50

E1020 730	Laboratory Equipment, L.F.	COST PER L.F.		
		MAT.	INST.	TOTAL
0110	Laboratory equipment, cabinets, wall, open	149	51	200
0120	Base, drawer units	470	57	527
0130	Fume hoods, not incl. HVAC, economy	665	190	855
0140	Deluxe incl. fixtures	1,975	425	2,400

E1020 810	Medical Equipment, EACH	COST EACH		
		MAT.	INST.	TOTAL
0100	Dental equipment, central suction system, economy	1,700	520	2,220
0110	Compressor-air, deluxe	7,925	1,050	8,975
0120	Chair, hydraulic, economy	2,825	1,050	3,875
0130	Deluxe	15,900	2,100	18,000
0140	Drill console with accessories, economy	1,725	330	2,055
0150	Deluxe	5,925	330	6,255
0160	X-ray unit, portable	2,400	131	2,531
0170	Panoramic unit	15,700	875	16,575
0300	Medical equipment, autopsy table, standard	9,750	625	10,375
0310	Deluxe	19,400	1,050	20,450
0320	Incubators, economy	3,750		3,750
0330	Deluxe	22,100		22,100
0700	Station, scrub-surgical, single, economy	4,500	208	4,708
0710	Dietary, medium, with ice	17,600		17,600
0720	Sterilizers, general purpose, single door, 20" x 20" x 28"	16,800		16,800
0730	Floor loading, double door, 28" x 67" x 52"	224,000		224,000
0740	Surgery tables, standard	15,200	940	16,140
0750	Deluxe	34,600	1,325	35,925
0770	Tables, standard, with base cabinets, economy	1,425	340	1,765
0780	Deluxe	4,250	510	4,760
0790	X-ray, mobile, economy	14,700		14,700
0800	Stationary, deluxe	239,500		239,500

E10 Equipment

E1030   Vehicular Equipment

E1030 110	Vehicular Service Equipment, EACH	COST EACH		
		MAT.	INST.	TOTAL
0110	Automotive equipment, compressors, electric, 1-1/2 H.P., std. controls	440	955	1,395
0120	5 H.P., dual controls	2,400	1,425	3,825
0130	Hoists, single post, 4 ton capacity, swivel arms	5,475	3,600	9,075
0140	Dual post, 12 ton capacity, adjustable frame	9,400	9,575	18,975
0150	Lube equipment, 3 reel type, with pumps	8,700	2,875	11,575
0160	Product dispenser, 6 nozzles, w/vapor recovery, not incl. piping, installed	24,800		24,800
0800	Scales, dial type, built in floor, 5 ton capacity, 8' x 6' platform	8,000	3,075	11,075
0810	10 ton capacity, 9' x 7' platform	11,800	4,400	16,200
0820	Truck (including weigh bridge), 20 ton capacity, 24' x 10'	12,300	5,125	17,425

E1030 210	Parking Control Equipment, EACH	COST EACH		
		MAT.	INST.	TOTAL
0110	Parking equipment, automatic gates, 8 ft. arm, one way	4,125	1,050	5,175
0120	Traffic detectors, single treadle	2,625	485	3,110
0130	Booth for attendant, economy	6,525		6,525
0140	Deluxe	27,700		27,700
0150	Ticket printer/dispenser, rate computing	9,000	835	9,835
0160	Key station on pedestal	800	284	1,084

E1030 310	Loading Dock Equipment, EACH	COST EACH		
		MAT.	INST.	TOTAL
0110	Dock bumpers, rubber blocks, 4-1/2" thick, 10" high, 14" long	53.50	19.70	73.20
0120	6" thick, 20" high, 11" long	147	39.50	186.50
0130	Dock boards, H.D., 5' x 5', aluminum, 5000 lb. capacity	1,275		1,275
0140	16,000 lb. capacity	1,725		1,725
0150	Dock levelers, hydraulic, 7' x 8', 10 ton capacity	9,325	1,400	10,725
0160	Dock lifters, platform, 6' x 6', portable, 3000 lb. capacity	9,150		9,150
0170	Dock shelters, truck, scissor arms, economy	1,650	510	2,160
0180	Deluxe	2,150	1,025	3,175

431

E10 Equipment

E1090 Other Equipment

E1090 210	Solid Waste Handling Equipment, EACH	COST EACH		
		MAT.	INST.	TOTAL
0110	Waste handling, compactors, single bag, 250 lbs./hr., hand fed	9,025	600	9,625
0120	Heavy duty industrial, 5 C.Y. capacity	20,700	2,875	23,575
0130	Incinerator, electric, 100 lbs./hr., economy	13,000	2,900	15,900
0140	Gas, 2000 lbs./hr., deluxe	405,000	23,800	428,800
0150	Shredder, no baling, 35 tons/hr.	297,000		297,000
0160	Incl. baling, 50 tons/day	593,500		593,500

E1090 350	Food Service Equipment, EACH	COST EACH		
		MAT.	INST.	TOTAL
0110	Kitchen equipment, bake oven, single deck	5,125	140	5,265
0120	Broiler, without oven	3,900	140	4,040
0130	Commercial dish washer, semiautomatic, 50 racks/hr.	9,400	865	10,265
0140	Automatic, 275 racks/hr.	33,300	3,650	36,950
0150	Cooler, beverage, reach-in, 6 ft. long	4,975	187	5,162
0160	Food warmer, counter, 1.65 kw	740		740
0170	Fryers, with submerger, single	1,650	161	1,811
0180	Double	2,500	225	2,725
0185	Ice maker, 1000 lb. per day, with bin	5,475	1,125	6,600
0190	Kettles, steam jacketed, 20 gallons	7,825	246	8,071
0200	Range, restaurant type, burners, 2 ovens and 24" griddle	8,475	187	8,662
0210	Range hood, incl. carbon dioxide system, economy	4,725	375	5,100
0220	Deluxe	37,200	1,125	38,325

E1090 360	Food Service Equipment, S.F.	COST PER S.F.		
		MAT.	INST.	TOTAL
0110	Refrigerators, prefab, walk-in, 7'-6" high, 6' x 6'	154	18.70	172.70
0120	12' x 20'	97	9.35	106.35

E1090 410	Residential Equipment, EACH	COST EACH		
		MAT.	INST.	TOTAL
0110	Arch. equip., appliances, range, cook top, 4 burner, economy	330	97	427
0120	Built in, single oven 30" wide, economy	750	97	847
0130	Standing, single oven-21" wide, economy	360	81.50	441.50
0140	Double oven-30" wide, deluxe	1,775	81.50	1,856.50
0150	Compactor, residential, economy	585	102	687
0160	Deluxe	910	171	1,081
0170	Dish washer, built-in, 2 cycles, economy	234	300	534
0180	4 or more cycles, deluxe	1,150	605	1,755
0190	Garbage disposer, sink type, economy	79.50	121	200.50
0200	Deluxe	210	121	331
0210	Refrigerator, no frost, 10 to 12 C.F., economy	490	81.50	571.50
0220	21 to 29 C.F., deluxe	3,375	272	3,647
0300	Washing machine, automatic	1,125	625	1,750

E1090 610	School Equipment, EACH	COST EACH		
		MAT.	INST.	TOTAL
0110	School equipment, basketball backstops, wall mounted, wood, fixed	1,250	900	2,150
0120	Suspended type, electrically operated	5,350	1,725	7,075
0130	Bleachers-telescoping, manual operation, 15 tier, economy (per seat)	86	32	118
0140	Power operation, 30 tier, deluxe (per seat)	330	56.50	386.50
0150	Weight lifting gym, universal, economy	435	815	1,250
0160	Deluxe	16,200	1,625	17,825
0170	Scoreboards, basketball, 1 side, economy	2,550	770	3,320
0180	4 sides, deluxe	6,125	10,600	16,725
0800	Vocational shop equipment, benches, metal	495	205	700
0810	Wood	540	205	745

E10 Equipment

E1090 Other Equipment

E1090 610	School Equipment, EACH	COST EACH		
		MAT.	INST.	TOTAL
0820	Dust collector, not incl. ductwork, 6' diam.	4,550	540	5,090
0830	Planer, 13" x 6"	1,075	256	1,331

E1090 620	School Equipment, S.F.	COST PER S.F.		
		MAT.	INST.	TOTAL
0110	School equipment, gym mats, naugahyde cover, 2" thick	3.95		3.95
0120	Wrestling, 1" thick, heavy duty	6.05		6.05

E1090 810	Athletic, Recreational, and Therapeutic Equipment, EACH	COST EACH		
		MAT.	INST.	TOTAL
0050	Bowling alley with gutters	49,900	10,800	60,700
0060	Combo. table and ball rack	1,250		1,250
0070	Bowling alley automatic scorer	10,900		10,900
0110	Sauna, prefabricated, incl. heater and controls, 7' high, 6' x 4'	5,100	785	5,885
0120	10' x 12'	12,300	1,725	14,025
0130	Heaters, wall mounted, to 200 C.F.	665		665
0140	Floor standing, to 1000 C.F., 12500 W	2,775	194	2,969
0610	Shooting range incl. bullet traps, controls, separators, ceilings, economy	30,800	6,825	37,625
0620	Deluxe	49,500	11,500	61,000
0650	Sport court, squash, regulation, in existing building, economy			40,400
0660	Deluxe			45,000
0670	Racketball, regulation, in existing building, economy	43,800	9,725	53,525
0680	Deluxe	47,400	19,500	66,900
0700	Swimming pool equipment, diving stand, stainless steel, 1 meter	6,800	380	7,180
0710	3 meter	11,500	2,550	14,050
0720	Diving boards, 16 ft. long, aluminum	3,700	380	4,080
0730	Fiberglass	2,800	380	3,180
0740	Filter system, sand, incl. pump, 6000 gal./hr.	1,875	695	2,570
0750	Lights, underwater, 12 volt with transformer, 300W	289	585	874
0760	Slides, fiberglass with aluminum handrails & ladder, 6' high, straight	4,000	640	4,640
0780	12' high, straight with platform	13,700	855	14,555

E1090 820	Athletic, Recreational, and Therapeutic Equipment, S.F.	COST PER S.F.		
		MAT.	INST.	TOTAL
0110	Swimming pools, residential, vinyl liner, metal sides	12.75	7.15	19.90
0120	Concrete sides	15.25	13	28.25
0130	Gunite shell, plaster finish, 350 S.F.	28.50	26.50	55
0140	800 S.F.	23	15.55	38.55
0150	Motel, gunite shell, plaster finish	35	33.50	68.50
0160	Municipal, gunite shell, tile finish, formed gutters	135	58.50	193.50

E2010 Fixed Furnishings

E2010 310	Window Treatment, EACH	COST EACH		
		MAT.	INST.	TOTAL
0110	Furnishings, blinds, exterior, aluminum, louvered, 1'-4" wide x 3'-0" long	66.50	51	117.50
0120	1'-4" wide x 6'-8" long	134	57	191
0130	Hemlock, solid raised, 1-'4" wide x 3'-0" long	44.50	51	95.50
0140	1'-4" wide x 6'-9" long	72.50	57	129.50
0150	Polystyrene, louvered, 1'-3" wide x 3'-3" long	30.50	51	81.50
0160	1'-3" wide x 6'-8" long	51	57	108
0200	Interior, wood folding panels, louvered, 7" x 20" (per pair)	51.50	30	81.50
0210	18" x 40" (per pair)	130	30	160

E2010 320	Window Treatment, S.F.	COST PER S.F.		
		MAT.	INST.	TOTAL
0110	Furnishings, blinds-interior, venetian-aluminum, stock, 2" slats, economy	2.02	.87	2.89
0120	Custom, 1" slats, deluxe	7.85	1.16	9.01
0130	Vertical, PVC or cloth, T&B track, economy	3.97	1.11	5.08
0140	Deluxe	13.90	1.28	15.18
0150	Draperies, unlined, economy	3.14		3.14
0160	Lightproof, deluxe	9.40		9.40
0510	Shades, mylar, wood roller, single layer, non-reflective	7.35	.75	8.10
0520	Metal roller, triple layer, heat reflective	12.15	.75	12.90
0530	Vinyl, light weight, 4 ga.	.62	.75	1.37
0540	Heavyweight, 6 ga.	1.87	.75	2.62
0550	Vinyl coated cotton, lightproof decorator shades	2.37	.75	3.12
0560	Woven aluminum, 3/8" thick, light and fireproof	6.05	1.46	7.51

E2010 420	Fixed Floor Grilles and Mats, S.F.	COST PER S.F.		
		MAT.	INST.	TOTAL
0110	Floor mats, recessed, inlaid black rubber, 3/8" thick, solid	22	2.63	24.63
0120	Colors, 1/2" thick, perforated	26	2.63	28.63
0130	Link-including nosings, steel-galvanized, 3/8" thick	19.95	2.63	22.58
0140	Vinyl, in colors	23	2.63	25.63

E2010 510	Fixed Multiple Seating, EACH	COST EACH		
		MAT.	INST.	TOTAL
0110	Seating, painted steel, upholstered, economy	145	29.50	174.50
0120	Deluxe	465	36.50	501.50
0400	Seating, lecture hall, pedestal type, economy	182	46.50	228.50
0410	Deluxe	610	70.50	680.50
0500	Auditorium chair, veneer construction	193	46.50	239.50
0510	Fully upholstered, spring seat	232	46.50	278.50

E20 Furnishings

E2020 Moveable Furnishings

E2020 210	Furnishings/EACH	COST EACH		
		MAT.	INST.	TOTAL
0200	Hospital furniture, beds, manual, economy	830		830
0210	Deluxe	2,825		2,825
0220	All electric, economy	1,450		1,450
0230	Deluxe	4,750		4,750
0240	Patient wall systems, no utilities, economy, per room	1,200		1,200
0250	Deluxe, per room	2,200		2,200
0300	Hotel furnishings, standard room set, economy, per room	2,550		2,550
0310	Deluxe, per room	9,125		9,125
0500	Office furniture, standard employee set, economy, per person	600		600
0510	Deluxe, per person	2,300		2,300
0550	Posts, portable, pedestrian traffic control, economy	139		139
0560	Deluxe	355		355
0700	Restaurant furniture, booth, molded plastic, stub wall and 2 seats, economy	390	256	646
0710	Deluxe	1,325	340	1,665
0720	Upholstered seats, foursome, single-economy	715	102	817
0730	Foursome, double-deluxe	1,650	171	1,821

E2020 220	Furniture and Accessories, L.F.	COST PER L.F.		
		MAT.	INST.	TOTAL
0210	Dormitory furniture, desk top (built-in),laminated plastc, 24"deep, economy	42.50	20.50	63
0220	30" deep, deluxe	239	25.50	264.50
0230	Dressing unit, built-in, economy	239	85.50	324.50
0240	Deluxe	720	128	848
0310	Furnishings, cabinets, hospital, base, laminated plastic	271	102	373
0320	Stainless steel	500	102	602
0330	Counter top, laminated plastic, no backsplash	47	25.50	72.50
0340	Stainless steel	151	25.50	176.50
0350	Nurses station, door type, laminated plastic	315	102	417
0360	Stainless steel	600	102	702
0710	Restaurant furniture, bars, built-in, back bar	200	102	302
0720	Front bar	275	102	377
0910	Wardrobes & coatracks, standing, steel, single pedestal, 30" x 18" x 63"	108		108
0920	Double face rack, 39" x 26" x 70"	119		119
0930	Wall mounted rack, steel frame & shelves, 12" x 15" x 26"	77.50	7.30	84.80
0940	12" x 15" x 50"	47.50	3.81	51.31

F10 Special Construction

F1010 Special Structures

F1010 120	Air-Supported Structures, S.F.	COST PER S.F.		
		MAT.	INST.	TOTAL
0110	Air supported struc., polyester vinyl fabric, 24oz., warehouse, 5000 S.F.	27	.33	27.33
0120	50,000 S.F.	12.65	.26	12.91
0130	Tennis, 7,200 S.F.	25.50	.27	25.77
0140	24,000 S.F.	17.75	.27	18.02
0150	Woven polyethylene, 6 oz., shelter, 3,000 S.F.	7.25	.54	7.79
0160	24,000 S.F.	6.15	.27	6.42
0170	Teflon coated fiberglass, stadium cover, economy	62	.14	62.14
0180	Deluxe	73.50	.19	73.69
0190	Air supported storage tank covers, reinf. vinyl fabric, 12 oz., 400 S.F.	25	.46	25.46
0200	18,000 S.F.	7.95	.41	8.36

F1010 210	Pre-Engineered Structures, EACH	COST EACH		
		MAT.	INST.	TOTAL
0600	Radio towers, guyed, 40 lb. section, 50' high, 70 MPH basic wind speed	3,550	1,325	4,875
0610	90 lb. section, 400' high, wind load 70 MPH basic wind speed	46,200	14,700	60,900
0620	Self supporting, 60' high, 70 MPH basic wind speed	7,425	2,575	10,000
0630	190' high, wind load 90 MPH basic wind speed	44,200	10,300	54,500
0700	Shelters, aluminum frame, acrylic glazing, 8' high, 3' x 9'	3,675	1,150	4,825
0710	9' x 12'	8,675	1,800	10,475

F1010 320	Other Special Structures, S.F.	COST PER S.F.		
		MAT.	INST.	TOTAL
0110	Swimming pool enclosure, transluscent, freestanding, economy	16.25	5.10	21.35
0120	Deluxe	250	14.65	264.65
0510	Tension structures, steel frame, polyester vinyl fabric, 12,000 S.F.	14.90	2.40	17.30
0520	20,800 S.F.	14.75	2.17	16.92

F1010 330	Special Structures, EACH	COST EACH		
		MAT.	INST.	TOTAL
0110	Kiosks, round, 5' diam., 1/4" fiberglass wall, 8' high	6,375		6,375
0120	Rectangular, 5' x 9', 1" insulated dbl. wall fiberglass, 7'-6" high	11,000		11,000
0200	Silos, concrete stave, industrial, 12' diam., 35' high	46,600	36,900	83,500
0210	25' diam., 75' high	175,000	83,000	258,000
0220	Steel prefab, 30,000 gal., painted, economy	23,100	5,350	28,450
0230	Epoxy-lined, deluxe	47,900	10,700	58,600

F1010 340	Special Structures, S.F.	COST PER S.F.		
		MAT.	INST.	TOTAL
0110	Comfort stations, prefab, mobile on steel frame, economy	51		51
0120	Permanent on concrete slab, deluxe	315	43	358
0210	Domes, bulk storage, wood framing, wood decking, 50' diam.	42.50	1.65	44.15
0220	116' diam.	25.50	1.91	27.41
0230	Steel framing, metal decking, 150' diam.	30.50	11.05	41.55
0240	400' diam.	25	8.45	33.45
0250	Geodesic, wood framing, wood panels, 30' diam.	25	1.86	26.86
0260	60' diam.	20.50	1.16	21.66
0270	Aluminum framing, acrylic panels, 40' diam.			116
0280	Aluminum panels, 400' diam.			26
0310	Garden house, prefab, wood, shell only, 48 S.F.	33	5.10	38.10
0320	200 S.F.	48	21.50	69.50
0410	Greenhouse, shell-stock, residential, lean-to, 8'-6" long x 3'-10" wide	37	30	67
0420	Freestanding, 8'-6" long x 13'-6" wide	25.50	9.50	35
0430	Commercial-truss frame, under 2000 S.F., deluxe			33.50
0440	Over 5,000 S.F., economy			23.50
0450	Institutional-rigid frame, under 500 S.F., deluxe			84
0460	Over 2,000 S.F., economy			32.50
0510	Hangar, prefab, galv. roof and walls, bottom rolling doors, economy	14.15	4.39	18.54
0520	Electric bifolding doors, deluxe	16.10	5.70	21.80

F10 Special Construction

F1020 Integrated Construction

F1020 110	Integrated Construction, EACH	COST EACH		
		MAT.	INST.	TOTAL
0110	Integrated ceilings, radiant electric, 2' x 4' panel, manila finish	99	23.50	122.50
0120	ABS plastic finish	112	45	157

F1020 120	Integrated Construction, S.F.	COST PER S.F.		
		MAT.	INST.	TOTAL
0110	Integrated ceilings, Luminaire, suspended, 5' x 5' modules, 50% lighted	4.32	12.25	16.57
0120	100% lighted	6.45	22	28.45
0130	Dimensionaire, 2' x 4' module tile system, no air bar	2.61	2.05	4.66
0140	With air bar, deluxe	3.83	2.05	5.88
0220	Pedestal access floors, stl pnls, no stringers, vinyl covering, >6000 S.F.	11.45	1.60	13.05
0230	With stringers, high pressure laminate covering, under 6000 S.F.	10.65	1.71	12.36
0240	Carpet covering, under 6000 S.F.	14.05	1.71	15.76
0250	Aluminum panels, no stringers, no covering	35	2.05	37.05

F1020 250	Special Purpose Room, EACH	COST EACH		
		MAT.	INST.	TOTAL
0110	Portable booth, acoustical, 27 db 1000 hz., 15 S.F. floor	3,975		3,975
0120	55 S.F. flr.	8,175		8,175

F1020 260	Special Purpose Room, S.F.	COST PER S.F.		
		MAT.	INST.	TOTAL
0110	Anechoic chambers, 7' high, 100 cps cutoff, 25 S.F.			2,550
0120	200 cps cutoff, 100 S.F.			1,325
0130	Audiometric rooms, under 500 S.F.	56.50	21	77.50
0140	Over 500 S.F.	53.50	17.10	70.60
0300	Darkrooms, shell, not including door, 240 S.F., 8' high	29.50	8.55	38.05
0310	64 S.F., 12' high	75	16	91
0510	Music practice room, modular, perforated steel, under 500 S.F.	34.50	14.65	49.15
0520	Over 500 S.F.	29.50	12.80	42.30

F1020 330	Special Construction, L.F.	COST PER L.F.		
		MAT.	INST.	TOTAL
0110	Spec. const., air curtains, shipping & receiving, 8'high x 5'wide, economy	355	119	474
0120	20' high x 8' wide, heated, deluxe	1,125	297	1,422
0130	Customer entrance, 10' high x 5' wide, economy	330	119	449
0140	12' high x 4' wide, heated, deluxe	690	149	839

F10 Special Construction

F1030 Special Construction Systems

F1030 120	Sound, Vibration, and Seismic Construction, S.F.	COST PER S.F.		
		MAT.	INST.	TOTAL
0020	Special construction, acoustical, enclosure, 4" thick, 8 psf panels	35.50	21.50	57
0030	Reverb chamber, 4" thick, parallel walls	50	25.50	75.50
0110	Sound absorbing panels, 2'-6" x 8', painted metal	13.30	7.15	20.45
0120	Vinyl faced	10.35	6.40	16.75
0130	Flexible transparent curtain, clear	8	8.30	16.30
0140	With absorbing foam, 75% coverage	11.20	8.30	19.50
0150	Strip entrance, 2/3 overlap	8.25	13.20	21.45
0160	Full overlap	10.35	15.50	25.85
0200	Audio masking system, plenum mounted, over 10,000 S.F.	.79	.27	1.06
0210	Ceiling mounted, under 5,000 S.F.	1.38	.49	1.87

F1030 210	Radiation Protection, EACH	COST EACH		
		MAT.	INST.	TOTAL
0110	Shielding, lead x-ray protection, radiography room, 1/16" lead, economy	8,900	3,475	12,375
0120	Deluxe	10,700	5,800	16,500
0210	Deep therapy x-ray, 1/4" lead, economy	24,800	10,900	35,700
0220	Deluxe	30,600	14,500	45,100

F1030 220	Radiation Protection, S.F.	COST PER S.F.		
		MAT.	INST.	TOTAL
0110	Shielding, lead, gypsum board, 5/8" thick, 1/16" lead	8	6.05	14.05
0120	1/8" lead	15.90	6.90	22.80
0130	Lath, 1/16" thick	7.90	6.45	14.35
0140	1/8" thick	15.90	7.25	23.15
0150	Radio frequency, copper, prefab screen type, economy	34.50	5.70	40.20
0160	Deluxe	39	7.05	46.05

F1030 910	Other Special Construction Systems, EACH	COST EACH		
		MAT.	INST.	TOTAL
0110	Disappearing stairways, folding, pine, 8'-6" ceiling	133	128	261
0200	Fire escape, galvanized steel, 8'-0" to 10'-6" ceiling	1,675	1,025	2,700
0220	Automatic electric, wood, 8' to 9' ceiling	9,075	1,025	10,100
0300	Fireplace prefabricated, freestanding or wall hung, economy	1,325	395	1,720
0310	Deluxe	3,700	570	4,270
0320	Woodburning stoves, cast iron, economy	1,300	790	2,090
0330	Deluxe	3,150	1,275	4,425

F10 Special Construction

F1040 Special Facilities

F1040 210	Ice Rinks, EACH	COAT EACH		
		MAT.	INST.	TOTAL
0100	Ice skating rink, 85' x 200', 55° system, 5 mos., 100 ton			632,500
0110	90° system, 12 mos., 135 ton			660,000
0120	Dash boards, acrylic screens, polyethylene coated plywood	137,500	34,600	172,100
0130	Fiberglass and aluminum construction	137,500	34,600	172,100

F1040 510	Liquid & Gas Storage Tanks, EACH	COST EACH		
		MAT.	INST.	TOTAL
0100	Tanks, steel, ground level, 100,000 gal.			222,500
0110	10,000,000 gal.			3,997,500
0120	Elevated water, 50,000 gal.		174,000	384,500
0130	1,000,000 gal.		812,500	1,795,000
0150	Cypress wood, ground level, 3,000 gal.	10,100	10,200	20,300
0160	Redwood, ground level, 45,000 gal.	76,000	27,800	103,800

F10 Special Construction

F1040 Special Facilities

F1040 910	Special Construction, EACH	COST EACH		
		MAT.	INST.	TOTAL
0110	Special construction, bowling alley incl. pinsetter, scorer etc., economy	45,400	10,200	55,600
0120	Deluxe	57,500	11,400	68,900
0130	For automatic scorer, economy, add	9,050		9,050
0140	Deluxe, add	10,900		10,900
0200	Chimney, metal, high temp. steel jacket, factory lining, 24" diameter	11,600	4,675	16,275
0210	60" diameter	84,000	20,300	104,300
0220	Poured concrete, brick lining, 10' diam., 200' high			1,600,000
0230	20' diam., 500' high			7,000,000
0240	Radial brick, 3'-6" I.D., 75' high			268,000
0250	7' I.D., 175' high			800,500
0300	Control tower, modular, 12' x 10', incl. instrumentation, economy			737,000
0310	Deluxe			1,028,500
0400	Garage costs, residential, prefab, wood, single car economy	5,600	1,025	6,625
0410	Two car deluxe	12,900	2,050	14,950
0500	Hangars, prefab, galv. steel, bottom rolling doors, economy (per plane)	14,200	4,875	19,075
0510	Electrical bi-folding doors, deluxe (per plane)	22,500	6,700	29,200

G1030 Site Earthwork

G1030 805	Trenching Common Earth	COST PER L.F.		
		MAT.	INST.	TOTAL
1310	Trenching, common earth, no slope, 2' wide, 2' deep, 3/8 C.Y. bucket		2.32	2.32
1330	4' deep, 3/8 C.Y. bucket		4.56	4.56
1360	10' deep, 1 C.Y. bucket		10.35	10.35
1400	4' wide, 2' deep, 3/8 C.Y. bucket		4.79	4.79
1420	4' deep, 1/2 C.Y. bucket		8.07	8.07
1450	10' deep, 1 C.Y. bucket		22.05	22.05
1480	18' deep, 2-1/2 C.Y. bucket		35.20	35.20
1520	6' wide, 6' deep, 5/8 C.Y. bucket w/trench box		21.80	21.80
1540	10' deep, 1 C.Y. bucket		29.10	29.10
1570	20' deep, 3-1/2 C.Y. bucket		49.50	49.50
1640	8' wide, 12' deep, 1-1/2 C.Y. bucket w/trench box		41.15	41.15
1680	24' deep, 3-1/2 C.Y. bucket		78.50	78.50
1730	10' wide, 20' deep, 3-1/2 C.Y. bucket w/trench box		69	69
1740	24' deep, 3-1/2 C.Y. bucket		91	91
3500	1 to 1 slope, 2' wide, 2' deep, 3/8 C.Y. bucket		4.56	4.56
3540	4' deep, 3/8 C.Y. bucket		13.56	13.56
3600	10' deep, 1 C.Y. bucket		62.50	62.50
3800	4' wide, 2' deep, 3/8 C.Y. bucket		7.04	7.04
3840	4' deep, 1/2 C.Y. bucket		15.86	15.86
3900	10' deep, 1 C.Y. bucket		77.50	77.50
4030	6' wide, 6' deep, 5/8 C.Y. bucket w/trench box		43.05	43.05
4050	10' deep, 1 C.Y. bucket		69.50	69.50
4080	20' deep, 3-1/2 C.Y. bucket		209.50	209.50
4500	8' wide, 12' deep, 1-1/2 C.Y. bucket w/trench box		104	104
4650	24' deep, 3-1/2 C.Y. bucket		317	317
4800	10' wide, 20' deep, 3-1/2 C.Y. bucket w/trench box		212.50	212.50
4850	24' deep, 3-1/2 C.Y. bucket		337	337

G1030 806	Trenching Loam & Sandy Clay	COST PER L.F.		
		MAT.	INST.	TOTAL
1310	Trenching, loam & sandy clay, no slope, 2' wide, 2' deep, 3/8 C.Y. bucket		2.26	2.26

G1030 807	Trenching Sand & Gravel	COST PER L.F.		
		MAT.	INST.	TOTAL
1310	Trenching, sand & gravel, no slope, 2' wide, 2' deep, 3/8 C.Y. bucket		2.13	2.13
1320	3' deep, 3/8 C.Y. bucket		3.54	3.54
1330	4' deep, 3/8 C.Y. bucket		4.23	4.23
1340	6' deep, 3/8 C.Y. bucket		5.35	5.35
1350	8' deep, 1/2 C.Y. bucket		7.05	7.05
1360	10' deep, 1 C.Y. bucket		7.65	7.65
1400	4' wide, 2' deep, 3/8 C.Y. bucket		4.41	4.41
1410	3' deep, 3/8 C.Y. bucket		6.50	6.50
1420	4' deep, 1/2 C.Y. bucket		7.45	7.45
1430	6' deep, 1/2 C.Y. bucket		11.50	11.50
1440	8' deep, 1/2 C.Y. bucket		16.65	16.65
1450	10' deep, 1 C.Y. bucket		16.60	16.60
1460	12' deep, 1 C.Y. bucket		21	21
1470	15' deep, 1-1/2 C.Y. bucket		24.50	24.50
1480	18' deep, 2-1/2 C.Y. bucket		27.50	27.50
1520	6' wide, 6' deep, 5/8 C.Y. bucket w/trench box		20	20
1530	8' deep, 3/4 C.Y. bucket		26.50	26.50
1540	10' deep, 1 C.Y. bucket		26	26
1550	12' deep, 1-1/2 C.Y. bucket		29	29
1560	16' deep, 2 C.Y. bucket		37.50	37.50
1570	20' deep, 3-1/2 C.Y. bucket		45	45
1580	24' deep, 3-1/2 C.Y. bucket		55.50	55.50

G1030 Site Earthwork

G1030 807	Trenching Sand & Gravel	COST PER L.F.		
		MAT.	INST.	TOTAL
1640	8' wide, 12' deep, 1-1/2 C.Y. bucket w/trench box		38.50	38.50
1650	15' deep, 1-1/2 C.Y. bucket		48	48
1660	18' deep, 2-1/2 C.Y. bucket		55	55
1680	24' deep, 3-1/2 C.Y. bucket		73	73
1730	10' wide, 20' deep, 3-1/2 C.Y. bucket w/trench box		73	73
1740	24' deep, 3-1/2 C.Y. bucket		91.50	91.50
1780	12' wide, 20' deep, 3-1/2 C.Y. bucket w/trench box		87.50	87.50
1790	25' deep, bucket		114	114
1800	1/2:1 slope, 2' wide, 2' deep, 3/8 CY bk		3.18	3.18
1810	3' deep, 3/8 C.Y. bucket		5.55	5.55
1820	4' deep, 3/8 C.Y. bucket		8.40	8.40
1840	6' deep, 3/8 C.Y. bucket		13.25	13.25
1860	8' deep, 1/2 C.Y. bucket		21	21
1880	10' deep, 1 C.Y. bucket		27	27
2300	4' wide, 2' deep, 3/8 C.Y. bucket		5.45	5.45
2310	3' deep, 3/8 C.Y. bucket		8.85	8.85
2320	4' deep, 1/2 C.Y. bucket		11.10	11.10
2340	6' deep, 1/2 C.Y. bucket		20	20
2360	8' deep, 1/2 C.Y. bucket		33.50	33.50
2380	10' deep, 1 C.Y. bucket		37.50	37.50
2400	12' deep, 1 C.Y. bucket		63	63
2430	15' deep, 1-1/2 C.Y. bucket		71	71
2460	18' deep, 2-1/2 C.Y. bucket		110	110
2840	6' wide, 6' deep, 5/8 C.Y. bucket w/trench box		30	30
2860	8' deep, 3/4 C.Y. bucket		44	44
2880	10' deep, 1 C.Y. bucket		48	48
2900	12' deep, 1-1/2 C.Y. bucket		58.50	58.50
2940	16' deep, 2 C.Y. bucket		88	88
2980	20' deep, 3-1/2 C.Y. bucket		121	121
3020	24' deep, 3-1/2 C.Y. bucket		168	168
3100	8' wide, 12' deep, 1-1/4 C.Y. bucket w/trench box		68	68
3120	15' deep, 1-1/2 C.Y. bucket		94	94
3140	18' deep, 2-1/2 C.Y. bucket		119	119
3180	24' deep, 3-1/2 C.Y. bucket		186	186
3270	10' wide, 20' deep, 3-1/2 C.Y. bucket w/trench box		149	149
3280	24' deep, 3-1/2 C.Y. bucket		204	204
3370	12' wide, 20' deep, 3-1/2 C.Y. bucket w/trench box		166	166
3380	25' deep, bucket		236	236
3500	1:1 slope, 2' wide, 2' deep, 3/8 CY bk		5.75	5.75
3520	3' deep, 3/8 C.Y. bucket		7.90	7.90
3540	4' deep, 3/8 C.Y. bucket		12.55	12.55
3560	6' deep, 3/8 C.Y. bucket		13.25	13.25
3580	8' deep, 1/2 C.Y. bucket		35	35
3600	10' deep, 1 C.Y. bucket		46	46
3800	4' wide, 2' deep, 3/8 C.Y. bucket		6.50	6.50
3820	3' deep, 3/8 C.Y. bucket		11.15	11.15
3840	4' deep, 1/2 C.Y. bucket		14.80	14.80
3860	6' deep, 1/2 C.Y. bucket		28.50	28.50
3880	8' deep, 1/2 C.Y. bucket		50	50
3900	10' deep, 1 C.Y. bucket		58.50	58.50
3920	12' deep, 1 C.Y. bucket		84	84
3940	15' deep, 1-1/2 C.Y. bucket		117	117
3960	18' deep, 2-1/2 C.Y. bucket		151	151
4030	6' wide, 6' deep, 5/8 C.Y. bucket w/trench box		40	40
4040	8' deep, 3/4 C.Y. bucket		62	62
4050	10' deep, 1 C.Y. bucket		69.50	69.50
4060	12' deep, 1-1/2 C.Y. bucket		88.50	88.50
4070	16' deep, 2 C.Y. bucket		139	139

G1030 Site Earthwork

G1030 807	Trenching Sand & Gravel	COST PER L.F.		
		MAT.	INST.	TOTAL
4080	20' deep, 3-1/2 C.Y. bucket		197	197
4090	24' deep, 3-1/2 C.Y. bucket		281	281
4500	8' wide, 12' deep, 1-1/2 C.Y. bucket w/trench box		98	98
4550	15' deep, 1-1/2 C.Y. bucket		140	140
4600	18' deep, 2-1/2 C.Y. bucket		183	183
4650	24' deep, 3-1/2 C.Y. bucket		298	298
4800	10' wide, 20' deep, 3-1/2 C.Y. bucket w/trench box		225	225
4850	24' deep, 3-1/2 C.Y. bucket		315	315
4950	12' wide, 20' deep, 3-1/2 C.Y. bucket w/trench box		239	239
4980	25' deep, bucket		360	360

G1030 815	Pipe Bedding	COST PER L.F.		
		MAT.	INST.	TOTAL
1440	Pipe bedding, side slope 0 to 1, 1' wide, pipe size 6" diameter	.78	.63	1.41
1460	2' wide, pipe size 8" diameter	1.72	1.40	3.12
1500	Pipe size 12" diameter	1.81	1.47	3.28
1600	4' wide, pipe size 20" diameter	4.50	3.65	8.15
1660	Pipe size 30" diameter	4.77	3.87	8.64
1680	6' wide, pipe size 32" diameter	8.35	6.75	15.10
1740	8' wide, pipe size 60" diameter	13.90	11.30	25.20
1780	12' wide, pipe size 84" diameter	27.50	22	49.50

G20 Site Improvements

G2010 Roadways

G2010 210	Roadway Pavement	COST PER L.F.		
		MAT.	INST.	TOTAL
1500	Roadways, bituminous conc. paving, 2-1/2" thick, 20' wide	62.50	57.50	120
1580	30' wide	80.50	65	145.50
1800	3" thick, 20' wide	67	56	123
1880	30' wide	87	65.50	152.50
2100	4" thick, 20' wide	74.50	56	130.50
2180	30' wide	98	65.50	163.50
2400	5" thick, 20' wide	91	56.50	147.50
2480	30' wide	123	66.50	189.50
3000	6" thick, 20' wide	89.50	58	147.50
3080	30' wide	121	69	190
3300	8" thick 20' wide	116	59	175
3380	30' wide	161	70	231
3600	12" thick 20' wide	146	62.50	208.50
3700	32' wide	218	78.50	296.50

G2020 210	Parking Lots	COST PER CAR		
		MAT.	INST.	TOTAL
1500	Parking lot, 90° angle parking, 3" bituminous paving, 6" gravel base	860	380	1,240
1540	10" gravel base	975	440	1,415
1560	4" bituminous paving, 6" gravel base	1,000	375	1,375
1600	10" gravel base	1,125	440	1,565
1620	6" bituminous paving, 6" gravel base	1,325	395	1,720
1660	10" gravel base	1,450	455	1,905
1800	60° angle parking, 3" bituminous paving, 6" gravel base	860	380	1,240
1840	10" gravel base	975	440	1,415
1860	4" bituminous paving, 6" gravel base	1,000	375	1,375
1900	10" gravel base	1,125	440	1,565
1920	6" bituminous paving, 6" gravel base	1,325	395	1,720
1960	10" gravel base	1,450	455	1,905
2200	45° angle parking, 3" bituminous paving, 6" gravel base	880	385	1,265
2240	10" gravel base	1,000	455	1,455
2260	4" bituminous paving, 6" gravel base	1,025	385	1,410
2300	10" gravel base	1,150	450	1,600
2320	6" bituminous paving, 6" gravel base	1,350	405	1,755
2360	10" gravel base	1,475	470	1,945

G2040 810	Flagpoles	COST EACH		
		MAT.	INST.	TOTAL
0110	Flagpoles, on grade, aluminum, tapered, 20' high	1,075	620	1,695
0120	70' high	8,750	1,525	10,275
0130	Fiberglass, tapered, 23' high	1,200	620	1,820
0140	59' high	5,275	1,350	6,625
0150	Concrete, internal halyard, 20' high	1,725	490	2,215
0160	100' high	17,100	1,225	18,325

G2040 950	Other Site Development, EACH	COST EACH		
		MAT.	INST.	TOTAL
0110	Grandstands, permanent, closed deck, steel, economy (per seat)			60.50
0120	Deluxe (per seat)			121
0130	Composite design, economy (per seat)			143
0140	Deluxe (per seat)			286

G3030 Storm Sewer

G3030 210	Manholes & Catch Basins	COST PER EACH		
		MAT.	INST.	TOTAL
1920	Manhole/catch basin, brick, 4' I.D. riser, 4' deep	1,075	1,600	2,675
1980	10' deep	2,425	3,775	6,200
3200	Block, 4' I.D. riser, 4' deep	995	1,300	2,295
3260	10' deep	2,000	3,125	5,125
4620	Concrete, cast-in-place, 4' I.D. riser, 4' deep	1,125	2,225	3,350
4680	10' deep	2,500	5,300	7,800
5820	Concrete, precast, 4' I.D. riser, 4' deep	1,100	1,175	2,275
5880	10' deep	2,350	2,725	5,075
6200	6' I.D. riser, 4' deep	2,775	1,750	4,525
6260	10' deep	5,275	4,100	9,375

G4020 Site Lighting

Light Pole

G4020 210	Light Pole (Installed)	COST EACH		
		MAT.	INST.	TOTAL
0200	Light pole, aluminum, 20' high, 1 arm bracket	1,175	1,050	2,225
0240	2 arm brackets	1,300	1,050	2,350
0280	3 arm brackets	1,425	1,100	2,525
0320	4 arm brackets	1,550	1,100	2,650
0360	30' high, 1 arm bracket	2,100	1,325	3,425
0400	2 arm brackets	2,225	1,325	3,550
0440	3 arm brackets	2,350	1,375	3,725
0480	4 arm brackets	2,475	1,375	3,850
0680	40' high, 1 arm bracket	2,575	1,775	4,350
0720	2 arm brackets	2,700	1,775	4,475
0760	3 arm brackets	2,825	1,825	4,650
0800	4 arm brackets	2,950	1,825	4,775
0840	Steel, 20' high, 1 arm bracket	1,425	1,125	2,550
0880	2 arm brackets	1,525	1,125	2,650
0920	3 arm brackets	1,550	1,150	2,700
0960	4 arm brackets	1,675	1,150	2,825
1000	30' high, 1 arm bracket	1,625	1,425	3,050
1040	2 arm brackets	1,725	1,425	3,150
1080	3 arm brackets	1,750	1,450	3,200
1120	4 arm brackets	1,875	1,450	3,325
1320	40' high, 1 arm bracket	2,125	1,925	4,050
1360	2 arm brackets	2,225	1,925	4,150
1400	3 arm brackets	2,250	1,975	4,225
1440	4 arm brackets	2,375	1,975	4,350

Reference Section

All the reference information is in one section making it easy to find what you need to know . . . and easy to use the book on a daily basis. This section is visually identified by a vertical gray bar on the edge of pages.

In this Reference Section, we've included General Conditions; Location Factors for adjusting costs to the region you are in; Historical Cost Indexes for cost comparisons over time; a Glossary; an explanation of all Abbreviations used in the book; and sample Estimating Forms.

Table of Contents

General Conditions, Overhead & Profit

The total building costs in the Commercial/Industrial/Institutional section include a 10% allowance for general conditions and a 15% allowance for the general contractor's overhead and profit and contingencies.

General contractor overhead includes indirect costs such as permits, Workers' Compensation, insurances, supervision and bonding fees. Overhead will vary with the size of project, the contractor's operating procedures and location. Profits will vary with economic activity and local conditions.

Contingencies provide for unforeseen construction difficulties which include material shortages and weather. In all situations, the appraiser should give consideration to possible adjustment of the 25% factor used in developing the Commercial/Industrial/Institutional models.

Architectural Fees

Tabulated below are typical percentage fees by project size, for good professional architectural service. Fees may vary from those listed depending upon degree of design difficulty and economic conditions in any particular area.

Rates can be interpolated horizontally and vertically. Various portions of the same project requiring different rates should be adjusted proportionately. For alterations, add 50% to the fee for the first $500,000 of project cost and add 25% to the fee for project cost over $500,000.

Architectural fees tabulated below include Engineering fees.

Insurance Exclusions

Many insurance companies exclude from coverage such items as architect's fees, excavation, foundations below grade, underground piping and site preparation. Since exclusions vary among insurance companies, it is recommended that for greatest accuracy each exclusion be priced separately using the unit-in-place section.

As a rule of thumb, exclusions can be calculated at 9% of total building cost plus the appropriate allowance for architect's fees.

Building Types	Total Project Size in Thousands of Dollars						
	100	250	500	1,000	5,000	10,000	50,000
Factories, garages, warehouses, repetitive housing	9.0%	8.0%	7.0%	6.2%	5.3%	4.9%	4.5%
Apartments, banks, schools, libraries, offices, municipal buildings	12.2	12.3	9.2	8.0	7.0	6.6	6.2
Churches, hospitals, homes, laboratories, museums, research	15.0	13.6	12.7	11.9	9.5	8.8	8.0
Memorials, monumental work, decorative furnishings	—	16.0	14.5	13.1	10.0	9.0	8.3

Location Factors

Costs shown in *RSMeans Square Foot Costs* are based on national averages for materials and installation. To adjust these costs to a specific location, simply multiply the base cost by the factor for that city.

The data is arranged alphabetically by state and postal zip code numbers. For a city not listed, use the factor for a nearby city with similar economic characteristics.

STATE/ZIP	CITY	Residential	Commercial
ALABAMA			
350-352	Birmingham	.86	.88
354	Tuscaloosa	.77	.80
355	Jasper	.72	.80
356	Decatur	.77	.81
357-358	Huntsville	.83	.85
359	Gadsden	.74	.80
360-361	Montgomery	.75	.80
362	Anniston	.77	.83
363	Dothan	.75	.77
364	Evergreen	.73	.78
365-366	Mobile	.81	.84
367	Selma	.72	.77
368	Phenix City	.73	.79
369	Butler	.73	.77
ALASKA			
995-996	Anchorage	1.26	1.21
997	Fairbanks	1.27	1.21
998	Juneau	1.25	1.20
999	Ketchikan	1.29	1.27
ARIZONA			
850,853	Phoenix	.86	.89
851,852	Mesa/Tempe	.81	.84
855	Globe	.77	.82
856-857	Tucson	.84	.86
859	Show Low	.79	.83
860	Flagstaff	.84	.88
863	Prescott	.77	.82
864	Kingman	.81	.84
865	Chambers	.77	.82
ARKANSAS			
716	Pine Bluff	.79	.82
717	Camden	.67	.71
718	Texarkana	.72	.74
719	Hot Springs	.68	.72
720-722	Little Rock	.83	.84
723	West Memphis	.77	.79
724	Jonesboro	.76	.80
725	Batesville	.73	.75
726	Harrison	.75	.76
727	Fayetteville	.70	.75
728	Russellville	.75	.76
729	Fort Smith	.76	.80
CALIFORNIA			
900-902	Los Angeles	1.08	1.08
903-905	Inglewood	1.01	1.00
906-908	Long Beach	1.00	1.01
910-912	Pasadena	1.00	1.01
913-916	Van Nuys	1.03	1.03
917-918	Alhambra	1.04	1.02
919-921	San Diego	1.03	1.05
922	Palm Springs	1.00	1.00
923-924	San Bernardino	1.01	.99
925	Riverside	1.07	1.05
926-927	Santa Ana	1.02	1.01
928	Anaheim	1.07	1.06
930	Oxnard	1.08	1.06
931	Santa Barbara	1.07	1.06
932-933	Bakersfield	1.06	1.05
934	San Luis Obispo	1.03	1.02
935	Mojave	1.01	1.00
936-938	Fresno	1.09	1.06
939	Salinas	1.11	1.08
940-941	San Francisco	1.26	1.24
942,956-958	Sacramento	1.12	1.08
943	Palo Alto	1.14	1.10
944	San Mateo	1.22	1.15
945	Vallejo	1.15	1.11
946	Oakland	1.22	1.16
947	Berkeley	1.20	1.13
948	Richmond	1.23	1.13
949	San Rafael	1.21	1.14
950	Santa Cruz	1.14	1.11
951	San Jose	1.21	1.17
952	Stockton	1.08	1.06
953	Modesto	1.08	1.06

STATE/ZIP	CITY	Residential	Commercial
CALIFORNIA (CONT'D)			
954	Santa Rosa	1.16	1.12
955	Eureka	1.10	1.06
959	Marysville	1.09	1.06
960	Redding	1.09	1.06
961	Susanville	1.09	1.06
COLORADO			
800-802	Denver	.92	.94
803	Boulder	.89	.90
804	Golden	.87	.91
805	Fort Collins	.86	.90
806	Greeley	.76	.84
807	Fort Morgan	.89	.90
808-809	Colorado Springs	.87	.91
810	Pueblo	.89	.91
811	Alamosa	.85	.89
812	Salida	.87	.90
813	Durango	.87	.90
814	Montrose	.84	.88
815	Grand Junction	.88	.90
816	Glenwood Springs	.87	.90
CONNECTICUT			
060	New Britain	1.09	1.08
061	Hartford	1.09	1.09
062	Willimantic	1.10	1.08
063	New London	1.09	1.06
064	Meriden	1.09	1.07
065	New Haven	1.10	1.09
066	Bridgeport	1.10	1.09
067	Waterbury	1.10	1.09
068	Norwalk	1.10	1.09
069	Stamford	1.11	1.12
D.C.			
200-205	Washington	.96	.99
DELAWARE			
197	Newark	1.03	1.04
198	Wilmington	1.04	1.04
199	Dover	1.02	1.04
FLORIDA			
320,322	Jacksonville	.81	.84
321	Daytona Beach	.88	.88
323	Tallahassee	.78	.81
324	Panama City	.75	.79
325	Pensacola	.82	.86
326,344	Gainesville	.81	.86
327-328,347	Orlando	.89	.89
329	Melbourne	.88	.91
330-332,340	Miami	.88	.91
333	Fort Lauderdale	.87	.89
334,349	West Palm Beach	.87	.86
335-336,346	Tampa	.92	.92
337	St. Petersburg	.79	.84
338	Lakeland	.89	.91
339,341	Fort Myers	.87	.87
342	Sarasota	.90	.89
GEORGIA			
300-303,399	Atlanta	.88	.89
304	Statesboro	.71	.78
305	Gainesville	.77	.81
306	Athens	.76	.82
307	Dalton	.75	.81
308-309	Augusta	.82	.83
310-312	Macon	.80	.83
313-314	Savannah	.80	.82
315	Waycross	.79	.81
316	Valdosta	.74	.81
317,398	Albany	.77	.82
318-319	Columbus	.81	.83
HAWAII			
967	Hilo	1.19	1.15
968	Honolulu	1.21	1.17

Location Factors

STATE/ZIP	CITY	Residential	Commercial
STATES & POSS.			
969	Guam	.94	1.03
IDAHO			
832	Pocatello	.84	.88
833	Twin Falls	.71	.80
834	Idaho Falls	.73	.81
835	Lewiston	.94	.95
836-837	Boise	.84	.88
838	Coeur d'Alene	.92	.94
ILLINOIS			
600-603	North Suburban	1.20	1.14
604	Joliet	1.22	1.15
605	South Suburban	1.20	1.14
606-608	Chicago	1.22	1.16
609	Kankakee	1.12	1.08
610-611	Rockford	1.11	1.09
612	Rock Island	.98	.97
613	La Salle	1.09	1.06
614	Galesburg	1.03	1.00
615-616	Peoria	1.05	1.03
617	Bloomington	1.03	1.01
618-619	Champaign	1.04	1.03
620-622	East St. Louis	1.02	1.00
623	Quincy	1.02	.97
624	Effingham	1.03	.99
625	Decatur	1.03	1.00
626-627	Springfield	1.02	1.01
628	Centralia	1.02	.98
629	Carbondale	.99	.96
INDIANA			
460	Anderson	.90	.90
461-462	Indianapolis	.92	.92
463-464	Gary	1.04	1.02
465-466	South Bend	.90	.91
467-468	Fort Wayne	.89	.88
469	Kokomo	.90	.88
470	Lawrenceburg	.87	.87
471	New Albany	.86	.85
472	Columbus	.92	.89
473	Muncie	.91	.90
474	Bloomington	.92	.90
475	Washington	.90	.89
476-477	Evansville	.90	.91
478	Terre Haute	.91	.92
479	Lafayette	.92	.90
IOWA			
500-503,509	Des Moines	.86	.88
504	Mason City	.74	.80
505	Fort Dodge	.74	.78
506-507	Waterloo	.76	.80
508	Creston	.78	.80
510-511	Sioux City	.82	.85
512	Sibley	.70	.75
513	Spencer	.72	.75
514	Carroll	.72	.75
515	Council Bluffs	.80	.88
516	Shenandoah	.72	.75
520	Dubuque	.83	.87
521	Decorah	.73	.75
522-524	Cedar Rapids	.90	.90
525	Ottumwa	.81	.84
526	Burlington	.83	.84
527-528	Davenport	.93	.93
KANSAS			
660-662	Kansas City	.98	.97
664-666	Topeka	.77	.83
667	Fort Scott	.85	.84
668	Emporia	.73	.79
669	Belleville	.77	.81
670-672	Wichita	.77	.82
673	Independence	.82	.82
674	Salina	.75	.82
675	Hutchinson	.77	.78
676	Hays	.79	.81
677	Colby	.81	.82
678	Dodge City	.79	.83
679	Liberal	.78	.81
KENTUCKY			
400-402	Louisville	.90	.91
403-405	Lexington	.87	.87

STATE/ZIP	CITY	Residential	Commercial
KENTUCKY (CONT'D)			
406	Frankfort	.83	.87
407-409	Corbin	.74	.80
410	Covington	.96	.95
411-412	Ashland	.93	.95
413-414	Campton	.75	.80
415-416	Pikeville	.82	.87
417-418	Hazard	.70	.77
420	Paducah	.88	.89
421-422	Bowling Green	.88	.89
423	Owensboro	.85	.89
424	Henderson	.90	.90
425-426	Somerset	.74	.81
427	Elizabethtown	.86	.87
LOUISIANA			
700-701	New Orleans	.85	.88
703	Thibodaux	.81	.83
704	Hammond	.76	.80
705	Lafayette	.79	.81
706	Lake Charles	.81	.83
707-708	Baton Rouge	.83	.85
710-711	Shreveport	.77	.80
712	Monroe	.72	.78
713-714	Alexandria	.73	.78
MAINE			
039	Kittery	.91	.89
040-041	Portland	.90	.90
042	Lewiston	.90	.89
043	Augusta	.93	.91
044	Bangor	.89	.89
045	Bath	.91	.89
046	Machias	.91	.89
047	Houlton	.92	.89
048	Rockland	.92	.88
049	Waterville	.91	.89
MARYLAND			
206	Waldorf	.83	.86
207-208	College Park	.90	.93
209	Silver Spring	.85	.89
210-212	Baltimore	.90	.92
214	Annapolis	.83	.90
215	Cumberland	.87	.88
216	Easton	.66	.72
217	Hagerstown	.86	.89
218	Salisbury	.72	.75
219	Elkton	.85	.81
MASSACHUSETTS			
010-011	Springfield	1.05	1.03
012	Pittsfield	1.05	1.02
013	Greenfield	1.03	1.01
014	Fitchburg	1.11	1.07
015-016	Worcester	1.13	1.10
017	Framingham	1.15	1.10
018	Lowell	1.15	1.12
019	Lawrence	1.15	1.12
020-022, 024	Boston	1.21	1.17
023	Brockton	1.15	1.11
025	Buzzards Bay	1.12	1.06
026	Hyannis	1.12	1.08
027	New Bedford	1.14	1.09
MICHIGAN			
480,483	Royal Oak	1.01	.98
481	Ann Arbor	1.02	1.00
482	Detroit	1.06	1.03
484-485	Flint	.97	.97
486	Saginaw	.94	.95
487	Bay City	.94	.94
488-489	Lansing	.96	.97
490	Battle Creek	.94	.93
491	Kalamazoo	.93	.92
492	Jackson	.93	.94
493,495	Grand Rapids	.84	.87
494	Muskegon	.90	.90
496	Traverse City	.84	.87
497	Gaylord	.81	.85
498-499	Iron mountain	.91	.93
MINNESOTA			
550-551	Saint Paul	1.14	1.10
553-555	Minneapolis	1.16	1.12
556-558	Duluth	1.10	1.06

STATE/ZIP	CITY	Residential	Commercial
MINNESOTA (CONT'd)			
559	Rochester	1.06	1.03
560	Mankato	1.02	1.00
561	Windom	.82	.87
562	Willmar	.84	.90
563	St. Cloud	1.10	1.09
564	Brainerd	1.00	1.01
565	Detroit Lakes	.95	.98
566	Bemidji	.97	.99
567	Thief River Falls	.96	.97
MISSISSIPPI			
386	Clarksdale	.76	.79
387	Greenville	.83	.86
388	Tupelo	.77	.81
389	Greenwood	.78	.80
390-392	Jackson	.83	.84
393	Meridian	.82	.84
394	Laurel	.78	.81
395	Biloxi	.80	.82
396	Mccomb	.76	.78
397	Columbus	.77	.80
MISSOURI			
630-631	St. Louis	1.02	1.02
633	Bowling Green	.93	.92
634	Hannibal	.84	.87
635	Kirksville	.79	.87
636	Flat River	.92	.93
637	Cape Girardeau	.86	.92
638	Sikeston	.81	.86
639	Poplar Bluff	.81	.86
640-641	Kansas City	1.03	1.03
644-645	St. Joseph	.91	.93
646	Chillicothe	.84	.82
647	Harrisonville	.94	.95
648	Joplin	.81	.83
650-651	Jefferson City	.86	.91
652	Columbia	.89	.93
653	Sedalia	.83	.88
654-655	Rolla	.88	.86
656-658	Springfield	.86	.89
MONTANA			
590-591	Billings	.87	.89
592	Wolf Point	.82	.87
593	Miles City	.85	.86
594	Great Falls	.87	.90
595	Havre	.82	.88
596	Helena	.86	.88
597	Butte	.86	.89
598	Missoula	.85	.88
599	Kalispell	.83	.86
NEBRASKA			
680-681	Omaha	.89	.89
683-685	Lincoln	.85	.87
686	Columbus	.85	.86
687	Norfolk	.90	.89
688	Grand Island	.90	.90
689	Hastings	.92	.90
690	Mccook	.83	.86
691	North Platte	.90	.90
692	Valentine	.84	.86
693	Alliance	.83	.86
NEVADA			
889-891	Las Vegas	1.05	1.07
893	Ely	.84	.86
894-895	Reno	.92	.96
897	Carson City	.92	.95
898	Elko	.88	.88
NEW HAMPSHIRE			
030	Nashua	.94	.94
031	Manchester	.94	.95
032-033	Concord	.92	.92
034	Keene	.73	.76
035	Littleton	.79	.79
036	Charleston	.72	.75
037	Claremont	.73	.75
038	Portsmouth	.94	.94

STATE/ZIP	CITY	Residential	Commercial
NEW JERSEY			
070-071	Newark	1.12	1.12
072	Elizabeth	1.14	1.11
073	Jersey City	1.11	1.10
074-075	Paterson	1.12	1.11
076	Hackensack	1.11	1.10
077	Long Branch	1.11	1.09
078	Dover	1.11	1.10
079	Summit	1.12	1.10
080,083	Vineland	1.09	1.08
081	Camden	1.10	1.10
082,084	Atlantic City	1.12	1.09
085-086	Trenton	1.10	1.10
087	Point Pleasant	1.09	1.09
088-089	New Brunswick	1.12	1.10
NEW MEXICO			
870-872	Albuquerque	.84	.89
873	Gallup	.84	.89
874	Farmington	.84	.89
875	Santa Fe	.85	.90
877	Las Vegas	.84	.88
878	Socorro	.84	.88
879	Truth/Consequences	.83	.85
880	Las Cruces	.82	.84
881	Clovis	.84	.87
882	Roswell	.84	.88
883	Carrizozo	.84	.89
884	Tucumcari	.85	.88
NEW YORK			
100-102	New York	1.39	1.33
103	Staten Island	1.32	1.29
104	Bronx	1.33	1.28
105	Mount Vernon	1.16	1.16
106	White Plains	1.19	1.16
107	Yonkers	1.20	1.19
108	New Rochelle	1.20	1.16
109	Suffern	1.16	1.12
110	Queens	1.32	1.29
111	Long Island City	1.34	1.30
112	Brooklyn	1.36	1.30
113	Flushing	1.34	1.29
114	Jamaica	1.33	1.28
115,117,118	Hicksville	1.21	1.21
116	Far Rockaway	1.33	1.30
119	Riverhead	1.22	1.22
120-122	Albany	.97	.98
123	Schenectady	.98	.98
124	Kingston	1.06	1.09
125-126	Poughkeepsie	1.20	1.15
127	Monticello	1.08	1.10
128	Glens Falls	.88	.92
129	Plattsburgh	.94	.93
130-132	Syracuse	.96	.97
133-135	Utica	.93	.94
136	Watertown	.93	.96
137-139	Binghamton	.94	.95
140-142	Buffalo	1.04	1.01
143	Niagara Falls	1.01	.99
144-146	Rochester	.98	.99
147	Jamestown	.90	.92
148-149	Elmira	.88	.93
NORTH CAROLINA			
270,272-274	Greensboro	.81	.77
271	Winston-Salem	.81	.77
275-276	Raleigh	.83	.78
277	Durham	.82	.78
278	Rocky Mount	.72	.72
279	Elizabeth City	.73	.73
280	Gastonia	.82	.76
281-282	Charlotte	.83	.78
283	Fayetteville	.81	.78
284	Wilmington	.79	.75
285	Kinston	.72	.71
286	Hickory	.77	.72
287-288	Asheville	.80	.75
289	Murphy	.71	.69
NORTH DAKOTA			
580-581	Fargo	.76	.83
582	Grand Forks	.73	.80
583	Devils Lake	.76	.80
584	Jamestown	.71	.77
585	Bismarck	.76	.83

455

STATE/ZIP	CITY	Residential	Commercial
NORTH DAKOTA (CONT'D)			
586	Dickinson	.74	.82
587	Minot	.79	.85
588	Williston	.74	.81
OHIO			
430-432	Columbus	.94	.94
433	Marion	.88	.88
434-436	Toledo	.99	.97
437-438	Zanesville	.87	.88
439	Steubenville	.93	.93
440	Lorain	.96	.95
441	Cleveland	1.00	.99
442-443	Akron	.96	.96
444-445	Youngstown	.94	.94
446-447	Canton	.92	.92
448-449	Mansfield	.92	.92
450	Hamilton	.91	.90
451-452	Cincinnati	.91	.91
453-454	Dayton	.90	.90
455	Springfield	.92	.90
456	Chillicothe	.94	.91
457	Athens	.88	.88
458	Lima	.89	.90
OKLAHOMA			
730-731	Oklahoma City	.78	.82
734	Ardmore	.77	.80
735	Lawton	.80	.82
736	Clinton	.76	.80
737	Enid	.76	.81
738	Woodward	.75	.80
739	Guymon	.65	.67
740-741	Tulsa	.76	.79
743	Miami	.80	.80
744	Muskogee	.70	.72
745	Mcalester	.73	.75
746	Ponca City	.76	.79
747	Durant	.76	.79
748	Shawnee	.75	.79
749	Poteau	.77	.79
OREGON			
970-972	Portland	.98	.99
973	Salem	.96	.98
974	Eugene	.97	.98
975	Medford	.96	.98
976	Klamath Falls	.96	.98
977	Bend	.98	.98
978	Pendleton	.96	.95
979	Vale	.94	.90
PENNSYLVANIA			
150-152	Pittsburgh	.98	1.00
153	Washington	.95	.98
154	Uniontown	.92	.97
155	Bedford	.88	.94
156	Greensburg	.95	.98
157	Indiana	.92	.97
158	Dubois	.90	.96
159	Johnstown	.90	.95
160	Butler	.94	.96
161	New Castle	.93	.95
162	Kittanning	.95	.97
163	Oil City	.90	.94
164-165	Erie	.93	.94
166	Altoona	.88	.93
167	Bradford	.90	.95
168	State College	.91	.93
169	Wellsboro	.92	.95
170-171	Harrisburg	.95	.97
172	Chambersburg	.89	.93
173-174	York	.93	.96
175-176	Lancaster	.92	.92
177	Williamsport	.85	.88
178	Sunbury	.93	.95
179	Pottsville	.93	.96
180	Lehigh Valley	1.04	1.04
181	Allentown	1.05	1.04
182	Hazleton	.93	.96
183	Stroudsburg	.93	1.00
184-185	Scranton	.98	1.00
186-187	Wilkes-Barre	.95	.96
188	Montrose	.92	.96
189	Doylestown	1.08	1.07

STATE/ZIP	CITY	Residential	Commercial
PENNSYLVANIA (CONT'D)			
190-191	Philadelphia	1.17	1.15
193	Westchester	1.13	1.10
194	Norristown	1.12	1.12
195-196	Reading	.98	1.00
PUERTO RICO			
009	San Juan	.73	.79
RHODE ISLAND			
028	Newport	1.08	1.05
029	Providence	1.08	1.06
SOUTH CAROLINA			
290-292	Columbia	.82	.78
293	Spartanburg	.82	.76
294	Charleston	.85	.81
295	Florence	.78	.76
296	Greenville	.81	.76
297	Rock Hill	.80	.75
298	Aiken	.99	.86
299	Beaufort	.80	.74
SOUTH DAKOTA			
570-571	Sioux Falls	.77	.81
572	Watertown	.74	.78
573	Mitchell	.75	.77
574	Aberdeen	.76	.80
575	Pierre	.75	.79
576	Mobridge	.74	.78
577	Rapid City	.76	.80
TENNESSEE			
370-372	Nashville	.84	.87
373-374	Chattanooga	.81	.85
375,380-381	Memphis	.81	.86
376	Johnson City	.70	.79
377-379	Knoxville	.74	.80
382	Mckenzie	.71	.78
383	Jackson	.73	.80
384	Columbia	.72	.81
385	Cookeville	.70	.81
TEXAS			
750	Mckinney	.74	.78
751	Waxahachie	.76	.80
752-753	Dallas	.83	.85
754	Greenville	.70	.73
755	Texarkana	.74	.78
756	Longview	.69	.73
757	Tyler	.75	.80
758	Palestine	.68	.71
759	Lufkin	.72	.74
760-761	Fort Worth	.82	.83
762	Denton	.75	.76
763	Wichita Falls	.78	.80
764	Eastland	.72	.72
765	Temple	.75	.76
766-767	Waco	.78	.80
768	Brownwood	.68	.72
769	San Angelo	.72	.76
770-772	Houston	.85	.87
773	Huntsville	.68	.72
774	Wharton	.70	.75
775	Galveston	.83	.85
776-777	Beaumont	.82	.82
778	Bryan	.74	.81
779	Victoria	.73	.76
780	Laredo	.72	.76
781-782	San Antonio	.78	.81
783-784	Corpus Christi	.77	.78
785	Mc Allen	.75	.76
786-787	Austin	.78	.80
788	Del Rio	.66	.69
789	Giddings	.69	.71
790-791	Amarillo	.79	.82
792	Childress	.77	.78
793-794	Lubbock	.76	.80
795-796	Abilene	.75	.78
797	Midland	.76	.78
798-799,885	El Paso	.75	.78
UTAH			
840-841	Salt Lake City	.79	.87
842,844	Ogden	.77	.84
843	Logan	.77	.85

Location Factors

STATE/ZIP	CITY	Residential	Commercial
UTAH (CONT'd)			
845	Price	.68	.76
846-847	Provo	.78	.85
VERMONT			
050	White River Jct.	.74	.79
051	Bellows Falls	.78	.83
052	Bennington	.79	.83
053	Brattleboro	.80	.86
054	Burlington	.80	.86
056	Montpelier	.81	.84
057	Rutland	.80	.85
058	St. Johnsbury	.76	.80
059	Guildhall	.76	.79
VIRGINIA			
220-221	Fairfax	1.03	.92
222	Arlington	1.04	.92
223	Alexandria	1.07	.94
224-225	Fredericksburg	.93	.88
226	Winchester	.89	.84
227	Culpeper	.98	.86
228	Harrisonburg	.88	.84
229	Charlottesville	.89	.84
230-232	Richmond	.97	.87
233-235	Norfolk	.99	.87
236	Newport News	.98	.86
237	Portsmouth	.91	.85
238	Petersburg	.95	.86
239	Farmville	.87	.79
240-241	Roanoke	.95	.84
242	Bristol	.84	.80
243	Pulaski	.81	.78
244	Staunton	.89	.82
245	Lynchburg	.94	.84
246	Grundy	.82	.78
WASHINGTON			
980-981,987	Seattle	1.03	1.05
982	Everett	1.03	1.00
983-984	Tacoma	1.01	1.02
985	Olympia	1.00	1.01
986	Vancouver	.94	.99
988	Wenatchee	.89	.93
989	Yakima	.95	.96
990-992	Spokane	.96	.93
993	Richland	.95	.95
994	Clarkston	.94	.92
WEST VIRGINIA			
247-248	Bluefield	.88	.88
249	Lewisburg	.89	.92
250-253	Charleston	.95	.95
254	Martinsburg	.88	.90
255-257	Huntington	.97	.96
258-259	Beckley	.90	.92
260	Wheeling	.92	.95
261	Parkersburg	.90	.94
262	Buckhannon	.91	.94
263-264	Clarksburg	.90	.94
265	Morgantown	.92	.94
266	Gassaway	.91	.94
267	Romney	.89	.91
268	Petersburg	.91	.92
WISCONSIN			
530,532	Milwaukee	1.08	1.03
531	Kenosha	1.01	.99
534	Racine	1.01	1.00
535	Beloit	.97	.96
537	Madison	.98	.98
538	Lancaster	.95	.92
539	Portage	.94	.93
540	New Richmond	.97	.93
541-543	Green Bay	.99	.95
544	Wausau	.94	.92
545	Rhinelander	.93	.93
546	La Crosse	.93	.93
547	Eau Claire	.96	.95
548	Superior	.96	.95
549	Oshkosh	.93	.92
WYOMING			
820	Cheyenne	.80	.84
821	Yellowstone Nat. Pk.	.73	.79
822	Wheatland	.73	.80

STATE/ZIP	CITY	Residential	Commercial
WYOMING (CONT'd)			
823	Rawlins	.73	.81
824	Worland	.72	.79
825	Riverton	.72	.79
826	Casper	.75	.81
827	Newcastle	.72	.79
828	Sheridan	.77	.82
829-831	Rock Springs	.76	.81
CANADIAN FACTORS (reflect Canadian currency)			
ALBERTA			
	Calgary	1.11	1.10
	Edmonton	1.10	1.11
	Fort McMurray	1.12	1.10
	Lethbridge	1.09	1.06
	Lloydminster	1.04	1.02
	Medicine Hat	1.04	1.02
	Red Deer	1.04	1.02
BRITISH COLUMBIA			
	Kamloops	1.02	1.03
	Prince George	1.03	1.04
	Vancouver	1.04	1.07
	Victoria	.98	.99
MANITOBA			
	Brandon	1.00	.96
	Portage la Prairie	.99	.96
	Winnipeg	.99	1.01
NEW BRUNSWICK			
	Bathurst	.92	.92
	Dalhousie	.92	.92
	Fredericton	.99	.95
	Moncton	.93	.92
	Newcastle	.92	.92
	St. John	.99	.95
NEWFOUNDLAND			
	Corner Brook	.94	.95
	St Johns	.95	.96
NORTHWEST TERRITORIES			
	Yellowknife	1.05	1.04
NOVA SCOTIA			
	Bridgewater	.95	.96
	Dartmouth	.95	.97
	Halifax	.98	.99
	New Glasgow	.95	.96
	Sydney	.94	.94
	Truro	.95	.96
	Yarmouth	.95	.95
ONTARIO			
	Barrie	1.10	1.05
	Brantford	1.12	1.06
	Cornwall	1.11	1.05
	Hamilton	1.13	1.09
	Kingston	1.11	1.05
	Kitchener	1.07	1.02
	London	1.11	1.07
	North Bay	1.08	1.04
	Oshawa	1.10	1.06
	Ottawa	1.12	1.08
	Owen Sound	1.09	1.05
	Peterborough	1.09	1.05
	Sarnia	1.12	1.07
	Sault Ste Marie	1.05	1.02
	St. Catharines	1.07	1.02
	Sudbury	1.05	1.02
	Thunder Bay	1.09	1.02
	Timmins	1.08	1.04
	Toronto	1.15	1.11
	Windsor	1.09	1.02
PRINCE EDWARD ISLAND			
	Charlottetown	.90	.92
	Summerside	.90	.92
QUEBEC			
	Cap-de-la-Madeleine	1.10	1.01
	Charlesbourg	1.10	1.01
	Chicoutimi	1.13	1.03
	Gatineau	1.09	1.00

Location Factors

STATE/ZIP	CITY	Residential	Commercial
QUEBEC (CONT'D)			
	Granby	1.09	
	Hull	1.09	1.00
	Joliette	1.10	1.01
	Laval	1.09	1.00
	Montreal	1.15	1.07
	Quebec	1.15	1.06
	Rimouski	1.13	1.03
	Rouyn-Noranda	1.09	1.00
	Saint Hyacinthe	1.09	1.00
	Sherbrooke	1.09	1.01
	Sorel	1.10	1.01
	St Jerome	1.09	1.00
	Trois Rivieres	1.10	1.01
SASKATCHEWAN			
	Moose Jaw	.92	.92
	Prince Albert	.91	.91
	Regina	.94	.93
	Saskatoon	.92	.92
YUKON			
	Whitehorse	.91	.92

Historical Cost Indexes

The following tables are the estimated Historical Cost Indexes based on a 30-city national average with a base of 100 on January 1, 1993.

The indexes may be used to:

1. Estimate and compare construction costs for different years in the same city.
2. Estimate and compare construction costs in different cities for the same year.
3. Estimate and compare construction costs in different cities for different years.
4. Compare construction trends in any city with the national average.

EXAMPLES

1. Estimate and compare construction costs for different years in the same city.

   A. To estimate the construction cost of a building in Lexington, KY in 1970, knowing that it cost $915,000 in 2010.

   $$\text{Index Lexington, KY in 1970} = 26.9$$
   $$\text{Index Lexington, KY in 2009} = 159.0$$

   $$\frac{\text{Index 1970}}{\text{Index 2010}} \quad x \quad \text{Cost 2009} \quad = \quad \text{Cost 1970}$$

   $$\frac{26.9}{159.0} \quad x \quad \$915,000 \quad = \quad \$155,000$$

   Construction Cost in Lexington, KY in 1970 = $155,000

   B. To estimate the current construction cost of a building in Boston, MA that was built in 1980 for $900,000.

   $$\text{Index Boston, MA in 1980} = 64.0$$
   $$\text{Index Boston, MA in 2010} = 214.3$$

   $$\frac{\text{Index 2010}}{\text{Index 1980}} \quad x \quad \text{Cost 1980} \quad = \quad \text{Cost 2010}$$

   $$\frac{214.3}{64.0} \quad x \quad \$900,000 \quad = \quad \$3,013,500$$

   Construction Cost in Boston in 2010 = $3,013,500

2. Estimate and compare construction costs in different cities for the same year.

   To compare the construction cost of a building in Topeka, KS in 2010 with the known cost of $800,000 in Baltimore, MD in 2010

   $$\text{Index Topeka, KS in 2010} = 152.5$$
   $$\text{Index Baltimore, MD in 2010} = 168.6$$

   $$\frac{\text{Index Topeka}}{\text{Index Baltimore}} \quad x \quad \text{Cost Baltimore} = \text{Cost Topeka}$$

   $$\frac{152.5}{168.6} \quad x \quad \$800,000 \quad = \$723,500$$

   Construction Cost in Topeka in 2010 = $723,500

3. Estimate and compare construction costs in different cities for different years.

   To compare the construction cost of a building in Detroit, MI in 2010 with the known construction cost of $5,000,000 for the same building in San Francisco, CA in 1980.

   $$\text{Index Detroit, MI in 2010} = 188.9$$
   $$\text{Index San Francisco, CA in 1980} = 75.2$$

   $$\frac{\text{Index Detroit 2010}}{\text{Index San Francisco 1980}} \quad x \quad \text{Cost San Francisco 1980} = \text{Cost Detroit 2010}$$

   $$\frac{188.9}{75.2} \quad x \quad \$5,000,000 \quad = \quad \$12,560,000$$

   Construction Cost in Detroit in 2010 = $12,560,000

4. Compare construction trends in any city with the national average.

   To compare the construction cost in Las Vegas, NV from 1975 to 2010 with the increase in the National Average during the same time period.

   Index Las Vegas, NV for 1975 = 42.8   For 2010 = 195.0
   Index 30 City Average for 1975 = 43.7   For 2010 = 182.8

   A. National Average escalation $= \dfrac{\text{Index} - \text{30 City 2010}}{\text{Index} - \text{30 City 1975}}$
   From 1975 to 2010

   $$= \frac{182.8}{43.7}$$

   National Average escalation
   From 1975 to 2010   = 4.18 or increased by 418%

   B. Escalation for Las Vegas, NV $= \dfrac{\text{Index Las Vegas, NV 2010}}{\text{Index Las Vegas, NV 1975}}$
   From 1975 to 2010

   $$= \frac{195.0}{42.8}$$

   Las Vegas escalation
   From 1975 to 2010   = 4.56 or increased by 456%

Conclusion: Construction costs in Las Vegas are higher than National average costs and increased at a greater rate from 1975 to 2010 than the National Average.

Historical Cost Indexes

Year	National 30 City Average	Alabama Birming-ham	Alabama Hunts-ville	Alabama Mobile	Alabama Mont-gomery	Alabama Tusca-loosa	Alaska Anchor-age	Arizona Phoenix	Arizona Tuscon	Arkansas Fort Smith	Arkansas Little Rock	California Anaheim	California Bakers-field	California Fresno	California Los Angeles	California Oxnard
Jan 2010	182.8E	160.2E	154.6E	153.5E	146.3E	146.7E	220.7E	162.2E	156.6E	146.3E	153.4E	194.1E	191.1E	194.3E	197.4E	194.0E
2009	182.5	162.7	157.2	155.6	148.8	149.4	222.9	161.3	156.8	149.2	156.3	194.5	192.3	195.0	196.6	194.5
2008	171.0	150.3	146.9	143.7	138.4	138.8	210.8	152.2	148.4	138.6	145.4	182.7	179.8	183.2	184.7	182.6
2007	165.0	146.9	143.3	140.3	134.8	135.4	206.8	147.7	143.1	134.8	141.7	175.1	173.7	176.7	177.5	176.2
2006	156.2	135.7	133.6	126.7	124.0	122.2	196.4	137.9	134.8	123.2	127.2	166.8	164.9	169.8	167.3	167.4
2005	146.7	127.9	125.4	119.3	116.6	114.6	185.6	128.5	124.1	115.4	119.4	156.5	153.0	157.9	157.1	156.4
2004	132.8	115.9	112.9	107.0	104.9	102.9	167.0	116.5	113.6	103.8	108.6	142.1	139.5	143.8	142.0	140.9
2003	129.7	113.1	110.7	104.8	102.6	100.8	163.5	113.9	110.6	101.8	105.8	139.4	137.2	142.1	139.6	138.7
2002	126.7	110.0	103.4	103.6	101.7	99.7	159.5	113.3	110.2	100.4	102.1	136.5	133.4	136.0	136.4	136.3
2001	122.2	106.0	100.5	100.6	98.3	95.9	152.6	109.0	106.4	97.3	98.7	132.5	129.3	132.8	132.4	132.4
2000	118.9	104.1	98.9	99.1	94.2	94.6	148.3	106.9	104.9	94.4	95.7	129.4	125.2	129.4	129.9	129.7
1999	116.6	101.2	97.4	97.7	92.5	92.8	145.9	105.5	103.3	93.1	94.4	127.9	123.6	126.9	128.7	128.2
1998	113.6	96.2	94.0	94.8	90.3	89.8	143.8	102.1	101.0	90.6	91.4	125.2	120.3	123.9	125.8	124.7
1997	111.5	94.6	92.4	93.3	88.8	88.3	142.0	101.8	100.7	89.3	90.1	124.0	119.1	122.3	124.6	123.5
1996	108.9	90.9	91.4	92.3	87.5	86.5	140.4	98.3	97.4	86.5	86.7	121.7	117.2	120.2	122.4	121.5
1995	105.6	87.8	88.0	88.8	84.5	83.5	138.0	96.1	95.5	85.0	85.6	120.1	115.7	117.5	120.9	120.0
1994	103.0	85.9	86.1	86.8	82.6	81.7	134.0	93.7	93.1	82.4	83.9	118.1	113.6	114.5	119.0	117.5
1993	100.0	82.8	82.7	86.1	82.1	78.7	132.0	90.9	90.9	80.9	82.4	115.0	111.3	112.9	115.6	115.4
1992	97.9	81.7	81.5	84.9	80.9	77.6	128.6	88.8	89.4	79.7	81.2	113.5	108.1	110.3	113.7	113.7
1991	95.7	80.5	78.9	83.9	79.8	76.6	127.4	88.1	88.5	78.3	80.0	111.0	105.8	108.2	110.9	111.4
1990	93.2	79.4	77.6	82.7	78.4	75.2	125.8	86.4	87.0	77.1	77.9	107.7	102.7	103.3	107.5	107.4
1989	91.0	77.5	76.1	81.1	76.8	73.7	123.4	85.3	85.6	75.6	76.4	105.6	101.0	101.7	105.3	105.4
1985	81.8	71.1	70.5	72.7	70.9	67.9	116.0	78.1	77.7	69.6	71.0	95.4	92.2	92.6	94.6	96.6
1980	60.7	55.2	54.1	56.8	56.8	54.0	91.4	63.7	62.4	53.1	55.8	68.7	69.4	68.7	67.4	69.9
1975	43.7	40.0	40.9	41.8	40.1	37.8	57.3	44.5	45.2	39.0	38.7	47.6	46.6	47.7	48.3	47.0
1970	27.8	24.1	25.1	25.8	25.4	24.2	43.0	27.2	28.5	24.3	22.3	31.0	30.8	31.1	29.0	31.0
1965	21.5	19.6	19.3	19.6	19.5	18.7	34.9	21.8	22.0	18.7	18.5	23.9	23.7	24.0	22.7	23.9
1960	19.5	17.9	17.5	17.8	17.7	16.9	31.7	19.9	20.0	17.0	16.8	21.7	21.5	21.8	20.6	21.7
1955	16.3	14.8	14.7	14.9	14.9	14.2	26.6	16.7	16.7	14.3	14.5	18.2	18.1	18.3	17.3	18.2
1950	13.5	12.2	12.1	12.3	12.3	11.7	21.9	13.8	13.8	11.8	11.6	15.1	15.0	15.1	14.3	15.1
1945	8.6	7.8	7.7	7.8	7.8	7.5	14.0	8.8	8.8	7.5	7.4	9.6	9.5	9.6	9.1	9.6
1940	6.6	6.0	6.0	6.1	6.0	5.8	10.8	6.8	6.8	5.8	5.7	7.4	7.4	7.4	7.0	7.4

Year	National 30 City Average	California River-side	California Sacra-mento	California San Diego	California San Francisco	California Santa Barbara	California Stockton	California Vallejo	Colorado Colorado Springs	Colorado Denver	Colorado Pueblo	Connecticut Bridge-Port	Connecticut Bristol	Connecticut Hartford	Connecticut New Britain	Connecticut New Haven
Jan 2010	182.8E	192.4E	197.9E	191.2E	226.3E	193.1E	193.8E	202.0E	167.2E	172.3E	165.7E	199.2E	197.8E	198.7E	197.5E	199.6E
2009	182.5	193.1	198.1	191.6	224.9	193.6	194.5	202.7	168.4	172.0	166.7	199.0	197.7	198.7	197.4	199.3
2008	171.0	181.3	186.0	179.9	210.6	181.5	183.2	191.6	158.8	161.9	157.9	185.8	184.8	185.7	184.5	186.1
2007	165.0	174.7	179.1	173.6	201.1	175.2	178.2	185.9	152.6	155.9	152.3	179.6	178.3	179.6	178.0	179.5
2006	156.2	166.0	172.2	164.0	191.2	166.4	171.0	177.9	146.1	149.2	145.0	169.5	168.0	169.5	167.7	169.8
2005	146.7	155.4	161.1	153.8	179.7	155.7	160.0	167.2	138.3	141.1	136.4	160.8	159.4	159.6	159.1	160.8
2004	132.8	141.0	147.2	139.0	163.6	141.4	144.2	148.6	125.6	127.2	123.5	143.6	142.9	142.4	142.7	144.3
2003	129.7	138.7	144.9	136.6	162.1	139.2	141.8	146.4	123.1	123.7	120.7	141.3	140.2	140.3	140.0	141.6
2002	126.7	135.3	138.4	134.2	157.9	135.9	136.9	143.8	118.7	121.3	116.7	133.8	133.6	133.1	133.3	134.8
2001	122.2	131.4	135.5	129.7	151.8	131.5	134.1	140.2	113.3	117.2	113.1	128.6	128.5	128.5	128.3	128.6
2000	118.9	128.0	131.5	127.1	146.9	128.7	130.7	137.1	109.8	111.8	108.8	122.7	122.6	122.9	122.4	122.7
1999	116.6	126.5	129.4	124.9	145.1	127.2	127.9	135.3	107.0	109.1	107.1	121.1	121.3	121.2	121.1	121.3
1998	113.6	123.7	125.9	121.3	141.9	123.7	124.7	132.5	103.3	106.5	103.7	119.1	119.4	120.0	119.7	120.0
1997	111.5	122.6	124.7	120.3	139.2	122.4	123.3	130.6	101.1	104.4	102.0	119.2	119.5	119.9	119.7	120.0
1996	108.9	120.6	122.4	118.4	136.8	120.5	121.4	128.1	98.5	101.4	99.7	117.4	117.6	117.9	117.8	118.1
1995	105.6	119.2	119.5	115.4	133.8	119.0	119.0	122.5	96.1	98.9	96.8	116.0	116.5	116.9	116.3	116.5
1994	103.0	116.6	114.8	113.4	131.4	116.4	115.5	119.7	94.2	95.9	94.6	114.3	115.0	115.3	114.7	114.8
1993	100.0	114.3	112.5	111.3	129.6	114.3	115.2	117.5	92.1	93.8	92.6	108.7	107.8	108.4	106.3	106.1
1992	97.9	111.8	110.8	109.5	127.9	112.2	112.7	115.2	90.6	91.5	90.9	106.7	106.0	106.6	104.5	104.4
1991	95.7	109.4	108.5	107.7	125.8	110.0	111.0	113.4	88.9	90.4	89.4	97.6	97.3	98.0	97.3	97.9
1990	93.2	107.0	104.9	105.6	121.8	106.4	105.4	111.4	86.9	88.8	88.8	96.3	95.9	96.6	95.9	96.5
1989	91.0	105.2	103.1	103.6	119.2	104.6	103.7	109.8	85.6	88.7	87.4	94.5	94.2	94.9	94.2	94.6
1985	81.8	94.8	92.2	94.2	106.2	93.2	93.7	97.0	79.6	81.0	80.4	86.1	86.2	87.2	86.1	86.3
1980	60.7	68.7	71.3	68.1	75.2	71.1	71.2	71.9	60.7	60.9	59.5	61.5	60.7	61.9	60.7	61.4
1975	43.7	47.3	49.1	47.7	49.8	46.1	47.6	46.5	43.1	42.7	42.5	45.2	45.5	46.0	45.2	46.0
1970	27.8	31.0	32.1	30.7	31.6	31.3	31.8	31.8	27.6	26.1	27.3	29.2	28.2	29.6	28.2	29.3
1965	21.5	23.9	24.8	24.0	23.7	24.1	24.5	24.5	21.3	20.9	21.0	22.4	21.7	22.6	21.7	23.2
1960	19.5	21.7	22.5	21.7	21.5	21.9	22.3	22.2	19.3	19.0	19.1	20.0	19.8	20.0	19.8	20.0
1955	16.3	18.2	18.9	18.2	18.0	18.4	18.7	18.6	16.2	15.9	16.0	16.8	16.6	16.8	16.6	16.8
1950	13.5	15.0	15.6	15.0	14.9	15.2	15.4	15.4	13.4	13.2	13.2	13.9	13.7	13.8	13.7	13.9
1945	8.6	9.6	10.0	9.6	9.5	9.7	9.9	9.8	8.5	8.4	8.4	8.8	8.7	8.8	8.7	8.9
1940	6.6	7.4	7.7	7.4	7.3	7.5	7.6	7.6	6.6	6.5	6.5	6.8	6.8	6.8	6.8	6.8

Year	National 30 City Average	Connecticut Norwalk	Connecticut Stamford	Connecticut Waterbury	Delaware Wilmington	D.C. Washington	Florida Fort Lauderdale	Florida Jacksonville	Florida Miami	Florida Orlando	Florida Tallahassee	Florida Tampa	Georgia Albany	Georgia Atlanta	Georgia Columbus	Georgia Macon
Jan 2010	182.8E	199.1E	204.7E	198.6E	189.9E	180.6E	162.1E	153.6E	166.3E	163.4E	147.9E	168.5E	149.0E	162.2E	152.6E	150.9E
2009	182.5	198.8	204.3	198.4	188.9	181.4	160.7	152.0	165.2	163.9	145.1	165.1	152.0	164.5	155.6	153.5
2008	171.0	185.2	189.2	185.3	177.5	170.4	149.6	142.5	152.9	153.0	135.3	155.6	140.4	153.2	143.9	142.5
2007	165.0	178.6	183.3	179.1	173.4	163.1	145.6	139.0	148.9	148.1	131.7	152.0	135.8	148.0	140.1	138.1
2006	156.2	169.7	173.7	169.1	158.3	153.0	136.0	126.5	136.7	134.3	118.4	136.6	123.6	139.9	126.3	124.3
2005	146.7	160.9	164.1	160.5	149.6	142.7	126.0	119.2	127.1	126.0	110.7	127.7	115.7	131.7	118.9	116.1
2004	132.8	143.6	146.5	143.2	136.1	126.8	114.8	107.8	115.6	113.2	100.6	116.6	102.8	119.6	102.0	105.0
2003	129.7	140.6	145.1	140.8	133.0	124.6	109.0	105.6	110.5	107.8	98.0	103.5	100.5	115.8	99.1	102.7
2002	126.7	133.5	136.2	133.9	129.4	120.3	107.1	104.6	107.4	106.7	97.3	102.8	99.7	113.0	98.3	101.9
2001	122.2	128.2	131.5	128.8	124.8	115.9	104.3	100.8	104.6	103.8	94.8	100.3	96.4	109.1	95.5	99.0
2000	118.9	121.5	126.4	123.2	117.4	113.8	102.4	99.0	101.8	101.2	93.6	98.9	94.9	106.1	94.1	97.0
1999	116.6	120.0	122.0	121.2	116.6	111.5	101.4	98.0	100.8	100.1	92.5	97.9	93.5	102.9	92.8	95.8
1998	113.6	118.8	120.8	120.1	112.3	109.6	99.4	96.2	99.0	98.4	90.9	96.3	91.4	100.8	90.4	93.3
1997	111.5	118.9	120.9	120.2	110.7	106.4	98.6	95.4	98.2	97.2	89.9	95.5	89.9	98.5	88.7	91.6
1996	108.9	117.0	119.0	118.4	108.6	105.4	95.9	92.6	95.9	95.1	88.0	93.6	86.6	94.3	83.9	88.3
1995	105.6	115.5	117.9	117.3	106.1	102.3	94.0	90.9	93.7	93.5	86.5	92.2	85.0	92.0	82.6	86.8
1994	103.0	113.9	116.4	115.8	105.0	99.6	92.2	88.9	91.8	91.5	84.5	90.2	82.3	89.6	80.6	83.7
1993	100.0	108.8	110.6	104.8	101.5	96.3	87.4	86.1	87.1	88.5	82.1	87.7	79.5	85.7	77.8	80.9
1992	97.9	107.2	109.0	103.1	100.3	94.7	85.7	84.0	85.3	87.1	80.8	86.2	78.2	84.3	76.5	79.6
1991	95.7	100.6	103.2	96.5	94.5	92.9	85.1	82.8	85.2	85.5	79.7	86.3	76.3	82.6	75.4	78.4
1990	93.2	96.3	98.9	95.1	92.5	90.4	83.9	81.1	84.0	82.9	78.4	85.0	75.0	80.4	74.0	76.9
1989	91.0	94.4	97.0	93.4	89.5	87.5	82.4	79.7	82.5	81.6	76.9	83.5	73.4	78.6	72.4	75.3
1985	81.8	85.3	86.8	85.6	81.1	78.8	76.7	73.4	78.3	73.9	70.6	77.3	66.9	70.3	66.1	68.1
1980	60.7	60.7	60.9	62.3	58.5	59.6	55.3	55.8	56.5	56.7	53.5	57.2	51.7	54.0	51.1	51.3
1975	43.7	44.7	45.0	46.3	42.9	43.7	42.1	40.3	43.2	41.5	38.1	41.3	37.5	38.4	36.2	36.5
1970	27.8	28.1	28.2	28.9	27.0	26.3	25.7	22.8	27.0	26.2	24.5	24.2	23.8	25.2	22.8	23.4
1965	21.5	21.6	21.7	22.2	20.9	21.8	19.8	17.4	19.3	20.2	18.9	18.6	18.3	19.8	17.6	18.0
1960	19.5	19.7	19.7	20.2	18.9	19.4	18.0	15.8	17.6	18.3	17.2	16.9	16.7	17.1	16.0	16.4
1955	16.3	16.5	16.5	17.0	15.9	16.3	15.1	13.2	14.7	15.4	14.4	14.1	14.0	14.4	13.4	13.7
1950	13.5	13.6	13.7	14.0	13.1	13.4	12.5	11.0	12.2	12.7	11.9	11.7	11.5	11.9	11.0	11.3
1945	8.6	8.7	8.7	8.9	8.4	8.6	7.9	7.0	7.8	8.1	7.6	7.5	7.4	7.6	7.0	7.2
▼ 1940	6.6	6.7	6.7	6.9	6.4	6.6	6.1	5.4	6.0	6.2	5.8	5.7	5.7	5.8	5.5	5.6

Year	National 30 City Average	Georgia Savannah	Hawaii Honolulu	Idaho Boise	Idaho Pocatello	Illinois Chicago	Illinois Decatur	Illinois Joliet	Illinois Peoria	Illinois Rockford	Illinois Springfield	Indiana Anderson	Indiana Evansville	Indiana Fort Wayne	Indiana Gary	Indiana Indianapolis
Jan 2010	182.8E	149.9E	214.5E	160.4E	161.0E	212.1E	182.8E	208.9E	187.4E	198.6E	184.1E	163.9E	167.1E	161.4E	185.4E	170.4E
2009	182.5	152.8	218.6	162.6	163.4	209.0	182.2	205.9	186.2	196.5	184.0	164.9	166.9	163.0	185.2	170.3
2008	171.0	141.1	205.5	153.2	152.8	195.7	168.7	185.5	172.2	180.2	168.5	152.4	155.8	151.0	169.0	159.2
2007	165.0	136.9	200.4	148.0	148.1	186.5	160.5	177.7	166.0	175.0	159.2	149.0	152.6	147.7	165.4	154.1
2006	156.2	124.0	191.9	141.6	141.2	177.8	153.8	168.7	155.2	163.3	152.8	139.8	144.4	139.1	154.5	144.6
2005	146.7	116.9	181.2	133.8	132.5	164.4	145.0	158.3	146.5	151.2	145.4	131.4	136.2	131.0	146.4	137.4
2004	132.8	105.6	161.2	122.0	120.5	149.3	129.3	145.1	133.9	138.1	130.3	120.9	123.5	120.1	132.2	125.0
2003	129.7	103.0	159.4	120.2	118.7	146.2	127.4	143.5	132.4	137.3	128.4	119.7	121.4	118.2	131.4	122.5
2002	126.7	102.0	157.2	118.3	117.1	141.2	123.8	138.5	129.6	132.9	125.3	117.5	119.0	116.4	129.1	120.5
2001	122.2	99.0	150.0	114.3	113.4	135.8	120.1	133.7	124.3	127.8	119.8	113.4	115.6	112.1	123.4	116.4
2000	118.9	97.5	144.8	112.9	112.1	131.2	115.1	124.6	119.0	122.2	116.2	109.8	111.5	108.4	117.8	113.2
1999	116.6	96.0	143.0	110.2	109.6	129.6	113.0	122.6	116.4	120.7	113.8	107.1	109.3	106.8	112.8	110.6
1998	113.6	93.7	140.4	107.4	107.0	125.2	110.1	119.8	113.8	115.5	111.1	105.0	107.2	104.5	111.1	108.0
1997	111.5	92.0	139.8	104.6	104.7	121.3	107.8	117.5	111.5	113.1	108.9	101.9	104.4	101.8	110.3	105.2
1996	108.9	88.6	134.5	102.2	102.1	118.8	106.6	116.2	109.3	111.5	106.5	100.0	102.1	99.9	107.5	102.7
1995	105.6	87.4	130.3	99.5	98.2	114.2	98.5	110.5	102.3	103.6	98.1	96.4	97.2	95.0	100.7	100.1
1994	103.0	85.3	124.0	94.8	95.0	111.3	97.3	108.9	100.9	102.2	97.0	93.6	95.8	93.6	99.1	97.1
1993	100.0	82.0	122.0	92.2	92.1	107.6	95.6	106.8	98.9	99.6	95.2	91.2	94.3	91.5	96.7	93.9
1992	97.9	80.8	120.0	91.0	91.0	104.3	94.4	104.2	97.3	98.2	94.0	89.5	92.9	89.9	95.0	91.5
1991	95.7	79.5	106.1	89.5	89.4	100.9	92.3	100.0	95.9	95.8	91.5	87.8	91.4	88.3	93.3	89.1
1990	93.2	77.9	104.7	88.2	88.1	98.4	90.9	98.4	93.7	94.0	90.1	84.6	89.3	83.4	88.4	87.1
1989	91.0	76.0	102.8	86.6	86.5	93.7	89.4	92.8	91.6	92.1	88.6	82.3	87.8	81.7	86.5	85.1
1985	81.8	68.9	94.7	78.0	78.0	82.4	81.9	83.4	83.7	83.0	81.5	75.2	79.8	75.0	77.8	77.1
1980	60.7	52.2	68.9	60.3	59.5	62.8	62.3	63.4	64.5	61.6	61.1	56.5	59.0	56.7	59.8	57.9
1975	43.7	36.9	44.6	40.8	40.5	45.7	43.1	44.5	44.7	42.7	42.4	39.5	41.7	39.9	41.9	40.6
1970	27.8	21.0	30.4	26.7	26.6	29.1	28.0	28.6	29.0	27.5	27.6	25.4	26.4	25.5	27.1	26.2
1965	21.5	16.4	21.8	20.6	20.5	22.7	21.5	22.1	22.4	21.2	21.3	19.5	20.4	19.7	20.8	20.7
1960	19.5	14.9	19.8	18.7	18.6	20.2	19.6	20.0	20.3	19.2	19.3	17.7	18.7	17.9	18.9	18.4
1955	16.3	12.5	16.6	15.7	15.6	16.9	16.4	16.8	17.0	16.1	16.2	14.9	15.7	15.0	15.9	15.5
1950	13.5	10.3	13.7	13.0	12.9	14.0	13.6	13.9	14.0	13.3	13.4	12.3	12.9	12.4	13.1	12.8
1945	8.6	6.6	8.8	8.3	8.2	8.9	8.6	8.9	9.0	8.5	8.6	7.8	8.3	7.9	8.4	8.1
▼ 1940	6.6	5.1	6.8	6.4	6.3	6.9	6.7	6.8	6.9	6.5	6.6	6.0	6.4	6.1	6.5	6.3

461

Historical Cost Indexes

Year	National 30 City Average	Indiana Muncie	Indiana South Bend	Indiana Terre Haute	Iowa Cedar Rapids	Iowa Davenport	Iowa Des Moines	Iowa Sioux City	Iowa Waterloo	Kansas Topeka	Kansas Wichita	Kentucky Lexington	Kentucky Louisville	Louisiana Baton Rouge	Louisiana Lake Charles	Louisiana New Orleans
Jan 2010	182.8E	163.8E	165.4E	168.8E	164.2E	170.4E	160.3E	154.5E	145.7E	152.5E	150.3E	159.0E	165.5E	154.5E	152.2E	160.4E
2009	182.5	164.5	167.1	168.0	165.9	172.0	162.1	156.3	147.6	154.5	152.4	160.7	167.5	156.9	154.4	161.9
2008	171.0	153.0	153.1	155.9	157.3	163.4	152.4	147.8	139.2	144.9	143.1	152.0	156.8	144.0	141.5	149.3
2007	165.0	149.7	149.5	152.7	152.4	157.6	148.9	144.0	135.0	141.0	138.7	147.6	150.9	139.4	138.0	145.2
2006	156.2	141.2	141.3	144.0	145.8	151.1	142.9	137.3	128.4	134.7	132.9	129.0	143.0	129.4	129.8	135.8
2005	146.7	131.0	133.8	135.2	135.7	142.1	133.5	129.4	120.0	125.9	125.6	121.5	135.1	120.8	117.7	126.2
2004	132.8	120.0	120.6	123.0	122.7	128.6	121.6	116.2	108.3	112.6	113.4	110.1	120.3	105.6	109.0	115.3
2003	129.7	118.7	119.2	121.7	120.9	126.4	120.2	114.6	105.9	109.4	111.3	107.8	118.7	103.3	106.1	112.4
2002	126.7	116.7	117.1	119.9	116.0	122.1	116.7	111.4	103.7	107.8	109.6	106.2	116.4	102.5	105.0	110.6
2001	122.2	112.8	111.6	115.5	112.5	117.5	113.2	107.7	100.6	104.3	105.2	103.3	112.7	99.4	101.3	104.6
2000	118.9	109.1	106.9	110.8	108.8	112.6	108.9	99.0	98.1	101.0	101.1	101.4	109.3	97.8	99.7	102.1
1999	116.6	105.7	103.7	107.9	104.1	109.2	107.6	96.7	96.4	98.7	99.3	99.7	106.6	96.1	97.6	99.5
1998	113.6	103.6	102.3	106.0	102.5	106.8	103.8	95.1	94.8	97.4	97.4	97.4	101.5	94.7	96.1	97.7
1997	111.5	101.3	101.2	104.6	101.2	105.6	102.5	93.8	93.5	96.3	96.2	96.1	99.9	93.3	96.8	96.2
1996	108.9	99.4	99.1	102.1	99.1	102.0	99.1	90.9	90.7	93.6	93.8	94.4	98.3	91.5	94.9	94.3
1995	105.6	95.6	94.6	97.2	95.4	96.6	94.4	88.2	88.9	91.1	91.0	91.6	94.6	89.6	92.7	91.6
1994	103.0	93.2	93.0	95.8	93.0	92.3	92.4	85.9	86.8	89.4	88.1	89.6	92.2	87.7	89.9	89.5
1993	100.0	91.0	90.7	94.4	91.1	90.5	90.7	84.1	85.2	87.4	86.2	87.3	89.4	86.4	88.5	87.8
1992	97.9	89.4	89.3	93.1	89.8	89.3	89.4	82.9	83.8	86.2	85.0	85.8	88.0	85.2	87.3	86.6
1991	95.7	87.7	87.5	91.7	88.5	88.0	88.4	81.8	81.6	84.4	83.9	84.1	84.4	83.2	85.3	85.8
1990	93.2	83.9	85.1	89.1	87.1	86.5	86.9	80.4	80.3	83.2	82.6	82.8	82.6	82.0	84.1	84.5
1989	91.0	81.9	83.5	86.7	85.3	84.9	85.3	79.0	78.8	81.6	81.2	81.3	80.1	80.6	82.7	83.4
1985	81.8	75.2	76.0	79.2	75.9	77.6	77.1	72.1	72.3	75.0	74.7	75.5	74.5	74.9	78.5	78.2
1980	60.7	56.1	58.3	59.7	62.7	59.6	61.8	59.3	57.7	58.9	58.0	59.3	59.8	59.1	60.0	57.2
1975	43.7	39.2	40.3	41.9	43.1	41.0	43.1	41.0	40.0	42.0	43.1	42.7	42.5	40.4	40.6	41.5
1970	27.8	25.3	26.2	27.0	28.2	26.9	27.6	26.6	25.9	27.0	25.5	26.9	25.9	25.0	26.7	27.2
1965	21.5	19.5	20.2	20.9	21.7	20.8	21.7	20.5	20.0	20.8	19.6	20.7	20.3	19.4	20.6	20.4
1960	19.5	17.8	18.3	18.9	19.7	18.8	19.5	18.6	18.2	18.9	17.8	18.8	18.4	17.6	18.7	18.5
1955	16.3	14.9	15.4	15.9	16.6	15.8	16.3	15.6	15.2	15.8	15.0	15.8	15.4	14.8	15.7	15.6
1950	13.5	12.3	12.7	13.1	13.7	13.1	13.5	12.9	12.6	13.1	12.4	13.0	12.8	12.2	13.0	12.8
1945	8.6	7.8	8.1	8.4	8.7	8.3	8.6	8.2	8.0	8.3	7.9	8.3	8.1	7.8	8.3	8.2
1940	6.6	6.0	6.3	6.5	6.7	6.4	6.6	6.3	6.2	6.4	6.1	6.4	6.3	6.0	6.4	6.3

Year	National 30 City Average	Louisiana Shreveport	Maine Lewiston	Maine Portland	Maryland Baltimore	Massachusetts Boston	Massachusetts Brockton	Massachusetts Fall River	Massachusetts Lawrence	Massachusetts Lowell	Massachusetts New Bedford	Massachusetts Pittsfield	Massachusetts Springfield	Massachusetts Worcester	Michigan Ann Arbor	Michigan Dearborn
Jan 2010	182.8E	146.0E	162.5E	164.8E	168.6E	214.3E	201.4E	199.8E	204.1E	204.3E	199.1E	187.0E	188.6E	200.2E	181.2E	188.2E
2009	182.5	147.6	159.4	161.8	169.1	211.8	199.2	197.6	201.9	201.9	196.9	185.1	187.1	198.2	179.2	186.9
2008	171.0	135.9	149.1	150.9	157.7	198.6	186.4	185.1	188.7	189.1	184.5	171.3	173.7	184.5	169.6	176.4
2007	165.0	132.4	146.4	148.2	152.5	191.8	180.4	179.7	182.0	181.7	179.1	166.8	169.1	180.4	166.8	172.5
2006	156.2	125.2	140.2	139.8	144.9	180.4	171.6	170.4	172.3	173.5	170.4	157.3	159.7	171.6	161.1	165.5
2005	146.7	117.4	131.7	131.4	135.8	169.6	161.3	159.6	162.2	162.3	159.6	147.6	150.7	158.8	147.5	155.8
2004	132.8	105.7	119.8	119.4	121.4	154.1	144.4	143.6	146.4	146.7	143.6	131.9	136.5	143.2	135.9	142.1
2003	129.7	103.9	117.7	117.3	118.1	150.2	139.6	140.6	143.3	143.6	140.5	128.9	133.8	139.0	134.1	139.0
2002	126.7	102.1	117.5	117.1	115.6	145.6	136.0	135.0	136.9	137.3	134.9	124.7	128.3	134.9	131.4	134.3
2001	122.2	98.3	114.6	114.3	111.7	140.9	132.1	132.1	133.1	132.9	131.3	120.3	124.5	130.1	126.8	129.8
2000	118.9	96.2	105.7	105.4	107.7	138.9	129.4	128.4	129.7	130.3	128.3	116.7	121.0	127.3	124.3	125.8
1999	116.6	94.8	105.0	104.7	106.4	136.2	126.7	126.3	127.3	127.4	126.2	115.0	118.8	123.8	117.5	122.7
1998	113.6	92.0	102.8	102.5	104.1	132.8	125.0	124.7	125.0	125.3	124.6	114.2	117.1	122.6	116.0	119.7
1997	111.5	90.1	101.7	101.4	102.2	132.1	124.4	124.7	125.1	125.6	124.7	114.6	117.4	122.9	113.5	117.9
1996	108.9	88.5	99.5	99.2	99.6	128.7	121.6	122.2	121.7	122.1	122.2	112.7	115.0	120.2	112.9	116.2
1995	105.6	86.7	96.8	96.5	96.1	128.6	119.6	117.7	120.5	119.6	117.2	110.7	112.7	114.8	106.1	110.5
1994	103.0	85.0	95.2	94.9	94.4	124.9	114.2	113.7	116.8	116.9	111.6	109.2	111.1	112.9	105.0	109.0
1993	100.0	83.6	93.0	93.0	93.1	121.1	111.7	111.5	114.9	114.3	109.4	107.1	108.9	110.3	102.8	105.8
1992	97.9	82.3	91.7	91.7	90.9	118.0	110.0	109.8	110.9	110.2	107.7	105.7	107.2	108.6	101.5	103.2
1991	95.7	81.6	89.8	89.9	89.1	115.2	105.9	104.7	107.3	105.5	105.1	100.6	103.2	105.7	95.1	98.3
1990	93.2	80.3	88.5	88.6	85.6	110.9	103.7	102.9	105.2	102.8	102.6	98.7	101.4	103.2	93.2	96.5
1989	91.0	78.8	86.7	86.7	83.5	107.2	100.4	99.7	102.9	99.7	99.8	94.7	96.4	99.1	89.6	94.6
1985	81.8	73.6	76.7	77.0	72.7	92.8	88.8	88.7	89.4	88.2	88.6	85.0	85.6	86.7	80.1	82.7
1980	60.7	58.7	57.3	58.5	53.6	64.0	63.7	64.1	63.4	62.7	63.1	61.8	62.0	62.3	62.9	64.0
1975	43.7	40.5	41.9	42.1	39.8	46.6	45.8	45.7	46.2	45.7	46.1	45.5	45.8	46.0	44.1	44.9
1970	27.8	26.4	26.5	25.8	25.1	29.2	29.1	29.2	29.1	28.9	29.0	28.6	28.5	28.7	28.5	28.9
1965	21.5	20.3	20.4	19.4	20.2	23.0	22.5	22.5	22.5	22.2	22.4	22.0	22.4	22.1	22.0	22.3
1960	19.5	18.5	18.6	17.6	17.5	20.5	20.4	20.4	20.4	20.2	20.3	20.0	20.1	20.1	20.0	20.2
1955	16.3	15.5	15.6	14.7	14.7	17.2	17.1	17.1	17.1	16.9	17.1	16.8	16.9	16.8	16.7	16.9
1950	13.5	12.8	12.9	12.2	12.1	14.2	14.1	14.1	14.1	14.0	14.1	13.8	14.0	13.9	13.8	14.0
1945	8.6	8.1	8.2	7.8	7.7	9.1	9.0	9.0	9.0	8.9	9.0	8.9	8.9	8.9	8.8	8.9
1940	6.6	6.3	6.3	6.0	6.0	7.0	6.9	7.0	7.0	6.9	6.9	6.8	6.9	6.8	6.8	6.9

Historical Cost Indexes

Year	National 30 City Average	Michigan Detroit	Flint	Grand Rapids	Kala-mazoo	Lansing	Sagi-naw	Minnesota Duluth	Minne-apolis	Roches-ter	Mississippi Biloxi	Jackson	Missouri Kansas City	St. Joseph	St. Louis	Spring-field
Jan 2010	182.8E	188.9E	177.4E	158.4E	168.1E	175.6E	170.6E	193.0E	203.7E	188.1E	150.1E	153.9E	188.2E	170.5E	187.3E	162.5E
2009	182.5	187.8	176.0	156.7	166.3	173.7	168.6	191.8	203.1	188.0	152.8	156.6	185.6	172.3	187.2	163.7
2008	171.0	177.0	164.4	140.2	155.7	161.6	158.7	177.4	190.6	175.0	141.8	147.1	175.5	164.2	176.2	152.9
2007	165.0	172.9	161.7	137.2	152.6	158.7	156.1	174.2	184.5	171.4	137.7	131.2	169.0	157.8	170.6	145.3
2006	156.2	165.8	153.8	131.1	146.0	152.5	150.6	166.2	173.9	161.9	123.3	117.1	162.0	152.4	159.3	139.4
2005	146.7	156.2	145.1	123.1	134.9	142.0	140.6	157.3	164.6	152.8	116.0	109.9	151.3	143.1	149.4	131.7
2004	132.8	141.9	129.6	112.8	122.5	129.7	127.6	139.1	150.1	137.0	105.3	99.3	135.1	126.9	135.5	116.3
2003	129.7	138.7	127.9	110.9	120.8	127.7	126.0	136.7	146.5	134.5	101.4	96.7	131.9	124.8	133.3	113.9
2002	126.7	134.2	126.9	107.4	119.9	125.3	124.4	131.9	139.5	130.0	101.0	96.3	128.4	122.7	129.6	112.5
2001	122.2	129.4	122.4	104.3	115.2	120.1	119.7	131.4	136.1	124.9	98.7	93.9	121.8	114.7	125.5	106.7
2000	118.9	125.3	119.8	102.7	111.6	117.6	115.9	124.1	131.1	120.1	97.2	92.9	118.2	111.9	122.4	104.3
1999	116.6	122.6	113.7	100.9	106.1	111.5	110.1	120.3	126.5	117.1	95.7	91.8	114.9	106.1	119.8	101.1
1998	113.6	119.5	112.3	99.7	104.9	110.1	108.7	117.7	124.6	115.5	92.1	89.5	108.2	104.2	115.9	98.9
1997	111.5	117.6	110.8	98.9	104.2	108.7	107.5	115.9	121.9	114.1	90.9	88.3	106.4	102.4	113.2	97.3
1996	108.9	116.0	109.7	93.9	103.5	107.8	106.9	115.0	120.4	113.5	87.1	85.4	103.2	99.9	110.1	94.6
1995	105.6	110.1	104.1	91.1	96.4	98.3	102.3	100.3	111.9	102.5	84.2	83.5	99.7	96.1	106.3	89.5
1994	103.0	108.5	102.9	89.7	95.0	97.1	101.1	99.1	109.3	101.1	82.4	81.7	97.3	92.8	103.2	88.2
1993	100.0	105.4	100.7	87.7	92.7	95.3	98.8	99.8	106.7	99.9	80.0	79.4	94.5	90.9	99.9	86.2
1992	97.9	102.9	99.4	86.4	91.4	94.0	97.4	98.5	105.3	98.6	78.8	78.2	92.9	89.5	98.6	84.9
1991	95.7	97.6	93.6	85.3	90.0	92.1	90.3	96.0	102.9	97.3	77.6	76.9	90.9	88.1	96.4	83.9
1990	93.2	96.0	91.7	84.0	87.4	90.2	88.9	94.3	100.5	95.6	76.4	75.6	89.5	86.8	94.0	82.6
1989	91.0	94.0	90.2	82.4	85.8	88.6	87.4	92.5	97.4	93.7	75.0	74.1	87.3	85.1	91.8	80.9
1985	81.8	81.6	80.7	75.8	79.4	80.0	80.5	85.2	87.9	86.5	69.4	68.5	78.2	77.3	80.8	73.2
1980	60.7	62.5	63.1	59.8	61.7	60.3	63.3	63.4	64.1	64.8	54.3	54.3	59.1	60.4	59.7	56.7
1975	43.7	45.8	44.9	41.8	43.7	44.1	44.8	45.2	46.0	45.3	37.9	37.7	42.7	45.4	44.8	41.1
1970	27.8	29.7	28.5	26.7	28.1	27.8	28.7	28.9	29.5	28.9	24.4	21.3	25.3	28.3	28.4	26.0
1965	21.5	22.1	21.9	20.6	21.7	21.5	22.2	22.3	23.4	22.3	18.8	16.4	20.6	21.8	21.8	20.0
1960	19.5	20.1	20.0	18.7	19.7	19.5	20.1	20.3	20.5	20.2	17.1	14.9	19.0	19.8	19.5	18.2
1955	16.3	16.8	16.7	15.7	16.5	16.3	16.9	17.0	17.2	17.0	14.3	12.5	16.0	16.6	16.3	15.2
1950	13.5	13.9	13.8	13.0	13.6	13.5	13.9	14.0	14.2	14.0	11.8	10.3	13.2	13.7	13.5	12.6
1945	8.6	8.9	8.8	8.3	8.7	8.6	8.9	8.9	9.0	8.9	7.6	6.6	8.4	8.8	8.6	8.0
1940	6.6	6.8	6.8	6.4	6.7	6.6	6.9	6.9	7.0	6.9	5.8	5.1	6.5	6.7	6.7	6.2

Year	National 30 City Average	Montana Billings	Great Falls	Nebraska Lincoln	Omaha	Nevada Las Vegas	Reno	New Hampshire Man-chester	Nashua	New Jersey Camden	Jersey City	Newark	Pater-son	Trenton	NM Albu-querque	NY Albany
Jan 2010	182.8E	163.0E	163.8E	159.7E	163.0E	195.0E	175.1E	172.8E	172.2E	200.3E	201.6E	204.3E	203.2E	200.7E	162.3E	178.5E
2009	182.5	165.3	166.0	161.7	165.0	191.5	176.5	174.0	173.5	196.8	200.5	203.6	202.0	198.6	163.0	178.1
2008	171.0	153.1	154.6	152.0	154.8	176.2	167.0	162.5	161.9	184.0	187.3	189.6	188.7	185.6	152.6	166.0
2007	165.0	147.7	147.8	147.7	150.3	166.7	161.4	159.3	158.7	179.3	183.3	185.2	184.6	181.7	146.7	159.4
2006	156.2	140.5	140.8	132.4	140.7	160.0	154.5	146.7	146.4	167.7	171.0	173.6	172.2	169.9	140.1	151.6
2005	146.7	131.9	132.4	124.1	132.8	149.5	145.4	136.8	136.5	159.0	162.4	163.9	163.0	161.5	130.4	142.2
2004	132.8	118.2	117.9	112.1	119.4	137.1	130.5	124.2	123.9	143.4	145.4	147.2	146.6	146.0	118.3	128.6
2003	129.7	115.8	115.8	110.2	117.3	133.8	128.3	122.4	122.2	142.0	144.3	145.7	145.0	143.0	116.4	126.8
2002	126.7	114.6	114.5	108.2	115.0	131.9	126.4	119.9	119.6	135.9	138.5	140.2	140.1	137.6	114.6	122.6
2001	122.2	117.5	117.7	101.3	111.6	127.8	122.5	116.2	116.2	133.4	136.7	136.9	136.8	136.0	111.4	119.2
2000	118.9	113.7	113.9	98.8	107.0	125.8	118.2	111.9	111.9	128.4	130.5	132.6	132.4	130.6	109.0	116.5
1999	116.6	112.1	112.7	96.6	104.8	121.9	114.4	109.6	109.6	125.3	128.8	131.6	129.5	129.6	106.7	114.6
1998	113.6	109.7	109.1	94.9	101.2	118.1	111.4	110.1	110.0	124.3	127.7	128.9	128.5	127.9	103.8	113.0
1997	111.5	107.4	107.6	93.3	99.5	114.6	109.8	108.6	108.6	121.9	125.1	126.4	126.4	125.2	100.8	110.0
1996	108.9	108.0	107.5	91.4	97.4	111.8	108.9	106.5	106.4	119.1	122.5	122.5	123.8	121.8	98.6	108.3
1995	105.6	104.7	104.9	85.6	93.4	108.5	105.0	100.9	100.8	107.0	112.2	111.9	112.1	111.2	96.3	103.6
1994	103.0	100.2	100.7	84.3	91.0	105.3	102.2	97.8	97.7	105.4	110.6	110.1	110.6	108.2	93.4	102.4
1993	100.0	97.9	97.8	82.3	88.7	102.8	99.9	95.4	95.4	103.7	109.0	108.2	109.0	106.4	90.1	99.8
1992	97.9	95.5	96.4	81.1	87.4	99.4	98.4	90.4	90.4	102.0	107.5	107.0	107.8	103.9	87.5	98.5
1991	95.7	94.2	95.1	80.1	86.4	97.5	95.8	87.8	87.8	95.0	98.5	95.6	100.0	96.6	86.2	96.1
1990	93.2	92.9	93.8	78.8	85.0	96.3	94.5	86.3	86.3	93.0	93.5	93.8	97.2	94.5	84.9	93.2
1989	91.0	91.3	92.2	77.3	83.6	94.8	92.3	84.8	84.8	90.4	91.4	91.8	95.5	91.9	83.3	88.4
1985	81.8	83.9	84.3	71.5	77.5	87.6	85.0	78.1	78.1	81.3	83.3	83.5	84.5	82.8	76.7	79.5
1980	60.7	63.9	64.6	58.5	63.5	64.6	63.0	56.6	56.0	58.6	60.6	60.1	60.0	58.9	59.0	59.5
1975	43.7	43.1	43.8	40.9	43.2	42.8	41.9	41.3	40.8	42.3	43.4	44.2	43.9	43.8	40.3	43.9
1970	27.8	28.5	28.9	26.4	26.8	29.4	28.0	26.2	25.6	27.2	27.8	29.0	27.8	27.4	26.4	28.3
1965	21.5	22.0	22.3	20.3	20.6	22.4	21.6	20.6	19.7	20.9	21.4	23.8	21.4	21.3	20.6	22.3
1960	19.5	20.0	20.3	18.5	18.7	20.2	19.6	18.0	17.9	19.0	19.4	19.4	19.4	19.2	18.5	19.3
1955	16.3	16.7	17.0	15.5	15.7	16.9	16.4	15.1	15.0	16.0	16.3	16.3	16.3	16.1	15.6	16.2
1950	13.5	13.9	14.0	12.8	13.0	14.0	13.6	12.5	12.4	13.2	13.5	13.5	13.5	13.3	12.9	13.4
1945	8.6	8.8	9.0	8.1	8.3	8.9	8.7	8.0	7.9	8.4	8.6	8.6	8.6	8.5	8.2	8.5
1940	6.6	6.8	6.9	6.3	6.4	6.9	6.7	6.2	6.1	6.5	6.6	6.6	6.6	6.6	6.3	6.6

Historical Cost Indexes

| Year | National 30 City Average | New York | | | | | | | | North Carolina | | | | | N. Dakota | Ohio |
		Bing-hamton	Buffalo	New York	Roches-ter	Schen-ectady	Syracuse	Utica	Yonkers	Charlotte	Durham	Greens-boro	Raleigh	Winston-Salem	Fargo	Akron
Jan 2010	182.8E	173.1E	185.3E	243.5E	180.5E	178.8E	176.9E	171.3E	217.8E	142.3E	143.0E	141.2E	142.3E	139.9E	152.0E	175.7E
2009	182.5	172.9	184.4	239.9	179.6	178.8	176.9	171.3	217.5	145.1	145.8	144.0	145.1	142.7	154.3	175.7
2008	171.0	160.7	173.9	226.8	168.0	167.2	165.9	161.6	202.1	135.8	136.6	134.6	135.3	133.8	145.0	165.4
2007	165.0	155.7	168.6	215.2	163.5	159.8	159.8	155.0	196.5	132.8	133.7	131.7	132.1	131.0	139.8	159.0
2006	156.2	147.5	159.2	204.5	155.2	151.1	150.9	146.4	186.0	125.1	124.6	123.8	124.2	123.0	133.2	152.8
2005	146.7	137.5	149.4	194.0	147.3	142.2	141.9	136.9	176.1	110.5	112.1	112.1	112.1	111.1	125.3	144.4
2004	132.8	123.5	136.1	177.7	132.0	128.4	127.3	124.4	161.8	98.9	100.0	100.1	100.3	99.4	113.0	131.9
2003	129.7	121.8	132.8	173.4	130.3	126.7	124.7	121.7	160.0	96.2	97.5	97.5	97.8	96.8	110.6	129.6
2002	126.7	119.0	128.5	170.1	127.1	123.0	121.8	118.4	154.8	94.8	96.1	96.1	96.3	95.5	106.3	127.2
2001	122.2	116.0	125.2	164.4	123.1	120.1	118.6	115.3	151.4	91.5	92.9	92.9	93.3	92.3	103.1	123.5
2000	118.9	112.4	122.3	159.2	120.0	117.5	115.1	112.4	144.8	90.3	91.6	91.6	91.9	91.0	97.9	117.8
1999	116.6	108.6	120.2	155.9	116.8	114.7	113.7	108.5	140.6	89.3	90.2	90.3	90.5	90.1	96.8	116.0
1998	113.6	109.0	119.1	154.4	117.2	114.2	113.6	108.6	141.4	88.2	89.1	89.2	89.4	89.0	95.2	113.1
1997	111.5	107.2	115.7	150.3	115.2	111.2	111.2	110.6	138.8	86.8	87.7	87.8	87.9	87.6	93.5	110.6
1996	108.9	105.5	114.0	148.0	113.3	109.5	108.2	105.2	137.2	85.0	85.9	86.0	86.1	85.8	91.7	107.9
1995	105.6	99.3	110.1	140.7	106.6	104.6	104.0	97.7	129.4	81.8	82.6	82.6	82.7	82.6	88.4	103.4
1994	103.0	98.1	107.2	137.1	105.2	103.5	102.3	96.5	128.1	80.3	81.1	81.0	81.2	81.1	87.0	102.2
1993	100.0	95.7	102.2	133.3	102.0	100.8	99.6	94.1	126.3	78.2	78.9	78.9	78.9	78.8	85.8	100.3
1992	97.9	93.7	100.1	128.2	99.3	99.3	98.1	90.5	123.9	77.1	77.7	77.8	77.8	77.6	83.1	98.0
1991	95.7	89.5	96.8	124.4	96.0	96.7	95.5	88.7	121.5	76.0	76.6	76.8	76.7	76.6	82.6	96.8
1990	93.2	87.0	94.1	118.1	94.6	93.9	91.0	85.6	111.4	74.8	75.4	75.5	75.5	75.3	81.3	94.6
1989	91.0	85.4	91.8	114.8	92.6	89.6	88.4	84.1	102.1	73.3	74.0	74.0	74.0	73.8	79.7	92.9
1985	81.8	77.5	83.2	94.9	81.6	80.1	81.0	77.0	92.8	66.9	67.6	67.7	67.6	67.5	73.4	86.8
1980	60.7	58.0	60.6	66.0	60.5	60.3	61.6	58.5	65.9	51.1	52.2	52.5	51.7	50.8	57.4	62.3
1975	43.7	42.5	45.1	49.5	44.8	43.7	44.8	42.7	47.1	36.1	37.0	37.0	37.3	36.1	39.3	44.7
1970	27.8	27.0	28.9	32.8	29.2	27.8	28.5	26.9	30.0	20.9	23.7	23.6	23.4	23.0	25.8	28.3
1965	21.5	20.8	22.2	25.5	22.8	21.4	21.9	20.7	23.1	16.0	18.2	18.2	18.0	17.7	19.9	21.8
1960	19.5	18.9	19.9	21.5	19.7	19.5	19.9	18.8	21.0	14.4	16.6	16.6	16.4	16.1	18.1	19.9
1955	16.3	15.8	16.7	18.1	16.5	16.3	16.7	15.8	17.6	12.1	13.9	13.9	13.7	13.5	15.2	16.6
1950	13.5	13.1	13.8	14.9	13.6	13.5	13.8	13.0	14.5	10.0	11.5	11.5	11.4	11.2	12.5	13.7
1945	8.6	8.3	8.8	9.5	8.7	8.6	8.8	8.3	9.3	6.4	7.3	7.3	7.2	7.1	8.0	8.8
1940	6.6	6.4	6.8	7.4	6.7	6.7	6.8	6.4	7.2	4.9	5.6	5.6	5.6	5.5	6.2	6.8

| Year | National 30 City Average | Ohio | | | | | | | | | Oklahoma | | | Oregon | | PA |
		Canton	Cincin-nati	Cleve-land	Colum-bus	Dayton	Lorain	Spring-field	Toledo	Youngs-town	Lawton	Oklahoma City	Tulsa	Eugene	Port-land	Allen-town
Jan 2010	182.8E	168.4E	168.4E	181.9E	171.5E	165.3E	174.0E	166.0E	178.5E	172.8E	149.7E	150.4E	143.9E	179.1E	181.4E	189.5E
2009	182.5	168.4	168.1	181.6	171.2	165.3	174.0	166.0	178.5	172.8	152.5	153.3	146.8	181.3	183.6	188.3
2008	171.0	158.4	159.0	170.6	160.2	157.5	163.9	157.5	168.2	161.5	141.3	140.6	136.2	172.3	174.7	175.7
2007	165.0	153.2	152.4	164.9	155.1	149.4	158.1	150.3	161.6	156.7	136.2	135.6	132.5	168.5	169.5	169.4
2006	156.2	147.0	144.9	156.9	146.9	142.9	151.8	143.4	154.5	150.3	129.5	129.5	125.7	160.5	161.7	160.1
2005	146.7	137.7	136.8	147.9	138.0	134.5	143.3	134.8	145.5	141.1	121.6	121.6	117.8	150.9	152.2	150.2
2004	132.8	125.6	124.3	135.6	126.2	121.1	131.5	122.0	132.8	129.3	109.3	109.4	107.3	136.8	137.8	133.9
2003	129.7	123.7	121.3	132.7	123.7	119.4	126.1	120.0	130.4	126.2	107.5	107.8	105.0	134.2	135.9	130.3
2002	126.7	120.9	119.2	130.0	120.9	117.5	124.3	118.0	128.4	123.6	106.3	106.0	104.1	132.2	133.9	128.1
2001	122.2	117.7	115.9	125.9	117.1	114.0	120.4	114.7	123.8	119.9	102.0	102.1	99.7	129.8	131.2	123.5
2000	118.9	112.7	110.1	121.3	112.5	109.0	115.0	109.2	115.7	114.1	98.7	98.9	97.5	126.3	127.4	119.9
1999	116.6	110.6	107.9	118.7	109.5	107.1	112.0	106.4	113.6	111.9	97.5	97.4	96.2	120.9	124.3	117.8
1998	113.6	108.5	105.0	114.8	106.8	104.5	109.4	104.1	111.2	109.0	94.3	94.5	94.5	120.0	122.2	115.7
1997	111.5	106.4	102.9	112.9	104.7	102.6	107.4	102.2	108.5	107.2	92.9	93.2	93.0	118.1	119.6	113.9
1996	108.9	103.2	100.2	110.1	101.1	98.8	103.8	98.4	105.1	104.1	90.9	91.3	91.2	114.4	116.1	112.0
1995	105.6	98.8	97.1	106.4	99.1	94.6	97.1	92.0	100.6	100.1	85.3	88.0	89.0	112.2	114.3	108.3
1994	103.0	97.7	95.0	104.8	95.4	93.0	96.0	90.1	99.3	98.9	84.0	86.5	87.6	107.6	109.2	106.4
1993	100.0	95.9	92.3	101.9	93.9	90.6	93.9	87.7	98.3	97.2	81.3	83.9	84.8	107.2	108.8	103.6
1992	97.9	94.5	90.6	98.7	92.6	89.2	92.2	86.4	97.1	96.0	80.1	82.7	83.4	101.1	102.6	101.6
1991	95.7	93.4	88.6	97.2	90.6	87.7	91.1	84.9	94.2	92.7	78.3	80.7	81.4	99.5	101.1	98.9
1990	93.2	92.1	86.7	95.1	88.1	85.9	89.4	83.4	92.9	91.8	77.1	79.9	80.0	98.0	99.4	95.0
1989	91.0	90.3	84.6	93.6	85.8	82.6	87.9	81.0	91.4	88.4	75.8	78.5	78.4	95.2	97.0	92.4
1985	81.8	83.7	78.2	86.2	77.7	76.3	80.7	74.3	84.7	82.7	71.2	73.9	73.8	88.7	90.1	81.9
1980	60.7	60.6	59.8	61.0	58.6	56.6	60.8	56.2	64.0	61.5	52.2	55.5	57.2	68.9	68.4	58.7
1975	43.7	44.0	43.7	44.6	42.2	40.5	42.7	40.0	44.9	44.5	38.6	38.6	39.2	44.6	44.9	42.3
1970	27.8	27.8	28.2	29.6	26.7	27.0	27.5	25.4	29.1	29.6	24.1	23.4	24.1	30.4	30.0	27.3
1965	21.5	21.5	20.7	21.1	20.2	20.0	21.1	19.6	21.3	21.0	18.5	17.8	20.6	23.4	22.6	21.0
1960	19.5	19.5	19.3	19.6	18.7	18.1	19.2	17.8	19.4	19.1	16.9	16.1	18.1	21.2	21.1	19.1
1955	16.3	16.3	16.2	16.5	15.7	15.2	16.1	14.9	16.2	16.0	14.1	13.5	15.1	17.8	17.7	16.0
1950	13.5	13.5	13.3	13.6	13.0	12.5	13.3	12.3	13.4	13.2	11.7	11.2	12.5	14.7	14.6	13.2
1945	8.6	8.6	8.5	8.7	8.3	8.0	8.5	7.8	8.6	8.4	7.4	7.1	8.0	9.4	9.3	8.4
1940	6.6	6.7	6.5	6.7	6.4	6.2	6.5	6.1	6.6	6.5	5.7	5.5	6.1	7.2	7.2	6.5

Historical Cost Indexes

Year	National 30 City Average	Pennsylvania Erie	Harrisburg	Philadelphia	Pittsburgh	Reading	Scranton	RI Providence	South Carolina Charleston	Columbia	South Dakota Rapid City	Sioux Falls	Tennessee Chattanooga	Knoxville	Memphis	Nashville
Jan 2010	182.8E	171.9E	177.2E	210.7E	182.1E	181.9E	182.5E	193.4E	148.3E	141.7E	145.7E	148.5E	154.7E	146.7E	156.2E	159.2E
2009	182.5	171.9	176.5	209.5	181.7	180.7	181.3	192.9	150.8	144.3	148.2	151.1	155.7	148.0	157.2	159.8
2008	171.0	161.0	165.7	196.2	169.2	169.3	167.9	178.2	142.1	134.8	138.4	141.3	136.7	133.4	146.0	147.5
2007	165.0	156.0	159.4	188.4	163.2	164.9	163.1	174.4	139.3	131.7	129.5	133.0	133.7	130.6	143.1	144.4
2006	156.2	148.7	150.7	177.8	155.5	155.0	153.9	163.6	122.4	118.7	122.8	126.6	127.0	123.9	136.5	135.5
2005	146.7	140.6	141.5	166.4	146.6	145.5	142.9	155.6	110.2	109.6	115.2	118.5	116.0	115.1	128.6	127.9
2004	132.8	127.0	126.2	148.4	133.1	128.7	129.0	138.8	98.9	98.4	103.9	107.3	105.0	104.5	115.6	115.8
2003	129.7	124.5	123.8	145.5	130.5	126.3	126.0	135.9	96.1	95.7	101.3	104.0	102.8	102.3	111.1	110.9
2002	126.7	122.1	121.1	142.0	127.8	123.4	124.2	131.1	94.8	93.9	100.2	103.2	102.2	101.2	107.6	109.2
2001	122.2	118.9	118.4	136.9	124.1	120.4	121.1	127.8	92.1	91.3	96.8	99.7	98.2	96.5	102.6	104.3
2000	118.9	114.6	114.0	132.1	120.9	115.9	117.7	122.8	89.9	89.1	94.2	98.0	96.9	95.0	100.8	100.8
1999	116.6	113.8	112.8	129.8	119.6	114.8	116.3	121.6	89.0	88.1	92.4	95.8	95.9	93.4	99.7	98.6
1998	113.6	109.8	110.5	126.6	117.1	112.5	113.7	120.3	88.0	87.2	90.7	93.6	94.9	92.4	97.2	96.8
1997	111.5	108.5	108.8	123.3	113.9	110.8	112.3	118.9	86.5	85.6	89.3	92.1	93.6	90.8	96.1	94.9
1996	108.9	106.8	105.7	120.3	110.8	108.7	109.9	117.2	84.9	84.1	87.2	90.0	91.6	88.8	94.0	91.9
1995	105.6	99.8	100.6	117.1	106.3	103.6	103.8	111.1	82.7	82.2	84.0	84.7	89.2	86.0	91.2	87.6
1994	103.0	98.2	99.4	115.2	103.7	102.2	102.6	109.6	81.2	80.6	82.7	83.1	87.4	84.1	89.0	84.8
1993	100.0	94.9	96.6	107.4	99.0	98.6	99.8	108.2	78.4	77.9	81.2	81.9	85.1	82.3	86.8	81.9
1992	97.9	92.1	94.8	105.3	96.5	96.7	97.8	106.9	77.1	76.8	80.0	80.7	83.6	77.7	85.4	80.7
1991	95.7	90.5	92.5	101.7	93.2	94.1	94.8	96.1	75.0	75.8	78.7	79.7	81.9	76.6	83.0	79.1
1990	93.2	88.5	89.5	98.5	91.2	90.8	91.3	94.1	73.7	74.5	77.2	78.4	79.9	75.1	81.3	77.1
1989	91.0	86.6	86.5	94.2	89.4	87.9	88.8	92.4	72.1	72.8	75.7	77.0	78.5	73.7	81.2	74.8
1985	81.8	79.6	77.2	82.2	81.5	77.7	80.0	83.0	65.9	66.9	69.7	71.2	72.5	67.7	74.3	66.7
1980	60.7	59.4	56.9	58.7	61.3	57.7	58.4	59.2	50.6	51.1	54.8	57.1	55.0	51.6	55.9	53.1
1975	43.7	43.4	42.2	44.5	44.5	43.6	41.9	42.7	35.0	36.4	37.7	39.6	39.1	37.3	40.7	37.0
1970	27.8	27.5	25.8	27.5	28.7	27.1	27.0	27.3	22.8	22.9	24.8	25.8	24.9	22.0	23.0	22.8
1965	21.5	21.1	20.1	21.7	22.4	20.9	20.8	21.9	17.6	17.6	19.1	19.9	19.2	17.1	18.3	17.4
1960	19.5	19.1	18.6	19.4	19.7	19.0	18.9	19.1	16.0	16.0	17.3	18.1	17.4	15.5	16.6	15.8
1955	16.3	16.1	15.6	16.3	16.5	15.9	15.9	16.0	13.4	13.4	14.5	15.1	14.6	13.0	13.9	13.3
1950	13.5	13.3	12.9	13.5	13.6	13.2	13.1	13.2	11.1	11.1	12.0	12.5	12.1	10.7	11.5	10.9
1945	8.6	8.5	8.2	8.6	8.7	8.4	8.3	8.4	7.1	7.1	7.7	8.0	7.7	6.8	7.3	7.0
1940	6.6	6.5	6.3	6.6	6.7	6.4	6.5	6.5	5.4	5.5	5.9	6.2	6.0	5.3	5.6	5.4

Year	National 30 City Average	Texas Abilene	Amarillo	Austin	Beaumont	Corpus Christi	Dallas	El Paso	Fort Worth	Houston	Lubbock	Odessa	San Antonio	Waco	Wichita Falls	Utah Ogden
Jan 2010	182.8E	143.5E	147.5E	146.6E	150.2E	142.1E	155.9E	141.3E	150.4E	161.0E	145.0E	139.5E	151.0E	146.7E	146.1E	153.0E
2009	182.5	143.5	147.3	146.4	149.8	141.7	155.5	141.1	150.1	160.9	144.6	139.1	150.8	146.7	146.1	155.1
2008	171.0	132.4	137.0	137.5	140.3	133.2	144.1	130.7	138.3	149.1	134.7	128.9	141.0	136.1	136.1	144.8
2007	165.0	128.5	132.2	131.6	137.1	128.8	138.6	126.2	134.8	146.0	130.2	125.0	136.4	132.1	132.1	140.0
2006	156.2	121.6	125.7	125.5	130.1	122.0	131.1	120.3	127.6	138.2	123.2	118.3	129.8	125.2	125.3	133.4
2005	146.7	113.8	117.3	117.9	121.3	114.4	123.7	112.5	119.4	129.0	115.5	110.5	121.3	116.1	117.3	126.0
2004	132.8	102.9	106.3	105.7	108.5	102.9	112.0	101.5	108.4	115.9	104.7	99.9	108.4	104.8	105.1	115.2
2003	129.7	100.5	104.4	103.9	106.1	100.4	109.3	99.5	105.5	113.4	102.3	97.6	104.8	102.7	103.0	112.7
2002	126.7	99.9	101.9	102.4	104.5	98.9	107.9	98.7	104.6	111.5	100.5	95.9	103.7	100.5	101.4	110.8
2001	122.2	93.4	98.4	99.8	102.3	96.6	103.8	95.5	100.9	107.8	97.6	93.4	100.5	97.8	97.3	107.7
2000	118.9	93.4	98.1	99.1	101.6	96.9	102.7	92.4	99.9	106.0	97.6	93.4	99.4	97.3	96.9	104.6
1999	116.6	91.8	94.5	96.0	99.7	94.0	101.0	90.7	97.6	104.6	96.0	92.1	98.0	94.8	95.5	103.3
1998	113.6	89.8	92.7	94.2	97.9	91.8	97.9	88.4	94.5	101.3	93.3	90.1	94.8	92.7	92.9	98.5
1997	111.5	88.4	91.3	92.8	96.8	90.3	96.1	87.0	93.3	100.1	91.9	88.8	93.4	91.4	91.5	96.1
1996	108.9	86.8	89.6	90.9	95.3	88.5	94.1	86.7	91.5	97.9	90.2	87.1	92.3	89.7	89.9	94.1
1995	105.6	85.2	87.4	89.3	93.7	87.4	91.4	85.2	89.5	95.9	88.4	85.6	88.9	86.4	86.8	92.2
1994	103.0	83.3	85.5	87.0	91.8	84.6	89.6	82.2	87.4	93.4	87.0	83.9	87.0	84.9	85.4	89.4
1993	100.0	81.4	83.4	84.9	90.0	82.8	87.8	80.2	85.5	91.1	84.8	81.9	85.0	83.0	83.5	87.1
1992	97.9	80.3	82.3	83.8	88.9	81.6	86.2	79.0	84.1	89.8	83.6	80.7	83.9	81.8	82.4	85.1
1991	95.7	79.2	81.1	82.8	87.8	80.6	85.9	78.0	83.3	87.9	82.7	79.8	83.3	80.6	81.5	84.1
1990	93.2	78.0	80.1	81.3	86.5	79.3	84.5	76.7	82.1	85.4	81.5	78.6	80.7	79.6	80.3	83.4
1989	91.0	76.6	78.7	79.7	85.1	77.8	82.5	75.2	80.6	84.0	80.0	77.0	78.7	78.1	78.7	81.9
1985	81.8	71.1	72.5	74.5	79.3	72.3	77.6	69.4	75.1	79.6	74.0	71.2	73.9	71.7	73.3	75.2
1980	60.7	53.4	55.2	54.5	57.6	54.5	57.9	53.1	57.0	59.4	55.6	57.2	55.0	54.9	55.4	62.2
1975	43.7	37.6	39.0	39.0	39.6	38.1	40.7	38.0	40.4	41.2	38.9	37.9	39.0	38.6	38.0	40.0
1970	27.8	24.5	24.9	24.9	25.7	24.5	25.5	23.7	25.9	25.4	25.1	24.6	23.3	24.8	24.5	26.8
1965	21.5	18.9	19.2	19.2	19.9	18.9	19.9	19.0	19.9	20.0	19.4	19.0	18.5	19.2	18.9	20.6
1960	19.5	17.1	17.4	17.4	18.1	17.1	18.2	17.0	18.1	18.2	17.6	17.3	16.8	17.4	17.2	18.8
1955	16.3	14.4	14.6	14.6	15.1	14.4	15.3	14.3	15.2	15.2	14.8	14.5	14.1	14.6	14.4	15.7
1950	13.5	11.9	12.1	12.1	12.5	11.9	12.6	11.8	12.5	12.6	12.2	12.0	11.6	12.1	11.9	13.0
1945	8.6	7.6	7.7	7.7	8.0	7.6	8.0	7.5	8.0	8.0	7.8	7.6	7.4	7.7	7.6	8.3
1940	6.6	5.9	5.9	5.9	6.1	5.8	6.2	5.8	6.2	6.2	6.0	5.9	5.7	5.9	5.8	6.4

Historical Cost Indexes

Year	National 30 City Average	Utah Salt Lake City	Vermont Burlington	Vermont Rutland	Virginia Alexandria	Virginia Newport News	Virginia Norfolk	Virginia Richmond	Virginia Roanoke	Washington Seattle	Washington Spokane	Washington Tacoma	West Virginia Charleston	West Virginia Huntington	Wisconsin Green Bay	Wisconsin Kenosha
Jan 2010	182.8E	160.4E	157.0E	155.5E	171.2E	158.0E	159.5E	158.5E	152.8E	191.2E	171.0E	186.8E	173.8E	176.3E	174.1E	181.2E
2009	182.5	160.4	158.3	156.7	172.5	160.0	161.9	160.7	155.6	188.5	173.0	186.3	174.6	177.4	175.5	182.5
2008	171.0	149.7	147.2	145.8	161.8	150.8	150.8	150.9	145.9	176.9	162.4	174.9	162.8	165.7	165.6	172.2
2007	165.0	144.6	144.4	143.2	155.0	146.2	146.1	147.3	142.9	171.4	156.7	168.7	158.7	160.1	159.4	164.8
2006	156.2	137.7	131.7	130.7	146.2	133.7	134.6	134.8	129.7	162.9	150.1	160.7	149.4	151.4	152.7	157.1
2005	146.7	129.4	124.6	123.8	136.5	123.5	124.4	125.4	112.2	153.9	141.9	151.5	140.8	141.0	144.4	148.3
2004	132.8	117.8	113.0	112.3	121.5	108.9	110.2	110.9	99.7	138.0	127.6	134.5	124.8	125.6	128.9	132.4
2003	129.7	116.0	110.7	110.1	119.5	104.8	106.2	108.6	97.0	134.9	125.9	133.0	123.3	123.6	127.4	130.9
2002	126.7	113.7	109.0	108.4	115.1	102.9	104.1	106.6	95.2	132.7	123.9	131.4	121.2	120.9	123.0	127.3
2001	122.2	109.1	105.7	105.2	110.8	99.8	100.3	102.9	92.1	127.9	120.3	125.7	114.6	117.5	119.1	123.6
2000	118.9	106.5	98.9	98.3	108.1	96.5	97.6	100.2	90.7	124.6	118.3	122.9	111.5	114.4	114.6	119.1
1999	116.6	104.5	98.2	97.7	106.1	95.6	96.5	98.8	89.8	123.3	116.7	121.6	110.6	113.4	112.1	115.8
1998	113.6	99.5	97.8	97.3	104.1	93.7	93.9	97.0	88.3	119.4	114.3	118.3	106.7	109.0	109.5	112.9
1997	111.5	97.2	96.6	96.3	101.2	91.6	91.7	92.9	86.9	118.1	111.7	117.2	105.3	107.7	105.6	109.1
1996	108.9	94.9	95.1	94.8	99.7	90.2	90.4	91.6	85.5	115.2	109.2	114.3	103.1	104.8	103.8	106.4
1995	105.6	93.1	91.1	90.8	96.3	86.0	86.4	87.8	82.8	113.7	107.4	112.8	95.8	97.2	97.6	97.9
1994	103.0	90.2	89.5	89.3	93.9	84.6	84.8	86.3	81.4	109.9	104.0	108.3	94.3	95.3	96.3	96.2
1993	100.0	87.9	87.6	87.6	91.6	82.9	83.0	84.3	79.5	107.3	103.9	106.7	92.6	93.5	94.0	94.3
1992	97.9	86.0	86.1	86.1	90.1	81.0	81.6	82.0	78.3	105.1	101.4	103.7	91.4	92.3	92.0	92.1
1991	95.7	84.9	84.2	84.2	88.2	77.6	77.9	79.8	77.3	102.2	100.0	102.2	89.7	88.6	88.6	89.8
1990	93.2	84.3	83.0	82.9	86.1	76.3	76.7	77.6	76.1	100.1	98.5	100.5	86.1	86.8	86.7	87.8
1989	91.0	82.8	81.4	81.3	83.4	74.9	75.2	76.0	74.2	96.1	96.8	98.4	84.5	84.8	84.3	85.6
1985	81.8	75.9	74.8	74.9	75.1	68.7	68.8	69.5	67.2	88.3	89.0	91.2	77.7	77.7	76.7	77.4
1980	60.7	57.0	55.3	58.3	57.3	52.5	52.4	54.3	51.3	67.9	66.3	66.7	57.7	58.3	58.6	58.3
1975	43.7	40.1	41.8	43.9	41.7	37.2	36.9	37.1	37.1	44.9	44.4	44.5	41.0	40.0	40.9	40.5
1970	27.8	26.1	25.4	26.8	26.2	23.9	21.5	22.0	23.7	28.8	29.3	29.6	26.1	25.8	26.4	26.5
1965	21.5	20.0	19.8	20.6	20.2	18.4	17.1	17.2	18.3	22.4	22.5	22.8	20.1	19.9	20.3	20.4
1960	19.5	18.4	18.0	18.8	18.4	16.7	15.4	15.6	16.6	20.4	20.8	20.8	18.3	18.1	18.4	18.6
1955	16.3	15.4	15.1	15.7	15.4	14.0	12.9	13.1	13.9	17.1	17.4	17.4	15.4	15.2	15.5	15.6
1950	13.5	12.7	12.4	13.0	12.7	11.6	10.7	10.8	11.5	14.1	14.4	14.4	12.7	12.5	12.8	12.9
1945	8.6	8.1	7.9	8.3	8.1	7.4	6.8	6.9	7.3	9.0	9.2	9.2	8.1	8.0	8.1	8.2
1940	6.6	6.3	6.1	6.4	6.2	5.7	5.3	5.3	5.7	7.0	7.1	7.1	6.2	6.2	6.3	6.3

Year	National 30 City Average	Wisconsin Madison	Wisconsin Milwaukee	Wisconsin Racine	Wyoming Cheyenne	Canada Calgary	Canada Edmonton	Canada Hamilton	Canada London	Canada Montreal	Canada Ottawa	Canada Quebec	Canada Toronto	Canada Vancouver	Canada Winnipeg
Jan 2010	182.8E	178.7E	188.7E	181.8E	153.1E	203.6E	204.3E	201.3E	197.2E	197.3E	198.4E	196.0E	204.0E	197.2E	186.4E
2009	182.5	180.0	187.3	182.8	155.2	205.4	206.6	203.0	198.9	198.8	200.1	197.4	205.6	199.6	188.5
2008	171.0	168.4	176.3	173.0	146.5	190.2	191.0	194.3	188.6	186.7	188.0	186.8	194.6	184.9	174.4
2007	165.0	160.7	168.9	164.7	141.3	183.1	184.4	186.8	182.4	181.7	182.4	182.0	187.7	180.6	170.1
2006	156.2	152.8	158.8	157.2	128.0	163.6	164.8	169.0	164.9	158.6	163.7	159.0	170.6	169.3	155.9
2005	146.7	145.2	148.4	147.9	118.9	154.4	155.7	160.0	156.2	149.2	155.1	150.1	162.6	159.4	146.9
2004	132.8	129.7	134.2	132.7	104.7	138.8	139.4	142.6	140.4	134.5	141.0	135.7	146.0	141.7	129.7
2003	129.7	128.4	131.1	131.5	102.6	133.6	134.2	139.1	137.0	130.2	137.4	131.4	142.3	137.0	124.4
2002	126.7	124.5	128.2	126.5	101.7	122.5	122.2	136.3	134.1	127.2	134.9	128.4	139.8	134.6	121.4
2001	122.2	120.6	123.9	123.3	99.0	117.5	117.4	131.5	129.1	124.4	130.2	125.5	134.7	130.2	117.2
2000	118.9	116.6	120.5	118.7	98.1	115.9	115.8	130.0	127.5	122.8	128.8	124.1	133.1	128.4	115.6
1999	116.6	115.9	117.4	115.5	96.9	115.3	115.2	128.1	125.6	120.8	126.8	121.9	131.2	127.1	115.2
1998	113.6	110.8	113.4	112.7	95.4	112.5	112.4	126.3	123.9	119.0	124.7	119.6	128.5	123.8	113.7
1997	111.5	106.1	110.1	109.3	93.2	110.7	110.6	124.4	121.9	114.6	122.8	115.4	125.7	121.9	111.4
1996	108.9	104.4	107.1	106.5	91.1	109.1	109.0	122.6	120.2	112.9	121.0	113.6	123.9	119.0	109.7
1995	105.6	96.5	103.9	97.8	87.6	107.4	107.4	119.9	117.5	110.8	118.2	111.5	121.6	116.2	107.6
1994	103.0	94.5	100.6	96.1	85.4	106.7	106.6	116.5	114.2	109.5	115.0	110.2	117.8	115.1	105.5
1993	100.0	91.3	96.7	93.8	82.9	104.7	104.5	113.7	111.9	106.9	112.0	107.0	114.9	109.2	102.8
1992	97.9	89.2	93.9	91.7	81.7	103.4	103.2	112.4	110.7	104.2	110.8	103.6	113.7	108.0	101.6
1991	95.7	86.2	91.6	89.3	80.3	102.1	102.0	108.2	106.7	101.8	106.9	100.4	109.0	106.8	98.6
1990	93.2	84.3	88.9	87.3	79.1	98.0	97.1	103.8	101.4	99.0	102.7	96.8	104.6	103.2	95.1
1989	91.0	81.8	86.4	85.0	77.8	95.6	94.7	98.2	96.5	94.9	98.5	92.5	98.5	97.6	92.9
1985	81.8	74.3	77.4	77.0	72.3	90.2	89.1	85.8	84.9	82.4	83.9	79.9	86.5	89.8	83.4
1980	60.7	56.8	58.8	58.1	56.9	64.9	63.3	63.7	61.9	59.2	60.9	59.3	60.9	65.0	61.7
1975	43.7	40.7	43.3	40.7	40.6	42.2	41.6	42.9	41.6	39.7	41.5	39.0	42.2	42.4	39.2
1970	27.8	26.5	29.4	26.5	26.0	28.9	28.6	28.5	27.8	25.6	27.6	26.0	25.6	26.0	23.1
1965	21.5	20.6	21.8	20.4	20.0	22.3	22.0	22.0	21.4	18.7	21.2	20.1	19.4	20.5	17.5
1960	19.5	18.1	19.0	18.6	18.2	20.2	20.0	20.0	19.5	17.0	19.3	18.2	17.6	18.6	15.8
1955	16.3	15.2	15.9	15.6	15.2	17.0	16.8	16.7	16.3	14.3	16.2	15.3	14.8	15.5	13.3
1950	13.5	13.2	13.2	12.9	12.6	14.0	13.9	13.8	13.5	11.8	13.4	12.6	12.2	12.8	10.9
1945	8.6	8.0	8.4	8.2	8.0	9.0	8.8	8.8	8.6	7.5	8.5	8.0	7.8	8.2	7.0
1940	6.6	6.2	6.5	6.3	6.2	6.9	6.8	6.8	6.6	5.8	6.6	6.2	6.0	6.3	5.4

Accent lighting–Fixtures or directional beams of light arranged to bring attention to an object or area.

Acoustical material–A material fabricated for the sole purpose of absorbing sound.

Addition–An expansion to an existing structure generally in the form of a room, floor(s) or wing(s). An increase in the floor area or volume of a structure.

Aggregate–Materials such as sand, gravel, stone, vermiculite, perlite and fly ash (slag) which are essential components in the production of concrete, mortar or plaster.

Air conditioning system–An air treatment to control the temperature, humidity and cleanliness of air and to provide for its distribution throughout the structure.

Air curtain–A stream of air directed downward to prevent the loss of hot or cool air from the structure and inhibit the entrance of dust and insects. Generally installed at loading platforms.

Anodized aluminum–Aluminum treated by an electrolytic process to produce an oxide film that is corrosion resistant.

Arcade–A covered passageway between buildings often with shops and offices on one or both sides.

Ashlar–A square-cut building stone or a wall constructed of cut stone.

Asphalt paper–A sheet paper material either coated or saturated with asphalt to increase its strength and resistance to water.

Asphalt shingles–Shingles manufactured from saturated roofing felts, coated with asphalt and topped with mineral granules to prevent weathering.

Assessment ratio–The ratio between the market value and assessed valuation of a property.

Back-up–That portion of a masonry wall behind the exterior facing-usually load bearing.

Balcony–A platform projecting from a building either supported from below or cantilevered. Usually protected by railing.

Bay–Structural component consisting of beams and columns occurring consistently throughout the structure.

Beam–A horizontal structural framing member that transfers loads to vertical members (columns) or bearing walls.

Bid–An offer to perform work described in a contract at a specified price.

Board foot–A unit or measure equal in volume to a board one foot long, one foot wide and one inch thick.

Booster pump–A supplemental pump installed to increase or maintain adequate pressure in the system.

Bowstring roof–A roof supported by trusses fabricated in the shape of a bow and tied together by a straight member.

Brick veneer–A facing of brick laid against, but not bonded to, a wall.

Bridging–A method of bracing joists for stiffness, stability and load distribution.

Broom finish–A method of finishing concrete by lightly dragging a broom over freshly placed concrete.

Built-up roofing–Installed on flat or extremely low-pitched roofs, a roof covering composed of plies or laminations of saturated roofing felts alternated with layers of coal tar pitch or asphalt and surfaced with a layer of gravel or slag in a thick coat of asphalt and finished with a capping sheet.

Caisson–A watertight chamber to expedite work on foundations or structure below water level.

Cantilever–A structural member supported only at one end.

Carport–An automobile shelter having one or more sides open to the weather.

Casement window–A vertical opening window having one side fixed.

Caulking–A resilient compound generally having a silicone or rubber base used to prevent infiltration of water or outside air.

Cavity wall–An exterior wall usually of masonry having an inner and outer wall separated by a continuous air space for thermal insulation.

Cellular concrete–A lightweight concrete consisting of cement mixed with gas-producing materials which, in turn, forms bubbles resulting in good insulating qualities.

Chattel mortgage–A secured interest in a property as collateral for payment of a note.

Clapboard–A wood siding used as exterior covering in frame construction. It is applied horizontally with grain running lengthwise. The thickest section of the board is on the bottom.

Clean room–An environmentally controlled room usually found in medical facilities or precision manufacturing spaces where it is essential to eliminate dust, lint or pathogens.

Close studding–A method of construction whereby studs are spaced close together and the intervening spaces are plastered.

Cluster housing–A closely grouped series of houses resulting in a high density land use.

Cofferdam–A watertight enclosure used for foundation construction in waterfront areas. Water is pumped out of the cofferdam allowing free access to the work area.

Collar joint–Joint between the collar beam and roof rafters.

Column–A vertical structural member supporting horizontal members (beams) along the direction of its longitudinal axis.

Combination door (windows)–Door (windows) having interchangeable screens and glass for seasonal use.

Common area–Spaces either inside or outside the building designated for use by the occupant of the building but not the general public.

Common wall–A wall used jointly by two dwelling units.

Compound wall–A wall constructed of more than one material.

Concrete–A composite material consisting of sand, coarse aggregate (gravel, stone or slag), cement and water that when mixed and allowed to harden forms a hard stone-like material.

Concrete floor hardener–A mixture of chemicals applied to the surface of concrete to produce a dense, wear-resistant bearing surface.

Conduit–A tube or pipe used to protect electric wiring.

Coping–The top cover or capping on a wall.

Craneway–Steel or concrete column and beam supports and rails on which a crane travels.

Curtain wall–A non-bearing exterior wall not supported by beams or girders of the steel frame.

Dampproofing–Coating of a surface to prevent the passage of water or moisture.

Dead load–Total weight of all structural components plus permanently attached fixtures and equipment.

Decibel–Unit of acoustical measurement.

Decorative block–A concrete masonry unit having a special treatment on its face.

Deed restriction–A restriction on the use of a property as set forth in the deed.

Depreciation–A loss in property value caused by physical aging, functional or economic obsolescence.

Direct heating–Heating of spaces by means of exposed heated surfaces (stove, fire, radiators, etc.).

Distributed load–A load that acts evenly over a structural member.

Dock bumper–A resilient material attached to a loading dock to absorb the impact of trucks backing in.

Dome–A curved roof shape spanning an area.

Dormer–A structure projecting from a sloping roof.

Double-hung window–A window having two vertical sliding sashes-one covering the upper section and one covering the lower.

Downspout–A vertical pipe for carrying rain water from the roof to the ground.

Drain tile–A hollow tile used to drain water-soaked soil.

Dressed size–In lumber about 3/8″ - 1/2″ less than nominal size after planing and sawing.

Drop panel–The depressed surface on the bottom side of a flat concrete slab which surrounds a column.

Dry pipe sprinkler system–A sprinkler system that is activated with water only in case of fire. Used in areas susceptible to freezing or to avoid the hazards of leaking pipes.

Drywall–An interior wall constructed of gypsum board, plywood or wood paneling. No water is required for application.

Duct–Pipe for transmitting warm or cold air. A pipe containing electrical cable or wires.

Dwelling–A structure designed as living quarters for one or more families.

Easement–A right of way or free access over land owned by another.

Eave–The lower edge of a sloping roof that overhangs the side wall of the structure.

Economic rent–That rent on a property sufficient to pay all operating costs exclusive of services and utilities.

Economic life–The term during which a structure is expected to be profitable. Generally shorter than the physical life of the structure.

Economic obsolescence–Loss in value due to unfavorable economic influences occurring from outside the structure itself.

Effective age–The age a structure appears to be based on observed physical condition determined by degree of maintenance and repair.

Elevator–A mechanism for vertical transport of personnel or freight equipped with car or platform.

Ell–A secondary wing or addition to a structure at right angles to the main structure.

Eminent domain–The right of the state to take private property for public use.

Envelope–The shape of a building indicating volume.

Equity–Value of an owner's interest calculated by subtracting outstanding mortgages and expenses from the value of the property.

Escheat–The assumption of ownership by the state, of property whose owner cannot be determined.

Facade–The exterior face of a building sometimes decorated with elaborate detail.

Face brick–Brick manufactured to present an attractive appearance.

Facing block–A concrete masonry unit having a decorative exterior finish.

Feasibility study–A detailed evaluation of a proposed project to determine its financial potential.

Felt paper–Paper sheathing used on walls as an insulator and to prevent infiltration of moisture.

Fenestration–Pertaining to the density and arrangement of windows in a structure.

Fiberboard–A building material composed of wood fiber compressed with a binding agent, produced in sheet form.

Fiberglass–Fine spun filaments of glass processed into various densities to produce thermal and acoustical insulation.

Field house–A long structure used for indoor athletic events.

Finish floor–The top or wearing surface of the floor system.

Fireproofing–The use of fire-resistant materials for the protection of structural members to ensure structural integrity in the event of fire.

Fire stop–A material or member used to seal an opening to prevent the spread of fire.

Flashing–A thin, impervious material such as copper or sheet metal used to prevent air or water penetration.

Float finish–A concrete finish accomplished by using a flat tool with a handle on the back.

Floor drain–An opening installed in a floor for removing excess water into a plumbing system.

Floor load–The live load the floor system has been designed for and which may be applied safely.

Flue–A heat-resistant enclosed passage in a chimney to remove gaseous products of combustion from a fireplace or boiler.

Footing–The lowest portion of the foundation wall that transmits the load directly to the soil.

Foundation–Below grade wall system that supports the structure.

Foyer–A portion of the structure that serves as a transitional space between the interior and exterior.

Frontage–That portion of a lot that is placed facing a street, public way or body of water.

Frost action–The effects of freezing and thawing on materials and the resultant structural damage.

Functional obsolescence–An inadequacy caused by outmoded design, dated construction materials or over or undersized areas, all of which cause excessive operating costs.

Furring strips–Wood or metal channels used for attaching gypsum or metal lath to masonry walls as a finish or for leveling purposes.

Gambrel roof (mansard)–A roof system having two pitches on each side.

Garden apartment–Two or three story apartment building with common outside areas.

General contractor–The prime contractor responsible for work on the construction site.

Girder–A principal horizontal supporting member.

Girt–A horizontal framing member for adding rigid support to columns which also acts as a support for sheathing or siding.

Grade beam–That portion of the foundation system that directly supports the exterior walls of the building.

Gross area–The total enclosed floor area of a building.

Ground area–The area computed by the exterior dimensions of the structure.

Ground floor–The floor of a building in closest proximity to the ground.

Grout–A mortar containing a high water content used to fill joints and cavities in masonry work.

Gutter–A wood or sheet metal conduit set along the building eaves used to channel rainwater to leaders or downspouts.

Gunite–A concrete mix placed pneumatically.

Gypsum–Hydrated calcium sulphate used as both a retarder in Portland cement and as a major ingredient in plaster of Paris. Used in sheets as a substitute for plaster.

Hall–A passageway providing access to various parts of a building-a large room for entertainment or assembly.

Hangar–A structure for the storage or repair of aircraft.

Head room–The distance between the top of the finished floor to the bottom of the finished ceiling.

Hearth–The floor of a fireplace and the adjacent area of fireproof material.

Heat pump–A mechanical device for providing either heating or air conditioning.

Hip roof–A roof whose four sides meet at a common point with no gabled ends.

I-beam–A structural member having a cross-section resembling the letter "I".

Insulation–Material used to reduce the effects of heat, cold or sound.

Jack rafter–An unusually short rafter generally found in hip roofs.

Jalousie–Adjustable glass louvers which pivot simultaneously in a common frame.

Jamb–The vertical member on either side of a door frame or window frame.

Glossary

Joist–Parallel beams of timber, concrete or steel used to support floor and ceiling.

Junction box–A box that protects splices or joints in electrical wiring.

Kalamein door–Solid core wood doors clad with galvanized sheet metal.

Kip–A unit of weight equal to 1,000 pounds.

Lally column–A concrete-filled steel pipe used as a vertical support.

Laminated beam–A beam built up by gluing together several pieces of timber.

Landing–A platform between flights of stairs.

Lath–Strips of wood or metal used as a base for plaster.

Lead-lined door, sheetrock–Doors or sheetrock internally lined with sheet lead to provide protection from X-ray radiation.

Lean-to–A small shed or building addition with a single pitched roof attached to the exterior wall of the main building.

Lintel–A horizontal framing member used to carry a load over a wall opening.

Live load–The moving or movable load on a structure composed of furnishings, equipment or personnel weight.

Load bearing partition–A partition that supports a load in addition to its own weight.

Loading dock leveler–An adjustable platform for handling off-on loading of trucks.

Loft building–A commercial/industrial type building containing large, open unpartitioned floor areas.

Louver window–A window composed of a series of sloping, overlapping blades or slats that may be adjusted to admit varying degrees of air or light.

Main beam–The principal load bearing beam used to transmit loads directly to columns.

Mansard roof–A roof having a double pitch on all four sides, the lower level having the steeper pitch.

Mansion–An extremely large and imposing residence.

Masonry–The utilization of brick, stone, concrete or block for walls and other building components.

Mastic–A sealant or adhesive compound generally used as either a binding agent or floor finish.

Membrane fireproofing–A coating of metal lath and plaster to provide resistance to fire and heat.

Mesh reinforcement–An arrangement of wire tied or welded at their intersection to provide strength and resistance to cracking.

Metal lath–A diamond-shaped metallic base for plaster.

Mezzanine–A low story situated between two main floors.

Mill construction–A heavy timber construction that achieves fire resistance by using large wood structural members, noncombustible bearing and non-bearing walls and omitting concealed spaces under floors and roof.

Mixed occupancy–Two or more classes of occupancy in a single structure.

Molded brick–A specially shaped brick used for decorative purposes.

Monolithic concrete–Concrete poured in a continuous process so there are no joints.

Movable partition–A non-load bearing demountable partition that can be relocated and can be either ceiling height or partial height.

Moving walk–A continually moving horizontal passenger carrying device.

Multi-zone system–An air conditioning system that is capable of handling several individual zones simultaneously.

Net floor area–The usable, occupied area of a structure excluding stairwells, elevator shafts and wall thicknesses.

Non-combustible construction–Construction in which the walls, partitions and the structural members are of non-combustible materials.

Nurses call system–An electrically operated system for use by patients or personnel for summoning a nurse.

Occupancy rate–The number of persons per room, per dwelling unit, etc.

One-way joist connection–A framing system for floors and roofs in a concrete building consisting of a series of parallel joists supported by girders between columns.

Open web joist–A lightweight prefabricated metal chord truss.

Parapet–A low guarding wall at the point of a drop. That portion of the exterior wall that extends above the roof line.

Parging–A thin coat of mortar applied to a masonry wall used primarily for waterproofing purposes.

Parquet–Inlaid wood flooring set in a simple design.

Peaked roof–A roof of two or more slopes that rises to a peak.

Penthouse–A structure occupying usually half the roof area and used to house elevator, HVAC equipment, or other mechanical or electrical systems.

Pier–A column designed to support a concentrated load.

Pilaster–A column usually formed of the same material and integral with, but projecting from, a wall.

Pile–A concrete, wood or steel column usually less than 2′ in diameter which is driven or otherwise introduced into the soil to carry a vertical load or provide vertical support.

Pitched roof–A roof having one or more surfaces with a pitch greater than 10°.

Plenum–The space between the suspended ceiling and the floor above.

Plywood–Structural wood made of three or more layers of veneer and bonded together with glue.

Porch–A structure attached to a building to provide shelter for an entranceway or to serve as a semi-enclosed space.

Post and beam framing–A structural framing system where beams rest on posts rather than bearing walls.

Post-tensioned concrete–Concrete that has the reinforcing tendons tensioned after the concrete has set.

Pre-cast concrete–Concrete structural components fabricated at a location other than in-place.

Pre-stressed concrete–Concrete that has the reinforcing tendons tensioned prior to the concrete setting.

Purlins–A horizontal structural member supporting the roof deck and resting on the trusses, girders, beams or rafters.

Rafters–Structural members supporting the roof deck and covering.

Resilient flooring–A manufactured interior floor covering material in either sheet or tile form, that is resilient.

Ridge–The horizontal line at the junction of two sloping roof edges.

Rigid frame–A structural framing system in which all columns and beams are rigidly connected; there are no hinged joints.

Roof drain–A drain designed to accept rainwater and discharge it into a leader or downspout.

Rough floor–A layer of boards or plywood nailed to the floor joists that serves as a base for the finished floor.

Sanitary sewer–A sewer line designed to carry only liquid or waterborne waste from the structure to a central treatment plant.

Sawtooth roof–Roof shape found primarily in industrial roofs that creates the appearance of the teeth of a saw.

Semi-rigid frame–A structural system wherein the columns and beams are so attached that there is some flexibility at the joints.

Septic tank–A covered tank in which waste matter is decomposed by natural bacterial action.

Service elevator–A combination passenger and freight elevator.

Sheathing–The first covering of exterior studs by boards, plywood or particle board.

Shoring–Temporary bracing for structural components during construction.

Span–The horizontal distance between supports.

Specification–A detailed list of materials and requirements for construction of a building.

Storm sewer–A system of pipes used to carry rainwater or surface waters.

Stucco–A cement plaster used to cover exterior wall surfaces usually applied over a wood or metal lath base.

Stud–A vertical wooden structural component.

Subfloor–A system of boards, plywood or particle board laid over the floor joists to form a base.

Superstructure–That portion of the structure above the foundation or ground level.

Terrazzo–A durable floor finish made of small chips of colored stone or marble, embedded in concrete, then polished to a high sheen.

Tile–A thin piece of fired clay, stone or concrete used for floor, roof or wall finishes.

Unit cost–The cost per unit of measurement.

Vault–A room especially designed for storage.

Veneer–A thin surface layer covering a base of common material.

Vent–An opening serving as an outlet for air.

Wainscot–The lower portion of an interior wall whose surface differs from that of the upper wall.

Wall bearing construction–A structural system where the weight of the floors and roof are carried directly by the masonry walls rather than the structural framing system.

Waterproofing–Any of a number of materials applied to various surfaces to prevent the infiltration of water.

Weatherstrip–A thin strip of metal, wood or felt used to cover the joint between the door or window sash and the jamb, casing or sill to keep out air, dust, rain, etc.

Wing–A building section or addition projecting from the main structure.

X-ray protection–Lead encased in sheetrock or plaster to prevent the escape of radiation.

A	Area Square Feet; Ampere	BTUH	BTU per Hour	Cwt.	100 Pounds
ABS	Acrylonitrile Butadiene Stryrene; Asbestos Bonded Steel	B.U.R.	Built-up Roofing	C.W.X.	Cool White Deluxe
A.C.	Alternating Current; Air-Conditioning; Asbestos Cement; Plywood Grade A & C	BX	Interlocked Armored Cable	C.Y.	Cubic Yard (27 cubic feet)
		°C	degree centegrade	C.Y./Hr.	Cubic Yard per Hour
		c	Conductivity, Copper Sweat	Cyl.	Cylinder
		C	Hundred; Centigrade	d	Penny (nail size)
A.C.I.	American Concrete Institute	C/C	Center to Center, Cedar on Cedar	D	Deep; Depth; Discharge
AD	Plywood, Grade A & D	C-C	Center to Center	Dis., Disch.	Discharge
Addit.	Additional	Cab.	Cabinet	Db.	Decibel
Adj.	Adjustable	Cair.	Air Tool Laborer	Dbl.	Double
af	Audio-frequency	Calc	Calculated	DC	Direct Current
A.G.A.	American Gas Association	Cap.	Capacity	DDC	Direct Digital Control
Agg.	Aggregate	Carp.	Carpenter	Demob.	Demobilization
A.H.	Ampere Hours	C.B.	Circuit Breaker	d.f.u.	Drainage Fixture Units
A hr.	Ampere-hour	C.C.A.	Chromate Copper Arsenate	D.H.	Double Hung
A.H.U.	Air Handling Unit	C.C.F.	Hundred Cubic Feet	DHW	Domestic Hot Water
A.I.A.	American Institute of Architects	cd	Candela	DI	Ductile Iron
AIC	Ampere Interrupting Capacity	cd/sf	Candela per Square Foot	Diag.	Diagonal
Allow.	Allowance	CD	Grade of Plywood Face & Back	Diam., Dia	Diameter
alt.	Altitude	CDX	Plywood, Grade C & D, exterior glue	Distrib.	Distribution
Alum.	Aluminum			Div.	Division
a.m.	Ante Meridiem	Cefi.	Cement Finisher	Dk.	Deck
Amp.	Ampere	Cem.	Cement	D.L.	Dead Load; Diesel
Anod.	Anodized	CF	Hundred Feet	DLH	Deep Long Span Bar Joist
Approx.	Approximate	C.F.	Cubic Feet	Do.	Ditto
Apt.	Apartment	CFM	Cubic Feet per Minute	Dp.	Depth
Asb.	Asbestos	c.g.	Center of Gravity	D.P.S.T.	Double Pole, Single Throw
A.S.B.C.	American Standard Building Code	CHW	Chilled Water; Commercial Hot Water	Dr.	Drive
Asbe.	Asbestos Worker			Drink.	Drinking
ASCE.	American Society of Civil Engineers	C.I.	Cast Iron	D.S.	Double Strength
A.S.H.R.A.E.	American Society of Heating, Refrig. & AC Engineers	C.I.P.	Cast in Place	D.S.A.	Double Strength A Grade
		Circ.	Circuit	D.S.B.	Double Strength B Grade
A.S.M.E.	American Society of Mechanical Engineers	C.L.	Carload Lot	Dty.	Duty
		Clab.	Common Laborer	DWV	Drain Waste Vent
A.S.T.M.	American Society for Testing and Materials	Clam	Common maintenance laborer	DX	Deluxe White, Direct Expansion
		C.L.F.	Hundred Linear Feet	dyn	Dyne
Attchmt.	Attachment	CLF	Current Limiting Fuse	e	Eccentricity
Avg., Ave.	Average	CLP	Cross Linked Polyethylene	E	Equipment Only; East
A.W.G.	American Wire Gauge	cm	Centimeter	Ea.	Each
AWWA	American Water Works Assoc.	CMP	Corr. Metal Pipe	E.B.	Encased Burial
Bbl.	Barrel	C.M.U.	Concrete Masonry Unit	Econ.	Economy
B&B	Grade B and Better; Balled & Burlapped	CN	Change Notice	E.C.Y	Embankment Cubic Yards
		Col.	Column	EDP	Electronic Data Processing
B.&S.	Bell and Spigot	CO_2	Carbon Dioxide	EIFS	Exterior Insulation Finish System
B.&W.	Black and White	Comb.	Combination	E.D.R.	Equiv. Direct Radiation
b.c.c.	Body-centered Cubic	Compr.	Compressor	Eq.	Equation
B.C.Y.	Bank Cubic Yards	Conc.	Concrete	EL	elevation
BE	Bevel End	Cont.	Continuous; Continued, Container	Elec.	Electrician; Electrical
B.F.	Board Feet	Corr.	Corrugated	Elev.	Elevator; Elevating
Bg. cem.	Bag of Cement	Cos	Cosine	EMT	Electrical Metallic Conduit; Thin Wall Conduit
BHP	Boiler Horsepower; Brake Horsepower	Cot	Cotangent		
		Cov.	Cover	Eng.	Engine, Engineered
B.I.	Black Iron	C/P	Cedar on Paneling	EPDM	Ethylene Propylene Diene Monomer
Bit., Bitum.	Bituminous	CPA	Control Point Adjustment		
Bit., Conc.	Bituminous Concrete	Cplg.	Coupling	EPS	Expanded Polystyrene
Bk.	Backed	C.P.M.	Critical Path Method	Eqhv.	Equip. Oper., Heavy
Bkrs.	Breakers	CPVC	Chlorinated Polyvinyl Chloride	Eqlt.	Equip. Oper., Light
Bldg.	Building	C.Pr.	Hundred Pair	Eqmd.	Equip. Oper., Medium
Blk.	Block	CRC	Cold Rolled Channel	Eqmm.	Equip. Oper., Master Mechanic
Bm.	Beam	Creos.	Creosote	Eqol.	Equip. Oper., Oilers
Boil.	Boilermaker	Crpt.	Carpet & Linoleum Layer	Equip.	Equipment
B.P.M.	Blows per Minute	CRT	Cathode-ray Tube	ERW	Electric Resistance Welded
BR	Bedroom	CS	Carbon Steel, Constant Shear Bar Joist	E.S.	Energy Saver
Brg.	Bearing			Est.	Estimated
Brhe.	Bricklayer Helper	Csc	Cosecant	esu	Electrostatic Units
Bric.	Bricklayer	C.S.F.	Hundred Square Feet	E.W.	Each Way
Brk.	Brick	CSI	Construction Specifications Institute	EWT	Entering Water Temperature
Brng.	Bearing			Excav.	Excavation
Brs.	Brass	C.T.	Current Transformer	Exp.	Expansion, Exposure
Brz.	Bronze	CTS	Copper Tube Size	Ext.	Exterior
Bsn.	Basin	Cu	Copper, Cubic	Extru.	Extrusion
Btr.	Better	Cu. Ft.	Cubic Foot	f.	Fiber stress
Btu	British Thermal Unit	cw	Continuous Wave	F	Fahrenheit; Female; Fill
		C.W.	Cool White; Cold Water	Fab.	Fabricated

FBGS	Fiberglass	H.P.	Horsepower; High Pressure	LE	Lead Equivalent
F.C.	Footcandles	H.P.F.	High Power Factor	LED	Light Emitting Diode
f.c.c.	Face-centered Cubic	Hr.	Hour	L.F.	Linear Foot
f'c.	Compressive Stress in Concrete; Extreme Compressive Stress	Hrs./Day	Hours per Day	L.F. Nose	Linear Foot of Stair Nosing
F.E.	Front End	HSC	High Short Circuit	L.F. Rsr	Linear Foot of Stair Riser
FEP	Fluorinated Ethylene Propylene (Teflon)	Ht.	Height	Lg.	Long; Length; Large
F.G.	Flat Grain	Htg.	Heating	L & H	Light and Heat
F.H.A.	Federal Housing Administration	Htrs.	Heaters	LH	Long Span Bar Joist
Fig.	Figure	HVAC	Heating, Ventilation & Air-Conditioning	L.H.	Labor Hours
Fin.	Finished	Hvy.	Heavy	L.L.	Live Load
Fixt.	Fixture	HW	Hot Water	L.L.D.	Lamp Lumen Depreciation
Fl. Oz.	Fluid Ounces	Hyd.;Hydr.	Hydraulic	lm	Lumen
Flr.	Floor	Hz.	Hertz (cycles)	lm/sf	Lumen per Square Foot
F.M.	Frequency Modulation; Factory Mutual	I.	Moment of Inertia	lm/W	Lumen per Watt
Fmg.	Framing	IBC	International Building Code	L.O.A.	Length Over All
Fdn.	Foundation	I.C.	Interrupting Capacity	log	Logarithm
Fori.	Foreman, Inside	ID	Inside Diameter	L-O-L	Lateralolet
Foro.	Foreman, Outside	I.D.	Inside Dimension; Identification	long.	longitude
Fount.	Fountain	I.F.	Inside Frosted	L.P.	Liquefied Petroleum; Low Pressure
fpm	Feet per Minute	I.M.C.	Intermediate Metal Conduit	L.P.F.	Low Power Factor
FPT	Female Pipe Thread	In.	Inch	LR	Long Radius
Fr.	Frame	Incan.	Incandescent	L.S.	Lump Sum
F.R.	Fire Rating	Incl.	Included; Including	Lt.	Light
FRK	Foil Reinforced Kraft	Int.	Interior	Lt. Ga.	Light Gauge
FRP	Fiberglass Reinforced Plastic	Inst.	Installation	L.T.L.	Less than Truckload Lot
FS	Forged Steel	Insul.	Insulation/Insulated	Lt. Wt.	Lightweight
FSC	Cast Body; Cast Switch Box	I.P.	Iron Pipe	L.V.	Low Voltage
Ft.	Foot; Feet	I.P.S.	Iron Pipe Size	M	Thousand; Material; Male; Light Wall Copper Tubing
Ftng.	Fitting	I.P.T.	Iron Pipe Threaded		
Ftg.	Footing	I.W.	Indirect Waste	M²CA	Meters Squared Contact Area
Ft lb.	Foot Pound	J	Joule	m/hr.; M.H.	Man-hour
Furn.	Furniture	J.I.C.	Joint Industrial Council	mA	Milliampere
FVNR	Full Voltage Non-Reversing	K	Thousand; Thousand Pounds; Heavy Wall Copper Tubing, Kelvin	Mach.	Machine
FXM	Female by Male			Mag. Str.	Magnetic Starter
Fy.	Minimum Yield Stress of Steel	K.A.H.	Thousand Amp. Hours	Maint.	Maintenance
g	Gram	KCMIL	Thousand Circular Mils	Marb.	Marble Setter
G	Gauss	KD	Knock Down	Mat; Mat'l.	Material
Ga.	Gauge	K.D.A.T.	Kiln Dried After Treatment	Max.	Maximum
Gal., gal.	Gallon	kg	Kilogram	MBF	Thousand Board Feet
gpm, GPM	Gallon per Minute	kG	Kilogauss	MBH	Thousand BTU's per hr.
Galv.	Galvanized	kgf	Kilogram Force	MC	Metal Clad Cable
Gen.	General	kHz	Kilohertz	M.C.F.	Thousand Cubic Feet
G.F.I.	Ground Fault Interrupter	Kip.	1000 Pounds	M.C.F.M.	Thousand Cubic Feet per Minute
Glaz.	Glazier	KJ	Kiljoule	M.C.M.	Thousand Circular Mils
GPD	Gallons per Day	K.L.	Effective Length Factor	M.C.P.	Motor Circuit Protector
GPH	Gallons per Hour	K.L.F.	Kips per Linear Foot	MD	Medium Duty
GPM	Gallons per Minute	Km	Kilometer	M.D.O.	Medium Density Overlaid
GR	Grade	K.S.F.	Kips per Square Foot	Med.	Medium
Gran.	Granular	K.S.I.	Kips per Square Inch	MF	Thousand Feet
Grnd.	Ground	kV	Kilovolt	M.F.B.M.	Thousand Feet Board Measure
H	High Henry	kVA	Kilovolt Ampere	Mfg.	Manufacturing
H.C.	High Capacity	K.V.A.R.	Kilovar (Reactance)	Mfrs.	Manufacturers
H.D.	Heavy Duty; High Density	KW	Kilowatt	mg	Milligram
H.D.O.	High Density Overlaid	KWh	Kilowatt-hour	MGD	Million Gallons per Day
H.D.P.E.	high density polyethelene	L	Labor Only; Length; Long; Medium Wall Copper Tubing	MGPH	Thousand Gallons per Hour
Hdr.	Header			MH, M.H.	Manhole; Metal Halide; Man-Hour
Hdwe.	Hardware	Lab.	Labor	MHz	Megahertz
Help.	Helper Average	lat	Latitude	Mi.	Mile
HEPA	High Efficiency Particulate Air Filter	Lath.	Lather	MI	Malleable Iron; Mineral Insulated
		Lav.	Lavatory	mm	Millimeter
Hg	Mercury	lb.; #	Pound	Mill.	Millwright
HIC	High Interrupting Capacity	L.B.	Load Bearing; L Conduit Body	Min., min.	Minimum, minute
HM	Hollow Metal	L. & E.	Labor & Equipment	Misc.	Miscellaneous
HMWPE	high molecular weight polyethylene	lb./hr.	Pounds per Hour	ml	Milliliter, Mainline
		lb./L.F.	Pounds per Linear Foot	M.L.F.	Thousand Linear Feet
H.O.	High Output	lbf/sq.in.	Pound-force per Square Inch	Mo.	Month
Horiz.	Horizontal	L.C.L.	Less than Carload Lot	Mobil.	Mobilization
		L.C.Y.	Loose Cubic Yard	Mog.	Mogul Base
		Ld.	Load	MPH	Miles per Hour
				MPT	Male Pipe Thread

MRT	Mile Round Trip	Pl.	Plate	S.F.C.A.	Square Foot Contact Area
ms	Millisecond	Plah.	Plasterer Helper	S.F. Flr.	Square Foot of Floor
M.S.F.	Thousand Square Feet	Plas.	Plasterer	S.F.G.	Square Foot of Ground
Mstz.	Mosaic & Terrazzo Worker	Pluh.	Plumbers Helper	S.F. Hor.	Square Foot Horizontal
M.S.Y.	Thousand Square Yards	Plum.	Plumber	S.F.R.	Square Feet of Radiation
Mtd., mtd.	Mounted	Ply.	Plywood	S.F. Shlf.	Square Foot of Shelf
Mthe.	Mosaic & Terrazzo Helper	p.m.	Post Meridiem	S4S	Surface 4 Sides
Mtng.	Mounting	Pntd.	Painted	Shee.	Sheet Metal Worker
Mult.	Multi; Multiply	Pord.	Painter, Ordinary	Sin.	Sine
M.V.A.	Million Volt Amperes	pp	Pages	Skwk.	Skilled Worker
M.V.A.R.	Million Volt Amperes Reactance	PP, PPL	Polypropylene	SL	Saran Lined
MV	Megavolt	P.P.M.	Parts per Million	S.L.	Slimline
MW	Megawatt	Pr.	Pair	Sldr.	Solder
MXM	Male by Male	P.E.S.B.	Pre-engineered Steel Building	SLH	Super Long Span Bar Joist
MYD	Thousand Yards	Prefab.	Prefabricated	S.N.	Solid Neutral
N	Natural; North	Prefin.	Prefinished	S-O-L	Socketolet
nA	Nanoampere	Prop.	Propelled	sp	Standpipe
NA	Not Available; Not Applicable	PSF, psf	Pounds per Square Foot	S.P.	Static Pressure; Single Pole; Self-Propelled
N.B.C.	National Building Code	PSI, psi	Pounds per Square Inch		
NC	Normally Closed	PSIG	Pounds per Square Inch Gauge	Spri.	Sprinkler Installer
N.E.M.A.	National Electrical Manufacturers Assoc.	PSP	Plastic Sewer Pipe	spwg	Static Pressure Water Gauge
		Pspr.	Painter, Spray	S.P.D.T.	Single Pole, Double Throw
NEHB	Bolted Circuit Breaker to 600V.	Psst.	Painter, Structural Steel	SPF	Spruce Pine Fir
N.L.B.	Non-Load-Bearing	P.T.	Potential Transformer	S.P.S.T.	Single Pole, Single Throw
NM	Non-Metallic Cable	P. & T.	Pressure & Temperature	SPT	Standard Pipe Thread
nm	Nanometer	Ptd.	Painted	Sq.	Square; 100 Square Feet
No.	Number	Ptns.	Partitions	Sq. Hd.	Square Head
NO	Normally Open	Pu	Ultimate Load	Sq. In.	Square Inch
N.O.C.	Not Otherwise Classified	PVC	Polyvinyl Chloride	S.S.	Single Strength; Stainless Steel
Nose.	Nosing	Pvmt.	Pavement	S.S.B.	Single Strength B Grade
N.P.T.	National Pipe Thread	Pwr.	Power	sst, ss	Stainless Steel
NQOD	Combination Plug-on/Bolt on Circuit Breaker to 240V.	Q	Quantity Heat Flow	Sswk.	Structural Steel Worker
		Qt.	Quart	Sswl.	Structural Steel Welder
N.R.C.	Noise Reduction Coefficient/Nuclear Regulator Commission	Quan., Qty.	Quantity	St.;Stl.	Steel
		Q.C.	Quick Coupling	S.T.C.	Sound Transmission Coefficient
N.R.S.	Non Rising Stem	r	Radius of Gyration	Std.	Standard
ns	Nanosecond	R	Resistance	Stg.	Staging
nW	Nanowatt	R.C.P.	Reinforced Concrete Pipe	STK	Select Tight Knot
OB	Opposing Blade	Rect.	Rectangle	STP	Standard Temperature & Pressure
OC	On Center	Reg.	Regular	Stpi.	Steamfitter, Pipefitter
OD	Outside Diameter	Reinf.	Reinforced	Str.	Strength; Starter; Straight
O.D.	Outside Dimension	Req'd.	Required	Strd.	Stranded
ODS	Overhead Distribution System	Res.	Resistant	Struct.	Structural
O.G.	Ogee	Resi.	Residential	Sty.	Story
O.H.	Overhead	Rgh.	Rough	Subj.	Subject
O&P	Overhead and Profit	RGS	Rigid Galvanized Steel	Subs.	Subcontractors
Oper.	Operator	R.H.W.	Rubber, Heat & Water Resistant; Residential Hot Water	Surf.	Surface
Opng.	Opening			Sw.	Switch
Orna.	Ornamental	rms	Root Mean Square	Swbd.	Switchboard
OSB	Oriented Strand Board	Rnd.	Round	S.Y.	Square Yard
O.S.&Y.	Outside Screw and Yoke	Rodm.	Rodman	Syn.	Synthetic
Ovhd.	Overhead	Rofc.	Roofer, Composition	S.Y.P.	Southern Yellow Pine
OWG	Oil, Water or Gas	Rofp.	Roofer, Precast	Sys.	System
Oz.	Ounce	Rohe.	Roofer Helpers (Composition)	t.	Thickness
P.	Pole; Applied Load; Projection	Rots.	Roofer, Tile & Slate	T	Temperature; Ton
p.	Page	R.O.W.	Right of Way	Tan	Tangent
Pape.	Paperhanger	RPM	Revolutions per Minute	T.C.	Terra Cotta
P.A.P.R.	Powered Air Purifying Respirator	R.S.	Rapid Start	T & C	Threaded and Coupled
PAR	Parabolic Reflector	Rsr	Riser	T.D.	Temperature Difference
Pc., Pcs.	Piece, Pieces	RT	Round Trip	Tdd	Telecommunications Device for the Deaf
P.C.	Portland Cement; Power Connector	S.	Suction; Single Entrance; South		
P.C.F.	Pounds per Cubic Foot	SC	Screw Cover	T.E.M.	Transmission Electron Microscopy
P.C.M.	Phase Contrast Microscopy	SCFM	Standard Cubic Feet per Minute	TFE	Tetrafluoroethylene (Teflon)
P.E.	Professional Engineer; Porcelain Enamel; Polyethylene; Plain End	Scaf.	Scaffold	T. & G.	Tongue & Groove; Tar & Gravel
		Sch., Sched.	Schedule		
		S.C.R.	Modular Brick	Th., Thk.	Thick
Perf.	Perforated	S.D.	Sound Deadening	Thn.	Thin
PEX	Cross linked polyethylene	S.D.R.	Standard Dimension Ratio	Thrded	Threaded
Ph.	Phase	S.E.	Surfaced Edge	Tilf.	Tile Layer, Floor
P.I.	Pressure Injected	Sel.	Select	Tilh.	Tile Layer, Helper
Pile.	Pile Driver	S.E.R., S.E.U.	Service Entrance Cable	THHN	Nylon Jacketed Wire
Pkg.	Package	S.F.	Square Foot	THW.	Insulated Strand Wire

THWN	Nylon Jacketed Wire	USP	United States Primed	Wrck.	Wrecker
T.L.	Truckload	UTP	Unshielded Twisted Pair	W.S.P.	Water, Steam, Petroleum
T.M.	Track Mounted	V	Volt	WT., Wt.	Weight
Tot.	Total	V.A.	Volt Amperes	WWF	Welded Wire Fabric
T-O-L	Threadolet	V.C.T.	Vinyl Composition Tile	XFER	Transfer
T.S.	Trigger Start	VAV	Variable Air Volume	XFMR	Transformer
Tr.	Trade	VC	Veneer Core	XHD	Extra Heavy Duty
Transf.	Transformer	Vent.	Ventilation	XHHW, XLPE	Cross-Linked Polyethylene Wire
Trhv.	Truck Driver, Heavy	Vert.	Vertical		Insulation
Trlr	Trailer	V.F.	Vinyl Faced	XLP	Cross-linked Polyethylene
Trlt.	Truck Driver, Light	V.G.	Vertical Grain	Y	Wye
TTY	Teletypewriter	V.H.F.	Very High Frequency	yd	Yard
TV	Television	VHO	Very High Output	yr	Year
T.W.	Thermoplastic Water Resistant	Vib.	Vibrating	Δ	Delta
	Wire	V.L.F.	Vertical Linear Foot	%	Percent
UCI	Uniform Construction Index	Vol.	Volume	~	Approximately
UF	Underground Feeder	VRP	Vinyl Reinforced Polyester	Ø	Phase; diameter
UGND	Underground Feeder	W	Wire; Watt; Wide; West	@	At
U.H.F.	Ultra High Frequency	w/	With	#	Pound; Number
U.I.	United Inch	W.C.	Water Column; Water Closet	<	Less Than
U.L.	Underwriters Laboratory	W.F.	Wide Flange	>	Greater Than
Uld.	unloading	W.G.	Water Gauge	Z	zone
Unfin.	Unfinished	Wldg.	Welding		
URD	Underground Residential	W. Mile	Wire Mile		
	Distribution	W-O-L	Weldolet		
US	United States	W.R.	Water Resistant		

RESIDENTIAL
COST ESTIMATE

OWNER'S NAME: _____ APPRAISER: _____

RESIDENCE ADDRESS: _____ PROJECT: _____

CITY, STATE, ZIP CODE: _____ DATE: _____

CLASS OF CONSTRUCTION
- ☐ ECONOMY
- ☐ AVERAGE
- ☐ CUSTOM
- ☐ LUXURY

RESIDENCE TYPE
- ☐ 1 STORY
- ☐ 1-1/2 STORY
- ☐ 2 STORY
- ☐ 2-1/2 STORY
- ☐ 3 STORY
- ☐ BI-LEVEL
- ☐ TRI-LEVEL

CONFIGURATION
- ☐ DETACHED
- ☐ TOWN/ROW HOUSE
- ☐ SEMI-DETACHED

OCCUPANCY
- ☐ ONE STORY
- ☐ TWO FAMILY
- ☐ THREE FAMILY
- ☐ OTHER _____

EXTERIOR WALL SYSTEM
- ☐ WOOD SIDING—WOOD FRAME
- ☐ BRICK VENEER—WOOD FRAME
- ☐ STUCCO ON WOOD FRAME
- ☐ PAINTED CONCRETE BLOCK
- ☐ SOLID MASONRY (AVERAGE & CUSTOM)
- ☐ STONE VENEER—WOOD FRAME
- ☐ SOLID BRICK (LUXURY)
- ☐ SOLID STONE (LUXURY)

*LIVING AREA (Main Building)		*LIVING AREA (Wing or Ell) ( )		*LIVING AREA (Wing or Ell) ( )	
First Level	_____ S.F.	First Level	_____ S.F.	First Level	_____ S.F.
Second Level	_____ S.F.	Second Level	_____ S.F.	Second Level	_____ S.F.
Third Level	_____ S.F.	Third Level	_____ S.F.	Third Level	_____ S.F.
Total	_____ S.F.	Total	_____ S.F.	Total	_____ S.F.

*Basement Area is not part of living area.

MAIN BUILDING	COSTS PER S.F. LIVING AREA
Cost per Square Foot of Living Area, from Page _____	$
Basement Addition: _____ % Finished, _____ % Unfinished	+
Roof Cover Adjustment: _____ Type, Page _____ (Add or Deduct)	( )
Central Air Conditioning: ☐ Separate Ducts ☐ Heating Ducts, Page _____	+
Heating System Adjustment: _____ Type, Page _____ (Add or Deduct)	( )
Main Building: Adjusted Cost per S.F. of Living Area	$

MAIN BUILDING TOTAL COST

$ /S.F.	x	S.F.	x	Town/Row House Multiplier (Use 1 for Detached)	=	$ TOTAL COST
Cost per S.F. Living Area		Living Area				

WING OR ELL ( ) _____ STORY	COSTS PER S.F. LIVING AREA
Cost per Square Foot of Living Area, from Page _____	$
Basement Addition: _____ % Finished, _____ % Unfinished	+
Roof Cover Adjustment: _____ Type, Page _____ (Add or Deduct)	( )
Central Air Conditioning: ☐ Separate Ducts ☐ Heating Ducts, Page _____	+
Heating System Adjustment: _____ Type, Page _____ (Add or Deduct)	( )
Wing or Ell ( ): Adjusted Cost per S.F. of Living Area	$

WING OR ELL ( ) TOTAL COST

$ /S.F.	x	S.F.	=	$ TOTAL COST
Cost per S.F. Living Area		Living Area		

WING OR ELL ( ) _____ STORY	COSTS PER S.F. LIVING AREA
Cost per Square Foot of Living Area, from Page _____	$
Basement Addition: _____ % Finished, _____ % Unfinished	+
Roof Cover Adjustment: _____ Type, Page _____ (Add or Deduct)	( )
Central Air Conditioning: ☐ Separate Ducts ☐ Heating Ducts, Page _____	+
Heating System Adjustment: _____ Type, Page _____ (Add or Deduct)	( )
Wing or Ell ( ): Adjusted Cost per S.F. of Living Area	$

WING OR ELL ( ) TOTAL COST

$ /S.F.	x	S.F.	=	$ TOTAL COST
Cost per S.F. Living Area		Living Area		

TOTAL THIS PAGE _____

Page 1 of 2

475

FORMS

RESIDENTIAL
COST ESTIMATE

	QUANTITY	UNIT COST	
Total Page 1			$
Additional Bathrooms: _____ Full, _____ Half			
Finished Attic: _____ Ft. x _____ Ft.	S.F.		+
Breezeway: ☐ Open ☐ Enclosed _____ Ft. x _____ Ft.	S.F.		+
Covered Porch: ☐ Open ☐ Enclosed _____ Ft. x _____ Ft.	S.F.		+
Fireplace: ☐ Interior Chimney ☐ Exterior Chimney ☐ No. of Flues ☐ Additional Fireplaces			+
Appliances:			+
Kitchen Cabinets Adjustment: (+/−)			
☐ Garage ☐ Carport: _____ Car(s) Description _____ (+/−)			
Miscellaneous:			+

ADJUSTED TOTAL BUILDING COST $ _____

REPLACEMENT COST	
ADJUSTED TOTAL BUILDING COST	$ _____
Site Improvements	
(A) Paving & Sidewalks	$ _____
(B) Landscaping	$ _____
(C) Fences	$ _____
(D) Swimming Pool	$ _____
(E) Miscellaneous	$ _____
TOTAL	$ _____
Location Factor	x _____
Location Replacement Cost	$ _____
Depreciation	−$ _____
LOCAL DEPRECIATED COST	$ _____

INSURANCE COST	
ADJUSTED TOTAL BUILDING COST	$ _____
Insurance Exclusions	
(A) Footings, Site Work, Underground Piping	−$ _____
(B) Architects' Fees	−$ _____
Total Building Cost Less Exclusion	$ _____
Location Factor	x _____
LOCAL INSURABLE REPLACEMENT COST	$ _____

SKETCH AND ADDITIONAL CALCULATIONS

Page 2 of 2

476

CII APPRAISAL

1. SUBJECT PROPERTY:

5. DATE:

6. APPRAISER:

2. BUILDING:

3. ADDRESS:

4. BUILDING USE:

7. YEAR BUILT:

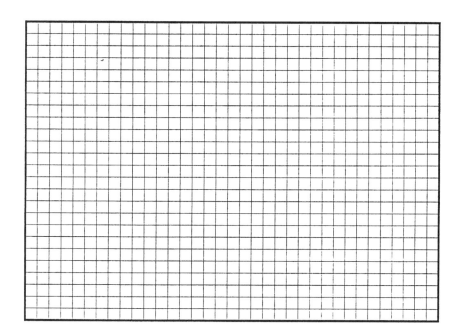

8. EXTERIOR WALL CONSTRUCTION:

9. FRAME:

10. GROUND FLOOR AREA: _____ S.F. 11. GROSS FLOOR AREA (EXCL. BASEMENT): _____ S.F.

12. NUMBER OF STORIES: _____ 13. STORY HEIGHT: _____

14. PERIMETER: _____ L.F. 15. BASEMENT AREA: _____ S.F.

16. GENERAL COMMENTS:

	NO.	DESCRIPTION		UNIT	UNIT COST	NEW SF COST	MODEL SF COST	+/- CHANGE
A SUBSTRUCTURE								
A	1010	Standard Foundations	Bay size:	S.F. Gnd.				
A	1030	Slab on Grade	Material:	Thickness:	S.F. Slab			
A	2010	Basement Excavation	Depth:	Area:	S.F. Gnd.			
A	2020	Basement Walls		L.F. Walls				
B SHELL								
	B10 Superstructure							
B	1010	Floor Construction	Elevated floors:	S.F. Floor				
				S.F. Floor				
B	1020	Roof Construction		S.F. Roof				
	B20 Exterior Enclosure							
B	2010	Exterior walls	Material:	Thickness: *% of wall*	S.F. Walls			
			Material:	Thickness: *% of wall*	S.F. Walls			
B	2020	Exterior Windows	Type:	*% of wall*	S.F. Wind.			
			Type:	*% of wall* Each	S.F. Wind.			
B	2030	Exterior Doors	Type:	Number:	Each			
			Type:	Number:	Each			
	B30 Roofing							
B	3010	Roof Coverings	Material:	S.F. Roof				
			Material:	S.F. Roof				
B	3020	Roof Openings		S.F. Opng.				
C INTERIORS								
C	1010	Partitions:	Material:	Density:	S.F. Part.			
			Material:	Density:				
C	1020	Interior Doors	Type:	Number:	Each			
C	1030	Fittings		Each				
C	2010	Stair Construction		Flight				
C	3010	Wall Finishes	Material:	% of Wall	S.F. Walls			
			Material:	S.F. Walls				
C	3020	Floor Finishes	Material:	S.F. Floor				
			Material:	S.F. Floor				
			Material:	S.F. Floor				
C	3030	Ceiling Finishes	Material:	S.F. Ceil.				

NO.	SYSTEM/COMPONENT	DESCRIPTION	UNIT	UNIT COST	NEW S.F. COST	MODEL S.F. COST	+/- CHANGE
D SERVICES							
D10 Conveying							
D 1010	Elevators & Lifts	Type: Capacity: Stops:	Each				
D 1020	Escalators & Moving Walks	Type:	Each				
D20 Plumbing							
D 2010	Plumbing		Each				
D 2020	Domestic Water Distribution		S.F. Floor				
D 2040	Rain Water Drainage		S.F. Roof				
D30 HVAC							
D 3010	Energy Supply		Each				
D 3020	Heat Generating Systems	Type:	S.F. Floor				
D 3030	Cooling Generating Systems	Type:	S.F. Floor				
D 3090	Other HVAC Sys. & Equipment		Each				
D40 Fire Protection							
D 4010	Sprinklers		S.F. Floor				
D 4020	Standpipes		S.F. Floor				
D50 Electrical							
D 5010	Electrical Service/Distribution		S.F. Floor				
D 5020	Lighting & Branch Wiring		S.F. Floor				
D 5030	Communications & Security		S.F. Floor				
D 5090	Other Electrical Systems		S.F. Floor				
E EQUIPMENT & FURNISHINGS							
E 1010	Commercial Equipment		Each				
E 1020	Institutional Equipment		Each				
E 1030	Vehicular Equipment		Each				
E 1090	Other Equipment		Each				
F SPECIAL CONSTRUCTION							
F 1020	Integrated Construction		S.F.				
F 1040	Special Facilities		S.F.				
G BUILDING SITEWORK							

ITEM		Total Change	
17	Model 020 total _____ $ _____		
18	Adjusted S.F. cost ___ $ _____ item 17 +/- changes		
19	Building area - from item 11 _____ S.F. x adjusted S.F. cost...	$...	
20	Basement area - from item 15 _____ S.F. x S.F. cost $	$..	
21	Base building sub-total - item 20 + item 19	$.....	
22	Miscellaneous addition (quality, etc.)	$......	
23	Sub-total - item 22 + 21	$.......	
24	General conditions -25 % of item 23	$......	
25	Sub-total - item 24 + item 23...	$......	
26	Architects fees _____ % of item 25	$	
27	Sub-total - item 26 + item 27	$......	
28	Location modifier	x....	
29	Local replacement cost - item 28 x item 27	$.....	
30	Depreciation _____ % of item 29	$	
31	Depreciated local replacement cost - item 29 less item 30	$.....	
32	Exclusions	$.......	
33	Net depreciated replacement cost - item 31 less item 32	$.....	

Base Cost per square foot floor area: *(from square foot table)*

Specify source:

Page: _____ , Model # _____ , Area _____ S.F.,	
Exterior wall _____ , Frame _____	

Adjustments for exterior wall variation:

Size Adjustment:
   (Interpolate)

Height Adjustment: _____   +   _____   =   _____

Adjusted Base Cost per square foot:

Building Cost $ _____   x   _____   =   _____
                                     Adjusted Base Cost   *Floor Area*
                                     per square foot

Basement Cost $ _____   x   _____   =   _____
                                     Basement Cost   *Basement Area*

Lump Sum Additions

TOTAL BUILDING COST *(Sum of above costs)*   _____

Modifications: *(complexity, workmanship, size)*   +/- _____ %   _____

Location Modifier: City _____ Date _____   x _____

     Local cost of replacement   _____

Less depreciation: _____   - _____

     Local cost of replacement less depreciation $ _____

481

Notes

484

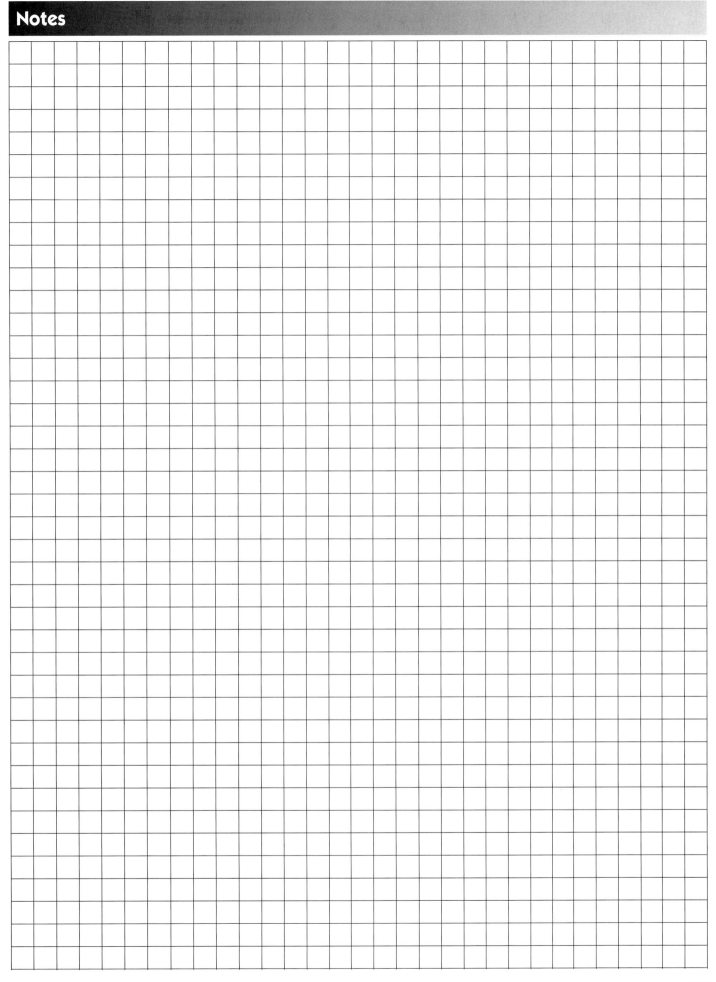

For more information visit the RSMeans websit at www.rsmeans.com

Reed Construction Data/RSMeans— a tradition of excellence in construction cost information and services since 1942

Book Selection Guide

The following table provides definitive information on the content of each cost data publication. The number of lines of data provided in each unit price or assemblies division, as well as the number of reference tables and crews, is listed for each book. The presence of other elements such as equipment rental costs, historical cost indexes, city cost indexes, square foot models, or cross-referenced indexes is also indicated. You can use the table to hel select the RSMeans book that has the quantity and type of information you most need in your work.

Unit Cost Divisions	Building Construction Costs	Mechanical	Electrical	Repair & Remodel	Square Foot	Site Work Landsc.	Assemblies	Interior	Concrete Masonry	Open Shop	Heavy Construction	Light Commercial	Facilities Construction	Plumbing	Residentia
1	553	372	385	474		489		301	437	552	483	220	967	386	157
2	737	264	85	676		964		358	215	734	706	434	1154	288	242
3	1506	213	201	850		1357		277	1837	1506	1497	347	1422	182	359
4	897	24	0	697		699		572	1105	873	592	489	1145	0	405
5	1851	201	156	958		795		1021	758	1838	1050	826	1841	276	722
6	1841	67	69	1429		449		1160	299	1820	384	1530	1453	22	2043
7	1438	209	125	1462		524		546	466	1436	0	1143	1500	218	905
8	2048	61	10	2118		294		1737	694	1974	0	1527	2320	0	1397
9	1783	72	26	1598		268		1875	367	1737	15	1511	2027	54	1331
10	986	17	10	597		224		800	157	986	29	481	1072	231	214
11	1020	214	163	487		126		865	30	1005	0	231	1034	173	110
12	547	0	2	303		248		1691	12	536	19	349	1736	23	317
13	733	119	111	247		344		255	69	732	256	80	743	74	79
14	280	36	0	229		28		259	0	279	0	12	299	21	6
21	78	0	16	30		0		250	0	78	0	60	368	373	0
22	1148	7031	155	1106		1466		622	20	1103	1648	823	6940	8763	634
23	1185	7032	609	774		158		679	38	1105	110	660	5089	1869	418
26	1246	493	9875	1003		739		1053	55	1234	523	1080	9811	438	559
27	67	0	210	34		9		59	0	67	35	48	212	0	4
28	86	58	107	61		0		68	0	71	0	40	122	44	25
31	1477	735	608	798		2793		7	1208	1421	2865	600	1538	660	606
32	811	114	0	675		3967		359	274	784	1481	373	1443	226	422
33	519	1025	465	201		1967		0	225	518	1890	117	1553	1196	115
34	109	0	20	4		144		0	31	64	122	0	129	0	0
35	18	0	0	0		166		0	0	18	281	0	83	0	0
41	52	0	0	24		7		22	0	52	30	0	59	14	0
44	98	90	0	0		32		0	0	23	32	0	98	105	0
Totals	23114	18447	13408	16835		18257		14836	8297	22546	14048	12981	46158	15636	11070

Assembly Divisions	Building Construction Costs	Mechanical	Elec-trical	Repair & Remodel	Square Foot	Site Work Landscape	Assemblies	Interior	Concrete Masonry	Open Shop	Heavy Construction	Light Commercial	Facilities Construction	Plumbing	Asm Div	Residentia
A		15	0	182	150	530	598	0	536		570	154	24	0	1	37
B		0	0	809	2471	0	5577	329	1914		368	2031	143	0	2	21
C		0	0	635	865	0	1235	1567	146		0	765	247	0	3	58
D		1066	811	694	1793	0	2422	767	0		0	1271	1025	1078	4	86
E		0	0	85	260	0	297	5	0		0	257	5	0	5	39
F		0	0	0	123	0	124	0	0		0	123	3	0	6	35
G		527	294	311	202	2366	668	0	535		1194	200	293	685	7	30
															8	76
															9	8
															10	
															11	
															12	
Totals		1608	1105	2716	5864	2896	10921	2668	3131		2132	4801	1740	1763		393

Reference Section	Building Construction Costs	Mechanical	Electrical	Repair & Remodel	Square Foot	Site Work Landscape	Assemblies	Interior	Concrete Masonry	Open Shop	Heavy Construction	Light Commercial	Facilities Construction	Plumbing	Residentia
Reference Tables	yes	yes	yes	yes	no	yes	yes	yes	yes	yes	yes	yes	yes	yes	yes
Models					107							46			28
Crews	487	487	487	467		487		487	487	465	487	465	467	487	465
Equipment Rental Costs	yes	yes	yes	yes		yes		yes	yes	yes	yes	yes	yes	yes	yes
Historical Cost Indexes	yes	yes	yes	yes	yes	yes	yes	yes	yes	yes	yes	yes	yes	yes	no
City Cost Indexes	yes	yes	yes	yes	yes	yes	yes	yes	yes	yes	yes	yes	yes	yes	yes

Annual Cost Guides

**For more information
visit the RSMeans website
at www.rsmeans.com**

RSMeans Building Construction Cost Data 2010

ffers you unchallenged unit price
liability in an easy-to-use format.
hether used for verifying complete,
ished estimates or for periodic checks,
supplies more cost facts better and
ster than any comparable source. Over
,000 unit prices have been updated
2010. The City Cost Indexes and
cation Factors cover over 930 areas,
indexing to any project location in
orth America. Order and get *RSMeans
uarterly Update Service* FREE. You'll
ve year-long access to the RSMeans
timating **HOTLINE** FREE with your
bscription. Expert assistance when
ing RSMeans data is just a phone call
vay.

69.95 per copy
ailable Sept. 2009
talog no. 60010

**Unit prices organized according to
the latest MasterFormat!**

RSMeans Mechanical Cost Data 2010
- **HVAC**
- **Controls**

Total unit and systems price guidance for
mechanical construction. . . materials,
parts, fittings, and complete labor cost
information. Includes prices for piping,
heating, air conditioning, ventilation,
and all related construction.

Plus new 2010 unit costs for:

- Thousands of installed HVAC/controls
  assemblies components
- "On-site" Location Factors for over
  930 cities and towns in the U.S. and
  Canada
- Crews, labor, and equipment

$169.95 per copy
Available Oct. 2009
Catalog no. 60020

RSMeans Plumbing Cost Data 2010

omprehensive unit prices and
semblies for plumbing, irrigation
stems, commercial and residential fire
otection, point-of-use water heaters,
d the latest approved materials. This
blication and its companion, *RSMeans
echanical Cost Data*, provide full-
ige cost estimating coverage for all the
echanical trades.

ow contains updated costs for potable
ter and radiant heat systems and high
iciency fixtures.

59.95 per copy
ailable Oct. 2009
talog no. 60210

**Unit prices organized according to
the latest MasterFormat!**

RSMeans Facilities Construction Cost Data 2010

For the maintenance and construction of
commercial, industrial, municipal, and
institutional properties. Costs are shown
for new and remodeling construction and
are broken down into materials, labor,
equipment, and overhead and profit.
Special emphasis is given to sections
on mechanical, electrical, furnishings,
site work, building maintenance, finish
work, and demolition.

More than 43,000 unit costs, plus
assemblies costs and a comprehensive
Reference Section are included.

$419.95 per copy
Available Nov. 2009
Catalog no. 60200

Annual Cost Guides

RSMeans Square Foot Costs 2010

It's Accurate and Easy To Use!

- **Updated price information** based on nationwide figures from suppliers, estimators, labor experts, and contractors

- "How-to-Use" sections, with **clear examples** of commercial, residential, industrial, and institutional structures

- Realistic graphics, offering true-to-life illustrations of building projects

- Extensive information on using square foot cost data, including sample estimates and alternate pricing methods

$184.95 per copy
Available Oct. 2009
Catalog no. 60050

RSMeans Repair & Remodeling Cost Data 2010

Commercial/Residential

Use this valuable tool to estimate commercial and residential renovation and remodeling.

Includes: New costs for hundreds of unique methods, materials, and conditions that only come up in repair and remodeling, PLUS:

- Unit costs for over 15,000 constructi components

- Installed costs for over 90 assemblies

- Over 930 "on-site" localization facto for the U.S. and Canada.

Unit prices organized according to the latest MasterFormat!

$144.95 per copy
Available Oct. 2009
Catalog no. 60040

RSMeans Electrical Cost Data 2010

Pricing information for every part of electrical cost planning. More than 13,000 unit and systems costs with design tables; clear specifications and drawings; engineering guides; illustrated estimating procedures; complete labor-hour and materials costs for better scheduling and procurement; and the latest electrical products and construction methods.

- A variety of special electrical systems including cathodic protection

- Costs for maintenance, demolition, HVAC/mechanical, specialties, equipment, and more

Unit prices organized according to the latest MasterFormat!

$169.95 per copy
Available Oct. 2009
Catalog no. 60030

RSMeans Electrical Change Order Cost Data 2010

RSMeans Electrical Change Order Cos Data provides you with electrical unit prices exclusively for pricing change orders—based on the recent, direct experience of contractors and suppliers Analyze and check your own change order estimates against the experience others have had doing the same work. It also covers productivity analysis and change order cost justifications. With useful information for calculating the effects of change orders and dealing wi their administration.

$169.95 per copy
Available Dec. 2009
Catalog no. 60230

Annual Cost Guides

**For more information
visit the RSMeans website
at www.rsmeans.com**

RSMeans Assemblies Cost Data 2010

RSMeans Assemblies Cost Data takes the guesswork out of preliminary or conceptual estimates. Now you don't have to try to calculate the assembled cost by working up individual component costs. We've done all the work for you.

Presents detailed illustrations, descriptions, specifications, and costs for every conceivable building assembly—240 types in all—arranged in the easy-to-use UNIFORMAT II system. Each illustrated "assembled" cost includes a complete grouping of materials and associated installation costs, including the installing contractor's overhead and profit.

$279.95 per copy
Available Sept. 2009
Catalog no. 60060

RSMeans Open Shop Building Construction Cost Data 2010

The latest costs for accurate budgeting and estimating of new commercial and residential construction. . . renovation work. . . change orders. . . cost engineering.

RSMeans Open Shop "BCCD" will assist you to:
• Develop benchmark prices for change orders
• Plug gaps in preliminary estimates and budgets
• Estimate complex projects
• Substantiate invoices on contracts
• Price ADA-related renovations

Unit prices organized according to the latest MasterFormat!

$169.95 per copy
Available Dec. 2009
Catalog no. 60150

RSMeans Residential Cost Data 2010

Contains square foot costs for 30 basic home models with the look of today, plus hundreds of custom additions and modifications you can quote right off the page. With costs for the 100 residential systems you're most likely to use in the year ahead. Complete with blank estimating forms, sample estimates, and step-by-step instructions.

Now contains line items for cultured stone and block, PVC trim lumber, and TPO roofing.

$144.95 per copy
Available Oct. 2009
Catalog no. 60170

**Unit prices organized according to
the latest MasterFormat!**

RSMeans Site Work & Landscape Cost Data 2010

Includes unit and assemblies costs for earthwork, sewerage, piped utilities, site improvements, drainage, paving, trees & shrubs, street openings/repairs, underground tanks, and more. Contains 57 tables of assemblies costs for accurate conceptual estimates.

Includes:
• Estimating for infrastructure improvements
• Environmentally-oriented construction
• ADA-mandated handicapped access
• Hazardous waste line items

$169.95 per copy
Available Nov. 2009
Catalog no. 60280

RSMeans Facilities Maintenance & Repair Cost Data 2010

RSMeans Facilities Maintenance & Repair Cost Data gives you a complete system to manage and plan your facility repair and maintenance costs and budget efficiently. Guidelines for auditing a facility and developing an annual maintenance plan. Budgeting is included, along with reference tables on cost and management, and information on frequency and productivity of maintenance operations.

The only nationally recognized source of maintenance and repair costs. Developed in cooperation with the Civil Engineering Research Laboratory (CERL) of the Army Corps of Engineers.

$379.95 per copy
Available Nov. 2009
Catalog no. 60300

RSMeans Light Commercial Cost Data 2010

Specifically addresses the light commercial market, which is a specialized niche in the construction industry. Aids you, the owner/designer/contractor, in preparing all types of estimates—from budgets to detailed bids. Includes new advances in methods and materials.

Assemblies Section allows you to evaluate alternatives in the early stages of design/planning.

Over 11,000 unit costs ensure that you have the prices you need. . . when you need them.

Unit prices organized according to the latest MasterFormat !

$144.95 per copy
Available Nov. 2009
Catalog no. 60180

For more information
visit the RSMeans website
at www.rsmeans.com

Annual Cost Guides

RSMeans Concrete & Masonry Cost Data 2010

Provides you with cost facts for virtually all concrete/masonry estimating needs, from complicated formwork to various sizes and face finishes of brick and block—all in great detail. The comprehensive Unit Price Section contains more than 7,500 selected entries. Also contains an Assemblies [Cost] Section, and a detailed Reference Section that supplements the cost data.

Unit prices organized according to the latest MasterFormat !

$154.95 per copy
Available Dec. 2009
Catalog no. 60110

RSMeans Labor Rates for the Construction Industry 2010

Complete information for estimating labor costs, making comparisons, and negotiating wage rates by trade for over 300 U.S. and Canadian cities. With 46 construction trades listed by local union number in each city, and historical wage rates included for comparison. Each city chart lists the county and is alphabetically arranged with handy visual flip tabs for quick reference.

$379.95 per copy
Available Dec. 2009
Catalog no. 60120

RSMeans Construction Cost Indexes 2010

What materials and labor costs will change unexpectedly this year? By how much?

- Breakdowns for 316 major cities
- National averages for 30 key cities
- Expanded five major city indexes
- Historical construction cost indexes

$362.00 per year (subscription)
Catalog no. 50140

$90.50 individual quarters
Catalog no. 60140 A,B,C,D

RSMeans Interior Cost Data 2010

Provides you with prices and guidance needed to make accurate interior work estimates. Contains costs on materials, equipment, hardware, custom installations, furnishings, and labor costs. . . for new and remodel commercial and industrial interior construction, including updated information on office furnishings, and reference information.

Unit prices organized according to the latest MasterFormat !

$169.95 per copy
Available Nov. 2009
Catalog no. 60090

RSMeans Heavy Construction Cost Data 2010

A comprehensive guide to heavy construction costs. Includes costs for highly specialized projects such as tunnels, dams, highways, airports, and waterways. Information on labor rates, equipment, and materials costs is included. Features unit price costs, systems costs, and numerous reference tables for costs and design.

$169.95 per copy
Available Nov. 2009
Catalog no. 60160

Unit prices organized according to the latest MasterFormat!

Reference Books

For more information
visit the RSMeans website
at www.rsmeans.com

Value Engineering: Practical Applications

For Design, Construction, Maintenance & Operations

by Alphonse Dell'Isola, PE

A tool for immediate application—for engineers, architects, facility managers, owners, and contractors. Includes making the case for VE—the management briefing; integrating VE into planning, budgeting, and design; conducting life cycle costing; using VE methodology in design review and consultant selection; case studies; VE workbook; and a life cycle costing program on disk.

$79.95 per copy
Over 450 pages, illustrated, softcover
Catalog no. 67319A

Facilities Operations & Engineering Reference

by the Association for Facilities Engineering and RSMeans

An all-in-one technical reference for planning and managing facility projects and solving day-to-day operations problems. Selected as the official certified plant engineer reference, this handbook covers financial analysis, maintenance, HVAC and energy efficiency, and more.

$54.98 per copy
Over 700 pages, illustrated, hardcover
Catalog no. 67318

The Building Professional's Guide to Contract Documents

3rd Edition

by Waller S. Poage, AIA, CSI, CVS

A comprehensive reference for owners, design professionals, contractors, and students

Structure your documents for maximum efficiency.
Effectively communicate construction requirements.
Understand the roles and responsibilities of construction professionals.
Improve methods of project delivery.

$64.95 per copy, 400 pages
Diagrams and construction forms, hardcover
Catalog no. 67261A

Complete Book of Framing

by Scot Simpson

This straightforward, easy-to-learn method will help framers, carpenters, and handy homeowners build their skills in rough carpentry and framing. Shows how to frame all the parts of a house: floors, walls, roofs, door & window openings, and stairs—with hundreds of color photographs and drawings that show every detail.

The book gives beginners all the basics they need to go from zero framing knowledge to a journeyman level.

$29.95 per copy
352 pages, softcover
Catalog no. 67353

Cost Planning & Estimating for Facilities Maintenance

In this unique book, a team of facilities management authorities shares their expertise on:

Evaluating and budgeting maintenance operations
Maintaining and repairing key building components
Applying *RSMeans Facilities Maintenance & Repair Cost Data* to your estimating

Covers special maintenance requirements of the ten major building types.

$99.95 per copy
Over 475 pages, hardcover
Catalog no. 67314

Life Cycle Costing for Facilities

by Alphonse Dell'Isola and Dr. Steven Kirk

Guidance for achieving higher quality design and construction projects at lower costs! Cost-cutting efforts often sacrifice quality to yield the cheapest product. Life cycle costing enables building designers and owners to achieve both. The authors of this book show how LCC can work for a variety of projects — from roads to HVAC upgrades to different types of buildings.

$99.95 per copy
450 pages, hardcover
Catalog no. 67341

Planning & Managing Interior Projects 2nd Edition

by Carol E. Farren, CFM

Expert guidance on managing renovation & relocation projects

This book guides you through every step in relocating to a new space or renovating an old one. From initial meeting through design and construction, to post-project administration, it helps you get the most for your company or client. Includes sample forms, spec lists, agreements, drawings, and much more!

$74.98 per copy
400 pages, softcover
Catalog no. 67245A

Construction Business Management

by Nick Ganaway

Only 43% of construction firms stay in business after four years. Make sure your company thrives with valuable guidance from a pro with 25 years of success as a commercial contractor. Find out what it takes to build all aspects of a business that is profitable, enjoyable, and enduring. With a bonus chapter on retail construction.

$49.95 per copy
200 pages, softcover
Catalog no. 67352

Interior Home Improvement Costs 9th Edition

Updated estimates for the most popular remodeling and repair projects—from small, do-it-yourself jobs to major renovations and new construction. Includes: kitchens & baths; new living space from your attic, basement, or garage; new floors, paint, and wallpaper; tearing out or building new walls; closets, stairs, and fireplaces; new energy-saving improvements, home theaters, and more!

$24.95 per copy
250 pages, illustrated, softcover
Catalog no. 67308E

Exterior Home Improvement Costs 9th Edition

Updated estimates for the most popular remodeling and repair projects—from small, do-it-yourself jobs, to major renovations and new construction. Includes: curb appeal projects—landscaping, patios, porches, driveways, and walkways; new windows and doors; decks, greenhouses, an sunrooms; room additions and garages; roofing, siding, and painting; "green" improvements to save energy & water.

$24.95 per copy
Over 275 pages, illustrated, softcover
Catalog no. 67309E

Builder's Essentials: Plan Reading & Material Takeoff

For Residential and Light Commercial Construction

by Wayne J. DelPico

A valuable tool for understanding plans and specs, and accurately calculating material quantities. Step-by-step instructions and takeoff procedures based on a full set of working drawings.

$35.95 per copy
Over 420 pages, softcover
Catalog no. 67307

Means Unit Price Estimating Methods
New 4th Edition

This new edition includes up-to-date cost data and estimating examples, updated to reflect changes to the CSI numbering system and new features of RSMeans cost data. It describes the most productive, universally accepted ways to estimate, and uses checklists and forms to illustrate shortcuts and timesavers. A model estimate demonstrates procedures. A new chapter explores computer estimating alternatives.

$59.95 per copy
Over 350 pages, illustrated, softcover
Catalog no. 67303B

Total Productive Facilities Management

by Richard W. Sievert, Jr.

Today, facilities are viewed as strategic resources. . . elevating the facility manager to the role of asset manager supporting the organization's overall business goals. Now, Richard Sievert Jr., in this well-articulated guidebook, sets forth a new operational standard for the facility manager's emerging role. . . a comprehensive program for managing facilities as a true profit center.

$29.98 per copy
275 pages, softcover
Catalog no. 67321

Green Building: Project Planning & Cost Estimating 2nd Edition

This new edition has been completely updated with the latest in green building technologies, design concepts, standards, and costs. Now includes a 2010 Green Building *CostWorks* CD with more than 300 green building assemblies and over 5,000 unit price line items for sustainable building. The new edition is also full-color with all new case studies—plus a new chapter on deconstruction, a key aspect of green building.

$129.95 per copy
350 pages, softcover
Catalog no. 67338A

Concrete Repair and Maintenance Illustrated

by Peter Emmons

Hundreds of illustrations show users how to analyze, repair, clean, and maintain concrete structures for optimal performance and cost effectiveness. From parking garages to roads and bridges to structural concrete, this comprehensive book describes the causes, effects, and remedies for concrete wear and failure. Invaluable for planning jobs, selecting materials, and training employees, this book is a must-have for concrete specialists, general contractors, facility managers, civil and structural engineers, and architects.

$69.95 per copy
300 pages, illustrated, softcover
Catalog no. 67146

Reference Books

**For more information
visit the RSMeans website
at www.rsmeans.com**

Means Illustrated Construction Dictionary Condensed, 2nd Edition

e best portable dictionary for office or field use—an
ential tool for contractors, architects, insurance and
l estate personnel, facility managers, homeowners, and
one who needs quick, clear definitions for construction
ns. The second edition has been further enhanced with
ates and hundreds of new terms and illustrations . . .
eeping with the most recent developments in the
struction industry.

w with a quick-reference Spanish section. Includes tools
equipment, materials, tasks, and more!

.95 per copy
r 500 pages, softcover
alog no. 67282A

Means Repair & Remodeling Estimating 4th Edition
By Edward B. Wetherill and RSMeans

This important reference focuses on the unique problems
of estimating renovations of existing structures, and helps
you determine the true costs of remodeling through careful
evaluation of architectural details and a site visit.

New section on disaster restoration costs.

$69.95 per copy
Over 450 pages, illustrated, hardcover
Catalog no. 67265B

acilities Planning & elocation
David D. Owen

omplete system for planning space needs and managing
cations. Includes step-by-step manual, over 50 forms,
extensive reference section on materials and
iishings.

v lower price and user-friendly format.

.98 per copy
r 450 pages, softcover
alog no. 67301

Means Square Foot & UNIFORMAT Assemblies Estimating Methods 3rd Edition

Develop realistic square foot and assemblies costs for
budgeting and construction funding. The new edition
features updated guidance on square foot and assemblies
estimating using UNIFORMAT II. An essential reference for
anyone who performs conceptual estimates.

$69.95 per copy
Over 300 pages, illustrated, softcover
Catalog no. 67145B

eans Electrical Estimating ethods 3rd Edition

anded edition includes sample estimates and cost
rmation in keeping with the latest version of the CSI
sterFormat and UNIFORMAT II. Complete coverage
ber optic and uninterruptible power supply electrical
tems, broken down by components, and explained in
il. Includes a new chapter on computerized estimating
:hods. A practical companion to *RSMeans Electrical
t Data.*

)5 per copy
r 325 pages, hardcover
alog no. 67230B

Means Mechanical Estimating Methods 4th Edition

Completely updated, this guide assists you in making a
review of plans, specs, and bid packages, with suggestions
for takeoff procedures, listings, substitutions, and pre-bid
scheduling for all components of HVAC. Includes suggestions
for budgeting labor and equipment usage. Compares materials
and construction methods to allow you to select the best
option.

$64.95 per copy
Over 350 pages, illustrated, softcover
Catalog no. 67294B

eans ADA Compliance Pricing uide 2nd Edition
Adaptive Environments and Means

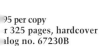

npletely updated and revised to the new 2004 *Americans
h Disabilities Act Accessibility Guidelines,*
book features more than 70 of the most commonly
ded modifications for ADA compliance. Projects range
n installing ramps and walkways, widening
rways and entryways, and installing and refitting
ators, to relocating light switches and signage.

95 per copy
r 350 pages, illustrated, softcover
log no. 67310A

Project Scheduling & Management for Construction 3rd Edition
by David R. Pierce, Jr.

A comprehensive yet easy-to-follow guide to construction
project scheduling and control—from vital project
management principles through the latest scheduling,
tracking, and controlling techniques. The author is a
leading authority on scheduling, with years of field and
teaching experience at leading academic institutions. Spend
a few hours with this book and come away with a solid
understanding of this essential management topic.

$64.95 per copy
Over 300 pages, illustrated, hardcover
Catalog no. 67247B

For more information
visit the RSMeans website
at www.rsmeans.com

Reference Books

The Practice of Cost Segregation Analysis

by Bruce A. Desrosiers and Wayne J. DelPico

This expert guide walks you through the practice of cost segregation analysis, which enables property owners to defer taxes and benefit from "accelerated cost recovery" through depreciation deductions on assets that are properly identified and classified.

With a glossary of terms, sample cost segregation estimates for various building types, key information resources, and updates via a dedicated website, this book is a critical resource for anyone involved in cost segregation analysis.

$99.95 per copy
Over 225 pages
Catalog no. 67345

Preventive Maintenance for Multi-Family Housing

by John C. Maciha

Prepared by one of the nation's leading experts on multi-family housing.

This complete PM system for apartment and condominium communities features expert guidance, checklists for buildings and grounds maintenance tasks and their frequencies, a reusable wall chart to track maintenance, and a dedicated website featuring customizable electronic forms. A must-have for anyone involved with multi-family housing maintenance and upkeep.

$89.95 per copy
225 pages
Catalog no. 67346

How to Estimate with Means Data & CostWorks

New 3rd Edition

by RSMeans and Saleh A. Mubarak, Ph.D.

New 3rd Edition—fully updated with new chapters, plus new CD with updated *CostWorks* cost data and MasterFormat organization. Includes all major construction items—with more than 300 exercises and two sets of plans that show how to estimate for a broad range of construction items and systems—including general conditions and equipment costs.

$59.95 per copy
272 pages, softcover
Includes CostWorks CD
Catalog no. 67324B

Job Order Contracting

Expediting Construction Project Delivery

by Allen Henderson

Expert guidance to help you implement JOC—fast becoming the preferred project delivery method for repair and renovation, minor new construction, and maintenance projects in the public sector and in many states and municipalities. The author, a leading JOC expert and practitioner, shows how to:

• Establish a JOC program

• Evaluate proposals and award contracts

• Handle general requirements and estimating

• Partner for maximum benefits

$89.95 per copy
192 pages, illustrated, hardcover
Catalog no. 67348

Builder's Essentials: Estimating Building Costs

For the Residential & Light Commercial Contractor

by Wayne J. DelPico

Step-by-step estimating methods for residential and light commercial contractors. Includes a detailed look at every construction specialty—explaining all the components, takeoff units, and labor needed for well-organized, complete estimates. Covers correctly interpreting plans and specifications, and developing accurate and complete labor and material costs.

$29.95 per copy
Over 400 pages, illustrated, softcover
Catalog no. 67343

Building & Renovating Schools

This all-inclusive guide covers every step of the school construction process—from initial planning, needs assessment, and design, right through moving into the new facility. A must-have resource for anyone concerned with new school construction or renovation. With square foot cost models for elementary, middle, and high school facilities, and real-life case studies of recently completed school projects.

The contributors to this book—architects, construction project managers, contractors, and estimators who specialize in school construction—provide start-to-finish, expert guidance on the process.

$69.98 per copy
Over 425 pages, hardcover
Catalog no. 67342

Reference Books

**For more information
visit the RSMeans website
at www.rsmeans.com**

The Homeowner's Guide to Mold By Michael Pugliese

Expert guidance to protect your health and your home.

Mold, whether caused by leaks, humidity or flooding, is a real health and financial issue—for homeowners and contractors. This full-color book explains:
- Construction and maintenance practices to prevent mold
- How to inspect for and remove mold
- Mold remediation procedures and costs
- What to do after a flood
- How to deal with insurance companies

$...95 per copy
... pages, softcover
Catalog no. 67344

Means Illustrated Construction Dictionary
Unabridged 3rd Edition, with CD-ROM

Long regarded as the industry's finest, *Means Illustrated Construction Dictionary* is now even better. With the addition of over 1,000 new terms and hundreds of new illustrations, it is the clear choice for the most comprehensive and current information. The companion CD-ROM that comes with this new edition adds many extra features: larger graphics, expanded definitions, and links to both CSI MasterFormat numbers and product information.

$99.95 per copy
Over 790 pages, illustrated, hardcover
Catalog no. 67292A

Means Landscape Estimating Methods
New 5th Edition

Answers questions about preparing competitive landscape construction estimates, with up-to-date cost estimates and the new MasterFormat classification system. Expanded and revised to address the latest materials and methods, including new coverage on approaches to green building. Includes:
- Step-by-step explanation of the estimating process
- Sample forms and worksheets that save time and prevent errors

$...95 per copy
Over 350 pages, softcover
Catalog no. 67295C

Means Plumbing Estimating Methods 3rd Edition
by Joseph Galeno and Sheldon Greene

Updated and revised! This practical guide walks you through a plumbing estimate, from basic materials and installation methods through change order analysis. *Plumbing Estimating Methods* covers residential, commercial, industrial, and medical systems, and features sample takeoff and estimate forms and detailed illustrations of systems and components.

$29.98 per copy
330+ pages, softcover
Catalog no. 67283B

Understanding & Negotiating Construction Contracts
by Kit Werremeyer

Take advantage of the author's 30 years' experience in small-to-large (including international) construction projects. Learn how to identify, understand, and evaluate high risk terms and conditions typically found in all construction contracts—then negotiate to lower or eliminate the risk, improve terms of payment, and reduce exposure to claims and disputes. The author avoids "legalese" and gives real-life examples from actual projects.

$...95 per copy
... pages, softcover
Catalog no. 67350

The Gypsum Construction Handbook (Spanish)

The Gypsum Construction Handbook, Spanish edition, is an illustrated, comprehensive, and authoritative guide on all facets of gypsum construction. You'll find comprehensive guidance on available products, installation methods, fire- and sound-rated construction information, illustrated framing-to-finish application instructions, estimating and planning information, and more.

$29.95 per copy
Over 450 pages, softcover
Catalog No. 67357S

Means Spanish/English Construction Dictionary 2nd Edition
by RSMeans and the International Code Council

This expanded edition features thousands of the most common words and useful phrases in the construction industry with easy-to-follow pronunciations in both Spanish and English. Over 800 new terms, phrases, and illustrations have been added. It also features a new stand-alone "Safety & Emergencies" section, with colored pages for quick access.

Unique to this dictionary are the systems illustrations showing the relationship of components in the most common building systems for all major trades.

$...95 per copy
Over 400 pages
Catalog no. 67327A

How Your House Works
by Charlie Wing

A must-have reference for every homeowner, handyman, and contractor—for repair, remodeling, and new construction. This book uncovers the mysteries behind just about every major appliance and building element in your house—from electrical, heating and AC, to plumbing, framing, foundations and appliances. Clear, "exploded" drawings show exactly how things should be put together and how they function—what to check if they don't work, and what you can do that might save you having to call in a professional.

$21.95 per copy
160 pages, softcover
Catalog no. 67351

**For more information
visit the RSMeans website
at www.rsmeans.com**

Professional Development

Means CostWorks® CD Training

This one-day course helps users become more familiar with the functionality of *Means CostWorks* program. Each menu, icon, screen, and function found in the program is explained in depth. Time is devoted to hands-on estimating exercises.

Some of what you'll learn:
- Searching the database utilizing all navigation methods
- Exporting RSMeans Data to your preferred spreadsheet format
- Viewing crews, assembly components, and much more!
- Automatically regionalizing the database

You are required to bring your own laptop computer for this course.

When you register for this course you will receive an outline for your laptop requirements.

Also offering web training for CostWorks CD!

Unit Price Estimating

This interactive *two-day* seminar teaches attendees how to interpret project information and process it into final, detailed estimates with the greatest accuracy level.

The most important credential an estimator can take to the job is the ability to visualize construction, and estimate accurately.

Some of what you'll learn:
- Interpreting the design in terms of cost
- The most detailed, time-tested methodology for accurate pricing
- Key cost drivers—material, labor, equipment, staging, and subcontracts
- Understanding direct and indirect costs for accurate job cost accounting and change order management

Who should attend: Corporate and government estimators and purchasers, architects, engineers... and others needing to produce accurate project estimates.

Square Foot & Assemblies Estimating

This *two-day* course teaches attendees how to quickly deliver accurate square foot estimates using limited budget and design information.

Some of what you'll learn:
- How square foot costing gets the estimate done faster
- Taking advantage of a "systems" or "assemblies" format
- The RSMeans "building assemblies/square foot cost approach"
- How to create a reliable preliminary systems estimate using bare-bones design information

Who should attend: Facilities managers, facilities engineers, estimators, planners, developers, construction finance professionals... and others needing to make quick, accurate construction cost estimates at commercial, government, educational, and medical facilities.

Repair & Remodeling Estimating

This *two-day* seminar emphasizes all the underlying considerations unique to repair/remodeling estimating and presents the correct methods for generating accurate, reliable R&R project costs using the unit price and assemblies methods.

Some of what you'll learn:
- Estimating considerations—like labor-hours, building code compliance, working within existing structures, purchasing materials in smaller quantities, unforeseen deficiencies
- Identifying problems and providing solutions to estimating building alterations
- Rules for factoring in minimum labor costs, accurate productivity estimates, and allowances for project contingencies
- R&R estimating examples calculated using unit price and assemblies data

Who should attend: Facilities managers, plant engineers, architects, contractors, estimators, builders... and others who are concerned with the proper preparation and/or evaluation of repair and remodeling estimates.

Mechanical & Electrical Estimating

This *two-day* course teaches attendees how to prepare more accurate and complete mechanical/electrical estimates, avoiding the pitfalls of omission and double-counting, while understanding the composition and rationale within the RSMeans Mechanical/Electrical database.

Some of what you'll learn:
- The unique way mechanical and electrical systems are interrelated
- M&E estimates–conceptual, planning, budgeting, and bidding stages
- Order of magnitude, square foot, assemblies, and unit price estimating
- Comparative cost analysis of equipment and design alternatives

Who should attend: Architects, engineers, facilities managers, mechanical and electrical contractors... and others needing a highly reliable method for developing, understanding, and evaluating mechanical and electrical contracts.

Green Building: Planning & Construction

In this *two-day* course, learn about tailoring a building and its placement on the site to the local climate, site conditions, culture, and community. Includes information to reduce resource consumption and augment resource supply.

Some of what you'll learn:
- Green technologies, materials, systems, and standards
- Modeling tools and how to use them
- Cost vs. value of green products over their life cycle
- Low-cost green strategies, and economic incentives and funding
- Health, comfort, and productivity

Who should attend: Contractors, project managers building owners, building officials, healthcare and insurance professionals.

Facilities Maintenance & Repair Estimating

This *two-day* course teaches attendees how to plan, budget, and estimate the cost of ongoing and preventive maintenance and repair for existing buildings and grounds.

Some of what you'll learn:
- The most financially favorable maintenance, repair, and replacement scheduling and estimating
- Auditing and value engineering facilities
- Preventive planning and facilities upgrading
- Determining both in-house and contract-out service costs
- Annual, asset-protecting M&R plan

Who should attend: Facility managers, maintenance supervisors, buildings and grounds superintendents, plant managers, planners, estimators... and others involved in facilities planning and budgeting.

Practical Project Management for Construction Professionals

In this *two-day* course, acquire the essential knowledge and develop the skills to effectively and efficiently execute the day-to-day responsibilities of the construction project manager.

Covers:
- General conditions of the construction contract
- Contract modifications: change orders and construction change directives
- Negotiations with subcontractors and vendors
- Effective writing: notification and communications
- Dispute resolution: claims and liens

Who should attend: Architects, engineers, owner's representatives, project managers.

Facilities Construction Estimating

In this *two-day* course, professionals working in facilities management can get help with their daily challenges to establish budgets for all phase of a project

Some of what you'll learn:
- Determine the full scope of a project
- Identify the scope of risks & opportunities
- Creative solutions to estimating issues
- Organizing estimates for presentation & discussion
- Special techniques for repair/remodel and maintenance projects
- Negotiating project change orders

Who should attend: Facility managers, engineers, contractors, facility trades-people, planners & project managers

Professional Development

For more information
visit the RSMeans website
at www.rsmeans.com

Scheduling and Project Management

This *two-day* course teaches attendees the most current and proven scheduling and management techniques needed to bring projects in on time and on budget.

Some of what you'll learn:
- Crucial phases of planning and scheduling
- How to establish project priorities and develop realistic schedules and management techniques
- Critical Path and Precedence Methods
- Special emphasis on cost control

Who should attend: Construction project managers, supervisors, engineers, estimators, contractors... and others who want to improve their project planning, scheduling, and management skills.

Assessing Scope of Work for Facility Construction Estimating

This *two-day* practical training program addresses the vital importance of understanding the SCOPE of projects in order to produce accurate cost estimates for facility repair and remodeling.

Some of what you'll learn:
- Discussions of site visits, plans/specs, record drawings of facilities, and site-specific lists
- Review of CSI divisions, including means, methods, materials, and the challenges of scoping each topic
- Exercises in SCOPE identification and SCOPE writing for accurate estimating of projects
- Hands-on exercises that require SCOPE, take-off, and pricing

Who should attend: Corporate and government estimators, planners, facility managers, and others who need to produce accurate project estimates.

Site Work & Heavy Construction Estimating

This *two-day* course teaches attendees how to estimate earthwork, site utilities, foundations, and site improvements, using the assemblies and the unit price methods.

Some of what you'll learn:
- Basic site work and heavy construction estimating skills
- Estimating foundations, utilities, earthwork, and site improvements
- Correct equipment usage, quality control, and site investigation for estimating purposes

Who should attend: Project managers, design engineers, estimators, and contractors doing site work and heavy construction.

2010 RSMeans Seminar Schedule

Note: call for exact dates and details.

Location	Dates
Las Vegas, NV	March
Washington, DC	April
Denver, CO	May
San Francisco, CA	June
Washington, DC	September
Dallas, TX	September
Las Vegas, NV	October
Atlantic City, NJ	October
Orlando, FL	November
San Diego, CA	December

2010 Regional Seminar Schedule

Location	Dates	Location	Dates
Seattle, WA	January and August	St. Louis, MO	August
Dallas/Ft. Worth, TX	January	Houston, TX	October
Austin, TX	February	Colorado Springs, CO	October
Anchorage, AK	March and September	Baltimore, MD	November
Oklahoma City, OK	March	Sacramento, CA	November
Phoenix, AZ	March	San Antonio, TX	December
Aiea, HI	May	Raleigh, NC	December
Kansas City, MO	May		
Bethesda, MD	June		
Columbus, GA	June		
El Segundo, CA	August		

1-800-334-3509, ext. 5115

For more information
visit the RSMeans website
at www.rsmeans.com

Professional Development

Registration Information

Register early... Save up to $100! Register 30 days before the start date of a seminar and save $100 off your total fee. *Note: This discount can be applied only once per order. It cannot be applied to team discount registrations or any other special offer.*

How to register Register by phone today! The RSMeans toll-free number for making reservations is **1-800-334-3509, ext. 5115.**

Individual seminar registration fee - $935. *Means CostWorks®* **training registration fee - $375.** To register by mail, complete the registration form and return, with your full fee, to: RSMeans Seminars, 63 Smiths Lane, Kingston, MA 02364.

Government pricing All federal government employees save off the regular seminar price. Other promotional discounts cannot be combined with the government discount.

Team discount program for two to four seminar registrations. Call for pricing: 1-800-334-3509, ext. 5115

Multiple course discounts When signing up for two or more courses, call for pricing.

Refund policy Cancellations will be accepted up to ten business days prior to the seminar start. There are no refunds for cancellations received later than ten working days prior to the first day of the seminar. A $150 processing fee will be applied for all cancellations. Written notice of cancellation is required. Substitutions can be made at any time before the session starts. **No-shows are subject to the full seminar fee.**

AACE approved courses Many seminars described and offered here have been approved for 14 hours (1.4 recertification credits) of credit by the AACE

International Certification Board toward meeting the continuing education requirements for recertification as a Certified Cost Engineer/Certified Cost Consultant.

AIA Continuing Education We are registered with the AIA Continuing Education System (AIA/CES) and are committed to developing quality learning activities in accordance with the CES criteria. Many seminars meet the AIA/CES criteria for Quality Level 2. AIA members may receive (14) learning units (LUs) for each two-day RSMeans course.

Daily course schedule The first day of each seminar session begins at 8:30 a.m. and ends at 4:30 p.m. The second day begins at 8:00 a.m. and ends at 4:00 p.m. Participants are urged to bring a hand-held calculator, since many actual problems will be worked out in each session.

Continental breakfast Your registration includes the cost of a continental breakfast, and a morning and afternoon refreshment break. These informal segments allow you to discuss topics of mutual interest with other seminar attendees. (You are free to make your own lunch and dinner arrangements.)

Hotel/transportation arrangements RSMeans arranges to hold a block of rooms at most host hotels. To take advantage of special group rates when making your reservation, be sure to mention that you are attending the RSMeans seminar. You are, of course, free to stay at the lodging place of your choice. **(Hotel reservations and transportation arrangements should be made directly by seminar attendees.)**

Important Class sizes are limited, so please register as soon as possible.

Note: Pricing subject to change.

Registration Form

ADDS-1000

Call 1-800-334-3509, ext. 5115 to register or FAX this form 1-800-632-6732. Visit our Web site: www.rsmeans.com

Please register the following people for the RSMeans construction seminars as shown here. We understand that we must make our own hotel reservations if overnight stays are necessary.

☐ Full payment of $_____ enclosed.

☐ Bill me

Please print name of registrant(s)

(To appear on certificate of completion)

_____

_____

_____

_____

P.O. #:_____
GOVERNMENT AGENCIES MUST SUPPLY PURCHASE ORDER NUMBER OR TRAINING FORM.

Firm name_____

Address_____

City/State/Zip_____

Telephone no._____ Fax no._____

E-mail address_____

Charge registration(s) to: ☐ MasterCard ☐ VISA ☐ American Express

Account no._____ Exp. date_____

Cardholder's signature_____

Seminar name_____

Seminar City_____

Please mail check to: RSMeans Seminars, 63 Smiths Lane, P.O. Box 800, Kingston, MA 02364 USA

CONSTRUCTION COST INFORMATION... PRODUCTS & SERVICES

503

**For more information
visit the RSMeans website
at www.rsmeans.com**

New Titles

The Gypsum Construction Handbook by USG

An invaluable reference of construction procedures for gypsum drywall, cement board, veneer plaster, and conventional plaster. This new edition includes the newest product developments, installation methods, fire- and sound-rated construction information, illustrated framing-to-finish application instructions, estimating and planning information, and more. Great for architects and engineers; contractors, builders, and dealers; apprentices and training programs; building inspectors and code officials; and anyone interested in gypsum construction. Features information on tools and safety practices, a glossary of construction terms, and a list of important agencies and associations. Also available in Spanish.

**$29.95
Over 575 pages, illustrated, softcover
Catalog no. 67357A**

BIM Handbook
A Guide to Building Information Modeling for Owners, Managers, Designers, Engineers and Contractors
by Chuck Eastman, Paul Teicholz, Rafael Sacks, Kathleen Liston

Building Information Modeling (BIM) is a new approach to design, construction, and facility management where a digital depiction of the building process is used to facilitate the exchange of information in digital format.

The BIM Handbook provides an in-depth understanding of BIM technologies, the business and organizational issues associated with its implementation, and the profound advantages that effective use of BIM can provide to all members of a project team. Also includes a rich set of informative BIM case studies.

**$85.00
504 pages
Catalog no. 67356**

Green Home Improvement
by Daniel D. Chiras, PhD

With energy costs rising and environmental awareness increasing, many people are looking to make their homes greener. This book makes it easy, with 65 projects and actual costs and projected savings that will help homeowners prioritize their green improvements.

Projects range from simple water savers that cost only a few dollars, to bigger-ticket items such as efficient appliances and HVAC systems, green roofing, flooring, landscaping, and much more. With color photos and cost estimates, each project compares options and describes the work involved, the benefits, and the savings.

**$24.95
320 pages, illustrated, softcover
Catalog no. 67355**

RSMeans Estimating Handbook, 3ʳᵈ Edition

Widely used in the industry for tasks ranging from routine estimates to special cost analysis projects this handbook has been completely updated and reorganized with new and expanded technical information.

RSMeans Estimating Handbook will help construction professionals:

- Evaluate architectural plans and specifications
- Prepare accurate quantity takeoffs
- Compare design alternatives and costs
- Perform value engineering
- Double-check estimates and quotes
- Estimate change orders

**$99.95 per copy
Over 900 pages, hardcover
Catalog No. 67276B**

Universal Design Ideas for Style, Comfort & Safety
by RSMeans and Lexicon Consulting, Inc.

Incorporating universal design when building or remodeling helps people of any age and physical ability more fully and safely enjoy their living spaces. This book shows how universal design can be artfully blended into the most attractive home It discusses specialized products like adjustable countertops and chair lifts, as well as simple way to enhance a home's safety and comfort. With col photos and expert guidance, every area of the hon is covered. Includes budget estimates that give an idea how much projects will cost.

**$21.95
160 pages, illustrated, softcover
Catalog no. 67354**

Residential & Light Commercial Construction Standards, 3rd Edition, Updated
by RSMeans and contributing authors

This book provides authoritative requirements and recommendations compiled from leading professional associations, industry publications, a building code organizations. This all-in-one refere helps establish a standard for workmanship, quic resolve disputes, and avoid defect claims. Update third edition includes new coverage of green building, seismic, hurricane, and mold-resistant construction.

**$59.95
Over 550 pages, illustrated, softcover
Catalog no. 67322B**

2010 Order Form

Qty.	Book no.	COST ESTIMATING BOOKS	Unit Price	Total
	60060	Assemblies Cost Data 2010	$279.95	
	60010	Building Construction Cost Data 2010	169.95	
	60110	Concrete & Masonry Cost Data 2010	154.95	
	50140	Construction Cost Indexes 2010 (subscription)	362.00	
	60140A	Construction Cost Index–January 2010	90.50	
	60140B	Construction Cost Index–April 2010	90.50	
	60140C	Construction Cost Index–July 2010	90.50	
	60140D	Construction Cost Index–October 2010	90.50	
	60340	Contr. Pricing Guide: Resid. R & R Costs 2010	39.95	
	60330	Contr. Pricing Guide: Resid. Detailed 2010	39.95	
	60320	Contr. Pricing Guide: Resid. Sq. Ft. 2010	39.95	
	60230	Electrical Change Order Cost Data 2010	169.95	
	60030	Electrical Cost Data 2010	169.95	
	60200	Facilities Construction Cost Data 2010	419.95	
	60300	Facilities Maintenance & Repair Cost Data 2010	379.95	
	60160	Heavy Construction Cost Data 2010	169.95	
	60090	Interior Cost Data 2010	169.95	
	60120	Labor Rates for the Const. Industry 2010	379.95	
	60180	Light Commercial Cost Data 2010	144.95	
	60020	Mechanical Cost Data 2010	169.95	
	60150	Open Shop Building Const. Cost Data 2010	169.95	
	60210	Plumbing Cost Data 2010	169.95	
	60040	Repair & Remodeling Cost Data 2010	144.95	
	60170	Residential Cost Data 2010	144.95	
	60280	Site Work & Landscape Cost Data 2010	169.95	
	60050	Square Foot Costs 2010	184.95	
	62019	Yardsticks for Costing (2009)	154.95	
	62010	Yardsticks for Costing (2010)	169.95	
		REFERENCE BOOKS		
	67310A	ADA Compliance Pricing Guide, 2nd Ed.	79.95	
	67356	BIM Handbook	85.00	
	67330	Bldrs Essentials: Adv. Framing Methods	12.98	
	67329	Bldrs Essentials: Best Bus. Practices for Bldrs	29.95	
	67298A	Bldrs Essentials: Framing/Carpentry 2nd Ed.	15.98	
	67298AS	Bldrs Essentials: Framing/Carpentry Spanish	15.98	
	67307	Bldrs Essentials: Plan Reading & Takeoff	35.95	
	67342	Building & Renovating Schools	69.98	
	67339	Building Security: Strategies & Costs	44.98	
	67353	Complete Book of Framing	29.95	
	67146	Concrete Repair & Maintenance Illustrated	69.95	
	67352	Construction Business Management	49.98	
	67314	Cost Planning & Est. for Facil. Maint.	89.95	
	67328A	Designing & Building with the IBC, 2nd Ed.	99.95	
	67230B	Electrical Estimating Methods, 3rd Ed.	64.95	
	64777A	Environmental Remediation Est. Methods, 2nd Ed.	49.98	
	67343	Estimating Bldg. Costs for Resi. & Lt. Comm.	29.95	
	67276B	Estimating Handbook, 3rd Ed.	$99.95	
	67318	Facilities Operations & Engineering Reference	54.98	
	67301	Facilities Planning & Relocation	49.98	
	67338A	Green Building: Proj. Planning & Cost Est., 2nd Ed.	129.95	
	67355	Green Home Improvement	24.95	

Qty.	Book no.	REFERENCE BOOKS (Cont.)	Unit Price	Total
	67357A	The Gypsum Construction Handbook	29.95	
	67357S	The Gypsum Construction Handbook (Spanish)	29.95	
	67323	Historic Preservation: Proj. Planning & Est.	49.98	
	67308E	Home Improvement Costs–Int. Projects, 9th Ed.	24.95	
	67309E	Home Improvement Costs–Ext. Projects, 9th Ed.	24.95	
	67344	Homeowner's Guide to Mold	21.95	
	67324B	How to Est.w/Means Data & CostWorks, 3rd Ed.	59.95	
	67351	How Your House Works	21.95	
	67282A	Illustrated Const. Dictionary, Condensed, 2nd Ed.	59.95	
	67292A	Illustrated Const. Dictionary, w/CD-ROM, 3rd Ed.	99.95	
	67348	Job Order Contracting	89.95	
	67347	Kitchen & Bath Project Costs	19.98	
	67295C	Landscape Estimating Methods, 5th Ed.	64.95	
	67341	Life Cycle Costing for Facilities	99.95	
	67294B	Mechanical Estimating Methods, 4th Ed.	64.95	
	67245A	Planning & Managing Interior Projects, 2nd Ed.	34.98	
	67283B	Plumbing Estimating Methods, 3rd Ed.	29.98	
	67345	Practice of Cost Segregation Analysis	99.95	
	67337	Preventive Maint. for Higher Education Facilities	149.95	
	67346	Preventive Maint. for Multi-Family Housing	89.95	
	67326	Preventive Maint. Guidelines for School Facil.	149.95	
	67247B	Project Scheduling & Management for Constr. 3rd Ed.	64.95	
	67265B	Repair & Remodeling Estimating Methods, 4th Ed.	69.95	
	67322B	Resi. & Light Commercial Const. Stds., 3rd Ed.	59.95	
	67327A	Spanish/English Construction Dictionary, 2nd Ed.	23.95	
	67145B	Sq. Ft. & Assem. Estimating Methods, 3rd Ed.	69.95	
	67321	Total Productive Facilities Management	29.98	
	67350	Understanding and Negotiating Const. Contracts	69.95	
	67303B	Unit Price Estimating Methods, 4th Ed.	59.95	
	67354	Universal Design	21.95	
	67319A	Value Engineering: Practical Applications	79.95	

MA residents add 6.25% state sales tax

Shipping & Handling**

Total (U.S. Funds)*

Prices are subject to change and are for U.S. delivery only. *Canadian customers may call for current prices. **Shipping & handling charges: Add 7% of total order for check and credit card payments. Add 9% of total order for invoiced orders.

Send order to: **ADDV-1000**

Name (please print) _____

Company _____

☐ **Company**

☐ **Home Address** _____

_____

City/State/Zip _____

Phone # _____ P.O. # _____

(Must accompany all orders being billed)

ail to: RSMeans, P.O. Box 800, Kingston, MA 02364

505

RSMeans Project Cost Report

By filling out this report, your project data will contribute to the database that supports the RSMeans Project Cost Square Foot Data. When you fill out this form, RSMeans will provide a $30 discount off one of the RSMeans products advertised in the preceding pages. Please complete the form including all items where you have cost data, and all the items marked (✔).

$30.00 Discount per product for each report you submit.

Project Description (No remodeling projects, please.)

✔ Building Use (Office, School...) _____

✔ Address (City, State) _____

_____

✔ Total Building Area (SF) _____

✔ Ground Floor (SF) _____

✔ Frame (Wood, Steel...) _____

✔ Exterior Wall (Brick, Tilt-up...) _____

✔ Basement: (check one) ☐ Full ☐ Partial ☐ None

✔ Number Stories _____

✔ Floor-to-Floor Height _____

✔ Volume (C.F.) _____

% Air Conditioned _____ Tons _____

Total Project Cost $ _____

Owner _____

Architect _____

General Contractor _____

✔ Bid Date _____

✔ Typical Bay Size _____

✔ Capacity _____

✔ Labor Force: _____ % Union _____ % Non-Union

✔ Project Description (Circle one number in each line)

1. Economy 2. Average 3. Custom 4. Luxury
1. Square 2. Rectangular 3. Irregular 4. Very Irregular

Comments _____

_____

A	✔ General Conditions	$	
B	✔ Site Work	$	
C	✔ Concrete	$	
D	✔ Masonry	$	
E	✔ Metals	$	
F	✔ Wood & Plastics	$	
G	✔ Thermal & Moisture Protection	$	
GR	Roofing & Flashing	$	
H	✔ Doors and Windows	$	
J	✔ Finishes	$	
JP	Painting & Wall Covering	$	

K	✔ Specialties	$	
L	✔ Equipment	$	
M	✔ Furnishings	$	
N	✔ Special Construction	$	
P	✔ Conveying Systems	$	
Q	✔ Mechanical	$	
QP	Plumbing	$	
QB	HVAC	$	
R	✔ Electrical	$	
S	✔ Mech./Elec. Combined	$	

Please specify the RSMeans product you wish to receive. Complete the address information.

Product Name _____

Product Number _____

Your Name _____

Title _____

Company _____
  ☐ Company
  ☐ Home   Street Address _____

City, State, Zip _____

Return by mail or Fax 888-492-6770.

Method of Payment:

Credit Card # _____

Expiration Date _____

Check _____

Purchase Order _____

RSMeans
a division of Reed Construction Data
Square Foot Costs Department
P.O. Box 800
Kingston, MA 02364-3008